Birkhäuser

# Advances in Mathematical Fluid Mechanics



*Advances in Mathematical Fluid Mechanics* is a forum for the publication of high quality monographs, or collections of works, on the mathematical theory of fluid mechanics, with special regards to the Navier-Stokes equations. Its mathematical aims and scope are similar to those of the *Journal of Mathematical Fluid Mechanics*. In particular, mathematical aspects of computational methods and of applications to science and engineering are welcome as an important part of the theory. So also are works in related areas of mathematics that have a direct bearing on fluid mechanics.

The monographs and collections of works published here may be written in a more expository style than is usual for research journals, with the intention of reaching a wide audience. Collections of review articles will also be sought from time to time.

More information about this series at http://www.springer.com/series/5032

Herbert Amann • Yoshikazu Giga • Hideo Kozono •
Hisashi Okamoto • Masao Yamazaki
Editors

# Recent Developments of Mathematical Fluid Mechanics

Birkhäuser

*Editors*
Herbert Amann
Institut für Mathematik
University of Zürich
Zürich, Switzerland

Yoshikazu Giga
Graduate School of Mathematical Sciences
University of Tokyo
Tokyo, Japan

Hideo Kozono
Department of Mathematics
Waseda University
Tokyo, Japan

Hisashi Okamoto
Research Institute for Mathematical Sciences
Kyoto University Kitashirakawa
Kyoto, Japan

Masao Yamazaki
Department of Mathematics
Waseda University
Tokyo, Japan

ISSN 2297-0320 ISSN 2297-0339 (electronic)
Advances in Mathematical Fluid Mechanics
ISBN 978-3-0348-0938-2 ISBN 978-3-0348-0939-9 (eBook)
DOI 10.1007/978-3-0348-0939-9

Library of Congress Control Number: 2015960758

Mathematics Subject Classification (2010): 35XX, 76XX

Springer Basel Heidelberg New York Dordrecht London


Printed on acid-free paper

Springer Basel is part of Springer Science+Business Media (www.birkhauser-science.com)

# Preface

This volume collects research papers and survey articles of participants in the

"International Conference on Mathematical Fluid Dynamics on the Occasion of Yoshihiro Shibata's 60th Birthday",

which was held from March 5 to 9, 2013, in Nara, the former capital of Japan.

We thank all the participants, in particular the invited speakers, who contributed to this volume. Our thanks also go to the referees for their efficient work, and to the secretaries, the staff, and the students who helped us during the meeting.

The conference was supported, in part, by the Grant in Aid for Scientific Research of the Japan Society for the Promotion of Science (No. 24224003, 24224004, 24340025) and by the JSPS-DFG Japanese-German Graduate Externship.

Zürich, Switzerland — Herbert Amann
Tokyo, Japan — Yoshikazu Giga
Tokyo, Japan — Hideo Kozono
Kyoto, Japan — Hisashi Okamoto
Tokyo, Japan — Masao Yamazaki

# Contents

# The Work of Yoshihiro Shibata

**Herbert Amann, Yoshikazu Giga, Hisashi Okamoto, Hideo Kozono, and Masaso Yamazaki**

**Abstract** Introducing his research carrier, we address Prof. Yoshihiro Shibata's great contributions to the mathematical analysis. His out-standing influence to the mathematical society is also clarified.

**Keywords** $L_p - L_q$-estimates • Elastodynamical system • Free boundary value problems • Hyperbolic equations • Local and global well-posedness • Local energy decay • Maximal $L_p - L_q$ regularity • Navier-Stokes equations • Oseen flow • R-boundedness

Yoshihiro Shibata was born in Tokyo on August 28, 1952. After graduating from Azabu Gakuen High School in March 1971, he studied at Tokyo University of Education from April 1971 through March 1977 and at Tsukuba University from April 1977 through March 1978. Afterwards he got a position at the Mathematical Department of Tsukuba University. There he obtained, in October 1981, the degree of a Doctor of Sciences under the supervision of Matsumura Mutsuhide.

During that period he studied initial-boundary value problems for linear hyperbolic equations [68, 70, 71] and uniqueness criteria for general partial differential equations with constant coefficients [58, 69, 72, 73].

Then his interest evolved toward initial-boundary value problems for non-linear hyperbolic equations and systems. In [75] he succeeded to prove global

H. Amann
Institut für Mathematik, Universität Zürich, 8057 Zürich, Switzerland
e-mail: herbert.amann@math.uzh.ch

Y. Giga
Graduate School of Mathematical Sciences, University of Tokyo, 153-8914 Tokyo, Japan
e-mail: labgiga@ms.u-tokyo.ac.jp

H. Okamoto
Research Institute for Mathematical Sciences, Kyoto University, Kyoto 606-8502, Japan
e-mail: okamoto@kurims.kyoto-u.ac.jp

H. Kozono (✉) • M. Yamazaki
Department of Mathematics, Waseda University, Tokyo 169-8555, Japan
e-mail: kozono@waseda.jp; masao.yamazaki@waseda.jp



H. Amann et al. (eds.), *Recent Developments of Mathematical Fluid Mechanics*,
Advances in Mathematical Fluid Mechanics, DOI 10.1007/978-3-0348-0939-9_1

well-posedness results for nonlinear wave equations with damping terms in $N$-dimensional exterior domains ($N \geq 3$). The main step was to establish decay properties of solutions to the linearized equations in exterior domains. To achieve this, he studied the local energy decay near the boundary. In combination with decay estimates for the whole space, this furnishes the required estimates in the exterior domain. To derive the local energy decay, he constructed the solution operator to the corresponding resolvent problem near the origin. Since the latter belongs to the continuous spectrum, he restricted the domain of the solution operator to compactly supported data. In addition, he considered it in local $L_2$ spaces. Thus he constructed the solution operator in a topology which is weaker than the usual one.

He is the first mathematician who applied this method, which was originally used in scattering theory by Vainberg [11, 12] and Lax and Phillips [8], for example, to the study of nonlinear evolution equations. Moreover, in [110, 111, 113, 114], together with Yoshio Tsutsumi, he proved global well-posedness results for non-linear wave equations with Dirichlet as well as Neumann boundary conditions in $N$-dimensional exterior domains ($N \geq 3$). The main idea was again to establish decay properties of solutions of wave equations in exterior domains. To treat the high frequency part, they applied a device of B. Vainberg [12], based on the non-trapping condition. For the low frequency part, they applied the technique developed in [75].

In 1985 Yoshihiro Shibata spent 10 months at the University of Paris 6 as a young researcher, supported by the Ministry of Education, Science, and Culture. He participated in the Leray seminar which was then organized by J. Vaillant. During that time, he proved local well-posedness theorems for fully nonlinear wave equations of the form $\partial_t u = F(u)$, $t \geq 0$, with Neumann boundary conditions [77]. To attack this problem he introduced a simple linearization procedure: By differentiating the equation with respect to $t$ he obtained the system $\partial_t v + A(u)v = 0$, $F(u) = v$, where $A(u) = -DF(u)$. The first equation is linear hyperbolic in $v$, and the second one is a fully nonlinear elliptic equation in $u$. Then he solved this system by the usual iteration method. Prior to his paper, fully nonlinear hyperbolic equations were treated by employing Nash-Moser iteration. His contribution was to show that this technique can be avoided. Later, Tosio Kato [5] formulated Y. Shibata's method in an abstract framework.

The ideas of [77] were further exploited by Gen Nakamura and Yoshihiro Shibata in [63] to proved the local well-posedness for the elastodynamical system with Neumann type boundary conditions. In the series of papers [42, 78, 79], Y. Shibata provided the detailed proof of the local well-posedness for some fully nonlinear hyperbolic systems.

During the period in which he studied nonlinear hyperbolic equations and systems, he was invited to the Bulgarian Academy, in 1987, to work together with Vladimir Georgiev, to Fudan University, in 1988, to collaborate with Song Mu Zheng [109], and, in 1988 and 1989, to the University of Bonn as a member of SFB 256, then chaired by Rolf Leis, to interact with Reinhard Racke.

In the late 80s he shifted his interest toward thermoelastic and viscoelastic equations, motivated by the fact that these hyperbolic systems exhibit physically reasonable dissipation mechanisms. By combining spectral analysis and multiplicative techniques, he could establish several global well-posedness results for small initial data [40, 41, 57, 65, 66].

In the middle of 1980, Hideo Kozono asked Yoshihiro Shibata to apply his idea, obtained in [75], to the study of Navier-Stokes equations in exterior domains. This question was passed on to Hirokazu Iwashita, who then, under the supervision of Y. Shibata, proved the global well-posedness, for small initial data, of the Navier-Stokes equations in exterior domains [7]. This result is an extension of one by Kato [4], who considered the full space problem. The key idea consisted in proving $L_p$-$L_q$ decay estimates for the Stokes equations. For this, Iwashita first established the local energy decay by combining the method of [75] with the Bogovski technique to preserve the divergence free condition for the velocity field. Then he combined these findings with $L_p$-$L_q$ decay estimates for the whole space in order to get the required estimates in exterior domains of $\mathbb{R}^N$ with $N \geq 3$.

Starting in the middle of 90s, Y. Shibata's concentrated his interest on the study of the Navier-Stokes equations. The first result he obtained, was the global well-posedness, for small initial data, of the Oseen flow in a three-dimensional exterior domain [45, 91]. Using these results, Galdi, Heywood and Shibata [34] solved a problem, proposed by Finn [2], concerning the Navier-Stokes flow past a moving obstacle that is started from rest. Although this model cannot be handled in the $L_2$ framework, they succeeded to get a solution in $L_3$.

Together with Wakako Dan, Yoshihiro Shibata [20, 24] obtained the optimal decay rate for the non-stationary Stokes equations in 2-dimensional exterior domains. In this case, the fundamental solution of the Stokes resolvent problem has a logarithmic singularity at the origin, whereas it is smooth there if $N \geq 3$. Thus a new idea, essentially different from the ones used in the case $N \geq 3$, was needed. They employed a contradiction argument to eliminate the logarithmic singularity by reflection at the boundary.

In addition, Y. Shibata studied the Navier-Stokes equations describing the motion of a compressible viscous barotropic fluid in 3-dimensional exterior domains. In collaboration with Takayuki Kobayashi [9, 46], he established the optimal decay rate of the Matsumura-Nishida solutions [57]. For this they used $L_p$-$L_q$ decay estimates ($1 < p \leq 2 \leq q \leq \infty$) for the linearized equations in exterior domains, thus providing a new approach to the mathematical study of compressible viscous fluids in exterior domains.

In 1996, Y. Shibata organized the international workshop "Some applications of real analytic methods to the study of nonlinear partial differential equations in mathematical physics " at Tsukuba University. It was supported by the Mathematical Society of Japan as a Regional Workshop (RW) and co-organized by Hideo Kozono (Kozono and Shibata [48]). Afterwards, he acted as co-organizer of many further international conferences concerning Navier-Stokes equations.

In April 1997, Yoshihiro Shibata moved to Waseda University. From then on, his main interest concerns the study of incompressible and compressible viscous fluid flows. It is one of his achievements in the study of Navier-Stokes equations that he "reintroduced" the pressure term. In fact, starting with the seminal papers by Fujita and Kato [3, 6], the Japanese school, including K. Masuda, T. Miyakawa, H. Okamoto, Y. Giga, H. Kozono, M. Yamazaki..., contributed a lot to the study of the Navier-Stokes equations by means of the analytic semigroup approach. Thereby, the pressure term is eliminated with the help of the Helmholtz-Leray projection. In contrast, Y. Shibata, and H. Iwashita included the pressure in their studies, using the Bogovski technique to preserve the solenoidal condition for the velocity field. This approach is more in the spirit of PDEs than of functional analysis. It was successfully applied by H. Iwashita [7] to obtain the $L_p$-$L_q$ decay rate for the Stokes equation in exterior domains, by Kobayashi and Shibata [45] to the Oseen equation in exterior domains, and to the exterior problem for rotating obstacles by T. Hishida and Y. Shibata [37].

Following a suggestion of H. Amann in 2000, Y. Shibata turned to the investigation of free boundary value problems for Navier-Stokes equations by means of maximal regularity theory. This set of problems—in particular the drop and the ocean problem—had already attracted many researchers, predominantly V. A. Solonnikov and his students, T. Beale, T. Nishida, A. Tani, and their Japanese followers, who employed either an $L_2$ framework or Hölder theory.

In their paper [133], Yoshihiro Shibata and Senjo Shimizu proved maximal $L_p$-$L_q$ regularity for the Stokes equations with a free boundary condition in a bounded domain. This was based on the $\mathcal{R}$-boundedness of the solution operators in the half-space model problem, which was then used to obtain the desired result with the help of the Weis operator-valued Fourier multiplier theorem [13]. Y. Shibata got the idea to use $\mathcal{R}$-boundedness during his stay at the University of Konstanz in September 2004, on an invitation of Reinhard Racke, by reading the booklet by Denk, Hieber and Prüss [1].

More recently, he extended this approach to obtain maximal $L_p$-$L_q$ regularity for the non-stationary Stokes equations with free boundary conditions in very general unbounded domains [103]. Here, he applies the Fourier multiplier theorem of L. Weis to the inverse Laplace transform of the $\mathcal{R}$-bounded solution operator of the generalized Stokes resolvent problem, which acts simultaneously on the right hand side and on the non-homogeneous boundary data. This idea goes back to the work of Reiko Sakamoto [10], who is an academic mother of Yoshihiro Shibata. She used the Plancherel theorem to prove the $L_2$ well-posedness of the initial-boundary value problem for hyperbolic equations satisfying a uniform Shapiro-Lopatinski condition. The method of [103] was further extended to obtain maximal $L_p$-$L_q$ regularity for the linearized Navier-Stokes equations describing the barotropic motion of compressible viscous fluid flows [32, 33], and also for compressible-incompressible two phase flows [53, 107]. Maximal $L_p$-$L_q$ regularity with different values of $p$ and $q$ is needed to prove global well-posedness in unbounded domains. In fact, since, unlike in the bounded domain case, only polynomial decay is expected, one has to be able to choose $p$ large to guarantee global integrability in time (cf. Saito and Shibata [67]). At present, he is interested to get global well-posedness results in unbounded domains by combining maximal $L_p$-$L_q$ regularity with $L_p$-$L_q$ decay estimates.

Yoshihiro Shibata has been a steady source of new ideas, and he has influenced many researchers. He supervised more than 10 students who obtained a Doctor Degree (Doctor of Natural Science) either from Tsukuba University or Waseda University.

Between 2009 and 2014 Y. Shibata coordinated, together with Matthias Hieber, the Japanese-German Graduate Externship Program for doctor course students between Waseda University and TU Darmstadt. It concerns the theory of mathematical fluid dynamics and is supported by JSPS and DFG. (Since 2014 this program is coordinated by Hideo Kozono and Matthias Hieber.) Also since 2009, he is the team leader of a project in the JST Mathematics Project organized by Yasumasa Nishiura. In 2014 he became the project leader of the Mathematics and Physics Unit of the Top Global University Project of Waseda University, supported by the Ministry of Education, Culture, Sports, Science, and Technology. Together with Hideo Kozono, Tohru Ozawa, Hiroaki Yoshimura, Shinichi Oishi and four professors from the physics department he recently started a new study group and a new education system in the doctor course of the Department of Sciences and Engineering of Waseda University. Its main subject is multi scale analysis, modeling and simulation with interdisciplinary and international background.

## References

1. R. Denk, M. Hieber, J. Pruss, $\mathcal{R}$-boundedness, Fourier multipliers and problems of elliptic and parabolic type. Mem. Am. Math. Soc. **166**(788) (2003)
2. R. Finn, Stationary solutions of the Navier-Stokes equations. Am. Math. Soc. Proc. Symp. Appl. Math. **19**, 121–153 (1965)
3. H. Fujita, T. Kato, On the Navier-Stokes initial value problem I. Arch. Ration. Mech. Anal. **16**, 269–315 (1964)
4. T. Kato, Strong $L^p$ solutions of the Navier-Stokes equation in $\mathbb{R}^m$ with applications to weak solutions. Math. Z. **187**, 471–480 (1984)
5. T. Kato, *Abstract Differential Equations and Nonlinear Mixed Problems*. Accademia Nazionale Dei Lincei Scuola Normale Superiore, Lezionei Fermiane, Pisa, 1985
6. T. Kato, H. Fujita, On the non-stationary Navier-Stokes system. Rend. Sem. Math. Univ. Padova **32**, 243–260 (1962)
7. H. Iwashita, $L_q$-$L_r$ estimates for solutions of the non-stationary Stokes equations in an exterior domain and the Navier-Stokes initial value problems in $L_q$ spaces. Math. Ann. **285**, 265–288 (1989)
8. P.D. Lax, R.S. Phillips, Scattering theory, in *Pure and Applied Mathematics*, vol. 26 (Academic, New York/London, 1967)
9. A. Matsumura, T. Nishida, Initial-boundary value problems for the equations of motion of compressible viscous and heat-conductive fluids. Commun. Math. Phys. **89**(4), 445–464 (1983)
10. R. Sakamoto, Mixed problems for hyperbolic equations I, II. J. Math Kyoto Univ. **10**, 349–373/403–417 (1970)
11. B. Vainberg, On the analytic properties of the resolvent for a certain class of operator pencils, Mat. Sb. (N.S.) **77** (1968); Math. USSR Sb. **6**, 241–273 (1968) [English translation]

12. B. Vainberg, On the short wave asymptotic behaviour of solutions of stationary problems and asymptotic behaviour as $t \to \infty$ of solutions of non-stationary problem. Ushpekhi Mat. Nauk. **30**, 3–55 (1975); Russ. Math. Surveys **30**, 1–58 (1975)
13. L. Weis, Operator-valued Fourier multiplier theorems and maximal $L_p$-regularity. Math. Ann. **319**, 735–758 (2001)

## List of Publications of Yoshihiro Shibata

14. T. Abe, Y. Shibata, On the Stokes and Navier-Stokes flows between parallel planes, in *Harmonic Analysis and Nonlinear Partial Differential Equations* (Japanese) (1235) (Sūrikaisekikenkyūsho Kōkyūroku, Kyoto, 2001), pp. 160–191
15. T. Abe, Y. Shibata, On a generalized resolvent estimate of the Stokes equation on an infinite layer, Part 2 $\lambda = 0$ case. J. Math. Fluid Mech. **5**(3), 245–274 (2003)
16. T. Abe, Y. Shibata, On a generalized resolvent estimate of the Stokes equation on an infinite layer, Part 1 $|\lambda| > 0$ case. J. Math. Soc. Jpn. **55**(2), 469–497 (2003)
17. T. Akiyama, H. Kasai, Y. Shibata, M. Tsutsumi, On a resolvent estimate of a system of Laplace operators with perfect wall condition. Funkcial. Ekvaj. **47**(3), 361–394 (2004)
18. T. Akiyama, Y. Shibata, On an $L_p$ approach to the stationary and non-stationary problems to the Ginzburg-Landau-Maxwell equations. J. Differ. Equ. **243**(1), 1–23 (2007)
19. P. D'Ancona, Y. Shibata, On global solvability of nonlinear viscoelastic equations in the analytic category. Math. Meth. Appl. Sci. **17**(6), 477–486 (1994)
20. W. Dan, Y. Shibata, On the $L_q$–$L_r$ estimate of the Stokes semigroup in a two dimensional exterior domain. J. Math. Soc. Jpn. **51**(1), 181–207 (1999)
21. W. Dan, T. Kobayashi, Y. Shibata, On the local energy decay approach to some fluid flow in an exterior domain, in *Recent Topics on Mathematical Theory of Viscous Incompressible Fluid* (Tsukuba, 1996), Lecture Notes Numerical Application Analysis, vol. 16 (Kinokuniya, Tokyo, 1998), pp. 1–51
22. W. Dan, Y Shibata, On a local energy decay of solutions of a dissipative wave equation. Funkcial. Ekvaj. **38**(3), 545–568 (1995)
23. W. Dan, Y. Shibata, On the $L_p$-$L_q$ estimates of the Stokes semigroup in a two-dimensional exterior domain, in *Nonlinear Evolution Equations and Their Applications* (Japanese) (Kyoto, 1996) (1009) (Sūrikaisekikenkyūsho Kōkyūroku, 1997), pp. 79–99
24. W. Dan, Y. Shibata, Remark on the $L_q$–$L_\infty$ estimate of the Stokes semigroup in a two dimensional exterior domain. Pacific J. Math. **189**(2), 223–239 (1999)
25. R. Denk, R. Racke, Y. Shibata, $L_p$ theory for the linear thermoelastic plate equations in bounded and exterior domains. Adv. Differ. Equ. **14**(7–8), 685–715 (2009)
26. R. Denk, R. Racke, Y. Shibata, Local energy decay estimate of solutions to the thermoelastic plate equations in two- and three- dimensional exterior domains. Z. Anal. Anwend. **29**(1), 21–62 (2010)
27. Y. Enomoto, Y. Shibata, Local energy decay of solutions to the Oseen equation in the exterior domains. Indiana Univ. Math. J. **53**(5), 1291–1330 (2004)
28. Y. Enomoto, Y. Shibata, On the rate of decay of the Oseen semigroup in exterior domains and its application to Navier-Stokes equations. J. Math. Fluid Mech. **7**(3), 339–367 (2005)
29. Y. Enomoto, Y. Shibata, On a Stability theorem of the navier-stokes equation in an exterior domain. in *Hyperbolic Problems, Theory, Numerics and Applications I* (Yokohama Publisher, Yokohama, 2006), pp. 383–389
30. Y. Enomoto, Y. Shibata, On some decay properties of Stokes semigroup of compressible viscous fluid flow in a 2-dimensional exterior domain. J. Differ. Equ. **252**(12), 6214–6249 (2012)

31. Y. Enomoto, Y. Shibata, About compressible viscous fluid flow in a 2-Dimensional exterior domain, in *Spectral Theory, Mathematical System Theory, Evolution Equations, Differential and Difference Equations, Operator Theory: Advances and Applications*, vol. 221 (Birkhäuser/Springer Basel AG, Basel, 2012), pp. 305–321
32. Y. Enomoto, Y. Shibata, On the $\mathcal{R}$-sectoriality and the initial boundary value problem for the viscous compressible fluid flow. Funkcial. Ekvac. **56**(3), 441–505 (2013)
33. Y. Enomoto, L. von Below, Y. Shibata, On some free boundary problem for a compressible barotropic viscous fluid flow. Ann. Univ. Ferrara Sez. VII Sci. Mat., **60**(1), 55–89 (2014)
34. G.P. Galdi, J.G. Heywood, Y. Shibata, On the global existence and convergence to steady state of Navier–Stokes flow past an obstacle that is started from rest. Arch. Ration. Mech. Anal. **138**(4), 307–318 (1997)
35. D. Gotz, Y. Shibata, On the $\mathcal{R}$-boundedness of the solution operators in the study of the compressible viscous fluid flow with free boundary conditions. Asymptot. Anal. **90**(3–4), 207–236 (2014)
36. T. Hishida, Y. Shibata, Globally in time existence theorem for the Navier-Stokes flow in the exterior of a rotating obstacle. WSWAS Trans. Math. **5**(3), 303–307 (2006)
37. T. Hishida, Y. Shibata, $L_p$-$L_q$ estimate of the Stokes operator and Navier-Stokes flows in the exterior of a rotating obstacle. Arch. Ration. Mech. Anal. **193**(2), 339–421 (2009)
38. M. Hieber, Y. Shibata, The Fujita-Kato approach to the Navier-Stokes equations in the rotational framework. Math. Z. **265**(2), 481–491 (2010)
39. H. Iwashita, Y. Shibata, On the analyticity of spectral functions for some exterior boundary value problems. Glasnik Math. Ser. III **23**(43, 2), 291–313 (1988)
40. S. Kawashima, Y. Shibata, Global existence and exponential stability of small solutions to nonlinear viscoelasticity. Commun. Math. Phys. **148**(1), 189–208 (1992)
41. S. Kawashima, Y. Shibata, On the Neumann problem of one–dimensional nonlinear thermoelasticity with time–independent external force. Czechoslovak Math. J. **45**(120, 1), 39–67 (1995)
42. M. Kikuchi, Y. Shibata, On the mixed problem for some quasi–linear hyperbolic system with fully nonlinear boundary condition. J. Differ. Equ. **80**(1), 154–197 (1989)
43. T. Kobayashi, H. Pecher, Y. Shibata, On a global in time existence theorem of smooth solutions to nonlinear wave equation with viscosity. Math. Ann. **296**(2), 215–234 (1993)
44. T. Kobayashi, Y. Shibata, Exterior problems for the Navier-Stokes equations, in *Nonlinear Evolution Equations and their Applications*(Japanese) (913) (Sūrikaisekikenkyūsho Kōkyūroku, Kyoto, 1994/1995) pp. 185–190
45. T. Kobayashi, Y. Shibata, On the Oseen equation in exterior domains. Math. Ann. **310**(1), 1–45 (1998)
46. T. Kobayashi, Y. Shibata, Decay estimates of solutions for the equations of motion of compressible viscous and heat-conductive gases in an exterior domain in $\mathbf{R}^3$. Commun. Math. Phys. **200**(3), 621–659 (1999)
47. T. Kobayashi, Y. Shibata, Remark on the rate of decay of solutions to linearized compressible Navier-Stokes equations. Pacific J. Math. **207**(1), 199–234 (2002)
48. H. Kozono, Y. Shibata, Recent topics on mathematical theory of viscous incompressible fluid, in *Lecture Notes in Numerical and Applied Analysis*, vol. 16 (Kinokuniya, Tokyo, 1998)
49. T. Kubo, Y. Shibata, On some properties of solutions to the Stokes equation in the half-space and perturbed half-space, in *Dispersive Nonlinear Problems in Mathematical Physics*, Quad. Mat., Dept. Math., vol. 15 (Seconda Univ. Napoli, Caserta, 2004), pp. 149–220
50. T. Kubo, Y. Shibata, On the Stokes and Navier-Stokes equation in a perturbed half-space. Adv. Differ. Equ. **10**(6), 695–720 (2005)
51. T. Kubo, Y. Shibata, On the Stokes and Navier-Stokes flows in a perturbed half space, in *Regularity and Other Aspects of the Navier-Stokes Equations*. Banach Center Publications vol. 70 (Polish Acad. Sci., Warsaw, 2005), pp. 157–167
52. T. Kubo, Y. Shibata, $L^p$-$L^q$ estimate of the stokes semigroup and its application to navier-stokes equation in a perturbed half-space, in *Hyperbolic Problems, Theory, Numerics and Applications II* (Yokohama Publisher, Yokohama, 2006), pp. 125–132

53. T. Kubo, Y. Shibata, K. Soga, On the $\mathcal{R}$-boundedness for the two phase problem: compressible-incompressible model problem. Bound. Value Probl. **141**, 33 pp. (2014)
54. J. Prüss, Y. Shibata, S. Shimizu, G. Simonett, On well-posedness of incompressible two-phase flows with phase transitions: the case of equal densities. Evol. Equ. Control Theory **1**(1), 171–194 (2012)
55. A. Milani, Y. Shibata, On compatible regularizing data for second order hyperbolic initial–boundary value problems. Osaka J. Math **32**(2), 347–362 (1995)
56. A. Milani, Y. Shibata, On the strong well–posedness of quasilinear hyperbolic initial–boundary value problems. Funkcial. Ekvaj. **38**(3), 491–503 (1995)
57. J. Muñoz Rivera, Y. Shibata, A linear thermoelastic plate equation with Dirichlet boundary condition. Math. Methods Appl. Sci. **20**(11), 915–932 (1997)
58. M. Murata, Y. Shibata, Lower bounds at infinity of solutions of partial differential equations in the exterior of a proper cone. Israel J. Math. **31**(2), 193–203 (1978)
59. Y. Naito, Y. Shibata, On the $L_p$ analytic semigroup associated with the linear thermoelastic plate equations in the half-space. J. Math. Soc. Jpn. **61**(4), 971–1011 (2009)
60. Y. Naito, R. Racke, Y. Shibata, Low frequency expansion in thermoelasticity with second sound in three dimensions. J. Math. Soc. Jpn. **62**(4), 1289–1316 (2010)
61. G. Nakamura, Y. Shibata, On a local existence theorem for quasi–linear hyperbolic mixed problems with Neumann type boundary conditions, Proc. Japan Acad. Ser. A Math. Sci. **62**(4), 117–120 (1986)
62. G. Nakamura, Y. Shibata, K. Tanuma, Whispering gallery waves in a neighborhood of a higher order zero of the curvature of the boundary. Publ. RIMS Kyoto Univ. **25**(4), 605–629 (1989)
63. G. Nakamura, Y. Shibata, On a local existence theorem of Neumann problem for some quasi–linear hyperbolic systems of 2nd order. Math. Z. **202**(1), 1–64 (1989)
64. M. Okamura, Y. Shibata, N. Yamaguchi, A Stokes approximation of two dimensional exterior Oseen flow near the boundary, in *Asymptotic Analysis and Singularities–Hyperbolic and Dispersive PDEs and Fluid Mechanics*. Advanced Studies in Pure Mathematics, vol. 47 (Mathematical Society, Tokyo, 2007), pp. 273–289
65. R. Racke, Y. Shibata, Global smooth solution and asymptotic stability in one-dimensional nonlinear thermoelasticity. Arch. Ration. Mech. Anal. **116**(1), 1–34 (1991)
66. R. Racke, Y. Shibata, S. Mu Zheng, Global solvability and exponential stability in one–dimensional nonlinear thermoelasticity. Quart. Appl. Math. **51**(4), 751–763 (1993)
67. H. Saito, Y. Shibata, On the Stokes equations with surface tension and gravity in $\mathbb{R}^N_+$. J. Math. Soc. Jpn. (to appear)
68. Y. Shibata, A characterization of the hyperbolic mixed problems in a quarter space for differential operators with constant coefficients. Publ. RIMS Kyoto Univ. **15**, 357–399 (1979)
69. Y. Shibata, Liouville type theorem for a system $\{P(D), B_j(D), j = 1, \ldots, p\}$ of differential operators with constant coefficients in a half–space. Publ. RIMS Kyoto Univ. **16**, 61–104 (1980)
70. Y. Shibata, $\mathcal{E}$–well posedness of mixed initial–boundary value problems with constant coefficients in a quarter space. J. D'Analyse Math. **37**, 32–45 (1980)
71. Y. Shibata, $\mathcal{E}$–well posedness of mixed initial–boundary value problem with constant coefficients in a quarter– space II. Proc. Jpn. Acad. Ser. A. **56**(7), 318–320 (1980)
72. Y. Shibata, Lower bounds at infinity of solutions of differential equations with constant coefficients in unbounded domains, in *Singularities in Boundary Value Problems*, ed. by H.G. Garnir. Proceedings of NATO Advanced Institute, Maratea, 1980. NATO Advanced Study Institute Series. Series C: Mathematical and Physical Sciences, vol. 65 (Reidel, Dordrecht, 1981), pp. 213–234
73. Y. Shibata, Lower bounds of solutions of general boundary value problems for differential operators with constant coefficients in a half–space, Japan. J. Math. (N.S.) **8**(2), 343–382 (1982)

74. Y. Shibata, On the global existence of classical solutions of mixed problem for some second order non–linear hyperbolic operators with dissipative term in the interior domain. Funkcial. Ekvac. **25**(3), 303–345 (1982)
75. Y. Shibata, On the global existence of classical solutions of second order fully nonlinear hyperbolic equations with first order dissipation in the exterior domain. Tsukuba J. Math. **7**(1), 1–68 (1983)
76. Y. Shibata, On a local existence theorem for quasilinear hyperbolic mixed problems with Neumann type boundary conditions, in *Hyperbolic Equations* (Padua, 1985) Pitman Research Notes in Mathematics Series, vol. 158 (Longman Scientific & Technical, Harlow, 1987), pp. 282–286
77. Y. Shibata, On a local existence theorem of Neumann problem for some quasi–linear hyperbolic equations, in *Calcul d'operateurs et fronts d'ondes*, ed. by J. Vaillant, Travaux en Cours, vol. 29 (Hermann, Paris, 1988), pp. 133–167
78. Y. Shibata, On the Neumann problem for some linear hyperbolic systems of second order. Tsukuba J. Math. **12**(1), 149–209 (1988)
79. Y. Shibata, On the Neumann problem for some linear hyperbolic systems of 2nd order with coefficients in Sobolev spaces. Tsukuba J. Math. **13**(2), 283–352 (1989)
80. Y. Shibata, On one-dimensional nonlinear thermoelasticity, in *Nonlinear Hyperbolic Equations and Field Theory* (Lake Como, 1990), Pitmann Res. Notes Math. Ser., vol. 253 (Longman Scientific & Technical, Harlow, 1992), pp. 178–184
81. Y. Shibata, Neumann problem for one-dimensional nonlinear thermoelasticity, in *Partial Differential Equations*, Warsaw, 1990, Parts 1, 2, vol. 27 (Banach Center Publications/Polish Academy of Sciences, Warsaw, 1992), pp. 457–480
82. Y. Shibata, Global in time solvability of the initial boundary value problem for some nonlinear dissipative evolution equations. Comment. Math. Univ. Carol. **34**(2), 295–312 (1993)
83. Y. Shibata, Neumann problem of one-dimensional nonlinear thermoelastic equations, in *Mathematical Analysis of Phenomena in Fluid and Plasma Dynamics* Kyoto, 1992. Surikaisekikenkyusho Kokyuroku (Japanese), vol. 824 (1993), pp. 283–296
84. Y. Shibata, On the exponential decay of the energy of a linear thermoelastic plate. Math. Appl. Comput. **13**(2), 81–102 (1994)
85. Y. Shibata, Global in time existence of small solutions of nonlinear thermoviscoelastic equations. Math. Methods Appl. Sci. **18**(11), 871–895 (1995)
86. Y. Shibata, On a linear thermoelastic plate equation, in *Nonlinear Evolution Equations and Their Applications* (898) (Japanese) (Sūrikaisekikenkyūsho Kōkyūroku, Kyoto ,1993/1995), pp. 149–154
87. Y. Shibata, An initial-boundary value problem for some hyperbolic-parabolic coupled system, in *Nonlinear Waves* (Sapporo. 1995), GAKUTO International Series Mathematical Sciences Application, vol. 10 (Gakkōtosho, Tokyo, 1997), pp. 447–450
88. Y. Shibata, An exterior initial-boundary value problem for the Navier-Stokes equation, in *Nonlinear waves (Sapporo, 1995)*, GAKUTO International Series Mathematical Sciences Application, vol. 10 (Gakkōtosho, Tokyo, 1997), pp. 431–446
89. Y. Shibata, On the decay estimate of the Stokes semigroup in a two dimensional exterior domain. *Navier-Stokes Equations and Related Nonlinear Problems*(Palanga, 1997) (VSP, Utrecht, 1998), pp. 315–330
90. Y. Shibata, *On a Decay Rate of Solutions to One-Dimensional Thermoelastic Equations on a Half Line; Linear Part*, ed. by S. Kawashima, T. Yanagisawa. Advances in Nonlinear Partial Differential Equations and Stochastics (World Scientific, Singapore, 1998), pp. 198–291
91. Y. Shibata, On an exterior initial-boundary value problem for Navier-Stokes equations. Quart. Appl. Math. **57**(1), 117–155 (1999)
92. Y. Shibata, Global solutions of nonlinear evolution equations and their stability. Sūgaku, **51**(1), 1–17 (1999) (in Japanese).
93. Y. Shibata, On the rate of decay of solutions to linear viscoelastic equation. Math. Methods Appl. Sci. **23**(3), 203–226 (2000)

94. Y. Shibata, On a stability theorem of the Navier-Stokes equation in a three dimensional exterior domain, in *Tosio Kato's Method and Principle for Evolution Equations in Mathematical Physics*, Sapporo, 2001 (1234) (Sūrikaisekikenkyūsho Kōkyūroku, 2001) pp. 146–172
95. Y. Shibata, On some stability theorems about viscous fluid flow, Quaderni del Seminario Matematico di Brescia (2003)
96. Y. Shibata, Time-global solutions of nonlinear evolution equations and their stability [translation of Sūgaku **51**(1), 1–17 (1999)], in *Selected Papers on Analysis and Differential Equations*. American Mathematical Society Translations: Series 2, vol. 211 (American Mathematical Society, Providence, RI, 2003), pp. 87–105
97. Y. Shibata, On some stability theorem of the steady flow of compressible viscous fluid with respect to the initial disturbance, in *Hyperbolic Problems and Related Topics, Graduate Series Analysis* (International Press, Somerville, MA, 2003), pp. 341–357
98. Y. Shibata, On the Oseen semigroup with rotating effect. in *Functional Analysis and Evolution Equations* (Birkhäuser, Basel, 2008), pp. 595–611
99. Y. Shibata, A stability theorem of the Navier-Stokes flow past a rotating body. in *Parabolic and Navier-Stokes Equations. Part 2*, vol. 81 (Banach Center Publication, Polish Academy of Science Institute of Mathematics, Warsaw, 2008), pp. 441–455
100. Y. Shibata, On a $C^0$ semigroup associated with a modified Oseen equation with rotating effect, in *Advances in Mathematical Fluid Mechanics*, (Springer, Berlin, 2010), pp. 513–551
101. Y. Shibata, Generalized resolvent estimates of the Stokes equations with first order boundary condition in a general domain. J. Math. Fluid Mech. **15**(1), 1–40 (2013)
102. Y. Shibata, On the $\mathcal{R}$-boundedness of solution operators for the weak Dirichlet-Neumann problem, in *RIMS Kōkyūroku 1875*, Mathematical Analysis of Incompressible Flow (4–6 Feb 2013), ed. by T. Hishida (RIMS, Kyoto University, Kyoto), pp. 1–18
103. Y. Shibata, On the $\mathcal{R}$-boundedness of solution operators for the Stokes equations with free boundary condition. Differ. Integr. Equ. **27**(3–4), 313–368 (2014)
104. Y. Shibata, On some free boundary problem of the Navier-Stokes equations in the maximal $L_p$-$L_q$ regularity class. J. Differ. Equ. **258**, 4127–4155 (2015)
105. Y. Shibata, On the global well-posedness of some free boundary problem for a compressible barotropic viscous fluid flow, in *The Contemporary Mathematics Series of the American Mathematical Society: Recent Advances in PDEs and Applications*, Levico Terme, 17–21 February 2014 [to celebrate the 70th Birthday of Professor Hugo Beirao da Veiga]
106. Y. Shibata, Local well-posedness of compressible-incompressible two-phase flows with phase transitions, To appear in the Proceedings of Levico Conf. on fluid Dyn and Electromagnetism. arXiv:submit/1156284 [math. AP] 10 Jan 2015
107. Y. Shibata, On the $\mathcal{R}$-boundedness for the two phase problem with phase transition: compressible-incompressible model problem. Funk. Ekvaj. (to appear)
108. Y. Shibata, W. Dan, On a local energy decay of solutions of a dissipative wave equation, in *Mathematical Analysis of Phenomena in Fluid and Plasma Dynamics* (Japanese) (862) (Sūrikaisekikenkyūsho Kōkyūroku, Kyoto, 1993/1994) pp. 181–190
109. Y. Shibata, S. Mu Zheng, On some nonlinear hyperbolic system with damping boundary condition. Nonlinear Anal. TMA **17**(3), 233–266 (1991)
110. Y. Shibata, Y. Tsutsumi, Global existence theorem for nonlinear wave equations in exterior domain, in *Recent Topics in Nonlinear PDE* (Hiroshima, 1983), North-Holland Mathematics Studies vol. 98 (North-Holland, Amsterdam, 1984) pp. 155–196
111. Y. Shibata, Y. Tsutsumi, Global existence theorem for nonlinear wave equation in exterior domain. Proc. Japan Acad. A Mat. Sci. **60**(1), 14–17 (1984)
112. Y. Shibata, Y. Tsutsumi, Local existence of $C^\infty$–solution for the initial–boundary value problem of fully nonlinear wave equation. Proc. Japan Acad. Ser. A Math. Sci. **60**(5), 149–152 (1984)
113. Y. Shibata, Y. Tsutsumi, On a global existence theorem of Neumann problem for some quasilinear hyperbolic equations, in *Recent Topics in Nonlinear PDE, II* (Sendai, 1984), North-Holland Mathematics Studies, vol. 128 (North-Holland, Amsterdam, 1985), pp. 175–228

114. Y. Shibata, Y. Tsutsumi, On a global existence theorem of small amplitude solutions for nonlinear wave equations in an exterior domain. Math. Z. **191**(2), 165–199 (1986)
115. Y. Shibata, Y. Tsutsumi, Local existence of solutions for the initial boundary value problem of fully nonlinear wave equation. Nonlinear Anal. TMA **11**(3), 335–365 (1987)
116. Y. Shibata, S.Shimizu, A decay property of the Fourier transform and its application to the Stokes problem. J. Math. Fluid Mech. **3**(3), 213–230 (2001)
117. Y. Shibata, S. Shimizu, On a resolvent estimate of the interface problem for the Stokes system in a bounded domain, *Harmonic Analysis and Nonlinear Partial Differential Equations* (Japanese) (1235) (Sūrikaisekikenkyūsho Kōkyūroku, Kyoto, 2001), pp. 132–159
118. Y. Shibata, S. Shimizu, On the $L_p$ and Schauder estimates of solutions to elastostatic interface problems, in *Proceedings of the Fourth International Conference on Functional Analysis and Approximation Theory*, Potenza, 2000, vol. II; Rend. Circ. Mat. Palermo (2) Suppl. **68**(Part II), 821–835 (2002)
119. Y. Shibata, S. Shimizu, On a resolvent estimate of the interface problem for the Stokes system in a bounded domain. J. Differ. Equ. **191**(2), 408–444 (2003)
120. P. Secchi, Y. Shibata, On the decay of solutions to the 2D Neumann exterior problem for the wave equation. J. Differ. Equ. **194**(1), 221–236 (2003)
121. Y. Shibata, S. Shimizu, On a resolvent estimate for the Stokes system with Neumann boundary condition. Differ. Integr. Equ. **16**(4), 385–426 (2003)
122. Y. Shibata, S. Shimizu, Applications of the Fourier transform to some resolvent estimates for the Stokes system, in *Progress in Analysis*, vols. I, II (Berlin, 2001) (World Science Publisher, River Edge, 2003), pp. 125–134
123. Y. Shibata, S. Shimizu, $L_p$–$L_q$ maximal regularity and viscous incompressible flows with free surface. Proc. Japan Acad. Ser A, Math. Sci. **81**(9), 151–155 (2005)
124. Y. Shibata, S. Shimizu, On the stokes equation with neumann boundary condition, in *Regularity and other Aspects of the Navier-Stokes Equations*. Banach Center Publication, vol. 70 (Polish Academy of Sciences, Warsaw, 2005), pp. 239–250
125. Y. Shibata, S. Shimizu, On a Free Boundary Problem for the Navier-Stokes Equations. Differ. Integr. Equ. **20**(3), 241–276 (2007)
126. Y Shibata, S. Shimizu, $L_p$-$L_q$ maximal regularity of the Neumann problem for the Stokes equations in a bounded domain, in *Asymptotic Analysis and Singularities–Hyperbolic and Dispersive PDEs and Fluid Mechanics*. Advanced Studies in Pure Mathematics, vol. 47 (Mathematical Society of Japan, Tokyo, 2007), pp. 349–362.
127. Y. Shibata, S. Shimizu, Decay properties of the Stokes semigroup in exterior domains with Neumann boundary condition. J. Math. Soc. Jpn. **59**(1), 1–34 (2007)
128. Y. Shibata, S. Shimizu, Report on a local in time solvability of free surface problems for the Navier-Stokes equations with surface tension. Appl. Anal. **90**(1), 201–214 (2011)
129. Y. Shibata, S. Shimizu, $L_p$-$L_q$ regularity for the two-phase Stokes equations; model problems. J. Differ. Equ. **251**(2), 373–419 (2011)
130. Y. Shibata, S. Shimizu, On the maximal $L_p$-$L_q$ regularity of the Stokes problem with first order boundary condition; model problems, J. Math. Soc. Jpn. **64**(2), 561–626 (2012)
131. Y. Shibata, H. Soga, Scattering theory for the elastic wave equation. Publ. Res. Inst. Math. Sci. **25**(6), 861–887 (1989)
132. Y. Shibata, K. Tanaka, On a resolvent problem for the linearized system from the dynamical system describing the compressible viscous fluid motion. Math. Methods Appl. Sci. **27**(13), 1579–1606 (2004)
133. Y. Shibata, S. Shimizu, On the $L_p$-$L_q$ maximal regularity of the Neumann problem for the Stokes equations in a bounded domain. J. Reine Angew. Math. **615**, 157–209 (2008)
134. Y. Shibata, M. Yamazaki, On some stability theorem of the Navier-Stokes equation in the three dimensional exterior domain, in *Mathematical Analysis of Liquids and Gases* (Japanese) (Kyoto, 1999) (1146) (Sūrikaisekikenkyūsho Kōkyūroku, 2000), pp. 73–91
135. Y. Shibata, N. Yamaguchi, Global existence of strong solutions to the micropolar fluid system, in *Hyperbolic Problems, Theory, Numerics and Applications II* (Yokohama Publisher, Yokohama, 2006), pp. 305–312

136. K. Tanaka, Y. Shibata, On the steady flow of compressible viscous fluid and its stability with respect to initial disturbance. J. Math. Soc. Jpn. **55**(3), 797–826 (2003)
137. Y. Shibata, R. Shimada, On a generalized resolvent estimate for the Stokes system with Robin boundary condition. J. Math. Soc. Jpn. **59**(2), 469–519 (2007)
138. Y. Shibata, R. Shimada, On the Stokes equation with Robin boundary condition, in *Asymptotic Analysis and Singularities–Hyperbolic and Dispersive PDEs and Fluid Mechanics*. Advanced Studies in Pure Mathematics, vol. 47 (Mathematical Society of Japan, Tokyo, 2007) pp. 341–348
139. M. Yamazaki, Y. Shibata, Uniform estimates in the velocity at infinity for stationary solutions to the Navier-Stokes exterior problem. Japanese J. Math. (N.S.) **31**(2), 225–279 (2005)
140. T. Hishida, Y. Shibata, Decay estimates of the Stokes flow around a rotating obstacle, in *Kyoto Conference on the Navier-Stokes Equations and Their Applications*. RIMS Kokyūroku Bessatsu, vol. B1 pp. 167–186 (Res. Inst. Math. Sci. (RIMS), Kyoto, 2007)
141. Y. Shibata, S. Shimizu, Free boundary problems for a viscous incompressible fluid, in *Kyoto Conference on the Navier-Stokes Equations and Their Applications*. RIMS Kokyūroku Bessatsu, vol. B1 (Res. Inst. Math. Sci. (RIMS), Kyoto, 2007), pp. 343–358
142. Y. Shibata, S. Shimizu, On a resolvent estimate of the Stokes system in a half space arising from a free boundary problem for the Navier-Stokes equations. Mathematische Nachrichten **282**(3), 482–499 (2009)
143. Y. Shibata, K. Tanaka, Rate of convergence of non-stationary flow to the steady flow of compressible viscous fluid. Comput. Math. Appl. **53**(3–4), 605–623 (2007)

# Existence of Weak Solutions for a Diffuse Interface Model of Power-Law Type Two-Phase Flows

**Helmut Abels, Lars Diening, and Yutaka Terasawa**

*Dedicated to Professor Yoshihiro Shibata's 60th birthday*

**Abstract** We first review results about existence of generalized or weak solutions for Newtonian and power-law type two-phase flows. Then we state a recent result by the authors about existence of weak solutions for diffuse interface model of power-law type two-phase flows and give a sketch of its proof. The latter part is a summary of Abels et al. (Nonlinear Anal Real World Appl 15:149–157, 2014).

**Keywords** Cahn-Hilliard equation • Diffuse interface model • Free boundary value problems • Lipschitz truncation • Power-law type fluids • Sharp interface model • Two-phase flow

## 1 Introduction

This is a review paper focusing on existence of generalized or weak solutions for Newtonian and power-law type two-phase flows. We also state a recent result by the authors in [5] and give a sketch of its proof.

We consider the flow of two macroscopically immiscible, incompressible Newtonian and power-law type fluids. There are models of two types concerning two-phase flows of Newtonian and power-law type fluids. One type are sharp interface models and the other are diffuse interface models. For sharp interface models, a partial mixing of the fluids is not taken into account. One *classical* sharp interface model is "Two-phase flows with surface tension (without phase transition)". For two-phase

H. Abels
Fakultät für Mathematik, Universität Regensburg, 93040 Regensburg, Germany
e-mail: helmut.abels@mathematik.uni-regensburg.de

L. Diening
Mathematisches Institut, LMU München, 80333 München, Germany
e-mail: diening@math.lmu.de

Y. Terasawa (✉)
Graduate School of Mathematics, Nagoya University, Nagoya 464-8602, Japan
e-mail: yutaka@math.nagoya-u.ac.jp



H. Amann et al. (eds.), *Recent Developments of Mathematical Fluid Mechanics*, Advances in Mathematical Fluid Mechanics, DOI 10.1007/978-3-0348-0939-9_2

flows of Newtonian fluid with surface tension, there are several studies concerning the existence of classical solution of the equations, see, e.g. [14–16, 32]. They treat local in time existence of solutions for arbitrary initial data. Especially in [32], space-time analyticity of solutions and interfaces are shown, together with a complete maximal regularity result of its linearized problem. For global in time behavior of the solutions, see [25]. For two-phase flows of power-law type fluid with surface tension, the global in time existence of a generalized solution, or a measure-valued solution for arbitrary initial data was proven by Abels [1] and Plotnikov [31]. Concerning results about "Two-phase flow of Newtonian fluids with phase transition", see, e.g. [33, 34].

In diffuse interface models, a partial mixing of the fluids is taken into account in contrast to sharp interface models. This has the advantage that flows beyond the occurrence of topological singularities e.g. due to droplet collision or pinch-off can be described. One well-known model in the case that both densities are the same is the so-called "model H". It leads to the following system of Navier-Stokes/Cahn-Hilliard type:

$$\rho\partial_t \mathbf{v} + \rho\mathbf{v}\cdot\nabla\mathbf{v} - \operatorname{div}\mathbf{S}(c,\mathbf{D}\mathbf{v}) + \nabla p = -\kappa\operatorname{div}(\nabla c\otimes\nabla c), \tag{1}$$

$$\operatorname{div}\mathbf{v} = 0, \tag{2}$$

$$\partial_t c + \mathbf{v}\cdot\nabla c = m\Delta\mu, \tag{3}$$

$$\mu = \kappa^{-1}\phi(c) - \kappa\Delta c \tag{4}$$

in $Q_T = \Omega\times(0,T)$, where $\Omega\subseteq\mathbb{R}^d$, $d\geq 2$, is a bounded domain and $T\in(0,\infty)$. Here $\mathbf{v}$ is the mean velocity, $\mathbf{D}\mathbf{v} = \frac{1}{2}(\nabla\mathbf{v}+\nabla\mathbf{v}^T)$, $p$ is the pressure, $c$ is an order parameter related to the concentration of the fluids e.g. the concentration difference or the concentration of one component, and $\rho$ is the density of the fluids, which is assumed to be constant. Moreover, $\mathbf{S}(c,\mathbf{D}\mathbf{v})$ is the viscous part of the stress tensor of the mixture to be specified below, $\kappa>0$ is a (small) parameter, which is related to the "thickness" of the interfacial region, $\Phi\colon\mathbb{R}\to\mathbb{R}$ is a homogeneous free energy density and $\phi=\Phi'$ and $\mu$ is the chemical potential. Capillary forces due to surface tension are modeled by an extra contribution $\kappa\nabla c\otimes\nabla c := \kappa\nabla c(\nabla c)^T$ in the stress tensor leading to the term on the right-hand side of (1). Moreover, we note that in the modeling, diffusion of the fluid components is taken into account. Therefore $m\Delta\mu$ is appearing in (3), where $m>0$ is a constant mobility coefficient.

We close the system by imposing the following boundary and initial conditions

$$\mathbf{v}|_{\partial\Omega} = 0 \qquad \text{on } \partial\Omega\times(0,T), \tag{5}$$

$$\mathbf{n}\cdot\nabla c|_{\partial\Omega} = \mathbf{n}\cdot\nabla\mu|_{\partial\Omega} = 0 \qquad \text{on } \partial\Omega\times(0,\infty), \tag{6}$$

$$(\mathbf{v},c)|_{t=0} = (\mathbf{v}_0,c_0) \qquad \text{in } \Omega. \tag{7}$$

Here $\mathbf{n}$ denotes the exterior normal at $\partial\Omega$. The boundary conditions are the most common in the mathematics literature.

In the case of Newtonian fluids, i.e., $\mathbf{S}(c, \mathbf{Dv}) = \nu(c)\mathbf{Dv}$ for some positive viscosity coefficient $\nu(c)$, the model was first discussed by Hohenberg and Halperin [22]. Later it was derived in the framework of rational continuum mechanics by Gurtin et al. [21]. The latter derivation can be easily modified to include a suitable non-Newtonian behavior of the fluids. If e.g. $\mathbf{S}(c, \mathbf{Dv})$ is chosen such that $\mathbf{S}(c, \mathbf{Dv}) : \mathbf{Dv} \geq 0$, the local dissipation inequality, which yields thermodynamical consistency, remains valid. For results on existence of weak and strong solutions in the case of Newtonian fluids we refer to Starovoitov [36], Boyer [7], and Abels [2]. First analytic results for the system (1)–(4) for Non-Newtonian fluids of power-law type were obtained by Kim et al. [23]. The authors proved existence of weak solutions if $q \geq \frac{3d+2}{d+2}$, $d = 2, 3$, where $q$ is the power describing the growth of the stress tensor with respect to $\mathbf{Dv}$. For this range of $q$ monotone operator techniques can be applied. Moreover, in the case $d = 3$ and $2 \leq q < \frac{11}{5}$ the authors prove existence of measure-valued solutions. Grasselli and Pražák [20] discussed the longtime behavior of solutions of (1)–(4) in the case $q \geq \frac{3d+2}{d-2}$, $d = 2, 3$ assuming periodic boundary conditions and a regular free energy density. For the same range of $q$ results on existence of weak solutions with a singular free energy density $\Phi$ and the longtime behavior were obtained by Bosia [6] in the case of a bounded domain in $\mathbb{R}^3$.

Let us review results on the Cahn–Hilliard equation. The Cahn-Hilliard equation is the following equation:

$$\partial_t c = m\Delta\mu, \tag{8}$$

$$\mu = \kappa^{-1}\phi(c) - \kappa\Delta c \tag{9}$$

in $Q_T$ with the boundary condition

$$\mathbf{n} \cdot \nabla c|_{\partial\Omega} = \mathbf{n} \cdot \nabla\mu|_{\partial\Omega} = 0 \qquad \text{on } \partial\Omega \times (0, \infty),$$

where $m, \kappa > 0$ and $\phi = \Phi'$. For $\Phi(s) = (s^2 - 1)^2$, Elliott and Zheng [19] proved the existence of global solutions to (8)–(9) in an $L^2$-setting. Nicolaenko et al. [30] proved the existence of an attractor of (8)–(9) in an $L^2$- setting with the same $\Phi$. Results on convergence of solutions to steady states using the Łojasiewicz-Simon inequality with the same $\Phi$ can be found in Rybka and Hoffmann [35].

For the singular potential satisfying *Assumption* 2.1 stated in Sect. 2, Elliot and Luckhaus [18] first proved the existence and uniqueness of solutions in the case of multi-component mixture. Debussche and Dettori [13] gave another proof in the case of two-component mixture and proved the existence of an attractor and estimated its dimension. Abels and Wilke [4] proved the existence and uniqueness of a solution by solving an abstract Cauchy problem for a suitable Lipschitz perturbation of a suitable monotone operator.

Concerning the relation between sharp interface models and diffuse interface models, see, e.g. [3, 12].

Next we review existence results of weak solutions for single power-law type fluid equations. Ladyzhenskaya [26] is the first to have investigated the Non-Newtonian fluid equations mathematically. She proved the existence of weak solutions for power-law type fluid equations for $p \geq \frac{11}{5}$ when $d = 3$. She also proved the uniqueness of weak solutions for $p \geq \frac{5}{2}$ when $d = 3$. Lions [27] treated the $p$-Laplacian case and proved the existence of weak solutions when $p > \frac{3d+2}{d+2}$ and the uniqueness of weak solutions when $p \geq \frac{d+2}{2}$ where $d \geq 2$. Later Målék et al. [29] proved the existence of weak solutions when $p \geq \frac{3d}{d+2}$, where $d \geq 2$, in periodic case. Wolf [37] proved the existence of weak solutions when $p > \frac{2d+2}{d+2}$ in an arbitrary domain using $L^\infty$ truncation method and a careful decomposition of the pressure which is needed since the $L^\infty$ truncation used does not preserve the divergence freeness of a velocity field. Then Diening et al. [17] proved existence of weak solutions when $q > \frac{2d}{d+2}$, $d \geq 2$ using parabolic Lipschitz truncation method and a decomposition of the pressure as similar to [37]. Recently a solenoidal parabolic Lipschitz truncation method, which keeps divergence free velocity fields divergence free, was developed by Breit et al. [9] and using that, a shorter proof of the existence result of weak solutions in [17] was given. For Serrin type uniqueness theorem and the existence of the global attractor with autonomous external force in the 3D case, see [10]. Recently, Bulíček et al. [11] proved existence of weak solutions for Implicitly constituted Incompressible fluids, which relate symmetric gradient of fluids to stress tensor of fluids in an implicit way. For a survey of results before 2006 concerning power-law fluids, see [28].

A recent contribution by the authors [5] employed a solenoidal parabolic Lipschitz truncation method developed in [9] to prove existence of weak solutions to (1)–(7) if $\mathbf{S}(c, \mathbf{D}\mathbf{v})$ is of power law type with an exponent $q > \frac{2d}{d+2}$. In Sect. 2, we review that existence result concerning weak solutions of (1)–(7). In Sect. 3, we give a sketch of its proof. For more details on its proof, we refer to [5].

## 2 Existence of Weak Solutions for the Diffuse Interface Model

We use standard notations. The usual Lebesgue spaces with respect to the Lebesgue measure are denoted by $L^p(M)$, $1 \leq p \leq \infty$, for some measurable $M \subset \mathbb{R}^N$. Moreover, $L^p(M; X)$ denotes its Banach space-valued variant and $L^p(0, T; X) = L^p((0, T); X)$. The standard $L^p$-Sobolev space is denoted by $W^m_p(\Omega)$. $W^m_{p,0}(\Omega)$ is the closure of $C_0^\infty(\Omega)$ in $W^m_p(\Omega)$ and $H^m(\Omega) = W^m_2(\Omega), H^m_0(\Omega) = W^m_{2,0}(\Omega)$. Finally $L^2_\sigma(\Omega)$ is the closure of divergence free $C_0^\infty(\Omega)$-vector fields in $L^2(\Omega)^d$.

For simplicity we assume that $\kappa = \rho = 1$ in (1)–(7) , but all results are true for general (fixed) $\kappa, \rho > 0$. Moreover, we assume:

*Assumption 2.1* Let $\Omega \subset \mathbb{R}^d$, $d = 2, 3$, be a bounded domain with $C^3$-boundary and let $\Phi \in C([a,b]) \cap C^2((a,b))$ be such that $\phi = \Phi'$ satisfies

$$\lim_{s\to a} \phi(s) = -\infty, \qquad \lim_{s\to b} \phi(s) = \infty, \qquad \phi'(s) \geq -\alpha$$

for some $\alpha \geq 0$. Let $m > 0$ and let $\mathbf{S}\colon [a,b] \times \mathbb{R}^{d\times d} \to \mathbb{R}^{d\times d}$ be such that

$$|\mathbf{S}(c,\mathbf{M})| \leq C(|\operatorname{sym}(\mathbf{M})|^{q-1} + 1) \tag{10}$$

$$|\mathbf{S}(c_1,\mathbf{M}) - \mathbf{S}(c_2,\mathbf{M})| \leq C|c_1 - c_2|(|\operatorname{sym}(\mathbf{M})|^{q-1} + 1) \tag{11}$$

$$\mathbf{S}(c,\mathbf{M}) : \mathbf{M} \geq \omega|\operatorname{sym}(\mathbf{M})|^q - C_1 \tag{12}$$

for all $\mathbf{M} \in \mathbb{R}^{d\times d}$, $c, c_1, c_2 \in [a,b]$, and some $C, C_1, \omega > 0$, $q \in (\frac{2d}{d+2}, \infty)$. Here $\operatorname{sym}(\mathbf{M}) = \frac{1}{2}(\mathbf{M} + \mathbf{M}^T)$. Moreover, we assume that $\mathbf{S}(c,\cdot)\colon \mathbb{R}^{d\times d}_{sym} \to \mathbb{R}^{d\times d}_{sym}$ is strictly monotone for every $c \in [a,b]$.

Let $\mathbf{v} \in L^q(0,T;W^1_{q,0}(\Omega)^d) \cap L^\infty(0,T;L^2_\sigma(\Omega))$, $c \in L^\infty(0,T;H^1(\Omega)) \cap L^2(0,T;H^2(\Omega))$ with $\Phi(c) \in L^2(\Omega\times(0,T))$, and $\mu \in L^2(0,T;H^1(\Omega))$, where $0 < T < \infty$. Then $(\mathbf{v}, c, \mu)$ is a weak solution of the system (1)–(7) if for any $\boldsymbol{\varphi} \in C^\infty(\overline{Q_T})^d$ with $\operatorname{div}\boldsymbol{\varphi} = 0$ and $\operatorname{supp}(\boldsymbol{\varphi}) \subset\subset \Omega \times [0,T)$ the following equations hold true:

$$\begin{aligned} -\int_{Q_T} \mathbf{v}\cdot\partial_t\boldsymbol{\varphi}\, d(x,t) - \int_{Q_T} \mathbf{v}\otimes\mathbf{v} : \mathbf{D}\boldsymbol{\varphi}\, d(x,t) + \int_{Q_T} \mathbf{S}(c,\mathbf{D}\mathbf{v}) : \mathbf{D}\boldsymbol{\varphi}\, d(x,t) \\ = \kappa \int_{Q_T} \nabla c\otimes\nabla c : \mathbf{D}\boldsymbol{\varphi}\, d(x,t) + \int_\Omega \mathbf{v}_0\cdot\boldsymbol{\varphi}(0)\, dx \end{aligned} \tag{13}$$

and for every $\psi \in C^\infty(\overline{Q_T})$ with $\operatorname{supp}(\psi) \subset\subset \overline{\Omega}\times[0,T)$

$$\begin{aligned} -\int_{Q_T} c\partial_t\psi\, d(x,t) - \int_\Omega c_0\psi(0)\, dx + \int_{Q_T} (\mathbf{v}\cdot\nabla c)\psi\, d(x,t) \\ = -m\int_{Q_T} \nabla\mu\cdot\nabla\psi\, d(x,t), \end{aligned} \tag{14}$$

$$\mu = \phi(c) - \Delta c \quad \text{in } \Omega\times(0,T), \tag{15}$$

$$\mathbf{n}\cdot\nabla c|_{\partial\Omega} = 0 \quad \text{on } \partial\Omega\times(0,T). \tag{16}$$

**Theorem 2.2** *Let Assumption 2.1 hold true and let* $0 < T < \infty$. *Then for any* $\mathbf{v}_0 \in L^2_\sigma(\Omega)$ *and* $c_0 \in H^1(\Omega)$ *with* $c_0(x) \in [a,b]$ *almost everywhere there exists a weak solution* $\mathbf{v} \in L^q(0,T;W^1_{q,0}(\Omega)) \cap L^\infty(0,T;L^2_\sigma(\Omega))$, $c \in L^\infty(0,T;H^1(\Omega)) \cap L^2(0,T;H^2(\Omega))$ *with* $\Phi(c) \in L^2(\Omega\times(0,T))$, *and* $\mu \in L^2(0,T;H^1(\Omega))$ *in the sense above.*

The following theorem, which is a summary of Theorem 2.16 and Corollary 2.17 of [9], is important for the proof of our theorem.

**Theorem 2.3** *Let $I_0$ be an open time interval, let $B_0$ be a ball in $\mathbb{R}^d$, and let $Q_0 := I_0 \times B_0$. Let $q, \sigma \in (1, \infty)$ with $q, q' > \sigma > 1$, where $q' = \frac{q}{q-1}$. Let $\zeta \in C_0^\infty(\frac{1}{6}Q_0)$ with $\chi_{\frac{1}{8}Q_0} \leq \zeta \leq \chi_{\frac{1}{6}Q_0}$. Let $\mathbf{u}_m$ and $\mathbf{G}_m$ satisfy $\partial_t \mathbf{u}_m = -\operatorname{div} \mathbf{G}_m$ in the sense of distributions $\mathcal{D}'_{\mathrm{div}}(Q_0)$, where $\mathcal{D}_{\mathrm{div}} = \{\varphi \in C_0^\infty(Q_0)^d : \operatorname{div} \varphi = 0\}$. Assume that $\mathbf{u}_m$ is a weak null sequence in $L^q(I_0; W_q^1(B_0))$, a strong null sequence in $L^\sigma(Q_0)$ and bounded in $L^\infty(I_0, L^\sigma(B_0))$. Further assume that $\mathbf{G}_m = \mathbf{G}_{1,m} + \mathbf{G}_{2,m}$ such that $\mathbf{G}_{1,m}$ is a weak null sequence in $L^{q'}(Q_0)$ and $\mathbf{G}_{2,m}$ converges strongly to zero in $L^\sigma(Q_0)$. Then there exists a double sequence of open sets $\mathcal{O}_{m,k}$, $k, m \in \mathbb{N}$, with $\limsup_{m\to\infty} |\mathcal{O}_{m,k}| \leq C\, 2^{-k} 2^{-q 2^k}$ for all $k \in \mathbb{N}$ such that for every $\mathbf{K} \in L^{q'}(\frac{1}{6}Q_0)$*

$$\limsup_{m\to\infty} \left| \int ((\mathbf{G}_{1,m} + \mathbf{K}) : \nabla \mathbf{u}_m) \zeta \chi_{\mathcal{O}_{m,k}^c}\, d(x,t) \right| \leq C\, 2^{-k/q}.$$

The proof of Theorem 2.3 is based on a solenoidal parabolic Lipschitz truncation. In particular, the sets $\mathcal{O}_{m,k}$ are level sets of suitable maximal operators defined by $\mathbf{u}_m$ and $\mathbf{G}_m$. It is a modification of the parabolic Lipschitz truncation of [17, 24]. The advantage of Theorem 2.3 is that the pressure can be completely avoided by using a solenoidal Lipschitz truncation. See [8] for a solenoidal Lipschitz truncation in the stationary case.

## 3 Sketch of Proof of Theorem 2.2

In order to approximate (1)–(7) we consider

$$\begin{aligned}
\partial_t \mathbf{v} + \operatorname{div}(\eta_\varepsilon(\mathbf{v})\mathbf{v} \otimes \mathbf{v}) - \operatorname{div} \mathbf{S}(c, \mathbf{D}\mathbf{v}) + \nabla p & \\
= -\Psi_\varepsilon(\operatorname{div}(\nabla c \otimes \nabla c)) & \quad \text{in } \Omega \times (0,T), && (17)\\
\operatorname{div} \mathbf{v} = 0, & \quad \text{in } \Omega \times (0,T), && (18)\\
\partial_t c + (\Psi_\varepsilon \mathbf{v}) \cdot \nabla c = m \Delta \mu, & \quad \text{in } \Omega \times (0,T), && (19)\\
\mu = \phi(c) - \Delta c. & \quad \text{in } \Omega \times (0,T) && (20)
\end{aligned}$$

together with (5)–(7), where $\Psi_\varepsilon \mathbf{w} = P_\sigma(\psi_\varepsilon * \mathbf{w})|_\Omega$, $\psi_\varepsilon(x) = \varepsilon^{-d}\psi(x/\varepsilon)$, $\varepsilon > 0$, is a usual smoothing kernel such that $\psi(-x) = \psi(x)$ for all $x \in \mathbb{R}^d$, $\mathbf{w}$ is extended by 0 outside of $\Omega$, and $P_\sigma$ is the Helmholtz projection. Moreover, $\eta_\varepsilon(\mathbf{s}) = \eta(\varepsilon|\mathbf{s}|^2)$ for all $\mathbf{s} \in \mathbb{R}^d$, $\varepsilon > 0$ with some $\eta \in C_0^\infty(\mathbb{R})$ with $\eta(0) = 1$.

We can construct weak solutions of the approximate system (17)–(20) together with (5)–(7) in an appropriate function space with the aid of the Leray-Schauder theorem. The solutions satisfy the energy identity

$$\frac{1}{2}\|\mathbf{v}(t)\|_{L^2(\Omega)}^2 + E_{mix}(c(t)) + \int_0^t \int_\Omega \mathbf{S}(c, \mathbf{D}\mathbf{v}) : \mathbf{D}\mathbf{v}\, dx\, d\tau$$
$$+ \int_0^t \int_\Omega m|\nabla\mu|^2\, dx\, d\tau = \frac{1}{2}\|\mathbf{v}_0\|_{L^2(\Omega)}^2 + E_{mix}(c_0) \tag{21}$$

for almost every $t \in (0, T)$, where

$$E_{mix}(c) = \int_\Omega \frac{|\nabla c|^2}{2}\, dx + \int_\Omega \Phi(c)\, dx.$$

We use those solutions to construct a weak solution of the original system (1)–(7).

In the following the solutions of the approximate system are denoted by $(\mathbf{v}_\varepsilon, c_\varepsilon, \mu_\varepsilon)$ for $\varepsilon > 0$. Using a priori estimates for solutions of the approximate system based on (21), we can conclude for a suitable subsequence $\varepsilon_i \to_{i\to\infty} 0$ that

$$\begin{aligned}
\mathbf{D}\mathbf{v}_{\varepsilon_i} &\to \mathbf{D}\mathbf{v} \quad \text{weakly in } L^q(Q_T)^{d\times d},\\
\mathbf{v}_{\varepsilon_i} &\to \mathbf{v} \quad \text{weakly in } L^{q\frac{d+2}{d}}(Q_T)^d,\\
\mathbf{S}(c_{\varepsilon_i}, \mathbf{D}\mathbf{v}_{\varepsilon_i}) &\to \tilde{\mathbf{S}} \quad \text{weakly in } L^{q'}(Q_T)^{d\times d},\\
\mathbf{v}_{\varepsilon_i} \otimes \mathbf{v}_{\varepsilon_i} \eta_{\varepsilon_i}(\mathbf{v}_{\varepsilon_i}) &\to \tilde{\mathbf{H}} \quad \text{weakly in } L^{q\frac{d+2}{2d}}(Q_T).
\end{aligned} \tag{22}$$

Moreover, from suitable a priori estimates of a solution of Cahn-Hilliard equations with convection term, we have for a suitable subsequence

$$\begin{aligned}
c_{\varepsilon_i} &\to_{i\to\infty} c \quad && \text{in } L^4(0, T; W_4^1(\Omega)) \cap L^2(0, T; W_r^2(\Omega)),\\
\mu_{\varepsilon_i} &\to_{i\to\infty} \mu \quad && \text{in } L^2(0, T; H^1(\Omega)),
\end{aligned} \tag{23}$$

where $(c, \mu)$ solve (14)–(16). Hence it only remains to prove (13). To this end let $\mathbf{K}_\varepsilon \in L^2(Q_T)^{d\times d}$ be such that

$$\int_{Q_T} \mathbf{K}_\varepsilon : \mathbf{D}\boldsymbol{\varphi}\, d(x,t) = \int_{Q_T} \nabla c_\varepsilon \otimes \nabla c_\varepsilon : \mathbf{D}\Psi_\varepsilon(\boldsymbol{\varphi})\, d(x,t)$$
$$- \int_{Q_T} \nabla c \otimes \nabla c : \mathbf{D}\boldsymbol{\varphi}\, d(x,t) \tag{24}$$

for all $\varphi \in L^2(0,T;H_0^1(\Omega)^d)$ and

$$\|\mathbf{K}_\varepsilon\|_{L^2(Q_T)} \leq C\|\operatorname{div}(\nabla c \otimes \nabla c) - \Psi_\varepsilon \operatorname{div}(\nabla c_\varepsilon \otimes \nabla c_\varepsilon)\|_{L^2(0,T;H_0^{-1})}$$

for some $C > 0$. We can assume that $\mathbf{K}_\varepsilon$ is pointwise a symmetric matrix. Then

$$\mathbf{K}_{\varepsilon_i} \to 0 \text{ strongly in } L^2(Q_T)^{d\times d},$$

due to (23).

Since $q > \frac{2d}{d+2}$, there exists some $\sigma_0 > 1$ such that $q\frac{d+2}{2d} > \sigma_0 > 1$. Hence, due to (22) we have for some $\varepsilon_i \to_{i\to\infty} 0$,

$$\mathbf{v}_{\varepsilon_i} \to \mathbf{v} \qquad \text{strongly in } L^{2\sigma_0}(Q_T)^d \tag{25}$$

$$\text{and } \mathbf{v}_{\varepsilon_i} \otimes \mathbf{v}_{\varepsilon_i}\eta_{\varepsilon_i}(|\mathbf{v}_\varepsilon|) \to \mathbf{v}\otimes\mathbf{v} \text{ strongly in } L^{\sigma_0}(Q_T)^{d\times d}. \tag{26}$$

Therefore $\tilde{\mathbf{H}} = \mathbf{v}\otimes\mathbf{v}$.

Then taking the limit of the weak form of the approximate system along the subsequence $\varepsilon_i$, we obtain the following limit equation:

$$\begin{aligned} -\int_{Q_T} \mathbf{v}\cdot\partial_t\boldsymbol{\varphi}\, d(x,t) + \int_{Q_T} (\tilde{\mathbf{S}} - \mathbf{v}\otimes\mathbf{v}) : \mathbf{D}\boldsymbol{\varphi}\, d(x,t) \\ = \int_{Q_T} \nabla c\otimes\nabla c : \mathbf{D}\boldsymbol{\varphi}\, d(x,t) + \int_\Omega \mathbf{v}_0\cdot\boldsymbol{\varphi}(0)\, dx. \end{aligned} \tag{27}$$

for all $\boldsymbol{\varphi} \in C^\infty(\overline{Q_T})^d$ with $\operatorname{div}\boldsymbol{\varphi} = 0$ and $\operatorname{supp}(\boldsymbol{\varphi}) \subset\subset \Omega\times[0,T)$.

By subtracting the above equation from the weak form of the approximate equations, we have the following equation:

$$\begin{aligned} &-\int_{Q_T} (\mathbf{v}_\varepsilon - \mathbf{v})\cdot\partial_t\boldsymbol{\varphi}\, d(x,t) + \int_{Q_T} \left(\mathbf{S}(c_\varepsilon, \mathbf{D}\mathbf{v}_\varepsilon) - \tilde{\mathbf{S}}\right) : \mathbf{D}\boldsymbol{\varphi}\, d(x,t) \\ = &\int_{Q_T} (\mathbf{v}_\varepsilon\otimes\mathbf{v}_\varepsilon\eta_\varepsilon(\mathbf{v}_\varepsilon) - \mathbf{v}\otimes\mathbf{v}) : \mathbf{D}\boldsymbol{\varphi}\, d(x,t) + \int_{Q_T} \mathbf{K}_\varepsilon : \mathbf{D}\boldsymbol{\varphi}\, d(x,t). \end{aligned}$$

Defining $\mathbf{u}_\varepsilon := \mathbf{v}_\varepsilon - \mathbf{v}$ we can write this as

$$\int_{Q_T} \mathbf{u}_{\varepsilon_i}\cdot\partial_t\boldsymbol{\varphi}\, d(x,t) = \int_{Q_T} \mathbf{H}_{\varepsilon_i} : \nabla\boldsymbol{\varphi}\, d(x,t) \tag{28}$$

for any $\varphi \in C^\infty(Q_T)^d$ with $\operatorname{div}\varphi = 0$ and $\operatorname{supp}(\varphi) \subset\subset \Omega \times [0,T)$, where $\mathbf{H}_i := \mathbf{H}_{1,i} + \mathbf{H}_{2,i}$ with

$$\mathbf{H}_{1,i} := \mathbf{S}(c_{\varepsilon_i}, \mathbf{D}\mathbf{v}_{\varepsilon_i}) - \tilde{\mathbf{S}},$$
$$\mathbf{H}_{2,i} := \mathbf{v}_{\varepsilon_i} \otimes \mathbf{v}_{\varepsilon_i} \eta_{\varepsilon_i}(\mathbf{v}_{\varepsilon_i}) - \mathbf{v} \otimes \mathbf{v} - \mathbf{K}_{\varepsilon_i}$$

Then we have the following convergences for suitable $\varepsilon_i \to_{i\to\infty} 0$

$$\mathbf{u}_{\varepsilon_i} \to 0 \text{ weakly in } L^q(0,t;V_q(\Omega)), \tag{29}$$
$$\mathbf{u}_{\varepsilon_i} \to 0 \text{ strongly in } L^{2\sigma_0}(Q_T), \tag{30}$$
$$\mathbf{u}_{\varepsilon_i} \to 0 \ *\text{-weakly in } L^\infty(0,T;L^2(\Omega)), \tag{31}$$
$$\mathbf{H}_{1,i} \to 0 \text{ weakly in } L^{q'}(Q_T), \tag{32}$$
$$\mathbf{H}_{2,i} \to 0 \text{ strongly in } L^{\sigma_0}(Q_T) \tag{33}$$

for some $1 < \sigma_0 < \min(q,q')$. Let $I_0 \subseteq (0,T)$ be a time interval, $B_0 \subset \mathbb{R}^d$ be a ball such that $Q_0 := I_0 \times B_0 \subset\subset Q_T$. Let $\zeta \in C_0^\infty(\frac{1}{6}Q_0)$ with $\chi_{\frac{1}{8}Q_0} \le \zeta \le \chi_{\frac{1}{6}Q_0}$. Here for $c > 0$ $cQ_0$ is defined as $cI_0 \times cB_0$, where $cI_0$ and $cB_0$ have the same center as $I_0$ and $B_0$ respectively and have $c$ times length and $c$ times radius respectively. Then we can apply Theorem 2.3 with $\mathbf{K} = \tilde{\mathbf{S}} - \mathbf{S}(c, \mathbf{D}\mathbf{v})$ to obtain

$$\limsup_{i\to\infty} \left| \int ((\mathbf{H}_{1,i} + \tilde{\mathbf{S}} - \mathbf{S}(c,\mathbf{D}\mathbf{v})) : \nabla(\mathbf{v}_{\varepsilon_i} - \mathbf{v}))\zeta \chi_{\mathcal{O}_{i,k}^c}\, d(x,t) \right| \le c\, 2^{-k/q}.$$

In other words

$$\limsup_{i\to\infty} \left| \int \Big( (\mathbf{S}(c_{\varepsilon_i}, \mathbf{D}\mathbf{v}_{\varepsilon_i}) - \mathbf{S}(c,\mathbf{D}\mathbf{v})) : \mathbf{D}(\mathbf{v}_{\varepsilon_i} - \mathbf{v})\Big)\zeta \chi_{\mathcal{O}_{i,k}^c}\, d(x,t) \right| \le c\, 2^{-k/q}.$$

Using this estimate, Hölder's inequality, (23) and *Assumption* 2.1, we have

$$\limsup_{i\to\infty} \left| \int \Big( (\mathbf{S}(c, \mathbf{D}\mathbf{v}_{\varepsilon_i}) - \mathbf{S}(c,\mathbf{D}\mathbf{v})) : \mathbf{D}(\mathbf{v}_{\varepsilon_i} - \mathbf{v})\Big)^{\frac{1}{2}} \zeta\, d(x,t) \right| \le c\, 2^{-\frac{k}{2q}}.$$

For $k \to \infty$ the right hand side converges to zero. Now the monotonicity of $\mathbf{S}$ can be used to prove $\tilde{\mathbf{S}} = \mathbf{S}(c, \mathbf{D}\mathbf{v})$. Hence we have (13). This ends the proof.

**Acknowledgements** This work was supported by the SPP 1506 "Transport Processes at Fluidic Interfaces" of the German Science Foundation (DFG) through the grant AB 285/4-1. Moreover, T. was supported by JSPS Research Fellowships for Young Scientists and by FMSP, a JSPS Program for Leading Graduate Schools in the University of Tokyo. The supports are gratefully acknowledged.

## References


1. H. Abels, On generalized solutions of two-phase flows for viscous incompressible fluids. Interfaces Free Bound. **9**(1), 31–65 (2007)
2. H. Abels, On a diffuse interface model for two-phase flows of viscous, incompressible fluids with matched densities. Arch. Ration. Mech. Anal. **194**(2), 463–506 (2009)
3. H. Abels, M. Röger, Existence of weak solutions for a non-classical sharp interface model for a two-phase flow of viscous, incompressible fluids. Ann. Inst. H. Poincaré Anal. Non Linéaire **26**(6), 2403–2424 (2009)
4. H. Abels, M. Wilke, Convergence to equilibrium for the Cahn-Hilliard equation with a logarithmic free energy. Nonlinear Anal. **67**, 3176–3193 (2007)
5. H. Abels, L. Diening, Y. Terasawa, Existence of weak solutions for a diffuse interface model of non-Newtonian two-phase flows. Nonlinear Anal. Real World Appl. **15**, 149–157 (2014)
6. S. Bosia, Analysis of a Cahn-Hilliard-Ladyzhenskaya system with singular potential. J. Math. Anal. Appl. **397**(1), 307–321 (2013)
7. F. Boyer, Mathematical study of multi-phase flow under shear through order parameter formulation. Asymptot. Anal. **20**(2), 175–212 (1999)
8. D. Breit, L. Diening, M. Fuchs, Solenoidal Lipschitz truncation and applications in fluid. J. Differ. Equ. **253**(6), 1910–1942 (2012)
9. D. Breit, L. Diening, S. Schwarzacher, Solenoidal Lipschitz truncation for parabolic PDE's. M3AS **23**(14), 2671–2700 (2014)
10. M. Bulíček, F. Ettwein, P. Kaplický, D. Pražák, Dimension of the attractor for 3D flow of non-Newtonian fluid. Commun. Pure Appl. Anal. **8**(5), 1503–1520 (2009)
11. M. Bulíček, P. Gwiazda, J. Málek, A. Świerczevska-Gwiazda, On Unsteady Flows of Implicitly Constituted Incompressible Fluids. SIAM J. Math. Anal. **44**(4), 2756–2801 (2012)
12. X. Chen, Global asymptotic limit of solutions of the Cahn-Hilliard equation. J. Differ. Geom. **44**, 262–311 (1996)
13. A. Debussche, L. Dettori, On the Cahn-Hilliard equation with a logarithmic free energy. Nonlinear Anal. **24**(10), 1491–1514 (1995)
14. I.V. Denisova, A priori estimates for the solution of the linear non stationary problem connected with the motion of a drop in a liquid medium. (Russian) Trudy Mat. Inst. Steklov **188**, 3–21 (1990). [translation in Proc. Steklov Inst. Math. **3**, 1–24 (1991)]
15. I.V. Denisova, Problem of the motion of two viscous incompressible fluids separated by a closed free interface. Mathematical problems for the Navier-Stokes equations (Centro, 1993). Acta Appl. Math. **37**, 31–40 (1994)
16. I.V. Denisova, V.A. Solonnikov, Classical solvability of the problem of the motion of two viscous incompressible fluids. (Russian) Algebra i Analiz. **7**(5), 101–142 (1995). [translation in St. Petersburg Math. J. **7**(5), 755–786 (1996)]
17. L. Diening, M. Růžička, J. Wolf, Existence of weak solutions for unsteady motions of generalized Newtonian fluids. Annali della Scuola Normale Superiore di Pisa Classe di scienze (5) **9**(1), 1–46 (2010)
18. C.M. Elliott, S. Luckhaus, A generalized equation for phase separation of a multi-component mixture with interfacial free energy. preprint SFB 256 Bonn No. 195, 1991
19. C.M. Elliott, S. Zheng, On the Cahn-Hilliard equation. Arch. Ration. Mech. Anal. **96**(4), 339–357 (1986)
20. M. Grasselli, D. Pražák, Longtime behavior of a diffuse interface model for binary fluid mixtures with shear dependent viscosity. Interfaces Free Bound. **13**(4), 507–530 (2011)
21. M.E. Gurtin, D. Polignone, J. Viñals, Two-phase binary fluids and immiscible fluids described by an order parameter. Math. Models Methods Appl. Sci. **6**(6), 815–831 (1996)
22. P.C. Hohenberg, B.I. Halperin, Theory of dynamic critical phenomena. Rev. Mod. Phys. **49**, 435–479 (1977)
23. N. Kim, L. Consiglieri, J.F. Rodrigues, On non-Newtonian incompressible fluids with phase transitions. Math. Methods Appl. Sci. **29**(13), 1523–1541 (2006)

24. J. Kinnunen, J.L. Lewis, Very weak solutions of parabolic systems of $p$-Laplacian type. Ark. Mat. **40**(1), 105–132 (2002)
25. M. Köhne, J. Prüss, M. Wilke, Qualitative behaviour of solutions for the two-phase Navier-Stokes equations with surface tension. Math. Ann. **356**(2), 737–792 (2013)
26. O.A. Ladyzhenskaya, Sur de nouvelles équation dans la dynamique de fluides visqueux et leur resolution globale. Troudi Mat. Inst. Stekloff **CII**, 85–104 (1967)
27. J.-L. Lions, *Quelques Méthodes de Résolution des Problèmes Aux Limites Non Linéaires*. (Dunod, France, 1969)
28. J. Málek, K.R. Rajagopal, Mathematical issues concerning the Navier-Stokes equations and some of its generalizations, in *Handbook Differential Equation: Evolutionary equations*, vol. II (Elsevier/NorthHolland, Amsterdam, 2005), pp. 371–459
29. J. Málek, J. Nečas, M. Růžička, On the non-Newtonian incompressible fluids. M3AS **3**(1), 35–63 (1993)
30. B. Nicolaenko, B. Scheurer, R. Temam, Some global dynamical properties of a class of pattern formation equations. Commun. Partial Differ. Equ. **14**(2), 245–297 (1989)
31. P.I. Plotnikov, Generalized solutions to a free boundary problem of motion of a non-Newetonian fluid. Siberian Math. J. **34**(4), 704–716 (1993)
32. J. Prüss, G. Simonett, On the two-phase Navier-Stokes equations with surface tension. Interfaces Free Bound. **12**(3), 311–34 (2010)
33. J. Prüss, G. Simonett, R. Zacher, Qualitative behavior of incompressible two-phase flows with phase transition: the case of equal densities. Interfaces Free Bound. **15**(4), 405–428 (2013)
34. J. Prüss, S. Shimizu, M. Wilke, Qualitative behavior of incompressible two-phase flows with phase transitions: the case of non-equal densities. Commun. Partial Differ. Equ. **39**(7), 1236–1283 (2014)
35. P. Rybka, K-H. Hoffmann, Convergence of solutions to Cahn-Hilliard equations. Commun. Partial Differ. Equ. **24**(5–6), 1055–1077 (1999)
36. V.N. Starovoĭtov, On the motion of a two-component fluid in the presence of capillary forces. Mat. Zametki **62**(2), 293–305 (1997)
37. J. Wolf, Existence of weak solutions to the equations of non-stationary motion of non-Newtonian fluids with shear rate dependent viscosity. J. Math. Fluid Mech. **9**(1), 104–138 (2007)

# Stationary Solutions for a Navier-Stokes/Cahn-Hilliard System with Singular Free Energies

**Helmut Abels and Josef Weber**

*Dedicated to Yoshihiro Shibata on the occasion of his 60th birthday*

**Abstract** We consider a stationary Navier-Stokes/Cahn-Hilliard type system. The system describes a so-called diffuse interface model for the two-phase flow of two macroscopically immiscible incompressible viscous fluids in the case of matched densities, also known as Model H. We prove existence of weak solutions for the stationary system for general exterior forces and singular free energies, which ensure that the order parameter stays in the physical reasonable interval. To this end we reduce the system to an abstract differential inclusion and apply the theory of multi-valued pseudo-monotone operators.

**Keywords** Cahn-Hilliard equation • Diffuse interface model • Navier-Stokes equation • Two-phase flow

## 1 Introduction and Main Result

We consider the following stationary system of Navier-Stokes/Cahn-Hilliard type

$$\mathbf{v}\cdot\nabla\mathbf{v} - \operatorname{div}(2\nu(c)\mathbf{D}\mathbf{v}) + \nabla p = \mu\nabla c + \tilde{f} \qquad \text{in } \Omega, \tag{1}$$

$$\operatorname{div}\ \mathbf{v} = 0 \qquad \text{in } \Omega, \tag{2}$$

$$\mathbf{v}\cdot\nabla c = \operatorname{div}(m(c)\nabla\mu) \qquad \text{in } \Omega, \tag{3}$$

$$\mu = -\varepsilon\Delta c + \varepsilon^{-1}f'(c) \qquad \text{in } \Omega, \tag{4}$$

$$\mathbf{v}_{|\partial\Omega} = 0 \qquad \text{on } \partial\Omega, \tag{5}$$

$$\mathbf{n}\cdot\nabla c_{|\partial\Omega} = \mathbf{n}\cdot\nabla\mu_{|\partial\Omega} = 0 \qquad \text{on } \partial\Omega. \tag{6}$$

H. Abels (✉) • J. Weber
Fakultät für Mathematik, Universität Regensburg, Universitätstr. 31, 93053 Regensburg, Germany
e-mail: helmut.abels@mathematik.uni-regensburg.de;
josef.weber@mathematik.uni-regensburg.de



H. Amann et al. (eds.), *Recent Developments of Mathematical Fluid Mechanics*, Advances in Mathematical Fluid Mechanics, DOI 10.1007/978-3-0348-0939-9_3

Moreover we prescribe

$$\frac{1}{|\Omega|}\int_\Omega c(x)dx = \bar{m} \in (-1,1). \tag{7}$$

Here $\mathbf{v}$ is the mean velocity, $\mathbf{D}\mathbf{v} = \frac{1}{2}(\nabla\mathbf{v} + \nabla\mathbf{v}^T)$, $p$ is the pressure, $c$ is an order parameter related to the concentration of the fluids (e.g. the concentration difference or the concentration of one component), $\tilde{\boldsymbol{f}}$ is an exterior force density, $\Omega$ is a bounded domain with sufficiently smooth boundary, and $\mathbf{n}$ is the outer unit normal to $\partial\Omega$. Moreover, $\nu(c) > 0$ is the viscosity of the mixture, $\varepsilon > 0$ is a (small) parameter, which will be related to the "thickness" of the interfacial region, and $f$ is a homogeneous free energy density specified below.

This system arises as a stationary version of the so-called "model H", cf. Hohenberg and Halperin [7] and Gurtin et al. [6]:

$$\begin{aligned}
\partial_t\mathbf{v} + \mathbf{v}\cdot\nabla\mathbf{v} - \operatorname{div}(\nu(c)\mathbf{D}\mathbf{v}) + \nabla p &= \mu\nabla c + \tilde{\boldsymbol{f}} && \text{in } \Omega\times(0,\infty),\\
\operatorname{div}\mathbf{v} &= 0 && \text{in } \Omega\times(0,\infty),\\
\partial_t c + \mathbf{v}\cdot\nabla c &= \operatorname{div}(m(c)\nabla\mu) && \text{in } \Omega\times(0,\infty),\\
\mu &= \varepsilon^{-1}f'(c) - \varepsilon\Delta c && \text{in } \Omega\times(0,\infty)
\end{aligned}$$

together with the same boundary conditions as above. The system describes the motion of two macroscopically immiscible, viscous, incompressible Newtonian fluids. The model takes a partial mixing on a small length scale measured by a parameter $\varepsilon > 0$ into account. Therefore the classical sharp interface between both fluids is replaced by an interfacial region and an order parameter related to the concentration difference of both fluids is introduced. Here it is assumed that the densities of both components as well as the density of the mixture are constant and for simplicity equal to one. In the following we set $\varepsilon = 1$ for simplicity. But all results hold true for general $\varepsilon > 0$. For the instationary system existence of weak solutions and well-posedness were obtained by Starovoitov [10], Boyer [4], Liu and Shen [8], and Abels [1]. Moreover, we refer to Abels, Depner and Garcke [3] for further references and results on similar models in the case of different densities.

In the following we will consider a class of singular free energies, which is motivated by the homogeneous free energy of the so-called regular solution models used by Cahn and Hilliard [5]:

$$\Psi(c) = \frac{\theta}{2}\left((1+c)\ln(1+c) + (1-c)\ln(1-c)\right) - \frac{\theta_c}{2}c^2, \quad c\in[-1,1],$$

where $0 < \theta < \theta_c$. Mathematically, these singular free energies ensure that the order parameter stays in the physically reasonable interval, which is $[-1,1]$ if $c$ is the difference of volume fractions of both fluids. But this leads to singular terms in the equation for the chemical potential. In order to deal with these terms we apply techniques, which were developed in Abels and Wilke [2].

More precisely we work with the following assumption:

**Assumption 1.1** *Let $\Omega \subseteq \mathbb{R}^d$, $d = 2, 3$, be a bounded domain with $C^3$-boundary, $\nu\colon [a,b] \to (0,\infty)$ be continuously differentiable, $m\colon [a,b] \to (0,\infty)$ be continuous, and $f \in C^0([a,b]) \cap C^2((a,b))$ such that*

$$\lim_{s\to a} f'(s) = -\infty, \quad \lim_{s\to b} f'(s) = +\infty, \tag{8}$$

$$f''(s) \geq -\kappa \tag{9}$$

*hold for all $s \in (a,b)$ and some constant $\kappa \geq 0$.*

Let $f$ be as in the previous assumption. Then we can conclude

$$f(s) = f_0(s) - \kappa\,\frac{s^2}{2} \tag{10}$$

with $f_0$ convex and $\kappa \in \mathbb{R}$ as in Assumption 1.1. Moreover we can deduce $\lim_{s\to a} f_0'(s) = -\infty$ and $\lim_{s\to b} f_0'(s) = +\infty$.

Weak solutions of the stationary Navier-Stokes/Cahn-Hilliard equations are defined as follows:

**Definition 1.2 (Weak Solution)** A triple $(\mathbf{v},\mu,c)$ with $\mathbf{v} \in H_0^1(\Omega)^d \cap L_\sigma^2(\Omega)$, $\mu \in H^1(\Omega)$ and $c \in H^1(\Omega) \cap L_{(\bar{m})}^2(\Omega)$ is called weak solution for the Eqs. (*1*)–(*6*) under the constraint (*7*) if

$$\int_\Omega (\mathbf{v}\cdot\nabla\mathbf{v})\cdot\boldsymbol{\varphi}dx + \int_\Omega 2\nu(c)\mathbf{D}\mathbf{v} : \mathbf{D}\boldsymbol{\varphi}dx = \int_\Omega \mu\,\nabla c\cdot\boldsymbol{\varphi}dx + \int_\Omega \tilde{\boldsymbol{f}}\cdot\boldsymbol{\varphi}dx \tag{11}$$

holds for all $\boldsymbol{\varphi} \in C_{0,\sigma}^\infty(\Omega)$ and if

$$\int_\Omega (\mathbf{v}\cdot\nabla c)\cdot\psi\,dx = -\int_\Omega m(c)\nabla\mu\cdot\nabla\psi\,dx \tag{12}$$

$$\int_\Omega \mu\,\psi\,dx = \int_\Omega \nabla c\cdot\nabla\psi\,dx + \int_\Omega f'(c)\,\psi\,dx \tag{13}$$

holds for all $\psi \in C^\infty(\overline{\Omega})$ and if $f'(c) \in L^2(\Omega)$.

Now our main result is the following:

**Theorem 1.3 (Existence of Weak Solutions)** *Let $\Omega$, $\nu$, and $f\colon [a,b] \to \mathbb{R}$ be as in Assumption* 1.1. *Then for all $\tilde{\boldsymbol{f}} \in L^2(\Omega)^d$ and $\bar{m} \in (-1,1)$ there exists a weak solution in the sense of Definition* 1.2.

## 2 Preliminaries and Notation

Throughout the paper the usual Lebesgue spaces with respect to the Lebesgue measure are denoted by $L^q(M)$, $1 \leq q \leq \infty$, for some measurable $M \subset \mathbb{R}^N$. The standard $L^q$-Sobolev space is denoted by $W^m_q(\Omega)$. $W^m_{q,0}(\Omega)$ is the closure of $C_0^\infty(\Omega)$ in $W^m_q(\Omega)$ and $H^m(\Omega) = W^m_2(\Omega), H^m_0(\Omega) = W^m_{2,0}(\Omega)$. Furthermore we use the notation

$$L^2_{(\bar{m})}(\Omega) = \left\{ f \in L^2(\Omega) : \frac{1}{|\Omega|} \int_\Omega f(x)dx = \bar{m} \right\},$$

$H^1_{(0)}(\Omega) = H^1(\Omega) \cap L^2_{(0)}(\Omega)$ and $H^{-1}_{(0)}(\Omega) := H^1_{(0)}(\Omega)'$. Finally $L^2_\sigma(\Omega)$ is the closure of divergence free $C_0^\infty(\Omega)$-vector fields in $L^2(\Omega)^d$. For a Banach space $X$, $X'$ denotes its duality space and $\langle .,. \rangle_{X',X}$ its duality product. If $H$ is a Hilbert space, $(.,.)_H$ denotes its inner product.

The projection $P_0 : L^1(\Omega) \to L^1_{(0)}(\Omega)$ on the $L^1$-space with mean value 0 is defined by

$$P_0 f := f - \frac{1}{|\Omega|} \int_\Omega f(x)dx$$

for every $f \in L^1(\Omega)$.

## 3 Subdifferential of $F$ and $\tilde{F}$

Let $f$ be as in Assumption 1.1. In this section, we study the functionals

$$F(c) := \frac{1}{2} \int_\Omega |\nabla c(x)|^2 dx + \int_\Omega f_0(c(x))dx \tag{14}$$

$$\tilde{F}(c) := \frac{1}{4} \int_\Omega |\nabla c(x)|^2 dx + \int_\Omega f_0(c(x))dx \tag{15}$$

where we assume $c \in L^2_{(0)}(\Omega)$, use the decomposition (10) of $f$ and extend $f_0$ to $\mathbb{R}$ by $f_0(x) = +\infty$ for all $x \notin [a,b]$. Then

$$\operatorname{dom}(F) = \operatorname{dom}(\tilde{F}) = \{c \in H^1(\Omega) \cap L^2_{(0)}(\Omega) : f_0(c) \in L^1(\Omega)\}.$$

For $c \notin \operatorname{dom}(F)$, we set $F(c) = +\infty$ and $\tilde{F}(c) = +\infty$.

In the following, we will collect some statements on the subgradients and subdifferentials of these two functionals $F$ and $\tilde{F}$. Therefore we define:

**Definition 3.1 (Subgradient, Subdifferential)** Let $X$ be a real Banach space and $f : X \to (-\infty, \infty]$ a functional on $X$. A linear functional $u' \in X'$ is called subgradient of $f$ at $u$ if $f(u) \neq +\infty$ and

$$f(v) \geq f(u) + \langle u', v - u\rangle_{X',X} \tag{16}$$

holds for all $v \in X$.

The set of all subgradients of $f$ at $u$ is called subdifferential $\partial f(u)$ of $f$ at $u$.

**Notation** For a functional $F : L^2_{(0)}(\Omega) \to (-\infty, +\infty]$, its subgradient maps from $L^2_{(0)}(\Omega)$ to $\mathcal{P}(L^2_{(0)}(\Omega)')$. Since we can also consider $F$ as a functional from $H^1_{(0)}(\Omega)$ to $(-\infty, +\infty]$, we have to distinguish between its different subgradients. Therefore we write $\partial_X f$ for the subgradient of a functional $f : X \to (-\infty, +\infty]$, where $X$ is a real Banach space.

In our specific case, $\partial_{L^2_{(0)}(\Omega)} F$ respectively $\partial_{L^2_{(0)}} F$ is the subgradient of the functional $F : L^2_{(0)}(\Omega) \to (-\infty, +\infty]$ and $\partial_{H^1_{(0)}(\Omega)} F$ respectively $\partial_{H^1_{(0)}} F$ is the subgradient of the functional $F : H^1_{(0)}(\Omega) \to (-\infty, +\infty]$.

First of all, we collect some statements on the subdifferentials of $F$ and $\tilde{F}$:

**Lemma 3.2** *The functionals $F, \tilde{F} : L^2_{(0)}(\Omega) \to \mathbb{R} \cup \{+\infty\}$ defined as in* (14) *and* (15) *are lower semi-continuous and convex with $dom(F), dom(\tilde{F}) \neq \emptyset$.*

The proof of this Lemma can be found in [2].

**Proposition 3.3** *The subgradients $\partial_{L^2_{(0)}} F$ and $\partial_{L^2_{(0)}} \tilde{F}$ are maximal monotone operators.*

*Proof* Because of Lemma 3.2, $F, \tilde{F} : L^2_{(0)}(\Omega) \to \mathbb{R} \cup \{+\infty\}$ are lower semi-continuous functionals with $F, \tilde{F} \not\equiv +\infty$.

The Theorem of Rockafeller, cf. Proposition 32.17 in [11], yields that the operators $\partial_{L^2_{(0)}} F, \partial_{L^2_{(0)}} \tilde{F} : L^2_{(0)} \to \mathcal{P}((L^2_{(0)}(\Omega))')$ are maximal monotone. □

**Proposition 3.4** *The subgradients $\partial_{H^1_{(0)}} F$ and $\partial_{H^1_{(0)}} \tilde{F}$ are maximal monotone operators.*

*Proof* Let $(c_k)_{k\in\mathbb{N}} \subseteq H^1_{(0)}(\Omega)$ with $c_k \to c$ in $H^1_{(0)}(\Omega)$. Then it holds $c_k \to c$ in $L^2(\Omega)$ and therefore $\liminf_{k\to\infty} F(c_k) \geq F(c)$ and $\liminf_{k\to\infty} \tilde{F}(c_k) \geq \tilde{F}(c)$ by Lemma 3.2. Thus the Theorem of Rockafeller, cf. Proposition 32.17 in [11], yields the statement. □

**Theorem 3.5** *Let $\Omega \subseteq \mathbb{R}^d$, $d = 2, 3$ be a bounded domain with $C^3$-boundary and $F$ defined as in* (14)*. Moreover let $f_0'(x) = +\infty$ for all $x \notin (a, b)$. Then*

$$\mathcal{D}(\partial_{L^2_{(0)}} F) = \{c \in H^2(\Omega) \cap L^2_{(0)}(\Omega) | f_0'(c) \in L^2(\Omega), \\ f_0''(c)|\nabla c|^2 \in L^1(\Omega),\ \partial_{\mathbf{n}} c_{|\partial\Omega} = 0\} \tag{17}$$

*and*

$$\partial_{L^2_{(0)}} F(c) = -\Delta c + P_0 f_0'(c). \tag{18}$$

*Furthermore it holds*

$$\|c\|^2_{H^2(\Omega)} + \|f_0'(c)\|^2_{L^2(\Omega)} + \int_\Omega f_0''(c(x))|\nabla c(x)|^2 dx \\ \leq C\left(\|\partial_{L^2_{(0)}} F(c)\|^2_{L^2(\Omega)} + \|c\|^2_{L^2(\Omega)} + 1\right)$$

*for some constant $C > 0$ independent of $c \in \mathcal{D}(\partial_{L^2_{(0)}} F)$.*

The proof can be found in [2], cf. Theorem 4.3.

*Remark 3.6* Here we identify $L^2_{(0)}(\Omega)'$ with $L^2_{(0)}(\Omega)$ in the standard way. But we will not identify $H^{-1}_{(0)}(\Omega)$ with $H^1_{(0)}(\Omega)$ via the Riesz isomorphism.

Equation (18) provides a characterization of $\partial_{L^2_{(0)}} F$. Later, we will need relations between $\partial_{H^1_{(0)}} F$ and $\partial_{L^2_{(0)}} F$. The two following lemmata provide such relations:

**Lemma 3.7** *Let $F$ be as in* (14)*. Then for all $c \in \mathcal{D}(\partial_{L^2_{(0)}} F)$ we have $\partial_{L^2_{(0)}} F(c) \subseteq \partial_{H^1_{(0)}} F(c)$.*

*Proof* Let $c \in \mathcal{D}(\partial_{L^2_{(0)}} F)$ and $\{w\} = \partial_{L^2_{(0)}} F(c)$. By definition it holds

$$\langle w, c' - c\rangle_{L^2(\Omega)', L^2(\Omega)} = (w, c' - c)_{L^2(\Omega)} \leq F(c') - F(c) \quad \forall\, c' \in L^2_{(0)}(\Omega).$$

Thus we can conclude

$$\langle w, c' - c\rangle_{H^{-1}_{(0)}(\Omega), H^1_{(0)}(\Omega)} \leq F(c') - F(c) \quad \forall\, c' \in H^1_{(0)}(\Omega).$$

This yields $\{w\} \subseteq \partial_{H^1_{(0)}} F(c)$. □

**Lemma 3.8** *Let $w \in \partial_{H^1_{(0)}} F(c)$ and $c \in \mathcal{D}(\partial_{H^1_{(0)}} F)$. Suppose $w \in L^2_{(0)}(\Omega)$. Then $c \in \mathcal{D}(\partial_{L^2_{(0)}} F)$ and*

$$w = \partial_{L^2_{(0)}} F(c) = -\Delta c + P_0 f_0'(c).$$

*Proof* Let $w \in \partial_{H^1_{(0)}} F(c)$. Then

$$\langle w, c' - c \rangle_{H^{-1}_{(0)}(\Omega), H^1_{(0)}(\Omega)} \leq F(c') - F(c) \quad \forall\, c' \in H^1_{(0)}(\Omega).$$

Since $w \in L^2_{(0)}(\Omega)$, we can conclude

$$(w, c' - c)_{L^2(\Omega)} \leq F(c') - F(c) \quad \forall\, c' \in L^2_{(0)}(\Omega).$$

The last conclusion holds since $\mathcal{D}(F) \subseteq H^1_{(0)}(\Omega)$ and $F(c') = +\infty$ for $c' \notin \mathcal{D}(F)$. This yields $w \in \partial_{L^2_{(0)}} F(c)$. Due to (18), we can identify $\{w\}$ and $w$, where $\{w\} = \partial_{L^2_{(0)}} F(c)$. Hence

$$w = \partial_{L^2_{(0)}} F(c) = -\Delta c + P_0 f_0'(c)$$

□

Finally we state the main result about maximal monotone operators on which the proof of Theorem 1.3 is based:

**Theorem 3.9 (Browder)**

1. *Let $C$ be a nonempty, closed and convex subset of a real and reflexive Banach space $X$.*
2. *Let $\mathcal{A} : C \to \mathcal{P}(X')$ be a maximal monotone operator.*
3. *Let $\mathcal{B} : C \to X'$ be a pseudo-monotone, bounded and demi-continuous mapping.*
4. *If the set $C$ is unbounded, then the operator $\mathcal{B}$ is $\mathcal{A}$-coercive with respect to $b \in X'$, i.e., there exists an element $u_0 \in C \cap \mathcal{D}(\mathcal{A})$ and $r > 0$ such that*

$$\langle \mathcal{B}u, u - u_0 \rangle_{X',X} \; > \; \langle b, u - u_0 \rangle_{X',X}$$

*for all $u \in C$ with $||u||_X > r$.*

*Then the problem*

$$b \in \mathcal{A}u + \mathcal{B}u \tag{19}$$

*has a solution $u \in \mathcal{D}(\mathcal{A}) \cap C$.*

The proof of this theorem can be found in [9], cf. Theorem 3.42, or [11], cf. Theorem 32.A.

# 4 Existence of Weak Solutions

## 4.1 Reformulation of the Problem

When working with Definition 1.2, some difficulties arise in the analysis of our problem. The most obvious one is that $L^2_{(\bar{m})}(\Omega)$ is no vector space for $\bar{m} \neq 0$. But w.l.o.g. we can assume $\bar{m} = 0$ in Eq. (7). Otherwise replace $c$ by $c_0 := c - \bar{m}$ and $f$ by $f_{\bar{m}}$ with $f_{\bar{m}}(x) := f(x + \bar{m})$ for all $x \in \mathbb{R}$. Note that this implies $0 = \bar{m} \in (a, b)$.

Moreover we will need that $\mu$ has mean value 0 in the following. Otherwise there will be some difficulties with the coercivity of the operator. But as we will see, we can reduce to this case. To this end we consider the following equations:

$$\int_\Omega (\mathbf{v} \cdot \nabla \mathbf{v}) \cdot \boldsymbol{\varphi} dx + \int_\Omega 2\nu(c)\mathbf{D}\mathbf{v} : \mathbf{D}\boldsymbol{\varphi} dx = \int_\Omega \mu_0 \nabla c \cdot \boldsymbol{\varphi} dx + \int_\Omega \tilde{\boldsymbol{f}} \cdot \boldsymbol{\varphi} dx \tag{20}$$

for all $\boldsymbol{\varphi} \in C^\infty_{0,\sigma}(\Omega)$ and

$$\int_\Omega (\mathbf{v} \cdot \nabla c) \cdot \psi dx = - \int_\Omega m(c) \nabla \mu_0 \cdot \nabla \psi dx \tag{21}$$

$$\int_\Omega \mu_0 \psi dx = \int_\Omega \nabla c \cdot \nabla \psi dx + \int_\Omega P_0 f'(c) \psi dx \tag{22}$$

for all $\psi \in C^\infty(\overline{\Omega})$, where $\mathbf{v} \in H^1_0(\Omega)^d \cap L^2_\sigma(\Omega)$, $\mu_0 \in H^1_{(0)}(\Omega)$, $c \in H^1_{(0)}(\Omega)$ and $f$ as in Assumption 1.1.

**Lemma 4.1** *Let* $(\mathbf{v}, \mu_0, c) \in (H^1_0(\Omega)^d \cap L^2_\sigma(\Omega)) \times H^1_{(0)}(\Omega) \times H^1_{(0)}(\Omega)$ *be a solution of the Eqs.* (20)–(22)*. Then* $(\mathbf{v}, \mu, c)$ *is a weak solution of the stationary Navier-Stokes/Cahn-Hilliard equations in the sense of Definition* 1.2*, where* $\mu = \mu_0 + \frac{1}{|\Omega|} \int_\Omega f'(c) dx$.

*Proof* Let $(\mathbf{v}, \mu_0, c) \in (H^1_0(\Omega)^d \cap L^2_\sigma(\Omega)) \times H^1_{(0)}(\Omega) \times H^1_{(0)}(\Omega)$ be a solution of the Eqs. (20)–(22). Since Eq. (20) is fulfilled, we conclude

$$\int_\Omega (\mathbf{v} \cdot \nabla \mathbf{v}) \cdot \boldsymbol{\varphi} dx + \int_\Omega 2\nu(c)\mathbf{D}\mathbf{v} : \mathbf{D}\boldsymbol{\varphi} dx$$

$$= \int_\Omega \mu_0 \nabla c \cdot \boldsymbol{\varphi} dx + \int_\Omega \overline{\mu}\, \nabla c \cdot \boldsymbol{\varphi} dx + \int_\Omega \tilde{\boldsymbol{f}} \cdot \boldsymbol{\varphi} dx$$

for all $\boldsymbol{\varphi} \in C^\infty_{0,\sigma}(\Omega)$, where $\overline{\mu} := \frac{1}{|\Omega|} \int_\Omega f'(c) dy$.

But this already yields Eq. (11). Equation (12) holds due to Eq. (21). It remains to show that Eq. (13) is fulfilled. Because of Eq. (22) holds, we can deduce

$$\int_\Omega \mu\psi dx = \int_\Omega \nabla c \cdot \nabla\psi dx + \int_\Omega (P_0 f'(c) + \overline{\mu})\psi dx.$$

Using the definitions of $\mu$ and $P_0$ yields Eq. (13) and therefore the lemma is proved. □

In Sect. 4.3 we will prove the existence of a solution for the Eqs. (20)–(22) and thus deduce the existence of a solution for the stationary Navier-Stokes/Cahn-Hilliard equations with the help of the previous lemma.

## 4.2 Abstract Reformulation

First of all, we define the following spaces:

$$X := (H_0^1(\Omega)^d \cap L^2_\sigma(\Omega)) \times H^1_{(0)}(\Omega) \times H^1_{(0)}(\Omega) \equiv X_1 \times X_2 \times X_3 \tag{23}$$

$$C := X_1 \times X_2 \times \{c \in X_3 : c(x) \in [a,b] \text{ a.e.}\} \equiv X_1 \times X_2 \times C_3 \tag{24}$$

Moreover we define a multivalued mapping $\mathcal{A} : C \to \mathcal{P}(X')$ by

$$\mathcal{A}(\mathbf{v}, \mu_0, c) := \begin{cases} \left\{ \begin{pmatrix} 0 \\ 0 \\ \partial_{H^1_{(0)}} \tilde{F}(c) \end{pmatrix} \right\} & \text{if } (\mathbf{v}, \mu_0, c) \in \mathcal{D}(\mathcal{A}), \\ \emptyset & \text{else} \end{cases} \tag{25}$$

where $\mathcal{D}(\mathcal{A}) := X_1 \times X_2 \times \mathcal{D}(\partial_{H^1_{(0)}} \tilde{F})$ and $\tilde{F}$ is defined as in (15), and another operator $\mathcal{B} : C \to X'$ by

$$\begin{aligned} &\langle \mathcal{B}(\mathbf{v}, \mu_0, c), (\boldsymbol{\varphi}, \psi, \phi) \rangle_{X',X} \\ &:= \int_\Omega (\mathbf{v}\cdot\nabla\mathbf{v})\cdot\boldsymbol{\varphi} dx + \int_\Omega 2\nu(c)\mathbf{D}\mathbf{v} : \mathbf{D}\boldsymbol{\varphi} dx - \int_\Omega \mu_0 \nabla c\cdot\boldsymbol{\varphi} dx \\ &+ \int_\Omega m(c)\nabla\mu_0\cdot\nabla\psi dx + \int_\Omega (\mathbf{v}\cdot\nabla c)\psi dx - \int_\Omega \mu_0 \phi dx \\ &- \int_\Omega P_0(\kappa\, c)\phi\, dx + \frac{1}{2}\int_\Omega \nabla c\cdot\nabla\phi dx \end{aligned} \tag{26}$$

for all $(\mathbf{v}, \mu_0, c) \in C$ and $(\boldsymbol{\varphi}, \psi, \phi) \in X$.

In addition, $b \in X'$ is given by

$$\langle b, (\boldsymbol{\varphi}, \psi, \phi)\rangle_{X',X} := \int_\Omega \tilde{\boldsymbol{f}} \cdot \boldsymbol{\varphi} dx$$

for all $(\boldsymbol{\varphi}, \psi, \phi) \in X$.

Because of the Eqs. (25) and (26), the problem $b \in \mathcal{A}(\mathbf{v}, \mu_0, c) + \mathcal{B}(\mathbf{v}, \mu_0, c)$ with $(\mathbf{v}, \mu_0, c) \in C \cap \mathcal{D}(\mathcal{A})$ can be written as

$$\begin{pmatrix} 0 \\ 0 \\ \partial_{H^1_{(0)}} \tilde{F}(c) \end{pmatrix} + \begin{pmatrix} \mathbf{v} \cdot \nabla \mathbf{v} - \operatorname{div}(2\nu(c)\mathbf{D}\mathbf{v}) - \mu_0 \nabla c \\ -\operatorname{div}_N(m(c)\nabla \mu_0) + \mathbf{v} \cdot \nabla c \\ -\mu_0 - P_0(\kappa\, c) - \frac{1}{2}\Delta_N c \end{pmatrix} = \begin{pmatrix} \tilde{\boldsymbol{f}} \\ 0 \\ 0 \end{pmatrix} \tag{27}$$

in $X' \cong X_1' \times X_2' \times X_3'$, where $-\Delta_N : H^1_{(0)}(\Omega) \to H^{-1}_{(0)}(\Omega)$ is defined as

$$\langle -\Delta_N u, \psi \rangle_{H^{-1}_{(0)}(\Omega), H^1_{(0)}(\Omega)} := \int_\Omega \nabla u \cdot \nabla \psi dx$$

for all $\psi \in H^1_{(0)}(\Omega)$ and

$$\langle -\operatorname{div}_N(m(c)\nabla \mu_0), \psi \rangle_{H^{-1}_{(0)}(\Omega), H^1_{(0)}(\Omega)} := \int_\Omega m(c)\nabla \mu_0 \cdot \nabla \psi dx$$

for all $\psi \in H^1_{(0)}(\Omega)$.

First of all, we suppose that the problem $b \in \mathcal{A}(\mathbf{v}, \mu_0, c) + \mathcal{B}(\mathbf{v}, \mu_0, c)$ has a solution. Then we show that this solution is a weak solution to the Navier-Stokes/Cahn-Hilliard equations. The existence of such a solution will be proved in the next section.

**Lemma 4.2** *Let $(\mathbf{v}, \mu_0, c) \in C \cap \mathcal{D}(\mathcal{A})$ be such that $b \in \mathcal{A}(\mathbf{v}, \mu_0, c) + \mathcal{B}(\mathbf{v}, \mu_0, c)$ holds. Then $(\mathbf{v}, \mu_0, c)$ is a weak solution of the stationary Navier-Stokes/Cahn-Hilliard equations in the sense of Definition* 1.2.

*Proof* Let $(\mathbf{v}, \mu_0, c) \in C \cap \mathcal{D}(\mathcal{A})$ be a triple such that $b \in \mathcal{A}(\mathbf{v}, \mu_0, c) + \mathcal{B}(\mathbf{v}, \mu_0, c)$ holds. Because of Eq. (27) we see that $w \in \partial_{H^1_{(0)}} \tilde{F}(c)$ exists, i.e.

$$\langle w, c' - c \rangle_{H^{-1}_{(0)}(\Omega), H^1_{(0)}(\Omega)} \le \tilde{F}(c') - \tilde{F}(c) \quad \text{for all } c' \in H^1_{(0)}(\Omega).$$

Thus we can conclude

$$\langle w, c' - c \rangle_{H^{-1}_{(0)}(\Omega), H^1_{(0)}(\Omega)} + \frac{1}{2}(\nabla c, \nabla c' - \nabla c)_{L^2(\Omega)}$$

$$\leq \tilde{F}(c') - \tilde{F}(c) + \frac{1}{2}(\nabla c, \nabla c' - \nabla c)_{L^2(\Omega)}$$
$$\leq \tilde{F}(c') - \tilde{F}(c) + \frac{1}{4}\|\nabla c'\|^2_{L^2(\Omega)} - \frac{1}{4}\|\nabla c\|^2_{L^2(\Omega)} - \frac{1}{4}\|\nabla(c - c')\|^2_{L^2(\Omega)}$$

for all $c' \in H^1_{(0)}(\Omega)$. Altogether we obtain

$$\langle w, c' - c\rangle_{H^{-1}_{(0)}(\Omega), H^1_{(0)}(\Omega)} + \frac{1}{2}(\nabla c, \nabla c' - \nabla c)_{L^2(\Omega)} \leq F(c') - F(c)$$

for all $c' \in H^1_{(0)}(\Omega)$. Thus $\tilde{w} \in \partial_{H^1_{(0)}} F(c)$, where $\tilde{w} \in H^{-1}_{(0)}(\Omega)$ is defined by

$$\langle \tilde{w}, \psi\rangle_{H^{-1}_{(0)}(\Omega), H^1_{(0)}(\Omega)} := \langle w, \psi\rangle_{H^{-1}_{(0)}(\Omega), H^1_{(0)}(\Omega)} + \frac{1}{2}(\nabla c, \nabla \psi)_{L^2(\Omega)} \tag{28}$$

for all $\psi \in H^1_{(0)}(\Omega)$. Since $(\mathbf{v}, \mu_0, c) \in C \cap \mathcal{D}(\mathcal{A})$ fulfills Eq. (27) by assumption, it holds

$$\mu_0 + P_0(\kappa\, c) + \frac{1}{2}\Delta_N c = w = \tilde{w} + \frac{1}{2}\Delta_N c \quad \text{in } X_3'.$$

Thus we obtain $\mu_0 + P_0(\kappa\, c) = \tilde{w}$ in $X_3'$. From $\mu_0 + P_0(\kappa\, c) \in L^2_{(0)}(\Omega)$ and $c \in \mathcal{D}(\partial_{H^1_{(0)}} F)$ we can deduce with Lemma 3.8 that it holds $\tilde{w} = -\Delta_N c + P_0 f_0'(c)$ in $X_3'$. In particular this yields $f_0'(c) \in L^2(\Omega)$. Altogether

$$\mu_0 + P_0(\kappa\, c) = -\Delta_N c + P_0 f_0'(c) \quad \text{in } X_3'.$$

Thus it holds in $X_3'$:

$$\mu_0 + P_0(\kappa\, c) = -\Delta_N c + P_0 f_0'(c)$$

which implies

$$\mu_0 = -\Delta_N c + P_0(f_0'(c) - \kappa\, c) = -\Delta_N c + P_0(f'(c)).$$

This means that Eq. (22) holds for all $\psi \in C^\infty(\overline{\Omega}) \cap L^2_{(0)}(\Omega)$. That Eq. (20) holds for all $\boldsymbol{\varphi} \in C^\infty_{0,\sigma}(\Omega)$ and (21) holds for all $\psi \in C^\infty(\overline{\Omega}) \cap L^2_{(0)}(\Omega)$ is a direct consequence of (26) and (27). Finally, Lemma 4.1 implies the statement. □

## 4.3 Proof of Existence of Weak Solutions

In Lemma 4.2 we considered the existence of a triple $(\mathbf{v}, \mu_0, c) \in C \cap \mathcal{D}(\mathcal{A})$ which fulfills $b \in \mathcal{A}(\mathbf{v}, \mu_0, c) + \mathcal{B}(\mathbf{v}, \mu_0, c)$. Then we showed that such a triple is a weak solution to the stationary Navier-Stokes/Cahn-Hilliard equations in the sense of Definition 1.2. Now it remains to show that there really exists such a triple.

*Proof of Theorem 1.3* According to Lemma 4.2 it is sufficient to show the existence of a solution $u \in C \cap \mathcal{D}(\mathcal{A})$ to the differential inclusion $b \in \mathcal{A}u + \mathcal{B}u$. To prove the existence of such a solution $u$, we will apply Theorem 3.9 and show that the conditions (1) – (4) are satisfied.

**To (1):** Obviously $C$ is non-empty and closed. Moreover it is easy to show that $C$ is convex. Since $X$ is a real, reflexive Banach space, $C$ and $X$ fulfill condition (1).
**To (2):** First we show that $\mathcal{D}(\partial_{H^1_{(0)}} \tilde{F}) \subseteq C_3$. To this end let $c \in \mathcal{D}(\partial_{H^1_{(0)}} \tilde{F})$. Then $\tilde{F}(c) \neq +\infty$ since $\mathcal{D}(\partial_{H^1_{(0)}} \tilde{F}) \subseteq \operatorname{dom}(\tilde{F})$. This yields $c(x) \in [a, b]$ a.e. since $f_0(x) = +\infty$ for $x \notin [a, b]$. Thus we obtain $c \in C_3$.

Because of Proposition 3.4, the operator $\partial_{H^1_{(0)}} \tilde{F} : H^1_{(0)}(\Omega) \to \mathcal{P}(H^{-1}_{(0)}(\Omega))$ is maximal monotone. Since $\mathcal{A}$ does not depend on $\mathbf{v}$ and $\mu_0$, the operator $\mathcal{A} : X \to \mathcal{P}(X')$ is maximal monotone. Due to $\mathcal{D}(\mathcal{A}) = (X_1 \times X_2 \times \mathcal{D}(\partial_{H^1_{(0)}} \tilde{F})) \subseteq (X_1 \times X_2 \times C_3) = C \subseteq X$, the operator $\mathcal{A} : C \to \mathcal{P}(X')$ is maximal monotone as well.

**To (3):** We split $\mathcal{B}$ into several operators $(\mathcal{B}_i)_{i\in\{1,\dots,8\}}$ with $\mathcal{B} = \mathcal{B}_1 + \ldots + \mathcal{B}_8$ and will show that all operators $\mathcal{B}_i : C \to X'$, $i \in \{1, \ldots, 8\}$, are pseudo-monotone. Moreover we will show that each operator is locally bounded and demi-continuous as well. Because of (26) we define

$$\langle \mathcal{B}_1(\mathbf{v}, \mu_0, c), (\boldsymbol{\varphi}, \psi, \phi)\rangle_{X',X} := \int_\Omega (\mathbf{v} \cdot \nabla \mathbf{v}) \cdot \boldsymbol{\varphi} dx,$$

$$\langle \mathcal{B}_2(\mathbf{v}, \mu_0, c), (\boldsymbol{\varphi}, \psi, \phi)\rangle_{X',X} := \int_\Omega 2\nu(c)\mathbf{D}\mathbf{v} : \nabla \boldsymbol{\varphi} dx,$$

$$\langle \mathcal{B}_3(\mathbf{v}, \mu_0, c), (\boldsymbol{\varphi}, \psi, \phi)\rangle_{X',X} := -\int_\Omega \mu_0 \nabla c \cdot \boldsymbol{\varphi} dx,$$

$$\langle \mathcal{B}_4(\mathbf{v}, \mu_0, c), (\boldsymbol{\varphi}, \psi, \phi)\rangle_{X',X} := \int_\Omega m(c)\nabla \mu_0 \cdot \nabla \psi dx,$$

$$\langle \mathcal{B}_5(\mathbf{v}, \mu_0, c), (\boldsymbol{\varphi}, \psi, \phi)\rangle_{X',X} := \int_\Omega (\mathbf{v} \cdot \nabla c)\psi dx,$$

$$\langle \mathcal{B}_6(\mathbf{v}, \mu_0, c), (\boldsymbol{\varphi}, \psi, \phi)\rangle_{X',X} := -\int_\Omega \mu\, \phi dx,$$

$$\langle \mathcal{B}_7(\mathbf{v}, \mu_0, c), (\boldsymbol{\varphi}, \psi, \phi)\rangle_{X',X} := -\int_\Omega P_0(\kappa\, c)\phi dx,$$

$$\langle \mathcal{B}_8(\mathbf{v}, \mu_0, c), (\boldsymbol{\varphi}, \psi, \phi)\rangle_{X',X} := \frac{1}{2}\int_\Omega \nabla c \cdot \nabla \phi dx.$$

for all $(\boldsymbol{\varphi}, \psi, \phi) \in X$.

**Pseudo-Monotonicity** Using the compact embeddings $H^1(\Omega) \hookrightarrow L^4(\Omega)$ and $L^{\frac{4}{3}}(\Omega) \hookrightarrow H^{-1}(\Omega)$, it is easy to show that the operators $\mathcal{B}_1, \mathcal{B}_3, \mathcal{B}_5, \mathcal{B}_6, \mathcal{B}_7$ and $\mathcal{B}_8$ are completely continuous and therefore pseudo-monotone. Thus it remains to show the pseudo-monotonicity of the operators $\mathcal{B}_2$ and $\mathcal{B}_4$. In the following, we only prove the pseudo-monotonicity of the operator $\mathcal{B}_2$ since the proof for the operator $\mathcal{B}_4$ is almost the same.

To prove the pseudo-monotonicity of $\mathcal{B}_2$, we define another operator $\tilde{\mathcal{B}}_2 : X \times X \to X'$ by

$$\langle \tilde{\mathcal{B}}_2(u_1, u_2), (\boldsymbol{\varphi}, \psi, \phi)\rangle_{X',X} := \int_\Omega 2\nu(c_1)\mathbf{D}\mathbf{v_2} : \mathbf{D}\boldsymbol{\varphi} dx,$$

where $u_i = (\mathbf{v_i}, \mu_{0_i}, c_i) \in X$, $i = 1, 2$, respectively $u = (\mathbf{v}, \mu_0, c) \in X$. Then $\tilde{\mathcal{B}}_2(u, \cdot)$ is hemi-continuous for every fixed $u \in X$. Moreover $\tilde{\mathcal{B}}_2(u, \cdot)$ is monotone since

$$\langle \tilde{\mathcal{B}}_2(u, u_1 - u_2), u_1 - u_2\rangle_{X',X} = \int_\Omega 2\nu(c_1)\mathbf{D}(\mathbf{v_1} - \mathbf{v_2}) : \mathbf{D}(\mathbf{v_1} - \mathbf{v_2})dx \geq 0.$$

Furthermore $\tilde{\mathcal{B}}_2(\cdot, u)$ is completely continuous for all $u \in X$. To this end we consider a sequence $(u_k)_{k\in\mathbb{N}} \subseteq X$ such that $u_k \rightharpoonup u$ in $X$ and $\tilde{u} = (\tilde{\mathbf{v}}, \tilde{\mu_0}, \tilde{c}) \in X$. Then it holds

$$\begin{aligned}\langle \tilde{\mathcal{B}_2}(u_k, \tilde{u}) - \tilde{\mathcal{B}_2}(u, \tilde{u}), (\boldsymbol{\varphi}, \psi, \phi)\rangle_{X',X} &= \int_\Omega 2\Big(\nu(c_k)\mathbf{D}\tilde{\mathbf{v}} - \nu(c)\mathbf{D}\tilde{\mathbf{v}}\Big) : \mathbf{D}\boldsymbol{\varphi} dx \\ &\leq 2||\nu(c_k)\mathbf{D}\tilde{\mathbf{v}} - \nu(c)\mathbf{D}\tilde{\mathbf{v}}||_{L^2(\Omega)}||\boldsymbol{\varphi}||_{H^1(\Omega)}.\end{aligned}$$

To verify $||\nu(c_k)\boldsymbol{D}\tilde{\boldsymbol{v}} - \nu(c)\boldsymbol{D}\tilde{\boldsymbol{v}}||_{L^2(\Omega)} \to 0$, we define

$$(Fc)(x) := f(x, c(x)) := \nu(c(x)) \cdot \mathbf{D}\tilde{\mathbf{v}}(x).$$

Then Lemma 1.19 in [9] yields that $F : L^2(\Omega) \to L^2(\Omega)$ is continuous and bounded. Since $c_k \to c$ in $L^2(\Omega)$, this yields that $\tilde{\mathcal{B}}_2(\cdot, u)$ is completely continuous for all $u \in X$.

Due to Lemma 5.1 below we can conclude that $\mathcal{B}_2$ is pseudo-monotone.

**Boundedness** Since the local boundedness is obvious for every operator $\mathcal{B}_i$, $i = 1, \ldots, 8$, the operator $\mathcal{B}$ is locally bounded as well.

**Demi-Continuity** Since the operators $\mathcal{B}_1, \ldots, \mathcal{B}_8$ are pseudo-monotone and locally bounded, they are also demi-continuous, cf. Chap. 3, Lemma 2.6 in [9], or Proposition 27.7 in [11]. Thus $\mathcal{B}$ is also demi-continuous.

**To (4):** Obviously, the set $C$ is unbounded. We choose $u_0 \equiv (0,0,0) \in C \cap \mathcal{D}(\mathcal{A})$. Then we have to show that there is some $r > 0$ such that

$$< \mathcal{B}(u) - b, u >_{X',X} > 0 \quad \text{for all } u \in C \text{ such that } ||u||_X > r \tag{29}$$

For the operator $\mathcal{B}$ we can write

$$\begin{aligned}\langle \mathcal{B}(u), u\rangle_{X',X} - \langle b, u\rangle_{X',X} &= \int_\Omega 2\nu(c)\mathbf{D}\mathbf{v} : \mathbf{D}\mathbf{v}dx + \int_\Omega m(c)\nabla\mu \cdot \nabla\mu dx \\ &- \int_\Omega \mu c\, dx - \int_\Omega P_0(\kappa\, c)cdx + \frac{1}{2}\int_\Omega \nabla c \cdot \nabla c dx - \int_\Omega \tilde{\boldsymbol{f}} \cdot \mathbf{v}dx.\end{aligned} \tag{30}$$

Using $|c(x)| \leq \max(|a|, |b|)$ a.e. and the fact that the mean value of $\mu_0$ is 0, we can derive the following inequalities:

$$\int_\Omega 2\nu(c)\mathbf{D}\mathbf{v} : \mathbf{D}\mathbf{v}dx \geq \tilde{C}_1\, ||\mathbf{v}||^2_{H^1(\Omega)}$$

$$\int_\Omega m(c)\nabla\mu \cdot \nabla\mu dx \geq \tilde{C}_2 ||\mu||^2_{H^1(\Omega)}$$

$$\int_\Omega \mu c\, dx \leq \tilde{C}_3\, ||\mu||_{H^1(\Omega)}$$

$$\int_\Omega P_0(\kappa\, c)c\, dx = \kappa\, ||c||^2_{L^2(\Omega)} \leq \tilde{C}_4$$

$$\frac{1}{2}\int_\Omega \nabla c \cdot \nabla c dx \geq \tilde{C}_5\, ||c||^2_{H^1(\Omega)}$$

$$\int_\Omega \tilde{\boldsymbol{f}} \cdot \mathbf{v}dx \leq \frac{1}{2}\left(\frac{1}{\tilde{C}_1}||\tilde{\boldsymbol{f}}||^2_{L^2(\Omega)} + \tilde{C}_1||\mathbf{v}||^2_{H^1(\Omega)}\right)$$

for some constants $\tilde{C}_1, \tilde{C}_2, \tilde{C}_3, \tilde{C}_4, \tilde{C}_5 > 0$. With the help of these inequalities, (30) can be estimated as

$$\begin{aligned}\langle \mathcal{B}(u), u\rangle_{X',X} - \langle b, u\rangle_{X',X} &\geq \tilde{C}_1||\mathbf{v}||^2_{H^1(\Omega)} + \tilde{C}_2||\mu||^2_{H^1(\Omega)} - \tilde{C}_3||\mu||_{H^1(\Omega)} - \tilde{C}_4 \\ &\quad + \tilde{C}_5\, ||c||^2_{H^1(\Omega)} - \frac{1}{2\tilde{C}_1}||\tilde{f}||^2_{L^2(\Omega)} - \frac{\tilde{C}_1}{2}||\mathbf{v}||^2_{H^1(\Omega)} \\ &= \frac{\tilde{C}_1}{2}||\mathbf{v}||^2_{H^1(\Omega)} + \underbrace{||\mu||_{H^1(\Omega)}(\tilde{C}_2||\mu||_{H^1(\Omega)} - \tilde{C}_3)}_{=:h(||\mu||_{H^1(\Omega)})} \\ &\quad + \tilde{C}_5\, ||c||^2_{H^1(\Omega)} - \tilde{C}_4 - \frac{1}{2\tilde{C}_1}||\tilde{f}||^2_{L^2(\Omega)}.\end{aligned}$$

Since $\tilde{f} \in L^2(\Omega)^d$ is fixed, $\tilde{C}_4 + \frac{1}{2\tilde{C}_1}||\tilde{f}||^2_{L^2(\Omega)} \leq M_1$ for a constant $M_1 > 0$ large enough. Moreover we deduce

$$\lim_{x\to+\infty} h(x) = +\infty.$$

Thus we can see that the condition (29) is fulfilled for all $(\mathbf{v}, \mu_0, c) \in C$ with $||(\mathbf{v}, \mu_0, c)||_X$ large enough. This yields the theorem. □

# Appendix

**Lemma 5.1** *Let $X$ be a real, reflexive Banach space and $\tilde{\mathcal{A}} : X \times X \to X'$ such that for all $u \in X$:*

1. *$\tilde{\mathcal{A}}(u, .) : X \to X'$ is monotone and hemi-continuous.*
2. *$\tilde{\mathcal{A}}(., u) : X \to X'$ is completely continuous.*

*Then the operator $\mathcal{A} : X \to X'$ defined by $\mathcal{A}(u) = \tilde{\mathcal{A}}(u, u)$ is pseudo-monotone.*

*Proof* Let $(u_k)_{k\in\mathbb{N}} \subseteq X$ be such that $u_k \rightharpoonup u$ in $X$ and

$$\limsup_{k\to\infty}\langle \mathcal{A}(u_k), u_k - u\rangle_{X',X} \leq 0.$$

We have to show that

$$\liminf_{k\to\infty}\langle \mathcal{A}(u_k), u_k - w\rangle_{X',X} \geq \langle \mathcal{A}(u), u - w\rangle_{X',X}$$

for all $w \in X$. Since $\tilde{\mathcal{A}}(u,.)$ is monotone,

$$\langle \tilde{\mathcal{A}}(u_k,u_k) - \tilde{\mathcal{A}}(u_k,w_t), u_k - w_t\rangle_{X',X} \geq 0,$$

where we choose $w_t := (1-t)u + tv$ for $0 < t < 1$ and $v \in X$ arbitrary.

Now we use that $u_k - w_t = (u_k - u) + t(u - v)$ and thus we get two terms. For the first one we use

$$\liminf_{k\to\infty} \langle \tilde{\mathcal{A}}(u_k,u_k) - \tilde{\mathcal{A}}(u_k,w_t), u_k - u\rangle_{X',X} = \liminf_{k\to\infty} \langle \mathcal{A}(u_k), u_k - u\rangle_{X',X},$$

where we used that $\tilde{\mathcal{A}}(\cdot,w_t)$ is completely continuous for every $0 < t < 1$.

This means that for the remaining terms it holds

$$\liminf_{k\to\infty} \left( \langle \tilde{\mathcal{A}}(u_k,u_k) - \tilde{\mathcal{A}}(u_k,w_t), t(u-v)\rangle_{X',X} + \langle \mathcal{A}(u_k), u_k - u\rangle_{X',X} \right) \geq 0.$$

Since $\limsup_{k\to\infty} \langle \mathcal{A}(u_k), u_k - u\rangle_{X',X} \leq 0$ and $0 < t < 1$, we can conclude

$$\liminf_{k\to\infty} \left( \langle \tilde{\mathcal{A}}(u_k,u_k) - \tilde{\mathcal{A}}(u_k,w_t), u-v\rangle_{X',X} + \langle \mathcal{A}(u_k), u_k - u\rangle_{X',X} \right) \geq 0.$$

This yields

$$\liminf_{k\to\infty} \left( \langle \mathcal{A}(u_k), u_k - v\rangle_{X',X} - \langle \tilde{\mathcal{A}}(u_k,w_t), u-v\rangle_{X',X} \right) \geq 0$$

Using that $\tilde{\mathcal{A}}(u_k,\cdot)$ is hemi-continuous for every $k \in \mathbb{N}$ and $\tilde{\mathcal{A}}(\cdot,u)$ is completely continuous for every $u \in X$ yields the lemma. □

## References

1. H. Abels, On a diffuse interface model for two-phase flows of viscous, incompressible fluids with matched densities. Arch. Ration. Mech. Anal. **194**(2), 463–506 (2009)
2. H. Abels, M. Wilke, Convergence to equilibrium for the Cahn-Hilliard equation with a logarithmic free energy. Nonlinear Anal. **67**(11), 3176–3193 (2007)
3. H. Abels, D. Depner, H. Garcke, Existence of weak solutions for a diffuse interface model for two-phase flows of incompressible fluids with different densities. J. Math. Fluid Mech. **15**(3), 453–480 (2013)
4. F. Boyer, Mathematical study of multi-phase flow under shear through order parameter formulation. Asymptot. Anal. **20**(2), 175–212 (1999). ISSN 0921–7134
5. J.W. Cahn, J.E. Hilliard, Free energy of a nonuniform system. I. Interfacial energy. J. Chem. Phys. **28**(2), 258–267 (1958)
6. M.E. Gurtin, D.Polignone, J. Viñals, Two-phase binary fluids and immiscible fluids described by an order parameter. Math. Models Methods Appl. Sci. **6**(6), 815–831 (1996). ISSN 0218–2025

7. P.C. Hohenberg, B.I. Halperin, Theory of dynamic critical phenomena. Rev. Mod. Phys. **49**, 435–479 (1977)
8. C. Liu, J. Shen, A phase field model for the mixture of two incompressible fluids and its approximation by a Fourier-spectral method. Phys. D **179**(3–4), 211–228 (2003). ISSN 0167–2789
9. M. Růžička, *Nichtlineare Funktionalanalysis. Eine Einführung* (Springer, Berlin, 2004)
10. V.N. Starovoĭtov, On the motion of a two-component fluid in the presence of capillary forces. Mat. Zametki **62**(2), 293–305 (1997). ISSN 0025–567X
11. E. Zeidler, *Nonlinear Functional Analysis and Its Applications. II/B* (Springer, New York, 1990), pp. i–xvi and 469–1202. ISBN 0–387–97167–X

# Parabolic Equations on Uniformly Regular Riemannian Manifolds and Degenerate Initial Boundary Value Problems

**Herbert Amann**

*Dedicated to Professor Yoshihiro Shibata on the occasion of his sixtieth birthday*

**Abstract** In this work there is established an optimal existence and regularity theory for second order linear parabolic differential equations on a large class of noncompact Riemannian manifolds. Then it is shown that it provides a general unifying approach to problems with strong degeneracies in the interior or at the boundary.

**Keywords** Degenerate equations • Linear parabolic boundary value problems • Noncompact Riemannian manifolds • Weighted Sobolev spaces

## 1 Introduction

This paper is devoted to second order initial boundary value problems for linear parabolic equations on a wide class of noncompact Riemannian manifolds, termed 'uniformly regular'. Important examples are complete Riemannian manifolds with no boundary and bounded geometry.[1] In this setting there is already a rich theory for linear parabolic equations—predominantly heat equations—based on kernel estimates. Our main interest concerns, however, noncompact Riemannian manifolds with boundary for which very little is known so far (see the following sections for references). Prototypes of such cases are $m$-dimensional Riemannian submanifolds of $\mathbb{R}^n$ with compact boundary or funnel-like ends (cf. Examples 3.5).

[1]Precise definitions of and notations for all terms used in this introduction without further explanation are found in the following sections and the Appendix.

H. Amann (✉)
Math. Institut, Universität Zürich, Winterthurerstr. 190, CH 8057 Zürich, Switzerland
e-mail: herbert.amann@math.uzh.ch

 
H. Amann et al. (eds.), *Recent Developments of Mathematical Fluid Mechanics*,
Advances in Mathematical Fluid Mechanics, DOI 10.1007/978-3-0348-0939-9_4

In order to give the flavor of our main results we consider in this introduction a simplified version of the general problem. Namely, we restrict ourselves to autonomous equations with homogeneous boundary conditions.

Let $M = (M, g)$ be a Riemannian manifold. We set

$$\mathcal{A}u := -\operatorname{div}(a \bullet \operatorname{grad} u), \tag{1}$$

with $a$ being a symmetric positive definite $(1, 1)$-tensor field on $M$ which is bounded and has bounded and continuous first order (covariant) derivatives. This is expressed by saying that $\mathcal{A}$ is a regular uniformly strongly elliptic differential operator.

We assume that $\partial_0 M$ is open and closed in $\partial M$ and $\partial_1 M := \partial M \setminus \partial_0 M$. Then we put

$$\mathcal{B}_0 u := u \text{ on } \partial_0 M, \quad \mathcal{B}_1 u := (\nu \,|\, a \bullet \operatorname{grad} u) \text{ on } \partial_1 M,$$

where these operators are understood in the sense of traces and $\nu$ is the inward pointing unit normal vector field on $\partial_1 M$. Thus $\mathcal{B} := (\mathcal{B}_0, \mathcal{B}_1)$ is the Dirichlet boundary operator on $\partial_0 M$ and the Neumann operator on $\partial_1 M$.

Throughout this paper, $0 < T < \infty$ and $J := [0, T]$. We write $M_T$ for the space time cylinder $M \times J$. Moreover, $\partial = \partial_t$ is the 'time derivative', $\partial M_T := \partial M \times J$ the lateral boundary, and $M_0 = M \times \{0\}$ the 'initial surface' of $M_T$. Then we consider the problem

$$\partial u + \mathcal{A}u = f \text{ on } M_T, \quad \mathcal{B}u = 0 \text{ on } \partial M_T, \quad u = u_0 \text{ on } M_0. \tag{2}$$

The last equation is to be understood as $\gamma_0 u = u_0$ with the 'initial' trace operator $\gamma_0$.

Of course, $\partial_0 M$ or $\partial_1 M$ or both may be empty. In such a situation obvious interpretations and modifications are to be applied.

We are interested in a strong $L_p$-theory for (2). To describe it we have to introduce (fractional order) Sobolev spaces. We always assume that $1 < p < \infty$. The Sobolev space $W_p^k(M)$ is defined for $k \in \mathbb{N}$ to be the completion of $\mathcal{D}(M)$, the space of smooth functions with compact support, in $L_{1,\mathrm{loc}}(M)$ with respect to the norm

$$u \mapsto \Big( \sum_{j=0}^{k} \big\| \, |\nabla^j u|_{g_j^0} \big\|_{L_p(M)}^p \Big)^{1/p}. \tag{3}$$

Here $\nabla = \nabla_g$ is the Levi-Civita covariant derivative and $|\cdot|_{g_j^0}$ the $(0, j)$-tensor norm naturally induced by $g$. Thus $W_p^0(M) = L_p(M)$. Moreover,

$$W_{p,\mathcal{B}}^2(M) := \big\{ u \in W_p^2(M) \; ; \; \mathcal{B}u = 0 \big\}.$$

We also need the space $W_p^{2-2/p}(M)$ which is defined for $p \neq 2$ by real interpolation:

$$W_p^{2-2/p}(M) := \big(L_p(M), W_p^2(M)\big)_{1-1/p,p}.$$

Then

$$W_{p,\mathcal{B}}^{2-2/p}(M) := \begin{cases} \{u \in W_p^{2-2/p}(M)\ ;\ \mathcal{B}u = 0\}, & 3 < p < \infty, \\ \{u \in W_p^{2-2/p}(M)\ ;\ \mathcal{B}_0 u = 0\}, & 3/2 < p < 3, \\ W_p^{2-2/p}(M), & 1 < p < 3/2. \end{cases} \tag{4}$$

We set

$$A := \mathcal{A} \,|\, W_{p,\mathcal{B}}^2(M), \tag{5}$$

considered as an unbounded linear operator in $L_p(M)$ with domain $W_{p,\mathcal{B}}^2(M)$. Then (2) can be expressed as an initial value problem for the evolution equation

$$\dot{u} + Au = f \text{ on } J, \quad u(0) = u_0 \tag{6}$$

in $L_p(M)$.

Now we are ready to formulate our main result in the present model setting. It is a special case of Theorem 3.4

**Theorem 1.1** *Let $M$ be a uniformly regular Riemannian manifold and let $p \notin \{3/2, 3\}$. Suppose that $\mathcal{A}$ is regular and uniformly strongly elliptic. Then* (2) *has for each*

$$(f, u_0) \in L_p\big(J, L_p(M)\big) \times W_{p,\mathcal{B}}^{2-2/p}(M)$$

*a unique solution*

$$u \in L_p\big(J, W_{p,\mathcal{B}}^2(M)\big) \cap W_p^1\big(J, L_p(M)\big).$$

*The map $(f, u_0) \mapsto u$ is linear and continuous.*

*Equivalently: $-A$ generates an analytic semigroup on $L_p(M)$ and has the property of maximal regularity.*

The finite time interval $J$ can be replaced by $\mathbb{R}^+$, provided we impose the additional assumption that the spectrum of $A$ is contained in $[\operatorname{Re} z \geq \gamma]$ for some $\gamma > 0$. This can always be achieved by replacing $A$ by $A + \omega$ for a sufficiently large $\omega > 0$.

On the surface, this theorem looks exactly the same as the very classical existence and uniqueness theorem for second order parabolic equations on open subsets of $\mathbb{R}^m$ with smooth compact boundary (e.g., Ladyzhenskaya et al. [27, Chap. IV] and Denk et al. [17]). However, it is in fact a rather deep-rooted vast generalization thereof since it applies to any uniformly regular Riemannian manifold.

Closely related to uniformly regular Riemannian manifolds are 'singular Riemannian manifolds' which are characterized by a 'singularity function' $\rho \in C^\infty\big(M,(0,\infty)\big)$. More precisely, let $M = (M,g)$ be a Riemannian manifold and consider the conformal metric $\hat{g} := g/\rho^2$ on $M$. Then the basic requirement for $M$ to be a singular Riemannian manifold is that $\hat{M} := (M,\hat{g})$ be a uniformly regular Riemannian manifold. In Examples 4.1 we present some important instances of singular Riemannian manifolds, most notably the class of $m$-dimensional Riemannian submanifolds of $\mathbb{R}^n$ with finitely many cuspidal singularities.

By considering parabolic equations on singular Riemannian manifolds we are naturally led to study degenerate parabolic equations in weighted Sobolev spaces. To be more precise, we now assume that $M = (M,g)$ is a singular Riemannian manifold and $\rho \in C^\infty\big(M,(0,\infty)\big)$ is a singularity function for it. Then $\mathcal{A}$ is said to be a $\rho$-regular uniformly $\rho$-elliptic differential operator if $\rho^{-2}a$ is symmetric, uniformly positive definite, and $\rho^{-2}a$ and $\rho^{-1}\nabla a$ are bounded and continuous. Note that this means that $\mathcal{A}$ is no longer uniformly strongly elliptic but that the ellipticity condition degenerates if $\rho$ tends to zero (or to infinity).

For $\lambda \in \mathbb{R}$ and $k \in \mathbb{N}$ we define the weighted Sobolev space $W_p^{k,\lambda}(M;\rho)$ to be the completion of $\mathcal{D}(M)$ in $L_{1,\mathrm{loc}}(M)$ with respect to the norm

$$u \mapsto \Big(\sum_{j=0}^{k} \big\| \rho^{\lambda+j}\, |\nabla^j u|_{g_j^0} \big\|_{L_p(M)}^p\Big)^{1/p}.$$

Then

$$W_p^{0,\lambda}(M;\rho) = L_p^\lambda(M;\rho) := \big\{\, u \in L_{p,\mathrm{loc}}(M)\ ;\ \rho^\lambda u \in L_p(M)\,\big\}.$$

If $p \neq 2$, then

$$W_p^{2-2/p,\lambda}(M;\rho) := \big(L_p^\lambda(M;\rho), W_p^{2,\lambda}(M;\rho)\big)_{1-1/p,p}.$$

Furthermore, $W_{p,\mathcal{B}}^{2,\lambda}(M;\rho)$ and $W_{p,\mathcal{B}}^{2-2/p,\lambda}(M;\rho)$ are defined analogously to $W_{p,\mathcal{B}}^2(M)$ and $W_{p,\mathcal{B}}^{2-2/p}(M)$, resp. Lastly, $W_p^2(M;\rho) := W_p^{2,0}(M;\rho)$, etc. Note that $L_p^0(M;\rho) = L_p(M)$.

Using this we can now formulate our main result for degenerate parabolic equations in the present setting.

**Theorem 1.2** *Let $M$ be a singular Riemannian manifold, $\rho$ a singularity function for it, and $p \notin \{3/2, 3\}$. Suppose $\mathcal{A}$ is $\rho$-regular and $\rho$-uniformly strongly elliptic. Then* (2) *has for each*

$$(f, u_0) \in L_p\big(J, L_p(M)\big) \times W_p^{2-2/p,2/p}(M;\rho)$$

*a unique solution*

$$u \in L_p\big(J, W^2_{p,\mathcal{B}}(M;\rho)\big) \cap W^1_p\big(J, L_p(M)\big).$$

*The map* $(f, u_0) \mapsto u$ *is linear and continuous.*

*Equivalently: Set* $A := \mathcal{A} \,|\, W^2_{p,\mathcal{B}}(M;\rho)$. *Then* $-A$ *generates a strongly continuous analytic semigroup on* $L_p(M)$ *and has the property of maximal regularity.*

This is a particular instance of Theorem 5.2 and its corollary, both of which apply to general weighted spaces, that is, to $\lambda \neq 0$ as well.

We should like to point out that we impose minimal regularity requirements on $a$ (within the framework of continuous coefficients). This allows to use Theorems 1.1 and 1.2 (and the more general results below) as a basis for the study of quasilinear equations along well-established lines (e.g., [1, 3]). For the sake of brevity we do not give details in this paper.

It should also be noted that only the behavior of $\rho$ near zero and infinity is of importance. In other words, if $\tilde{\rho} \in C^\infty\big(M,(0,\infty)\big)$ satisfies $\tilde{\rho} \sim \rho$, that is, $\rho/c \leq \tilde{\rho} \leq c\rho$ for some $c \geq 1$, then Theorem 1.2 remains valid with $\rho$ replaced by $\tilde{\rho}$. In particular, $W^2_{p,\mathcal{B}}(M;\tilde{\rho})$ equals $W^2_{p,\mathcal{B}}(M;\rho)$ except for equivalent norms.

Now we illustrate the strength of our results by means of relatively simple examples. For this we assume that $\Omega$ is a smooth open subset of $\mathbb{R}^m$ with a compact smooth boundary, that is, $\bar{\Omega}$ is a smooth $m$-dimensional submanifold of $\mathbb{R}^m$. We also assume that

> $\mathbf{\Gamma}$ is a finite family of compact connected smooth submanifolds $\Gamma$ of $\mathbb{R}^m$ without boundary and dimension $\ell_\Gamma \leq m-1$ such that the following applies:
>
> (i) if $\ell_\Gamma = m-1$ and $\Gamma \cap \partial\Omega \neq \emptyset$, then $\Gamma \subset \partial\Omega$.
>
> (ii) if $1 \leq \ell_\Gamma \leq m-2$, then $\Gamma \subset \Omega$.

Then $M := \bar{\Omega} \setminus \bigcup\{\,\Gamma \,;\, \Gamma \in \mathbf{\Gamma}\,\}$, endowed with the Euclidean metric, is an $m$-dimensional Riemannian submanifold of $\mathbb{R}^m$ whose boundary $\partial M$ equals $\partial\Omega \setminus \bigcup\{\,\Gamma \,;\, \Gamma \in \mathbf{\Gamma}\,\}$.

For each $\Gamma \in \mathbf{\Gamma}$ and $x \in M$ we denote by $\delta_\Gamma(x)$ the (Euclidean) distance from $x$ to $\Gamma$. Then $\delta_\Gamma$ is, sufficiently close to $\Gamma$, a well-defined strictly positive smooth function. If $M$ contains a neighborhood of infinity in $\mathbb{R}^m$, that is, if $\Omega$ is an exterior domain, then we put $\delta_\infty(x) := |x|$ with the Euclidean norm $|\cdot|$ in $\mathbb{R}^m$. We also fix $\alpha_\Gamma \geq 1$ and $\alpha_\infty \in (-\infty, 0)$. Then we choose a function $\rho \in C^\infty\big(M,(0,\infty)\big)$ satisfying $\rho \sim \delta_\Gamma^{\alpha_\Gamma}$ near $\Gamma \in \mathbf{\Gamma}$, $\rho \sim \delta_\infty^{\alpha_\infty}$ near infinity if $\Omega$ is an exterior domain, and $\rho \sim \mathbf{1}$ away from the 'singularity set' $\mathcal{S}(M) := \bigcup\{\,\Gamma \,;\, \Gamma \in \mathbf{\Gamma}\,\}$ and infinity. Then $M$ is a singular Riemannian manifold characterized by the singularity function $\rho$. Indeed, see Examples 4.1, each $\Gamma \in \mathbf{\Gamma}$ is an $(\alpha_\Gamma, \ell_\Gamma)$-wedge and $\{\, x \in M \,;\, |x| > R\,\}$ is for sufficiently large $R > 1$ diffeomorphic to an infinite $\alpha_\infty$-cusp (over $S^{m-1}$ in $\mathbb{R}^{m+1}$) if $\Omega$ is an exterior domain. Thus Theorem 1.2 applies to this situation.

Next we consider some particularly simple subcases which have been treated before in the literature.

(a) Suppose $\Omega$ is bounded and $\mathcal{S}(M) = \partial\Omega$. Thus $\rho \sim \delta^\alpha$ for some $\alpha \geq 1$, where $\delta$ is the distance to $\partial\Omega$. In this situation it is shown by Vespri [34] that $A$ generates an analytic semigroup on $L_p(\Omega) = L_p(M)$. Recently, Fornaro et al. [19] have given a new proof for this generation theorem.
(b) Let $\Omega$ be bounded and $\ell_\Gamma = 0$ for each $\Gamma \in \boldsymbol{\Gamma}$. Then $\mathcal{S}(M)$ consists of finitely many one-point sets $\{x_0\}, \ldots, \{x_k\}$ lying either in $\Omega$ or on $\partial\Omega$. We set $\delta_j(x) := |x - x_j|$ for $0 \leq j \leq k$ and $x \in M = \bar{\Omega} \setminus \bigcup_{j=0}^k \{x_j\}$. Assume $\alpha_j \geq 1$ for $0 \leq j \leq k$. Then Theorem 1.2 implies that, given any $\rho \in C^\infty\big(M, (0,\infty)\big)$ satisfying $\rho \sim \delta_j^{\alpha_j}$ near $x_j$ and $\rho \sim 1$ otherwise, $-A = -\mathcal{A} \,|\, W_{p,\mathcal{B}}^2(M;\rho)$ generates a strongly continuous analytic semigroup on $L_p(M) = L_p(\Omega)$ and has the property of maximal regularity.

The only paper known to the author treating the problem of semigroup generation by parabolic equations with strong degeneracies at isolated points is the recent publication of Fragnelli et al. [20]. These authors consider the case where $\Omega = (0,1)$ and $\mathcal{S}(M) = \{x_0\} \subset \Omega$ and show that $-A$ generates an analytic semigroup on $L_2(\Omega)$.

In none of the above papers it is shown that the maximal regularity property prevails. Furthermore, the proofs given there depend significantly on the fact that second order equations are being considered. In contrast, our approach does not depend on the particular structure of the problem but applies equally well to systems and higher order equations (cf. Amann [7]).

Observe that the preceding examples show that a given Riemannian manifold can possess uncountably many non-equivalent singular structures. This is related to and sheds new light on the non-uniqueness results observed by Pozio et al. [30]. Thus, besides being rather general and widely applicable, our approach to highly degenerate parabolic problems via Riemannian manifolds leads to a deeper understanding of such problems as well.

In the next section we give the precise definition of a uniformly regular Riemannian manifold. Then we formulate our main result, Theorem 3.1, in the setting of second order equations and trace it back to the much more general propositions in [7]. Note that, besides allowing lower order terms, we prove an optimal regularity theorem in the presence of nonhomogeneous boundary conditions. In addition, we show that we get classical solutions if we impose slightly stronger regularity assumptions on the data.

Singular Riemannian manifolds are precisely defined in Sect. 4 and basic examples are presented. Furthermore, weighted function spaces are introduced and their interrelation with non-weighted Sobolev-Slobodeckii spaces on uniformly regular manifolds is established.

Section 5 contains our main theorem for second order degenerate parabolic problems involving lower order terms and nonhomogeneous boundary conditions. We attract the reader's attention to Theorem 5.2 where it is shown that problems with homogeneous boundary conditions give rise to generators of analytic semigroups possessing the property of maximal regularity in general weighted spaces $L_p^\lambda(M;\rho)$ for any $\lambda \in \mathbb{R}$. This generalizes results by Barbu et al. [15], for example, where the case $M = \Omega$, with $\Omega$ a bounded domain on $\mathbb{R}^m$, $\mathcal{S}(M) = \partial\Omega$, and $\lambda = -1$ is considered (see also [20]).

For the reader's convenience there is included an Appendix in which some basic facts on tensor bundles over Riemannian manifolds are listed.

## 2 Function Spaces and Uniformly Regular Manifolds

By a *manifold* we always mean a smooth, that is, $C^\infty$ manifold with (possibly empty) boundary such that its underlying topological space is separable and metrizable. Thus, in the context of manifolds, we work in the smooth category. A manifold does not need to be connected, but all connected components are of the same dimension.

Let $M = (M,g)$ be a Riemannian manifold with boundary $\partial M$ and volume measure $dv$. The metric $g$ on $TM$ gives rise to a vector bundle metric on the tensor bundle $V_\tau^\sigma := T_\tau^\sigma M$ for $\sigma,\tau \in \mathbb{N}$, which we denote by $g_\sigma^\tau$ (see the Appendix for more details). In particular, $g_1^0 = g$ and $g_0^1 = g^*$, the adjoint (or contravariant) metric on the cotangent bundle $T^*M$.

For $k \in \mathbb{N}$ the vector space of all $(\sigma,\tau)$-tensor fields of class $C^k$, that is, of all $C^k$ sections of $V_\tau^\sigma$, is denoted by $C^k(V_\tau^\sigma)$. The Levi-Civita covariant derivative, $\nabla$, satisfies $\nabla^k a \in C(V_{\tau+k}^\sigma)$ for $a \in C^k(V_\tau^\sigma)$, where $\nabla^0 a := a$. We write $\mathcal{D}(V_\tau^\sigma)$ for the space of all smooth sections with compact support in $M$ (which may meet the boundary). As usual, $C^k(M)$ stands for $C^k(V_0^0)$, etc.

We fix $\sigma$ and $\tau$ and set $V := V_\tau^\sigma$. Then, given $k \in \mathbb{N}$, we denote by $W_p^k(V)$ the Sobolev space of order $k$, defined to be the completion of $\mathcal{D}(V)$ in $L_{1,\mathrm{loc}}(V) = L_{1,\mathrm{loc}}(V,dv)$ with respect to the norm

$$u \mapsto \Big(\sum_{j=0}^{k} \big\| \, |\nabla^j u|_{g_\sigma^{\tau+j}} \big\|_{L_p(V)}^p\Big)^{1/p}.$$

Thus $W_p^0(V) = L_p(V)$.

We also need fractional order Sobolev spaces, namely the Slobodeckii spaces $W_p^s(V)$, for $s \in \mathbb{R}^+ \setminus \mathbb{N}$. If $k < s < k+1$ with $k \in \mathbb{N}$, then

$$W_p^s(V) := \big(W_p^k(V), W_p^{k+1}(V)\big)_{s-k,p}, \tag{7}$$

which is the interpolation space between $W_p^k(V)$ and $W_p^{k+1}(V)$ obtained by means of the real interpolation method with exponent $s-k$ and integrability parameter $p$.

We denote by $B(V)$ the space of all bounded sections of $V$. It is a Banach space with the norm $u \mapsto \|u\|_\infty := \big\| \, |u| \, \big\|_\infty$, where $\|\cdot\|_\infty$ is the maximum norm.

Moreover, $BC(V) := B(V) \cap C(V)$ is a closed linear subspace thereof. For $k \in \mathbb{N}$ we write $BC^k(V)$ for the linear subspace of $C^k(V)$ consisting of all $u$ satisfying $\nabla^j u \in B(V^\sigma_{\tau+j})$ for $0 \leq j \leq k$. It is a Banach space with the obvious norm. Moreover, $BC^\infty(V) := \bigcap_k BC^k(V)$ and $bc^k(V)$ is the closure of $BC^\infty(V)$ in $BC^k(V)$. Now we define *Besov-Hölder spaces* $B^s_\infty(V)$ for $s > 0$ by

$$B^s_\infty(V) := \begin{cases} \big(bc^k(V), bc^{k+1}(V)\big)_{s-k,\infty}, & k < s < k+1, \\ \big(bc^k(V), bc^{k+2}(V)\big)_{1/2,\infty}, & s = k+1, \end{cases} \tag{8}$$

where $k \in \mathbb{N}$.

Besides these isotropic spaces we also need anisotropic versions adapted to parabolic problems. Anisotropic Sobolev-Slobodeckii spaces are introduced for $s \in \mathbb{R}^+ = [0, \infty)$ by

$$W_p^{(s,s/2)}(V \times J) := L_p\big(J, W_p^s(V)\big) \cap W_p^{s/2}\big(J, L_p(V)\big).$$

The second space on the right is a standard Sobolev-Slobodeckii space of Banach space valued distributions on $\mathring{J}$. Of course, $W_p^{(0,0)}(V \times J)$ is naturally identified with $L_p(V \times J) = L_p\big(V \times J, dvdt\big)$.

Analogously, we define anisotropic Besov-Hölder spaces for $s > 0$ by

$$B_\infty^{(s,s/2)}(V \times J) := B\big(J, B^s_\infty(V)\big) \cap B^{s/2}_\infty\big(J, B(V)\big).$$

Here the second space on the right is a standard Hölder space $C^{s/2}\big(J, B(V)\big)$ if $s \notin 2\mathbb{N}$, and a Zygmund space for $s \in 2\mathbb{N}$, of Banach space valued functions on $J$ (see Lunardi [28], for example). For $k \in \mathbb{N}^\times := \mathbb{N} \setminus \{0\}$ we put

$$BC^{(k,k/2)}(V \times J) := C\big(J, BC^k(V)\big) \cap C^{k/2}\big(J, B(V)\big),$$

recalling that $J$ is compact.

Although Sobolev-Slobodeckii spaces, respectively Besov-Hölder spaces, are well-defined for each $s \in \mathbb{R}^+$, respectively $s > 0$, they are not too useful on general Riemannian manifolds since, for example, the fundamental Sobolev type embedding theorems may not hold in general. Even more importantly, there may be no characterization by local coordinates. For this reason we restrict ourselves to the class of *uniformly regular Riemannian manifolds*. Loosely speaking, $M$ is a uniformly regular Riemannian manifold if its differentiable structure is induced by an atlas $\mathfrak{K}$ of finite multiplicity whose coordinate patches are all of comparable size, such that $\mathfrak{K}$ can be uniformly shrunk to an atlas for $M$, and the family of all charts in $\mathfrak{K}$ which intersect $\partial M$ induces an atlas of the same type for $\partial M$. In particular, $\partial M = (\partial M, \mathring{g})$, where $\mathring{g}$ is the Riemannian metric induced by $g$ on $\partial M$, is a uniformly regular $(m-1)$-dimensional Riemannian manifold if $M$ is uniformly regular (see Example 4.1(b)).

For the precise definition of a uniformly regular Riemannian manifold we introduce some notation and conventions. By $c$ we denote constants $\geq 1$ whose numerical value may vary from occurrence to occurrence; but $c$ is always independent of the free variables in a given formula, unless a dependence is explicitly indicated.

We denote by $\mathbb{H}^m$ the closed right half-space $\mathbb{R}^+ \times \mathbb{R}^{m-1}$ in $\mathbb{R}^m$, where $\mathbb{R}^0 = \{0\}$. The Euclidean metric on $\mathbb{R}^m$, $(dx^1)^2 + \cdots + (dx^m)^2$, is denoted by $g_m$. The same symbol is used for its restriction to an open subset $U$ of $\mathbb{R}^m$ or $\mathbb{H}^m$, that is, for $\iota^* g_m$, where $\iota : U \hookrightarrow \mathbb{R}^m$ is the natural embedding. Here and below, we employ standard definitions of pull-back and push-forward operations.

On the space of all nonnegative functions, defined on some nonempty set whose specific form will be clear in any given situation, we introduce an equivalence relation $\sim$ by setting $f \sim g$ iff there exists $c \geq 1$ such that $f/c \leq g \leq cf$. Inequalities between vector bundle metrics have to be understood in the sense of quadratic forms. By $\mathbf{1}$ we denote the constant function $s \mapsto 1$, whose domain will always be clear from the context.

We set $Q := (-1, 1) \subset \mathbb{R}$. If $\kappa$ is a local chart for an $m$-dimensional manifold $M$, then we write $U_\kappa$ for the corresponding coordinate patch $\operatorname{dom}(\kappa)$. A local chart $\kappa$ is *normalized* (at $q$) if $\kappa(U_\kappa) = Q^m$ whenever $U_\kappa \subset \mathring{M}$, the interior of $M$, whereas $\kappa(U_\kappa) = Q^m \cap \mathbb{H}^m$ if $U_\kappa \cap \partial M \neq \emptyset$ (and $\kappa(q) = 0$). We put $Q^m_\kappa := \kappa(U_\kappa)$ if $\kappa$ is normalized.

An atlas $\mathfrak{K}$ for $M$ has *finite multiplicity* if there exists $k \in \mathbb{N}$ such that any intersection of more than $k$ coordinate patches is empty. In this case

$$\mathfrak{N}(\kappa) := \{\, \tilde{\kappa} \in \mathfrak{K} \,;\ U_{\tilde{\kappa}} \cap U_\kappa \neq \emptyset \,\}$$

has cardinality $\leq k$ for each $\kappa \in \mathfrak{K}$. An atlas is *uniformly shrinkable* if it consists of normalized charts and there exists $r \in (0, 1)$ such that the family $\{\kappa^{-1}(rQ^m_\kappa) \,;\ \kappa \in \mathfrak{K}\}$ is a cover of $M$. We put $\mathfrak{K}_S := \{\, \kappa \in \mathfrak{K} \,;\ U_\kappa \cap S \neq \emptyset \,\}$ for any nonempty subset $S$ of $M$.

Given an open subset $X$ of $\mathbb{R}^m$ or $\mathbb{H}^m$ and a Banach space $\mathcal{X}$, we write $\|\cdot\|_{k,\infty}$ for the usual norm of $BC^k(X, \mathcal{X})$, the Banach space of all $u \in C^k(X, \mathcal{X})$ such that $|\partial^\alpha u|_{\mathcal{X}}$ is uniformly bounded for $\alpha \in \mathbb{N}^m$ of length at most $k$.

An atlas $\mathfrak{K}$ for $M$ is *uniformly regular* if

$$\begin{aligned} &\text{(i)} \quad \mathfrak{K} \text{ is uniformly shrinkable and has finite multiplicity.} \\ &\text{(ii)} \quad \|\tilde{\kappa} \circ \kappa^{-1}\|_{k,\infty} \leq c(k), \quad \kappa, \tilde{\kappa} \in \mathfrak{K}, \quad k \in \mathbb{N}. \end{aligned} \tag{9}$$

In (ii) and in similar situations it is understood that only $\kappa, \tilde{\kappa} \in \mathfrak{K}$ with $U_\kappa \cap U_{\tilde{\kappa}} \neq \emptyset$ are being considered. Two uniformly regular atlases $\mathfrak{K}$ and $\tilde{\mathfrak{K}}$ are *equivalent*, $\mathfrak{K} \approx \tilde{\mathfrak{K}}$, if

$$\begin{aligned} &\text{(i)} \quad \operatorname{card}\{\, \tilde{\kappa} \in \tilde{\mathfrak{K}} \,;\ U_{\tilde{\kappa}} \cap U_\kappa \neq \emptyset \,\} \leq c, \quad \kappa \in \mathfrak{K}; \\ &\text{(ii)} \quad \|\tilde{\kappa} \circ \kappa^{-1}\|_{k,\infty} + \|\kappa \circ \tilde{\kappa}^{-1}\|_{k,\infty} \leq c(k), \quad \kappa \in \mathfrak{K}, \quad \tilde{\kappa} \in \tilde{\mathfrak{K}}, \quad k \in \mathbb{N}. \end{aligned} \tag{10}$$

A uniformly regular structure is a maximal family of equivalent uniformly regular atlases. A **uniformly regular manifold** is a manifold endowed with a uniformly regular structure. Clearly, on such a manifold all local charts, atlases, etc. under consideration belong to its uniformly regular structure. An $m$-dimensional Riemannian manifold $(M, g)$ is a **uniformly regular Riemannian manifold** if

$$\begin{aligned} &\text{(i)} && M \text{ is uniformly regular;} \\ &\text{(ii)} && \kappa_* g \sim g_m, \quad \kappa \in \mathfrak{K}; \\ &\text{(iii)} && \|\kappa_* g\|_{k,\infty} \le c(k), \quad \kappa \in \mathfrak{K}, \quad k \in \mathbb{N}, \end{aligned} \tag{11}$$

for some uniformly regular atlas $\mathfrak{K}$ for $M$.

Let $M$ be a uniformly regular Riemannian manifold. Then the Sobolev-Slobodeckii and Besov-Hölder space scales possess all the properties known to hold in the case of the $m$-dimensional Euclidean space or half-space. In other words, there are embedding, interpolation, and trace theorems for $W_p^s(M)$ and $W_p^{(s,s/2)}(M \times J)$ which are completely analogous to the corresponding theorems for the classical Sobolev-Slobodeckii spaces. In particular, the anisotropic Sobolev-Morrey type embedding theorem

$$W_p^{(s,s/2)}(V \times J) \hookrightarrow B_\infty^{(t,t/2)}(V \times J), \qquad s + (m+2)/p > t > 0, \tag{12}$$

is valid. In addition, $W_p^s(V)$ and $B_\infty^s(V)$ can be characterized by means of local coordinates, similarly as in the case of compact manifolds.

The spaces $BC^k(V)$ and $BC^{(k,k/2)}(V)$ do not belong to either one of these scales. However, they can be arbitrarily well approximated by Besov-Hölder spaces. In fact, given $k \in \mathbb{N}^\times$,

$$B_\infty^{(s_1,s_1/2)}(V \times J) \hookrightarrow BC^{(k,k/2)}(V \times J) \hookrightarrow B_\infty^{(s_0,s_0/2)}(V \times J) \tag{13}$$

for $0 < s_0 < k < s_1$. Note that this implies a corresponding assertion for the isotropic spaces $BC^k(V)$ and $B_\infty^s(V)$, since $BC^k(V)$ is naturally identified with the closed linear subspace of $BC^{(k,k/2)}(V)$ of all 'time-independent' functions therein, etc.

Proofs, further results, references to related research, and many more details—in particular spaces of sections of general uniformly regular vector bundles over $M$—are found in the earlier work [5, 6] of the author (also see [7], [8], as well as [4]).

## 3 Parabolic Problems on Uniformly Regular Riemannian Manifolds

Let $M = (M, g)$ be a uniformly regular Riemannian manifold. We consider parabolic initial boundary value problems of the form

$$\partial u + \mathcal{A}u = f \text{ on } M_T, \quad \mathcal{B}u = h \text{ on } \partial M_T, \quad u = u_0 \text{ on } M_0. \tag{14}$$

In order to reduce the technical apparatus to a minimum we restrict ourselves to the important class of second order divergence form problems. Thus we fix $\delta \in C\big(\partial M, \{0, 1\}\big)$ and set $\partial_j M := \delta^{-1}(j)$ for $j \in \{0, 1\}$. Then $\partial M = \partial_0 M \cup \partial_1 M$ and $\partial_0 M \cap \partial_1 M = \emptyset$. We assume that $(\mathcal{A}, \mathcal{B})$ is of the form

$$\mathcal{A}u := -\operatorname{div}(a \bullet \operatorname{grad} u) + (\vec{a} \,|\, \operatorname{grad} u) + a_0 u \tag{15}$$

and

$$\mathcal{B}u := \begin{cases} \mathcal{B}_0 u & \text{on } \partial_0 M_T, \\ \mathcal{B}_1 u & \text{on } \partial_1 M_T, \end{cases}$$

where

$$\mathcal{B}_0 u := \gamma u, \quad \mathcal{B}_1 u := \big(\nu \,\big|\, \gamma(a \bullet \operatorname{grad} u)\big) + b_0 \gamma u.$$

Here $(\cdot|\cdot) = (\cdot|\cdot)_g := g(\cdot, \cdot)$, $\nu$ is the (inward pointing) unit normal on $\partial M$, $\gamma$ the trace map for $\partial M$, and $\bullet$ denotes complete contraction (see the Appendix). More precisely, $\mathcal{B}_0 u = (\gamma u) \,|\, \partial_0 M$, etc. We suppose $a \in C^1(T_1^1 M \times J)$, $\vec{a}$ is a time-dependent vector field, $a_0$ a function on $M_T$, and $b_0$ one on $\partial_1 M_T$. In local coordinates,

$$\mathcal{A}u = -\frac{1}{\sqrt{g}}\, \partial_i\big(\sqrt{g}\, a^i_j g^{jk} \partial_k u\big) + a^i \partial_i u + a_0 u.$$

Hence $\mathcal{B}$ is the Dirichlet boundary operator on $\partial_0 M$ and the Neumann or a Robin boundary operator on $\partial_1 M$. Note that either $\partial_0 M$ or $\partial_1 M$ may be empty. We also allow $M$ to be a manifold without boundary. In this case it is understood throughout the whole paper that all statements, assumptions, and formulas referring explicitly or implicitly to $\partial M$ are to be unconsidered. For example, problem (14) reduces to the Cauchy problem

$$\partial u + \mathcal{A}u = f \text{ on } M_T, \quad u = u_0 \text{ on } M_0$$

if $\partial M = \emptyset$.

A function $u$ satisfying (14) is a *strong $L_p$ solution* if it belongs to $W_p^{(2,1)}(M_T)$, and a *classical solution* if it is a member of $BC^{(2,1)}(M_T)$.

The differential operator $\mathcal{A}$ is *uniformly strongly elliptic on $M_T$* if $a(\cdot,t)$ is symmetric and uniformly positive definite, uniformly with respect to $t \in J$. Clearly, the latter means that there exists a constant $\varepsilon > 0$ such that

$$\big(a(q,t) \cdot X \,\big|\, X\big)_{g(q)} \geq \varepsilon\, |X|^2_{g(q)}, \qquad X \in T_qM, \quad q \in M, \quad t \in J.$$

For a concise formulation of the main result we introduce for $s \geq 0$ the boundary data spaces

$$\begin{aligned} &W_p^{(s+2-\delta-1/p)(1,1/2)}(\partial M_T) \\ &\qquad := W_p^{(s+2-1/p)(1,1/2)}(\partial_0 M_T) \times W_p^{(s+1-1/p)(1,1/2)}(\partial_1 M_T), \end{aligned}$$

whose general point is written $h = (h_0, h_1)$. Obvious interpretations apply if either $\partial_0 M$ or $\partial_1 M$ is empty. The *total data spaces* are then

$$\begin{aligned} &\mathcal{W}_p^{(s+2,(s+2)/2)}(M_T) \\ &\qquad := W_p^{(s,s/2)}(M_T) \times W_p^{(s+2-\delta-1/p)(1,1/2)}(\partial M_T) \times W_p^{s+2-2/p}(M_0) \end{aligned}$$

for $s \geq 0$.

Given Banach spaces $E$ and $F$, we denote by $\mathcal{L}(E,F)$ the Banach space of bounded linear operators from $E$ into $F$. We write $\mathcal{L}\mathrm{is}(E,F)$ for the subset of all bijections in $\mathcal{L}(E,F)$. Banach's homomorphism theorem guarantees that $A^{-1} \in \mathcal{L}(F,E)$ if $A \in \mathcal{L}\mathrm{is}(E,F)$.

Now we can formulate the main existence and uniqueness theorem for problem (14).

**Theorem 3.1** *Let $M$ be a uniformly regular Riemannian manifold and let $p \notin \{3/2, 3\}$. Suppose*

$$\begin{aligned} a \in BC^{(1,1/2)}(T_1^1 M \times J), &\quad \vec{a} \in L_\infty(TM \times J), \\ a_0 \in L_\infty(M_T), &\quad b_0 \in BC^{(1,1/2)}(\partial_1 M_T), \end{aligned} \tag{16}$$

*and $\mathcal{A}$ is uniformly strongly elliptic. Denote by $\mathcal{W}_{p,\mathrm{cc}}^{(2,1)}(M_T)$ the vector space of all $(f,h,u_0) \in \mathcal{W}_p^{(2,1)}(M_T)$ satisfying the compatibility conditions of order zero:*

$$\begin{aligned} \gamma u_0 = h_0(\cdot,0) \ \text{ on } \partial_0 M \quad &\text{if } 3/2 < p < 3, \\ \mathcal{B}(\cdot,0) u_0 = h(\cdot,0) \ \text{ on } \partial M \quad &\text{if } p > 3. \end{aligned} \tag{17}$$

*Then* $\mathcal{W}_{p,cc}^{(2,1)}(M_T)$ *is closed in* $\mathcal{W}_p^{(2,1)}(M_T)$ *and*

$$(\partial + \mathcal{A},\ \mathcal{B}, \gamma_0) \in \mathcal{L}\mathrm{is}\big(W_p^{(2,1)}(M_T), \mathcal{W}_{p,cc}^{(2,1)}(M_T)\big). \tag{18}$$

**Supplement** Suppose $0 < s < \bar{s} < 1 + 3/p$ with $s \neq 3/p$ and

$$a \in B_\infty^{(1+\bar{s})(1,1/2)}(T_1^1 M \times J), \quad \vec{a} \in B_\infty^{(\bar{s},\bar{s}/2)}(TM \times J), \tag{19}$$
$$a_0 \in B_\infty^{(\bar{s},\bar{s}/2)}(M_T), \quad b_0 \in B_\infty^{(1+\bar{s})(1,1/2)}(\partial_1 M_T).$$

Let $\mathcal{W}_{p,cc}^{(s+2)(1,1/2)}(M_T)$ be the linear subspace of $\mathcal{W}_p^{(s+2)(1,1/2)}(M_T)$ of all $(f, h, u_0)$ satisfying, in addition to (17), the first order compatibility condition

$$\partial h_0(\cdot, 0) + \gamma\mathcal{A}(\cdot, 0)u_0 = \gamma f(\cdot, 0) \text{ on } \partial_0 M \quad \text{if } s > 2/p. \tag{20}$$

Then $\mathcal{W}_{p,cc}^{(s+2)(1,1/2)}(M_T)$ is closed and

$$(\partial + \mathcal{A},\ \mathcal{B}, \gamma_0) \in \mathcal{L}\mathrm{is}\big(W_p^{(s+2)(1,1/2)}(M_T), \mathcal{W}_{p,cc}^{(s+2)(1,1/2)}(M_T)\big).$$

*Proof* We set $a_2 := -a^\sharp$ and $a_1 := \vec{a} - \mathrm{div}(a^\sharp)$, using the notations of the Appendix. Then we get from (75)

$$\mathcal{A} = a_2 \bullet \nabla^2 + a_1 \bullet \nabla + a_0. \tag{21}$$

We let $\nu_\flat$ be the unit conormal vector field $g_\flat \nu$ on $\partial M$ and set $b_1 := \nu_\flat \bullet \gamma a^\sharp$. Then

$$\mathcal{B}_1 = b_1 \bullet \gamma\nabla + b_0\gamma. \tag{22}$$

By means of the characterization of $BC^k(T^*M)$ by local coordinates referred to in the preceding section one verifies

$$\nu_\flat \in BC^\infty(T^*M). \tag{23}$$

Let (16) be satisfied. Then it is obvious that

$$a_2 \in BC^{(1,1/2)}(T_0^2 M \times J), \quad a_1 \in L_\infty(TM \times J), \quad a_0 \in L_\infty(M_T).$$

Furthermore, (23) implies

$$b_1 \in BC^{(1,1/2)}\big((TM)_{|\partial_1 M} \times J\big).$$

If (19) applies, then

$$a_2 \in B_\infty^{(1+\bar{s})(1,1/2)}(T_0^2M \times J), \quad a_1 \in B_\infty^{(\bar{s},\bar{s}/2)}(TM \times J),$$
$$a_0 \in B_\infty^{(\bar{s},\bar{s}/2)}(M_T),$$

and, once more by (23) and the point-wise multiplier result [5, Theorem 14.3],

$$b_1 \in B_\infty^{(1+\bar{s})(1,1/2)}\big((TM)_{|\partial_1 M} \times J\big).$$

This shows that $(\mathcal{A}, \mathcal{B})$ satisfies in either case the regularity assumptions of the main theorem of [7]. Since the uniform strong ellipticity of $\mathcal{A}$ implies that $(\partial + \mathcal{A}, \mathcal{B})$ is a uniformly strongly parabolic boundary value problem the assertion is a very particular consequence of the latter theorem. □

**Corollary 3.2** *Let* (16) *be satisfied. Then the initial boundary value problem* (14) *has for each* $(f, h, u_0) \in \mathcal{W}_{p,\mathrm{cc}}^{(2,1)}(M_T)$ *a unique strong* $L_p$ *solution* $u$ *on* $M_T$. *Suppose* $(m+2)/p < s < 1 + 3/p$ *with* $s \neq 3/p$, (19) *applies, and* $(f, h, u_0) \in \mathcal{W}_{p,\mathrm{cc}}^{(s+2)(1,1/2)}(M_T)$, *then* $u$ *is a classical solution.*

*Proof* The first assertion is clear and the second one follows from (12). □

*Remarks 3.3*

(a) If $\delta = 0$ (Dirichlet boundary value problem), then $p = 3$ is admissible as well. If $\delta = \mathbf{1}$ (Neumann or Robin boundary conditions), then $p = 3/2$ can be admitted also. Similarly, (20) is vacuous if $\delta = 1$.
(b) If all data are smooth and the compatibility conditions of all orders are satisfied, then $u$ is a smooth solution on $M_T$.
(c) We refer to [7] for higher order problems and operators acting on sections of general uniformly regular vector bundles over $M$.
(d) Theorem 3.1 is the basis for establishing results on the existence, uniqueness, and continuous dependence on the data of solutions of quasilinear parabolic problems of the form

$$\partial u + \mathcal{A}(u)u = F(u) \text{ on } M_T, \quad \mathcal{B}(u)u = H(u) \text{ on } \partial M_T, \quad u = u_0 \text{ on } M_0.$$

Such results are obtained by (more or less obvious) modifications of the proofs in [3]. This is to be carried out somewhere else. □

Of course, Theorem 3.1 applies in particular to autonomous problems. To simplify the presentation we restrict ourselves to the setting of strong $L_p$ solutions. Then (16) reduces to

$$\begin{aligned} &a \in BC^1(T_1^1M), && \vec{a} \in L_\infty(TM), \\ &a_0 \in L_\infty(M), && b_0 \in BC^1\big((TM)_{|\partial_1 M}\big). \end{aligned} \tag{24}$$

Of particular importance is the case of homogeneous boundary value problems.

Theorem 3.1 guarantees $\mathcal{A} \in \mathcal{L}\big(W_p^2(M), L_p(M)\big)$. Hence $A$, the restriction of $\mathcal{A}$ to $W_{p,\mathcal{B}}^2(M)$, is a well-defined element of $\mathcal{L}\big(W_{p,\mathcal{B}}^2(M), L_p(M)\big)$. Moreover, $A$ is closed (cf. [1, Lemma I.1.1.2]) and densely defined (since $\mathcal{D}(\mathring{M})$ is a subset of $W_{p,\mathcal{B}}^2(M)$ and $\mathcal{D}(\mathring{M})$ is dense in $L_p(M)$). By means of $A$ we can reformulate the autonomous homogeneous initial boundary value problem (2) as the evolution equation (6). This is made precise by the next theorem for which we rely on semigroup theory and maximal regularity (see Amann [1, Chap. III] and [2], Denk et al. [17], or Kunstmann et al. [26], for example, for information on these concepts).

**Theorem 3.4** *Let $M$ be a uniformly regular Riemannian manifold and let $p \notin \{3/2, 3\}$. Suppose $\mathcal{A}$ is autonomous, uniformly strongly elliptic, and conditions* (24) *are satisfied. Then $-A$ generates a strongly continuous analytic semigroup on $L_p(M)$ and has the property of maximal regularity, that is to say, $(\partial + A,\ \gamma_0)$ belongs to*

$$\mathcal{L}\mathrm{is}\big(L_p\big(J, W_{p,\mathcal{B}}^2(M)\big) \cap W_p^1\big(J, L_p(M)\big), L_p(M_T) \times W_{p,\mathcal{B}}^{2-2/p}(M)\big). \tag{25}$$

*Proof* In the present setting $\mathcal{W}_{p,\mathrm{cc}}^{(2,1)}(M_T) = L_p(M_T) \times W_{p,\mathcal{B}}^{2-2/p}(M)$. Hence (25) is a reformulation of (18). Now the semigroup assertion follows from a result of Dore [18]. □

To indicate the power of these theorems we need to know examples of uniformly regular Riemannian manifolds. This problem is dealt with in [9] where proofs for the following claims are found.

*Examples 3.5*

(a) Every compact Riemannian manifold is a uniformly regular Riemannian manifold.
(b) An $m$-dimensional Riemannian submanifold of $\mathbb{R}^m$ possessing a compact boundary is a uniformly regular Riemannian manifold.
(c) $\mathbb{R}^m = (\mathbb{R}^m, g_m)$ and $\mathbb{H}^m = (\mathbb{H}^m, g_m)$ are uniformly regular Riemannian manifolds.
(d) Let $\tilde{M} = (\tilde{M}, \tilde{g})$ be a Riemannian manifold and $\varphi : (M, g) \to (\tilde{M}, \tilde{g})$ an isometry. Then $M$ is a uniformly regular Riemannian manifold iff $\tilde{M}$ is one.
(e) A Riemannian manifold has *bounded geometry* if it has no boundary, a positive injectivity radius, and all covariant derivatives of the curvature tensor are bounded. Every complete Riemannian manifold with bounded geometry is a uniformly regular Riemannian manifold.
(f) Suppose $S \subset U \subset M$, where $S$ is closed and $U$ is open in $M$. An atlas $\mathfrak{K}$ for $U$ is *uniformly regular on* $S$ if (9) holds with $\mathfrak{K}$ replaced by $\mathfrak{K}_S$. Two uniformly regular atlases $\mathfrak{K}$ and $\tilde{\mathfrak{K}}$ for $U$ on $S$ are equivalent if (10) applies to $\mathfrak{K}_S$ and $\tilde{\mathfrak{K}}_S$. This defines a uniformly regular structure for $U$ *on* $S$. Then $U$ is *uniformly regular on* $S$ if it is endowed with a uniformly regular structure on $S$. Lastly, $U$ is a *uniformly regular Riemannian manifold on* $S$ if (11) is satisfied for $U$ and $\mathfrak{K}_S$, where $\mathfrak{K}$ is a uniformly regular atlas for $U$ on $S$.

Let $S_j \subset U_j \subset M$ and suppose $U_j$ is a uniformly regular Riemannian manifold on $S_j$ for $0 \le j \le \ell$. Let $\mathfrak{K}_j$ be a uniformly regular atlas for $U_j$ on $S_j$. Assume

($\alpha$) $\|\kappa_i \circ \kappa_j^{-1}\|_{k,\infty} \le c(k)$, $(\kappa_i, \kappa_j) \in \mathfrak{K}_{i,S_i} \times \mathfrak{K}_{j,S_j}$, $0 \le i,j \le \ell$, $k \in \mathbb{N}$;

($\beta$) $M = S_0 \cup \cdots \cup S_\ell$.

Then $\mathfrak{K} := \mathfrak{K}_0 \cup \cdots \cup \mathfrak{K}_\ell$ is a uniformly regular atlas for $M$ and $M$ is a uniformly regular Riemannian manifold. It is obtained by *patching together the uniformly regular pieces $U_j$ on $S_j$.*

(g) Assume $d \ge m$ and $B$ is an $(m-1)$-dimensional compact submanifold of $\mathbb{R}^{d-1}$. For a nonempty subinterval $I$ of $(1,\infty)$ and $0 \le \alpha \le 1$ we set

$$F_\alpha(I,B) := \{ (t, t^\alpha y) \; ; \; t \in I, \; y \in B \} \subset \mathbb{R} \times \mathbb{R}^{d-1} = \mathbb{R}^d,$$

Then $F_\alpha(B) := F_\alpha\big((1,\infty), B\big)$, endowed with the Riemannian metric induced by $\mathbb{R}^d$, that is, by $g_d$, is an $m$-dimensional Riemannian submanifold of $\mathbb{R}^d$ with boundary $F_\alpha(\partial B)$, where $F_\alpha(\emptyset) := \emptyset$. It is called *$\alpha$-funnel* in $\mathbb{R}^d$. Note that a 0-funnel is a cylinder and a 1-funnel a (blunt) cone over (the basis) $B$.

Let $F = F_\alpha(B)$ be an $\alpha$-funnel in $\mathbb{R}^d$ and set $S := F_\alpha\big([2,\infty), B\big)$. Then $F$ is an $m$-dimensional uniformly regular Riemannian manifold on $S$.

(h) Suppose $U$ is open in $M$ and $F = F_\alpha(B)$ an $m$-dimensional $\alpha$-funnel in $\mathbb{R}^d$. Set $F(I) := F_\alpha(I,B)$. Assume $\varphi : U \to F$ is a diffeomorphism such that $S := \varphi^{-1}\big(F[2,\infty)\big)$ satisfies

($\alpha$) $(\overline{U \setminus S}) \cap S = \varphi^{-1}\big(F(\{2\})\big)$;

($\beta$) $\varphi_*(g \,|\, S) \sim g_F \,|\, F[2,\infty)$.

Then $U$ is a uniformly regular Riemannian manifold on $S$, and $M$ is said to have an *$(\alpha, B)$-funnel-like end in $U$ with representation $\varphi$.*

(i) Let $U_0, \ldots, U_\ell$ be open in $M$. Suppose

($\alpha$) $U_i \cap U_j = \emptyset$, $1 \le i < j \le \ell$;

($\beta$) $M$ has an $(\alpha_j, B_j)$-funnel-like end in $U_j$ with representation $\varphi_j$ for $j \ge 1$;

($\gamma$) $\varphi_j(U_0 \cap U_j) = F_j(1,4)$, $j \ge 1$;

($\delta$) $S_0 := U_0 \setminus \bigcup_{j=1}^{\ell} \varphi_j^{-1}\big(F_j(3,\infty)\big)$ is compact.

Then $M$ is a uniformly regular Riemannian manifold, a *Riemannian manifold with finitely many funnel-like ends*. It is obtained by patching together the uniformly regular pieces $U_j$ on $S_j$ for $0 \le j \le \ell$. □

The most elementary situation in which Theorem 3.1 applies is the case in which $M$ is compact. If, notably, $M$ is the closure of a smooth bounded open subset of $\mathbb{R}^m$, then our theorem reduces essentially to a well-known classical result (e.g., [27]).

More recently, Grubb [22] has established a general $L_p$ theory for parabolic pseudo-differential boundary value problems acting on sections of vector bundles (also see Sect. IV.4.1 in [23]). It applies to a class of noncompact manifolds, called 'admissible' and being introduced in Grubb and Kokholm [24]. It is a subclass of the above family of manifolds with funnel-like ends, namely a family of manifolds with conical ends. Of course, aside from the requirements on the manifold, differential boundary value problems of the form considered in the present paper constitute a very particular subcase of Grubb's general class. However, in order to apply the results of Grubb [22] to (14) we have to require that $(\mathcal{A}, \mathcal{B})$ has $C^\infty$ coefficients. In contrast, we impose in essence minimal regularity assumptions on $(\mathcal{A}, \mathcal{B})$. This is important for the study of quasilinear equations on the basis of the linear theorems proved here.

Now we suppose that $M$ is a noncompact uniformly regular Riemannian manifold not belonging to the class of manifolds with funnel-like ends. This is the case, in particular, if $M$ has no boundary, is complete, and has bounded geometry. There is a tremendous amount of literature on heat equations for such manifolds, most of which is an $L_2$ theory and is concerned with kernel estimates and spectral theory (see, for example, Davies [16] or Grigor'yan [21] and the references therein). There are a few papers dealing with (semilinear) parabolic equations on noncompact complete Riemannian manifolds under various curvature assumptions which are based on heat kernel estimates (e.g., Zhang [35, 36], Mazzucato and Nistor [29], Punzo [31, 32], Bandle et al. [14]). In all these papers either the top-order part is the Laplace-Beltrami operator or smooth leading order coefficients are required.

Except for a recent paper by Shao and Simonett [33], the author is not aware of any result on parabolic equations on noncompact manifolds which do not rely on heat kernel techniques, leave alone noncompact manifolds with noncompact boundary. In [33] the authors, building on [5] and [6], establish a Hölder space existence theorem for autonomous nonlinear parabolic equations on uniformly regular manifolds without boundary. As an application they show that the solutions of the Yamabe flow instantaneously regularize and become real analytic in space and time.

A prototypical example to which our results apply is furnished by an $m$-dimensional Riemannian submanifold $M_{\mathsf{H}} = (M_{\mathsf{H}}, g_{\mathsf{H}})$ of the hyperbolic space $\mathsf{H}^m$ represented by the Poincaré model. More specifically, we denote by $\mathbb{B}^m$ the open unit ball in $\mathbb{R}^m$ with closure $\bar{\mathbb{B}}^m$ and boundary $S^{m-1}$, the $(m-1)$-sphere. Then $\mathsf{H} = \mathsf{H}^m = (\mathbb{B}^m, g_{\mathsf{H}})$, where $g_{\mathsf{H}} = 4g_m/(1-|x|^2)^2$ for $x \in \mathbb{B}^m$. If $\partial M_{\mathsf{H}}$ is not compact, then we assume that its closure in $\bar{\mathbb{B}}^m$ intersects $S^{m-1}$ transversally and that this intersection is the boundary of an $(m-1)$-dimensional Riemannian submanifold of $S^{m-1}$. Informally expressed this means, in particular, that $M_{\mathsf{H}}$ 'does not collapse at infinity'.

Writing $\operatorname{div}_{\mathsf{H}}$ for $\operatorname{div}_{g_{\mathsf{H}}}$, etc., problem (14) is on $M_{\mathsf{H}} \times J$ given by

$$\mathcal{A}_{\mathsf{H}} u := -\operatorname{div}_{\mathsf{H}}(a \cdot \operatorname{grad}_{\mathsf{H}} u) + (\vec{a} \,|\, \operatorname{grad}_{\mathsf{H}} u)_{\mathsf{H}} + a_0 u$$

and

$$\mathcal{B}_{\mathsf{H}} u := \begin{cases} \gamma u & \text{on} \quad \partial_0 M_{\mathsf{H}}, \\ \left(\nu_{\partial M_{\mathsf{H}}} \,\middle|\, \gamma(a \cdot \operatorname{grad}_{\mathsf{H}} u)\right)_{\mathsf{H}} & \text{on} \quad \partial_1 M_{\mathsf{H}}. \end{cases}$$

Using the fact that $g_{\mathsf{H}}$ is conformal to $g_m$ we can express $\mathcal{A}_{\mathsf{H}}$ and $\mathcal{B}_{\mathsf{H}}$ in terms of $g_m$, that is, as differential operators on $M := (M_{\mathsf{H}}, g_m)$. In fact, writing $\operatorname{div} = \operatorname{div}_{g_m}$, etc., we find with $\rho(x) := (1 - |x|^2)/2$

$$\operatorname{div}_{\mathsf{H}}(a \bullet \operatorname{grad}_{\mathsf{H}} u) = \rho^m \partial_i(\rho^{2-m} a^i_j g^{jk} \partial_k u) = \rho^m \operatorname{div}(\rho^{2-m} a \bullet \operatorname{grad} u) \tag{26}$$

$$= \operatorname{div}(\rho^2 a \bullet \operatorname{grad} u) - m(\rho a \bullet \operatorname{grad} \rho \,|\, \operatorname{grad} u) \tag{27}$$

and $(\vec{a} \,|\, \operatorname{grad}_{\mathsf{H}} u)_{\mathsf{H}} = (\vec{a} \,|\, \operatorname{grad} u)$. Moreover, $\nu_{\partial M_{\mathsf{H}}} = \rho\nu$ and, consequently,

$$\big(\nu_{\mathsf{H}} \,\big|\, \gamma(a \bullet \operatorname{grad}_{\mathsf{H}} u)\big)_{\mathsf{H}} = \rho\big(\nu \,\big|\, \gamma(a \bullet \operatorname{grad} u)\big).$$

This shows that the initial boundary value problem

$$\partial_t u + \mathcal{A}_{\mathsf{H}} u = f \text{ on } M_{\mathsf{H}} \times J, \quad \mathcal{B}_{\mathsf{H}} u = h \text{ on } \partial M_{\mathsf{H}} \times J, \quad u = u_0 \text{ on } M_{\mathsf{H}} \times \{0\}$$

can be seen as a degenerate initial boundary value problem on the 'underlying' Euclidean manifold $M$. Note that $M$ is not a uniformly regular Riemannian manifold, even if $\partial M = \emptyset$, that is, $M = \mathbb{B}^m$, since it cannot be covered by an atlas $\mathfrak{K}$ whose coordinate patches are uniformly comparable in size and such that a uniform shrinking of $\mathfrak{K}$ is still an atlas.

## 4 Singular Riemannian Manifolds and Weighted Function Spaces

Generalizing the preceding example we are led to the concept of singular Riemannian manifolds. Informally speaking, such a manifold is characterized by a singularity function $\rho \in C^\infty\big(M, (0,\infty)\big)$ such that the conformal metric $\hat{g} := g/\rho^2$ gives rise to a uniformly regular Riemannian manifold $\hat{M} := (M, \hat{g})$. To be precise:

Let $M$ be an $m$-dimensional uniformly regular manifold. A pair $(\rho, \mathfrak{K})$ is a *singularity datum* for $M$ if $\rho \in C^\infty\big(M, (0,\infty)\big)$ and $\mathfrak{K}$ is a uniformly regular atlas such that

$$\begin{aligned} &\text{(i)} \quad \|\kappa_* \rho\|_{k,\infty} \le c(k)\rho_\kappa, \quad \kappa \in \mathfrak{K}, \ \ k \in \mathbb{N}, \\ &\qquad\ \text{where } \rho_\kappa := \kappa_*\rho(0) = \rho\big(\kappa^{-1}(0)\big); \\ &\text{(ii)} \quad \rho \,|\, U_\kappa \sim \rho_\kappa, \quad \kappa \in \mathfrak{K}. \end{aligned} \tag{28}$$

Two singularity data $(\rho, \mathfrak{K})$ and $(\tilde{\rho}, \tilde{\mathfrak{K}})$ are *equivalent*, $(\rho, \mathfrak{K}) \approx (\tilde{\rho}, \tilde{\mathfrak{K}})$, if $\rho \sim \tilde{\rho}$ and $\mathfrak{K} \approx \tilde{\mathfrak{K}}$.

A *singularity structure*, $\mathfrak{S}(M)$, for $M$ is a maximal family of equivalent singularity data. A *singularity function* for $M$ is a $\rho \in C^\infty\big(M, (0,\infty)\big)$ such that there exists an atlas $\mathfrak{K}$ with $(\rho, \mathfrak{K}) \in \mathfrak{S}(M)$. The set of all singularity functions is the *singularity type* of $M$. It is convenient to denote it by $[\![\rho]\!]$, where $\rho$ is one of its representatives.

A **singular Riemannian manifold of type** $[\![\rho]\!]$ is a Riemannian manifold $(M, g)$ such that

$$\begin{aligned} &\text{(i)} \quad M \text{ is uniformly regular and endowed with} \\ &\qquad \text{a singularity structure } \mathfrak{S}(M) \text{ of singularity type } [\![\rho]\!]; \\ &\text{(ii)} \quad (M, g/\rho^2) \text{ is a uniformly regular Riemannian manifold.} \end{aligned} \tag{29}$$

This definition is independent of the particular choice of $\rho$ in the following sense: Let $(\tilde{\rho}, \tilde{\mathfrak{K}}) \approx (\rho, \mathfrak{K})$. Then it follows from (11)(ii), (iii) and (28) that $(M, g/\tilde{\rho}^2)$ is a uniformly regular Riemannian manifold and $g/\tilde{\rho}^2 \sim g/\rho^2$. In [7] it is shown that (28)(i) is equivalent to

$$d \log \rho \in BC^{\infty}(T^*\hat{M}). \tag{30}$$

In [9] there is carried out a detailed study of singular Riemannian manifolds. We refer the reader to that paper for proofs of the following examples.

*Examples 4.1*

(a) A uniformly regular Riemannian manifold is singular of type $[\![\mathbf{1}]\!]$, and conversely.

(b) Let $\Gamma$ be a union of connected components of $\partial M$ and $m \geq 2$. We endow $\Gamma$ with the induced Riemannian metric $\mathring{g} := \mathring{\iota}^* g$, where $\mathring{\iota} : \Gamma \hookrightarrow M$ is the natural embedding. Let $\mathfrak{K}$ be a uniformly regular atlas for $M$. For $\kappa \in \mathfrak{K}_\Gamma$ we set $U_{\mathring{\kappa}} := \partial U_\kappa := U_\kappa \cap \partial M = U_\kappa \cap \Gamma$ and $\mathring{\kappa} := \iota_0 \circ (\mathring{\iota}^* \kappa) : U_{\mathring{\kappa}} \to \mathbb{R}^{m-1}$ with $\iota_0 : \{0\} \times \mathbb{R}^{m-1} \to \mathbb{R}^{m-1}$, $(0, x') \mapsto x'$. Then $\mathring{\mathfrak{K}} := \{\, \mathring{\kappa} \ ; \ \kappa \in \mathfrak{K}_\Gamma \,\}$ is a uniformly regular atlas for $\Gamma$, the one *induced by* $\mathfrak{K}$.

Suppose $(M, g)$ is a singular Riemannian manifold of type $[\![\rho]\!]$. We set $\mathring{\rho} := \mathring{\iota}^* \rho = \rho \,|\, \Gamma$. Then $(\mathring{\rho}, \mathring{\mathfrak{K}})$ is a singularity datum for $\Gamma$. Thus it defines a singular structure $\mathring{\mathfrak{S}}(\Gamma)$ for $\Gamma$, the one *induced by* $\mathfrak{S}(M)$. Furthermore, $(\Gamma, \mathring{g}/\mathring{\rho}^2)$ is a singular Riemannian manifold of type $[\![\mathring{\rho}]\!]$ (and dimension $m-1$). It is always understood that $\Gamma$ is endowed with the singular structure induced by the one of $M$.

(c) Let $(\tilde{M}, \tilde{g})$ be a Riemannian manifold and $f : M \to \tilde{M}$ an isometric diffeomorphism, that is, $\tilde{g} = f_* g$. Suppose $(M, g)$ is singular of type $[\![\rho]\!]$ with singularity datum $(\rho, \mathfrak{K})$. Set $f_*\mathfrak{K} := \{ f_*\kappa \ ; \ \kappa \in \mathfrak{K} \}$. Then the pair $(f_*\rho, f_*\mathfrak{K})$ is a singularity datum for $(\tilde{M}, \tilde{g})$ and the latter is a singular Riemannian manifold of type $[\![f_*\rho]\!]$.

(d) Suppose $S \subset U \subset M$, where $S$ is closed and $U$ is open in $M$. Assume $\rho \in C^{\infty}\big(U, (0, \infty)\big)$ and $\mathfrak{K}$ is a uniformly regular atlas for $U$ on $S$ such that (28) holds for $\mathfrak{K}_S$. Then $(\rho, \mathfrak{K})$ is a singularity structure for $U$ *on* $S$. Two such singularity structures $(\rho, \mathfrak{K})$ and $(\tilde{\rho}, \tilde{\mathfrak{K}})$ are equivalent *on* $S$ if $\rho \sim \tilde{\rho}$ and $\mathfrak{K}$ and $\tilde{\mathfrak{K}}$ are equivalent on $S$. This defines a singularity structure *for* $U$ *on* $S$ of type $[\![\rho]\!]$. Then $U$ is a *singular Riemannian manifold on* $S$ *of type* $[\![\rho]\!]$ if it is endowed with a singularity structure on $S$ of type $[\![\rho]\!]$ and $(U, g/\rho^2)$ is a uniformly regular Riemannian manifold on $S$.

Assume $S_j \subset U_j \subset M$ and $U_j$ is a uniformly regular Riemannian manifold on $S_j$ of type $[\![\rho_j]\!]$ for $0 \le j \le \ell$. Let $(\rho_j, \mathfrak{K}_j)$ be a singularity structure for $U_j$ on $S_j$ and assume that $(\alpha)$ and $(\beta)$ of Example 3.5(f) apply and

$$\rho_i \,|\, (S_i \cap S_j) \sim \rho_j \,|\, (S_i \cap S_j), \qquad 0 \le i < j \le \ell.$$

Then there exists $\rho \in C^\infty\big(M, (0,\infty)\big)$ such that $\rho \,|\, S_j \sim \rho_j \,|\, S_j$ for $0 \le j \le \ell$ and $(\rho, \mathfrak{K})$ is a singularity datum for $M$, where $\mathfrak{K} = \mathfrak{K}_0 \cup \cdots \cup \mathfrak{K}_\ell$. Furthermore, $M$ is a singular Riemannian manifold of type $[\![\rho]\!]$. It is said to be obtained by *patching together the singular Riemannian manifolds $U_j$ on $S_j$ of type $[\![\rho_j]\!]$*.

(e) Let $d \ge 2$ and suppose $B$ is a $b$-dimensional compact Riemannian submanifold of $\mathbb{R}^{d-1}$. For a nonempty subinterval $I$ of $(0,1)$ and $\alpha \ge 1$ we set

$$C^d_\alpha(I,B) := \big\{ (t, t^\alpha y) \; ; \; t \in I, \; y \in B \big\} \subset \mathbb{R} \times \mathbb{R}^{d-1} = \mathbb{R}^d.$$

We endow $C^d_\alpha(B) := C^d_\alpha\big((0,1), B\big)$ with the metric induced by $\mathbb{R}^d$. It is called *model $\alpha$-cusp over* (the base) $B$ in $\mathbb{R}^d$. Note that a 1-cusp is a cone.

For $\ell \in \mathbb{N}$ we set $C^d_{\alpha,\ell}(I,B) := C^d_\alpha(I,B)$ if $\ell = 0$, and

$$C^d_{\alpha,\ell}(I,B) := C^d_\alpha(I,B) \times IQ^\ell, \qquad \ell > 0,$$

where $IQ^\ell = \{ tz \in \mathbb{R}^\ell \; ; \; t \in I, \; z \in Q^\ell \}$. Then $C^d_{\alpha,\ell}(B) := C^d_{\alpha,\ell}\big((0,1), B\big)$ is a $(1+b+\ell)$-dimensional Riemannian submanifold of $\mathbb{R}^d \times \mathbb{R}^\ell = \mathbb{R}^{d+\ell}$, a *model $(\alpha,\ell)$-wedge over $B$*, also called *model $\ell$-wedge over $C^d_\alpha(B)$*. Thus every model cusp is a model wedge, a 0-wedge. Every $(\alpha,\ell)$-wedge $C = C^d_{\alpha,\ell}(B)$ is a singular Riemannian manifold on $S := C^d_{\alpha,\ell}\big((0,3/4], B\big)$ of type $[\![R_\alpha]\!]$, where the *cusp characteristic* $R_\alpha$ is defined by $R_\alpha(x) := t^\alpha$ for $x = (t,y,z) \in C$ with $y \in Q^\ell$.

(f) Let $U$ be open in $M$ and set $C := C^d_{\alpha,\ell}(B)$ and $S := C^d_{\alpha,\ell}\big((0,3/4], B\big)$ with $\ell := m - 1 - b \ge 0$. Suppose that $\varphi : U \to C$ is a diffeomorphism such that $(\varphi_* g) \,|\, S \sim g_C \,|\, S$. Then $U$ is a singular Riemannian manifold on $\varphi^{-1}(S)$ of type $[\![\varphi^* R_\alpha]\!]$. It is said to be an *$(\alpha,\ell)$-wedge represented by $\varphi$*.

(g) Let $d \ge m$ and let $B$ be an $(m-1)$-dimensional compact submanifold of $\mathbb{R}^{d-1}$. For a nonempty subinterval $I$ of $(1,\infty)$ and $\alpha < 0$ we set

$$K^d_\alpha(I,B) := \big\{ (t, t^\alpha y) \; ; \; t \in I, \; y \in B \big\} \subset \mathbb{R} \times \mathbb{R}^{d-1} = \mathbb{R}^d.$$

Then $K^d_\alpha(B) := K^d_\alpha\big((1,\infty), B\big)$ is considered as an $m$-dimensional Riemannian submanifold of $\mathbb{R}^d$, an *infinite $\alpha$-cusp over $B$* in $\mathbb{R}^d$. Its cusp characteristic $R_\alpha$ is given by $R_\alpha(x) := t^\alpha$ for $x = (t, t^\alpha y) \in K^d_\alpha(B)$. It holds that $K^d_\alpha(B)$ is a singular Riemannian manifold on $K^d_\alpha\big([2,\infty), B\big)$ of type $[\![R_\alpha]\!]$.

(h) Let $U$ be open in $M$ and let $K := K_\alpha^d(B)$ be an infinite $\alpha$-cusp over $B$ in $\mathbb{R}^d$. Set $S := K_\alpha^d\big([2,\infty),B\big)$. Let $\varphi : U \to K$ be a diffeomorphism satisfying

($\alpha$) $\overline{U\backslash\varphi^{-1}(S)} \cap S = \varphi^{-1}\big(K_\alpha^d(\{2\},B)\big)$;

($\beta$) $(\varphi_* g)\,|\,S \sim g_K\,|\,S$.

Then $U$ is a singular Riemannian manifold on $\varphi^{-1}(S)$ of type $[\![\varphi^* R_\alpha]\!]$. Furthermore, $M$ is said to have in $U$ an *infinite $\alpha$-cusp* (more precisely: *$(\alpha,B)$-cusp) represented by $\varphi$*.

(i) Assume that $M$ is an $m$-dimensional Riemannian submanifold for $\mathbb{R}^n$ for some $n \geq m$. Then $\mathcal{S}(M) := \bar{M}\backslash M$, where $\bar{M}$ is the closure of $M$ in $\mathbb{R}^n$, is the *singularity set* of $M$. It is independent of $n$ since the closure of $M$ in $\mathbb{R}^{\tilde{n}}$ with $\tilde{n} > n$ and $\mathbb{R}^n = \mathbb{R}^n \times \{0\} \subset \mathbb{R}^{\tilde{n}}$ equals $\bar{M}$ also.

Suppose $\Sigma$ is a connected component of $\mathcal{S}(M)$ with the following properties:

($\alpha$) it is an $\ell$-dimensional compact Riemannian submanifold of $\mathbb{R}^n$ without boundary, where $\ell \in \{0,\dots,m-1\}$;

($\beta$) there exist $\alpha \geq 1$, a compact $(m-\ell-1)$-dimensional Riemannian submanifold $B$ of $\mathbb{R}^d$ with $d \geq m-\ell$, and for each $p \in \Sigma$ a normalized chart $\Phi_p$ for $\mathbb{R}^n$ at $p$ such that, setting $V_p := \operatorname{dom}(\Phi_h)$,

$$\Phi_p(M \cap V_p) = C^d_{(\alpha,\ell)}(B) \times \{0\} \subset \mathbb{R}^{d+\ell} \times \mathbb{R}^{n-d-\ell} = \mathbb{R}^n$$

and

$$\Phi_p(\Sigma \cap V_p) = \big(\{0\} \times Q^\ell\big) \times \{0\};$$

($\gamma$) $U_p := M \cap V_p$ is an $(\alpha,\ell)$-wedge represented by $\varphi_p := \Phi_p\,|\,U_p$.

Then $M$ is said to possess a *smooth cuspidal singularity of type* $(\alpha,\ell)$ (more precisely: $(\alpha,\ell,B)$) *near* $\Sigma$.

Let $\Sigma \subset \mathcal{S}(M)$ and assume $M$ has a smooth cuspidal singularity of type $(\alpha,\ell)$ near $\Sigma$. Also assume that there exist relatively compact open neighborhoods $V$ and $W$ of $\Sigma$ in $\mathbb{R}^n$ with $\bar{W} \subset V$ possessing the following properties: set $U := V \cap M$ and $S := \bar{W} \cap M$. Then there is $\rho_\alpha \in C^\infty\big(U,(0,\infty)\big)$ such that

$$\rho_\alpha\,|\,(S \cap U_p) \sim \varphi_p^* R_\alpha\,|\,(S \cap U_p), \qquad p \in \Sigma,$$

and $U$ is on $S$ a singular Riemannian manifold of type $[\![\rho_\alpha]\!]$. Loosely speaking: $\Sigma$ is then said to be a *smooth* $(\alpha,\ell)$*-wedge*, more precisely, a *smooth* $(\alpha,\ell,B)$*-wedge*. It is a *smooth $\alpha$-cusp*, respectively $(\alpha,B)$*-cusp*, if $\ell = 0$. Note that near every smooth $(\alpha,S^{m-1})$-cusp $M$ looks locally like $\mathbb{R}^m\backslash\{0\}$ near 0. (In this case we choose $d = m+1$.)

(j) Let $M$ be an $m$-dimensional Riemannian submanifold of $\mathbb{R}^n$ for some $n \geq m$. Suppose:

($\alpha$) $\mathcal{S}(M)$ is compact and for each connected component $\Sigma$ of $\mathcal{S}(M)$ there exist $\alpha_\Sigma \geq 1$, $\ell_\Sigma \in \{0,\dots,m-1\}$, and a compact $(m-\ell-1)$-dimensional submanifold $B_\Sigma$ of $\mathbb{R}^d$, where $d \geq m-\ell$, such that $M$ has a smooth cuspidal singularity of type $(\alpha_\Sigma, \ell_\Sigma, B_\Sigma)$ near $\Sigma$;

($\beta$) there are $k \in \mathbb{N}$ and for each $i \in \{1,\dots,k\}$ an open subset $U_i$ of $M$, $\alpha_i \in (-\infty, 0)$, an $(m-1)$-dimensional compact Riemannian submanifold $B_i$ of $\mathbb{R}^d$ for some $d \geq m$, and a diffeomorphism $\varphi_i : U_j \to K^d_{\alpha_i}(B_j)$ such that $M$ has in $U_i$ an infinite $(\alpha_i, B_i)$-cusp represented by $\varphi_i$;

($\gamma$) $U_i \cap U_j = \emptyset$ for $1 \leq i < j \leq k$, $M \backslash (U_1 \cup \cdots \cup U_k)$ is relatively compact in $\mathbb{R}^n$, and $\mathcal{S}(M) \backslash (U_1 \cup \cdots \cup U_k) = \mathcal{S}(M)$.

Then $M$ is said to be a *manifold with cuspidal singularities*. Note that $M$ is relatively compact in $\mathbb{R}^n$ if $k = 0$.

Every manifold with cuspidal singularities is a singular Riemannian manifold of type $[\![\rho]\!]$, where $\rho \sim \rho_\alpha$ near a smooth $(\alpha, \ell)$-wedge $\Sigma \subset \mathcal{S}(M)$, $\rho \sim \varphi_i^* R_{\alpha_i}$ on $U_i$, $1 \leq i \leq k$, and $\rho \sim \mathbf{1}$ away from the singularities. □

The qualifier 'smooth' in the preceding definitions refers to the fact that the bases of the cusps are uniformly regular. If they are singular Riemannian manifolds themselves then we get manifolds with cuspidal corners of various orders. For this we refer to [9] as well.

Ammann et al. [11] introduce a class of noncompact Riemannian manifolds, termed Lie manifolds, in order to establish regularity properties of solutions to elliptic boundary value problems on polyhedral domains; also see Ammann et al. [12], Ammann and Nistor [10], and the references therein, as well as the survey by Bacuta et al. [13]. These authors use a desingularization technique by which they introduce conformal metrics $g/\rho^2$, where $\rho$ is the distance to the singular set.

Let $M = (M, g)$ be a singular Riemannian manifold of type $[\![\rho]\!]$. Then we can apply the results of Sects. 2 and 3 to $\hat{M}$, where we have to use $\hat{\nabla} := \nabla_{\hat{g}}$, of course. Fortunately, since $\hat{g}$ is conformal to $g$ we can express all spaces and operators in terms of $g$, so that $\hat{M}$ does not appear in the final results.

Specifically, set $V := V^\sigma_\tau$ and let $\lambda \in \mathbb{R}$. Then the weighted Sobolev space $W^{k,\lambda}_p(V;\rho)$ is for $k \in \mathbb{N}$ the completion of $\mathcal{D}(V)$ in $L_{1,\mathrm{loc}}(V)$ with respect to the norm

$$u \mapsto \Big( \sum_{j=0}^{k} \big\| \rho^{\lambda+j+\tau-\sigma} \, |\nabla^j u|_{g^{\tau+j}_\sigma} \big\|^p_{L_p(V)} \Big)^{1/p}.$$

Weighted Slobodeckii spaces $W^{s,\lambda}_p(V;\rho)$ are for $k < s < k+1$ again defined by interpolation, that is, by replacing $W^\ell_p(V)$ in (7) by $W^{\ell,\lambda}_p(V;\rho)$ for $\ell \in \{k, k+1\}$.

Analogously, $B^\lambda(V;\rho)$ is the vector space of all sections $u$ of $V$ such that $\rho^{\lambda+\tau-\sigma} \, |u|_{g^\tau_\sigma} \in B(M)$. The norm $u \mapsto \big\| \rho^{\lambda+\tau-\sigma} \, |u|_{g^\tau_\sigma} \big\|_\infty$ makes it a Banach space. If

$k \in \mathbb{N}$, then $B^{k,\lambda}(V;\rho)$ is the Banach space of all $u \in C^k(V)$ for which

$$\max_{0 \le j \le k} \left\| \rho^{\lambda+j+\tau-\sigma} \, |\nabla^j u|_{g_\sigma^{\tau+j}} \right\|_\infty$$

is finite, endowed with this norm. Furthermore, $bc^{k,\lambda}(V;\rho)$ is the closure of $BC^{\infty,\lambda}(V;\rho) := \bigcap_k BC^k(V;\rho)$ in $BC^{k,\lambda}(V;\rho)$. Then weighted Besov-Hölder spaces $B_\infty^{s,\lambda}(V;\rho)$ are defined by interpolation in complete analogy to (8).

Weighted $L_2$ Sobolev spaces of this type have been introduced by Kondrat'ev [25] in the study of elliptic boundary value problems on domains with singular points. Since then they have been used by numerous authors, predominantly in an Euclidean $L_2$ setting. A detailed study of the $L_p$ case and references are found in [6].

We set

$$W_p^s(V;\rho) := W_p^{s,0}(V;\rho), \quad BC^k(V;\rho) := BC^{k,0}(V;\rho),$$
$$B_\infty^s(V;\rho) := B_\infty^{s,0}(V;\rho).$$

In [7] it is proved that

$$W_p^s(\hat{V}) \doteq W_p^{s,-m/p}(V;\rho), \quad BC^k(\hat{V}) \doteq BC^k(V;\rho), \qquad (31)$$
$$B_\infty^s(\hat{V}) \doteq B_\infty^s(V;\rho),$$

where $\doteq$ means 'equal except for equivalent norms' and $\hat{V} := T_\tau^\sigma \hat{M}$. In [7] it is also shown that $(u \mapsto \rho^\lambda u)$ belongs to

$$\mathcal{L}\mathrm{is}\big(W_p^{s,\lambda'+\lambda}(V;\rho), W_p^{s,\lambda'}(V;\rho)\big) \cap \mathcal{L}\,\mathrm{is}\,\big(B_\infty^{s,\lambda'+\lambda}(V;\rho), B_\infty^{s,\lambda'}(V;\rho)\big) \qquad (32)$$

for $\lambda, \lambda' \in \mathbb{R}$, and $(u \mapsto \rho^\lambda u)^{-1} = (v \mapsto \rho^{-\lambda} v)$. Thus it suffices to study the spaces $W_p^s(\hat{V})$ and $B_\infty^s(\hat{V})$ since by this isomorphism and by (31) we can transfer all properties from $W_p^s(\hat{V})$ onto $W_p^{s,\lambda}(V;\rho)$ and from $B_\infty^s(\hat{V})$ onto $B_\infty^{s,\lambda}(V;\rho)$. Alternatively, we can refer directly to [6].

Anisotropic weighted Sobolev-Slobodeckii spaces are defined for $s \ge 0$ by

$$W_p^{(s,s/2),\lambda}(V \times J;\rho) := L_p\big(J, W_p^{s,\lambda}(V;\rho)\big) \cap W_p^{s/2}\big(J, L_p^\lambda(V;\rho)\big).$$

Analogously, we introduce anisotropic Besov-Hölder spaces for $s > 0$ by

$$B_\infty^{(s,s/2),\lambda}(V;\rho)) := B\big(J, B_\infty^{s,\lambda}(V;\rho)\big) \cap B_\infty^{s/2}\big(J, B^\lambda(V;\rho)\big).$$

Again, we omit the superscript $\lambda$ if it equals zero. It is obvious from the above that all embedding, interpolation, and trace theorems, etc. proved in [5] carry over to the present setting using natural adaptions. It is also clear that (31) implies

$$W_p^{(s,s/2)}(\hat{V}) \doteq W_p^{(s,s/2),-m/p}(V;\rho). \qquad (33)$$

Furthermore,

$$(u \mapsto \rho^\lambda u) \in \mathcal{L}\mathrm{is}\big(W_p^{(s,s/2),\lambda'+\lambda}(V;\rho), W_p^{(s,s/2),\lambda'}(V;\rho)\big) \tag{34}$$

for $\lambda, \lambda' \in \mathbb{R}$ is a consequence of (32).

## 5 Degenerate Parabolic Problems

In this section we study problem (14) in the case of singular Riemannian manifolds. It turns out that in this situation Theorem 3.1 leads to an isomorphism theorem for degenerate parabolic initial boundary value problems on weighted Sobolev spaces.

Let $M = (M, g)$ be a singular Riemannian manifold of type $[\![\rho]\!]$ and $\lambda \in \mathbb{R}$. Similarly as in Sect. 3, we introduce data spaces which are now weighted and $\lambda$-dependent. To simplify the presentation we restrict ourselves to the setting of strong $L_p$ solutions. Thus we put

$$\begin{aligned} &W_p^{(2-\delta-1/p)(1,1/2),\lambda+\delta+1/p}(\partial M_T;\mathring{\rho}) \\ &\quad := W_p^{(2-1/p)(1,1/2),\lambda+1/p}(\partial_0 M_T;\mathring{\rho}) \times W_p^{(1-1/p)(1,1/2),\lambda+1+1/p}(\partial_1 M_T;\mathring{\rho}) \end{aligned}$$

and

$$\begin{aligned} &\mathcal{W}_p^{(2,1),\lambda}(M_T;\rho) \\ &\quad := L_p^\lambda(M_T;\rho) \times W_p^{(2-\delta-1/p)(1,1/2),\lambda+\delta+1/p}(\partial M_T;\mathring{\rho}) \times W_p^{2-2/p,\lambda}(M;\rho). \end{aligned}$$

Similarly as before, $\mathcal{W}_{p,\mathrm{cc}}^{(2,1),\lambda}(M_T;\rho)$ is the linear subspace hereof consisting of all $(f, h, u_0)$ satisfying the compatibility conditions (17).

The differential operator (15) is *uniformly strongly $\rho$-elliptic* if $a(\cdot, t)$ is symmetric for $t \in J$ and there exists a constant $\varepsilon > 0$ such that

$$\big(a(q,t) \cdot X \,\big|\, X\big)_{g(q)} \ge \varepsilon\rho^2(q)\,|X|_{g(q)}^2, \qquad X \in T_qM, \quad q \in M, \quad t \in J. \tag{35}$$

Henceforth, we say that $(\mathcal{A}, \mathcal{B})$ is a *$\rho$-regular uniformly $\rho$-elliptic boundary value problem on $M_T$* if $\mathcal{A}$ is uniformly strongly $\rho$-elliptic,

$$a \in BC^{(1,1/2),-2}(T_1^1M \times J;\rho), \quad \vec{a} \in L_\infty(TM \times J;\rho), \tag{36}$$
$$a_0 \in L_\infty(M_T),$$

and

$$b_0 \in BC^{(1,1/2),-1}(\partial_1 M_T). \tag{37}$$

If $\rho = \mathbf{1}$, then $(\mathcal{A}, \mathcal{B})$ is simply called regularly uniformly elliptic. Note that the first part of (36) implies $|a|_{g_1^1} \le c\rho^2$. Using this and the symmetry of $a(\cdot, t)$ we see that (35) is equivalent to the existence of $\varepsilon \in (0, 1)$ with

$$\varepsilon\rho^2(q)\,|X|^2_{g(q)} \le \big(a(q,t) \bullet X \,\big|\, X\big)_{g(q)} \le \rho^2(q)\,|X|^2_{g(q)}/\varepsilon \tag{38}$$

for $X \in T_qM$, $q \in M$, and $t \in J$.

Now we can formulate the following isomorphism theorem for degenerate parabolic equations.

**Theorem 5.1** *Let $M$ be a singular Riemannian manifold of type $[\![\rho]\!]$ and $p \notin \{3/2, 3\}$. Suppose that $(\mathcal{A}, \mathcal{B})$ is a $\rho$-regular uniformly $\rho$-elliptic boundary value problem on $M_T$ and $\lambda \in \mathbb{R}$.*

*Then $\mathcal{W}^{(2,1),\lambda}_{p,\mathrm{cc}}(M_T; \rho)$ is closed and*

$$(\partial + \mathcal{A},\ \mathcal{B}, \gamma_0) \in \mathcal{L}\mathrm{is}\big(W^{(2,1),\lambda}_p(M_T;\rho), \mathcal{W}^{(2,1),\lambda}_{p,\mathrm{cc}}(M_T;\rho)\big).$$

The proof of this theorem is given later in this section. First we derive an analogue of Theorem 3.4. For this we define

$$W^{s,\lambda}_{p,\mathcal{B}}(M;\rho), \qquad s \in [0,2]\backslash\{1/p, 1+1/p\},$$

by replacing $W^s_p(M)$ in (4) by $W^{s,\lambda}_p(M;\rho)$.

**Theorem 5.2** *Let $M$ be a singular Riemannian manifold of type $[\![\rho]\!]$ and $p \notin \{3/2, 3\}$. Suppose $(\mathcal{A}, \mathcal{B})$ is an autonomous $\rho$-regular uniformly $\rho$-elliptic boundary value problem on $M_T$ and $\lambda \in \mathbb{R}$. Set $A^\lambda := \mathcal{A}\,|\,W^{2,\lambda}_{p,\mathcal{B}}(M;\rho)$, considered as an unbounded linear operator in $L^\lambda_p(M;\rho)$. Then $-A^\lambda$ generates a strongly continuous analytic semigroup on $L^\lambda_p(M;\rho)$ and has the property of maximal regularity, that is, $(\partial + \mathcal{A},\ \gamma)$ belongs to*

$$\mathcal{L}\mathrm{is}\big(L_p(J, W^{2,\lambda}_{p,\mathcal{B}}(M;\rho)) \cap W^1_p(J, L^\lambda_p(M;\rho)), L^\lambda_p(M;\rho) \times W^{2-2/p,\lambda}_{p,\mathcal{B}}(M)\big).$$

*Proof* This follows from Theorem 5.1 by the arguments which led from Theorem 3.1 to Theorem 3.4. □

**Corollary 5.3** *Set $A := A^0$. Then $-A$ generates a strongly continuous analytic semigroup on $L_p(M)$ and has the maximal regularity property on $L_p(M)$.*

In order to facilitate the proof of Theorem 5.1 we precede it with a technical lemma. In this connection we identify $\rho^\lambda$ with the point-wise multiplication operator $u \mapsto \rho^\lambda u$.

**Lemma 5.4** *Let $(\mathcal{A}, \mathcal{B})$ be a $\rho$-regular uniformly $\rho$-elliptic boundary value problem on $M_T$ and $\lambda \in \mathbb{R}$. Then there exists another such pair $(\mathcal{A}', \mathcal{B}')$ such that*

$$(\mathcal{A}, \mathcal{B}) \circ \rho^\lambda = \rho^\lambda \circ (\mathcal{A}', \mathcal{B}'). \tag{39}$$

*Proof*

(1) Note that

$$(\mathcal{A},\mathcal{B})\circ\rho^{\lambda}=\rho^{\lambda}\circ(\mathcal{A},\mathcal{B})+\big[(\mathcal{A},\mathcal{B}),\rho^{\lambda}\big] \tag{40}$$

where the commutator

$$\big[(\mathcal{A},\mathcal{B}),\rho^{\lambda}\big]u:=(\mathcal{A},\mathcal{B})(\rho^{\lambda}u)-\rho^{\lambda}(\mathcal{A},\mathcal{B})u$$

is given by $\big([\mathcal{A},\rho^{\lambda}],[\mathcal{B},\rho^{\lambda}]\big)$ with

$$[\mathcal{A},\rho^{\lambda}]u=-2(a\bullet\operatorname{grad}\rho^{\lambda}\mid\operatorname{grad}u)+\big((\vec{a}|\operatorname{grad}\rho^{\lambda})-\operatorname{div}(a\bullet\operatorname{grad}\rho^{\lambda})\big)u$$

and

$$[\mathcal{B},\rho^{\lambda}]u=\big(0,(\nu\,|\,\gamma(a\bullet\operatorname{grad}\rho^{\lambda}))u\big).$$

For abbreviation, $\mathrm{gl}:=\operatorname{grad}\log$. Then

$$\operatorname{grad}\rho^{\lambda}=\lambda\rho^{\lambda-1}\operatorname{grad}\rho=\rho^{\lambda}\lambda\,\mathrm{gl}\,\rho. \tag{41}$$

We set

$$\vec{a}':=-2\lambda a\bullet\mathrm{gl}\,\rho,\quad a_0':=\lambda\big((\vec{a}|\,\mathrm{gl}\,\rho)-\operatorname{div}(a\bullet\mathrm{gl}\,\rho)\big)-\lambda^2(a\bullet\mathrm{gl}\,\rho|\,\mathrm{gl}\,\rho),$$
$$b_0':=\lambda\big(\nu\,|\,\gamma(a\bullet\mathrm{gl}\,\rho)\big).$$

Moreover,

$$\mathcal{A}'u:=\mathcal{A}u+(\vec{a}'\,|\operatorname{grad}u)+a_0'u,\quad \mathcal{B}'u:=\mathcal{B}u+(0,b_0'\gamma u).$$

It follows from (40) and (41) that $(\mathcal{A}',\mathcal{B}')$ satisfies (39). Hence it remains to show that $(\vec{a}',a_0',b_0')$ possesses the same regularity properties as $(\vec{a},a_0,b_0)$.

(2) We know from (30) and (31) that

$$d\log\rho\in BC^1(T^*\hat{M})\doteq BC^1(T^*M;\rho).$$

Using this, $\operatorname{grad}=g^{\sharp}d$, and (71) it follows

$$\mathrm{gl}\,\rho=g^{\sharp}\,d\log\rho\in BC^{1,2}(TM;\rho). \tag{42}$$

It is now an easy consequence of this, the assumptions on $a$ and $\vec{a}$, and of (68) that

$$\vec{a}'\in L_\infty(TM\times J),\quad (\vec{a}|\,\mathrm{gl}\,\rho)\in L_\infty(M_T). \tag{43}$$

From (42), (72), and (73) we infer

$$\begin{aligned}|\operatorname{div}(a \bullet \operatorname{gl}\rho)| &= |\nabla(a \bullet \operatorname{gl}\rho)|_{g_1^1} \\ &\le \rho^{-1}\,|\nabla a|_{g_1^2}\,\rho\,|\operatorname{gl}\rho|_{g_1^0} + \rho^{-2}\,|a|_{g_1^1}\,\rho^2\,|\nabla\operatorname{gl}\rho|_{g_1^1}.\end{aligned}$$

This guarantees that the second summand of $a_0'$ belongs to $L_\infty(M_T)$. Similarly,

$$|(a \bullet \operatorname{gl}\rho \,|\, \operatorname{gl}\rho)| \le \rho^{-2}\,|a|_{g_1^1}\,(\rho\,|\operatorname{gl}\rho|_{g_1^0})^2$$

implies that the third summand lies in $L_\infty(M_T)$ as well. Hence, by the second part of (43),

$$a_0' \in L_\infty(M_T). \tag{44}$$

Let $\hat{\nu}$ be the unit normal vector field of $\partial\hat{M}$. In local coordinates,

$$\hat{\nu} = (\hat{g}_{11})^{-1/2}\partial/\partial x^1 = \rho(g_{11})^{-1/2}\partial/\partial x^1 = \rho\nu. \tag{45}$$

Thus $\nu \in BC^{1,1}(TM_{|\partial M};\rho)$. This implies for the conormal field

$$\nu_\flat = g_\flat\nu \in BC^{1,-1}(T^*M_{|\partial M};\rho). \tag{46}$$

As above, we derive from (42)

$$\gamma(a \bullet \operatorname{gl}\rho) \in BC^{(1,1/2)}(TM_{|\partial M};\rho).$$

Therefore, by (46),

$$b_0' = \lambda\nu_\flat \bullet \gamma(a \bullet \operatorname{gl}\rho) \in BC^{(1,1/2),-1}(\partial M_T;\rho). \tag{47}$$

Now (43), (44), and (47) imply the assertion. □

*Proof of Theorem 5.1*

(1) By the definitions of $\hat{g}$ and the gradient we get

$$\rho^{-2}(X\,|\,\operatorname{grad}_{\hat{g}} u) = (X\,|\,\operatorname{grad}_{\hat{g}} u)_{\hat{g}} = \langle du, X\rangle = (X\,|\,\operatorname{grad} u) \tag{48}$$

for any $C^1$ function $u$ and any vector field $X$ on $M$. From this we obtain

$$\operatorname{grad} u = \rho^{-2}\operatorname{grad}_{\hat{g}} u. \tag{49}$$

We also note that (65)–(67) imply

$$|\cdot|_{\hat{g}_\sigma^\tau} = \rho^{\tau-\sigma}\,|\cdot|_{g_\sigma^\tau}, \qquad \sigma,\tau \in \mathbb{N}. \tag{50}$$

We put $\hat{a} := \rho^{-2}a$. Then we infer from (50)

$$|\hat{a}|_{\hat{g}_1^1} = \rho^{-2} |a|_{g_1^1}. \tag{51}$$

Note that $\nabla \hat{a} = \rho^{-2}\nabla a - 2(d\log\rho)\bullet\hat{a}$ (cf. (41)). Hence, see (68),

$$\begin{aligned}\rho\,|\nabla\hat{a}|_{g_1^2} &\le \rho^{-1}\,|\nabla a|_{g_1^2} + 2\rho\,|d\log\rho|_{g_0^1}\,|\hat{a}|_{g_1^1}\\ &= \rho^{-1}\,|\nabla a|_{g_1^2} + 2\,|d\log\rho|_{\hat{g}_0^1}\,|\hat{a}|_{g_1^1},\end{aligned}$$

the last equality being a consequence of (50). From this, (30), (51), and the assumption on $a$ we deduce

$$\hat{a} \in BC^{(1,1/2)}(T_1^1 M \times J;\rho) \doteq BC^{(1,1/2)}(T_1^1 \hat{M} \times J). \tag{52}$$

By replacing the index $\mathsf{H}$ in (26) by $\hat{g}$ and using (48) and (49) we find

$$\begin{aligned}\operatorname{div}(a\bullet\operatorname{grad}u) &= \operatorname{div}(\rho^2\hat{a}\bullet\operatorname{grad}u)\\ &= \operatorname{div}_{\hat{g}}(\hat{a}\bullet\operatorname{grad}_{\hat{g}}u) + m(\rho\hat{a}\bullet\operatorname{grad}\rho\,|\,\operatorname{grad}u)\\ &= \operatorname{div}_{\hat{g}}(\hat{a}\bullet\operatorname{grad}_{\hat{g}}u) + (m\hat{a}\bullet\rho^{-1}\operatorname{grad}_{\hat{g}}\rho\,|\,\operatorname{grad}_{\hat{g}}u)_{\hat{g}}.\end{aligned} \tag{53}$$

Observe that

$$\rho^{-1}\operatorname{grad}_{\hat{g}}\rho = \rho^{-1}\hat{g}^\sharp\,d\rho = \hat{g}^\sharp\,d\log\rho.$$

Hence, by (68) and (71),

$$|\hat{a}\bullet\rho^{-1}\operatorname{grad}_{\hat{g}}\rho|_{\hat{g}} \le |\hat{a}|_{\hat{g}_1^1}\,|d\log\rho|_{\hat{g}_0^1}.$$

This, (52), and (30) imply

$$\hat{a}\bullet\rho^{-1}\operatorname{grad}_{\hat{g}}\rho \in L_\infty(T\hat{M}\times J). \tag{54}$$

Furthermore, by (49) and $\hat{g} = \rho^{-2}g$,

$$(\vec{a}\,|\,\operatorname{grad}u) = (\vec{a}\,|\,\rho^{-2}\operatorname{grad}_{\hat{g}}u) = (\vec{a}\,|\,\operatorname{grad}_{\hat{g}}u)_{\hat{g}}.$$

From (50) we derive

$$\|\,|\vec{a}|_{\hat{g}}\,\|_\infty = \|\rho^{-1}\,|\vec{a}|_g\,\|_\infty = \|\vec{a}\|_{L_\infty(TM\times J;\rho)}. \tag{55}$$

Thus it follows from (54) that

$$\hat{d} := m\hat{a}\bullet\rho^{-1}\operatorname{grad}_{\hat{g}}\rho + \vec{a} \in L_\infty(T\hat{M}\times J). \tag{56}$$

Now we put

$$\hat{\mathcal{A}}u := -\operatorname{div}_{\hat{g}}(\hat{a} \cdot \operatorname{grad}_{\hat{g}} u) + (\hat{d} \,|\, \operatorname{grad}_{\hat{g}} u)_{\hat{g}} + a_0 u.$$

Then (53) shows

$$\mathcal{A}u = \hat{\mathcal{A}}u, \qquad u \in W_p^{(2,1),-m/p}(M_T;\rho) \doteq W_p^{(2,1)}(\hat{M}_T), \tag{57}$$

the last equivalence being a consequence of (31).

(2) Recalling (45) and $\dot{\rho} = \rho \,|\, \partial M$, we find

$$\begin{aligned}\big(\nu \,|\, \gamma(a \cdot \operatorname{grad} u)\big) &= \big(\nu \,|\, \gamma(\hat{a} \cdot \operatorname{grad}_{\hat{g}} u)\big) \\ &= \dot{\rho}^2\big(\nu \,|\, \gamma(\hat{a} \cdot \operatorname{grad}_{\hat{g}} u)\big)_{\hat{g}} = \dot{\rho}\big(\hat{\nu} \,|\, \gamma(\hat{a} \cdot \operatorname{grad}_{\hat{g}} u)\big)_{\hat{g}}.\end{aligned} \tag{58}$$

We denote by $\dot{\nabla}$ the Levi-Civita connection of $\partial M$ and set $\hat{b}_0 := \dot{\rho}^{-1} b_0$. Then

$$\dot{\nabla}\hat{b}_0 = \dot{\rho}^{-1}\dot{\nabla} b_0 - (d \log \dot{\rho})\hat{b}_0. \tag{59}$$

Relation (50) implies

$$\dot{\rho}\,|d \log \dot{\rho}|_{\dot{g}_0^1} = |d \log \dot{\rho}|_{\hat{\dot{g}}_0^1}.$$

Since $b_0 \in BC^{(1,1/2),-1}(\partial_1 M_T;\rho)$ it holds

$$\begin{aligned}\|\hat{b}_0\|_{L_\infty(\partial_1 \hat{M}_T)} &= \|\dot{\rho}^{-1} b_0\|_{L_\infty(\partial_1 M_T)} < \infty, \\ &\big\| \, |\dot{\nabla} b_0|_{\dot{g}_0^1} \big\|_{L_\infty(\partial_1 M_T)} < \infty.\end{aligned} \tag{60}$$

Thus we get from (59), (60), Example 4.1(b), and (30) (applied to $\partial_1 M = \partial_1 \hat{M}$) that $\big\|\dot{\rho}\,|\dot{\nabla}\hat{b}_0|_{\dot{g}_0^1}\big\|_{L_\infty(\partial_1 M_T)}$ is finite. This implies

$$\hat{b}_0 \in BC^{(1,1/2)}(\partial_1 M_T;\rho) \doteq BC^{(1,1/2)}(\partial_1 \hat{M}_T). \tag{61}$$

Now we set

$$\hat{\mathcal{B}}_1 u := \big(\hat{\nu} \,|\, \gamma(\hat{a} \cdot \operatorname{grad}_{\hat{g}} u)\big)_{\hat{g}} + \hat{b}_0 \gamma u$$

and $\hat{\mathcal{B}} := (\mathcal{B}_0, \hat{\mathcal{B}}_1)$. Then we see from (52), (56), (61), and (35) that $(\hat{\mathcal{A}}, \hat{\mathcal{B}})$ is a regular uniformly elliptic boundary value problem on $\hat{M}_T$. Furthermore,

$$\mathcal{B}u = (\hat{\mathcal{B}}_0 u, \dot{\rho}\hat{\mathcal{B}}_1 u). \tag{62}$$

(3) Suppose $(f, h, u_0) \in \mathcal{W}_{p,cc}^{(2,1),-m/p}(M_T;\rho)$. Then, by (34),

$$\mathring{\rho}^{-1}h_1 \in W_p^{(1-1/p)(1,1/2),-(m-1)/p}(\partial_1 M_T;\mathring{\rho}).$$

More precisely, set $\hat{h} := (h_0, \mathring{\rho}^{-1}h_1)$. Then (34), (31), and (33) imply

$$\big((f,h,u_0) \mapsto (f,\hat{h},u_0)\big) \in \mathcal{L}\mathrm{is}\big(\mathcal{W}_{p,cc}^{(2,1),-m/p}(M_T;\rho), \mathcal{W}_{p,cc}^{(2,1)}(\hat{M}_T)\big).$$

In addition, we deduce from (57) and (62) that $u \in W_p^{(2,1)}(M_T;\rho)$ is a solution of (14) iff $u \in W_p^{(2,1)}(\hat{M}_T)$ and $(\partial + \hat{\mathcal{A}},\ \hat{\mathcal{B}}, \gamma_0)u = (f, \hat{h}, u_0)$. Now Theorem 3.1 implies the validity of the assertion if $\lambda = -m/p$.

(4) Let $\lambda \neq -m/p$. Lemma 5.4 guarantees the existence of a $\rho$-regular uniformly $\rho$-elliptic boundary value problem $(\mathcal{A}', \mathcal{B}')$ on $M_T$ such that

$$\rho^\lambda \circ (\mathcal{A},\mathcal{B}) = (\mathcal{A}',\mathcal{B}') \circ \rho^\lambda. \tag{63}$$

By (32) and (34) it follows that $(f,h,u_0) \in \mathcal{W}_{p,cc}^{(2,1),\lambda}(M_T;\rho)$ iff

$$(f',h',u_0') := \rho^{\lambda+m/p}(f,h,u_0) \in \mathcal{W}_{p,cc}^{(2,1),-m/p}(M_T;\rho)$$

and $u \in W_p^{(2,1),\lambda}(M_T;\rho)$ iff

$$u' := \rho^{\lambda+m/p}u \in W_p^{(2,1),-m/p}(M_T;\rho).$$

From (63) we get

$$(\mathcal{A},\mathcal{B},\gamma_0)u = (f,h,u_0) \iff (\mathcal{A}',\mathcal{B}',\gamma_0)u' = (f',h',u_0).$$

As the claim holds for $\lambda = -m/p$, the assertion follows. □

*Remarks 5.5*

(a) It is obvious from this proof that there is a straightforward parameter-dependent analogue of the supplement to Theorem 3.1 for degenerate parabolic problems.
(b) Remarks 3.3 apply in the present setting also. □

## Appendix: Tensor Bundles

Let $M$ be a manifold and $V = (V, \pi, M)$ a vector bundle of rank $n$ over it. For a nonempty subset $S$ of $M$ we denote by $V_{|S}$ the restriction $\pi^{-1}(S)$ of $V$ to $S$. If $S$ is a submanifold or a union of connected components of $\partial M$, then $V_{|S}$ is a vector bundle of rank $n$ over $S$. As usual, $V_p := V_{\{p\}}$ is the fibre $\pi^{-1}(p)$ of $V$ over $p$. By $\Gamma(S,V)$ we mean the $\mathbb{R}^S$ module of all sections of $V$ (**no** smoothness).

As usual, $TM$ and $T^*M$ are the tangent and cotangent bundles of $M$. Then $T^\sigma_\tau M := TM^{\otimes\sigma} \otimes T^*M^{\otimes\tau}$ is for $\sigma,\tau \in \mathbb{N}$ the $(\sigma,\tau)$-tensor bundle of $M$, that is, the vector bundle of all tensors on $M$ being contravariant of order $\sigma$ and covariant of order $\tau$. In particular, $T^1_0 M = TM$ and $T^0_1 M = T^*M$, as well as $T^0_0 M = M \times \mathbb{R}$.

For $\nu \in \mathbb{N}^\times$ we put $\mathbb{J}_\nu := \{1,\dots,m\}^\nu$. Then, given local coordinates $\kappa = (x^1,\dots,x^m)$ and setting

$$\frac{\partial}{\partial x^{(i)}} := \frac{\partial}{\partial x^{i_1}} \otimes \cdots \otimes \frac{\partial}{\partial x^{i_\sigma}}, \quad dx^{(j)} := dx^{j_1} \otimes \cdots \otimes dx^{j_\tau}$$

for $(i) = (i_1,\dots,i_\sigma) \in \mathbb{J}_\sigma$, $(j) \in \mathbb{J}_\tau$, the local representation of a $(\sigma,\tau)$-tensor field $a \in \Gamma(T^\sigma_\tau M)$ with respect to these coordinates is given by

$$a = a^{(i)}_{(j)} \frac{\partial}{\partial x^{(i)}} \otimes dx^{(j)}$$

with $a^{(i)}_{(j)} \in \mathbb{R}^{U_\kappa}$. We use the summation convention for (multi-)indices labeling coordinates or bases. Thus such a repeated index, which appears once as a superscript and once as a subscript, implies summation over its whole range.

Suppose $\sigma_1,\sigma_2,\tau_1,\tau_2 \in \mathbb{N}$. Then the *complete contraction*

$$\Gamma(T^{\sigma_2+\tau_1}_{\tau_2+\sigma_1}M) \times \Gamma(T^{\sigma_1}_{\tau_1}M) \to \Gamma(T^{\sigma_2}_{\tau_2}M), \quad (a,b) \mapsto a \bullet b$$

is defined as follows: Given $(i_k) \in \mathbb{J}_{\sigma_k}$ and $(j_k) \in \mathbb{J}_{\tau_k}$ for $k = 1,2$, we set

$$(i_2;j_1) := (i_{2,1},\dots,i_{2,\sigma_2},j_{1,1},\dots,j_{1,\tau_1}) \in \mathbb{J}_{\sigma_2+\tau_1}$$

etc., using obvious interpretations if $\min\{\sigma,\tau\} = 0$. Suppose $a \in \Gamma(T^{\sigma_2+\tau_1}_{\tau_2+\sigma_1}M)$ and $b \in \Gamma(T^{\sigma_1}_{\tau_1}M)$ are locally represented on $U_\kappa$ by

$$a = a^{(i_2;j_1)}_{(j_2;i_1)} \frac{\partial}{\partial x^{(i_2)}} \otimes \frac{\partial}{\partial x^{(j_1)}} \otimes dx^{(j_2)} \otimes dx^{(i_1)}, \quad b = b^{(i_1)}_{(j_1)} \frac{\partial}{\partial x^{(i_1)}} \otimes dx^{(j_1)}.$$

Then the local representation of $a \bullet b$ on $U_\kappa$ is given by

$$a^{(i_2;j_1)}_{(j_2;i_1)} b^{(i_1)}_{(j_1)} \frac{\partial}{\partial x^{(i_2)}} \otimes dx^{(j_2)}.$$

Let $g$ be a Riemannian metric on $TM$. We write $g_\flat : TM \to T^*M$ for the (fiber-wise defined) Riesz isomorphism. Thus $\langle g_\flat X, Y\rangle = g(X,Y)$ for $X,Y \in \Gamma(TM)$, where $\langle\cdot,\cdot\rangle : \Gamma(T^*M) \times \Gamma(TM) \to \mathbb{R}^M$ is the natural (fiber-wise defined) duality pairing. The inverse of $g_\flat$ is denoted by $g^\sharp$. Then $g^*$, the adjoint Riemannian metric on $T^*M$, is defined by $g^*(\alpha,\beta) := g(g^\sharp\alpha, g^\sharp\beta)$ for $\alpha,\beta \in \Gamma(T^*M)$. In local coordinates

$$g = g_{ij}\, dx^i \otimes dx^j, \quad g^* = g^{ij}\, \frac{\partial}{\partial x^i} \otimes \frac{\partial}{\partial x^j}, \tag{64}$$

$[g^{ij}]$ being the inverse of the $(m \times m)$-matrix $[g_{ij}]$.

The metric $g$ induces a vector bundle metric on $T^\sigma_\tau M$ which we denote by $g^\tau_\sigma$. In local coordinates

$$g^\tau_\sigma(a, b) = g_{(i)(j)} g^{(k)(\ell)} a^{(i)}_{(k)} b^{(j)}_{(\ell)}, \qquad a, b \in \Gamma(T^\sigma_\tau M), \tag{65}$$

where

$$g_{(i)(j)} := g_{i_1 j_1} \cdots g_{i_\sigma j_\sigma}, \quad g^{(k)(\ell)} := g^{k_1 \ell_1} \cdots g^{k_\tau \ell_\tau} \tag{66}$$

for $(i), (j) \in \mathbb{J}_\sigma$ and $(k), (\ell) \in \mathbb{J}_\tau$. Note $g^0_1 = g$ and $g^1_0 = g^*$ and $g^0_0(a, b) = ab$ for $a, b \in \Gamma(M \times \mathbb{R}) = \mathbb{R}^M$. Moreover,

$$|\cdot|_{g^\tau_\sigma} : \Gamma(T^\sigma_\tau M) \to (\mathbb{R}^+)^M, \quad a \mapsto \sqrt{g^\tau_\sigma(a, a)} \tag{67}$$

is the *vector bundle norm* on $T^\sigma_\tau M$ induced by $g$. It follows that the complete contraction satisfies

$$|a \bullet b|_{g^{\tau_2}_{\sigma_2}} \le |a|_{g^{\tau_2+\sigma_1}_{\sigma_2+\tau_1}} |b|_{g^{\tau_1}_{\sigma_1}}, \qquad a \in \Gamma(T^{\sigma_2+\tau_1}_{\tau_2+\sigma_1} M), \quad b \in \Gamma(T^{\sigma_1}_{\tau_1} M). \tag{68}$$

We define a vector bundle isomorphism $T^\sigma_{\tau+1} M \to T^{\sigma+1}_\tau M,\ a \mapsto a^\sharp$ by

$$a^\sharp(\alpha_1, \ldots, \alpha_\sigma, \alpha, X_1, \ldots, X_\tau) := a(\alpha_1, \ldots, \alpha_\sigma, X_1, \ldots, X_\tau, g^\sharp \alpha) \tag{69}$$

for $X_1, \ldots, X_\tau \in \Gamma(TM)$ and $\alpha, \alpha_1, \ldots, \alpha_\sigma \in \Gamma(T^*M)$. If $a^{(i)}_{(j;k)}$ with $(i) \in \mathbb{J}_\sigma$, $(j) \in \mathbb{J}_\tau$, and $k \in \mathbb{J}_1$ is the coefficient of $a$ in a local coordinate representation, then

$$(a^\sharp)^{(i;k)}_{(j)} = g^{k\ell} a^{(i)}_{(j;\ell)}. \tag{70}$$

This implies

$$|a^\sharp|_{g^\tau_{\sigma+1}} = |a|_{g^{\tau+1}_\sigma}. \tag{71}$$

The Levi-Civita connection on $TM$ is denoted by $\nabla = \nabla_g$. We use the same symbol for its natural extension to a metric connection on $T^\sigma_\tau M$. Then the corresponding covariant derivative is the linear map

$$\nabla : C^\infty(T^\sigma_\tau M) \to C^\infty(T^\sigma_{\tau+1} M), \quad a \mapsto \nabla a,$$

defined by $\langle \nabla a, b \otimes X \rangle := \langle \nabla_X a, b \rangle$ for $b \in C^\infty(T^\tau_\sigma M)$ and $X \in C^\infty(TM)$. It is a well-defined continuous linear map from $C^1(T^\sigma_\tau M)$ into $C(T^\sigma_{\tau+1} M)$, as follows from its local representation. For $k \in \mathbb{N}$ we define

$$\nabla^k : C^k(T^\sigma_\tau M) \to C(T^\sigma_{\tau+k} M), \quad a \mapsto \nabla^k a$$

by $\nabla^0 a := a$ and $\nabla^{k+1} := \nabla \circ \nabla^k$.

In local coordinates $\kappa = (x^1, \dots, x^m)$ the volume measure $dv = dv_g$ of $(M, g)$ is represented by $\kappa_* \, dv = \kappa_* \sqrt{g} \, dx$, where $\sqrt{g} := \big(\det[g_{ij}]\big)^{1/2}$ and $dx$ is the Lebesgue measure on $\mathbb{R}^m$.

The *contraction* $\mathsf{C} : T^{\sigma+1}_{\tau+1} M \to T^\sigma_\tau M$, $a \mapsto \mathsf{C}a$ is given in local coordinates by $(\mathsf{C}a)^{(i)}_{(j)} := a^{(i;k)}_{(j;k)}$. It follows

$$|\mathsf{C}a|_{g^\tau_\sigma} = |a|_{g^{\tau+1}_{\sigma+1}}. \tag{72}$$

Recall that the divergence of tensor fields is the map

$$\operatorname{div} = \operatorname{div}_g : C^1(T^{\sigma+1}_\tau M) \to C(T^\sigma_\tau M), \quad a \mapsto \operatorname{div} a := \mathsf{C}(\nabla a). \tag{73}$$

If $X$ is a $C^1$ vector field on $M$, then $\operatorname{div} X$ has the well-known local representation

$$\frac{1}{\sqrt{g}} \frac{\partial}{\partial x^i} \big(\sqrt{g} X^i\big), \qquad X = X^i \frac{\partial}{\partial x^i}. \tag{74}$$

The gradient, $\operatorname{grad} u = \operatorname{grad}_g u$, of a $C^1$ function $u$ is the continuous vector field $g^\sharp du$.

Suppose $a \in C^1(T^1_1 M)$. Then, in terms of covariant derivatives,

$$\operatorname{div}(a \operatorname{grad} u) = a^\sharp \bullet \nabla^2 u + \operatorname{div}(a^\sharp) \bullet \nabla u. \tag{75}$$

## References


1. H. Amann, *Linear and Quasilinear Parabolic Problems, Volume I: Abstract Linear Theory* (Birkhäuser, Basel, 1995)
2. H. Amann, Maximal regularity for nonautonomous evolution equations. Adv. Nonlinear Stud. **4**, 417–430 (2004)
3. H. Amann, Quasilinear parabolic problems via maximal regularity. Adv. Differ. Equ. **10**(10), 1081–1110 (2005)
4. H. Amann, *Anisotropic Function Spaces and Maximal Regularity for Parabolic Problems. Part 1: Function Spaces*. Jindřich Nečas Center for Mathematical Modeling, Lecture Notes, Prague, vol. 6, 2009.
5. H. Amann, Anisotropic function spaces on singular manifolds (2012). arXiv: 1204.0606
6. H. Amann, Function spaces on singular manifolds. Math. Nachr. **286**, 436–475 (2012)
7. H. Amann, Parabolic equations on noncompact Riemannian manifolds (in preparation)

8. H. Amann, Pseudodifferential boundary value problems (in preparation)
9. H. Amann, Uniformly regular and singular Riemannian manifolds, in *Elliptic and Parabolic Equations, Hannover, September 2013,* ed. by J. Escher, E. Schrohe, J. Seiler, C. Walker. Springer Proceedings in Mathematics & Statistics (2015), pp.1–43
10. B. Ammann, V. Nistor, Weighted Sobolev spaces and regularity for polyhedral domains. Comput. Methods Appl. Mech. Eng. **196**, 3650–3659 (2007)
11. B. Ammann, R. Lauter, V. Nistor, On the geometry of Riemannian manifolds with a Lie structure at infinity. Int. J. Math. Math. Sci. 161–193 (2004)
12. B. Ammann, A.D. Ionescu, V. Nistor, Sobolev spaces on Lie manifolds and regularity for polyhedral domains. Doc. Math. **11**, 161–206 (2006)
13. C. Bacuta, A.L. Mazzucato, V. Nistor, Anisotropic regularity and optimal rates of convergence for the finite element method on three dimensional polyhedral domains (2012). arXiv:1205.2128
14. C. Bandle, F. Punzo, A. Tesei, Existence and nonexistence of patterns on Riemannian manifolds. J. Math. Anal. Appl. **387**(1), 33–47 (2012)
15. V. Barbu, A. Favini, S. Romanelli, Degenerate evolution equations and regularity of their associated semigroups. Funkcial. Ekvac. **39**(3), 421–448 (1996)
16. E.B. Davies, *Heat Kernels and Spectral Theory* (Cambridge University Press, Cambridge, 1989)
17. R. Denk, M. Hieber, J. Prüss, $\mathcal{R}$-boundedness, Fourier multipliers and problems of elliptic and parabolic type. Mem. Am. Math. Soc. **166**(788) (2003)
18. G. Dore, $L^p$ regularity for abstract differential equations, in *Functional Analysis and Related Topics*, 1991 (Kyoto). Lecture Notes in Mathematics, vol. 1540 (Springer, Berlin, 1993), pp. 25–38
19. S. Fornaro, G. Metafune, D. Pallara, Analytic semigroups generated in $L^p$ by elliptic operators with high order degeneracy at the boundary. Note Math. **31**(1), 103–116 (2011)
20. G. Fragnelli, G. Ruiz Goldstein, J.A. Goldstein, S. Romanelli, Generators with interior degeneracy on spaces of $L^2$ type. Electron. J. Differ. Equ. **2012**(189), 1–30 (2012)
21. A. Grigor'yan, *Heat Kernel and Analysis on Manifolds* (American Mathematical Society, Providence, 2009)
22. G. Grubb, Parameter-elliptic and parabolic pseudodifferential boundary problems in global $L_p$ Sobolev spaces. Math. Z. **218**, 43–90 (1995)
23. G. Grubb, *Functional Calculus of Pseudodifferential Boundary Problems* (Birkhäuser, Boston, 1996)
24. G. Grubb, N.J. Kokholm, A global calculus of parameter-dependent pseudodifferential boundary problems in $L_p$ Sobolev spaces. Acta Math. **171**, 165–229 (1993)
25. V.A. Kondrat'ev, Boundary value problems for elliptic equations in domains with conical or angular points. Trudy Moskov. Mat. Obšč. **16**, 209–292 (1967)
26. P.C. Kunstmann, L. Weis, Maximal $L_p$-regularity for parabolic equations, Fourier multiplier theorems and $H^\infty$-functional calculus, in *Functional Analytic Methods for Evolution Equations*. Lecture Notes in Mathematics, vol. 1855 (Springer, Berlin, 2004), pp. 65–311
27. O.A. Ladyzhenskaya, V.A. Solonnikov, N.N. Ural'ceva, *Linear and Quasilinear Equations of Parabolic Type*. Translations of Mathematical Monographs (American Mathematical Society, Providence, 1968)
28. A. Lunardi, *Analytic Semigroups and Optimal Regularity in Parabolic Problems* ( Birkhäuser, Basel, 1995)
29. A.L. Mazzucato, V. Nistor, Mapping properties of heat kernels, maximal regularity, and semilinear parabolic equations on noncompact manifolds. J. Hyperbolic Differ. Equ. **3**(4), 599–629 (2006)
30. M.A. Pozio, F. Punzo, A. Tesei, Criteria for well-posedness of degenerate elliptic and parabolic problems. J. Math. Pures Appl. (9) **90**(4), 353–386 (2008)
31. F. Punzo, On well-posedness of semilinear parabolic and elliptic problems in the hyperbolic space. J. Differ. Equ. **251**(7), 1972–1989 (2011)

32. F. Punzo, Blow-up of solutions to semilinear parabolic equations on Riemannian manifolds with negative sectional curvature. J. Math. Anal. Appl. **387**(2), 815–827 (2012)
33. Y. Shao, G. Simonett, Continuous maximal regularity on uniformly regular Riemannian manifolds (2013). arXiv:1309.2041
34. V. Vespri, Analytic semigroups, degenerate elliptic operators and applications to nonlinear Cauchy problems. Ann. Math. Pura Appl. (4) **155**, 353–388 (2012)
35. Q.S. Zhang, Nonlinear parabolic problems on manifolds, and a nonexistence result for the noncompact Yamabe problem. Electron. Res. Announc. Am. Math. Soc. **3**, 45–51 (1997)
36. Q.S. Zhang, Semilinear parabolic problems on manifolds and applications to the non-compact Yamabe problem. Electron. J. Differ. Equ. **2000**(46), 1–30 (2000)

# A Generalization of Some Regularity Criteria to the Navier–Stokes Equations Involving One Velocity Component

**Šimon Axmann and Milan Pokorný**

*To Yoshihiro Shibata*

**Abstract** We present generalizations of results concerning conditional global regularity of weak Leray–Hopf solutions to incompressible Navier–Stokes equations presented by Zhou and Pokorný in articles (Pokorný, Electron J Differ Equ (11):1–8, 2003; Zhou, Methods Appl Anal 9(4):563–578, 2002; Zhou, J Math Pure Appl 84(11):1496–1514, 2005); see also Neustupa et al. (Quaderni di Matematica, vol. 10. Topics in Mathematical Fluid Mechanics, 2002, pp. 163–183) We are able to replace the condition on one velocity component (or its gradient) by a corresponding condition imposed on a projection of the velocity (or its gradient) onto a more general vector field. Comparing to our other recent results from Axmann and Pokorný (A note on regularity criteria for the solutions to Navier-Stokes equations involving one velocity component, in preparation), the conditions imposed on the projection are more restrictive here, however due to the technique used in Axmann and Pokorný (A note on regularity criteria for the solutions to Navier-Stokes equations involving one velocity component, in preparation), there appeared a specific additional restriction on geometrical properties of the reference field, which could be omitted here.

**Keywords** Global regularity • Incompressible Navier–Stokes equations • Regularity criteria

Š. Axmann • M. Pokorný (✉)
Charles University in Prague, Faculty of Mathematics and Physics, Sokolovská 83, 186 75 Praha 8, Czech Republic
e-mail: axmas7am@karlin.mff.cuni.cz; pokorny@karlin.mff.cuni.cz



H. Amann et al. (eds.), *Recent Developments of Mathematical Fluid Mechanics*, Advances in Mathematical Fluid Mechanics, DOI 10.1007/978-3-0348-0939-9_5

# 1 Introduction

We consider the Cauchy problem for the instationary incompressible Navier–Stokes equations in the full three space dimensions

$$\left.\begin{aligned} \frac{\partial \mathbf{v}}{\partial t} + \mathbf{v}\cdot\nabla\mathbf{v} - \nu\Delta\mathbf{v} + \nabla p = \mathbf{f} \\ div\mathbf{v} = 0 \end{aligned}\right\} \text{ in } (0,T)\times\mathbb{R}^3, \tag{1}$$

$$\mathbf{v}(0,\mathbf{x}) = \mathbf{v}_0(\mathbf{x}) \quad \text{in } \mathbb{R}^3,$$

where $\mathbf{v} : (0,T)\times\mathbb{R}^3 \to \mathbb{R}^3$ is the velocity field, $p : (0,T)\times\mathbb{R}^3 \to \mathbb{R}$ is the pressure, $\mathbf{f} : (0,T)\times\mathbb{R}^3 \to \mathbb{R}^3$ is the given density of external forces, and $\nu > 0$ is given kinematic viscosity. For the sake of simplicity, we set $\nu = 1$ and $\mathbf{f} \equiv \mathbf{0}$ in our further considerations. Indeed, the actual value of viscosity does not play any role. It would be also possible to formulate some suitable assumptions on the regularity of $\mathbf{f}$ in such a way that our main results remain true. However, it would lead to unnecessary technicalities which we prefer to omit here.

The mathematical theory of Navier–Stokes equations has long, interesting history (see e.g. [16]). In the celebrated works of Leray [11, 12] and Hopf [6] the existence of weak solutions to system (1) in space $\mathbf{v} \in L^2(0,T,(W^{1,2}_{div}(\mathbb{R}^3))) \cap L^\infty(0,T,(L^2(\mathbb{R}^3)))$ for any given $\mathbf{v}_0 \in L^2_{div}(\mathbb{R}^3)$ was proved; they satisfy energy inequality. Further, for $\mathbf{v}_0 \in W^{1,2}_{div}(\mathbb{R}^3)$ the existence of (possibly short) time interval $(0,T^*)$ such that there exists a unique strong solution in space $\mathbf{v} \in L^2(0,T^*,(W^{2,2}(\mathbb{R}^3))) \cap L^\infty(0,T^*,(W^{1,2}_{div}(\mathbb{R}^3)))$ was established (see [9]). The uniqueness and regularity of Leray–Hopf weak solutions is still a challenging open problem [10]. For overview of known results see e.g. [4].

On the other hand, there were established many criteria ensuring the smoothness of the solution under additional assumptions concerning the velocity and its components, the gradient of the velocity and its components, the pressure, the vorticity, or other quantities.

During the last decade, an interesting progress was achieved in the field of regularity criteria concerning only one velocity component. The very first result in this direction is criterion proved by Neustupa and Penel [13], which ensures the regularity for $v_3 \in L^\infty\left(0,T,\left(L^\infty(\mathbb{R}^3)\right)\right)$. Similar result for the gradient of one velocity component $\left(\nabla v_3 \in L^t\left(0,T,\left(L^s(\mathbb{R}^3)\right)\right), \frac{2}{t}+\frac{3}{s} \le 1,\ s \ge 3\right)$ is due to He [5]. These pioneering results were then improved by Neustupa et al. [14] $\left(v_3 \in L^t\left(0,T,\left(L^s(\mathbb{R}^3)\right)\right), \frac{2}{t}+\frac{3}{s} \le \frac{1}{2},\ s \ge 2\right.$, as local criterion for suitable weak solution), and Pokorný [15] $\left(\nabla v_3 \in L^t\left(0,T,\left(L^s(\mathbb{R}^3)\right)\right), \frac{2}{t}+\frac{3}{s} \le \frac{3}{2},\ s \ge 2.\right)$, observing the equation for vorticity; the same results were obtained also by Zhou [17, 18]. Further improvements were later done via several techniques by Kukavica and Ziane [8], Cao and Titi [2], and finally Zhou and Pokorný [19, 20]. Note that the results in [1] contain generalization of these criteria. However, due to the used technique (multiplicative Gagliardo–Nirenberg inequality which has to be generalized) we get additional geometrical restrictions on the field **b** which can be avoided in our present paper.

**Notation**
In the whole paper, the standard notation for Lebesgue spaces $L^p(\mathbb{R}^3)$ with the norm $\|\cdot\|_p$ will be used. For the sake of brevity, we will denote the norm on Bochner spaces $L^p\left(0, t, \left(L^q(\mathbb{R}^3)\right)\right)$ by $\|\cdot\|_{p,q}$, the length of the time interval will be everywhere clear from the context. We will also use the same notation for scalar spaces $X$ and their vector analogues $X^N$. All generic constants will be denoted by $C$, although its value may differ from line to line, or even in the same formula. We will use Einstein summation convention over repeated indices.

## 2 Main Results

As we have already mentioned above, our main goal is to generalize the results of Zhou and Pokorný from articles [15, 17, 18] (the latter proved originally for suitable weak solution in [14]).

**Theorem 1** *Let* $\mathbf{v}$ *be a weak Leray–Hopf solution to the Navier–Stokes equations corresponding to initial datum* $\mathbf{v}_0 \in W^{1,2}_{div}(\mathbb{R}^3)$. *Assume moreover that there exist* $\delta > 0$, *and a vector field* $\mathbf{b}(t,\mathbf{x}) : (0,T) \times \mathbb{R}^3 \mapsto \mathbb{R}^3$ *such that*

$$\nabla \mathbf{b} \in L^\infty\left(0, T, \left(L^\infty(\mathbb{R}^3)\right)\right),$$

$$\frac{\partial \mathbf{b}}{\partial t},\ \nabla^2 \mathbf{b} \in L^\infty\left(0, T, \left(L^3(\mathbb{R}^3)\right)\right),$$

*and* $|\mathbf{b}(t,\mathbf{x})| \geq \delta$ *such that the projection of the velocity*

$$v_{\mathbf{b}}(t,\mathbf{x}) := \mathbf{b}(t,\mathbf{x}) \cdot \mathbf{v}(t,\mathbf{x})$$

*satisfies either*

$$v_{\mathbf{b}}(t,\mathbf{x}) \in L^t\left(0, T, \left(L^s(\mathbb{R}^3)\right)\right),\ \frac{2}{t} + \frac{3}{s} \leq \frac{1}{2},\ 6 \leq s \leq \infty$$

*or*

$$\nabla v_{\mathbf{b}}(t,\mathbf{x}) \in L^t\left(0, T, \left(L^s(\mathbb{R}^3)\right)\right),\ \frac{2}{t} + \frac{3}{s} \leq \frac{3}{2},\ 2 \leq s \leq \infty.$$

*Then* $\mathbf{v}$ *is actually a strong solution to the Navier–Stokes equations in the interval* $[0, T]$.

Since the proofs of both cases have lot of similarities, we will prove them simultaneously. It is well known that there exists a unique strong solution to (1) on (possibly short) time interval $[0, T^*)$, we will work with this strong solution and show that actually $T^* \geq T$. The result concerning so-called weak-strong uniqueness will then yield the desired result. Let us denote space $Y(\tau) := L^\infty(0, \tau; L^2(\mathbb{R}^3)) \cap L^2(0, \tau; W^{1,2}(\mathbb{R}^3))$. Our first step in Lemma 1 will be to derive a suitable estimate of $\mathbf{b} \cdot \boldsymbol{\omega}$ in the norm of space $Y$, then we will test Eq. (1) by an analogue of the quantity $-\Delta \mathbf{v}$ and get the desired estimate of $\nabla \mathbf{v}$ using Lemma 1.

**Lemma 1** *Let* $\mathbf{v}$ *be a strong solution to the Navier–Stokes equations corresponding to the initial condition* $\mathbf{v}_0 \in W^{1,2}_{div}(\mathbb{R}^3)$. *Suppose that the assumptions of Theorem* 1 *are satisfied. Let* $0 < \tau < T^*$. *Then* $\omega_{\mathbf{b}} := \mathbf{b} \cdot \boldsymbol{\omega}$ *can be estimated on* $(0, \tau)$ *as follows*

$$\|\omega_{\mathbf{b}}\|^2_{\infty,2} + \|\nabla \omega_{\mathbf{b}}\|^2_{2,2} \leq \|\omega_{\mathbf{b}}(0)\|^2_2 + C(\tau) \left(1 + \|\nabla \mathbf{v}\|_{Y(\tau)}\right). \tag{2}$$

*In particular, if* $\tau \to 0^+$, *then* $C(\tau) \to 0$ *and if* $\tau \to (T^*)^-$, *then* $C(\tau)$ *remains bounded.*

## 3 Proof of the Lemma

By possible decreasing the value of $t$ we could easily achieve that $\frac{2}{t} + \frac{3}{s} = \frac{1}{2}$ (or $\frac{3}{2}$, respectively). Applying the curl operator on (1) we get

$$\frac{\partial \omega_i}{\partial t} + \mathbf{v} \cdot \nabla \omega_i = \boldsymbol{\omega} \cdot \nabla v_i + \Delta \omega_i, \qquad i = 1, 2, 3.$$

Multiplying these equations by $b_i(t, x)$

$$b_i \frac{\partial \omega_i}{\partial t} + b_i \mathbf{v} \cdot \nabla \omega_i = b_i (\boldsymbol{\omega} \cdot \nabla v_i) + (\Delta \omega_i) b_i,$$

summing up $\sum_{i=1}^{3}$, and multiplying the arisen equation by $\omega_{\mathbf{b}} = \sum_{i=1}^{3} b_i \omega_i$, we get four terms which could be rewritten in the following way

$$\begin{aligned}
\omega_{\mathbf{b}} b_i \frac{\partial \omega_i}{\partial t} &= \omega_{\mathbf{b}} \frac{\partial \omega_{\mathbf{b}}}{\partial t} - \omega_{\mathbf{b}} \frac{\partial b_i}{\partial t} \omega_i, \\
b_i (\mathbf{v} \cdot \nabla \omega_i) \omega_{\mathbf{b}} &= (\mathbf{v} \cdot \nabla \omega_{\mathbf{b}}) \omega_{\mathbf{b}} - (\mathbf{v} \cdot \nabla b_i) \omega_i \omega_{\mathbf{b}}, \\
b_i (\boldsymbol{\omega} \cdot \nabla v_i) \omega_{\mathbf{b}} &= (\boldsymbol{\omega} \cdot \nabla v_{\mathbf{b}}) \omega_{\mathbf{b}} - (\boldsymbol{\omega} \cdot \nabla b_i) v_i \omega_{\mathbf{b}}, \\
(\Delta \omega_i) b_i \omega_{\mathbf{b}} &= (\Delta \omega_{\mathbf{b}} - 2 \nabla b_i \cdot \nabla \omega_i - \omega_i \Delta b_i) \, \omega_{\mathbf{b}};
\end{aligned}$$

whence

$$\frac{\partial \omega_{\mathbf{b}}}{\partial t}\omega_{\mathbf{b}} + \mathbf{v}\cdot\nabla\omega_{\mathbf{b}}\omega_{\mathbf{b}} = \boldsymbol{\omega}\cdot\nabla v_{\mathbf{b}}\omega_{\mathbf{b}} + \Delta\omega_{\mathbf{b}}\omega_{\mathbf{b}}$$
$$\underbrace{+\,\omega_i\frac{\partial b_i}{\partial t}\omega_{\mathbf{b}} + \omega_i(\mathbf{v}\cdot\nabla b_i)\omega_{\mathbf{b}} - v_i(\boldsymbol{\omega}\cdot\nabla b_i)\omega_{\mathbf{b}} - \Delta b_i\omega_i\omega_{\mathbf{b}} - 2(\nabla\omega_i\cdot\nabla b_i)\omega_{\mathbf{b}}}_{=:\mathrm{I}}\,.$$

Integration over the whole $\mathbb{R}^3$ with integration by parts gives us

$$\frac{1}{2}\frac{\mathrm{d}}{\mathrm{d}t}\|\omega_{\mathbf{b}}\|_2^2 + \|\nabla\omega_{\mathbf{b}}\|_2^2 \le -\underbrace{\int_{\mathbb{R}^3}\mathbf{v}\cdot\nabla\omega_{\mathbf{b}}\omega_{\mathbf{b}}\mathrm{d}\mathbf{x}}_{=0} + \int_{\mathbb{R}^3}\boldsymbol{\omega}\cdot\nabla v_{\mathbf{b}}\omega_{\mathbf{b}}\mathrm{d}\mathbf{x} + \int_{\mathbb{R}^3} I\mathrm{d}\mathbf{x}.$$

The lower order terms $I$ could be easily estimated using again integration by parts and Hölder's inequality

$$\begin{aligned}\int_{\mathbb{R}^3}\Big|\omega_i\frac{\partial b_i}{\partial t}\omega_{\mathbf{b}} - v_i(\boldsymbol{\omega}\cdot\nabla b_i)\omega_{\mathbf{b}} + \Delta b_i\omega_i\omega_{\mathbf{b}}\\ + 2(\omega_i\nabla b_i)\cdot\nabla\omega_{\mathbf{b}} + \omega_i(\mathbf{v}\cdot\nabla b_i)\omega_{\mathbf{b}}\Big|\mathrm{d}\mathbf{x}\\ \le C_\varepsilon\|\nabla\mathbf{v}\|_2^2\left[\left\|\frac{\partial\mathbf{b}}{\partial t}\right\|_3^2 + \|\Delta\mathbf{b}\|_3^2 + 2\|\nabla\mathbf{b}\|_\infty^2\right] + 3\varepsilon\|\nabla\omega_{\mathbf{b}}\|_2^2\\ + 2\|\nabla\mathbf{b}\|_\infty\int_{\mathbb{R}^3}|\boldsymbol{\omega}|\,|\mathbf{v}|\,|\omega_{\mathbf{b}}|\,\mathrm{d}\mathbf{x}.\end{aligned}$$

In the last integral we will use Hölder's inequality, interpolation, and Young's inequality

$$\begin{aligned}\|\nabla\mathbf{b}\|_\infty\int_{\mathbb{R}^3}|\boldsymbol{\omega}|\,|\mathbf{v}|\,|\omega_{\mathbf{b}}|\,\mathrm{d}\mathbf{x} &\le C\|\nabla\mathbf{b}\|_\infty\|\nabla\mathbf{v}\|_2\|\mathbf{v}\|_3\|\omega_{\mathbf{b}}\|_6\\ &\le\varepsilon\|\nabla\omega_{\mathbf{b}}\|_2^2 + C_\varepsilon\|\nabla\mathbf{b}\|_\infty^2\|\nabla\mathbf{v}\|_2^2\|\mathbf{v}\|_3^2,\end{aligned}$$

where

$$\int_0^t\|\mathbf{v}\|_3^2\|\nabla\mathbf{v}\|_2^2\mathrm{d}\tau \le \|\nabla\mathbf{v}\|_{4,2}^2\|\mathbf{v}\|_{4,3}^2 \le \|\nabla\mathbf{v}\|_{2,2}\|\mathbf{v}\|_{4,3}^2\|\nabla\mathbf{v}\|_{\infty,2}\,.$$

Now, we will estimate the leading terms, distinguishing two considered cases.

1. Assume the projection $v_{\mathbf{b}}$ has better integrability properties. Then

$$\int_{\mathbb{R}^3} \boldsymbol{\omega} \cdot \nabla v_{\mathbf{b}} \omega_{\mathbf{b}} \mathrm{d}\mathbf{x} \leq \left| - \int_{\mathbb{R}^3} v_{\mathbf{b}} \boldsymbol{\omega} \cdot \nabla \omega_{\mathbf{b}} \mathrm{d}\mathbf{x} \right| \leq \varepsilon \left\| \nabla \omega_{\mathbf{b}} \right\|_2^2 + C_\varepsilon \int_{\mathbb{R}^3} |\boldsymbol{\omega}|^2 {v_{\mathbf{b}}}^2 \mathrm{d}\mathbf{x}$$

$$\leq \varepsilon \left\| \nabla \omega_{\mathbf{b}} \right\|_2^2 + 2C_\varepsilon \int_{\mathbb{R}^3} |\nabla \mathbf{v}|^2 {v_{\mathbf{b}}}^2 \mathrm{d}\mathbf{x}$$

$$\begin{aligned} \int_{\mathbb{R}^3} {v_{\mathbf{b}}}^2 |\nabla \mathbf{v}|^2 \, \mathrm{d}\mathbf{x} &\leq \|v_{\mathbf{b}}\|_s^2 \, \|\nabla \mathbf{v}\|_{\frac{2s}{s-2}}^2 \\ &\leq \|v_{\mathbf{b}}\|_s^2 \, \|\nabla \mathbf{v}\|_{\frac{2s}{s-4}} \, \|\nabla \mathbf{v}\|_2 \\ &\leq C \|v_{\mathbf{b}}\|_s^2 \, \|\nabla \mathbf{v}\|_2^{1-\frac{6}{s}} \, \|\Delta \mathbf{v}\|_2^{\frac{6}{s}} \, \|\nabla \mathbf{v}\|_2 \\ &= C \|v_{\mathbf{b}}\|_s^2 \|\nabla \mathbf{v}\|_2^{\frac{4}{t}} \|\Delta \mathbf{v}\|_2^{\frac{6}{s}} \|\nabla \mathbf{v}\|_2, \end{aligned}$$

(recall $\frac{2}{t} + \frac{3}{s} = \frac{1}{2}$) which gives using assumptions on $\mathbf{b}$

$$\frac{1}{2} \frac{\mathrm{d}}{\mathrm{d}t} \|\omega_{\mathbf{b}}\|_2^2 + \|\nabla \omega_{\mathbf{b}}\|_2^2 \leq C \|v_{\mathbf{b}}\|_s^2 \, \|\nabla \mathbf{v}\|_2^{\frac{4}{t}} \, \|\Delta \mathbf{v}\|_2^{\frac{6}{s}} \, \|\nabla \mathbf{v}\|_2 + C(\mathbf{b}) \, \|\nabla \mathbf{v}\|_2^2 \, (1 + \|\mathbf{v}\|_3^2).$$

Integrating over time interval $(0, \tau)$, with usage of Hölder's inequality then yields

$$\begin{aligned} \|\omega_{\mathbf{b}}(\tau)\|_2^2 + \int_0^\tau \|\nabla \omega_{\mathbf{b}}\|_2^2 \, \mathrm{d}\sigma & \\ \leq \|\omega_{\mathbf{b}}(0)\|_2^2 + C \int_0^\tau & \|v_{\mathbf{b}}\|_s^2 \, \|\nabla \mathbf{v}\|_2^{\frac{4}{t}} \, \|\Delta \mathbf{v}\|_2^{\frac{6}{s}} \, \|\nabla \mathbf{v}\|_2 \mathrm{d}\sigma \\ & + C(\mathbf{b}) \int_0^\tau (\|\nabla \mathbf{v}\|_2^2 + \|\mathbf{v}\|_3^2 \, \|\nabla \mathbf{v}\|_2^2) \mathrm{d}\sigma. \end{aligned}$$

Recall that from the energy inequality we have estimate of $\mathbf{v}$ in the spaces $L^2(0, T, (W^{1,2}_{div}(\mathbb{R}^3)))$ and $L^\infty(0, T, (L^2(\mathbb{R}^3)))$, hence due to the interpolation also in $L^4(0, T, (L^3(\mathbb{R}^3)))$. Thus

$$
\begin{aligned}
&\|\omega_{\mathbf{b}}(\tau)\|_2^2 + \int_0^\tau \|\nabla\omega_{\mathbf{b}}\|_2^2 \,\mathrm{d}\sigma \\
&\quad \le \|\omega_{\mathbf{b}}(0)\|_2^2 + C(\mathbf{b}) \left[ \|v_{\mathbf{b}}\|_{t,s}^2 \|\nabla \mathbf{v}\|_{\infty,2}^{\frac{4}{t}} \|\Delta \mathbf{v}\|_{2,2}^{\frac{6}{s}} + 1 \right] \|\nabla \mathbf{v}\|_{2,2}^{2} \\
&\qquad + C(\mathbf{b}) \|\nabla \mathbf{v}\|_{2,2} \|\mathbf{v}\|_{4,3}^2 \|\nabla \mathbf{v}\|_{\infty,2}
\end{aligned}
\tag{3}
$$

which yields the conclusion of Lemma in the first case as $\frac{4}{t} + \frac{6}{s} = 1$.

2. For a given $2 \le s \le \infty$, we will find $2 \le p \le 6$, $2 \le q \le 3$ such that $\frac{1}{s} + \frac{1}{p} + \frac{1}{q} = 1$. We will use gradually Hölder's inequality, interpolation, and Young's inequality to obtain

$$
\begin{aligned}
\int_{\mathbb{R}^3} \boldsymbol{\omega} \cdot \nabla v_{\mathbf{b}} \omega_{\mathbf{b}} \mathrm{d}\mathbf{x} &\le \|\nabla v_{\mathbf{b}}\|_s \|\omega_{\mathbf{b}}\|_p \|\boldsymbol{\omega}\|_q \\
&\le \|\nabla v_{\mathbf{b}}\|_s \|\omega_{\mathbf{b}}\|_2^{\frac{6-p}{2p}} \|\omega_{\mathbf{b}}\|_6^{\frac{3p-6}{2p}} \|\boldsymbol{\omega}\|_2^{\frac{6-q}{2q}} \|\boldsymbol{\omega}\|_6^{\frac{3q-6}{2q}} \\
&\le \varepsilon \|\nabla\omega_{\mathbf{b}}\|_2^2 + C \|\nabla v_{\mathbf{b}}\|_s^{\frac{4p}{6+p}} \|\boldsymbol{\omega}\|_2^{2\frac{p}{q}\frac{6-q}{6+p}} \|\boldsymbol{\omega}\|_6^{2\frac{p}{q}\frac{3q-6}{6+p}} \|\omega_{\mathbf{b}}\|_2^{2\frac{6-p}{6+p}} .
\end{aligned}
$$

Altogether we get

$$
\begin{aligned}
&\frac{\mathrm{d}}{\mathrm{d}t} \|\omega_{\mathbf{b}}(\tau)\|_2^2 + \|\nabla\omega_{\mathbf{b}}\|_2^2 \\
&\quad \le C \|\nabla v_{\mathbf{b}}\|_s^{\frac{4p}{6+p}} \|\boldsymbol{\omega}\|_2^{2\frac{p}{q}\frac{6-q}{6+p}} \|\boldsymbol{\omega}\|_6^{2\frac{p}{q}\frac{3q-6}{6+p}} \|\omega_{\mathbf{b}}\|_2^{2\frac{6-p}{6+p}} \\
&\qquad + C(\mathbf{b}) \|\nabla \mathbf{v}\|_2^2 (1 + \|\mathbf{v}\|_3^2),
\end{aligned}
$$

thus using generalized Gronwall inequality in the form of Theorem 2 from [3] gives us $\left(\frac{p-6}{p+6} + 1 = \frac{2p}{6+p}\right)$

$$
\begin{aligned}
\|\omega_{\mathbf{b}}(\tau)\|_2^{\frac{4p}{6+p}} &\le \|\omega_{\mathbf{b}}(0)\|_2^{\frac{4p}{6+p}} \\
&\quad + C \|\boldsymbol{\omega}\|_{\infty,2}^{4\frac{p}{q}\frac{3-q}{6+p}} \int_0^\tau \|\nabla v_{\mathbf{b}}\|_s^{\frac{4p}{6+p}} \|\boldsymbol{\omega}\|_2^{\frac{2p}{6+p}} \|\boldsymbol{\omega}\|_6^{2\frac{p}{q}\frac{3q-6}{6+p}} \,\mathrm{d}\sigma \\
&\quad + C\Big( \int_0^\tau \|\nabla \mathbf{v}\|_2^2 (1 + \|\mathbf{v}\|_3^2) \mathrm{d}\sigma \Big)^{\frac{2p}{6+p}} .
\end{aligned}
$$

Estimating the second term as above and using the Hölder inequality in the form

$$\int_0^\tau \|\nabla v_{\mathbf{b}}\|_s^{\frac{4p}{6+p}} \|\boldsymbol{\omega}\|_2^{\frac{2p}{6+p}} \|\boldsymbol{\omega}\|_6^{2\frac{p}{q}\frac{3q-6}{6+p}} \mathrm{d}\sigma \leq \|\nabla v_{\mathbf{b}}\|_{t,s}^{\frac{4p}{6+p}} \|\boldsymbol{\omega}\|_{2,2}^{\frac{2p}{6+p}} \|\boldsymbol{\omega}\|_{2,6}^{2\frac{p}{q}\frac{3q-6}{6+p}}$$

we conclude

$$\|\omega_{\mathbf{b}}\|_{\infty,2}^2 \leq C_1 + C_2 \|\boldsymbol{\omega}\|_{Y(\tau)} .$$

This finishes the proof of Lemma 1.

## 4 Proof of the Main Theorem

Without loss of generality we may assume that $|\mathbf{b}(\tau, \mathbf{x})| \equiv 1$. Then for every time $\sigma \in (0, T)$, and point $\mathbf{x} \in \mathbb{R}^3$, there exists at least one component $b_j(\sigma, \mathbf{x})$ such that $\left|b_j(\sigma, \mathbf{x})\right| > \frac{1}{2}$. Since vector field $\mathbf{b}(\cdot, \cdot)$ is continuous on $(0, T) \times \mathbb{R}^3$, the sets $\Omega_{r,\sigma} = \left\{\mathbf{x} \in \mathbb{R}^3 \mid b_r(\sigma, \mathbf{x}) > \frac{1}{2}\right\}$ a $\Omega_{r+3,\sigma} = \left\{\mathbf{x} \in \mathbb{R}^3 \mid b_r(\sigma, \mathbf{x}) < -\frac{1}{2}\right\}$, $r = 1,\ 2,\ 3$ compose at each particular time a covering of $\mathbb{R}^3$ by six open sets $\{\Omega_{r,\sigma}\}_{r=1}^6$. For simplicity, we set $b_{r+3} := -b_r$, $r = 1, 2, 3$. Using the partition of unity (see e.g. [7]) we get functions $\varphi_{r,\sigma} \in C_0^\infty(\Omega_{r,\sigma})$, $0 \leq \varphi_{r,\sigma}(\mathbf{x}) \leq 1$ such that $\sum_r \varphi_{r,\sigma} = 1$, and $|\nabla \varphi_{r,\sigma}| \leq C(\mathbf{b})$. We will multiply the following equivalent form of Navier–Stokes equations (1)

$$\frac{\partial \mathbf{v}}{\partial t} - \nu \Delta \mathbf{v} = -\boldsymbol{\omega} \times \mathbf{v} - \nabla \left(p + \frac{1}{2} |\mathbf{v}|^2\right) \tag{4}$$

by the test function $-\sum\limits_{r=1}^{6} \partial_l(\varphi_{r,\sigma} b_r \partial_l \mathbf{v})$. Let work with each term separately, for illustration only with $r = 1$:

$$\begin{aligned} -\int_{\Omega_{1,\sigma}} \frac{\partial \mathbf{v}}{\partial t} \cdot \partial_l(\varphi_{1,\sigma} b_1 \partial_l \mathbf{v}) \mathrm{d}\mathbf{x} &= \int_{\Omega_{1,\sigma}} \frac{\partial}{\partial t} \partial_l \mathbf{v} \cdot \varphi_{1,\sigma} b_1 \partial_l \mathbf{v} \mathrm{d}\mathbf{x} \\ &\geq \int_{\Omega_{1,\sigma}} \frac{1}{4} \varphi_{1,\sigma} \frac{\mathrm{d}}{\mathrm{d}t} |\nabla \mathbf{v}|^2 \mathrm{d}\mathbf{x} \end{aligned}$$

$$\int_{\Omega_{1,\sigma}} \Delta \mathbf{v} \cdot \partial_l(\varphi_{1,\sigma} b_1 \partial_l \mathbf{v}) d\mathbf{x} = \int_{\Omega_{1,\sigma}} \Delta \mathbf{v} \cdot \big(\varphi_{1,\sigma} b_1 \partial_l \partial_l \mathbf{v} + \partial_l(\varphi_{1,\sigma} b_1)\partial_l \mathbf{v}\big) d\mathbf{x}$$
$$\geq \int_{\Omega_{1,\sigma}} \frac{1}{2}\varphi_{1,\sigma} |\Delta \mathbf{v}|^2 d\mathbf{x} + \underbrace{\int_{\Omega_{1,\sigma}} \Delta \mathbf{v} \cdot \partial_l(\varphi_{1,\sigma} b_1)\partial_l \mathbf{v} d\mathbf{x}}_{=Z_1^1}$$

$$\int_{\Omega_{1,\sigma}} \partial_k \left(p + \frac{1}{2}|\mathbf{v}|^2\right) \partial_l(\varphi_{1,\sigma} b_1 \partial_l v_k) d\mathbf{x} = \int_{\Omega_{1,\sigma}} \partial_l \left(p + \frac{1}{2}|\mathbf{v}|^2\right) \partial_k(\varphi_{1,\sigma} b_1 \partial_l v_k) d\mathbf{x}$$
$$= \underbrace{\int_{\Omega_{1,\sigma}} \partial_l p \partial_k(\varphi_{1,\sigma} b_1)\partial_l v_k d\mathbf{x}}_{=Z_2^1} + \underbrace{\int_{\Omega_{1,\sigma}} \frac{1}{2}\partial_l |\mathbf{v}|^2 \partial_k(\varphi_{1,\sigma} b_1)\partial_l v_k d\mathbf{x}}_{=Z_3^1}$$

$$\int_{\Omega_{1,\sigma}} (\boldsymbol{\omega} \times \mathbf{v}) \cdot \partial_l (\varphi_{1,\sigma} b_1 \partial_l \mathbf{v}) d\mathbf{x}$$
$$= \int_{\Omega_{1,\sigma}} \varphi_{1,\sigma} b_1 (\boldsymbol{\omega} \times \mathbf{v}) \cdot \Delta \mathbf{v} d\mathbf{x} - \underbrace{\int_{\Omega_{1,\sigma}} (\boldsymbol{\omega} \times \mathbf{v}) \cdot \partial_l(\varphi_{1,\sigma} b_1)\partial_l \mathbf{v} d\mathbf{x}}_{=Z_4^1}$$

We rewrite the term in the first integral and get

$$\begin{aligned}
b_1(\boldsymbol{\omega} \times \mathbf{v}) \cdot \Delta \mathbf{v} = {} & b_1\omega_2 v_3 \Delta v_1 \; -b_1\omega_1 v_3 \Delta v_2 \; +b_1\omega_1 v_2 \Delta v_3 \\
& -b_1\omega_3 v_2 \Delta v_1 \; +b_1\omega_3 v_1 \Delta v_2 \; -b_1\omega_2 v_1 \Delta v_3 \\
& +b_2\omega_2 v_3 \Delta v_2 \; -b_2\omega_2 v_3 \Delta v_2 \\
& +b_3\omega_2 v_3 \Delta v_3 \qquad\qquad\qquad -b_3\omega_2 v_3 \Delta v_3 \\
& -b_2\omega_3 v_2 \Delta v_2 \; +b_2\omega_3 v_2 \Delta v_2 \\
& -b_3\omega_3 v_2 \Delta v_3 \qquad\qquad\qquad +b_3\omega_3 v_2 \Delta v_3 \\
& \qquad\qquad\qquad +b_3\omega_3 v_3 \Delta v_2 \; +b_2\omega_2 v_2 \Delta v_3 \\
& \qquad\qquad\qquad -b_3\omega_3 v_3 \Delta v_2 \; -b_2\omega_2 v_2 \Delta v_3 .
\end{aligned}$$

Thus,

$$\begin{aligned} b_1(\boldsymbol{\omega}\times\mathbf{v})\cdot\Delta\mathbf{v} =& \omega_2 v_3 \Delta v_{\mathbf{b}} - \omega_3 v_2 \Delta v_{\mathbf{b}} \\ &+\omega_{\mathbf{b}} v_2 \Delta v_3 - \omega_{\mathbf{b}} v_3 \Delta v_2 \\ &+\omega_3 v_{\mathbf{b}} \Delta v_2 - \omega_2 v_{\mathbf{b}} \Delta v_3 + \text{ lower order terms}(I_4^1,\ I_5^1). \end{aligned} \tag{5}$$

**Observation** The above mentioned equality holds true without additional lower order terms, if the vector $\mathbf{b}(\sigma,\cdot)$ is constant in space, otherwise we use the following identity

$$\begin{aligned} \int\limits_{\Omega_{1,\sigma}} \varphi_{1,\sigma}(\omega_l v_m) b_l \Delta v_n \mathrm{d}\mathbf{x} =& \int\limits_{\Omega_{1,\sigma}} \varphi_{1,\sigma}(\omega_l v_m)\Delta(b_l v_n)\mathrm{d}\mathbf{x} \\ &- \int\limits_{\Omega_{1,\sigma}} \varphi_{1,\sigma}\big((\omega_l v_m) v_n \Delta b_l + 2\varphi_{1,\sigma}\nabla v_n\cdot\nabla b_l(\omega_l v_m)\big)\mathrm{d}\mathbf{x}. \end{aligned}$$

Similarly, for $b_2$ we get

$$\begin{aligned} b_2(\boldsymbol{\omega}\times\mathbf{v})\cdot\Delta\mathbf{v} =& \omega_3 v_1 \Delta v_{\mathbf{b}} - \omega_1 v_3 \Delta v_{\mathbf{b}} \\ &+\omega_{\mathbf{b}} v_3 \Delta v_2 - \omega_{\mathbf{b}} v_1 \Delta v_3 \\ &+\omega_1 v_{\mathbf{b}} \Delta v_3 - \omega_3 v_{\mathbf{b}} \Delta v_2 + \text{ lower order terms.} \end{aligned} \tag{6}$$

For the term with $b_3$ we use (as above) the shifts $1 \mapsto 2$, $2 \mapsto 3$, $3 \mapsto 1$. The case $r = 4, 5, 6$ is trivial.

Summing up $\sum_{r=1}^{6}$, and using the definitions of $\varphi_{r,\sigma}$ we get (recall, we use summation convention)

$$\begin{aligned} &\frac{1}{4}\int_{\mathbb{R}^3} \frac{\mathrm{d}}{\mathrm{d}t}|\nabla\mathbf{v}|^2\,\mathrm{d}\mathbf{x} + \frac{1}{2}\int_{\mathbb{R}^3} |\Delta\mathbf{v}|^2\,\mathrm{d}\mathbf{x} \\ &\qquad \le \sum_{r=1}^{6}\big(|Z_1^r| + |Z_2^r| + |Z_3^r| + |Z_4^r| + CI_4^r + CI_5^r\big) \\ &\qquad\qquad + \sum_{i=1}^{3}\varepsilon_{ijk}(I_1^{jk} + I_2^{jk} + I_3^{jk}), \end{aligned} \tag{7}$$

where

$$
\begin{aligned}
I_1^{jk} &= \int_{\mathbb{R}^3} \omega_k v_{\mathbf{b}} \Delta v_j \mathrm{d}\mathbf{x}, & I_2^{jk} &= \int_{\mathbb{R}^3} \omega_j v_k \Delta v_{\mathbf{b}} \mathrm{d}\mathbf{x},\\
I_3^{jk} &= \int_{\mathbb{R}^3} \omega_{\mathbf{b}} v_j \Delta v_k \mathrm{d}\mathbf{x}, & I_4^{r} &= \int_{\mathbb{R}^3} |\boldsymbol{\omega}|\, |\mathbf{v}|^2\, |\Delta(\varphi_{r,\sigma}\mathbf{b})|\, \mathrm{d}\mathbf{x},\\
I_5^{r} &= 2\int_{\mathbb{R}^3} |\nabla\mathbf{v}|\, |\nabla(\varphi_{r,\sigma}\mathbf{b})|\, |\boldsymbol{\omega}|\, |\mathbf{v}|\, \mathrm{d}\mathbf{x}, & Z_1^{r} &= \int_{\mathbb{R}^3} \Delta\mathbf{v}\cdot \partial_l(\varphi_{r,\sigma} b_r)\partial_l \mathbf{v} \mathrm{d}\mathbf{x}, \qquad (8)\\
Z_2^{r} &= \int_{\mathbb{R}^3} \partial_l p \partial_k(\varphi_{r,\sigma} b_r)\partial_l v_k \mathrm{d}\mathbf{x}, & Z_3^{r} &= \int_{\mathbb{R}^3} \frac{1}{2}\partial_l\, |\mathbf{v}|^2\, \partial_k(\varphi_{r,\sigma} b_r)\partial_l v_k \mathrm{d}\mathbf{x},\\
Z_4^{r} &= \int_{\mathbb{R}^3} (\boldsymbol{\omega}\times\mathbf{v})\cdot \partial_l(\varphi_{r,\sigma} b_r)\partial_l \mathbf{v} \mathrm{d}\mathbf{x}
\end{aligned}
$$

(9)

and $\varepsilon_{ijk}$ is the Levi-Civita tensor, i.e. it is zero unless all indices are different, it is equal to $+1$ for a positive permutation of 123 and equal to $-1$ otherwise.

Now, we will estimate these integrals in order to finish the proof. At first, we will consider the case, in which we have the additional information about the projection $v_b$ itself; we will proceed quite analogously with [18].

$$
\begin{aligned}
\int_0^\tau \left|I_1^{jk}\right| \mathrm{d}\sigma &\le \int_0^\tau \int_{\mathbb{R}^3} \left|\omega_k v_{\mathbf{b}} \Delta v_j\right| \mathrm{d}\mathbf{x}\mathrm{d}\sigma \le \int_0^\tau \|v_{\mathbf{b}}\|_s \|\omega_k\|_{\frac{2s}{s-2}} \left\|\Delta v_j\right\|_2 \mathrm{d}\sigma\\
&\le C\int_0^\tau \|v_{\mathbf{b}}\|_s \|\omega_k\|_2^{\frac{s-3}{s}} \|\nabla\omega_k\|_2^{\frac{3}{s}} \left\|\Delta v_j\right\|_2 \mathrm{d}\sigma\\
&\le \varepsilon \|\Delta\mathbf{v}\|_{2,2}^2 + C_\varepsilon \int_0^\tau \|v_{\mathbf{b}}\|_s^{\frac{2s}{s-3}} \|\nabla\mathbf{v}\|_2^2 \,\mathrm{d}\sigma\\
&\le \varepsilon \|\Delta\mathbf{v}\|_{2,2}^2 + C_\varepsilon(T) \|v_{\mathbf{b}}\|_{t,s}^{\frac{2s}{s-3}} \|\nabla\mathbf{v}\|_{\infty,2}^2 \qquad \left(t \ge \frac{2s}{s-3}\right).
\end{aligned}
$$

Next

$$\begin{aligned} I_2^{jk} =&\varepsilon_{jmn} \int\limits_{\mathbb{R}^3} \partial_m v_n v_k \Delta v_{\mathbf{b}} \mathrm{d}\mathbf{x} \\ =&-\varepsilon_{jmn} \int\limits_{\mathbb{R}^3} \partial_m \partial_l v_n v_k \partial_l v_{\mathbf{b}} \mathrm{d}\mathbf{x} - \varepsilon_{jmn} \int\limits_{\mathbb{R}^3} \partial_m v_n \partial_l v_k \partial_l v_{\mathbf{b}} \mathrm{d}\mathbf{x} \\ =&\underbrace{-\varepsilon_{jmn} \int\limits_{\mathbb{R}^3} \partial_m \partial_l v_n v_k \partial_l v_{\mathbf{b}} \mathrm{d}\mathbf{x}}_{J_1^{jk}} + \underbrace{\varepsilon_{jmn} \int\limits_{\mathbb{R}^3} \partial_m \partial_l v_n \partial_l v_k v_{\mathbf{b}} \mathrm{d}\mathbf{x}}_{J_2^{jk}} \\ &+ \underbrace{\varepsilon_{jmn} \int\limits_{\mathbb{R}^3} \partial_l \partial_l v_k \partial_m v_n v_{\mathbf{b}} \mathrm{d}\mathbf{x}}_{J_3^{jk}}. \end{aligned}$$

Here,

$$\begin{aligned} \left|J_1^{jk}\right| \leq& \int\limits_{\mathbb{R}^3} \left|\varepsilon_{jmn} \partial_m \partial_l v_n v_k \partial_l v_{\mathbf{b}}\right| \mathrm{d}\mathbf{x} \leq \varepsilon \left\|\Delta \mathbf{v}\right\|_2^2 + C_\varepsilon \int\limits_{\mathbb{R}^3} v_k^2 (\partial_l v_{\mathbf{b}})^2 \mathrm{d}\mathbf{x} \\ \leq& \varepsilon \left\|\Delta \mathbf{v}\right\|_2^2 - C_\varepsilon \int\limits_{\mathbb{R}^3} v_{\mathbf{b}} \partial_l \partial_l v_{\mathbf{b}} v_k^2 \mathrm{d}\mathbf{x} - C_\varepsilon \int\limits_{\mathbb{R}^3} v_{\mathbf{b}} \partial_l v_{\mathbf{b}} \partial_l \left(v_k^2\right) \mathrm{d}\mathbf{x}. \end{aligned} \tag{10}$$

Let us estimate the first integral on the right hand side of (10):

$$\begin{aligned} \left| \int\limits_0^\tau \int\limits_{\mathbb{R}^3} v_{\mathbf{b}} \partial_l \partial_l v_{\mathbf{b}} v_k^2 \mathrm{d}\mathbf{x} \mathrm{d}\sigma \right| \leq& \underbrace{\int\limits_0^\tau \int\limits_{\mathbb{R}^3} |\mathbf{b}| \, |\Delta \mathbf{v}| \, |v_{\mathbf{b}}| \, |\mathbf{v}|^2 \, \mathrm{d}\mathbf{x} \mathrm{d}\sigma}_{J_{11}} \\ &+ \underbrace{2 \int\limits_0^\tau \int\limits_{\mathbb{R}^3} |\nabla \mathbf{b}| \, |\nabla \mathbf{v}| \, |v_{\mathbf{b}}| \, |\mathbf{v}|^2 \, \mathrm{d}\mathbf{x} \mathrm{d}\sigma}_{J_{12}} + \underbrace{\int\limits_0^\tau \int\limits_{\mathbb{R}^3} |\Delta \mathbf{b}| \, |v_{\mathbf{b}}| \, |\mathbf{v}|^3 \, \mathrm{d}\mathbf{x} \mathrm{d}\sigma}_{J_{13}}. \end{aligned}$$

Then

$$J_{11} \le \int_0^\tau \|\Delta \mathbf{v}\|_2 \|\nu_{\mathbf{b}}\|_s \|\mathbf{v}\|^2_{\frac{4s}{s-2}} \,\mathrm{d}\sigma$$

$$\le C \int_0^\tau \|\Delta \mathbf{v}\|_2 \|\nu_{\mathbf{b}}\|_s \|\mathbf{v}\|_{\frac{3s}{s-3}} \|\mathbf{v}\|_6 \,\mathrm{d}\sigma$$

$$\le C \|\Delta \mathbf{v}\|_{2,2} \|\nu_{\mathbf{b}}\|_{t,s} \|\mathbf{v}\|_{\frac{4s}{s+6},\frac{3s}{s-3}} \|\mathbf{v}\|_{\infty\, 6}\,.$$

As $\frac{2}{4s/(s+6)} + \frac{3}{3s/(s-3)} = \frac{3}{2}$, we can interpolate

$$\|\mathbf{v}\|_{\frac{4s}{s+6},\frac{3s}{s-3}} \le C(s) \|\mathbf{v}\|_{\infty,2}^{\frac{s-6}{2s}} \|\nabla \mathbf{v}\|_{2,2}^{\frac{s+6}{2s}}\,.$$

Thus, using Young's inequality and energy inequality we get

$$J_{11} \le \varepsilon \|\Delta \mathbf{v}\|^2_{2,2} + C_\varepsilon \|\nu_{\mathbf{b}}\|^2_{t,s} \|\nabla \mathbf{v}\|^2_{\infty,2}\,.$$

Note that for fixed $\varepsilon > 0$, $C_\varepsilon \to 0$ for $\tau \to 0$, uniformly for $s \in [6,\infty]$. The lower order terms $J_{12}$, and $J_{13}$ may be bounded as follows:

$$J_{12} \le \int_0^\tau \|\nu_{\mathbf{b}}\|_s \|\nabla \mathbf{v}\|_2 \|\nabla \mathbf{b}\|_\infty \|\mathbf{v}\|^2_{\frac{4s}{s-2}} \,\mathrm{d}\sigma$$

$$\le \int_0^\tau \|\nu_{\mathbf{b}}\|_s \|\nabla \mathbf{v}\|_2 \|\nabla \mathbf{b}\|_\infty \|\mathbf{v}\|_{\frac{3s}{s-3}} \|\mathbf{v}\|_6 \,\mathrm{d}\sigma$$

$$\le \|\nu_{\mathbf{b}}\|_{t,s} \|\nabla \mathbf{v}\|_{\infty,2} \|\nabla \mathbf{b}\|_{2,\infty} \|\mathbf{v}\|_{\frac{4s}{s+6},\frac{3s}{s-3}} \|\mathbf{v}\|_{\infty,6}$$

$$\le C \|\nu_{\mathbf{b}}\|_{t,s} \|\nabla \mathbf{v}\|^2_{\infty,2}\,,$$

$$J_{13} \le \int_0^\tau \|\nu_{\mathbf{b}}\|_s \|\mathbf{v}\|^2_6 \|\mathbf{v}\|_{\frac{3s}{s-3}} \left\|\nabla^2 \mathbf{b}\right\|_3 \mathrm{d}\sigma$$

$$\le \|\nu_{\mathbf{b}}\|_{t,s} \|\nabla \mathbf{v}\|^2_{\infty,2} \|\mathbf{v}\|_{\frac{4s}{s+6},\frac{3s}{s-3}} \left\|\nabla^2 \mathbf{b}\right\|_{2,3}$$

$$\le C \|\nu_{\mathbf{b}}\|_{t,s} \|\nabla \mathbf{v}\|^2_{\infty,2}\,,$$

where we have used Hölder's inequality $\left(\frac{1}{t} + \frac{1}{2} + \frac{s+6}{4s} = 1\right)$, the assumptions on $\mathbf{b}(\cdot,\cdot)$, and the fact that from energy inequality we have estimate of the norm of $\mathbf{v}$ in space $L^{\frac{4s}{s+6}}(0,T,(L^{\frac{3s}{s-3}}(\mathbb{R}^3))^3)$. Note that $C = C(\tau) \to 0$ for $\tau \to 0$ uniformly for $s \in [6,\infty]$.

Further, we will estimate the last integral from (10):

$$\left|\int_0^\tau \int_{\mathbb{R}^3} v_{\mathbf{b}} \partial_l v_{\mathbf{b}} \partial_l \left(v_k^2\right) \mathrm{d}\mathbf{x}\mathrm{d}\sigma\right| \le \underbrace{\int_0^\tau \int_{\mathbb{R}^3} |v_{\mathbf{b}}| \, |\nabla \mathbf{v}|^2 \, |\mathbf{v}| \, |\mathbf{b}| \, \mathrm{d}\mathbf{x}\mathrm{d}\sigma}_{=J_{14}}$$

$$+ \underbrace{\int_0^\tau \int_{\mathbb{R}^3} |v_{\mathbf{b}}| \, |\nabla \mathbf{b}| \, |\nabla \mathbf{v}| \, |\mathbf{v}|^2 \, \mathrm{d}\mathbf{x}\mathrm{d}\sigma}_{J_{15}=J_{12}}$$

$$J_{14} \le C \int_0^\tau \|v_{\mathbf{b}}\|_s \, \|\nabla \mathbf{v}\|_3^2 \, \|\mathbf{v}\|_{\frac{3s}{s-3}} \, \mathrm{d}\sigma \le C \, \|v_{\mathbf{b}}\|_{t,s} \, \|\nabla \mathbf{v}\|_{4,3}^2 \, \|\mathbf{v}\|_{\frac{2t}{t-2},\frac{3s}{s-3}}$$

The interpolation inequalities

$$\|\mathbf{v}\|_{\frac{2t}{t-2},\frac{3s}{s-3}} \le C \, \|\mathbf{v}\|_{\infty,2}^{2/t} \, \|\mathbf{v}\|_{2,6}^{(t-2)/t}, \text{ and } \|\nabla \mathbf{v}\|_{4,3} \le C \|\nabla \mathbf{v}\|_{\infty,2}^{\frac{1}{2}} \|\Delta \mathbf{v}\|_{2,2}^{\frac{1}{2}},$$

and Young's inequality yield

$$\left|\int_0^\tau \int_{\mathbb{R}^3} v_{\mathbf{b}} \partial_l v_{\mathbf{b}} \partial_l \left(v_k^2\right) \mathrm{d}\mathbf{x}\mathrm{d}\sigma\right| \le \varepsilon \, \|\Delta \mathbf{v}\|_{2,2}^2 + C_\varepsilon \, \|v_{\mathbf{b}}\|_{t,s}^2 \, \|\nabla \mathbf{v}\|_{\infty,2}^2 ,$$

which implies the bound on $J_1^{jk}$

$$\left|\int_0^\tau J_1^{jk} \mathrm{d}\sigma\right| \le 3\varepsilon \, \|\Delta \mathbf{v}\|_{2,2}^2 + C_\varepsilon \left(\|v_{\mathbf{b}}\|_{t,s}^2 + \|v_{\mathbf{b}}\|_{t,s}\right) \|\nabla \mathbf{v}\|_{\infty,2}^2 .$$

Further,

$$\int_0^\tau \left|J_2^{jk} + J_3^{jk}\right| \mathrm{d}\sigma \le \int_0^\tau \left(\int_{\mathbb{R}^3} |\partial_l \partial_m v_n \partial_l v_k v_{\mathbf{b}}| \, \mathrm{d}\mathbf{x} + \int_{\mathbb{R}^3} |\partial_l \partial_l v_k \partial_m v_n v_{\mathbf{b}}| \, \mathrm{d}\mathbf{x}\right) \mathrm{d}\sigma$$

$$\le 2 \int_0^\tau \|\Delta \mathbf{v}\|_2 \, \|v_{\mathbf{b}}\|_s \, \|\nabla \mathbf{v}\|_{\frac{2s}{s-2}} \, \mathrm{d}\sigma$$

$$\leq C \int_0^\tau \|v_{\mathbf{b}}\|_s \|\nabla \mathbf{v}\|_2^{\frac{s-3}{s}} \|\Delta \mathbf{v}\|_2^{\frac{3}{s}} \|\Delta \mathbf{v}\|_2 \mathrm{d}\sigma$$

$$\leq \varepsilon \|\Delta \mathbf{v}\|_{2,2}^2 + C_\varepsilon \|v_{\mathbf{b}}\|_{t,s} \|\nabla \mathbf{v}\|_{\infty,2}^2 ,$$

$$\int_0^\tau \left|I_3^{jk}\right| \mathrm{d}\sigma \leq \int_0^\tau \int_{\mathbb{R}^3} |\omega_{\mathbf{b}}| \left|v_j\right| |\Delta v_k| \, \mathrm{d}\mathbf{x}\mathrm{d}\sigma \leq \varepsilon \|\Delta \mathbf{v}\|_{2,2}^2 + C_\varepsilon \int_0^\tau \|\mathbf{v}\|_3^2 \|\omega_{\mathbf{b}}\|_6^2 \, \mathrm{d}\sigma$$

$$\leq \varepsilon \|\Delta \mathbf{v}\|_{2,2}^2 + C_\varepsilon \|\mathbf{v}\|_{\infty,3}^2 \|\omega_{\mathbf{b}}\|_{2,6}^2 .$$

Using $\|\mathbf{v}\|_{\infty,3}^2 \leq \|\mathbf{v}\|_{\infty,2} \|\mathbf{v}\|_{\infty,6} \leq C \|\nabla \mathbf{v}\|_{\infty,2}$, and the information which comes from (3), we get

$$\|\mathbf{v}\|_{\infty,3}^2 \|\omega_{\mathbf{b}}\|_{2,6}^2 \leq C \|\nabla \mathbf{v}\|_{\infty,2} (1 + \|\nabla \mathbf{v}\|_{Y(\tau)})$$

$$\leq C(\tau)(\|\nabla \mathbf{v}\|_{\infty,2} + \|\nabla \mathbf{v}\|_{\infty,2}^2 + \|\Delta \mathbf{v}\|_{2,2}^2) + C_0 \|\mathbf{v}_0\|_{1,2}^2 .$$

Recall that $C \to 0$ for $\tau \to 0^+$, uniformly for $s \in [6, \infty]$. It remains to deduce suitable estimates of the lower order terms with derivatives of $\mathbf{b}(\cdot, \cdot)$.

$$\int_0^\tau |I_4^r| \, \mathrm{d}\sigma \leq \int_0^\tau \int_{\mathbb{R}^3} |\boldsymbol{\omega}| \, |\mathbf{v}|^2 \, |\Delta(\varphi_{r,\sigma}\mathbf{b})| \, \mathrm{d}\mathbf{x}\mathrm{d}\sigma$$

$$\leq \int_0^\tau \|\nabla \mathbf{v}\|_6 \|\mathbf{v}\|_3 \|\mathbf{v}\|_6 \|\Delta(\varphi_{r,\sigma}\mathbf{b})\|_3 \, \mathrm{d}\sigma$$

$$\leq \varepsilon \|\Delta \mathbf{v}\|_{2,2}^2 + C_\varepsilon \int_0^\tau \|\nabla \mathbf{v}\|_2^2 \|\mathbf{v}\|_3^2 \|\Delta(\varphi_{r,\sigma}\mathbf{b})\|_3^2 \, \mathrm{d}\sigma$$

$$\leq \varepsilon \|\Delta \mathbf{v}\|_{2,2}^2 + C_\varepsilon \|\nabla \mathbf{v}\|_{\infty,2}^2 \|\mathbf{v}\|_{4,3}^2 \|\Delta(\varphi_{r,\sigma}\mathbf{b})\|_{4,3}^2$$

$$\int_0^\tau |Z_1^r| \, \mathrm{d}\sigma \leq \int_0^\tau \int_{\mathbb{R}^3} |\Delta \mathbf{v}| \, |\nabla(\varphi_{r,\sigma}\mathbf{b})| \, |\nabla \mathbf{v}| \, \mathrm{d}\mathbf{x}\mathrm{d}\sigma$$

$$\leq \varepsilon \|\Delta \mathbf{v}\|_{2,2} + C_\varepsilon \|\nabla \mathbf{v}\|_{\infty,2}^2 \|\nabla(\varphi_{r,\sigma}\mathbf{b})\|_{2,\infty}^2$$

$$
\begin{aligned}
&\int_0^\tau |I_5^r + Z_2^r + Z_3^r + Z_4^r| \, \mathrm{d}\sigma \\
&\quad \le \int_0^\tau \int_{\mathbb{R}^3} \Big( |\nabla \mathbf{v}| \, |\nabla(\varphi_{r,\sigma}\mathbf{b})| \, |\boldsymbol{\omega}| \, |\mathbf{v}| + |\nabla p| |\nabla(\varphi_{r,\sigma}\mathbf{b})| |\nabla \mathbf{v}| \\
&\qquad\qquad + |\nabla \mathbf{v}|^2 \, |\mathbf{v}| |\nabla(\varphi_{r,\sigma}\mathbf{b})| + |\boldsymbol{\omega}| \, |\mathbf{v}| \, |\nabla \mathbf{v}| \, |\nabla(\varphi_{r,\sigma}\mathbf{b})| \Big) \mathrm{d}\mathbf{x} \mathrm{d}\sigma \\
&\quad \le C \int_0^\tau \|\nabla \mathbf{v}\|_6 \|\nabla \mathbf{v}\|_2 \|\mathbf{v}\|_3 \left(\|\nabla \mathbf{b}\|_\infty + 1\right) \mathrm{d}\sigma \\
&\quad \le C \, \|\Delta \mathbf{v}\|_{2,2} \, \|\nabla \mathbf{v}\|_{\infty,2} \, \|\mathbf{v}\|_{4,3} \left( \|\nabla \mathbf{b}\|_{4,\infty} + (T^*)^{\frac{1}{4}} \right) \\
&\quad \le \varepsilon \, \|\Delta \mathbf{v}\|_{2,2}^2 + C_\varepsilon \, \|\nabla \mathbf{v}\|_{\infty,2}^2 \, \|\mathbf{v}\|_{4,3}^2
\end{aligned}
$$

Collecting all the above estimates together, we see that

$$
\|\nabla \mathbf{v}\|_{\infty,2}^2 + \|\Delta \mathbf{v}\|_{2,2}^2 \le C_0 \, \|\mathbf{v}_0\|_{1,2}^2 + C(\|\nabla \mathbf{v}\|_{\infty,2}^2 + \|\Delta \mathbf{v}\|_{2,2}^2 + 1), \tag{11}
$$

where $C \to 0$ for $\tau \to 0^+$, uniformly for $s \in [6, \infty]$. Therefore, taking $\tau$ sufficiently small, we get

$$
\|\nabla \mathbf{v}\|_{\infty,2}^2 + \|\Delta \mathbf{v}\|_{2,2}^2 \le 4C_0 \, \|\mathbf{v}_0\|_{1,2}^2 \, .
$$

Repeating the same estimates on $(\tau, 2\tau)$ we get that

$$
\|\nabla \mathbf{v}\|_{L^\infty(\tau,2\tau;L^2(\mathbb{R}^3))}^2 + \|\Delta \mathbf{v}\|_{L^2(\tau,2\tau;L^2(\mathbb{R}^3))}^2 \le 4C_0 \, \|\mathbf{v}(\tau)\|_{1,2}^2 \, .
$$

Therefore, after finite number of steps, we get that the regular solution exists on the whole time interval $(0, T)$.

Let us move to the case where we have information about the gradient of the projection and let us estimate all terms from (8). We start with the term $I_2^{jk}$:

$$
\begin{aligned}
I_2^{jk} &= \varepsilon_{jmn} \int_{\mathbb{R}^3} \partial_m v_n v_k \Delta v_{\mathbf{b}} \mathrm{d}\mathbf{x} \\
&= \underbrace{-\varepsilon_{jmn} \int_{\mathbb{R}^3} \partial_m \partial_l v_n v_k \partial_l v_{\mathbf{b}} \mathrm{d}\mathbf{x}}_{J_1^{jk}} \underbrace{-\varepsilon_{jmn} \int_{\mathbb{R}^3} \partial_m v_n \partial_l v_k \partial_l v_{\mathbf{b}} \mathrm{d}\mathbf{x}}_{J_2^{jk}}
\end{aligned}
$$

$$
\left| J_1^{jk} \right| \le \varepsilon \, \|\Delta \mathbf{v}\|_2^2 + C_\varepsilon \int_{\mathbb{R}^3} |\mathbf{v}|^2 \, |\nabla v_{\mathbf{b}}|^2 \, \mathrm{d}\mathbf{x} \tag{12}
$$

In the estimate of the right hand side of (12), we will distinguish between two possible cases. For $2 \le s \le 3$, we get

$$\begin{aligned}\int_{\mathbb{R}^3} |\mathbf{v}|^2 \, |\nabla v_{\mathbf{b}}|^2 \, \mathrm{d}\mathbf{x} &\le \|\nabla v_{\mathbf{b}}\|_s^2 \, \|\mathbf{v}\|_{\frac{2s}{s-2}}^2 \\ &\le \|\nabla v_{\mathbf{b}}\|_s^2 \, \|\nabla \mathbf{v}\|_2^{\frac{4s-6}{s}} \, \left\|\nabla^2 \mathbf{v}\right\|_2^{\frac{6-2s}{s}} \\ &\le \varepsilon \left\|\nabla^2 \mathbf{v}\right\|_2^2 + C_\varepsilon \, \|\nabla v_{\mathbf{b}}\|_s^{\frac{2s}{2s-3}} \, \|\nabla \mathbf{v}\|_2^2 ,\end{aligned}$$

while for $s > 3$, we will proceed in the following way

$$\int_{\mathbb{R}^3} |\mathbf{v}|^2 \, |\nabla v_{\mathbf{b}}|^2 \, \mathrm{d}\mathbf{x} \le \|\nabla v_{\mathbf{b}}\|_s^{\frac{6s}{5s-6}} \, \|\nabla v_{\mathbf{b}}\|_2^{\frac{4s-12}{5s-6}} \, \|\mathbf{v}\|_{\frac{2}{3}\frac{5s-6}{s-2}}^2 .$$

Further, due to $2 \le \frac{2}{3}\frac{5s-6}{s-2} \le 6$, we can interpolate

$$\|\mathbf{v}\|_{\frac{2}{3}\frac{5s-6}{s-2}}^2 \le C \|\mathbf{v}\|_2^{\frac{4s-12}{5s-6}} \, \|\nabla \mathbf{v}\|_2^{\frac{6s}{5s-6}} .$$

Moreover,

$$\|\nabla v_{\mathbf{b}}\|_2^{\frac{4s-12}{5s-6}} \le \left(\|\nabla \mathbf{v}\|_2 + \|\mathbf{v}\|_2 \, \|\nabla \mathbf{b}\|_\infty\right)^{\frac{4s-12}{5s-6}} ,$$

and using Young's inequality we have

$$\int_{\mathbb{R}^3} |\mathbf{v}|^2 \, |\nabla v_{\mathbf{b}}|^2 \, \mathrm{d}\mathbf{x} \le C \, \|\nabla v_{\mathbf{b}}\|_s^{\frac{6s}{5s-6}} \, \|\mathbf{v}\|_2^{\frac{4s-12}{5s-6}} \left(\|\nabla \mathbf{v}\|_2^2 + C_0(\mathbf{b})\right) .$$

The integrals $J_2^{jk}$, and $I_3^{jk}$ can be estimated in a straightforward way, analogously as in [15]

$$\left|J_2^{jk}\right| \le \|\nabla v_{\mathbf{b}}\|_s \, \|\nabla \mathbf{v}\|_{\frac{2s}{s-1}}^2 \le \varepsilon \left\|\nabla^2 \mathbf{v}\right\|_2^2 + C_\varepsilon \, \|\nabla v_{\mathbf{b}}\|_s^{\frac{2s}{2s-3}} \, \|\nabla \mathbf{v}\|_2^2 ;$$

$$\left|I_3^{jk}\right| \le \left\|\nabla^2 \mathbf{v}\right\|_2 \|\omega_{\mathbf{b}}\|_3 \|\mathbf{v}\|_6 \le \varepsilon \left\|\nabla^2 \mathbf{v}\right\|_2^2 + \varepsilon \|\omega_{\mathbf{b}}\|_3^4 + C_\varepsilon \|\nabla \mathbf{v}\|_2^4 .$$

We now return to the term $I_1^{jk}$. We have (below, $\delta_{ij}$ denotes the Kronecker symbol)

$$\begin{aligned}\varepsilon_{ijk}I_1^{jk} &= \varepsilon_{ijk}\int_{\mathbb{R}^3}\omega_k v_{\mathbf{b}}\Delta v_j \mathrm{d}\mathbf{x} = \varepsilon_{ijk}\varepsilon_{klm}\int_{\mathbb{R}^3}\partial_l v_m v_{\mathbf{b}}\Delta v_j \mathrm{d}\mathbf{x} \\ &= (\delta_{il}\delta_{jm} - \delta_{im}\delta_{jl})\int_{\mathbb{R}^3}\partial_l v_m v_{\mathbf{b}}\Delta v_j\mathrm{d}\mathbf{x} = \int_{\mathbb{R}^3}(\partial_i v_j v_{\mathbf{b}}\Delta v_j - \partial_j v_i v_{\mathbf{b}}\Delta v_j)\mathrm{d}\mathbf{x} \\ &= -\int_{\mathbb{R}^3}\partial_i v_j \partial_l v_{\mathbf{b}}\partial_l v_j \mathrm{d}\mathbf{x} + \frac{1}{2}\int_{\mathbb{R}^3}(\partial_l v_j)^2\partial_i v_{\mathbf{b}}\mathrm{d}\mathbf{x} + \int_{\mathbb{R}^3} v_i\partial_j v_{\mathbf{b}}\Delta v_j \mathrm{d}\mathbf{x}.\end{aligned}$$

Therefore these terms can be treated exactly as terms above coming from $I_2^{jk}$. Next, we have to estimate the lower order terms. Since they can be treated exactly as in the previous case (additional information about $v_{\mathbf{b}}$), we skip the details.

Altogether we get

$$\begin{aligned}\|\nabla\mathbf{v}\|_{\infty,2}^2 + \left\|\nabla^2\mathbf{v}\right\|_{2,2}^2 &\le \varepsilon\int_0^\tau \|\omega_{\mathbf{b}}\|_3^4 \mathrm{d}\sigma \\ &+ C_\varepsilon\int_0^\tau\Big\{g(s)\,\|\nabla v_{\mathbf{b}}\|_s^{\frac{2s}{2s-3}}\,C_0(\mathbf{b}) + \|\nabla v_{\mathbf{b}}\|_s^{\frac{6s}{5s-6}} \\ &+\|\Delta\mathbf{b}\|_6^2 + \|\mathbf{v}\|_3^4 + \|\nabla\mathbf{b}\|_\infty^4 + \|\nabla\mathbf{v}\|_2^2\Big\}\,\|\nabla\mathbf{v}\|_2^2\,\mathrm{d}\sigma,\end{aligned}$$

where $g(s) = 0$, for $2 \le s \le 3$, and $g(s) = 1$, for $s > 3$. Note that both $\frac{2s}{2s-3}$ and $\frac{6s}{5s-6}$ are less than $t = \frac{4s}{3s-6}$. Using the estimate from Lemma 1 we obtain

$$\int_0^\tau \|\omega_{\mathbf{b}}\|_3^4\mathrm{d}\sigma \le C(1 + \|\nabla\mathbf{v}\|_{Y(\tau)}^2).$$

Choosing $\varepsilon$ sufficiently small, we can use Gronwall's inequality in order to conclude that

$$\|\nabla\mathbf{v}\|_{\infty,2}^2 + \left\|\nabla^2\mathbf{v}\right\|_{2,2}^2 \le C(\mathbf{v}_0, \|\nabla v_{\mathbf{b}}\|_{t,s}).$$

As this inequality holds for any $\tau < T^*$ and $C$ is independent of $\tau$, the proof of Theorem 1 is complete.

**Acknowledgements** The work of the first author was supported by the grant SVV-2015-260226. The work of the second author was partially supported by the grant No. 201/09/0917 of the Grant Agency of the Czech Republic.

## References


1. Š. Axmann, M. Pokorný, A note on regularity criteria for the solutions to Navier-Stokes equations involving one velocity component (in preparation)
2. C. Cao, E.S. Titi, Regularity criteria for the three-dimensional Navier–Stokes equations. Indiana U. Math. J. **57**(6), 2643–2660 (2008)
3. H. El-Owaidy et al., On some new integral inequalities of Gronwall–Bellman type. Appl. Math. Comput. **106**, 289–303 (1999)
4. G.P. Galdi, An introduction to the Navier–Stokes initial–boundary value problem, in *Fundamental Directions in Mathematical Fluid Mechanics*. Advances in Mathematical Fluid Mechanics (Birkhäuser, Basel, 2000), pp. 1–70
5. C. He, Regularity for solutions to the Navier–Stokes equations with one velocity component regular. Electron. J. Differ. Equ. **2002**(29), 1–13 (2002)
6. E. Hopf, Über die Anfangswertaufgabe für die hydrodynamischen Grundgleichungen. Math. Nachr. **4**(1), 213–231 (1951)
7. A. Kufner, O. John, S. Fučík, *Function Spaces*, 1st edn. (Academia, Prague, 1977)
8. I. Kukavica, M. Ziane, One component regularity for the Navier–Stokes equations. Nonlinearity **19**(2), 453–469 (2006)
9. O.A. Ladyzhenskaya, *The Mathematical Theory of Viscous Incompressible Flow*, 2nd edn. (Gordon and Breach, New York/London/Paris, 1969)
10. O.A. Ladyzhenskaya, Sixth problem of the millenium: Navier–Stokes equations, existence and smoothness. Russ. Math. Surv. **58**(2), 251–286 (2003)
11. J. Leray, Étude de diverses équations intégrales non linéaires et de quelques problèmes que pose l'hydrodynamique. J. Math. Pure Appl. **12**(2), 1–82 (1933)
12. J. Leray, Sur le mouvement d'un liquide visqueux emplissant l'espace. Acta Math. **63**(1), 193–248 (1934)
13. J. Neustupa, P. Penel, Regularity of a suitable weak solution to the Navier–Stokes equations as a consequence of regularity of one velocity component, in *Applied Nonlinear Analysis*, ed. by A. Sequeira et al. (Kluwer Academic/Plenum Pulishers, New York, 1999), pp. 391–402
14. J. Neustupa, A. Novotný, P. Penel, An interior regularity of weak solution to the Navier–Stokes equations in dependence on one component of velocity, in *Quaderni di Matematica*, ed. by A.O. Eden. Topics in Mathematical Fluid Mechanics, vol. 10 (Dept. Math., Seconda Univ. Napoli, Caserta, 2002), pp. 163–183
15. M. Pokorný, On the result of He concerning the smoothness of solutions to the Navier–Stokes equations. Electron. J. Differ. Equ. **2003**(11), 1–8 (2003)
16. R. Temam, Some developments on Navier–Stokes equations in the second half of the 20th century, in *Development of Mathematics 1950–2000* (Birkhäuser, Basel, 2000), pp. 1049–1106
17. Y. Zhou, A new regularity criterion for the Navier–Stokes equations in terms of the gradient of one velocity component. Methods Appl. Anal. **9**(4), 563–578 (2002)
18. Y. Zhou, A new regularity criterion for weak solutions to the Navier–Stokes equations. J. Math. Pure Appl. **84**(11), 1496–1514 (2005)
19. Y. Zhou, M. Pokorný, On a regularity criterion for the Navier–Stokes equations involving gradient of one velocity component. J. Math. Phys. **50**(12), 123514 (2009)
20. Y. Zhou, M. Pokorný, On the regularity criterion of the Navier–Stokes equations via one velocity component. Nonlinearity **23**(5), 1097–1107 (2010)

# On the Singular $p$-Laplacian System Under Navier Slip Type Boundary Conditions: The Gradient-Symmetric Case

**H. Beirão da Veiga**

*Dedicated to Yoshihiro Shibata on the occasion of his 60th birthday*

**Abstract** We consider the $p$-Laplacian system of $N$ equations in $n$ space variables, $1 < p \leq 2$, under the homogeneous Navier slip boundary condition without friction. Here, the gradient of the velocity is replaced by the (more physical) symmetric gradient, and the classical non-slip boundary condition is replaced by the Navier slip boundary condition without friction. These combination of circumstances leads to some additional obstacles. We prove $W^{2,q}$ regularity, up to the boundary, under suitable assumptions on the couple $p,\ q$. The singular case $\mu = 0$ is covered.

**Keywords** Regularity up to the boundary • Singular p-Laplacian elliptic systems • Slip boundary conditions

**Mathematics Subject Classification (2010).** 35J57; 35J60; 35J75; 35J92

## 1 Introduction: The Main Result

In the sequel $\Omega$ is a bounded, open, connected, subset of $\mathbb{R}^n$, locally situated on one side of its boundary, a smooth manifold $\Gamma$. We denote by $\underline{n}$ the outer unit normal to $\partial\Omega$. For convenience, we assume that $\Omega$ has not axis of symmetry. The reason will be clear below.

By $D\,u = \nabla u + \nabla^T u$ we denote the symmetric gradient. So

$$D_{ij}(u) = \partial_i\, u_j + \partial_j\, u_i\,, \tag{1}$$

H. Beirão da Veiga (✉)
Dipartimento di Matematica, Università di Pisa, Via Buonarroti 1/C, 56127 Pisa, Italy
e-mail: bveiga@dma.unipi.it

 
H. Amann et al. (eds.), *Recent Developments of Mathematical Fluid Mechanics*, Advances in Mathematical Fluid Mechanics, DOI 10.1007/978-3-0348-0939-9_6

where $i, j = 1, 2, \ldots, n$. Often we simply write $D$, provided that the vector field under consideration follows from the context. Further, we denote by $\underline{t}(u)$ the Cauchy stress vector

$$\underline{t}(u) = (D u) \cdot \underline{n}\,.$$

So, $t_j = (\partial_i u_j + \partial_j u_i) n_i$, where (here and in the sequel) we use the summation convention on repeated indexes.

In these notes we consider the system

$$- \nabla \cdot \big( (\mu + |D u|^2)^{\frac{p-2}{2}} D u \big) = f \text{ in } \Omega\,, \tag{2}$$

where $u$ is an $N$-dimensional vector field, $\mu \geq 0$ is a given parameter (we are particularly interested in the singular case $\mu = 0$), and $f$ is a given vector field. The vector field $u$ is subjected to the Navier slip boundary condition without friction, see [22],

$$\begin{cases} \underline{u} \cdot \underline{n} = 0, \\ (\underline{t}(u))_\tau = 0\,, \end{cases} \tag{3}$$

where, in general, the subscript $\tau$ denotes "tangential component".

Equation (2) has been considered by mathematicians mostly with $D u$ replaced by $\nabla u$. It is worth noting that (2) satisfies the Stokes Principle (see [23, 26], p. 231), a significant physical requirement of isotropy, which does not hold if we replace $D u$ by $\nabla u$.

For a mathematical study of the above boundary condition see, for instance, [7, 25], where this boundary condition is associated to the linear Stokes problem.

By $L^p(\Omega)$ and $W^{m,p}(\Omega)$, $1 \leq p \leq \infty$, $m$ nonnegative integer, we denote the usual Lebesgue and Sobolev spaces, with the standard norms $\| \cdot \|_p$ and $\| \cdot \|_{m,p}$, respectively.

In notation concerning norms and functional spaces, we do not distinguish between scalar and vector fields. For instance $L^p(\Omega; \mathbb{R}^N) = [L^p(\Omega)]^N$, $N > 1$, is denoted simply by $L^p(\Omega)$. We define

$$V_p = V_p(\Omega) = \{ v \in W^{1,p}(\Omega) : v \cdot \underline{n} = 0 \text{ on } \Gamma \}.$$

The linear space $V_p(\Omega)$, endowed with one of the following norms

$$\big( \| v \|_p + \| D v \|_p \big)^{\frac{1}{p}}, \quad \big( \| v \|_p + \| \nabla v \|_p \big)^{\frac{1}{p}}, \quad \| \nabla v \|_p\,,$$

is a Banach space. The above norms are equivalent in $V_p(\Omega)$. Further, since we assume that the domain $\Omega$ has not axis of symmetry, $\| D v \|_p$ alone is also an equivalent norm. Without the above assumption on $\Omega$, the equivalence does not hold. For a quite complete discussion on this, and related, arguments, we refer to [7]. See in particular the Lemma 2.3, and related results, in this last reference. Note, in particular, that the space $\mathbb{V}^1_\tau(\Omega)$ considered in Ref. [7] coincides here with $V_2(\Omega)$, since the space $Z$, defined in the above reference, is here reduced to $\{0\}$, due to the absence of axis of symmetry in $\Omega$. Further, we remark that in [7] we appeal to (12) instead of appealing to the totally equivalent formulation (11).

Our main result is the following.

**Theorem 1.1** *Let* $1 < p \leq 2$ *be fixed. Assume that* $\mu \geq 0$, *and let* $f \in L^q(\Omega)$, *where* $q > n$. *Let* $C_q = C(q, \Omega)$ *be the constant that appears in the linear estimate* (14) *below. Assume that*

$$(2 - p)\, C_q < 1\,. \tag{4}$$

*Then, the weak solution* $u$ *to the problem* (2), (3) *belongs to* $W^{2,q}(\Omega)$. *Moreover, the following estimate holds*

$$\|u\|_{2,q} \leq C\left(\|f\|_q + \|f\|_q^{\frac{1}{p-1}}\right). \tag{5}$$

Roughly speaking, below we tried to follow the main lines of the proof given in [10]. However, the presence of the symmetric gradient in place of the gradient, and that of the Navier slip boundary condition in place of the non-slip boundary condition, lead to additional obstacles. In booth proofs, following [8], the very starting point is the analysis, and subtle utilization, of the inner structure of the equations.

*An open problem.* Let us now consider the absolute vorticity boundary condition

$$\begin{cases} \underline{u} \cdot \underline{n} = 0, \\ \underline{\omega}(u) \times \underline{n} = 0\,, \end{cases} \tag{6}$$

where $N = 3$. This conditions was introduced by Bardos in Ref. [5]. It became very popular, and has been studied by a large number of authors. On flat portions of the boundary it coincides with the boundary condition (3). So, at least at first glance, it seems very likely that proofs and results shown below for (3) also hold for (6), with simple adaptations. We are able to prove $\mu$-uniform a priori estimates in $W^{2,q}(\Omega)$, but we have not a suitable definition of "weak solution" to the above boundary value problem, even in the full gradient case.

Regularity of solutions for systems like (2) has received substantial attention from many authors. We refer, for instance, to [1, 16, 18, 20, 27, 28]. Other related results may be found in [3, 4, 11–13, 15, 21], and references therein. We refer readers particularly interested in studying new connections between different boundary value assumptions, to the recent challenging paper [6]. Finally, we refer the reader to [17], where the authors prove very interesting regularity results for the Stokes problem (insert, as usual, pressure and divergence free conditions in equation (2)), under the boundary value problem (3).

The plan of the paper is the following: In Sect. 2 we recall the existence and uniqueness result of weak solutions. In Sect. 3 we introduce an auxiliary linear problem and state (by appealing to well know classical results) the existence of $W^{2,q}(\Omega)$ solutions to this linear problem. In Sect. 4 we formulate the non-linear problem in a more explicit (formally equivalent) form, in which the non-linearities are (roughly speaking) concentrated in the right hand side (see Eq. (18) below). Further, we appeal to this last formulation to define "strong solution". In Sect. 5, by assuming $\mu > 0$, and by appealing to the result stated in Sect. 3 for the auxiliary linear problem, we show that the strong solutions introduced in Sect. 4 exist and belongs to $W^{2,q}(\Omega)$, for each $\mu > 0$. Moreover, the estimates obtained are independent of $\mu$. This last property allows us, in Sect. 6, to extend the regularity result to the singular case $\mu = 0$. This is accomplished by passing to the limit in the variational formulation (10), as $\mu$ tends to zero. Sections 5 and 6 follow arguments developed in [10].

## 2 Existence and Uniqueness of the Weak Solution

Existence and uniqueness of weak solutions follows from well know results. Let us recall some basic points. Set

$$B(D\,u) = (\mu + |D\,u|^2)^{\frac{p-2}{2}}\,. \tag{7}$$

By appealing to the identity $D_{ij}u\,D_{ij}v = 2\,D_{ij}u\,\partial_j v_i$, integration by parts shows that

$$\begin{aligned}\frac{1}{2}\int_\Omega B(D\,u)\,D\,u\cdot D\,v\,dx &= -\int_\Omega \nabla\cdot\big(B(D\,u)\,D(u)\big)\cdot v\,dx\\ &+\int_\Gamma B(D\,u)\,[\,(D\,u)\cdot v\cdot n\,]\,dS\,.\end{aligned} \tag{8}$$

Hence,

$$\begin{aligned}\frac{1}{2}\int_\Omega B(D\,u)\,D\,u\cdot D\,v\;dx &= -\int_\Omega \nabla\cdot\big(B(D\,u)\,D(u)\big)\cdot v\,dx\\ &+\int_\Gamma B(D\,u)\,(\underline{t}(u))_\tau\cdot v\,dS\,.\end{aligned} \tag{9}$$

Note that on $\Gamma$ one has $\underline{t}(u)\cdot v = (\underline{t}(u))_\tau\cdot v$, since $v\cdot n = 0$.

Identity (9) justifies the following definition.

**Definition 2.1** Let $f \in V_p'(\Omega)$. We say that $u$ is a *weak solution* of problem (2), (3) if $u \in V_p(\Omega)$ satisfies

$$\frac{1}{2}\int_\Omega B(D\,u)\cdot D\,u\cdot D\,v\,dx = \int_\Omega f\cdot v\,dx\,, \tag{10}$$

for all $v \in V_p(\Omega)$.

Existence and uniqueness of the weak solution, for each fixed $\mu \geq 0$, follows by appealing to the theory of monotone operators, see Lions, [19, Chap. 2, Theorem 2.1]. Recall that $\| D\,v \|_p$ is a norm in $V_p(\Omega)$. Further, take into account that, for $y \geq 0$,

$$(\mu + y^2)^{\frac{p-2}{2}}\,y^2 \geq c\,y^p - c\,, \quad \text{if} \quad c \leq \left(\frac{1}{1+\mu}\right)^{\frac{2-p}{2}}.$$

## 3 An Auxiliary Linear Problem

In this section we consider the following linear elliptic equation

$$-\nabla\cdot D\,u = F \ \text{ in } \Omega\,, \tag{11}$$

under the linear boundary condition (3). We remark that Eq. (11) may be written in the equivalent form

$$-\Delta\,u - \nabla(\nabla\cdot u) = F\,. \tag{12}$$

The following definition coincides with definition 2.1, if $p = 2$.

**Definition 3.1** Let $f \in V_2'(\Omega)$. We say that $u$ is a *weak solution* to the linear problem (11), (3) if $u \in V_2(\Omega)$ satisfies

$$\frac{1}{2}\int_\Omega D\,u\cdot D\,v\,dx = \int_\Omega F\cdot v\,dx\,, \tag{13}$$

for all $v \in V_2(\Omega)$.

Coerciveness of the bilinear form on the left hand side of (13) follows here by appealing to the fact that $\| D\,v \|$ is a norm in $V_2(\Omega)$, since we have assumed that $\Omega$ has not axis of symmetry. Hence, existence, uniqueness, and the standard estimate holds for the above problem.

In the next sections we appeal to the following regularity result.

**Theorem 3.1** *Consider the linear boundary value problem* (11), (3). *Assume that* $F \in L^q(\Omega)$, *for some* $q \geq 2$. *Then, the weak solution* $u$ *to the above linear problem belongs to* $W^{2,q}(\Omega)$. *Furthermore, there is a constant* $C_q = C_q(q, \Omega)$, *such that*

$$\| \nabla Du \|_q \leq C_q \| F \|_q . \tag{14}$$

The $W^{2,2}(\Omega)$ regularity of weak solutions claimed in the theorem may be proved, for instance, by following [25] or [7]. The reader may adapt the argument developed in [25, Sect. 4] or, alternatively, in [7, Sects. 3–7]. In this last reference a particularly complete proof of the $W^{2,2}(\Omega)$ regularity is given. Actually, in [7], we consider a much more involved problem. So it is a quite tedious, but mathematically straightforward exercise, to adapt the proof given in [7] to the simpler problem considered here. This is simply done by cutting out many pieces in the proof.

Further, as claimed in [25, Sect. 4], once we have proved $W^{2,2}(\Omega)$ existence and regularity, the deeper $W^{2,q}(\Omega)$ regularity result, for arbitrarily large exponents $q$, follows by appealing to the corresponding a priori estimates. For these estimates see the classical references [2] by Agmon et al. or [24] by Solonnikov. Since all the linear maps at stake are bounded, there is a constant $\tilde{C}_q$ such that

$$\| u \|_{2,q} \leq \tilde{C}_q \| F \|_q . \tag{15}$$

Finally, the point-wise estimate $|\nabla^2 u| \leq 3 |\nabla D u| \leq 6 |\nabla^2 u|$ shows that the norms $\| u \|_{2,q}$ and $\| \nabla Du \|_q$ are equivalent in $W^{2,q}(\Omega) \cap V_q(\Omega)$. So, (14) holds.

## 4 The Strong Solution: Definition

The main lines followed in this section have their starting point in some ideas introduced, in a more complex context, in [8].

Since

$$\nabla (\mu + |D u|^2)^{\frac{p-2}{2}} = \frac{p-2}{2} (\mu + |D u|^2)^{\frac{p-4}{2}} \nabla (|D u|^2),$$

straightforward calculations show that

$$\begin{aligned} &\nabla \cdot \left( (\mu + |D u|^2)^{\frac{p-2}{2}} D u \right) = (\mu + |D u|^2)^{\frac{p-2}{2}} \nabla \cdot (Du) \\ &+ (p-2) (\mu + |D u|^2)^{\frac{p-4}{2}} I(u) \end{aligned} \tag{16}$$

where, by definition,

$$I(u) = \frac{1}{2} \nabla (|D u|^2) \cdot D u = (Du : \nabla Du) \cdot Du .$$

The $j$ component of the vector field $I(u)$ is given by

$$I_j(u) = \sum_k \sum_{l,m} D_{lm}(\partial_k D_{lm}) D_{kj} .$$

By improving an argument already used in [10], we may prove (as in the proof of Lemma 3.4 in [9]) the algebraic relation

$$|I \cdot \xi| \le |D|^2 |\nabla D u| |\xi| ,$$

for each arbitrary vector field in $\xi \in \mathbb{R}^N$, where

$$|\nabla D u|^2 = \sum_{m,l,k} (\partial_k D_{m\,l})^2 .$$

Consequently, the pointwise estimate

$$|I(u)| \le |D|^2 |\nabla D u| \tag{17}$$

holds.

Finally, we introduce the notion of strong solution used in the next section.

**Definition 4.1** Assume that $\mu > 0$, and let $f \in L^q(\Omega)$ be given, $q > 1$. We say that $u \in W^{2,q}(\Omega)$ is a *strong solution* of problem (2), (3) if $u$ satisfies (3) in the trace sense and, moreover, the equation

$$-\nabla \cdot D u = (p-2)\, G(u) + (\mu + |D u|^2)^{\frac{2-p}{2}} f \tag{18}$$

holds almost everywhere in $\Omega$, where

$$G(v) = (\mu + |D v|^2)^{-1} I(v) .$$

Note that, by (17), $G(v) \le |\nabla D v|$ almost everywhere in $\Omega$, independently of $\mu$. So,

$$\| G(v) \|_q \le \|\nabla D v\|_q . \tag{19}$$

## 5 Existence of the Strong Solution for $\mu > 0$

In this and in the next section we prove the existence of a (unique) strong solution $u \in W^{2,q}(\Omega)$ of our problem. In this section the case $\mu > 0$ is considered. In the next section the result is extended to the singular case. In these two sections we follow the proofs given in [10]. For the reader's convenience, we show the main points. For more details, we refer to the original proof, in [10].

Fix $\mu > 0$, and let $f \in L^q(\Omega)$. Since $q > n$, there is a constant $\hat{C}(q,\Omega)$ such that

$$\| D\,v\|_\infty \le \hat{C}\,\|\nabla D\,v\,\|_q\,, \tag{20}$$

for all $v \in W^{2,q}(\Omega) \cap V_q(\Omega)$. Hence,

$$\|\,|Dv|^{2-p} f\|_q \le \|\,Dv\,\|_\infty^{2-p}\,\|f\|_q \le \hat{C}^{2-p}\,\|\nabla D\,v\,\|_q^{2-p}\,\|f\|_q\,. \tag{21}$$

Further, since $(a+b)^\alpha \le a^\alpha + b^\alpha$ for nonnegative $a$ and $b$, and $0 < \alpha < 1$, it follows that

$$(\mu + |D\,v|^2)^{\frac{2-p}{2}} \le \mu^{\frac{2-p}{2}} + |D\,v|^{2-p}\,. \tag{22}$$

From (21) and (22) we show that

$$\|\,(\mu + |\,D\,v\,|^2)^{\frac{2-p}{2}} f\|_q \le \mu^{\frac{2-p}{2}}\,\|f\,\|_q + \hat{C}^{2-p}\,\|\,\nabla\,Dv\,\|_q^{2-p}\,\|f\|_q\,. \tag{23}$$

Next we define the convex closed set

$$\mathbb{K} = \mathbb{K}(R) = \{v \in W^{2,q}(\Omega) \cap V_q(\Omega)\,:\,\|\nabla D\,v\|_q \le R\}\,, \tag{24}$$

and consider, for each $v \in \mathbb{K}$, the solution $u = T(v)$ to the problem

$$-\,\nabla \cdot D\,u = F(v) \equiv (p-2)\,G(v) + \,(\,\mu + |\,D\,v|^2\,)^{\frac{2-p}{2}} f\,, \tag{25}$$

under the boundary conditions (3). We want to prove the existence of a fixed point $T(u) = u \in \mathbb{K}$.

By appealing to Eqs. (14), (19), and (23), we obtain the estimate

$$\|\,\nabla\,Du\,\|_q \le C_q\{\,(2-p)\,\|\,\nabla\,Dv\,\|_q + \mu^{\frac{2-p}{2}}\,\|f\,\|_q + \hat{C}^{2-p}\,\|\,\nabla\,Dv\,\|_q^{2-p}\,\|f\|_q\,\}\,. \tag{26}$$

Next we show that if $\|\,\nabla\,Dv\|_q \le R$ then the corresponding solution $u = T(v)$ satisfies the same estimate, namely $\|\,\nabla\,Du\|_q \le R$. This shows that $T(\mathbb{K}) \subset \mathbb{K}$.

Since $v \in \mathbb{K}$, it follows that

$$\|\,\nabla\,Du\,\|_q \le \mu^{\frac{2-p}{2}}\,C_q\,\|f\|_q + (2-p)\,C_q R + C_q\,\hat{C}^{2-p}\,\|f\|_q\,R^{2-p}\,. \tag{27}$$

By assuming (4), we show that $u \in \mathbb{K}(R)$ if

$$[\,1 - (2-p)\,C_q\,]\,R \ge \mu^{\frac{2-p}{2}}\,C_q\,\|f\|_q + C_q\,\hat{C}^{2-p}\,\|f\|_q\,R^{2-p}\,.$$

This inequality is satisfied if, for instance, its left hand side is larger or equal to two times each of the two terms on the right hand side. This holds for

$$R = \frac{2}{\alpha}\,\mu^{\frac{2-p}{2}}\,C_q\,\|f\|_q + (\frac{2\,C_q\,\hat{C}^{2-p}}{\alpha})^{\frac{1}{p-1}}\,\|f\|_q^{\frac{1}{p-1}}\,, \tag{28}$$

where $\alpha = 1 - C_2(q)\,(2-p)$. Hence $\|\,\nabla\, Du\,\|_q \leq R$. The inclusion $T(\mathbb{K}) \subset \mathbb{K}$ follows. This is the main ingredient to prove the existence of a fixed point in $\mathbb{K}$, by appealing to the Theorem 3.2 in [10]. For more details we refer to the argument developed in this last reference.

The expression of $R$ shows that the uniform estimate (5) holds. Summarizing, we have shown that, for each positive $\mu$, the estimate

$$\|u^\mu\|_{2,q} \leq C\left(\|f\|_q + \|f\|_q^{\frac{1}{p-1}}\right) \tag{29}$$

holds, where $u^\mu$ denotes the strong solution related to the particular positive value $\mu$.

## 6 Existence of the Strong Solution for $\mu = 0$

Following [10], we appeal here to the uniformity with respect to $\mu > 0$ of (29), and to a compactness argument, to pass to the limit, as $\mu$ tends to zero, in the weak formulation (10). In this way we show that the weak solutions $u$ to the singular problem belong to $W^{2,q}(\Omega)$, and satisfy (5).

We start by recalling the definition of weak solution $u^\mu$ of problem (2), for $\mu \geq 0$, namely,

$$\int_\Omega \left(\mu + |\,D\,u^\mu|^2\right)^{\frac{p-2}{2}}\,D\,u^\mu\cdot D\,v\,dx = \int_\Omega f\cdot v\,dx\,, \tag{30}$$

for all $v \in V_p(\Omega)$. This condition is satisfied by the strong solutions $u^\mu$, for $\mu > 0$, constructed in the previous section. Since these solutions are uniformly bounded in $W^{2,q}(\Omega)$, suitable sub-sequences, which we continue to denote by $u^\mu$, weakly converge in $W^{2,q}(\Omega)$ to some $u$. Let us show, by passing to the limit in (30), that

$$\int_\Omega |\,D\,u|^{\frac{p-2}{2}}\,D\,u\cdot D\,v\,dx = \int_\Omega f\cdot v\,dx. \tag{31}$$

for all $v \in V_p(\Omega)$. Consequently, $u \in W^{2,q}(\Omega)$ is the solution (know to be unique) for $\mu = 0$. To prove this claim, we have to show that the left hand side of Eq. (30) converges to the left hand side of (31), for each fixed $v \in V_p(\Omega)$. Essentially, the proof followed in [10, Sect. 4] applies here. For the reader's convenience, we repeat the main argument.

Since $u^\mu \rightharpoonup u$ weakly in $W^{2,q}(\Omega)$, and $q > n$, strong convergence (of suitable subsequences) in $W^{1,s}(\Omega)$, for any $s$, follows. So, strong convergence in $W^{1,p}(\Omega)$ holds.

Write the integral on the left-hand side of (30) as

$$\int_\Omega \Big[ \big(\mu + |Du^\mu|^2\big)^{\frac{p-2}{2}} Du^\mu - \big(\mu + |Du|^2\big)^{\frac{p-2}{2}} Du \Big] \cdot Dv\, dx \tag{32}$$

$$+ \int_\Omega \big(\mu + |Du|^2\big)^{\frac{p-2}{2}} Du \cdot Dv\, dx,$$

and show that the first integral tends to zero, and the second integral tends to the left hand side of (31). The inequality

$$|(\mu + |A|)^{p-2} A - (\mu + |B|)^{p-2} B| \leq C\,|A - B|\,(\mu + |A| + |B|)^{p-2}, \tag{33}$$

where $C$ is independent of $\mu$ (see [14, Eq. (6.8)]), shows that the absolute value of the first integral in Eq. (32) is bounded by

$$C \int_\Omega (\mu + |Du| + |Du^\mu|)^{p-2}\, |Du - Du^\mu|\, |Dv|\, dx.$$

Since

$$(\mu + |Du| + |Du^\mu|)^{p-2}\, |Du - Du^\mu| \leq |Du - Du^\mu|^{p-1},$$

the absolute value of the first integral in Eq. (32) is bounded by

$$C\, \| Du^\mu - Du \|_p^{p-1}\, \| Dv \|_p,$$

which tends to zero with $\mu$.

Finally, by Lebesgue's dominated convergence theorem,

$$\lim_{\mu \to 0+} \int_\Omega \big(\mu + |Du|^2\big)^{\frac{p-2}{2}} Du \cdot Dv\, dx = \int_\Omega |Du|^{p-2} Du \cdot Dv\, dx.$$

## References

1. E. Acerbi, N. Fusco, Regularity for minimizers of nonquadratic functionals: the case $1 < p < 2$. J. Math. Anal. Appl. **140**, 115–135 (1989)
2. S. Agmon, A. Douglis, L. Nirenberg, Estimates near the boundary for solutions of elliptic partial differential equations satisfying general boundary conditions II. Commun. Pure Appl. Math. **17**, 35–92 (1964)
3. S. Antontsev, S. Shmarev, Elliptic equations and systems with nonstandard growth conditions: existence, uniqueness and localization properties for solutions. Nonlinear Anal. **65**, 728–761 (2006)

4. S. Antontsev, S. Shmarev, Anisotropic parabolic equations with variable nonlinearity. Publ. Mat. **53**, 355–399 (2009)
5. C. Bardos, Existence et unicité de la solution de l'équation de Euler en dimension deux. J. Math. Anal. Appl. **40**, 769–790 (1972)
6. C. Bardos, F. Golse, L. Paillard, The incompressible Euler limit of the Boltzmann equation with accommodation boundary condition. Commun. Math. Sci. **10**, 159–190 (2012)
7. H. Beirão da Veiga, Regularity for Stokes and generalized Stokes systems under non homogeneous slip type boundary conditions. Adv. Differ. Equ. **9**(9–10), 1079–1114 (2004)
8. H. Beirão da Veiga, On the regularity of flows with Ladyzhenskaya shear dependent viscosity and slip or non-slip boundary conditions. Commun. Pure Appl. Math. **58**, 552–577 (2005). (online: 15th June 2004, doi:10.1002/cpa.20036)
9. H. Beirão da Veiga, On the global regularity for singular $p$-systems under non-homogeneous Dirichlet boundary conditions. J. Math. Anal. Appl. **398**, 527–533 (2013). doi:10.1016/j.jmaa.2012.08.058
10. H. Beirão da Veiga, F. Crispo, On the global $W^{2,q}$ regularity for nonlinear $N$-systems of the p-Laplacian type in $n$ space variables. Nonlinear Anal. Theory Methods Appl. **75**, 4346–4354 (2012). doi:10.1016/j.na.2012.03.021
11. H. Beirão da Veiga, F. Crispo, On the global regularity for nonlinear systems of the p-Laplacian type. Discrete Continuous Dyn. Syst. Ser. S **6**, 1173–1191 (2013). arXiv:1008.3262v1 [math.AP]
12. E. DiBenedetto, $C^{1+\alpha}$ local regularity of weak solutions of degenerate elliptic equations. Nonlinear Anal. **7**, 827–850 (1983)
13. E. DiBenedetto, J. Manfredi, On the higher integrability of the gradient of weak solutions of certain degenerate elliptic systems. Am. J. Math. **115**, 1107–1134 (1993)
14. L. Diening, C. Ebmeyer, M. Růžička, Optimal convergence for the implicit space-time discretization of parabolic systems with $p$-structure. SIAM J. Numer. Anal. **45**, 457–472 (2007)
15. M. Fuchs, G. Seregin, *Variational Methods for Problems from Plasticity Theory and for Generalized Newtonian Fluids*. Lecture Notes in Mathematics, vol. 1749 (Springer, Berlin, 2000)
16. C. Hamburger, Regularity of differential forms minimizing degenerate elliptic functionals. J. Reine Angew. Math. **431**, 7–64 (1992)
17. P. Kaplický, J. Tichý, Boundary regularity of flows under perfect slip boundary conditions. Cent. Eur. J. Math. **11**, 1243–1263 (2013)
18. G.M. Lieberman, Boundary regularity for solutions of degenerate elliptic equations. Nonlinear Anal. **12**, 1203–1219 (1988)
19. J.-L. Lions, *Quelques Méthodes de Résolution des Problèmes aux Limites Non Linéaires* (Dunod, Paris, 1969)
20. W.B. Liu, J.W. Barrett, A remark on the regularity of the solutions of the $p$-Laplacian and its application to their finite element approximation. J. Math. Anal. Appl. **178**, 470–487 (1993)
21. P. Marcellini, G. Papi, Nonlinear elliptic systems with general growth. J. Differ. Equ. **221**, 412–443 (2006)
22. C.L.M. Navier, Mémoire sur les lois du mouvement des fluides. Mémoires de l'Académie Royale des Sciences de l'Institut de France **6**, 389–440 (1927)
23. J. Serrin, Mathematical principles of classical fluid mechanics, in *Handbuch der Physik*, Bd VIII/1, (Springer, Berlin 1959), pp. 125–263
24. V.A. Solonnikov, General boundary value problems for Douglis-Nirenberg elliptic systems. II, in *Proceedings of the Steklov Institute of Mathematics*, vol. 92 (1966)
25. V.A. Solonnikov, V.E. Ščadilov, On a boundary value problem for a stationary system of Navier-Stokes equations. Proc. Steklov Inst. Math. **125**, 186–199 (1973)
26. G. Stokes, Trans. Camb. Philos. Soc. **8**, 287, 75–129 (1845)
27. P. Tolksdorf, Regularity for a more general class of quasilinear elliptic equations. J. Differ. Equ. **51**, 126–150 (1984)
28. N.N. Ural'tseva, Degenerate quasilinear elliptic systems. Zap. NauK. Sem. LOMI **7**, 184–222 (1968) (in Russian)

# Thermodynamically Consistent Modeling for Dissolution/Growth of Bubbles in an Incompressible Solvent

**Dieter Bothe and Kohei Soga**

*Dedicated to Yoshihiro Shibata on the Occasion of his 60th Anniversary*

**Abstract** We derive mathematical models of the elementary process of dissolution/growth of bubbles in a liquid under pressure control. The modeling starts with a fully compressible version, both for the liquid and the gas phase so that the entropy principle can be easily evaluated. This yields a full PDE system for a compressible two-phase fluid with mass transfer of the gaseous species. Then the passage to an incompressible solvent in the liquid phase is discussed, where a carefully chosen equation of state for the liquid mixture pressure allows for a limit in which the solvent density is constant. We finally provide a simplification of the PDE system in case of a dilute solution.

**Keywords** Entropy principle • Incompressible limit • Mass transfer • Two-phase fluid system

## 1 Introduction

The process of dissolution or growth of gas bubbles in an ambient liquid phase is very common in many situations. In everyday life, we often see bubbles in carbonated mineral water, beer, champagne etc. In particular the dissolution of gases is of huge technological and industrial importance in the context of gas scrubbing. This is, for instance, relevant for $CO_2$ disposal, where gas from a combustion process is injected into a reactive liquid medium. Such processes are usually run under pressure control instead of volume control. Note that the latter is much more

D. Bothe (✉)
Technische Universität Darmstadt, Center of Smart Interfaces, 64287 Darmstadt, Germany
e-mail: bothe@csi.tu-darmstadt.de

K. Soga
Department of Mathematics, Faculty of Science and Technology, Keio University, Hiyoshi 3-14-1, Kohoku-ku, Yokohama 223-8522, Japan
e-mail: soga@math.keio.ac.jp



H. Amann et al. (eds.), *Recent Developments of Mathematical Fluid Mechanics*, Advances in Mathematical Fluid Mechanics, DOI 10.1007/978-3-0348-0939-9_7

common in the mathematical analysis of such mass transfer problems, since it allows for a fixed domain in which the mathematical model—usually in the form of a system of partial differential equations—holds. The massive impact of the external pressure is known from the above mentioned everyday life examples, but also can be seen in the medical context. This is the case with decompression sickness or caisson disease, where severe symptoms can be caused by bubble generation in the blood after a fast change of the ambient pressure. There is a large literature on experiments and numerical computation of dissolution/growth of bubbles in a liquid, e.g. Liger-Belair et al. [10], Sauzade and Cubaud [13], Takemura and Yabe [16]. A rigorous mathematical model is necessary for possible theoretical investigations and mathematical analysis on this topic.

Based on Continuum Physics, we derive a mathematical model of a two-phase fluid system of type liquid/gas, where both gas and liquid phases are composed of molecularly miscible constituents and the pressure is controlled via a free (upper) surface $\Gamma(t)$. The system consists of chemical components $A_1, \ldots, A_N$. The gas phase is denoted by $\Omega^+(t)$, the liquid phase by $\Omega^-(t)$ and the movable free interface by $\Sigma(t)$. See Fig. 1 below.

In common mathematical models for mass transfer from or to gas bubbles in a liquid phase, the transferred gas is treated as a dilute component in both phases. This allows to use a two-phase Navier-Stokes system together with advection-diffusion equations for passive scalars. If the bubble is composed of a pure gas, this is no longer possible since the dissolution then significantly changes the bubble volume. In this case a much more elaborate modeling is required for both of the bulk phases and the transmission condition at the interface. In particular, the two one-sided limits of the bulk velocities at the interface and the interface's own velocity need to be distinguished. Since such a more rigorous model accounts for the mass and

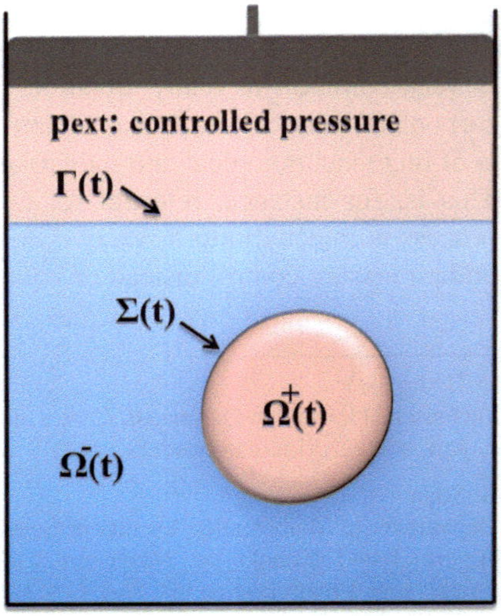


**Fig. 1** The two phase system under pressure control

volume of individual constituents, an incompressible model for the liquid phase will still lead to non-zero divergence of the barycentric velocity field. Moreover, a thermodynamically rigorous model needs to be developed for compressible bulk phases in the first place. Only then, an incompressible version may be derived as a limit, where the latter depends on the notion of incompressibility which is neither a priori clear nor unique in the mixture context.

The novel aspect in the present paper is the idea of an incompressible solvent (associated to $A_N$) carrying dissolved gas components which add their partial pressure to the total one like being ideal gases. The underlying mixture is supposed to be described by an equation of state according to

$$p = p_N^R + K(\frac{\rho_N}{\rho_N^R} - 1) + \sum_{k=1}^{N-1} \frac{\rho_k}{M_k} RT,$$

where $p_N^R$ is a reference pressure and $\rho_N^R$ a reference density for the solvent, while $K$ is the solvent bulk modulus. The incompressible limit will be attained (formally) by letting $K$ tend to infinity. This leads to the constraint

$$\rho_N \equiv \rho_N^R,$$

i.e. to a constant solvent density. Since the continuity equation for the solvent then reduces to

$$\nabla \cdot v_N = 0,$$

it makes sense to employ the solvent momentum balance instead of the one for the mixture. This is attractive, because it leads to a standard incompressible Navier-Stokes equation for the bulk liquid. Only the diffusive fluxes, which rely on the relative velocity to the barycentric one, become slightly more intricate, but only involving a simple linear relation.

The obtained PDE systems still comprise of a compressible gas phase model. Low Mach number approximation seems possible and will be given in a forthcoming paper. Note that the gas phase density in the incompressible limit will still be a function of time, determined by the dynamical mass transfer process.

## 2 Balance Equations

### (a) Mass balance

For simplicity, we assume that there are no chemical reactions (which could be easily added) and that there is no absorbed mass at the interface, i.e. $\rho_i^\Sigma \equiv 0$ for all $i = 1, \dots, N$. The partial mass balance in its integral form for a fixed control volume $V$ with the outer normal $n$ reads as

$$\frac{d}{dt}\int_V \rho_i \, dx = -\int_{\partial V} \rho_i v_i \cdot n \, do.$$

Using the two-phase transport and divergence theorems (see the Appendix), this implies

$$\int_{V\setminus\Sigma} \partial_t \rho_i \, dx - \int_{\Sigma_V} [\![\rho_i]\!] v^\Sigma \cdot n_\Sigma \, do = -\int_{V\setminus\Sigma} \nabla\cdot(\rho_i v_i)\, dx - \int_{\Sigma_V} [\![\rho_i v_i \cdot n_\Sigma]\!] \, do$$

with $\Sigma_V := \Sigma(t) \cap V$, the surface velocity $v^\Sigma$ and the surface unit normal $n_\Sigma$ pointing toward $\Omega^-$. Comparison of bulk and interface terms yields the local form

$$\begin{cases} \partial_t \rho_i + \nabla\cdot(\rho_i v_i) = 0 \ \text{ in } \Omega^+(t) \cup \Omega^-(t), \\ [\![\rho_i (v_i - v^\Sigma)\cdot n_\Sigma]\!] = 0 \ \text{ on } \Sigma(t). \end{cases} \tag{1}$$

Above, the bracket $[\![\cdot]\!]$ denotes the jump of a quantity across the interface (crossing $\Sigma$ in the direction opposite to $n_\Sigma$). The mixture is described by the total density $\rho$ and the barycentric velocity $v$, given by

$$\rho := \sum_{i=1}^N \rho_i, \quad \rho v := \sum_{i=1}^N \rho_i v_i.$$

As a consequence of (1), the mixture obeys the continuity equation

$$\begin{cases} \partial_t \rho + \nabla\cdot(\rho v) = 0 \ \text{ in } \Omega^+(t) \cup \Omega^-(t), \\ [\![\rho (v - v^\Sigma)\cdot n_\Sigma]\!] = 0 \ \text{ on } \Sigma(t). \end{cases} \tag{2}$$

Let $\dot m^\pm$ denote the one-sided limits $\rho^\pm(v^\pm - v^\Sigma)\cdot n_\Sigma$ on $\Sigma(t)$. Then the second equation in (2) becomes $\dot m^- = \dot m^+$, and hence $\dot m := \dot m^- = \dot m^+$ is well-defined. Similarly, we introduce $\dot m_i := \dot m_i^\pm = \rho_i^\pm(v_i^\pm - v^\Sigma)\cdot n_\Sigma$. We define diffusion velocities $u_i := v_i - v$, mass fractions $y_i := \rho_i/\rho$ and diffusion mass fluxes $j_i := \rho_i u_i = \rho_i(v_i - v)$. Then we have the following equivalent form of Eqs. (1):

$$\begin{cases} \partial_t \rho_i + \nabla\cdot(\rho_i v + j_i) = 0 \ \text{ in } \Omega^+(t) \cup \Omega^-(t), \\ [\![j_i \cdot n_\Sigma]\!] + [\![\rho_i (v - v^\Sigma)\cdot n_\Sigma]\!] = 0 \ \text{ at } \Sigma(t), \end{cases} \tag{3}$$

or

$$\begin{cases} \rho(\partial_t y_i + v\cdot\nabla y_i) + \nabla\cdot j_i = 0 \ \text{ in } \Omega^+(t) \cup \Omega^-(t), \\ [\![j_i \cdot n_\Sigma]\!] + \dot m [\![y_i]\!] = 0 \ \text{ at } \Sigma(t). \end{cases} \tag{4}$$

In the common models for mass transfer, the jump condition in (3) or (4) is simplified to read $[\![j_i \cdot n_\Sigma]\!] = 0$, assuming $\dot{m} = 0$ which means that the total phase change effect of the mass transfer is neglected; cf. Bothe and Fleckenstein [4] for an assessment of this approximation.

**(b) Momentum balance**

The mixture is to be described by a so-called class-I model, where we consider only a single (common) momentum balance. The integral form is

$$\frac{d}{dt}\int_V \rho v\, dx = -\int_{\partial V} \rho v(v\cdot n)\, do + \int_{\partial V} Sn\, do + \int_V \rho b\, dx + \int_{\partial \Sigma_V} S^\Sigma \nu\, ds$$

with the bulk stress tensor $S$, the surface stress tensor $S^\Sigma$ and the body force $\rho b$. Note that $\rho b = \sum_{k=1}^N \rho_k b_k$ with (possibly) individual body forces $b_k$, for instance due to forces in an electrical field. Here $\nu$ is the outer unit normal of the bounding curve $\partial\Sigma_V$ of $\Sigma_V$, being tangential to $\Sigma$. The transport and (surface) divergence theorems yield the local form

$$\begin{cases} \partial_t(\rho v) + \nabla\cdot(\rho v\otimes v - S) = \rho b & \text{in } \Omega^+(t)\cup\Omega^-(t), \\ \dot{m}[\![v]\!] - [\![Sn_\Sigma]\!] = \nabla_\Sigma\cdot S^\Sigma & \text{at } \Sigma(t). \end{cases} \tag{5}$$

We assume non-polar fluids, for which the balance of angular momentum has a simple form without body couples or surface couples. This is equivalent to the assumptions

$$S = S^\mathsf{T}, \quad S^\Sigma = (S^\Sigma)^\mathsf{T}.$$

This is a constitutive assumption which is made right away.

**(c) Energy balance**

The integral form of the total energy balance is

$$\begin{aligned} \frac{d}{dt}\Big[\int_V \rho(e+\frac{v^2}{2})\, dx + \int_{\Sigma_V} u^\Sigma\, do\Big] &= -\int_{\partial V}\rho(e+\frac{v^2}{2})v\cdot n\, do \\ &- \int_{\partial\Sigma_V} u^\Sigma v^\Sigma\cdot\nu\, ds - \int_{\partial V} q\cdot n\, do - \int_{\partial\Sigma_V} q^\Sigma\cdot\nu\, ds + \int_{\partial V} v\cdot Sn\, do \\ &+ \int_{\partial\Sigma_V} v^\Sigma\cdot S^\Sigma\nu\, ds + \int_V v\cdot\rho b\, dx + \int_V \sum_{k=1}^N j_k\cdot b_k\, dx \end{aligned}$$

with the specific internal energy of the bulk $e$ and the internal energy density of the surface $u^\Sigma$. After straightforward computations, the local form turns out as

$$\begin{cases} \partial_t \Big(\rho(e+\frac{v^2}{2})\Big) + \nabla\cdot\Big(\rho(e+\frac{v^2}{2})v + q\Big) \\ \qquad = \nabla\cdot(Sv) + \rho v\cdot b + \sum_{k=1}^N j_k\cdot b_k \quad \text{in } \Omega^+(t)\cup\Omega^-(t), \\ \frac{D^\Sigma u^\Sigma}{Dt} + u^\Sigma\nabla_\Sigma\cdot v^\Sigma + [\![\rho(e+\frac{v^2}{2})(v-v^\Sigma)\cdot n_\Sigma]\!] + [\![q\cdot n_\Sigma]\!] + \nabla_\Sigma\cdot q^\Sigma \\ \qquad = [\![Sv\cdot n_\Sigma]\!] + \nabla_\Sigma\cdot(S^\Sigma v^\Sigma) \quad \text{at } \Sigma(t). \end{cases}$$

Subtracting the balance of kinetic energy derived from (5), one obtains the balance of internal energy as

$$\begin{cases} \partial_t(\rho e) + \nabla\cdot(\rho e v + q) = \nabla v : S + \sum_{k=1}^N j_k\cdot b_k \\ \qquad\qquad \text{in } \Omega^+(t)\cup\Omega^-(t), \\ \frac{D^\Sigma u^\Sigma}{Dt} + u^\Sigma\nabla_\Sigma\cdot v^\Sigma + \dot m[\![e + \frac{(v-v^\Sigma)^2}{2} - \frac{1}{\rho}n_\Sigma\cdot S n_\Sigma]\!] \\ \quad + [\![q\cdot n_\Sigma]\!] + \nabla_\Sigma\cdot q^\Sigma - [\![(v-v^\Sigma)_\| \cdot S n_\Sigma]\!] = \nabla_\Sigma v^\Sigma : S^\Sigma \\ \qquad\qquad \text{at } \Sigma(t), \end{cases} \tag{6}$$

where $(v-v^\Sigma)_\|$ stands for the tangential projection of $(v-v^\Sigma)$ onto the local tangent plane to $\Sigma$, i.e. $(v-v^\Sigma)_\| = P_\Sigma(v-v^\Sigma)$ with the projection tensor $P_\Sigma = I - n_\Sigma\otimes n_\Sigma$. Later, we will use the constitutive relation $S^\Sigma = \gamma^\Sigma P_\Sigma$ with a scalar $\gamma^\Sigma$. Then we have $\nabla_\Sigma v^\Sigma : S^\Sigma = \gamma^\Sigma \nabla_\Sigma\cdot v^\Sigma$.

**(d) Entropy balance**

Let $\rho s$ denote the density of entropy in the bulk (i.e. $s$ is the specific entropy) and $\eta^\Sigma$ the area-density of interfacial entropy. The integral form of the entropy balance is

$$\frac{d}{dt}\Big[\int_V \rho s\,dx + \int_{\Sigma_V} \eta^\Sigma\,do\Big] = -\int_{\partial V} (\rho s v + \Phi)\cdot n\,do$$
$$- \int_{\partial\Sigma_V} (\eta^\Sigma v^\Sigma + \Phi^\Sigma)\cdot \nu\,ds + \int_V \xi\,dx + \int_{\Sigma_V} \xi^\Sigma\,do$$

with the bulk entropy flux $\Phi$ and the interfacial entropy flux $\Phi^\Sigma$. Hence we obtain the local form

$$\begin{cases} \partial_t(\rho s) + \nabla\cdot(\rho s v + \Phi) = \xi \quad \text{in } \Omega^+(t)\cup\Omega^-(t), \\ \frac{D^\Sigma \eta^\Sigma}{Dt} + \eta^\Sigma\nabla_\Sigma\cdot v^\Sigma + \nabla_\Sigma\cdot\Phi^\Sigma + \dot m[\![s]\!] + [\![\Phi\cdot n_\Sigma]\!] = \xi^\Sigma \quad \text{at } \Sigma(t). \end{cases} \tag{7}$$

# 3 Entropy Principle

If the entropy fluxes $\Phi$ and $\Phi^\Sigma$ in (7) are related to the primitive variables via constitutive relations in such a way that the following entropy principle holds, we speak of a thermodynamically consistent model.

**Entropy principle.** *The entropy flux* $(\Phi, \Phi^\Sigma)$ *is such that*

- *The entropy production is a sum of binary products of "fluxes" times "driving force", i.e.* $\xi = \sum_m F_m D_m$ *and* $\xi^\Sigma = \sum_{m'} F^\Sigma_{m'} D^\Sigma_{m'}$.
- $\xi \geq 0$, $\xi^\Sigma \geq 0$ *for any thermodynamical process.*
- $\xi \equiv 0$ *and* $\xi^\Sigma \equiv 0$ *characterizes equilibria of the system.*

This is a condensed form of the full entropy principle. For more details see Bothe and Dreyer [3], as well as Dreyer [8]. We consider the simplest class of isotropic fluids without mesoscopic forces. This corresponds to the choice of certain primitive variables in modeling the entropy of the material. We assume

$$\rho s = h(\rho e, \rho_1, \ldots, \rho_N), \quad \eta^\Sigma = h^\Sigma(u^\Sigma), \tag{8}$$

where $h$ and $h^\Sigma$ are concave functions. The concavity is required for thermodynamic stability properties of the mixture. Then we define the (absolute) temperature $T$, respectively $T^\Sigma$ of bulk and interface, as well as the bulk chemical potentials $\mu_i$ via

$$\frac{1}{T} := \frac{\partial h}{\partial(\rho e)}, \quad -\frac{\mu_i}{T} := \frac{\partial h}{\partial \rho_i}, \quad \frac{1}{T^\Sigma} := \frac{\partial h^\Sigma}{\partial u^\Sigma}. \tag{9}$$

Next, we compute $\xi$ and $\xi^\Sigma$ from (7)–(9), where we eliminate the time derivatives of $\rho_i$, $\rho e$, $u^\Sigma$ by means of the balance equations in (1), (6). This yields the following results.

**(a) Bulk entropy production**

$$\xi = \nabla \cdot (\Phi - \frac{q}{T} + \sum_{k=1}^N \frac{\mu_k j_k}{T}) - \frac{1}{T}(\rho e - \rho s T - \sum_{k=1}^N \rho_k \mu_k)\nabla \cdot v$$
$$+\frac{1}{T}\nabla v : S + q \cdot \nabla \frac{1}{T} - \sum_{k=1}^N j_k \cdot \left(\nabla \frac{\mu_k}{T} - \frac{b_k}{T}\right).$$

We choose the entropy flux as

$$\Phi = \frac{q}{T} - \sum_{k=1}^N \frac{\mu_k j_k}{T}$$

and determine further constitutive relations so that the entropy principle holds. We decompose the stress tensor $S$ as $S = -PI + S^\circ$ with the traceless part $S^\circ$ of $S$ and $P = -\frac{1}{3}\text{tr}(S)$. We decompose the pressure $P$ as $P = p + \Pi$, where $\Pi$ vanishes in equilibrium. This is important, since $\Pi$ can depend on $\nabla \cdot v$, while $p$ cannot. Hence $S$ is rewritten as

$$S = -(p + \Pi)I + S^\circ.$$

Introducing the Helmholtz free energy

$$\rho\psi = \rho\psi(T, \rho_1, \ldots, \rho_N) = \rho e - \rho s T,$$

we change from $\rho e$ as a primitive variable to $T$ (via Legendre transform with $\frac{\partial(\rho e)}{\partial(\rho s)} = T$). Then $\xi$ becomes

$$\xi = -\frac{1}{T}\left(\rho\psi + p - \sum_{k=1}^{N} \rho_k \mu_k\right) \nabla \cdot v - \frac{\Pi}{T} \nabla \cdot v + \frac{1}{T} \nabla v : S^\circ + q \cdot \nabla \frac{1}{T}$$
$$- \sum_{k=1}^{N} j_k \cdot \left(\nabla \frac{\mu_k}{T} - \frac{b_k}{T}\right).$$

Now, $\xi \geq 0$ for any thermodynamical process implies the Gibbs-Duhem relation

$$p = -\rho\psi + \sum_{i=k}^{N} \rho_k \mu_k.$$

Thus the entropy production in the bulk reduces to read

$$\xi = -\frac{\Pi}{T} \nabla \cdot v + \frac{1}{T} \nabla v : S^\circ + q \cdot \nabla \frac{1}{T} - \sum_{k=1}^{N} j_k \cdot \left(\nabla \frac{\mu_k}{T} - \frac{b_k}{T}\right).$$

Since $\xi \geq 0$ is required, the simplest closure is linear in the driving forces and such that a quadratic form is obtained. Note that the constraint $\sum_{k=1}^{N} j_k = 0$ has to be accounted for. Hence we eliminate $j_N$, which is chosen as $-\sum_{k=1}^{N-1} j_k$. For $D := \frac{1}{2}(\nabla v + (\nabla v)^{\mathsf{T}})$ and its traceless part $D^\circ$ we have $\nabla v : S^\circ = D^\circ : S^\circ$ and $\text{tr} D = \nabla \cdot v$. Then $\xi$ becomes

$$\xi = -\frac{\Pi}{T} \nabla \cdot v + \frac{1}{T} D^\circ : S^\circ + q \cdot \nabla \frac{1}{T} - \sum_{k=1}^{N-1} j_k \cdot \left(\nabla \frac{\mu_k - \mu_N}{T} - \frac{b_k - b_N}{T}\right). \tag{10}$$

Note that the viscous entropy production can be written as $\frac{1}{T}D : S^{irr}$, if we let $S^{irr} := -\Pi I + S^{\circ}$, i.e. $S^{irr}$ is the irreversible part of $S$ which produces entropy.

**(b) Interfacial entropy production**

We do not consider viscous surface dissipation, hence $S^{\Sigma} = \gamma^{\Sigma} P_{\Sigma}$. Then it follows from the second equation in (7) and the other balance equations that

$$\xi^{\Sigma} = \frac{1}{T^{\Sigma}}(\gamma^{\Sigma} - u^{\Sigma} + T^{\Sigma}\eta^{\Sigma})\nabla_{\Sigma} \cdot v^{\Sigma} + \nabla_{\Sigma} \cdot (\Phi^{\Sigma} - \frac{q^{\Sigma}}{T^{\Sigma}}) + q^{\Sigma} \cdot \nabla_{\Sigma}\frac{1}{T^{\Sigma}}$$
$$+[\![(\frac{1}{T} - \frac{1}{T^{\Sigma}})(\dot{m}sT + q \cdot n_{\Sigma})]\!] + \frac{1}{T^{\Sigma}}[\![(v - v^{\Sigma})_{\|} \cdot (S^{irr} n_{\Sigma})]\!]$$
$$-\sum_{k=1}^{N}[\![\frac{\mu_k j_k \cdot n_{\Sigma}}{T}]\!] - \frac{\dot{m}}{T^{\Sigma}}[\![\sum_{k=1}^{N} y_k \mu_k + \frac{(v - v^{\Sigma})^2}{2} - \frac{1}{\rho} n_{\Sigma} \cdot S^{irr} n_{\Sigma}]\!].$$

We choose the entropy flux as

$$\Phi^{\Sigma} = q^{\Sigma}/T^{\Sigma}$$

and obtain the surface Gibbs-Duhem equation

$$\gamma^{\Sigma} = u^{\Sigma} - T^{\Sigma}\eta^{\Sigma},$$

which shows that $\gamma^{\Sigma}$ is the interfacial free energy. For simplification, we assume from here on that there is no temperature jump at $\Sigma(t)$, i e.

$$[\![T]\!] = 0, \quad T|_{\Sigma} = T^{\Sigma}.$$

Then, with (3), we see that $\xi^{\Sigma}$ becomes

$$\xi^{\Sigma} = q^{\Sigma} \cdot \nabla_{\Sigma}\frac{1}{T} + \frac{1}{T}[\![(v - v^{\Sigma})_{\|} \cdot (S^{irr} n_{\Sigma})]\!] \tag{11}$$
$$-\frac{1}{T}\sum_{k=1}^{N} \dot{m}_k [\![\mu_k + \frac{(v - v^{\Sigma})^2}{2} - \frac{1}{\rho} n_{\Sigma} \cdot S^{irr} n_{\Sigma}]\!],$$

where $\dot{m}_i$ satisfies $[\![\dot{m}_i]\!] = 0$ for all $i = 1, \ldots, N$.

In the next section, we further determine appropriate constitutive relations such that the entropy principle holds. In addition, one needs a constitutive modeling for the Helmholtz free energy $\rho\psi$. This will be constructed from an equation of state for the pressure $p$ and from the chemical potentials $\mu_i$.

# 4 Constitutive Modeling

Constitutive relations can be derived from the entropy principle in (10) and (11). The standard closure is as follows (cf. de Groot and Mazur [7]; Slattery [15]; Hutter and Jöhnk [9]).

**(a) Bulk**

(B1) $\Pi = -\lambda \nabla \cdot v,\ \ \lambda = \lambda(T, \rho_i) \geq 0$ the bulk viscosity,
(B2) $S^\circ = 2\eta D^\circ,\ \ \eta = \eta(T, \rho_i) \geq 0$ the dynamic viscosity (Newton's law),
(B3) $q = \alpha \nabla \frac{1}{T},\ \ \alpha = \alpha(T, \rho_i) \geq 0$ the heat conductivity (Fourier's law),
(B4) $j_i = -\sum_{k=1}^{N-1} L_{ik} \left(\nabla \frac{\mu_k - \mu_N}{T} - \frac{b_k - b_N}{T}\right)$ with a positive (semi-)definite matrix $[L_{ik}] = [L_{ik}(T, \rho_1, \ldots, \rho_N)]$ of mobilities (Fick's law for multi-component mixture).

**(b) Interface**

(B5) $q^\Sigma = \alpha^\Sigma \nabla_\Sigma \frac{1}{T}, \alpha^\Sigma = \alpha^\Sigma(T) \geq 0$ the interfacial heat conductivity,
(B6) $[\![v_\parallel]\!] = 0,\ v_\parallel^\pm = v_\parallel^\Sigma$, i.e. continuous tangential velocities,
(B7) If $i \in I^\pm := \{i \,|\, A_i \text{ is only in } \Omega^\pm\}$, then $\dot{m}_i = 0$ (no transfer) and otherwise

$$[\![\mu_i]\!] = [\![\frac{1}{\rho} n_\Sigma \cdot S^{irr} n_\Sigma - \frac{(v - v^\Sigma)^2}{2}]\!],$$

or, more general but still neglecting mass transfer cross-effects,

(B7') $\dot{m}_i = -\beta_i [\![\mu_i + \dfrac{(v - v^\Sigma)^2}{2} - \dfrac{1}{\rho} n_\Sigma \cdot S^{irr} n_\Sigma]\!],\ \beta_i = \beta_i(T) \geq 0.$

Now we model the Helmholtz free energy $\rho\psi$, where we follow Example 2 in Bothe and Dreyer [3]. The free energy can be constructed from an equation of state for the pressure $p$ and from relations for the "chemical part" of the chemical potential $\mu_i$. We consider the gas phase as an ideal mixture of ideal gases and the liquid phase as a solution with $A_N$ as the solvent and $A_1, \ldots, A_{N-1}$ the solutes (i.e. dissolved components). We introduce the following notation:

$$c_i := \frac{\rho_i}{M_i} \text{ (molar density)}, \quad c := \sum_{i=1}^N c_i, \quad x_i := \frac{c_i}{c} \text{ (molar fraction)},$$

$$x' := (x_1, \ldots, x_{N-1}),$$

where $\sum_{k=1}^N x_k = 1$. We use $(\rho, x')$ as a set of primitive variables as well as $(\rho_1, \ldots, \rho_N)$. Note that $(\rho, x') \mapsto (\rho_1, \ldots, \rho_N)$ is one-to-one with the relations above and

$$\rho_i = \rho_i(\rho, x') := \frac{\rho M_i x_i}{M(x')}, \quad M(x') := \sum_{k=1}^N M_k x_k, \quad x_N := 1 - \sum_{k=1}^{N-1} x_k.$$

Each thermodynamic quantity $f$ is represented as

$$f = f(T, \rho_1, \ldots, \rho_N) = \tilde{f}(T, \rho, x'),$$

where we always suppose the above relations among $(T, \rho_1, \ldots, \rho_N)$, $(T, \rho, x')$ and $x_N$.

Now we model the pressure. In the gas phase $\Omega^+(t)$, we assume $p = \sum_{k=1}^N p_k$ with partial pressures $p_i$ according to the ideal gas law $p_i = \frac{\rho_i}{M_i}RT$, namely

$$p = p(T, \rho_1, \ldots, \rho_N) = \sum_{k=1}^{N} \frac{\rho_i}{M_i}RT = \tilde{p}(T, \rho, x') = \frac{\rho RT}{M(x')}, \tag{12}$$

where $\rho_i = 0$ means that $A_i$ does not exist in $\Omega^+(t)$.

In the liquid phase $\Omega^-(t)$, for the later passage to the incompressible case, we use

$$p_N = p_N^R + K(\frac{\rho_N}{\rho_N^R} - 1) \tag{13}$$

with a bulk modulus $K = \partial_{\rho_N} p_N(\rho_N^R)\rho_N^R > 0$ and reference quantities $p_N^R$ and $\rho_N^R$. Later we let $K \to \infty$, which leads to $\rho_N \equiv \rho_N^R$. Note that the "1" in (13) can be generalized to an appropriate function of temperature and composition, but then $\rho_N$ will not become constant in the incompressible limit. For all other species in the liquid, we assume that they behave as ideal gas components (in the solvent "matrix" instead of a gas volume), namely $p_i = \frac{\rho_i}{M_i}RT$ for all $i < N$. Hence we have

$$\begin{aligned} p = p(T, \rho_1, \ldots, \rho_N) &= p_N^R + K(\frac{\rho_N}{\rho_N^R} - 1) + \sum_{k=1}^{N-1} \frac{\rho_k}{M_k}RT \\ = \tilde{p}(T, \rho, x') &= p_N^R + K\left(\frac{\rho M_N x_N}{\rho_N^R M(x')} - 1\right) + \frac{\rho RT}{M(x')}\sum_{k=1}^{N-1} x_k, \end{aligned} \tag{14}$$

where $\rho_i = 0$ $(i < N)$ means that $A_i$ does not exist in $\Omega^-(t)$.

The full chemical potential cannot be modeled directly, but needs to be computed from a Helmholtz free energy function $\rho\psi$. The modeling of $\psi$ follows the concept laid out in Sect. 13 of Bothe and Dreyer [3] and employs a decomposition of $\psi$ into an "elastic" part $\psi^{el}$, which takes into account the mechanical (pressure) work, and a "thermal" part $\psi^{th}$ which accounts for the entropy of mixing.

We start with a fixed temperature $T$ and a reference pressure $p^R$. We have a reference density function $\rho^* = \rho^*(T, x')$ through the equation

$$\tilde{p}(T, \rho^*, x') = p^R.$$

We then define

$$\psi^{th}(T, x') := \tilde{\psi}(T, \rho^*(T, x'), x'), \quad \psi^{el}(T, \rho, x') := \tilde{\psi}(T, \rho, x') - \psi^{th}(T, x').$$

Note that $\psi^{el}(T, \rho^*, x') = 0$. From the Gibbs-Duhem relation, we obtain

$$\rho^* \psi^{th}(T, x') = p^R + \sum_{k=1}^{N} \rho_k(\rho^*, x') \mu_k^{th}(T, x'),$$

$$\mu_k^{th}(T, x') := \tilde{\mu}_k(T, \rho^*(T, x'), x').$$

The thermal part of the chemical potential needs to be modeled, where we only consider the case of ideal mixtures (only containing entropy of mixing), namely

$$\mu_i^{th}(T, x') = g_i(T, p^R) + \frac{RT}{M_i} \ln x_i, \quad i = 1, \ldots, N,$$

where $g_i$ denotes the Gibbs free energy of the pure component $A_i$ in the respective phase. Next we compute $\psi^{el}$ through the relation

$$\frac{\partial}{\partial \rho} \psi^{el}(T, \rho, x') = \frac{\tilde{p}(T, \rho, x')}{\rho^2},$$

inserting $\tilde{p}$ modeled in (12) and (14), respectively.

For the gas phase, we obtain

$$\psi^{el}(T, \rho, x') = \int_{\rho^*}^{\rho} \frac{\tilde{p}(T, \tilde{\rho}, x')}{\tilde{\rho}^2} \, d\tilde{\rho} = \frac{RT}{M(x')} \ln \frac{\rho}{\rho^*}.$$

Hence we have

$$\rho\psi = \rho\tilde{\psi}(T, \rho, x') = -p^R \frac{\rho}{\rho^*} + \frac{\rho}{\rho^*} \sum_{k=1}^{N} \rho_k(\rho^*, x') \left\{ g_k(T, p^R) + \frac{RT}{M_k} \ln x_k \right\}$$
$$+ \frac{\rho RT}{M(x')} \ln \frac{\rho}{\rho^*}.$$

In order to compute $\rho\psi(T,\rho_1 \dots, \rho_N) = \rho\tilde{\psi}(T,\rho,x'(\rho_1,\dots,\rho_N))$, we observe the following relations:

$$\frac{\tilde{p}(T,\rho,x')}{\tilde{p}(T,\rho^*,x')} = \frac{\rho^* RT/M(x')}{\rho RT/M(x')} = \frac{\rho^*}{\rho} = \frac{p}{p^R} = \frac{RT}{p^R}\sum_{k=1}^{N}\frac{\rho_k}{M_k},$$

$$M(x'(\rho_1 \dots, \rho_N)) = \frac{\rho}{c}, \quad x_i = \frac{\rho_i/M_i}{\sum_{k=1}^N \rho_k/M_k}, \tag{15}$$

$$\rho_i(\rho^*(T,x'(\tilde{\rho}_1 \dots, \tilde{\rho}_N)), x'(\tilde{\rho}_1 \dots, \tilde{\rho}_N)) = \rho^* \frac{M_i x_i}{M(x')} = \rho^* \frac{\tilde{\rho}_i}{\tilde{\rho}}. \tag{16}$$

Direct calculation yields

$$\begin{aligned}\rho\psi &= \rho\psi(T,\rho_1 \dots, \rho_N) = \rho\tilde{\psi}(T,\rho,x'(\rho_1,\dots,\rho_N))\\ &= -RT\sum_{k=1}^{N}\frac{\rho_k}{M_k} + \sum_{k=1}^{N}\rho_k\left(g_k(T,p^R) + \frac{RT}{M_k}\ln\frac{\rho_k}{M_k}\right)\\ &\quad +RT\left(\sum_{k=1}^{N}\frac{\rho_k}{M_k}\right)\ln\frac{RT}{p^R}.\end{aligned}$$

Hence we obtain for $i = 1,\dots,N$ the chemical potentials as

$$\begin{aligned}\mu_i = \mu_i(T,\rho_1,\dots,\rho_N) &:= \frac{\partial(\rho\psi(T,\rho_1,\dots,\rho_N))}{\partial\rho_i}\\ &= g_i(T,p^R) + \frac{RT}{M_i}\ln\left(\frac{\rho_i RT}{p^R M_i}\right).\end{aligned}$$

With the relation $\rho_i RT/M_i = (RT/M_i)(\rho M_i x_i/M(x')) = \tilde{p}(T,\rho,x')x_i$, we also obtain

$$\begin{aligned}\mu_i &= \tilde{\mu}_i(T,\rho,x')\\ &= g_i(T,p^R) + \frac{RT}{M_i}\ln\frac{\tilde{p}(T,\rho,x')}{p^R} + \frac{RT}{M_i}\ln x_i \quad \text{for } i = 1,\dots,N.\end{aligned} \tag{17}$$

This reproduces the formulas known from the thermodynamical literature; see, e.g., Müller [11].

For the liquid phase, we obtain

$$\begin{aligned}\psi^{el}(T,\rho,x') &= \int_{\rho^*}^{\rho} \frac{\tilde{p}(T,\tilde{\rho},x')}{\tilde{\rho}^2} d\tilde{\rho} \\ &= -(p_N^R - K)\left(\frac{1}{\rho} - \frac{1}{\rho^*}\right) + \frac{KM_N x_N}{\rho_N^R M(x')} \ln\frac{\rho}{\rho^*} \\ &\quad + \frac{RT}{M(x')}\left(\sum_{k=1}^{N-1} x_k\right) \ln\frac{\rho}{\rho^*}.\end{aligned}$$

Hence we have

$$\begin{aligned}\rho\psi = \rho\tilde{\psi}(T,\rho,x') &= -p^R\frac{\rho}{\rho^*} + \frac{\rho}{\rho^*}\sum_{k=1}^{N}\rho_k(\rho^*,x')\left(g_k(T,p^R) + \frac{RT}{M_k}\ln x_k\right) \\ &\quad + (p_N^R - K)\left(\frac{\rho}{\rho^*} - 1\right) + \frac{K\rho M_N x_N}{\rho_N^R M(x')}\ln\frac{\rho}{\rho^*} + \frac{\rho RT}{M(x')}\left(\sum_{k=1}^{N-1} x_k\right)\ln\frac{\rho}{\rho^*}.\end{aligned}$$

Solving $\tilde{p}(T,\rho^*,x') = p^R$ with (15), we get

$$\frac{\rho}{\rho^*} = \frac{\rho}{\rho^*(T,x'(\rho_1,\ldots,\rho_N))} = \left(K\frac{\rho_N}{\rho_N^R} + RT\sum_{k=1}^{N-1}\frac{\rho_i}{M_i}\right)(p^R - p_N^R + K)^{-1}.$$

Straightforward computation with (15) and (16) yields

$$\begin{aligned}\rho\psi &= \rho\psi(T,\rho_1\ldots,\rho_N) = \rho\tilde{\psi}(T,\rho,x'(\rho_1,\ldots,\rho_N)) \\ &= \left(RT\sum_{k=1}^{N-1}\frac{\rho_k}{M_k} + K\frac{\rho_N}{\rho_N^R}\right)\left(\ln\frac{\rho}{\rho^*} - 1\right) + K - p_N^R \\ &\quad + \sum_{k=1}^{N}\rho_k\left(g_k(T,p^R) + \frac{RT}{M_k}\ln\frac{\rho_k/M_k}{c}\right),\end{aligned}$$

where the above $\rho/\rho^*$ and $c$ still have to be plugged in. Therefore we obtain, for $i = 1,\ldots,N-1$,

$$\begin{aligned}\mu_i &= \mu_i(T,\rho_1,\ldots,\rho_N) := \frac{\partial(\rho\psi(T,\rho_1,\ldots,\rho_N))}{\partial\rho_i} \\ &= g_i(T,p^R) + \frac{RT}{M_i}\ln\frac{\rho_i/M_i}{c} \\ &\quad + \frac{RT}{M_i}\ln\left(\left(K\frac{\rho_N}{\rho_N^R} + RT\sum_{k=1}^{N-1}\frac{\rho_k}{M_k}\right)\frac{1}{p^R - p_N^R + K}\right). \qquad (18)\end{aligned}$$

For $i = N$, we obtain

$$\begin{aligned}\mu_N = \mu_N(T, \rho_1, \ldots, \rho_N) &:= \frac{\partial(\rho\psi(T, \rho_1, \ldots, \rho_N))}{\partial \rho_N} \\ &= g_N(T, p^R) + \frac{RT}{M_N} \ln \frac{\rho_N / M_N}{c} \\ &\quad + \frac{K}{\rho_N^R} \ln\left(\left(K\frac{\rho_N}{\rho_N^R} + RT\sum_{k=1}^{N-1} \frac{\rho_i}{M_i}\right)\frac{1}{p^R - p_N^R + K}\right). \end{aligned} \tag{19}$$

Let us sum up: Up to here, the balance equations (1), (5) and (6) with constitutive relations (B1)–(B7), where the chemical potentials are modeled via (17), (18) and (19), form—up to boundary and initial conditions—a thermodynamically consistent full PDE system for a two-phase gas/liquid multicomponent system with compressible liquid and gas phases and mass transfer. For the non-isothermal case, the temperature dependencies need to be specified and the internal energy balance is usually transformed into a temperature form, i.e. of heat equation type. In the isothermal case, it can be dropped.

## 5 Incompressible Limit

We discuss the passage to an incompressible limit for the liquid solvent. As $K \to \infty$, assuming that the pressure stays bounded, we get $\rho_N/\rho_N^R \to 1$. After a (formal) computation, the passage $K \to \infty$ yields $\mu_i \to \mu_i^\infty$, where

$$\begin{aligned}\mu_i^\infty &= g_i(T, p^R) + \frac{RT}{M_i} \ln x_i \quad \text{for } i < N. \\ \mu_N^\infty &= g_N(T, p^R) + \frac{p - p^R}{\rho_N^R} + \frac{RT}{M_N} \ln x_N.\end{aligned}$$

Note that for an incompressible pure substance $A_N$, the Gibbs free energy satisfies

$$g_N(T, p) = g_N(T, p^R) + \frac{p - p^R}{\rho_N}.$$

Hence we have

$$\mu_N^\infty = g_N(T, p) + \frac{RT}{M_N} \ln x_N.$$

Therefore, we obtain the usual formulas for the chemical potential in the limit of $K \to \infty$, except for the fact that the chemical potentials of the solutes do not depend on the pressure. This is not a priori clear. Below, the superscript "$\infty$" is dropped.

Note that $\rho_N$ is constant and $p = p_N + \sum_{k=1}^{N-1} \frac{\rho_k}{M_k} RT$ with $p_N$ a free primitive variable. In fact, $p_N$ acts as a Lagrange multiplier in the liquid phase to account for the constraint $\nabla \cdot v_N = 0$ which results from (1) for $i = N$. As mentioned in the introduction, we employ the solvent momentum balance in the liquid phase and couple it to the barycentric momentum balance in the gas phase. For this purpose we use the relation

$$v_N = v + u_N = v + \frac{j_N}{\rho_N} = v - \frac{1}{\rho_N} \sum_{k=1}^{N-1} j_k. \tag{20}$$

Then each mass balance equation in (1) is rewritten with $v_N$, instead of $v$, in the liquid phase. In particular, the mass transfer transmission conditions $[\![\dot{m}_i]\!] = 0$ become, for $i < N$,

$$\left(j_i^+ + \rho_i^+ (v^+ - v^\Sigma)\right) \cdot n_\Sigma = \left(J_i^- + \rho_i^- (v_N^- - v^\Sigma)\right) \cdot n_\Sigma \quad \text{on } \Sigma(t),$$

where

$$J_i := \rho_i (v_i - v_N) = j_i + \frac{\rho_i}{\rho_N} \sum_{k=1}^{N-1} j_k$$

is the diffusion flux relative to the solvent. For $i = N$, the transfer condition is rewritten to become a substitution for the second equation in (2) and reads as

$$j_N^+ \cdot n_\Sigma + \dot{m} y_N^+ \cdot n_\Sigma = \rho_N^- (v_N^- - v^\Sigma) \cdot n_\Sigma \quad \text{on } \Sigma(t). \tag{21}$$

If the solvent evaporation is neglected, i.e. $\dot{m}_N = 0$ and $\rho_N^+ = 0$, then (21) simplifies to

$$v^\Sigma \cdot n_\Sigma = v_N^- \cdot n_\Sigma \quad \text{on } \Sigma(t).$$

As for the momentum balance, the standard approach would be to employ the barycentric momentum balance (5). However, this would lead to a velocity field $v$ of non-zero divergence. As an interesting alternative which leads to a divergence free velocity field, we make use of the partial momentum balance for $A_N$. According to Bothe and Dreyer [3], the partial momentum balance for $A_i$ reads as

$$\rho_i (\partial_t v_i + v_i \cdot \nabla v_i) = -\rho_i \nabla \mu_i + \nabla \cdot S_i^{irr} + \rho_i b_i - T \sum_{k=1}^{N} f_{ik} \rho_i \rho_k (v_i - v_k), \tag{22}$$

where $S_i^{irr} = p_i I + S_i$ is the irreversible stress part of $S_i$ and $f_{ik} = f_{ki} > 0$ are friction coefficients governing the exchange of momentum between the constituents. Comparing (22) to the barycentric momentum balance in dimensionless form, it turns out that the difference of $\partial_t v_i + v_i \cdot \nabla v_i$ to the mixture acceleration $\partial_t v + v \cdot \nabla v$ is negligible against the remaining terms, if the characteristic speed of diffusion is small compared to $\sqrt{p/\rho}$ which is about the speed of sound in a gas. The latter is assumed to hold, in which case (22) can be replaced by

$$\rho_i(\partial_t v_i + v_i \cdot \nabla v_i) = -y_i \nabla p + y_i \lambda \nabla(\nabla \cdot v) + y_i \nabla \cdot S^\circ + \rho_i b.$$

Applied to the solvent ($i = N$), we obtain

$$\rho^-(\partial_t v_N^- + v_N^- \cdot \nabla v_N^-) = -\nabla p_N^- - RT \sum_{k=1}^{N-1} \nabla c_k^- + \lambda \nabla(\nabla \cdot v^-) + \nabla \cdot S^{\circ -} + \rho^- b^- \tag{23}$$

with the standard constraint $\nabla \cdot v_N^- \equiv 0$ in the incompressible limit, where the superscript "–" indicates that a quantity refers to the liquid phase. For the momentum transmission, the jump condition in (5) is rewritten with $v_N$ and $j_k$, namely

$$\frac{\rho^-}{\rho_N^-}(\dot{m}_N^- - j_N^- \cdot n_\Sigma)(v_N^- - \frac{j_N^-}{\rho_N^-}) - S^- n_\Sigma - (\dot{m}^+ v^+ - S^+ n_\Sigma) = \nabla_\Sigma \cdot S^\Sigma \quad \text{on } \Sigma(t). \tag{24}$$

In (23) and (24), $v^-$, $S^{\circ -} = 2\eta D^{\circ -}$ and $S^- = -p^- I + \lambda(\nabla \cdot v^-)I + 2\eta D^{\circ -}$ are to be rewritten with $v_N$ and $j_k$ instead of $v$ by means of (20).

In order to obtain more detailed information about the diffusive fluxes, we first compute $\nabla(\mu_i/T)$. Since we are finally interested in the isothermal case, we consider constant $T$ from here on. In the gas phase, with

$$\mu_i^+ = g_i^+(T, p^R) + \frac{RT}{M_i} \ln \frac{x_i^+ p^+}{p^R} = g_i^+(T, p^R) + \frac{RT}{M_i} \ln \frac{c_i^+ RT}{p^R}$$

for the assumed ideal gas mixture, we obtain the result

$$\nabla \frac{\mu_i^+}{T} = \frac{R}{M_i} \frac{\nabla c_i^+}{c_i^+} = \frac{R}{c_i^+} \nabla \rho_i^+,$$

where the superscript "+" indicates that a quantity refers to the gas phase. In the liquid phase, we obtain $\nabla(\mu_i^-/T)$ for $i < N$ and $i = N$ as

$$\nabla\frac{\mu_i^-}{T} = \frac{R}{M_i}\frac{\nabla x_i^-}{x_i^-},$$

$$\nabla\frac{\mu_N^-}{T} = \frac{\nabla p^-}{T\rho_N^R} + \frac{R}{M_N}\frac{\nabla x_N^-}{x_N^-} = \frac{\nabla p_N^-}{T\rho_N^R} + \frac{R}{\rho_N^R}\sum_{k=1}^{N-1}\nabla c_k^- + \frac{R}{M_N}\frac{\nabla x_N^-}{x_N^-}.$$

If these are inserted into the Fickean form of the diffusive mass fluxes, the (molar) mass densities in the denominator only cancel, if the dependence of the phenomenological coefficients $L_{ik}$ on $\rho_1, \ldots, \rho_N$ has a special structure. To incorporate such structural information, while keeping the derivation as rigorous as possible, we prefer to use the generalized Maxwell-Stefan equations as constitutive relations determining the diffusion fluxes. The Maxwell-Stefan equations read

$$-\sum_{k=1}^{N}\frac{x_k j_i^m - x_i j_k^m}{Đ_{ik}} = \frac{\rho_i}{RT}\nabla\mu_i - \frac{y_i}{RT}\nabla p - \frac{\rho_i}{RT}(b_i - b) \tag{25}$$

with an individual body force $b_i$ for $A_i$ and the molar mass fluxes

$$j_i^m := \frac{j_i}{M_i} = c_i(v_i - v).$$

For a rigorous derivation of (25) see Bothe and Dreyer [3]. There you also find the additional contribution $\nabla \cdot S_i - y_i \nabla \cdot S$ in the right-hand side of (25). The latter is not included in the classical form of the Maxwell-Stefan equations as given in, e.g., Taylor and Krishna [17] and Bird et al. [1]. For simplicity, we also neglect the effect of diffusion driven by viscous stress. In (25), the $Đ_{ik}$ are the so-called Maxwell-Stefan diffusivities, which are symmetric (cf. [3]). From measurements, one knows that the $Đ_{ik}$ depend only weakly on the composition (often as affine functions), in contrast to the Fickean diffusivities. We assume the $Đ_{ik}$ to be constant with $Đ_{ik} = Đ_{ki} > 0$. Note that the Maxwell-Stefan equations sum up to zero, and hence the N equations are not independent. Concerning the inversion of this equation system, see Bothe [2] and the references cited there.

From here on, we assume equal body forces $b_k \equiv b$ for all components. Insertion of the chemical potential gradients yields for the gas phase

$$-\sum_{k \neq i}\frac{x_k^+ j_i^{m+} - x_i^+ j_k^{m+}}{Đ_{ik}^+} = \nabla c_i^+ - \frac{y_i^+}{RT}\nabla p^+ = \nabla c_i^+ - y_i^+ \nabla c^+.$$

In the liquid phase, we obtain for $i < N$

$$-\sum_{k \neq i}\frac{x_k^- j_i^{m-} - x_i^- j_k^{m-}}{Đ_{ik}^-} = c^-\nabla x_i^- - \frac{y_i^-}{RT}\nabla p^-.$$

For $i = N$, we obtain

$$-\sum_{k \neq N} \frac{x_k^- j_N^{m-} - x_N^- j_k^{m-}}{Đ_{Nk}^-} = \frac{\nabla p^-}{RT} + c^- \nabla x_N^- - \frac{y_N^-}{RT} \nabla p^-.$$

We simplify the jump conditions of the chemical potential. Neglecting the viscous and the kinetic effect in (B7), we assume

$$[\![\mu_i]\!] = 0.$$

See Bothe and Fleckenstein [4] for an assessment of this approximation. For $i < N$, we have

$$\mu_i^+(T, \rho^+, x'^+) = g_i^+(T, p^R) + \frac{RT}{M_i} \ln \frac{p^+ x_i^+}{p^R} = g_i^+(T, p^R) + \frac{RT}{M_i} \ln \frac{p_i^+}{p^R},$$

$$\mu_i^-(T, \rho^-, x'^-) = g_i^-(T, p^R) + \frac{RT}{M_i} \ln x_i^-.$$

For given $T$, choose $p^R = p_i^R(T)$ so that $g_i^+(T, p_i^R(T)) = g_i^-(T, p_i^R(T))$ holds for each $i$ and for a planar interface. Then, neglecting curvature effects via the pressure jump, we obtain

$$\mu_i^+(T, \rho^+, x'^+) = \mu_i^-(T, \rho^-, x'^-) \Leftrightarrow \ln x_i^- = \ln \frac{p_i^+}{p_i^R(T)} = \ln \frac{\rho_i^+ RT}{M_i p_i^R(T)}$$

$$\Leftrightarrow x_i^- p_i^R(T) = c_i^+ RT.$$

This is a version of Henry's law. Thus we obtain the following PDE system for incompressible solvent and compressible gas phase, where we assume equal body forces and a continuous temperature field.

**Non-dilute solution with incompressible solvent:**

Gas phase:

$$\begin{cases} \partial_t c_i + \nabla \cdot (c_i v + j_i^m) = 0, \qquad i = 1, \dots, N, \\ -\sum_{k \neq i} \frac{x_k j_i^m - x_i j_k^m}{Đ_{ik}} = \nabla c_i - y_i \nabla c, \qquad i = 1, \dots, N, \\ \rho(\partial_t v + v \cdot \nabla v) + \nabla p = \lambda \nabla (\nabla \cdot v) + \eta \Delta v + \rho b, \\ p = cRT = RT \sum_{k=1}^N c_k, \quad \rho = \sum_{k=1}^N M_k c_k. \end{cases}$$

Liquid phase:

$$\begin{cases} \partial_t c_i + \nabla\cdot(c_i v_N + J_i^m) = 0, \qquad i = 1,\dots,N-1, \\ -\sum_{k\neq i} \frac{x_k j_i^m - x_i j_k^m}{Ð_{ik}} = c\nabla x_i - \frac{y_i}{RT}\nabla p, \qquad i = 1,\dots,N-1, \\ -\sum_{k\neq N} \frac{x_k j_N^m - x_N j_k^m}{Ð_{Nk}} = c\nabla x_N + \frac{1-y_N}{RT}\nabla p, \\ J_i^m = j_i^m + \frac{\rho_i}{\rho_N}\sum_{k=1}^{N-1} j_k^m,\ j_i^m = \frac{j_i}{M_i},\ p = p_N + \sum_{k=1}^{N-1} c_k RT, \\ \rho(\partial_t v_N + v_N\cdot\nabla v_N) + \nabla p_N = \lambda\nabla(\nabla\cdot v) + \nabla\cdot S^\circ + \rho b - RT\sum_{k=1}^{N-1}\nabla c_k, \\ \nabla\cdot v_N \equiv 0,\ \rho_N \equiv \rho_N^R, \end{cases}$$

where $v$ and $S^\circ = 2\eta D^\circ = \eta\big(\nabla v + (\nabla v)^{\mathsf{T}} - \frac{1}{3}(\nabla\cdot v)I\big)$ are to be rewritten with $v_N$ and $j_k$ through (20).

Interface:

$$\begin{cases} (j_i^{m+} + c_i^+(v^+ - v^\Sigma))\cdot n_\Sigma = (j_i^{m-} + c_i^-(v_N^- - v^\Sigma))\cdot n_\Sigma, \\ \qquad\qquad\qquad\qquad\qquad\qquad i = 1,\dots,N-1, \\ (j_N^+\cdot n_\Sigma + y_N^+\rho^+(v^+ - v^\Sigma))\cdot n_\Sigma = \rho_N^-(v_N^- - v^\Sigma)\cdot n_\Sigma \\ (\text{or } v^\Sigma\cdot n = v_N^-\cdot n_\Sigma \text{ in case of negligible evaporation of } A_N), \\ x_i^- p_i^R(T) = c_i^+ RT, \qquad i = 1,\dots,N-1, \\ \frac{\rho^-}{\rho_N^-}(\dot m_N^- - j_N^-\cdot n_\Sigma)(v_N^- - \frac{j_N^-}{\rho_N^-}) - S^- n_\Sigma - (\dot m^+ v^+ - S^+ n_\Sigma) \\ \qquad\qquad\qquad\qquad\qquad = \gamma^\Sigma\kappa_\Sigma n_\Sigma + \nabla_\Sigma\gamma^\Sigma, \end{cases}$$

where $\kappa_\Sigma = \nabla_\Sigma\cdot(-n_\Sigma)$ is the curvature, $S^\pm = -p^\pm I + \lambda(\nabla\cdot v)^\pm I + 2\eta D^{\circ\pm}$ with $D^\circ = \frac{1}{2}\big(\nabla v + (\nabla v)^{\mathsf{T}} - \frac{1}{3}(\nabla\cdot v)I\big)$ and $S^-$ is to be rewritten with $v_N$ and $j_k$ through (20).

## 6 Dilute Solution with Incompressible Solvent

We ignore bulk viscosities in both phases and assume $\dot m_N = 0$. Note that in the dilute case ($x_i \ll 1$ for $i < N$), we have $J_i^m \approx j_i^m$, $\rho \approx \rho_N$ and $S \approx S_N$ in the liquid phase. We obtain a simple Fick's law for $i < N$, namely

$$j_i^{m-} = -Ð_{Ni}^-\nabla c_i^-.$$

We may approximate $c^- \approx c_N^-$. Then Henry's law becomes

$$\frac{c_i^-}{c_i^+} = \frac{c_N^- RT}{p_i^R(T)} =: H_i.$$

Thus we obtain the following PDE system for a dilute solution with incompressible solvent and compressible gas phase, where we assume equal body forces and a continuous temperature field.

**Dilute solution with incompressible solvent:**

Gas phase:

$$\begin{cases} \partial_t c_i + \nabla \cdot (c_i v + j_i^m) = 0, \qquad i = 1, \ldots, N, \\ -\sum_{k \neq i} \frac{x_k j_i^m - x_i j_k^m}{Ð_{ik}} = \nabla c_i - y_i \nabla c, , \qquad i = 1, \ldots, N, \\ \rho(\partial_t v + v \cdot \nabla v) + RT \nabla c = \eta \Delta v + \rho b, \\ p = cRT = RT \sum_{k=1}^{N} c_k, \quad \rho = \sum_{k=1}^{N} M_k c_k. \end{cases}$$

Liquid phase:

$$\begin{cases} \partial_t c_i + \nabla \cdot (c_i v_N + j_i^m) = 0, \qquad i = 1, \ldots, N-1, \\ j_i^m = -Ð_{iN} \nabla c_i, \qquad i = 1, \ldots, N-1, \\ \rho_N(\partial_t v_N + v_N \cdot \nabla v_N) + \nabla p_N = \eta_N \Delta v_N + \rho_N b - RT \sum_{k=1}^{N-1} \nabla c_k, \\ \nabla \cdot v_N \equiv 0, \quad \rho_N \equiv \rho_N^R. \end{cases}$$

Interface:

$$\begin{cases} (j_i^{m+} + c_i^+(v^+ - v^\Sigma)) \cdot n_\Sigma = j_i^{m-} \cdot n_\Sigma, \qquad i = 1, \ldots, N-1, \\ v^\Sigma \cdot n = v_N^- \cdot n_\Sigma, \\ \frac{c_i^-}{c_i^+} = H_i, \\ (-j_N^- \cdot n_\Sigma)(v_N^- - \frac{j_N^-}{\rho_N^-}) - S_N^- n_\Sigma - (\dot{m}^+ v^+ - S^+ n_\Sigma) = \gamma^\Sigma \kappa_\Sigma n_\Sigma + \nabla_\Sigma \gamma^\Sigma, \end{cases}$$

where $S^+ = -p^+ I + 2\eta D^{\circ +}$ and $S_N^- = -p_N^- I + 2\eta_N D^{\circ -} = -p_N^- I + \eta_N^- (\nabla v_N + (\nabla v_N)^\mathsf{T})^-$.

# 7 Boundary Conditions

The mathematical model is to be complemented by appropriate boundary conditions at the fixed walls, called $\partial\Omega$, and at the free upper surface $\Gamma(t)$. Since the derivation of physically sound boundary conditions is a topic on its own (cf., e.g., Bothe et al. [6]), we rest content with the simplest reasonable choice.

### (a) Boundary conditions at fixed walls

The fixed walls are impermeable. Hence

$$v \cdot n_w = 0 \quad \text{and} \quad j_i \cdot n_w = 0 \quad \text{at } \partial\Omega,$$

where $n_w$ is the unit outer normal to the walls. This also implies

$$v_N \cdot n_w = 0 \quad \text{at } \partial\Omega.$$

In order to allow for a movable upper surface, the tangential velocities $v_{\|}$ and $v_{N\|}$ shall not be assumed to vanish. Instead, we assume a Navier slip condition of the form

$$v_{\|} + a(Sn_w)_{\|} = 0, \;\; v_{N\|} + a_N(S_N n_w)_{\|} = 0 \quad \text{at } \partial\Omega \text{ with } a, a_N \geq 0.$$

In the non-isothermal case, we add a Robin-condition for the temperature, i.e.

$$T + \beta \nabla T \cdot n_w = T_{ext} \quad \text{at } \partial\Omega$$

with $\beta \geq 0$.

### (b) Boundary conditions at the free upper surface

The Robin condition for the temperature can also be applied at the free surface. The other conditions are

$$v \cdot n_\Gamma = V_\Gamma \;\text{ and }\; (p_{ext} - p)n_\Gamma + S^{irr} n_\Gamma = \gamma^\Gamma \kappa_\Gamma n_\Gamma + \nabla_\Gamma \gamma^\Gamma \quad \text{on } \Gamma(t) \qquad (26)$$

with the outer unit normal $n_\Gamma$ on $\Gamma(t)$ and the curvature $\kappa_\Gamma = \nabla_\Gamma \cdot (-n_\Gamma)$. Let us note that, in the dilute solution limit and for a constant surface tension $\gamma^\Gamma$, the condition (26) becomes

$$v_N \cdot n_\Gamma = V_\Gamma,$$
$$p_{ext} - p_N + n_\Gamma \cdot S_N^{irr} n_\Gamma = \gamma_\Gamma \kappa_\Gamma \;\text{ and }\; n_\Gamma \times S_N^{irr} n_\Gamma = 0 \quad \text{on } \Gamma(t).$$

We assume the mixture composition to be given at $\Gamma(t)$ due to local chemical equilibrium with a large and well-mixed external gas phase. Hence

$$x_i^- = x_i^\Gamma \quad \text{on } \Gamma(t) \text{ for } i < N \text{ with } x_i^\Gamma \geq 0,$$

where we assume $\sum_{i=1}^{N-1} x_i^\Gamma < 1$.

**(c) Condition at the contact line**

The free surface $\Gamma(t)$ touches the fixed wall in a set of points which forms the so-called contact line $C$. The modeling of dynamic contact lines is, again, a topic on its own and we refer to Shikhmurzaev [14] and the references therein for detailed information. Here, in order to close the system in the simplest possible manner, we assume a fixed contact angle of $\pi/2$, i.e.

$$n_\Gamma \perp n_w \quad \text{on } C.$$

## Appendix

The derivation of the balance equations is based on standard two-phase transport and divergence theorems: Let $V$ denote an arbitrary fixed control volume with outer normal $n$. Then

$$\frac{d}{dt}\int_V \phi dx = \int_{V\setminus\Sigma} \partial_t \phi dx - \int_{\Sigma_V} [\![\phi]\!] v^\Sigma \cdot n_\Sigma do$$

with $\Sigma_V := \Sigma(t) \cap V$, the surface velocity (including tangential part) $v^\Sigma$ and the surface unit normal $n_\Sigma$. Here $[\![\phi]\!] := \lim_{h\to 0+}(\phi(x + hn_\Sigma) - \phi(x - hn_\Sigma))$ defined for $x \in \Sigma$. We also have

$$\int_{\partial V} f \cdot n do = \int_{V\setminus\Sigma} \nabla \cdot f dx + \int_{\Sigma_V} [\![f \cdot n_\Sigma]\!] do.$$

Since the internal energy and the entropy have surface contributions, we also need the surface transport theorem for $\phi^\Sigma$ defined on $\Sigma$. It states that

$$\frac{d}{dt}\int_{\Sigma_V} \phi^\Sigma do = \int_{\Sigma_V} \left(\frac{D^\Sigma \phi^\Sigma}{Dt} + \phi^\Sigma \nabla_\Sigma \cdot v^\Sigma\right) do - \int_{\partial\Sigma_V} \phi^\Sigma v^\Sigma \cdot \nu ds,$$

which—in this simple form—holds for all fixed $V$ such that its outer normal $n$ satisfies $n \perp n_\Sigma$ on $\Sigma_V$, and hence $n = \nu$, where $\nu$ is tangential to $\Sigma$ and normal to the bounding curve $\partial\Sigma_V$. We always choose such control volumes in the integral balances above. For more details and mathematical proofs see, e.g., Slattery [15], Romano and Marasco [12] or the Appendix in Bothe et al. [5].

**Acknowledgements** This work was supported by the DFG within scope of the IRTG 1529 "Mathematical Fluid Dynamics" and by the JSPS Japanese-German Graduate Externship. The authors gratefully acknowledge this support.

## References


1. R.B. Bird, W.E. Stewart, E.N. Lightfoot, *Transport Phenomena*, 2nd edn. (Wiley, New York 2007)
2. D. Bothe, On the Maxwell-Stefan equations to multicomponent diffusion, in *Progress in Nonlinear Differential Equations and Their Applications*, vol. 60, ed. by P. Guidotti, C. Walker et al. (Springer, Basel, 2011), pp. 81–93
3. D. Bothe, W. Dreyer, Continuum thermodynamics of chemically reacting fluid mixtures. Acta Mech. **226**, 1757–1805 (2015)
4. D. Bothe, S. Fleckenstein, A Volume-of-Fluid-based method for mass transfer processes at fluid particles. Chem. Eng. Sci. **101**, 283–302 (2013)
5. D. Bothe, J. Prüss, G. Simonett, Well-posedness of a two-phase flow with soluble surfactant, in *Progress in Nonlinear Differential Equations and Their Applications*, vol. 64, ed. by M. Chipot, J. Escher (Birkhäuser, Basel, 2005), pp. 37–61
6. D. Bothe, M. Köhne, J. Prüss, On a class of energy preserving boundary conditions for incompressible Newtonian flows. SIAM J. Math. Anal. **45**(6), 3768–3822 (2013)
7. S.R. de Groot, P. Mazur, *Non-equilibrium Thermodynamics* (Dover, New York, 1984)
8. W. Dreyer, On jump conditions at phase boundaries for ordered and disordered phases. WIAS Preprint No. 869 (2003)
9. K. Hutter, K. Jöhnk, *Continuum Methods of Physical Modeling* (Springer, Heidelberg, 2004)
10. G. Liger-Belair, M. Bourget, S. Villaume, P. Jeandet, H. Pron, G. Polidori, On the losses of dissolved $CO_2$ during champagne serving. J. Agric. Food Chem. **58**, 8768–8775 (2010)
11. I. Müller, *Thermodynamics* (Pitman, London, 1985)
12. A. Romano, A. Marasco, *Continuum Mechanics: Advanced Topics and Research Trends* (Birkhäuser, Basel, 2010)
13. M. Sauzade, T. Cubaud, Initial microfluidic dissolution regime of $CO_2$ bubbles in viscous oils. Phys. Rev. E **88**, 051001(R) (2013)
14. Y.D. Shikhmurzaev, *Capillary Flows with Forming Interfaces* (Chapman & Hall/CRC, Boca Raton, 2008)
15. J.C. Slattery, *Advanced Transport Phenomena* (Cambridge University Press, Cambridge, 1999)
16. F. Takemura, A. Yabe, Gas dissolution process of spherical rising gas bubbles. Chem. Eng. Sci. **53**(15), 2691–2699 (1998)
17. R. Taylor, R. Krishna, *Multicomponent Mass Transfer* (Wiley, New York 1993)

# On Unsteady Internal Flows of Bingham Fluids Subject to Threshold Slip on the Impermeable Boundary

**Miroslav Bulíček and Josef Málek**

*To Yoshihiro Shibata on the occasion of his 60th birthday.*

**Abstract** In the analysis of weak solutions relevant to evolutionary flows of incompressible fluids with *non-constant viscosity* or with *non-linear constitutive equation*, it is in general an open question whether a globally integrable pressure exists if the flows are subject to *no-slip* boundary conditions. Here we overcome this deficiency by considering *threshold* boundary conditions stating that the fluid adheres to the boundary until certain critical value for the wall shear stress is reached. Once the wall shear stress exceeds this critical value, the fluid slips. The main ingredient in our approach is to look at this type of activated, *stick-slip*, boundary condition as an *implicit constitutive equation* on the boundary.

We prove the long-time and large-data existence of weak solutions, with integrable pressure, to unsteady internal flows of Bingham and Navier-Stokes fluids subject to such threshold slip boundary conditions.

**Keywords** Bingham fluid • Implicit constitutive theory • Incompressible fluid • Integrable pressure • Long-time and large-data existence • Navier's slip • Navier-Stokes fluid • No-slip • Non-Newtonian fluid • Stick-slip • Threshold slip • Unsteady flow • Weak solution

## 1 Introduction

In the analysis of partial differential equations, especially in the theory of weak (distributional or variational) solutions the homogeneous, Dirichlet data are considered to be the simplest boundary conditions to deal with. It is however not the case when

This work was initiated with the support of GACR 201/09/0917 financed by the Czech Science Foundation and completed with the support of the ERC-CZ project LL1202 financed by the Ministry of Education, Youth and Sports of the Czech Republic.

M. Bulíček • J. Málek (✉)
Charles University in Prague, Faculty of Mathematics and Physics, Mathematical Institute, Sokolovská 83, 186 75 Prague 8, Czech Republic
e-mail: mbul8060@karlin.mff.cuni.cz; malek@karlin.mff.cuni.cz



H. Amann et al. (eds.), *Recent Developments of Mathematical Fluid Mechanics*, Advances in Mathematical Fluid Mechanics, DOI 10.1007/978-3-0348-0939-9_8

studying the flows of incompressible fluids, in particular, if the flows are varying with time. The no-slip boundary conditions are not compatible with the Helmholtz decomposition and this fact leads to difficulties in the analysis of such flows. Recall that the Helmholtz decomposition splits any vector field into the sum of the field that is divergence free and another field that is the gradient of a suitable function (solution of the Neumann problem for the Laplace equation). In the case of a no-slip boundary condition, while the velocity to which the Helmholtz decomposition is applied vanishes on the boundary, its divergence-free counterpart that comes from the Helmholtz decomposition is not zero on the boundary (in general merely its normal component vanishes). Consequently, up to some exceptional cases that will be specified below, it seems impossible (within the context of weak solutions) to identify the pressure field as an integrable function, belonging for example to $L^1(0,T;L^1(\Omega))$, see [35, 36]. This deficiency is however overcome if one requires that only the normal component of the velocity vanishes (thus the boundary is impermeable) and considers Navier's slip that relates the tangential components of the velocity to the projection of the normal stress to the tangent plane at the boundary (see [1, 2]). Another setting where the question of (global) integrability of the pressure is positively answered represents spatially periodic problems, see for example [34].

Not only is the question of the global integrability of the pressure interesting as a problem itself but it also appears to be a crucial step in the large-data analysis of unsteady flows for heat-conducting incompressible fluids [2, 3], one/two equation turbulence models [5], incompressible fluids with pressure and shear-rate dependent viscosity [1, 3, 26], and finally for the corresponding numerical methods and their analysis.

The above discussion about the integrability of the pressure is concerned with the concept of weak solutions.[1] In our opinion, such concept of solution is preferable for several reasons. Let us briefly mention a few of them.

From the point of view of fluid mechanics, a part of the governing system of equations comes from the balance equations that are based on the balancing the mass, linear and angular momentum, energy, and entropy over any subpart of the flow domain. Thus, the primary form of these balance equations has an integral representation and their classical formulation is derived under additional assumptions on the smoothness (regularity) of the quantities involved. Also, for internal flows, one controls the (internal, kinetic and potential) energy of the system and then also the rate of entropy (that are due to mechanical working, thermal heating, etc.) in terms of the data. These a priori bounds, although rather poor from the point of view of classical formulation, give sense to all terms in the integral (weak) formulation of the problem. On the level of mathematical analysis, weak solutions are in many cases the only concepts (see [19, 24]) that exist globally in time for any size of data measured in reasonable norms (frequently given by the a priori estimates). Furthermore, the concept of weak solution or the integral

[1]By *weak solution,* we mean here a broad class of generalized solutions such as suitable weak solution, renormalized solution, entropy solution, dissipative solution, etc.

formulation is a point of departure for several numerical approaches - spectral methods, finite volume methods, finite element methods are just the most known examples of methods used intensively in computer simulations.

In order to have globally integrable pressure for considered evolutionary problems (and approximate the no-slip boundary conditions as close as possible), we propose to replace no-slip by stick-slip boundary conditions. The stick-slip (or threshold slip) boundary conditions are usually characterized as follows: the fluid exhibits no-slip on the boundary if and only if the magnitude of the projections of the normal stress to the tangent plane is below a certain critical value, called *threshold*. Once the magnitude of the projection of the normal stress to the tangent plane exceeds this threshold, the fluid slips following the Navier's slip boundary conditions. Note that this type of boundary condition is observed especially (but not only) for certain non-Newtonian fluids (see [9, 16] for example) and it seems very natural in general. Having said this, we however take a different (purely mathematical) standpoint in this study. We consider the threshold slip as an auxiliary tool for incorporating a no-slip boundary condition into the formulation of the problem following the goal to prove long-time and large-data existence of weak solution, having, in particular, an integrable pressure. The threshold can have only purely auxiliary character and can be any (even very large) number. In some applications this threshold may not even be activated during the processes considered. Also, if one would be capable of proving that the weak solution to the relevant problem involving threshold slip not only exists but also possesses the velocity field that is continuous (up to the boundary), then the problem that is successfully solved in this way is also the problem with no-slip boundary condition provided that the threshold is chosen bigger that the $L^{\infty}$-norm of the velocity. We wish to emphasize that stick-slip boundary conditions lead to a weak formulation and function spaces that are different from those used for the pure no-slip boundary conditions.

Furthermore, we look at this type of activated boundary condition as an *implicit constitutive relation* on the boundary, which leads again to a formulation that is different from those used earlier—we do not need tools such as variational inequalities, penalty methods, etc. To the best of our knowledge, this is a novel feature of our approach.

For the sake of completeness, we wish to mention a couple of cases (and conditions) where the global integrability of the pressure is established even for no-slip boundary conditions for incompressible fluid models.

First, for fluids with constant viscosity the issue of obtaining an integrable pressure when analyzing unsteady flows subject to the homogeneous Dirichlet boundary condition is solved by means of maximal regularity theory developed for the evolutionary Stokes system (with homogeneous Dirichlet data), see [15, 32]. These results are based on inverting the "evolutionary Stokes operator", and finding the fundamental solution explicitly. Generalizations of the $L^p(0,T;L^q(\Omega))$ maximal regularity theory to more complicated operators seem to be open with the one exception, namely, evolutionary Stokes type problems with the viscosity being a Hölder continuous function of time and the spatial variables, see [33]. Note that these results require, in addition to a Hölder continuous viscosity, smooth domains (with the boundary that is at least of the class $C^2$) and sufficiently regular right-hand side and the initial value. In other words, the approach presented in this study can be useful for the evolutionary Navier-Stokes equations in situations when the right-hand side is only functional (and not an integrable function); also the boundary can be Lipschitz in some cases.

Second, in the analysis of time-dependent flows of incompressible generalized Newtonian fluids, it means the fluids where the relation between the deviatoric part of the Cauchy stress and the symmetric part of the velocity gradient is uniformly monotone, there are rather restrictive situations (e.g. the velocity itself has to be at least an admissible test function) in which it is possible to prove (see [20] and [4]) that the time derivative is an integrable function from which then follows easily (by a consequence of de Rham's theorem) that the pressure itself is integrable as well.

To conclude, except these two cases, the question of the existence of weak solution with an integrable pressure to the three-dimensional evolutionary problems with the no-slip boundary condition is unanswered.

In order to illustrate our approach (based on replacing no-slip by slip-stick boundary conditions) for interesting problems that generalize those for the incompressible Navier-Stokes equations, yet the basic function space setting is the same as that for the Navier-Stokes system, we consider Bingham fluids in this study. Bingham fluids represents a non-linear (it means non-Newtonian) fluid model described by the following constitutive relation: the fluid responses as a rigid body if and only if the magnitude of the deviatoric part of the Cauchy stress is below a certain critical value for the stress, called the yield stress; once the magnitude of the deviatoric part of the Cauchy stress exceeds the yield stress the fluid responses as a Navier-Stokes fluid. There is a recent observation [28, 29] that this dichotomy description can be included into the setting of implicit constitutive relations, and as recent studies [6, 7, 11, 17] show this leads to a new mathematical setting useful both for the existence, numerical analysis and computer simulation of the relevant problems. We also consider the Navier-Stokes equations with the viscosity dependent continuously on the time and spatial variables. Such a system can be viewed for example as a simplified model for fluids with the viscosity dependent on some quantity like the concentration or the temperature, where such a dependence is somehow known a priori. As mentioned above, if the viscosity is merely continuous the result of Solonnikov [33] is not applicable, hence the question of the existence of globally integrable pressure is open and worth investigating.

Although the most of the studies concerning analysis of flows of incompressible fluids described by the Stokes and Navier-Stokes equations concern no-slip boundary conditions, there are several studies analyzing (both steady and unsteady) flows subject to different type of boundary conditions. We mention several studies being aware of the fact that the list is incomplete. (Note that some studies concerning the analysis of problems with Navier's slip boundary conditions were cited above.)

Slip and leak boundary conditions for (mostly steady flows of) Stokes or Navier-Stokes fluids were studied in the works by Fujita [13, 14] focusing on the question of existence and uniqueness of weak solution. Its regularity properties are investigated in Saito [30], and strong solutions are constructed in a recent paper by Kashiwabara [18], where one can find further references to studies focused on numerical (mostly finite element) methods. Threshold slip boundary conditions were investigated in [22] and [23].

Bingham fluids subject to a nonlocal slip boundary condition were studied in Consiglieri [8]. We also refer to [31], and [27] for the analysis of steady flows of Bingham fluids with no-slip boundary conditions using regularity technique. Flows of second grade fluids subject to slip boundary conditions were investigated in [21].

The paper is organized in the following way. In Sect. 2 we describe the problem studied in this paper, specify the class of fluids considered, introduce the function spaces and formulate the results. In Sect. 3 we state and prove the convergence lemma that is suitable for the analysis of the problem. Finally, Sect. 4 contains the proof of the results performed via a suitable approximation scheme based on the truncation of the convective term and modification of the constitutive equations in the bulk and on the boundary in such a way that standard monotone operator theory is directly applicable to this kind of approximations. Then we derive the uniform estimates and take the limit. The assumptions of the convergence lemma are achieved by using a suitable truncation as a test function.

## 2 Formulation of the Problem and the Main Result

### 2.1 Formulation of the Problem

We consider the following problem: Given the length of the time interval $T > 0$, a connected bounded open set $\Omega \subset \mathbb{R}^d$, the density $\varrho^* \in (0, \infty)$, the density of the external body forces $\boldsymbol{b} : (0,T) \times \Omega \to \mathbb{R}^d$, the functions $\mathbb{G} : \mathbb{R}^{d\times d}_{\mathrm{sym}} \times \mathbb{R}^{d\times d}_{\mathrm{sym}} \to \mathbb{R}^{d\times d}_{\mathrm{sym}}$ and $\boldsymbol{g} : \mathbb{R}^d \times \mathbb{R}^d \to \mathbb{R}^d$, and the initial velocity $\boldsymbol{v}_0 : \Omega \to \mathbb{R}^d$, find $(\boldsymbol{v}, p, \mathbb{S}) : Q_T := (0,T) \times \Omega \to \mathbb{R}^d \times \mathbb{R} \times \mathbb{R}^{d\times d}_{\mathrm{sym}}$ and $\boldsymbol{s} : \Sigma_T := (0,T) \times \partial\Omega \to \mathbb{R}^d$ such that

$$
\begin{aligned}
\operatorname{div} \boldsymbol{v} &= 0 && \text{in } Q_T,\\
\varrho_* \left(\boldsymbol{v}_{,t} + \operatorname{div}(\boldsymbol{v} \otimes \boldsymbol{v})\right) - \operatorname{div} \mathbb{S} + \nabla p &= \varrho_* \boldsymbol{b} && \text{in } Q_T,\\
\mathbb{G}(\mathbb{S}, \mathbb{D}) &= \mathbb{O} && \text{in } Q_T,\\
\boldsymbol{v} \cdot \boldsymbol{n} &= 0 && \text{on } \Sigma_T,\\
\boldsymbol{g}(\boldsymbol{s}, \boldsymbol{v}_\tau) &= \boldsymbol{0} && \text{on } \Sigma_T,\\
\boldsymbol{v}(0, \cdot) &= \boldsymbol{v}_0 && \text{in } \Omega.
\end{aligned}
\tag{1}
$$

In this setting, $\boldsymbol{v}$ represents the velocity, $-p$ stands for the mean normal stress (the pressure), $\mathbb{S}$ is the deviatoric part of the Cauchy stress $\mathbb{T}$ (hence $\mathbb{T} := \mathbb{S} - p\mathbb{I}$) and $\boldsymbol{s}$ denotes the projection of the normal stress into the corresponding tangent plane, i.e., $\boldsymbol{s} := (\mathbb{T}\boldsymbol{n})_\tau$. Here, for any vector $\boldsymbol{z}$ defined on $\partial\Omega$ we set $\boldsymbol{z}_\tau := \boldsymbol{z} - (\boldsymbol{z} \cdot \boldsymbol{n})\boldsymbol{n}$, whereas

$\boldsymbol{n} : \partial\Omega \to \mathbb{R}^d$ denotes the normal vector. Note that $(\mathbb{T}\boldsymbol{n})_\tau = (\mathbb{S}\boldsymbol{n})_\tau$. Finally, the symbol $\mathbb{D}$ stands for the symmetric part of the velocity gradient $\nabla \boldsymbol{v}$ defined through $\mathbb{D} := \frac{1}{2}(\nabla \boldsymbol{v} + (\nabla \boldsymbol{v})^T)$; if needed we use $\mathbb{D}\boldsymbol{v}$ instead of $\mathbb{D}$.

Equation $(1)_1$ expresses the fact that the fluid is incompressible (and since the density $\varrho_*$ is constant the fluid is even homogeneous). The second equation $(1)_2$ is the balance of linear momentum. The third equation $(1)_3$, characterizing the response of the fluid one deals with, says that the symmetric part of the velocity gradient $\mathbb{D}$ (that generalizes the shear rate) and the deviatoric part of the Cauchy stress $\mathbb{S}$ are related[2]: while the Navier-Stokes fluid is characterized by a *linear* relation, namely,

$$\mathbb{S} = 2\nu_* \mathbb{D}, \tag{2}$$

here we consider general implicit relations opening the possibility of describing, in terms of the same quantities that enter into (2), various phenomena (such as stress thickening or stress thinning, shear thickening or shear thinning, activation phenomena such as the presence of the yield stress, and even normal stress differences) that the Navier-Stokes model (2) cannot capture.

Equation $(1)_4$ says that the boundary is impermeable, and Eq. $(1)_5$ describes the result of interactions of the fluid with the boundary. Equation $(1)_5$ states that the tangential components of the velocity are related to the $(\mathbb{T}\boldsymbol{n})_\tau$. If this relation is linear, which means that

$$-\boldsymbol{s} = \gamma_* \boldsymbol{v}_\tau \qquad \text{with } \gamma_* > 0, \tag{3}$$

we obtain Navier's slip boundary condition that can be viewed also as the bridge between no-slip (letting $\gamma_* \to +\infty$) and (perfect) slip (letting $\gamma_* \to 0$).

The last equation $(1)_6$ is the requirement on the initial datum for the velocity: given $\boldsymbol{v}_0$ should be divergence-free and have vanishing normal component on $\partial\Omega$, i.e., $\boldsymbol{v}_0$ should satisfy $(1)_1$ and $(1)_5$.

## 2.2 Assumptions on Constitutive Functions $\mathbb{G}$ and $\mathbf{g}$

Next, we shall specify what kind of implicit relations $(1)_3$ and $(1)_5$ we consider in this study. Concerning $(1)_3$ we make the following identification:

$$(\mathbb{S}, \mathbb{D}) \in \mathcal{A} \qquad \Longleftrightarrow \qquad \mathbb{G}(\mathbb{S}, \mathbb{D}) = \mathbb{O}, \tag{4}$$

[2]The reason why only $\mathbb{S}$ appears in the constitutive $(1)_3$ is due to the fact that the fluid is incompressible.

and we put the following assumptions on $\mathcal{A} : Q_T \to \mathbb{R}^{d\times d}_{\text{sym}} \times \mathbb{R}^{d\times d}_{\text{sym}}$:

**(A1)** *$\mathcal{A}$ contains the origin:* $(\mathbb{O}, \mathbb{O}) \in \mathcal{A}$.

**(A2)** *$\mathcal{A}$ is a monotone graph:* For any $(\mathbb{S}_1, \mathbb{D}_1), (\mathbb{S}_2, \mathbb{D}_2) \in \mathcal{A}$

$$(\mathbb{S}_1 - \mathbb{S}_2) \cdot (\mathbb{D}_1 - \mathbb{D}_2) \geq 0.$$

**(A3)** *$\mathcal{A}$ is a maximal monotone graph.* Let $(\mathbb{S}, \mathbb{D}) \in \mathbb{R}^{d\times d}_{\text{sym}} \times \mathbb{R}^{d\times d}_{\text{sym}}$.

$$\text{If } (\mathbb{S} - \tilde{\mathbb{S}}) \cdot (\mathbb{D} - \tilde{\mathbb{D}}) \geq 0 \text{ for all } (\tilde{\mathbb{S}}, \tilde{\mathbb{D}}) \in \mathcal{A}, \text{ then } (\mathbb{S}, \mathbb{D}) \in \mathcal{A}.$$

**(A4)** *$\mathcal{A}$ is a 2 graph:* There are $\alpha_* \in (0, 1]$ and $c_* \geq 0$ so that

$$\mathbb{S} \cdot \mathbb{D} \geq \alpha_* \left(|\mathbb{D}|^2 + |\mathbb{S}|^2\right) - c_* \qquad \text{for any} \quad (\mathbb{S}, \mathbb{D}) \in \mathcal{A}.$$

The last assumption can be relaxed and referring to the setting used in [7] and [6] one can replace **(A4)** by the condition

**(A4*)** $\mathcal{A}$ **is a** $\psi$ **graph:** There are $\alpha_* \in (0, 1]$ and $c_* \geq 0$ so that

$$\mathbb{S} \cdot \mathbb{D} \geq \alpha_* \left(\psi(|\mathbb{D}|) + \psi^*(|\mathbb{S}|)\right) - c_* \qquad \text{for any} \quad (\mathbb{S}, \mathbb{D}) \in \mathcal{A},$$

where $\psi$ is an Orlicz function (convex, even, continuous) and $\psi^*$ is its dual (conjugate) function.

As mentioned in the Introduction, we restrict ourselves in this study to two cases to which the assumptions **(A1)**–**(A4)** are appropriate. First, we consider

$$\mathbb{D} = \frac{1}{2\nu_*} \frac{(|\mathbb{S}| - \tau_*)^+}{|\mathbb{S}|} \mathbb{S} \quad \text{with } \tau_* \geq 0, \nu_* > 0, \tag{5}$$

where $x^+ = \max\{x, 0\}$. Setting $\tau_* = 0$ in (5) we obtain the Navier-Stokes fluid model (2). If $\tau_* > 0$ the above formula is equivalent to

$$|\mathbb{S}| \leq \tau_* \Leftrightarrow \mathbb{D} = \mathbb{O} \quad \text{and} \quad |\mathbb{S}| > \tau_* \Leftrightarrow \mathbb{S} = \frac{\tau_* \mathbb{D}}{|\mathbb{D}|} + 2\nu(|\mathbb{D}|^2)\mathbb{D}, \tag{6}$$

which is the form in which the response of Bingham fluids is usually described, see for example [12]. Note that (5) fulfills the assumptions **(A1)**–**(A4)** above (see Lemma 2.1 in [7] for details).

We are also interested in analyzing the model

$$\mathbb{D} = \frac{1}{2\mu(t, x)} \mathbb{S} \quad \text{with } 0 < \mu_* \leq \mu(t, x) \leq \mu^* \text{ in } Q_T, \tag{7}$$

that can be viewed as a simplified model for incompressible fluids where the viscosity $\nu$ changes with another quantity $z$ (such as temperature, concentration, electric or magnetic field) and this dependence is somehow known a priori; then $\mu(t,x) = \nu(z(t,x))$. It is easy to check that the linear constitutive equation (7) fulfills the assumptions **(A1)**–**(A4)**.

Referring to $(1)_5$, we introduce the graph $\mathcal{B}$ similarly as above through the following identification:

$$(\boldsymbol{s}, \boldsymbol{v}_\tau) \in \mathcal{B} \qquad \Longleftrightarrow \qquad \boldsymbol{g}(\boldsymbol{s}, \boldsymbol{v}_\tau) = \boldsymbol{0}, \tag{8}$$

and we require that

**(B1)** *$\mathcal{B}$ contains the origin.* $(\boldsymbol{0}, \boldsymbol{0}) \in \mathcal{B}$.
**(B2)** *$\mathcal{B}$ is a monotone graph.*

$$-(\boldsymbol{s}_1 - \boldsymbol{s}_2) \cdot (\boldsymbol{v}_\tau^1 - \boldsymbol{v}_\tau^2) \geq 0 \qquad \text{for all } (\boldsymbol{s}_1, \boldsymbol{v}_\tau^1), (\boldsymbol{s}_2, \boldsymbol{v}_\tau^2) \in \mathcal{B}.$$

**(B3)** *$\mathcal{B}$ is a maximal monotone graph.* Let $(\boldsymbol{s}, \boldsymbol{v}) \in \mathbb{R}^d \times \mathbb{R}^d$.

$$\text{If} \quad -(\bar{\boldsymbol{s}} - \boldsymbol{s}) \cdot (\bar{\boldsymbol{v}}_\tau - \boldsymbol{u}) \geq 0 \quad \text{for all } (\bar{\boldsymbol{s}}, \bar{\boldsymbol{v}}_\tau) \in \mathcal{B} \quad \text{then } (\boldsymbol{s}, \boldsymbol{v}) \in \mathcal{B}.$$

**(B4)** *$\mathcal{B}$ is a 2 graph.* There are $\beta_* > 0$ and $d_* \geq 0$ such that

$$-\boldsymbol{s} \cdot \boldsymbol{v}_\tau \geq \beta_*(|\boldsymbol{v}_\tau|^2 + |\boldsymbol{s}|^2) - d_* \qquad \text{for all } (\boldsymbol{s}, \boldsymbol{v}_\tau) \in \mathcal{B}.$$

Our main interest is to analyze the stick-slip (threshold slip) boundary conditions that are usually written in the form

$$\begin{aligned} |(\mathbb{S}\boldsymbol{n})_\tau| \leq \sigma_* &\Leftrightarrow \boldsymbol{v}_\tau = \boldsymbol{0}, \\ |(\mathbb{S}\boldsymbol{n})_\tau| > \sigma_* &\Leftrightarrow -(\mathbb{S}\boldsymbol{n})_\tau = \sigma_* \frac{\boldsymbol{v}_\tau}{|\boldsymbol{v}_\tau|} + \gamma_* \boldsymbol{v}_\tau, \end{aligned} \tag{9}$$

where $\sigma_* \geq 0$ and $\gamma_* \geq 0$.

Recalling the notation $\boldsymbol{s} := (\mathbb{S}\boldsymbol{n})_\tau$ we re-write (9) into the equivalent implicit equation $(1)_5$ that takes the form:

$$\gamma_* \boldsymbol{v}_\tau = -\frac{(|\boldsymbol{s}| - \sigma_*)^+}{|\boldsymbol{s}|}\boldsymbol{s} \qquad \text{with } \sigma_* \geq 0, \gamma_* \geq 0. \tag{10}$$

If $\sigma_* = 0$ and $\gamma_* > 0$, (10) represents Navier's slip boundary condition $\gamma_* \boldsymbol{v}_\tau + \boldsymbol{s} = \boldsymbol{0}$. If $\sigma_* = \gamma_* = 0$ then we end up with (perfect) slip. As above, Lemma 2.1 from [7] can be again used as the reference for verifying that (10) meets all the assumptions **(B1)**–**(B4)**. It is important to note that the no-slip boundary condition $\boldsymbol{v}_\tau = \boldsymbol{0}$ is excluded by **(B4)**.

## 2.3 A Priori Estimates and Function Spaces

Multiplying $(1)_2$ by $\boldsymbol{v}$, we obtain

$$\tfrac{1}{2}\tfrac{\partial|\boldsymbol{v}|^2}{\partial t} + \operatorname{div}(\tfrac{1}{2}|\boldsymbol{v}|^2\boldsymbol{v}) - \operatorname{div}(\mathbb{S}\boldsymbol{v}) + \mathbb{S}\cdot\mathbb{D} = -\operatorname{div}(p\boldsymbol{v}) + \boldsymbol{b}\cdot\boldsymbol{v}\,. \tag{11}$$

Since $\boldsymbol{v}\cdot\boldsymbol{n} = 0$, integrating (11) over $\Omega$ leads to

$$\frac{1}{2}\frac{d}{dt}\|\boldsymbol{v}\|_2^2 + \int_\Omega \mathbb{S}\cdot\mathbb{D}\,dx - \int_{\partial\Omega}(\mathbb{S}\boldsymbol{n})_{\boldsymbol{\tau}}\cdot\boldsymbol{v}_{\boldsymbol{\tau}}\,dS = \int_\Omega \boldsymbol{b}\cdot\boldsymbol{v}\,. \tag{12}$$

The simplest relations, namely (2) and (3), or the assumptions **(A4)** and **(B4)** then imply

$$\begin{aligned}\frac{1}{2}\frac{d}{dt}\|\boldsymbol{v}\|_2^2 &+ \alpha_*\left(\|\mathbb{S}\|_2^2 + \|\mathbb{D}\|_2^2\right) + \beta_*\left(\|\boldsymbol{s}\|_{2,\partial\Omega}^2 + \|\boldsymbol{v}_{\boldsymbol{\tau}}\|_{2,\partial\Omega}^2\right)\\ &\le c_*\int_\Omega 1 + d_*\int_{\partial\Omega} 1 + \int_\Omega \boldsymbol{b}\cdot\boldsymbol{v}\,.\end{aligned} \tag{13}$$

If $\boldsymbol{b}$ is given appropriately (for example $\boldsymbol{b} = \boldsymbol{0}$), the last inequality indicates clearly in what function spaces we should look for the solution.

We introduce briefly the function spaces needed below. For $q \in [1,\infty]$, $(L^q(\Omega), \|\cdot\|_q)$ and $(W^{1,q}(\Omega), \|\cdot\|_{1,q})$ denote the Lebesgue and Sobolev spaces. If $X$ is a Banach space of scalar functions then $X^d$ is the space of vector-valued functions having $d$ components, each of them belonging to $X$. Similarly, $X^{d\times d}_{\text{sym}}$ is the space of tensor-valued symmetric functions. Next, we set, for any $q \in [1,\infty)$,

$$L^q_{\boldsymbol{n},\text{div}} := \overline{\{\boldsymbol{v}\in\mathcal{D}(\Omega)^d;\ \operatorname{div}\boldsymbol{v} = 0\}}^{\|\cdot\|_q}.$$

The subspaces (and their duals) of vector-valued Sobolev functions from $W^{1,q}(\Omega)^d$ which have zero normal component on the boundary are defined in the following way:

$$\begin{aligned}W^{1,q}_{\boldsymbol{n}} &:= \{\boldsymbol{v}\in W^{1,q}(\Omega)^d;\ \boldsymbol{v}\cdot\boldsymbol{n} = 0 \text{ on } \partial\Omega\},\\ W^{1,q}_{\boldsymbol{n},\text{div}} &:= \{\boldsymbol{v}\in W^{1,q}_{\boldsymbol{n}};\ \operatorname{div}\boldsymbol{v} = 0 \text{ in } \Omega\},\\ W^{-1,q}_{\boldsymbol{n}} &:= \left(W^{1,q}_{\boldsymbol{n}}\right)^*, \qquad W^{-1,q}_{\boldsymbol{n},\text{div}} := \left(W^{1,q}_{\boldsymbol{n},\text{div}}\right)^*.\end{aligned}$$

By the Helmholtz decomposition we observe that

$$W^{1,q}_{\boldsymbol{n}} = W^{1,q}_{\boldsymbol{n},\text{div}} \oplus \{\nabla\varphi;\ \varphi\in W^{2,q}(\Omega),\ \nabla\varphi\cdot\boldsymbol{n} = 0 \text{ on } \partial\Omega\}.$$

Note that this does not hold for $(W^{1,2}_0(\Omega))^d$.

## 2.4 Main Result

**Definition 2.1** Let

$$v_0 \in L^2_{n,\mathrm{div}}, \quad b \in L^2(0,T;W^{-1,2}_n)\,. \tag{14}$$

We say $(p, v, \mathbb{S}, s)$ is weak solution to the problem (1) if

$$\begin{aligned}
&p \in L^1(Q_T),\\
&v \in C_{\mathrm{weak}}(0,T;L^2_{n,\mathrm{div}}) \cap L^2(0,T;W^{1,2}_{\mathbf{n},\mathrm{div}}),\\
&\mathbb{S} \in L^2(Q_T)^{d\times d},\\
&s \in L^2(\Sigma_T)^d,\\
&\lim_{t\to 0_+} \|v(t) - v_0\|_2^2 = 0,\\
&\langle v_{,t}, w\rangle + (\mathbb{S}, \mathbb{D}w) - (v \otimes v, \mathbb{D}w) + (s, w_\tau)_{\partial\Omega} = \langle b, w\rangle, + (p, \mathrm{div}\, w)\,,\\
&\qquad \text{for all } w \in W^{1,\infty}_n \text{ and a.a. } t \in (0,T),\\
&(\mathbb{S}(t,x), \mathbb{D}v(t,x)) \in \mathcal{A} \text{ for a.a. } (t,x) \in Q_T,\\
&(s(t,x), v_\tau(t,x)) \in \mathcal{B} \text{ for a.a. } (t,x) \in \Sigma_T.
\end{aligned}$$

We formulate first the following theorem for general $\mathbb{G}$ and $g$ fulfilling **(A1)–(A4)** and **(B1)–(B4)**.

**Theorem 2.2** *Let $\mathcal{A}$ satisfy the assumptions* **(A1)–(A4)** *and let $\mathcal{B}$ satisfy the assumptions* **(B1)–(B4)**. *Then, for any $\Omega \in \mathcal{C}^{1,1}$, $T \in (0,\infty)$ and for arbitrary $v_0$ and $b$ fulfilling* (14), *there exists a weak solution to the problem* (1).

In order to be completely explicit in the construction of the solution in this study we restrict ourselves to equations (5) (or (7)) and (10). We establish the following result.

**Theorem 2.3** *For any $\Omega \in \mathcal{C}^{1,1}$ and $T \in (0,\infty)$ and for arbitrary $v_0$ and $b$ fulfilling* (14) *there exist*

$$\begin{aligned}
&v \in L^\infty(0,T;L^2(\Omega)^3) \cap L^2(0,T;W^{1,2}_{n,\mathrm{div}}),\\
&\mathbb{S} \in L^2(Q_T)^{d\times d}_{sym}, \quad s \in L^2(0,T;L^2(\partial\Omega)^3),\\
&p_1 \in L^2(Q_T), \quad p_2 \in L^{\frac{d+2}{d+1}}(0,T;W^{1,\frac{d+2}{d+1}}(\Omega)),
\end{aligned}$$

*solving, for almost all time $t \in (0,T)$ and for all $w \in W^{1,2}_n$,*

$$\langle v_{,t}, w\rangle - \int_\Omega (v \otimes v)\cdot \nabla w + \int_\Omega \mathbb{S}\cdot \mathbb{D}w + \int_{\partial\Omega} s\cdot w = \int_\Omega (p_1 + p_2)\,\mathrm{div}\, w.$$

*Furthermore, $\mathbb{S}$ and $\mathbb{D}$ fulfill the equation* (5) *(or* (7)*) a.e. in $Q_T$, and $s$ and $v_\tau$ fulfill* (10) *a.e. in $\Sigma_T$.*

In Sect. 4 we prove Theorem 2.3. The proof of Theorem 2.2 follows then from the main theorem proved in [7] and the proof of Theorem 2.3.

## 3 Convergence Lemma

In this section, we establish a convergence lemma that we use to verify the validity of the nonlinear constitutive equations, such as (5) or (10), or more generally $(1)_3$ or $(1)_5$. The fact that the result is stated locally is useful in applications. Although the lemma is proved in [6], for the sake of completeness, we give a simplified proof here.

**Lemma 3.1** *Let $U$ be an arbitrary measurable set and let a graph $\mathcal{A}$ fulfill the assumptions* **(A2)–(A3)**. *Assume that, for some $r \in (1, \infty)$,*

$$(\mathbb{S}^n, \mathbb{D}^n) \in \mathcal{A} \qquad \textit{almost everywhere in } U, \tag{15}$$

$$\mathbb{D}^n \rightharpoonup \mathbb{D} \qquad \textit{weakly in } L^r(U)^{d\times d}, \tag{16}$$

$$\mathbb{S}^n \rightharpoonup \mathbb{S} \qquad \textit{weakly in } L^{r'}(U)^{d\times d}, \tag{17}$$

$$\limsup_{n\to\infty} \int_U \mathbb{S}^n \cdot \mathbb{D}^n \le \int_U \mathbb{S} \cdot \mathbb{D}. \tag{18}$$

*Then*

$$(\mathbb{S}, \mathbb{D}) \in \mathcal{A} \textit{ almost everywhere in } U.$$

*Proof* We start with observing that

$$\mathbb{S}^n \cdot \mathbb{D}^n \rightharpoonup \mathbb{S} \cdot \mathbb{D} \quad \text{weakly in } L^1(U). \tag{19}$$

To prove this, we first notice that, for arbitrary $n, m \in \mathbb{N}$ and for $(\mathbb{S}^n, \mathbb{D}^n), (\mathbb{S}^m, \mathbb{D}^m) \in \mathcal{A}$, **(A2)** implies that

$$0 \le (\mathbb{S}^n - \mathbb{S}^m) \cdot (\mathbb{D}^n - \mathbb{D}^m) \qquad \text{a.e. in } U.$$

Then, it follows from (16)–(18) that

$$\begin{aligned}
&\limsup_{n\to\infty} \limsup_{m\to\infty} \|(\mathbb{S}^n - \mathbb{S}^m) \cdot (\mathbb{D}^n - \mathbb{D}^m)\|_1 \\
&= \limsup_{n\to\infty} \limsup_{m\to\infty} \left[\int_U \mathbb{S}^n \cdot \mathbb{D}^n - \int_U \mathbb{S}^m \cdot \mathbb{D}^n - \int_U \mathbb{S}^n \cdot \mathbb{D}^m + \int_U \mathbb{S}^m \cdot \mathbb{D}^m\right] \\
&\le \limsup_{n\to\infty} \left[\int_U \mathbb{S}^n \cdot \mathbb{D}^n - \int_U \mathbb{S} \cdot \mathbb{D}^n - \int_U \mathbb{S}^n \cdot \mathbb{D} + \int_U \mathbb{S} \cdot \mathbb{D}\right] \\
&\le 0.
\end{aligned}$$

Since

$$\lim_{n\to\infty}\lim_{m\to\infty}\|(\mathbb{S}^n-\mathbb{S}^m)\cdot(\mathbb{D}^n-\mathbb{D}^m)\|_1=0,$$

we observe that

$$\lim_{n\to\infty}\lim_{m\to\infty}\int_U(\mathbb{S}^n-\mathbb{S}^m)\cdot(\mathbb{D}^n-\mathbb{D}^m)\,\varphi=0\qquad\text{for all }\varphi\in L^\infty(\Omega).\tag{20}$$

Setting

$$L:=\lim_{\ell\to\infty}\int_U(\mathbb{S}^\ell\cdot\mathbb{D}^\ell)\varphi,$$

we conclude from (20) that

$$\begin{aligned}0&=\lim_{n\to\infty}\lim_{m\to\infty}\left[\int_U(\mathbb{S}^n\cdot\mathbb{D}^n)\,\varphi-\int_U(\mathbb{S}^n\cdot\mathbb{D}^m)\,\varphi-\int_U(\mathbb{S}^m\cdot\mathbb{D}^n)\,\varphi+\int_U(\mathbb{S}^m\cdot\mathbb{D}^m)\,\varphi\right]\\&=2\left(L-\int_U(\mathbb{S}\cdot\mathbb{D})\,\varphi\right),\end{aligned}$$

which is (19).

Next, we take an arbitrary $(\mathbb{S}^*,\mathbb{D}^*)\in\mathcal{A}$ and arbitrary nonnegative $\varphi\in L^\infty(U)$. It then follows from **(A2)**, (16), (17) and (19) that

$$0\le\lim_{n\to\infty}\int_U(\mathbb{S}^n-\mathbb{S}^*)\cdot(\mathbb{D}^n-\mathbb{D}^*)\varphi=\int_U(\mathbb{S}-\mathbb{S}^*)\cdot(\mathbb{D}-\mathbb{D}^*)\varphi.$$

Since $\varphi\ge0$ is arbitrary we get

$$0\le(\mathbb{S}-\mathbb{S}^*)\cdot(\mathbb{D}-\mathbb{D}^*)\qquad\text{almost everywhere in }U.$$

Since $(\mathbb{S}^*,\mathbb{D}^*)\in\mathcal{A}$ is arbitrary, the property **(A3)** implies that

$$(\mathbb{S},\mathbb{D})\in\mathcal{A}\qquad\text{almost everywhere in }U.$$

□

While the first three assumptions (15)–(17) of the lemma are usually easily available, the key assumption (18) suggests that one needs some kind of energy equalities in order to establish its validity. In applications, when one needs to identify a constitutive equations of the type $(1)_3$ in $Q_T$, the verification of (18) is usually straightforward if $\boldsymbol{v}$, the solutions itself, is an admissible test function in the weak formulation of the problem. If $\boldsymbol{v}$ is not an admissible test function, then one uses some kind of truncation instead of $\boldsymbol{v}$. Here, in Sect. 4, we apply an $L^\infty$-*truncation method* as this is sufficient to analyze the problem (1) with the constitutive relations (5) and (10). More powerful methods are those based on *Lipschitz-truncation* of Bochner functions having values in Sobolev or Orlicz-Sobolev spaces, we refer to [10] and [7] both for details and the references to earlier studies.

Concerning the constitutive equations on the boundary, such as $(1)_5$ or (10), one needs to check that sequences $(\boldsymbol{s}^n, \boldsymbol{v}^n) \in \mathcal{B}$ satisfying, for some $q > 1$,

$$\boldsymbol{s}^n \rightharpoonup \boldsymbol{s} \qquad \text{weakly in } L^{q'}(\Sigma_T)^d,$$

$$\boldsymbol{v}^n \rightharpoonup \boldsymbol{v} \qquad \text{weakly in } L^{q}(\Sigma_T)^d,$$

fulfill also the condition

$$\limsup_{n\to\infty} \int_{\Sigma_T} \boldsymbol{s}^n \cdot \boldsymbol{v}^n \le \int_{\Sigma_T} \boldsymbol{s} \cdot \boldsymbol{v}. \tag{21}$$

Here, however, we do not need to refer to any kind of energy equality as we usually have

$$\boldsymbol{v}^n \to \boldsymbol{v} \qquad \text{strongly in } L^1(\Sigma_T).$$

Then, by Egorov theorem, for any $\varepsilon > 0$ there exists $U_\varepsilon$ such that $|\Sigma_T \setminus U_\varepsilon| \le \varepsilon$ and

$$\boldsymbol{v}^n \to \boldsymbol{v} \qquad \text{strongly in } L^\infty(U_\varepsilon),$$

which implies (21). Consequently, by Lemma 3.1 $(\boldsymbol{s}, \boldsymbol{v}) \in \mathcal{B}$ a.e. in $U_\varepsilon$. However, $\varepsilon$ is arbitrary and thus $(\boldsymbol{s}, \boldsymbol{v}) \in \mathcal{B}$ a.e. in $\Sigma_T$.

This argument suggests that the maximal monotone graph setting outlined by the assumptions **(B1)**–**(B4)** is not necessary as $\boldsymbol{v}^n$ is compact on $\Sigma_T$; we will include details in a forthcoming study.

# 4 Proof of the Main Result

We set for simplicity $\varrho_* = 2\nu_* = \gamma_* = 1$ in this section. The proof holds true for any $\varrho_* > 0$, $\nu_* > 0$ and $\gamma_* > 0$ without any essential change.

## *4.1 Approximations and Their Existence*

Let $n \in \mathbb{N}$. We first set

$$\mathbb{G}^n(\mathbb{S}, \mathbb{D}) := \mathbb{D} - \left( \frac{(|\mathbb{S}| - \tau_*)_+}{|\mathbb{S}|} + \frac{1}{n} \right) \mathbb{S}, \tag{22}$$

$$g^n(\boldsymbol{s},\boldsymbol{v}) := \boldsymbol{v} + \left(\frac{(|\boldsymbol{s}|-\sigma_*)_+}{|\boldsymbol{s}|} + \frac{1}{n}\right)\boldsymbol{s}, \tag{23}$$

and notice that the constitutive equations

$$\mathbb{G}^n(\mathbb{S},\mathbb{D}) = \mathbb{O} \quad \Longleftrightarrow \quad \mathbb{D} = \left(\frac{(|\mathbb{S}|-\tau_*)_+}{|\mathbb{S}|} + \frac{1}{n}\right)\mathbb{S}, \tag{24}$$

$$g^n(\boldsymbol{s},\boldsymbol{v}_\tau) = \boldsymbol{0} \quad \Longleftrightarrow \quad \boldsymbol{0} = \boldsymbol{v}_\tau + \left(\frac{(|\boldsymbol{s}|-\sigma_*)_+}{|\boldsymbol{s}|} + \frac{1}{n}\right)\boldsymbol{s} \tag{25}$$

imply

$$\mathbb{S} = \tilde{\mathbb{S}}^n(\mathbb{D}) := \begin{cases} n\mathbb{D} & \text{if} \quad |\mathbb{S}| \le \tau_* \\ \frac{n}{n+1}\left(\tau_* \frac{\mathbb{D}}{|\mathbb{D}|} + \mathbb{D}\right) & \text{if} \quad |\mathbb{S}| > \tau_* \end{cases},$$

and

$$-\boldsymbol{s} = \tilde{\boldsymbol{s}}^n(\boldsymbol{v}_\tau) := \begin{cases} n\boldsymbol{v}_\tau & \text{if} \quad |\boldsymbol{s}| \le \sigma_* \\ \frac{n}{n+1}\left(\sigma_* \frac{\boldsymbol{v}_\tau}{|\boldsymbol{v}_\tau|} + \boldsymbol{v}_\tau\right) & \text{if} \quad |\boldsymbol{s}| > \sigma_* \end{cases},$$

where $\tilde{\mathbb{S}}_n$ and $\tilde{\boldsymbol{s}}_n$ are continuous monotone functions with linear growth (at infinity).

Next, we consider smooth functions $G_n : \mathbb{R} \to \mathbb{R}$ with $|G_n'| \le \frac{2}{n}$ such that

$$G_n(s) := 1 \text{ for } s \le n, \quad G_n(s) = 0 \text{ for } s > 2n.$$

Then we introduce the approximate problem relevant to (1): for each $n \in \mathbb{N}$, find $(\boldsymbol{v}^n, p^n)$ satisfying

$$\begin{aligned} \operatorname{div} \boldsymbol{v}^n &= 0 && \text{in } Q_T, \\ \boldsymbol{v}^n_{,t} + \operatorname{div}((\boldsymbol{v}^n \otimes \boldsymbol{v}^n)G_n(|\boldsymbol{v}^n|)) - \operatorname{div} \tilde{\mathbb{S}}^n(\mathbb{D}\boldsymbol{v}^n) + \nabla p^n &= \boldsymbol{b} && \text{in } Q_T, \\ \boldsymbol{v}^n \cdot \boldsymbol{n} &= 0 && \text{on } \Sigma_T, \\ \boldsymbol{s} + \tilde{\boldsymbol{s}}^n(\boldsymbol{v}^n_\tau) &= \boldsymbol{0} && \text{on } \Sigma_T, \\ \boldsymbol{v}^n(0,\cdot) &= \boldsymbol{v}_0 && \text{in } \Omega. \end{aligned} \tag{26}$$

In what follows, we set

$$\mathbb{S}^n := \tilde{\mathbb{S}}^n(\mathbb{D}\boldsymbol{v}^n) \quad \text{and} \quad \boldsymbol{s}^n := \tilde{\boldsymbol{s}}^n(\boldsymbol{v}^n_\tau).$$

Due to the presence of $G_n$ that truncates the convective term and due to the properties of $\tilde{\mathbb{S}}^n$ and $\tilde{\boldsymbol{s}}^n$ that generate monotone operators, the existence of weak solution $\boldsymbol{v}^n \in$

$C([0,T];L^2_{\boldsymbol{n},\mathrm{div}}) \cap L^2(0,T;W^{1,2}_{\boldsymbol{n},\mathrm{div}})$ to the problem (26) that satisfies

$$\begin{aligned}\langle \boldsymbol{v}^n_{,t}, \tilde{\boldsymbol{w}}\rangle + (\mathbb{S}^n, \mathbb{D}\tilde{\boldsymbol{w}}) + (\mathrm{div}(\boldsymbol{v}^n \otimes \boldsymbol{v}^n)G_n(|\boldsymbol{v}^n|), \tilde{\boldsymbol{w}}) + (\boldsymbol{s}^n, \tilde{\boldsymbol{w}}_{\boldsymbol{\tau}})_{\partial\Omega} \\ - \langle \boldsymbol{b}, \tilde{\boldsymbol{w}}\rangle = 0 \qquad \text{for all } \tilde{\boldsymbol{w}} \in W^{1,2}_{\boldsymbol{n},\mathrm{div}} \text{ and a.a. } t \in (0,T),\end{aligned} \tag{27}$$

follows from standard monotone operator theory. For fixed $n \in \mathbb{N}$ a construction of the solution can be done via the Galerkin approximations taking as the basis in $W^{1,2}_{\boldsymbol{n},\mathrm{div}} \cap W^{d,2}(\Omega)^d$ the eigenfunctions of the following eigenvalue problem

$$\int_\Omega \nabla^{(d)}\boldsymbol{v} \cdot \nabla^{(d)}\boldsymbol{w} + \nabla\boldsymbol{v} \cdot \nabla\boldsymbol{w} = \lambda \int_\Omega \boldsymbol{v} \cdot \boldsymbol{w} \qquad \text{for all } \boldsymbol{w} \in W^{1,2}_{\boldsymbol{n},\mathrm{div}} \cap W^{d,2}(\Omega)^d.$$

We do not go into more details referring to [19, 25] or [1].

Having $\boldsymbol{v}^n$ we introduce $p^n$ as the solution of the following problem

$$\begin{aligned}-(\nabla p^n, \nabla z) &= (\mathbb{S}^n, \nabla^{(2)} z) + (\mathrm{div}(\boldsymbol{v}^n \otimes \boldsymbol{v}^n)G_n(|\boldsymbol{v}^n|), \nabla z) \\ &\quad + (\boldsymbol{s}^n, (\nabla z)_{\boldsymbol{\tau}})_{\partial\Omega} - \langle \boldsymbol{b}, \nabla z\rangle \\ &\quad \text{for all } z \in W^{2,2}(\Omega) \text{ with } \nabla z \cdot \mathbf{n} = 0 \text{ on } \partial\Omega \text{ and a.a. } t \in (0,T).\end{aligned} \tag{28}$$

In short, it means that $p^n := (-\Delta_N)^{-1} \operatorname{div}(\operatorname{div} \mathbb{S}^n + \operatorname{div}(\boldsymbol{v}^n \otimes \boldsymbol{v}^n)G_n(|\boldsymbol{v}^n|) - \boldsymbol{b})$, where $-\Delta_N$ denotes the Laplace operator together with homogeneous Neumann boundary conditions. We consider $\int_\Omega p^n = 0$ for a.a. $t \in (0,T)$.

Applying the Helmholtz decomposition to an arbitrary $\boldsymbol{w} \in W^{1,2}_{\boldsymbol{n}}$, i.e.,

$$\boldsymbol{w} = \tilde{\boldsymbol{w}} + \nabla z \text{ with } z \in W^{2,2}(\Omega), \operatorname{div}\tilde{\boldsymbol{w}} = 0, \tilde{\boldsymbol{w}} \cdot \boldsymbol{n} = \nabla z \cdot \boldsymbol{n} = 0 \text{ on } \partial\Omega, \tag{29}$$

we can write, using (27) and (28),

$$\begin{aligned}&(\mathbb{S}^n, \mathbb{D}\boldsymbol{w})) + (\mathrm{div}(\boldsymbol{v}^n \otimes \boldsymbol{v}^n)G_n(|\boldsymbol{v}^n|), \boldsymbol{w}) + (\boldsymbol{s}^n, \boldsymbol{w}_{\boldsymbol{\tau}})_{\partial\Omega} - \langle \boldsymbol{b}, \boldsymbol{w}\rangle \\ &\quad = (\mathbb{S}^n, \mathbb{D}\tilde{\boldsymbol{w}})) + (\mathrm{div}(\boldsymbol{v}^n \otimes \boldsymbol{v}^n)G_n(|\boldsymbol{v}^n|), \tilde{\boldsymbol{w}}) + (\boldsymbol{s}^n, \tilde{\boldsymbol{w}}_{\boldsymbol{\tau}})_{\partial\Omega} - \langle \boldsymbol{b}, \tilde{\boldsymbol{w}}\rangle \\ &\qquad + (\mathbb{S}^n, \nabla^{(2)} z) + (\mathrm{div}(\boldsymbol{v}^n \otimes \boldsymbol{v}^n)G_n(|\boldsymbol{v}^n|), \nabla z) + (\boldsymbol{s}^n, (\nabla z)_{\boldsymbol{\tau}})_{\partial\Omega} - \langle \boldsymbol{b}, \nabla z\rangle \\ &\quad = -\langle \boldsymbol{v}^n_{,t}, \tilde{\boldsymbol{w}}\rangle - (\nabla p, \nabla z) \\ &\quad = -\langle \boldsymbol{v}^n_{,t}, \tilde{\boldsymbol{w}} + \nabla z\rangle - (\nabla p, \tilde{\boldsymbol{w}} + \nabla z),\end{aligned}$$

which finally leads to the following weak formulation of the problem (26):

$$\begin{aligned}\langle \boldsymbol{v}^n_{,t}, \boldsymbol{w}\rangle + (\mathbb{S}^n, \mathbb{D}\boldsymbol{w}) + (\mathrm{div}(\boldsymbol{v}^n \otimes \boldsymbol{v}^n)G_n(|\boldsymbol{v}^n|), \boldsymbol{w}) + (\boldsymbol{s}^n, \boldsymbol{w}_{\boldsymbol{\tau}})_{\partial\Omega} \\ = (p^n, \operatorname{div}\boldsymbol{w}) + \langle \boldsymbol{b}, \boldsymbol{w}\rangle \text{ for all } \boldsymbol{w} \in W^{1,2}_{\boldsymbol{n}} \text{ and a.a. } t \in (0,T).\end{aligned} \tag{30}$$

## 4.2 A Priori Estimates

Taking $\boldsymbol{v}^n$ as a test function in the weak formulation of $(26)_2$ and noticing that the convective term vanishes we obtain

$$\frac{1}{2}\frac{d}{dt}\|\boldsymbol{v}^n\|_2^2 + \int_\Omega \tilde{\mathbb{S}}^n(\mathbb{D}\boldsymbol{v}^n)\cdot\mathbb{D}\boldsymbol{v}^n + \int_{\partial\Omega} \tilde{\boldsymbol{s}}^n(\boldsymbol{v}^n_{\boldsymbol{\tau}})\cdot\boldsymbol{v}^n_{\boldsymbol{\tau}} = \langle \boldsymbol{b}, \boldsymbol{v}^n\rangle 0,$$

which imply that

$$\begin{aligned}\sup_{t\in(0,T)} \|\boldsymbol{v}^n(t)\|_2^2 + \int_{Q_T} |\mathbb{S}^n|^2 + |\nabla\boldsymbol{v}^n|^2 + |\boldsymbol{v}^n|^{\frac{2(d+2)}{d}} + \int_{\Sigma_T} |\boldsymbol{s}^n|^2 + |\boldsymbol{v}^n_{\boldsymbol{\tau}}|^2 \\ \le C(\|\boldsymbol{v}_0\|_2, \|\boldsymbol{b}\|_{L^2(0,T;(W^{1,2}_{\boldsymbol{n},\mathrm{div}})^*)}) =: C_*,\end{aligned} \tag{31}$$

where $C_* > 0$ is a generic constant. In (31), we also incorporated the information coming from interpolating[3] $\boldsymbol{v}^n \in L^\infty(0,T;L^2(\Omega)^d)$ and $\boldsymbol{v}^n \in L^2(0,T;W^{1,2}(\Omega)^d)$.

Further, we introduce $p_2^n$ with zero mean value as a weak solution of the following problem (stated at each time level):

$$\int_\Omega \nabla p_2^n \cdot \nabla\varphi = -\int_\Omega \operatorname{div}((\boldsymbol{v}^n \otimes \boldsymbol{v}^n)G_n(|\boldsymbol{v}^n|))\cdot\nabla\varphi \quad \text{for all } \varphi \in W^{1,d+2}(\Omega).$$

Since

$$\int_{Q_T} |\operatorname{div}(\boldsymbol{v}^n \otimes \boldsymbol{v}^n)G_n(|\boldsymbol{v}^n|)|^{\frac{d+2}{d+1}} \le C_*,$$

we observe that

$$\int_0^T \|p_2^n\|_{1,\frac{d+2}{d+1}}^{\frac{d+2}{d+1}} \le C_*. \tag{32}$$

Next, we set $p_1^n := p^n - p_2^n$ and find $\varphi$ with zero mean value solving

$$-\Delta\varphi = p_1^n \text{ in } \Omega \quad \text{and} \quad \nabla\varphi\cdot\boldsymbol{n} = 0 \text{ on } \partial\Omega.$$

Then

$$\int_{Q_T} |\nabla^2\varphi|^2 + \int_{(0,T)\times\partial\Omega} |\nabla\varphi|^2 \le \int_{Q_T} |p_1^n|^2. \tag{33}$$

---

[3]We refer to standard interpolation inequality $\|z\|_{\frac{2(d+2)}{d}} \le \|z\|_2^{\frac{2}{d+2}}\|z\|_{\frac{2d}{d-2}}^{\frac{d}{d+2}}$ valid if $d \ge 3$, and $\|z\|_{\frac{2(d+2)}{d}} \le c\|z\|_2^{\frac{2}{d+2}}\|\nabla z\|_2^{\frac{d}{d+2}}$ that holds even if $d \ge 2$.

Taking then $\nabla\varphi$ for such $\phi$ as a test function in the weak formulation of $(26)_2$ we arrive at

$$\int_{Q_T} |p_1^n|^2 = -\int_{Q_T} \nabla p_1^n \cdot \nabla\varphi = \int_{Q_T} (\nabla p_2^n - \operatorname{div}(\boldsymbol{v}^n \otimes \boldsymbol{v}^n) G_n(|\boldsymbol{v}^n|)) \cdot \nabla\varphi +$$
$$\int_{Q_T} \mathbb{S}^n \cdot \nabla^2\varphi + \int_{(0,T)\times\partial\Omega} \boldsymbol{s}^n \cdot \nabla\varphi$$
$$= \int_{Q_T} \mathbb{S}^n \cdot \nabla^2\varphi + \int_{(0,T)\times\partial\Omega} \boldsymbol{s}^n \cdot \nabla\varphi \leq C_* \left(\int_{Q_T} |p_1^n|^2\right)^{\frac{1}{2}}.$$

Finally, from the weak formulation of $(26)_2$, we directly conclude that

$$\|\boldsymbol{v}^n_{,t}\|_{(L^2(0,T;W^{1,2}_{\boldsymbol{n}} \cap L^{d+2}(\Omega)^d))^*} \leq C_*. \tag{34}$$

### 4.3 Limit $n \to \infty$

The uniform estimates (31)–(34) together with the Aubin-Lions compactness lemma and the (compact) embedding of the Sobolev spaces into the space of traces suffice to obtain the following convergences (for possibly subsequences that we again denote as the original sequences):

$$\begin{aligned}
\boldsymbol{v}^n &\rightharpoonup \boldsymbol{v} && \text{weakly in } L^2(0,T;W^{1,2}_{\boldsymbol{n}}),\\
\mathbb{S}^n &\rightharpoonup \mathbb{S} && \text{weakly in } L^2(Q_T)^{d\times d},\\
\boldsymbol{s}^n &\rightharpoonup \boldsymbol{s} && \text{weakly in } L^2(0,T;L^2(\partial\Omega)^d),\\
\boldsymbol{v}^n &\to \boldsymbol{v} && \text{strongly in } L^2(Q_T)^d,\\
\boldsymbol{v}^n &\to \boldsymbol{v} && \text{strongly in } L^2(0,T;L^2(\partial\Omega)^d),\\
p_1^n &\rightharpoonup p_1 && \text{weakly in } L^2(Q_T),\\
p_2^n &\rightharpoonup p_2 && \text{weakly in } L^{\frac{d+2}{d+1}}(0,T;W^{1,\frac{d+2}{d+1}}(\Omega)),\\
\boldsymbol{v}^n_{,t} &\rightharpoonup \boldsymbol{v}_{,t} && \text{weakly in } (L^2(0,T;W^{1,2}_{\boldsymbol{n}} \cap L^{d+2}(\Omega)^d))^*.
\end{aligned}$$

It is easy to observe that $\boldsymbol{v}, p := p_1 + p_2, \mathbb{S}$ and $\boldsymbol{s}$ fulfill the weak formulation of the balance equation $(1)_2$. It remains to show the validity of the constitutive equations, i.e., to prove that $\boldsymbol{v}$, $\mathbb{S}$ and $\boldsymbol{s}$ fulfill (5) and (10). For the latter, we argue via the Egorov theorem, see the end of Sect. 3. In the remaining part of Sect. 4 we prove the validity of (5).

For this purpose, assume that a sequence $\{k^n\}_{n=1}^{\infty}$ is such that $0 < A \le k^n \le B < \infty$, where $A$, $B$ and $k^n$ will be specified later. Next, we take as a test function in the weak formulation of the $n$-th approximation the function

$$\boldsymbol{w}^n = T_{k^n}(\boldsymbol{v}^n - \boldsymbol{v}) := (\boldsymbol{v}^n - \boldsymbol{v}) \min\left\{1, \frac{k^n}{|\boldsymbol{v}^n - \boldsymbol{v}|}\right\}.$$

Then (see [1] for details)

$$\begin{aligned}
&\limsup_{n\to\infty} \int_{Q_T} \mathbb{S}^n \cdot \mathbb{D}\boldsymbol{w}^n - p_1^n \operatorname{div} \boldsymbol{w}^n \\
&\quad = \limsup_{n\to\infty} \Bigg[ \int_{Q_T} -\langle \boldsymbol{v}^n_{,t}, \boldsymbol{w}^n \rangle - \operatorname{div}(\boldsymbol{v}^n \otimes \boldsymbol{v}^n) G_n(|\boldsymbol{v}^n|) \cdot \boldsymbol{w}^n \qquad (35)\\
&\qquad\qquad \boldsymbol{b} \cdot \boldsymbol{w}^n - \nabla p_2^n \cdot \boldsymbol{w}^n + \int_{\Sigma_T} \boldsymbol{s}^n \cdot \boldsymbol{w}^n \Bigg] \le 0.
\end{aligned}$$

Let $\overline{\mathbb{S}} \in L^2(Q_T)$ be such that

$$\mathbb{D}\boldsymbol{v} = \frac{(|\overline{\mathbb{S}}| - \tau_*)_+}{|\overline{\mathbb{S}}|} \overline{\mathbb{S}}. \qquad (36)$$

Then it follows from (35) that

$$\limsup_{n\to\infty} \int_{Q_T} (\mathbb{S}^n - \overline{\mathbb{S}}) \cdot \mathbb{D}\boldsymbol{w}^n \le \limsup_{n\to\infty} \int_{|\boldsymbol{v}^n - \boldsymbol{v}| \ge k^n} \frac{k^n}{|\boldsymbol{v}^n - \boldsymbol{v}|} |p_1^n| (|\nabla \boldsymbol{v}^n| + |\nabla \boldsymbol{v}|),$$

which then leads to

$$\limsup_{n\to\infty} \int_{|\boldsymbol{v}^n - \boldsymbol{v}| < k^n} (\mathbb{S}^n - \overline{\mathbb{S}}) \cdot \mathbb{D}(\boldsymbol{v}^n - \boldsymbol{v}) \le \limsup_{n\to\infty} \int_{|\boldsymbol{v}^n - \boldsymbol{v}| \ge k^n} \frac{k^n}{|\boldsymbol{v}^n - \boldsymbol{v}|} I^n, \qquad (37)$$

where

$$I^n := C_*(|p_1^n|^2 + |\nabla \boldsymbol{v}^n|^2 + |\nabla \boldsymbol{v}|^2 + |\overline{\mathbb{S}}|^2 + 1) \text{ with } C_* > 0.$$

Note that $I^n$ is uniformly bounded in $L^1(Q_T)$.

Let $N \in \mathbb{N}$ be arbitrary. We fix $A = N$ and $B = N^{N+1}$ and define

$$Q_i^n := \{(t, x) \in Q_T; N^i \le |\boldsymbol{v}^n - \boldsymbol{v}| \le N^{i+1}\} \qquad i = 1, \dots, N.$$

Since

$$\sum_{i=1}^{N} \int_{Q_i^n} I^n \leq C_*, \tag{38}$$

there is, for each $n \in \mathbb{N}$, an index $i_n \in \{1, \ldots, N\}$ such that

$$\int_{Q_{i_n}^n} I^n \leq \frac{C_*}{N}. \tag{39}$$

At this point we return to (37) with $k^n := N^{i_n}$ and estimate the right-hand side in the following way:

$$\begin{aligned} \int_{|\boldsymbol{v}^n - \boldsymbol{v}| \geq N^{i_n}} \frac{k^n}{|\boldsymbol{v}^n - \boldsymbol{v}|} I^n &= \int_{N^{i_n+1} \geq |\boldsymbol{v}^n - \boldsymbol{v}| \geq N^{i_n}} \frac{N^{i_n}}{|\boldsymbol{v}^n - \boldsymbol{v}|} I^n \\ &\quad + \int_{|\boldsymbol{v}^n - \boldsymbol{v}| \geq N^{i_n+1}} \frac{N^{i_n}}{|\boldsymbol{v}^n - \boldsymbol{v}|} I^n \\ &\leq \int_{Q_{i_n}^n} I^n + \frac{1}{N} \int_{|\boldsymbol{v}^n - \boldsymbol{v}| \geq N^{i_n+1}} I^n \leq \frac{C_*}{N}, \end{aligned} \tag{40}$$

where we used (39) to bound $\int_{Q_{i_n}^n} I^n$ and (38) to note that the last integral in (40) is bounded by $C_*$. Inserting (24) and (36) into (37) and using (40) we obtain

$$\limsup_{n \to \infty} \int_{|\boldsymbol{v}^n - \boldsymbol{v}| \leq k^n} \left(\mathbb{S}^n - \overline{\mathbb{S}}\right) \cdot \left( \left( \frac{(\mathbb{S}^n - \tau_*)_+}{|\mathbb{S}^n|} + \frac{1}{n} \right) \mathbb{S}^n - \frac{(\overline{\mathbb{S}}| - \tau)_+}{|\overline{\mathbb{S}}|} \overline{\mathbb{S}} \right) \leq \frac{C_*}{N},$$

which leads to

$$\limsup_{n \to \infty} \int_{|\boldsymbol{v}^n - \boldsymbol{v}| \leq k^n} Z^n \leq \frac{C_*}{N} + \limsup \frac{1}{n} \int_{Q_T} |\mathbb{S}^n||\mathbb{S}^n - \overline{\mathbb{S}}| \leq \frac{C_*}{N}, \tag{41}$$

where

$$Z^n := \left(\mathbb{S}^n - \overline{\mathbb{S}}\right) \cdot \left( \frac{(\mathbb{S}^n - \tau_*)_+}{|\mathbb{S}^n|} \mathbb{S}^n - \frac{(\overline{\mathbb{S}}| - \tau_*)_+}{|\overline{\mathbb{S}}|} \overline{\mathbb{S}} \right) \geq 0.$$

Since $A = N$ and $k^n \geq N$ it follows from (41) that

$$\limsup_{n \to \infty} \int_{|\boldsymbol{v}^n - \boldsymbol{v}| \leq N} Z^n \leq \frac{C_*}{N}.$$

The rest of the argument is the same as in [7], pp. 2787–2788, and will be only briefly sketched. Splitting $Q_T$ into a union of $|\boldsymbol{v}^n - \boldsymbol{v}| \leq N$ and $|\boldsymbol{v}^n - \boldsymbol{v}| > N$ one concludes from above that

$$\limsup_{n\to\infty} \int_{Q_T} \sqrt{Z^n} \leq \frac{C_*}{N}.$$

Letting $N \to \infty$ this gives, at least for a subsequence,

$$Z^n \to 0 \qquad \text{a.e. in } Q_T.$$

Applying then a biting lemma (see [7] for details), one then concludes that

$$Z^n \to 0 \qquad \text{strongly in } L^1(Q_T \setminus E_j),$$

where $E_j \subset Q_T$ are such that $\lim_{j\to\infty} |E_j| = 0$. It follows from the definition of $Z^n$, where we use (36) and replace $\frac{(\mathbb{S}^n - \tau_*)_+}{|\mathbb{S}^n|}\mathbb{S}^n$ by $\mathbb{D}\boldsymbol{v}^n - \frac{1}{n}\mathbb{S}^n$, that

$$\limsup_{n\to\infty} \int_{Q_T\setminus E_j} \mathbb{S}^n \cdot (\mathbb{D}\boldsymbol{v}^n - \frac{1}{n}\mathbb{S}^n) = \int_{Q_T\setminus E_j} \mathbb{S} \cdot \mathbb{D}\boldsymbol{v}.$$

Since

$$(\mathbb{S}^n, \mathbb{D}\boldsymbol{v}^n - \frac{1}{n}\mathbb{S}^n) \in \mathcal{A},$$

$$\mathbb{S}^n \rightharpoonup \mathbb{S} \text{ weakly in } L^2(Q_T)^{d\times d},$$

$$\mathbb{D}\boldsymbol{v}^n - \frac{1}{n}\mathbb{S}^n \rightharpoonup \mathbb{D}\boldsymbol{v} \text{ weakly in } L^2(Q_T)^{d\times d},$$

by Lemma 3.1 and properties of $E_j$ we conclude $(\mathbb{S}, \mathbb{D}\boldsymbol{v}) \in \mathcal{A}$ a.e. in $Q_T$, it means that (5) is proved.

## 5 Concluding Remarks

The main point of this study was to point out that the threshold slip is a possible way for overcoming the difficulties connected with the analysis of unsteady flows subject to homogeneous Dirichlet boundary conditions (no-slip) in particular if one needs to have a globally integrable pressure. The setting presented here is influenced by the framework of implicitly constituted materials.

We wish to remark that for implicitly constituted fluids characterized by **(A1)**–**(A3)** and **(A4*)** with $\psi(s) \geq c_1 s^r$ and $r > 2d/(d+2)$, we can define the concept of weak solution with an integrable pressure and prove its large data existence (see [7] and a forthcoming study by the authors and Luisa Consiglieri for further details). Thus, we obtain a subject for further numerical analysis and computer simulations for large class of non-Newtonian fluids.

**Acknowledgements** The authors acknowledge the membership to the Nečas Center for Mathematical Modeling (NCMM) and to the Charles University center for mathematical modeling, applied analysis and computational mathematics (MathMAC).

## References

1. M. Bulíček, J. Málek, K.R. Rajagopal, Navier's slip and evolutionary Navier-Stokes-like systems with pressure and shear-rate dependent viscosity. Indiana Univ. Math. J. **56** 51–85 (2007)
2. M. Bulíček, E. Feireisl, J. Málek, A Navier-Stokes-Fourier system for incompressible fluids with temperature dependent material coefficients. Nonlinear Anal. Real World Appl. **10**, 992–1015 (2009)
3. M. Bulíček, J. Málek, K.R. Rajagopal, Mathematical analysis of unsteady flows of fluids with pressure, shear-rate, and temperature dependent material moduli that slip at solid boundaries. SIAM J. Math. Anal. **41**, 665–707 (2009)
4. M. Bulíček, F. Ettwein, P. Kaplický, D. Pražák, On uniqueness and time regularity of flows of power-law like non-Newtonian fluids. Math. Methods Appl. Sci. **33** 1995–2010 (2010)
5. M. Bulíček, R. Lewandowski, J. Málek, On evolutionary Navier-Stokes-Fourier type systems in three spatial dimensions. Comment. Math. Univ. Carol. **52**, 89–114 (2011)
6. M. Bulíček, P. Gwiazda, J. Málek, K.R. Rajagopal, A. Świerczewska-Gwiazda, On flows of fluids described by an implicit constitutive equation characterized by a maximal monotone graph, in *Mathematical Aspects of Fluid Mechanics*, ed. by J.C. Robinson, J.L. Rodrigo, W. Sadowski. London Mathematical Society Lecture Note Series, vol. 402 (Cambridge University Press, Cambridge, 2012), pp. 23–51
7. M. Bulíček, P. Gwiazda, J. Málek, A. Świerczewska-Gwiazda, On unsteady flows of implicitly constituted incompressible fluids. SIAM J. Math. Anal. **44**, 2756–2801 (2012)
8. L. Consiglieri, Existence for a class of non-Newtonian fluids with a nonlocal friction boundary condition. Acta Math. Sin. (Engl. Ser.) **22** 523–534 (2006)
9. M.M. Denn, Fifty years of non-newtonian fluid dynamics. AIChE J. **50**, 2335–2345 (2004)
10. L. Diening, M. Ružička, J. Wolf, Existence of weak solutions for unsteady motions of generalized Newtonian fluids. Ann. Sc. Norm. Super. Pisa Cl. Sci. **IX**, 1–46 (2010)
11. L. Diening, C. Kreuzer, E. Süli, Finite element approximation of steady flows of incompressible fluids with implicit power-law-like rheology. SIAM J. Numer. Anal. **51**, 984–1015 (2013)
12. G. Duvant, J.-L. Lions, *Inequalities in Mechanics and Physics* (Springer, Berlin, 1976)
13. H. Fujita, A mathematical analysis of motions of viscous incompressible fluid under leak or slip boundary conditions, *Mathematical Fluid Mechanics and Modeling* (Sūrikaisekikenkyūsho Kōkyūroku, Kyoto, 1994), pp. 199–216
14. H. Fujita, H. Kawarada, A. Sasamoto, Analytical and numerical approaches to stationary flow problems with leak and slip boundary conditions, in *Advances in Numerical Mathematics; Proceedings of the Second Japan-China Seminar on Numerical Mathematics(Tokyo, 1994)*, Lecture Notes Numerical Applied Analysis, vol. 14 (Kinokuniya, Tokyo, 1995), pp. 17–31
15. M. Giga, Y. Giga, H. Sohr, $L^p$ estimates for the Stokes system, in *Functional Analysis and Related Topics*, 1991 (Kyoto), Lecture Notes in Mathematics, vol. 1540 (Springer, Berlin, 1993), pp. 55–67
16. H. Hervet, L.Léger, Flow with slip at the wall: from simple to complex fluids. C. R. Phys. **4**, 241–249 (2003)

17. J. Hron, J. Málek, J. Stebel, K. Touška, A novel view of computations of steady flows of Bingham and Herschel-Bulkley fluids using implicit constitutive relations. Submitted for publication (2015)
18. T. Kashiwabara, On a strong solution of the non-stationary Navier-Stokes equations under slip or leak boundary conditions of friction type. J. Differ. Equ. **254**, 756–778 (2013)
19. O.A. Ladyzhenskaya, Modifications of the Navier-Stokes equations for large gradients of the velocities. Zapiski Naukhnych Seminarov Leningrad. Otdel. Mat. Inst. Steklov. (LOMI) **7**, 126–154 (1968)
20. O.A. Ladyzhenskaya, Attractors for the modifications of the three-dimensional Navier-Stokes equations. Philos. Trans. R. Soc. Lond. A **346**, 173–190 (1994)
21. C. Le Roux, Existence and uniqueness of the flow of second-grade fluids with slip boundary conditions. Arch. Ration. Mech. Anal. **148**, 309–356 (1999)
22. C. Le Roux, Steady Stokes flows with threshold slip boundary conditions. Math. Models Methods Appl. Sci. **15**, 1141–1168 (2005)
23. C. Le Roux, A. Tani, Steady solutions of the Navier-Stokes equations with threshold slip boundary conditions. Math. Methods Appl. Sci. **30**, 595–624 (2007)
24. J. Leray, Sur le mouvement d'un liquide visqueux emplissant l'espace. Acta Math. **63**, 193–248 (1934)
25. J.-L. Lions, *Quelques méthodes de Résolution des Problèmes Aux Limites Non Linéaires* (Paris, Dunod, 1969)
26. J. Málek, J. Nečas, K.R. Rajagopal, Global analysis of the flows of fluids with pressure-dependent viscosities. Arch. Ration. Mech. Anal. **165**, 243–269 (2002)
27. J. Málek, M. Ružička, V.V. Shelukhin, Herschel-Bulkley fluids: existence and regularity of steady flows. Math. Models Methods Appl. Sci. **15**, 1845–1861 (2005)
28. J. Málek, K.R. Rajagopal, Compressible generalized Newtonian fluids. Z. Angew. Math. Phys. **61**, 1097–1110 (2010)
29. K.R. Rajagopal, A.R. Srinivasa, On the thermodynamics of fluids defined by implicit constitutive relations. Z. Angew. Math. Phys. **59**, 715–729 (2008)
30. N. Saito, On the Stokes equation with the leak and slip boundary conditions of friction type: regularity of solutions. Publ. Res. Inst. Math. Sci. **40**, 345–383 (2004)
31. V.V. Shelukhin, Bingham viscoplastic as a limit of non-Newtonian fluids. J. Math. Fluid Mech. **4**, 109–127 (2002)
32. V.A. Solonnikov, Estimates for solutions of nonstationary system of Navier-Stokes equations. J. Soviet Math. **8**, 467–523 (1977)
33. V.A. Solonnikov, $L_p$-estimates for solutions to the initial boundary-value problem for the generalized Stokes system in a bounded domain. Function theory and partial differential equations. J. Math. Sci. (New York), **105**, 2448–2484 (2001)
34. R. Temam, *Navier-Stokes Equations and Nonlinear Functional Analysis*. CBMS-NSF Regional Conference Series in Applied Mathematics, vol. 66, 2nd edn. (Society for Industrial and Applied Mathematics (SIAM), Philadelphia, 1995)
35. R. Temam, *Navier-Stokes equations*, Theory and numerical analysis (AMS Chelsea Publishing, Providence, RI, 2001). Reprint of the 1984 edition
36. J. Wolf, Existence of weak solutions to the equations of nonstationary motion of non-Newtonian fluids with shear-dependent viscosity. J. Math. Fluid Mech. **9**, 104–138 (2007)

# Inhomogeneous Boundary Value Problems in Spaces of Higher Regularity

**Robert Denk and Tim Seger**

*Dedicated to Yoshihiro Shibata on occasion of his $60^{th}$ birthday*

**Abstract** Uniform a priori estimates for parameter-elliptic boundary value problems are well-known if the underlying basic space equals $L^p(\Omega)$. However, much less is known for the $W_p^s(\Omega)$-realization, $s > 0$, of a parameter-elliptic boundary value problem. We discuss a priori estimates and the generation of analytic semigroups for these realizations in various cases. The Banach scale method can be applied for homogeneous boundary conditions if the right-hand side satisfies certain compatibility conditions, while for the general case parameter-dependent norms are used. In particular, we obtain a resolvent estimate for the general situation where no analytic semigroup is generated.

**Keywords** Boundary value problems • Higher order Sobolev spaces • Parameter-ellipticity • Resolvent estimates

## 1 Introduction

For the treatment of nonlinear parabolic equations, a priori estimates in $L^p$-Sobolev spaces are an important step. Based on the theory of parameter-ellipticity, resolvent estimates have been established for a large class of equations, implying sectoriality of the corresponding operator or even maximal regularity for the non-stationary problem. For a boundary value problem in a domain $\Omega \subset \mathbb{R}^n$, the basic space is usually $L^p(\Omega)$. This leads to a solution in $W_p^m(\Omega)$ where $m$ denotes the order of the differential operator. For the boundary traces, one obtains non-integer Besov spaces.

The situation becomes more complicated and much less investigated if one is interested in spaces of higher regularity. Here we start with $W_p^s(\Omega)$, $s > 0$, as the basic space and expect the solution to be in $W_p^{m+s}(\Omega)$. Apart from its own interest, spaces of higher regularity naturally appear in mixed-order systems (Douglis-Nirenberg

R. Denk (✉) • T. Seger
Department of Mathematics and Statistics, University of Konstanz, 78457 Konstanz, Germany
e-mail: robert.denk@uni-konstanz.de; tim.2.seger@gmail.com

 
H. Amann et al. (eds.), *Recent Developments of Mathematical Fluid Mechanics*, Advances in Mathematical Fluid Mechanics, DOI 10.1007/978-3-0348-0939-9_9

systems). Inhomogeneous boundary conditions and non-standard boundary spaces also appear in transmission problems and coupled systems. As an example, we mention the two-phase Stokes equations where the normal component of the velocity jumps across the interface. In the paper [23], Y. Shibata and S. Shimizu have shown maximal $L^p$-$L^q$-regularity for this system, introducing a special function space adapted to the inhomogeneous jump conditions. The proofs in this and many other papers in fluid mechanics (see, e.g., Shibata [22]) are based on partial Fourier transform and careful estimates of the solution operators. In the present text, we essentially follow the same approach, however, aiming at uniform a priori estimates where the basic space is $W_p^s(\Omega)$ instead of $L^p(\Omega)$. We will restrict ourselves to scalar parameter-elliptic equations which can be seen as a first step in the direction of the Stokes system and general mixed-order systems.

Let us consider the boundary value problem

$$\begin{aligned} (A-\lambda)u &= f \quad \text{in } \Omega, \\ Bu &= g \quad \text{on } \partial\Omega \end{aligned} \tag{1}$$

in a bounded sufficiently smooth domain $\Omega \subset \mathbb{R}^n$. Here $A$ is a scalar differential operator of order $m \in 2\mathbb{N}$, and $B$ is a column of $\frac{m}{2}$ boundary operators, $B = (B_1, \dots, B_{m/2})^T$, with $\operatorname{ord} B_j = m_j < m$. Classical parameter-elliptic theory states that, under suitable ellipticity and smoothness conditions, a uniform a priori estimate for the solution $u$ holds. More precisely, we have

$$|||u|||_{m,p,\Omega} \le C\Big( \|f\|_{L^p(\Omega)} + \sum_{j=1}^{m/2} |||g_j|||_{m-m_j-\frac{1}{p},p,\partial\Omega} \Big). \tag{2}$$

Here for $s > 0$ the parameter-dependent norms $|||\cdot|||$ are defined by

$$|||u|||_{s,p,\Omega} := \|u\|_{W_p^s(\Omega)} + |\lambda|^{s/m} \|u\|_{L^p(\Omega)} \quad (s \ge 0)$$

(analogously for $|||\cdot|||_{s,p,\partial\Omega}$). For $s \ge 0$, $W_p^s(\Omega)$ stands for the standard Sobolev-Slobedeckii space. From the a priori estimate (2), we immediately obtain the resolvent estimate

$$\|\lambda(\lambda - A_B)^{-1}\|_{L(L^p(\Omega))} \le C \tag{3}$$

for the $L^p$-realization $A_B$ of the boundary value problem $(A, B)$. This unbounded operator in $L^p(\Omega)$ is defined by $D(A_B) := \{u \in W_p^m(\Omega) : Bu = 0\}$ and $A_B u := A(D)u$ $(u \in D(A_B))$. In particular, under suitable parabolicity assumptions, $A_B$ is sectorial and generates an analytic semigroup. In fact, $A_B$ is even $\mathcal{R}$-sectorial and therefore admits maximal $L^p$-regularity. A priori estimates of the form (2) are known since long; we refer to the classical works Agmon [1], Agranovich–Vishik [4], Geymonat–Grisvard [13], and Roitberg–Sheftel [20]. Concerning $\mathcal{R}$-sectoriality and maximal regularity, we mention Denk-Hieber-Prüss [10] and the references therein.

In spaces of higher regularity, however, the resolvent estimate (3) in general does not hold. In fact, it is easily seen (cf. Nesensohn [19]) that the Dirichlet-Laplacian $\Delta_D$ in $W_p^1(\Omega)$ with domain $D(\Delta_D) := \{u \in W_p^3(\Omega) : u|_{\partial\Omega} = 0\}$ does not generate an analytic semigroup; its resolvent decays like $|\lambda|^{-1/2-1/(2p)}$ as $|\lambda| \to \infty$. The paper Denk-Dreher [8] deals with resolvent estimates for mixed-order systems. Here conditions on the basic space $Y \subset W_p^s(\Omega)$ were formulated which are necessary and sufficient for a generation of an analytic semigroup. It was shown that additional conditions have to be included in the basic space; these conditions can be seen as compatibility relations. For scalar equations or systems with the same order in each component, the method of Banach scales developed by Amann in [5] can be applied and gives a rather complete answer to the question of generation of an analytic semigroup. We will comment on this in Sect. 2 below. Generation of analytic semigroups for parabolic equations was also studied by Guidetti in [18] (see also Guidetti [16]). Here in particular mixed-order systems were studied which arise by the reduction of a higher-order system (in time) to a first-order system. A priori estimates in parameter-dependent norms have been studied, e.g., by Faierman and his coauthors in [2, 9, 12]. We also remark that a particular case of an a priori estimate in $W_p^s(\Omega)$ was used in the second author's thesis [21] to obtain a compactness property needed for a Schauder-type fixed-point argument in the context of a nonlinear elliptic-parabolic system.

In the present paper, we will discuss uniform a priori estimates for the boundary value problem (1) with inhomogeneous boundary conditions. To avoid technicalities, we will restrict ourselves to the model problems in the whole space $\mathbb{R}^n$ and the half-space $\mathbb{R}^n_+$. Here the operators are assumed to have constant coefficients and no lower-order terms. The generalization to bounded sufficiently smooth domains and to variable coefficients by localization and partition of unity is quite standard, and we will not dwell on this.

In Sect. 2, we will study the whole space case and the case of homogeneous boundary conditions. Whereas in the whole space the a priori estimates leading to the generation of an analytic semigroup follows quite directly from Michlin's theorem, the case of homogeneous boundary conditions can be treated by the Banach scale method. In Sect. 3, we will consider the case of inhomogeneous boundary conditions and derive the main a priori estimates of the present text.

## 2 The Whole Space Case and the Case of Homogeneous Boundary Conditions

Let $A(D) = \sum_{|\alpha|=m} a_\alpha D^\alpha$ be a linear differential operator in $\mathbb{R}^n$, $n \geq 2$, of order $m \in 2\mathbb{N}$ with constant coefficients $a_\alpha \in \mathbb{C}$. Here and in the following, we set $D := -i\partial$ and use the standard multi-index notation $D^\alpha := (-i)^{|\alpha|}\partial_{x_1}^{\alpha_1} \ldots \partial_{x_n}^{\alpha_n}$. Let $\mathcal{L} \subset \mathbb{C}$

be a closed sector in the complex plane with vertex at the origin. Without loss of generality, we may assume that $\mathcal{L} = \overline{\Sigma}_\theta$ with $\Sigma_\theta := \{z \in \mathbb{C} \setminus \{0\} : |\arg z| < \theta\}$ for some $\theta \in (0, \pi]$. The operator $A(D)$ is called parameter-elliptic in $\overline{\Sigma}_\theta$ (see [4]) if $A(\xi) - \lambda \neq 0$ holds for all $(\xi, \lambda) \in (\mathbb{R}^n \times \overline{\Sigma}_\theta) \setminus \{0\}$. Here $A(\xi) := \sum_{|\alpha|=m} a_\alpha \xi^\alpha$ is the symbol of $A(D)$. If the latter condition is satisfied with $\theta \geq \frac{\pi}{2}$, the operator $A$ is called parabolic.

We will consider the realization of the operator $A(D)$ in different scales of Sobolev spaces. For $s \in \mathbb{R}$ and $p \in (1, \infty)$, we denote by $H_p^s(\mathbb{R}^n)$ and $B_{pp}^s(\mathbb{R}^n)$ the standard Bessel potential and Besov space, respectively. The Sobolev-Slobodeckii space $W_p^s(\mathbb{R}^n)$, $s \geq 0$, coincides with $H_p^s(\mathbb{R}^n)$ for $s \in \mathbb{N}_0$ and with $B_{pp}^s(\mathbb{R}^n)$ for $s \in (0, \infty) \setminus \mathbb{N}$. We recall that a closed linear operator $A: X \supset D(A) \to X$ in a complex Banach space $X$ is called sectorial if the domain and the range of $A$ are dense in $X$ and if there exists $\phi \in (0, \pi)$ such that $\rho(A) \supset \Sigma_\phi$ and the set $\{\lambda(\lambda - A)^{-1} : \lambda \in \Sigma_\phi\}$ is bounded in $L(X)$. In this case, the supremum over all angles satisfying this condition is called the spectral angle $\phi_A$ of $A$.

In the following, $C$ stands for a generic constant which may vary from inequality to inequality but is independent of the variables appearing in the inequality (and in particular independent of $\lambda$).

In the whole space, it is easily seen that the realization of the operator $A(D)$ is sectorial:

**Lemma 2.1** *Let $A(D)$ be parameter-elliptic in $\overline{\Sigma}_\theta$, and let $s \in \mathbb{R}$ and $p \in (1, \infty)$. Then for every $\lambda \in \overline{\Sigma}_\theta \setminus \{0\}$ and every $f \in H_p^s(\mathbb{R}^n)$, the equation $(A(D) - \lambda)u = f$ has a unique solution $u \in H_p^{m+s}(\mathbb{R}^n)$, and the a priori estimate*

$$\|u\|_{H_p^{m+s}(\mathbb{R}^n)} + |\lambda| \, \|u\|_{H_p^s(\mathbb{R}^n)} \leq C \|f\|_{H_p^s(\mathbb{R}^n)}$$

*holds. In particular, if $\theta \geq \frac{\pi}{2}$, the operator $A^{(s)}$ in $H_p^s(\mathbb{R}^n)$, defined by $D(A^{(s)}) := H_p^{m+s}(\mathbb{R}^n)$, $A^{(s)}u := A(D)u$ $(u \in D(A^{(s)}))$, is sectorial with spectral angle larger than $\frac{\pi}{2}$ and therefore generates an analytic semigroup.*

*The analog results hold when $H_p^s(\mathbb{R}^n)$ is replaced by the Besov space $B_{pp}^s(\mathbb{R}^n)$.*

*Proof* This essentially follows from more general results on the existence of a bounded $H^\infty$-calculus, see, e.g., Denk-Saal-Seiler [11]. However, the result can also easily be seen by an application of the Michlin's theorem. In fact, for each $\lambda \in \overline{\Sigma}_\theta \setminus \{0\}$, the unique solution $u$ can be written as $u = \mathscr{F}^{-1}(A(\xi) - \lambda)^{-1}\mathscr{F}f$ where $\mathscr{F}$ stands for the Fourier transform in $\mathbb{R}^n$. Now it is immediately seen that

$$m(\xi) := (1 + |\xi|^2)^{m/2}(A(\xi) - \lambda)^{-1}$$

satisfies the conditions of Michlin's theorem (see [24], Sect. 2.2.4). By this and the definition of the Bessel potential spaces, the results in $H_p^s(\mathbb{R}^n)$ follow.

The analog results for Besov spaces are obtained by real interpolation. □

We will now consider boundary value problems where we will again restrict ourselves to the model problem in the half-space $\mathbb{R}^n_+ := \{x \in \mathbb{R}^n : x_n > 0\}$. As before, let $A(D) = \sum_{|\alpha|=m} a_\alpha D^\alpha$, $m \in 2\mathbb{N}$, be a differential operator with constant coefficients and let $B_j(D) = \sum_{|\beta|=m_j} b_{j\beta} D^\beta$, $j = 1, \ldots, \frac{m}{2}$, be boundary operators with constant coefficients $b_{j\beta} \in \mathbb{C}$. We assume $m_j < m$ throughout this paper. Setting $B := (B_1, \ldots, B_{m/2})^T$, we consider the boundary value problem

$$\begin{aligned} (A(D) - \lambda)u &= f \quad \text{in } \mathbb{R}^n_+, \\ \gamma_0 B(D)u &= g \quad \text{on } \mathbb{R}^{n-1}. \end{aligned} \tag{4}$$

Here $\gamma_0 : u \mapsto u|_{\mathbb{R}^{n-1}}$ denotes the boundary trace operator. The boundary value problem $(A(D), B(D))$ is called parameter-elliptic in $\overline{\Sigma}_\theta$ if $A(D)$ is parameter-elliptic in $\overline{\Sigma}_\theta$ and if the following Shapiro-Lopatinskii condition holds:

For all $(\xi', \lambda) \in (\mathbb{R}^{n-1} \times \overline{\Sigma}_\theta) \setminus \{0\}$ and all $h = (h_1, \ldots, h_{m/2})^T \in \mathbb{C}^{m/2}$, the ordinary differential equation

$$\begin{aligned} (A(\xi', D_n) - \lambda)v(x_n) &= 0 \quad (x_n > 0), \\ B(\xi', D_n)v(0) &= h, \\ v(x_n) &\to 0 \quad (x_n \to \infty) \end{aligned}$$

has a unique solution. If these conditions are satisfied with $\theta \geq \frac{\pi}{2}$, the boundary value problem $(A(D), B(D))$ is called parabolic.

Throughout the following, we assume that the boundary value problem $(A(D), B(D))$ is parameter-elliptic in $\overline{\Sigma}_\theta$. We first discuss the case of homogeneous boundary conditions, i.e., we assume $g = 0$ in (4), so we discuss the $L^p(\mathbb{R}^n_+)$-realization of $(A(D), B(D))$ which is given by $D(A_B) := \{u \in W^m_p(\mathbb{R}^n_+) : \gamma_0 B(D)u = 0\}$ and $A_B u := A(D)u$ $(u \in D(A_B))$. For this, we apply the method of Banach scales (see [5], Chap. V). We recall the main definitions and results. Let $X$ be a Banach space, $\{\cdot, \cdot\}$ an exact interpolation functor, and $A : X \supset D(A) \to X$ be the generator of a $C_0$-semigroup. Then for $k \in \mathbb{N}_0$, the space $E_k$ is defined by $E_k := D(A^k)$, and the $E_k$-realization $A_k$ of $A_{k-1}$ is iteratively defined by

$$D(A_k) := \{u \in E_k \cap D(A_{k-1}) : A_{k-1}u \in E_k\}, \quad A_k u := A_{k-1}u \ (u \in D(A_k)).$$

Here we have set $E_0 := X$ and $A_0 := A$. For $s \in (0, \infty) \setminus \mathbb{N}$, we write $s = k + \theta$ with $k \in \mathbb{N}_0$ and $\theta \in (0, 1)$ and define the space $E_s := \{E_k, E_{k+1}\}_\theta$ and the operator $A_s$ as the $E_s$-realization of $A_k$, i.e.,

$$D(A_s) := \{u \in E_s \cap D(A_k) : A_k u \in E_s\}, \quad A_s u := A_k u \ (u \in D(A_k)).$$

*Remark 2.2* In the above situation, $[(E_s, A_s) : s \geq 0]$ defines a scale of Banach spaces in the sense of [5], Definition V.1.1. Moreover, the operator $A_s$ is again the generator of a $C_0$-semigroup in $E_s$ for all $s \geq 0$. This follows from [5],

Theorem V.2.1.3 and Corollary V.2.1.4, after introducing a shift, i.e. considering $\tilde{A} := A - \omega$ with $\omega > 0$ sufficiently large such that $0 \in \rho(\tilde{A})$. Moreover, we have $\rho(A_s) = \rho(A)$ for all $s \geq 0$, and the resolvent estimate carries over from $A$ to $A_s$, see [5], inequality (2.1.16) in Theorem 2.1.3. In particular, if $A : E_0 \supset E_1 \to E_0$ is sectorial with angle $\phi$ then the same holds for $A_s : E_s \supset E_{s+1} \to E_s$.

To apply the above abstract definitions to the boundary value problem, we introduce the following spaces (see Amann [6], Sect. 4.9):

**Definition 2.3** Assume that the $L^p$-realization $A_B$ of $(A(D), B(D))$ generates a $C_0$-semigroup. For $p \in (1, \infty)$ and $s \in [0, \infty) \setminus \{km + m_j + \frac{1}{p} : k \in \mathbb{N}_0, j = 1, \dots, \frac{m}{2}\}$, we define the space $W^s_{p;(A,B)}(\mathbb{R}^n_+)$ as the set of all $u \in W^s_p(\mathbb{R}^n_+)$ which satisfy $\gamma_0 B_j A^k u = 0$ for all $k \in \mathbb{N}_0$ and $j = 1, \dots, \frac{m}{2}$ with $s - mk - m_j > \frac{1}{p}$.

The Banach scale method gives the following result.

**Theorem 4** *Let the boundary value problem $(A(D), B(D))$ be parabolic, and let $p \in (1, \infty)$ and $s \in [0, \infty) \setminus \{km + m_j + \frac{1}{p} : k \in \mathbb{N}_0, j = 1, \dots, \frac{m}{2}\}$. Then for all $f \in W^s_{p;(A,B)}(\mathbb{R}^n_+)$ and all $\lambda \in \Sigma_{\frac{\pi}{2}} \setminus \{0\}$ the problem $(A(D) - \lambda)u = f$, $\gamma_0 B(D)u = 0$, has a unique solution $u \in W^{m+s}_{p;(A,B)}(\mathbb{R}^n_+)$, and the a priori estimate*

$$\|u\|_{W^{m+s}_p(\mathbb{R}^n_+)} + |\lambda| \, \|u\|_{W^s_p(\mathbb{R}^n_+)} \leq C \|f\|_{W^s_p(\mathbb{R}^n_+)} \quad (u \in W^{m+s}_{p;(A,B)}(\mathbb{R}^n_+)) \tag{5}$$

*holds. In particular, the $W^s_{p;(A,B)}(\mathbb{R}^n_+)$-realization $A^{(s)}_B$ given by*

$$D(A^{(s)}_B) := W^{m+s}_{p;(A,B)}(\mathbb{R}^n_+), \quad A^{(s)}_B u := A(D)u \ (u \in D(A^{(s)}_B))$$

*is sectorial with angle larger than $\frac{\pi}{2}$ and therefore generates an analytic semigroup in $W^s_{p;(A,B)}(\mathbb{R}^n_+)$.*

*Proof* We will apply the method of Banach scales as introduced above. We consider the $L^p$-realization $A_B$ and remark that it is well known that $A_B$ is sectorial with angle larger than $\frac{\pi}{2}$ (see, e.g., [2]).

(i) First we show that for each $k \in \mathbb{N}_0$, we have $W^{km}_{p;(A,B)}(\mathbb{R}^n_+) = D(A^k_B)$. By definition, the inclusion "$\subset$" is obvious. To show the converse inclusion, we have to prove that $E_k := D(A^k_B) := \{u \in D(A_B) : A^\ell u \in D(A_B) \ (\ell = 0, \dots, k-1)\}$ is contained in $W^{km}_{p;(A,B)}(\mathbb{R}^n_+)$.

Due to the definition of $D(A_B)$, we have $\gamma_0 B_j A^\ell u = 0$ for all $\ell = 0, \dots, k-1$ and all $j = 1, \dots, \frac{m}{2}$. Therefore, we only have to show that $D(A^k_B) \subset W^{km}_p(\mathbb{R}^n_+)$. This is done iteratively. As $u \in D(A^2_B)$ and $\gamma_0 B_j u = 0$ for all $j = 1, \dots, \frac{m}{2}$, we have $u \in W^m_p(\mathbb{R}^n_+)$ and $Au \in W^m_p(\mathbb{R}^n_+)$. For $\lambda_0 \in \rho(A_B)$, the boundary value problem $(A - \lambda_0, B)$ is regular elliptic in the sense of Triebel [24], Definition 5.2.1/4. By elliptic regularity, we obtain $u \in W^{2m}_p(\mathbb{R}^n_+)$. Replacing

$u$ by $Au$ and using $A^2u \in W_p^m(\mathbb{R}_+^n)$, we can now prove $Au \in W_p^{2m}(\mathbb{R}_+^n)$. An iteration gives $u \in W_p^{km}(\mathbb{R}_+^n)$.

(ii) We consider $(A_B)_{s/m}$ and the scale generated by $A_B$. By real interpolation, we have for $k \in \mathbb{N}_0$ and $\theta \in (0,1)$ the identities

$$E_{k+\theta} = (E_k, E_{k+1})_{\theta,p} = \Big(W^{km}_{p;(A,B)}(\mathbb{R}^n_+), W^{(k+1)m}_{p;(A,B)}(\mathbb{R}^n_+)\Big)_{\theta,p} = W^{(k-\theta)m}_{p;(A,B)}(\mathbb{R}^n_+).$$

Here the last equality was shown in Amann [6], Corollary 4.9.2, in a more general setting. Due to Amann [5], Theorem 2.1.3, we see that $(A_B)_{s/m}$ generates an analytic semigroup and that $D((A_B)_{s/m}) = E_{\frac{s}{m}+1} = W^{m+s}_{p;(A,B)}(\mathbb{R}^n_+)$. Therefore, $(A_B)_{s/m}$ coincides with the $W^s_{p;(A,B)}$-realization $A_B^{(s)}$. In particular, $\rho(A_B^{(s)}) = \rho(A_B) \supset \overline{\Sigma_{\frac{\pi}{2}}} \setminus \{0\}$, and the resolvent estimate (5) holds. □

We remark that the exceptional cases $s = km + m_j + \frac{1}{p}$ arise due to the real interpolation results, see the discussion in Amann [6, 7].

## 3 The Case of Inhomogeneous Boundary Conditions

Now we consider the boundary value problem (4) for $g \neq 0$, again restricting ourselves to the model problem in $\mathbb{R}^n_+$. For this, we will use an explicit representation of the solution. We start with the definition of the basic solutions. Throughout this section, we assume that $(A(D), B(D))$ is parameter-elliptic in a fixed sector $\overline{\Sigma_\theta}$.

**Lemma 3.1** *For each $(\xi', \lambda) \in (\mathbb{R}^{n-1} \times \overline{\Sigma_\theta}) \setminus \{0\}$ and $j = 1, \dots, \frac{m}{2}$, we define the basic solution $w_j = w_j(\xi', x_n, \lambda)$ as the unique stable solution of the ordinary differential equation*

$$\begin{aligned} (A(\xi', D_n) - \lambda) w_j(x_n) &= 0 \quad (x_n > 0), \\ B_k(\xi', D_n) w_j(0) &= \delta_{jk} \quad (k = 1, \dots, \frac{m}{2}). \end{aligned}$$

*Then $w_j$ can be written in the form*

$$w_j(\xi', x_n, \lambda) = \int_{\gamma(\xi',\lambda)} e^{ix_n\tau} (A(\xi', \tau) - \lambda)^{-1} N_j(\xi', \tau, \lambda) d\tau$$

*where $\gamma(\xi', \lambda)$ is a smooth contour in the upper complex half-plane which encloses all roots of the polynomial $\tau \mapsto A(\xi', \tau) - \lambda$ with positive imaginary part. The* *functions* $N_j$ *and* $w_j$ *are smooth with respect to their arguments and continuous for all* $(\xi', \lambda) \in (\mathbb{R}^{n-1} \times \overline{\Sigma}_\theta) \setminus \{0\}$ , *and we have the quasi-homogeneities*

$$N_j(\rho\xi', \rho\tau, \rho^m\lambda) = \rho^{m-m_j-1} N_j(\xi', \tau, \lambda),$$
$$w_j(\rho\xi', \tfrac{x_n}{\rho}, \rho^m\lambda) = \rho^{-m_j} w_j(\xi', x_n, \lambda)$$

*for* $\rho > 0$, $\xi' \in \mathbb{R}^{n-1} \setminus \{0\}$ *and* $\lambda \in \overline{\Sigma}_\theta$.

*Proof* These assertions are stated in Denk-Faierman-Möller [9], Lemma 2.5. See also Volevich [25] for an explicit construction of $N_j$. □

To define the solution operators, we will use a parameter-dependent extension operator $E_\lambda \colon W_p^{k-\frac{1}{p}}(\mathbb{R}^{n-1}) \to W_p^k(\mathbb{R}^n_+)$ given by

$$(E_\lambda g)(x', x_n) := (\mathscr{F}')^{-1} \exp\big( - (|\xi'| + |\lambda|^{1/m}) x_n \big) (\mathscr{F}' g)(\xi').$$

Here $\mathscr{F}'$ stands for the partial Fourier transform with respect to the first $n-1$ variables. This operator was studied in Grubb–Kokholm [15] and in Agranovich-Denk-Faierman [2] in connection with the parameter-dependent norms above. It was shown that for all $k \in \mathbb{N}$, the trace operator

$$\gamma_0 \colon \big( W_p^k(\mathbb{R}^n_+), \|\!|\cdot|\!\|_{k,p,\mathbb{R}^n_+} \big) \to \big( W_p^{k-\frac{1}{p}}(\mathbb{R}^{n-1}), \|\!|\cdot|\!\|_{k-\frac{1}{p},p,\mathbb{R}^{n-1}} \big)$$

is continuous and $E_\lambda$ is a continuous right-inverse to $\gamma_0$. Here and in the following, we call a linear operator continuous with respect to the parameter-dependent norms if for each $\lambda_0 > 0$ the norm of this operator can be estimated by a constant $C = C(\lambda_0)$ which does not depend on $\lambda$ for each $\lambda \in \overline{\Sigma}_\theta$ with $|\lambda| \geq \lambda_0$.

Let us now consider the boundary value problem

$$\begin{aligned} (A(D) - \lambda) u &= 0 \quad \text{in } \mathbb{R}^n_+, \\ \gamma_0 B(D) u &= g \quad \text{on } \mathbb{R}^{n-1} \end{aligned} \tag{6}$$

with $g \in \prod_{j=1}^{m/2} W_p^{m+k-m_j-\frac{1}{p}}(\mathbb{R}^{n-1})$. Following an idea from Volevich [25], we write $\mathscr{F}'u$ in the form

$$\begin{aligned} (\mathscr{F}'u)(\xi', x_n) &= \sum_{j=1}^{m/2} w_j(\xi', x_n, \lambda)(\mathscr{F}' g_j)(\xi') \\ &= -\sum_{j=1}^{m/2} \int_0^\infty \frac{\partial}{\partial y_n} \Big[ w_j(\xi', x_n + y_n, \lambda)(\mathscr{F}' E_\lambda g_j)(\xi', y_n) \Big] dy_n \\ &= \sum_{j=1}^{m/2} \mathscr{F}' \Big( T_j^{(1)}(\lambda) E_\lambda g_j + T_j^{(2)}(\lambda)(\partial_n E_\lambda g_j) \Big). \end{aligned}$$

Here the solution operators $T_j^{(1)}$, $T_j^{(2)}$ are given by

$$(T_j^{(1)}(\lambda)\varphi)(x) := -\int_0^\infty (\mathscr{F}')^{-1}(\partial_n w_j)(\xi', x_n + y_n, \lambda)(\mathscr{F}'\varphi)(\xi', y_n)dy_n,$$
$$(T_j^{(2)}(\lambda)\varphi)(x) := -\int_0^\infty (\mathscr{F}')^{-1} w_j(\xi', x_n + y_n, \lambda)(\mathscr{F}'\varphi)(\xi', y_n)dy_n.$$

**Lemma 3.2** *Let $k \in \mathbb{N}_0$ and $p \in (1, \infty)$. Then the operators*

$$T_j^{(1)}(\lambda)\colon \big(W_p^{m+k-m_j}(\mathbb{R}^n_+), \|\!|\cdot|\!\|_{m+k-m_j,p,\mathbb{R}^n_+}\big) \to \big(W_p^{m+k}(\mathbb{R}^n_+), \|\!|\cdot|\!\|_{m+k,p,\mathbb{R}^n_+}\big),$$
$$T_j^{(2)}(\lambda)\colon \big(W_p^{m+k-m_j-1}(\mathbb{R}^n_+), \|\!|\cdot|\!\|_{m+k-m_j-1,p,\mathbb{R}^n_+}\big) \to \big(W_p^{m+k}(\mathbb{R}^n_+), \|\!|\cdot|\!\|_{m+k,p,\mathbb{R}^n_+}\big),$$

*$j = 1, \dots, \frac{m}{2}$, are continuous with respect to the parameter-dependent norms.*

*Proof* We only consider $T_j^{(1)}$, the proof for $T_j^{(2)}$ essentially being the same. Let $\lambda_0 > 0$. We make use of the equivalence of norms

$$\|\!|u|\!\|_{m+k,p,\mathbb{R}^n_+} \approx \sum_{\ell=0}^{m+k} \sum_{|\alpha|=\ell} \big\| \lambda^{\frac{m+k-\ell}{m}} (D')^{\alpha'} \partial_n^{\alpha_n} u \big\|_{L^p(\mathbb{R}^n_+)}.$$

Therefore, we have to estimate

$$\begin{aligned}
&\|\!|T_j^{(1)}(\lambda)\varphi|\!\|^p_{m+k,p,\mathbb{R}^n_+} \\
&\le C \sum_{\ell=0}^{m+k} \sum_{|\alpha|=\ell} \int_0^\infty \Big\| \int_0^\infty (\mathscr{F}')^{-1} \lambda^{\frac{m+k-\ell}{m}} (\xi')^{\alpha'} (\partial_n^{\alpha_n+1} w_j)(\xi', x_n + y_n, \lambda) \\
&\qquad\qquad (\mathscr{F}'\varphi)(\xi', y_n)dy_n \Big\|^p_{L^p(\mathbb{R}^{n-1})} dx_n \\
&\le C \sum_{\ell=0}^{m+k} \sum_{|\alpha|=\ell} \int_0^\infty \Big( \int_0^\infty \Big\| (\mathscr{F}')^{-1} M_{j,\ell,\alpha}(\xi', x_n + y_n, \lambda) \\
&\qquad\qquad (\mathscr{F}'\tilde{\varphi})(\xi', y_n) \Big\|_{L^p(\mathbb{R}^{n-1})} dy_n \Big)^p dx_n.
\end{aligned}$$

Here we have defined

$$\tilde{\varphi} := (\mathscr{F}')^{-1}(|\xi'| + |\lambda|^{1/m})^{m+k-m_j} \mathscr{F}'\varphi \in L^p(\mathbb{R}^n_+)$$

and

$$M_{j,\ell,\alpha}(\xi', x_n, \lambda) := (|\xi'| + |\lambda|^{1/m})^{-m-k+m_j} \lambda^{\frac{m+k-\ell}{m}} (\xi')^{\alpha'} (\partial_n^{\alpha_n-1} w_j)(\xi', x_n, \lambda).$$

We will apply Michlin's theorem to the functions $M_{j,\ell,\alpha}$. For this, we abbreviate $\rho := \rho(\xi',\lambda) := |\xi'| + |\lambda|^{1/m}$ and use the homogeneities stated in Lemma 3.1. In the integral representation for the basic solutions $w_j$ in Lemma 3.1, we make the substitution $\tau \mapsto \tilde{\tau} = \frac{\tau}{\rho}$ and use the fact that $\gamma(\frac{\xi'}{\rho}, \frac{\lambda}{\rho^m})$ can be replaced by a contour $\tilde{\gamma}$ which is independent of $\xi'$ and $\lambda$. For $\beta' \in \mathbb{N}_0^{n-1}$, we obtain

$$\begin{aligned}
&\Big|(\xi')^{\beta'} D_{\xi'}^{\beta'} M_{j,\ell,\alpha}(\xi', x_n, \lambda)\Big| \\
&= \Big|(\xi')^{\beta'} \lambda^{\frac{m+k-\ell}{m}} \int_{\gamma(\xi',\lambda)} \tau^{\alpha_n+1} e^{i\tau x_n} D_{\xi'}^{\beta'} \Big(\rho^{-m-k+m_j} (\xi')^{\alpha'} \\
&\qquad\qquad (A(\xi',\tau) - \lambda)^{-1} N_j(\xi',\tau,\lambda)\Big) d\tau\Big| \\
&= \Big|\big(\tfrac{\xi'}{\rho}\big)^{\beta'} \big(\tfrac{\lambda}{\rho^m}\big)^{\frac{m+k-\ell}{m}} \int_{\gamma(\xi',\lambda)} \big(\tfrac{\tau}{\rho}\big)^{\alpha_n+1} e^{i\tau x_n} H_{j,\alpha',\beta'}\big(\tfrac{\xi'}{\rho}, \tfrac{\tau}{\rho}, \tfrac{\lambda}{\rho^m}\big) d\tau\Big| \\
&= \Big|\big(\tfrac{\xi'}{\rho}\big)^{\beta'} \big(\tfrac{\lambda}{\rho^m}\big)^{\frac{m+k-\ell}{m}} \int_{\tilde{\gamma}} \rho\, \tilde{\tau}^{\alpha_n+1} e^{i\rho\tilde{\tau} x_n} H_{j,\alpha',\beta'}\big(\tfrac{\xi'}{\rho}, \tilde{\tau}, \tfrac{\lambda}{\rho^m}\big) d\tilde{\tau}\Big| \\
&\le C\rho \exp(-C\rho x_n) \le \frac{C}{x_n}.
\end{aligned}$$

Here we have set

$$H_{j,\alpha',\beta'}(\xi',\tau,\lambda) := D_{\xi'}^{\beta'} \Big(\rho^{-m-k+m_j} (\xi')^{\alpha'} (A(\xi',\tau) - \lambda)^{-1} N_j(\xi',\tau,\lambda)\Big)$$

and used the fact that $H_{j,\alpha',\beta'}$ is quasi-homogeneous in its arguments of degree $|\alpha'| - m - k - 1 - |\beta'|$. The two inequalities follow by a compactness argument and by the elementary inequality $te^{-t} \le 1$ for $t \ge 0$.

Now an application of Michlin's theorem in $\mathbb{R}^{n-1}$ gives

$$\begin{aligned}
&\|T_j^{(1)}(\lambda)\varphi\|_{m+k,p,\mathbb{R}^n_+} \\
&\le C \sum_{\ell=0}^{m+k} \sum_{|\alpha|=\ell} \Big( \int_0^\infty \Big( \int_0^\infty \frac{\|\tilde{\varphi}(\cdot, y_n)\|_{L^p(\mathbb{R}^{n-1})}}{x_n + y_n} dy_n \Big)^p dx_n \Big)^{1/p} \\
&\le C \Big( \int_0^\infty \|\tilde{\varphi}(\cdot, y_n)\|^p_{L^p(\mathbb{R}^{n-1})} dy_n \Big)^{1/p} \\
&= C \|\tilde{\varphi}\|_{L^p(\mathbb{R}^n_+)}.
\end{aligned}$$

Here we have used the continuity of the Hilbert transform in $L^p(\mathbb{R}_+)$ for the second inequality. Finally, we have

$$\begin{aligned}\|\tilde{\varphi}\|_{L^p(\mathbb{R}^n_+)} &= \left\|(\mathscr{F}')^{-1}(|\xi'| + |\lambda|^{1/m})^{m+k-m_j}\mathscr{F}'\varphi\right\|_{L^p(\mathbb{R}^n_+)} \\ &\leq C\Big(\|\varphi\|_{m+k-m_j,p,\mathbb{R}^n_+} + |\lambda|^{\frac{m+k-m_j}{m}}\|\varphi\|_{L^p(\mathbb{R}^n_+)}\Big) \\ &\leq C|\!|\!|\varphi|\!|\!|_{m+k-m_j,p,\mathbb{R}^n_+},\end{aligned}$$

which shows the continuity of $T_j^{(1)}(\lambda)$. □

The last lemma is the main step in the proof of uniform a priori estimates with respect to parameter-dependent norms. We obtain the following result, cf. Agranovich [3], Theorem 3.2.1 for the case $p = 2$.

**Theorem 3** *Let $s \in [0,\infty)$ and $p \in (1,\infty)$ with $m + s - m_j - \frac{1}{p} \notin \mathbb{N}_0$ for all $j = 1,\dots,\frac{m}{2}$. Then for every $\lambda \in \overline{\Sigma_\theta} \setminus \{0\}$, all $f \in W_p^s(\mathbb{R}^n_+)$ and all $g \in \prod_{j=1}^{m/2} W_p^{m+s-m_j-\frac{1}{p}}(\mathbb{R}^n)$ there exists a unique solution $u \in W_p^{m+s}(\mathbb{R}^n_+)$ of (4). Moreover, the operator*

$$(A(D) - \lambda, \gamma_0 B(D)) : W_p^{m+s}(\mathbb{R}^n_+) \to W_p^s(\mathbb{R}^n_+) \times \prod_{j=1}^{m/2} W_p^{m+s-m_j-\frac{1}{p}}(\mathbb{R}^{n-1}) \tag{7}$$

*is an isomorphism of Banach spaces with respect to the parameter-dependent norms. In particular, for every $\lambda_0 > 0$ there exists a constant $C = C(\lambda_0)$ such that*

$$|\!|\!|u|\!|\!|_{m+s,p,\mathbb{R}^n_+} \leq C\Big(|\!|\!|f|\!|\!|_{s,p,\mathbb{R}^n_+} + \sum_{j=1}^{m/2} |\!|\!|g_j|\!|\!|_{m+s-m_j-\frac{1}{p},p,\mathbb{R}^{n-1}}\Big) \tag{8}$$

*holds for all $\lambda \in \overline{\Sigma_\theta}$ with $|\lambda| \geq \lambda_0$.*

*Proof*

(i) First we assume $s \in \mathbb{N}_0$. The case $s = 0$ is well-known, see, e.g., Agranovich-Denk-Faierman [2], Theorem 2.1. In particular, we already know unique solvability of (4) with the solution $u$ being at least in $W_p^m(\mathbb{R}^n_+)$. Moreover, the continuity of the operator in (7) with respect to the parameter-dependent norms is an immediate consequence of the continuity of the derivatives and of the trace operator. Therefore, we only have to show the a priori estimate (8) which also gives the smoothness of the solution.

Let $r_+ \in L(W_p^s(\mathbb{R}^n), W_p^s(\mathbb{R}^n_+))$, $r_+ f := f|_{\mathbb{R}^n_+}$, denote the operator of restriction onto $\mathbb{R}^n_+$. Using the fact that there exists a coretraction $e_+$ of $r_+$

(see Amann [6], Sect. 4.4) with $e_+ \in L(W_p^\ell(\mathbb{R}^n_+), W_p^\ell(\mathbb{R}^n))$ for all $\ell \in [0,s]$, we see that both $r_+$ and $e_+$ are continuous with respect to the parameter-dependent norms, too.

We write $u = r_+u_1 + u_2$, where $u_1$ is the unique solution of $(A(D)-\lambda)u_1 = e_+f$ in $\mathbb{R}^n$. By the explicit representation of $u_1$ (see the proof of Lemma 2.1), we see that

$$|||r_+u_1|||_{m+s,p,\mathbb{R}^n_+} \le |||u_1|||_{m+s,p,\mathbb{R}^n} \le C|||e_+f|||_{s,p,\mathbb{R}^n} \le C|||f|||_{s,p,\mathbb{R}^n_+}.$$

For $u_2$ we obtain the boundary value problem

$$\begin{aligned} (A(D)-\lambda)u_2 &= 0 \quad \text{in } \mathbb{R}^n_+, \\ \gamma_0 B(D)u_2 &= \tilde g \quad \text{on } \mathbb{R}^{n-1} \end{aligned}$$

with $\tilde g := g - \gamma_0 B(D) r_+ u_1$. By the continuity of $r_+$, $B(D)$ and $\gamma_0$ (with respect to the parameter-dependent norms), we see that

$$|||\tilde g_j|||_{m+s-m_j-\frac{1}{p},p,\mathbb{R}^{n-1}} \le C\Big(|||g_j|||_{m+s-m_j-\frac{1}{p},p,\mathbb{R}^{n-1}} + |||f|||_{s,p,\mathbb{R}^n_+}\Big). \tag{9}$$

Due to Lemma 3.2, we have

$$u_2 = \sum_{j=1}^{m/2} \Big(T_j^{(1)}(\lambda)E_\lambda \tilde g_j + T_j^{(2)}(\lambda)\partial_n E_\lambda \tilde g_j\Big).$$

Now the continuity of $E_\lambda$, $\partial_n$, and $T_j^{(1)}, T_j^{(2)}$ yields

$$|||u_2|||_{m+s,p,\mathbb{R}^n_+} \le C \sum_{j=1}^{m/2} |||\tilde g_j|||_{m+s-m_j-\frac{1}{p},p,\mathbb{R}^{n-1}}$$

which in connection with (9) gives the a priori estimate (8) and the proof for $s \in \mathbb{N}_0$.

(ii) For $s \in (0,\infty)\setminus\mathbb{N}$ with $m+s-m_j-\frac{1}{p} \notin \mathbb{N}_0$, the result follows by real interpolation. Here we use the fact that for $k \in \mathbb{N}_0$ and $\theta \in (0,1)$ we have

$$\big(W_p^k(\mathbb{R}^n_+), W_p^{k+1}(\mathbb{R}^n_+)\big)_{\theta,p} = W_p^{k+\theta}(\mathbb{R}^n_+)$$

uniformly in $\lambda$ with respect to the parameter-dependent norms, see Grubb-Kokholm [15], Theorem 1.1. □

A combination of Theorems 4 and 3 immediately yields the following result.

**Corollary 3.4** *Let $(A(D), B(D))$ be parabolic, and let $s \in [0, \infty)\backslash\{km+m_j-\frac{1}{p} : k \in \mathbb{N}, j = 1, \ldots, \frac{m}{2}\}$. Then for all $\lambda \in \overline{\Sigma}_{\frac{\pi}{2}} \setminus \{0\}$, $|\lambda| \geq \lambda_0 > 0$, and all $f \in W^s_{p;(A,B)}(\mathbb{R}^n_+)$ and $g \in \prod_{j=1}^{m/2} W_p^{m+s-m_j-\frac{1}{p}}(\mathbb{R}^{n-1})$ there exists a unique solution $u \in W_p^{m+s}(\mathbb{R}^n_+)$ of (4), and*

$$\|u\|_{W_p^{m+s}(\mathbb{R}^n_+)} + |\lambda|\, \|u\|_{W_p^s(\mathbb{R}^n_+)} \leq C\Big(\|f\|_{W_p^s(\mathbb{R}^n_+)} + \sum_{j=1}^{m/2} |||g_j|||_{m+s-m_j-\frac{1}{p},p,\mathbb{R}^{n-1}}\Big). \tag{10}$$

*In particular, this holds for all $f \in W_p^s(\mathbb{R}^n_+)$ if $s < m_j + \frac{1}{p}$ for all $j = 1, \ldots, \frac{m}{2}$.*

*Proof* We write $u = u_1 + u_2$, where $u_1$ solves $(A(D) - \lambda)u_1 = f$, $\gamma_0 B(D)u_1 = 0$ and $u_2$ solves $(A(D) - \lambda)u_2 = 0$, $\gamma_0 B(D)u_2 = g$, and apply Theorems 4 and 3, respectively. For the application of Theorem 3, we note that by the interpolation inequality, the left-hand side of (10) is not larger than a constant times $|||u|||_{m+s,p,\mathbb{R}^n_+}$. The last statement follows directly from the fact that for these $s$, the spaces $W_p^s(\mathbb{R}^n_+)$ and $W^s_{p;(A,B)}(\mathbb{R}^n_+)$ coincide. □

*Remark 3.5*

(a) The results of Theorem 3 and Corollary 3.4 have some connection to results by D. Guidetti [16–18]. In particular, an estimate similar to (10) can be found as Proposition 2.16 in [17] in the Besov space setting, under the additional assumption $\min_j m_j < s < \min_j m_j + \frac{1}{p}$. We remark that in [17] essentially the same parameter-dependent norms for $g_j$ appear as well as some compatibility conditions. Our assumption $f \in W^s_{p;(A,B)}(\mathbb{R}^n_+)$ enables us to consider a wider range for $s$. Apart from some restrictions on the smoothness parameter $s$, the results by Guidetti in the Besov space setting are quite general and include the generation of an analytic semigroup.
(b) On the right-hand side of (10) large powers of $\lambda$ may appear, although we only have $|\lambda|$ on the left-hand side. The following elementary example in $\mathbb{R}^1_+$ shows that even in the one-dimensional case this cannot be avoided: Consider the boundary value problem $\lambda u(x_n) - u''(x_n) = 0$ $(x_n > 0)$, $u(0) = g \in \mathbb{C}$. Then for $\lambda > 0$ the stable solution is $u(x_n) = \exp(-\sqrt{\lambda} x_n)g$, and a direct calculation shows that for $s \in \mathbb{N}_0$ we have

$$\|u\|_{W_p^{2+s}(\mathbb{R}_+)} \geq C|\lambda|^{\frac{2+s-1/p}{2}}|g|.$$

The power on the right-hand side is consistent with the exponent $(m + s - m_j - \frac{1}{p})/m$ appearing on the right-hand side of (10).

In some sense the parameter-dependent norms in Theorem 3 are natural for parameter-elliptic problems. However, they do not yield resolvent estimates as the parameter $\lambda$ appears on both sides. Moreover, on the left-hand side we have $|\lambda|^{(m+s)/m}$ instead of $|\lambda|$. We will now derive a resolvent estimate for the

$W_p^s(\mathbb{R}^n_+)$-realization. As it was shown in [8], for a decay like $|\lambda|^{-1}$ additional conditions on $f$ are necessary. The following result gives a general resolvent estimate.

**Theorem 6**

*(a) In the situation of Theorem 3, assume that*

$$\sigma := \max_{j=1,\dots,m/2} (s - m_j - \tfrac{1}{p}) > 0. \tag{11}$$

*Then for all $\lambda_0 > 0$ there exists a constant $C = C(\lambda_0) > 0$ such that*

$$\|u\|_{m+s,p,\mathbb{R}^n_+} + |\lambda|\,\|u\|_{s,p,\mathbb{R}^n_+} \le C\Big(|\lambda|^{\sigma/m}\|f\|_{s,p,\mathbb{R}^n_+} + \sum_{j=1}^{m/2} |\!|\!| g_j |\!|\!|_{m+s-m_j-\frac{1}{p},p,\mathbb{R}^{n-1}}\Big). \tag{12}$$

*(b) Let $(A(D), B(D))$ be parabolic, and define $\sigma$ as in (11). For $s \in [0,\infty) \setminus \{mk + m_j - \frac{1}{p} : k \in \mathbb{N}_0, j = 1,\dots,m/2\}$ define the unbounded operator $\tilde{A}_B^{(s)}$ in $W_p^s(\mathbb{R}^n_+)$ by*

$$D(\tilde{A}_B^{(s)}) := \{u \in W_p^{m+s}(\mathbb{R}^n_+) : \gamma_0 B(D)u = 0\},$$

$$\tilde{A}_B^{(s)} u := A(D)u \ (u \in D(\tilde{A}_B^{(s)})).$$

*Then $\rho(\tilde{A}_B^{(s)}) \supset \overline{\Sigma}_{\frac{\pi}{2}} \setminus \{0\}$, and for all $\lambda_0 > 0$ there exists $C = C(\lambda_0) > 0$ such that*

$$\big\|(\tilde{A}_B^{(s)} - \lambda)^{-1}\big\|_{L(W_p^s(\mathbb{R}^n_+))} \le C|\lambda|^{-1+\max\{0,\sigma/m\}} \quad (\lambda \in \overline{\Sigma}_{\frac{\pi}{2}},\ |\lambda| \ge \lambda_0).$$

*Proof*

(a) We write the solution $u$ in the form $u = u_1 + u_2 + u_3$. Here $u_1$ solves $(A(D) - \lambda)u_1 = 0$, $\gamma_0 B(D)u_1 = g$, while $u_2 := r_+\tilde{u}_2$ with $\tilde{u}_2$ being the solution of $(A(D) - \lambda)\tilde{u}_2 = e_+ f$ in $\mathbb{R}^n$. Due to Corollary 3, we obtain

$$\|u_1\|_{m+s,p,\mathbb{R}^n_+} + |\lambda|\,\|u_1\|_{s,p,\mathbb{R}^n_+} \le C\sum_{j=1}^{m/2} |\!|\!| g_j |\!|\!|_{m+s-m_j-\frac{1}{p},p,\mathbb{R}^{n-1}}.$$

Moreover, by Lemma 2.1, we have

$$\|u_2\|_{m+s,p,\mathbb{R}^n_+} + |\lambda|\,\|u_2\|_{s,p,\mathbb{R}^n_+} \le C\|f\|_{s,p,\mathbb{R}^n_+}. \tag{13}$$

The function $u_3$ is the solution of the boundary value problem

$$(A(D) - \lambda)u_3 = 0 \quad \text{in } \mathbb{R}^n_+,$$
$$\gamma_0 B(D)u_3 = -\gamma_0 B(D)u_2 \quad \text{on } \mathbb{R}^{n-1}.$$

We apply Corollary 3 again and get

$$\|u_3\|_{m+s,p,\mathbb{R}^n_+} + |\lambda|\, \|u_3\|_{s,p,\mathbb{R}^n_+} \le C \sum_{j=1}^{m/2} \||\gamma_0 B(D)u_2\||_{m+s-m_j-\frac{1}{p},p,\mathbb{R}^{n-1}}$$
$$= C \sum_{j=1}^{m/2} \Big( \|\gamma_0 B(D)u_2\|_{m+s-m_j-\frac{1}{p}} + |\lambda|^{\frac{m+s-m_j-1/p}{m}} \|\gamma_0 B(D)u_2\|_{L^p(\mathbb{R}^{n-1})} \Big).$$

We estimate the norms on the right-hand side for each $j$. First, we have

$$\|\gamma_0 B(D)u_2\|_{m+s-m_j-\frac{1}{p},p,\mathbb{R}^{n-1}} \le C\|B(D)u_2\|_{m+s-m_j,p,\mathbb{R}^n_+}$$
$$\le C\|u_2\|_{m+s,p,\mathbb{R}^n_+}$$

for all $j = 1, \dots, \frac{m}{2}$.

If $s > m_j + \frac{1}{p}$, we can estimate

$$\begin{aligned}
|\lambda|^{\frac{m+s-m_j-1/p}{m}} &\|\gamma_0 B(D)u_2\|_{L^p(\mathbb{R}^{n-1})} \\
&\le |\lambda|^{\frac{m+s-m_j-1/p}{m}} \|\gamma_0 B(D)u_2\|_{s-m_j-\frac{1}{p},p,\mathbb{R}^{n-1}} \\
&\le C|\lambda|^{\frac{m+s-m_j-1/p}{m}} \|B(D)u_2\|_{s-m_j,p,\mathbb{R}^n_+} \\
&\le C|\lambda|^{\frac{m+s-m_j-1/p}{m}} \|u_2\|_{s,p,\mathbb{R}^n_+} \\
&\le C|\lambda|^{\sigma/m}\, |\lambda|\, \|u_2\|_{s,p,\mathbb{R}^n_+} \\
&\le C|\lambda|^{\sigma/m}\, \|f\|_{s,p,\mathbb{R}^n_+}.
\end{aligned}$$

Here we applied (13) in the last step.

If $s < m_j + \frac{1}{p}$, we similarly write

$$\begin{aligned}
|\lambda|^{\frac{m+s-m_j-1/p}{m}} &\|\gamma_0 B(D)u_2\|_{L^p(\mathbb{R}^{n-1})} \\
&\le |\lambda|^{\frac{m+s-m_j-1/p}{m}} \|\gamma_0 B(D)u_2\|_{\sigma,p,\mathbb{R}^{n-1}} \\
&\le C|\lambda|^{\sigma/m}\, |\lambda|^{\frac{m+s-m_j-\sigma-1/p}{m}} \|u_2\|_{\sigma+m_j+\frac{1}{p},p,\mathbb{R}^n_+}
\end{aligned}$$

$$\leq C|\lambda|^{\sigma/m}\left(\|u_2\|_{m+s,p,\mathbb{R}^n_+} + |\lambda|\,\|u\|_{s,p,\mathbb{R}^n_+}\right)$$
$$\leq C|\lambda|^{\sigma/m}\|f\|_{s,p,\mathbb{R}^n_+},$$

where we used the interpolation inequality (see Grisvard [14], Theorem 1.4.3.3) for the third inequality. This finishes the proof of the a priori estimate (12) and of part (a).

(b) In the case $\sigma < 0$ the a priori estimate (and the fact that the operator $\tilde{A}_B^{(s)}$ is sectorial) follows from the last statement in Corollary 3.4. For $\sigma > 0$, we apply part (a) with $g = 0$. Note that the case $\sigma = 0$ is excluded. □

*Remark 3.7* For the Dirichlet-Laplacian $\Delta_D$, we have $m = 2, m_1 = 0$, and therefore $\sigma = s - \frac{1}{p}$. For the resolvent in $W_p^s(\mathbb{R}^n_+)$, we obtain from Theorem 6

$$\|(\lambda - \Delta_D)^{-1}\|_{L(W_p^s(\mathbb{R}^n_+))} \leq C|\lambda|^{-1+\frac{s}{2}-\frac{1}{2p}}.$$

For $s = 1$, we have a decay like $|\lambda|^{-1/2-1/(2p)}$. It was shown in [19] that this is the exact decay rate.

## References


1. S. Agmon, On the eigenfunctions and on the eigenvalues of general elliptic boundary value problems. Commun. Pure Appl. Math. **15**, 119–147 (1962)
2. M. Agranovich, R. Denk, M. Faierman, Weakly smooth nonselfadjoint spectral elliptic boundary problems, in *Spectral Theory, Microlocal Analysis, Singular Manifolds, Math. Top.*, vol. 14 (Akademie Verlag, Berlin, 1997), pp. 138–199
3. M.S. Agranovich, Elliptic boundary problems, in *Partial Differential Equations, IX, Encyclopaedia Mathematical Science*, vol. 79 (Springer, Berlin, 1997), pp. 1–144, 275–281. [Translated from the Russian by the author]
4. M.S. Agranovich, M.I. Vishik, Elliptic problems with a parameter and parabolic systems of general form. Russ. Math. Surv. **19**, 53–157 (1964)
5. H. Amann, *Linear and Quasilinear Parabolic Problems. Vol. I: Abstract linear theory, Monographs in Mathematics*, vol. 89 (Birkhäuser, Boston, 1995)
6. H. Amann, *Anisotropic function spaces and maximal regularity for parabolic problems. Part 1: Function spaces*, Jindřich Nečas Center for Mathematical Modeling Lecture Notes, vol. 6 (Matfyzpress, Prague, 2009)
7. H. Amann, Function spaces on singular manifolds. Math. Nachr. **286**(5–6), 436–475 (2013)
8. R. Denk, M. Dreher, Resolvent estimates for elliptic systems in function spaces of higher regularity. Electron. J. Differ. Equ. **109**, 12 (2011)
9. R. Denk, M. Faierman, M. Möller, An elliptic boundary problem for a system involving a discontinuous weight. Manuscripta Math. **108**(3), 289–317 (2002)
10. R. Denk, M. Hieber, J. Prüss, $\mathcal{R}$-boundedness, Fourier multipliers and problems of elliptic and parabolic type. Mem. Am. Math. Soc. **166**(788), viii+114 (2003)
11. R. Denk, J. Saal, J. Seiler, Bounded $H_\infty$-calculus for pseudo-differential Douglis-Nirenberg systems of mild regularity. Math. Nachr. **282**(3), 386–407 (2009)
12. M. Faierman, On the resolvent arising in a parameter-elliptic multi-order boundary problem. Math. Nachr. **285**(13), 1643–1670 (2012)

13. G. Geymonat, P. Grisvard, Alcuni risultati di teoria spettrale per i problemi ai limiti lineari ellittici. Rendiconti del Seminario Matematico della Università di Padova **38**, 121–173 (1967)
14. P. Grisvard, *Elliptic Problems in Nonsmooth Domains*, *Monographs and Studies in Mathematics*, vol. 24 (Pitman Advanced Publishing Program, Boston, MA, 1985)
15. G. Grubb, N.J. Kokholm, A global calculus of parameter-dependent pseudodifferential boundary problems in $L_p$ Sobolev spaces. Acta Math. **171**(2), 165–229 (1993)
16. D. Guidetti, A maximal regularity result with applications to parabolic problems with nonhomogeneous boundary conditions. Rendiconti del Seminario Matematico della Università di Padova **84**, 1–37 (1990)
17. D. Guidetti, On elliptic problems in Besov spaces. Math. Nachr. **152**, 247–275 (1991)
18. D. Guidetti, On boundary value problems for parabolic equations of higher order in time. J. Differ. Equ. **124**(1), 1–26 (1996)
19. M. Nesensohn, Randwertprobleme in Sobolevräumen höherer Ordnung. Diploma thesis, University of Konstanz, 2009
20. J.A. Roitberg, Z.G. Sheftel, Boundary value problems with a parameter for systems elliptic in the sense of Douglis-Nirenberg. Ukrain. Mat. Ž. **19**(1), 115–120 (1967)
21. T. Seger, Elliptic-parabolic systems with applications to lithium-ion battery models. Ph.D. thesis, University of Konstanz, 2013
22. Y. Shibata, Generalized resolvent estimates of the Stokes equations with first order boundary condition in a general domain. J. Math. Fluid Mech. **15**(1), 1–40 (2013)
23. Y. Shibata, S. Shimizu, On the maximal $L_p$-$L_q$ regularity of the Stokes problem with first order boundary condition; model problems. J. Math. Soc. Japan **64**(2), 561–626 (2012)
24. H. Triebel, *Interpolation Theory, Function Spaces, Differential Operators* (Deutscher Verlag der Wissenschaften, Berlin, 1978)
25. L.R. Volevich, Solvability of boundary problems for general elliptic systems. Am. Math. Soc. Translat. II. Ser. **67**, 182–225 (1968)

# Blow-Up Criterion for 3D Navier-Stokes Equations and Landau-Lifshitz System in a Bounded Domain

**Jishan Fan and Tohru Ozawa**

*Dedicated to Professor Yoshihro Shibata on the occasion of his sixtieth birthday*

**Abstract** In this paper we prove a blow-up criterion for the 3D Navier-Stokes equations in a bounded domain in terms of a BMO norm of vorticity. We will also prove a regularity criterion for the Landau-Lifshitz system in a bounded domain.

**Keywords** BMO-norm • Landau-Lifshitz • Navier-Stokes • Regularity criterion

## 1 Introduction

Let $\Omega \subset \mathbb{R}^3$ be a bounded, simply connected domain with smooth boundary $\partial\Omega$, and $\nu$ be the unit outward normal vector to $\partial\Omega$. We consider the regularity criterion of the Navier-Stokes equations:

$$\operatorname{div} u = 0, \tag{1}$$

$$\partial_t u + u \cdot \nabla u + \nabla \pi = \mu \Delta u \quad in \quad \Omega \times (0, \infty), \tag{2}$$

$$u \cdot \nu = 0, \operatorname{curl} u \times \nu = 0 \quad on \quad \partial\Omega \times (0, \infty), \tag{3}$$

$$u(\cdot, 0) = u_0 \quad in \quad \Omega \subset \mathbb{R}^3, \tag{4}$$

where $u$ and $\pi$ denote unknown velocity vector field and pressure scalar of the fluid, respectively. The physical constant $\mu \geq 0$ is the viscosity and we will take $\mu = 1$ for simplicity.

J. Fan (✉)
Department of Applied Mathematics, Nanjing Forestry University, Nanjing 210037, P.R. China
e-mail: fanjishan@njfu.com.cn

T. Ozawa
Department of Applied Physics, Waseda University, Tokyo 169-8555, Japan
e-mail: txozawa@waseda.jp



H. Amann et al. (eds.), *Recent Developments of Mathematical Fluid Mechanics*, Advances in Mathematical Fluid Mechanics, DOI 10.1007/978-3-0348-0939-9_10

It is well known that the problem has at least a global-in-time weak solution and a unique local-in-time strong solution [8, 13]. However, the regularity of weak solutions is still a very challenging open problem. On the other hand, the development of blow-up criteria is of great importance for both theoretical and practical purpose. When $\Omega := \mathbb{R}^3$, many criteria have been obtained, see [1, 3, 4, 6, 11, 12, 15, 16] and the references therein. When $\Omega$ is a bounded domain and the boundary condition is the homogeneous Dirichlet boundary condition, Giga [7] and Kim [10] proved a Serrin type regularity criterion.

When $\Omega$ is a bounded domain and the boundary condition is (3), Kang and Kim [9] prove some Serrin type regularity criteria.

The aim of this paper is to prove a new regularity criterion for the problem (1)–(4).

We will prove that

**Theorem 1.1** *Let* $u_0 \in H^3(\Omega)$ *and* $\operatorname{div} u_0 = 0$ *in* $\Omega$ *and* $u_0 \cdot \nu = 0, \operatorname{curl} u_0 \times \nu = 0$ *on* $\partial\Omega$. *Let* $u$ *be a unique local smooth solution to the problem* (1)–(4). *If* $u$ *satisfies*

$$\nabla u \in L^1(0, T; BMO(\Omega)) \text{ with } 0 < T < \infty, \tag{5}$$

*then the solution* $u$ *can be extended beyond* $T > 0$.

*Remark 1.2* When $\mu = 0$, (5) is proved in [14] for the Euler equations, which is simpler than the case $\mu > 0$, since we need to deal with the boundary integrals.

To prove Theorem 1.1, we will use the following logarithmic Sobolev inequality [14]:

$$\|\nabla u\|_{L^\infty} \leq C(1 + \|\nabla u\|_{BMO} \log(e + \|u\|_{W^{s,p}})) \text{ with } s > 1 + \frac{3}{p}. \tag{6}$$

Next, we consider the 3D Landau-Lifshitz system:

$$d_t - \Delta d = d|\nabla d|^2 + d \times \Delta d,\ |d| = 1 \ \ in \ \ \Omega \times (0, \infty), \tag{7}$$

$$\partial_\nu d = 0, \ \ on \ \ \partial\Omega \times (0, \infty), \tag{8}$$

$$d(\cdot, 0) = d_0, |d_0| = 1 \ \ in \ \ \Omega \subset \mathbb{R}^d. \tag{9}$$

Carbou and Fabrie [2] showed the existence and uniqueness of local smooth solutions. When $\Omega := \mathbb{R}^n$ $(n = 2, 3, 4)$, Fan and Ozawa [5] prove some regularity criteria. The aim of this paper is to prove a blow-up criterion for the problem (7)–(9) when $\Omega$ is a bounded domain. We will prove

**Theorem 1.3** *Let* $d_0 \in H^3(\Omega)$ *with* $|d_0| = 1$ *in* $\Omega$ *and* $\partial_\nu d_0 = 0$ *on* $\partial\Omega$. *Let* $d$ *be a local smooth solution to the problem* (7)–(9). *If* $d$ *satisfies*

$$\int_0^T \|\nabla d\|_{L^q}^{\frac{2q}{q-3}} dt < \infty \ \ with \ \ 3 < q \leq \infty \ \ and \ \ 0 < T < \infty, \tag{10}$$

*then the solution $d$ can be extended beyond $T > 0$.*

## 2 Proof of Theorem 1.1

This section is devoted to the proof of Theorem 1.1. We only need to establish some a priori estimates.

First, testing (2) by $u$ and using (1) and (3), we have the well-known energy inequality:

$$\frac{1}{2}\frac{d}{dt}\int u^2 dx + \int |\operatorname{curl} u|^2 dx \leq 0,$$

which gives

$$\|u\|_{L^\infty(0,T;L^2)} + \|u\|_{L^2(0,T;H^1)} \leq C. \tag{11}$$

Taking curl to (2) and using (1), we get the well-known equation for the vorticity $\omega := \operatorname{curl} u$ :

$$\partial_t \omega + u \cdot \nabla \omega - \omega \cdot \nabla u = \Delta \omega. \tag{12}$$

Testing (12) by $\omega$ and using (1) and (6), we see that

$$\begin{aligned}
&\frac{1}{2}\frac{d}{dt}\int |\omega|^2 dx + \int |\operatorname{curl} \omega|^2 dx \\
&= \int (\omega \cdot \nabla) u \cdot \omega dx \leq \|\nabla u\|_{L^\infty} \int \omega^2 dx \\
&\leq C(1 + \|\nabla u\|_{BMO}) \log(e + \|u\|_{H^3}) \int |\omega|^2 dx,
\end{aligned}$$

and therefore

$$\int |\omega|^2 dx + \int_{t_0}^t \|\operatorname{curl} \omega\|_{L^2}^2 d\tau \leq C(e + y)^{C_0 \epsilon} \tag{13}$$

provided that

$$\int_{t_0}^t \|\nabla u\|_{BMO} d\tau \leq \epsilon << 1 \tag{14}$$

and $y(t) := \sup_{[t_0,t]} \|u\|_{H^3}$ for any $0 < t_0 \leq t \leq T$ and $C_0$ is a absolute constant.

Applying $\partial_t$ to (2), we have

$$\partial_t^2 u + u \cdot \nabla \partial_t u + \nabla \partial_t \pi - \Delta \partial_t u = -\partial_t u \cdot \nabla u. \tag{15}$$

Testing (15) by $\partial_t u$, using (1), (3) and (6), we obtain

$$\begin{aligned}
&\frac{1}{2}\frac{d}{dt}\int |\partial_t u|^2 dx + \int |\mathrm{curl}\, \partial_t u|^2 dx \\
&= -\int \partial_t u \cdot \nabla u \cdot \partial_t u dx \leq \|\nabla u\|_{L^\infty} \int |\partial_t u|^2 dx \\
&\leq C(1 + \|\nabla u\|_{BMO}) \log(e + \|u\|_{H^3}) \int |\partial_t u|^2 dx,
\end{aligned}$$

which leads to

$$\int |\partial_t u|^2 dx + \int_{t_0}^t \int |\mathrm{curl}\, \partial_t u|^2 dx d\tau \leq C(e+y)^{C_0 \epsilon}. \tag{16}$$

On the other hand, thanks to the $H^2$-theory of the Stokes system, it follows from (2), (13) and (16) that

$$\begin{aligned}
\|u\|_{H^2} &\leq C\| - \Delta u + \nabla \pi\|_{L^2} \leq C\|\partial_t u + u \cdot \nabla u\|_{L^2} \\
&\leq C\|\partial_t u\|_{L^2} + C\|u\|_{L^6}\|\nabla u\|_{L^3} \\
&\leq C\|\partial_t u\|_{L^2} + C\|\nabla u\|_{L^2}^{\frac{3}{2}}\|u\|_{H^2}^{\frac{1}{2}} \\
&\leq \frac{1}{2}\|u\|_{H^2} + C\|\partial_t u\|_{L^2} + C\|\omega\|_{L^2}^3,
\end{aligned}$$

which implies

$$\|u\|_{H^2} \leq C(e+y)^{C_0 \epsilon}. \tag{17}$$

Testing (15) by $-\Delta \partial_t u + \nabla \partial_t \pi$, using (1), (3), (16) and (17), we have

$$\begin{aligned}
&\frac{1}{2}\frac{d}{dt}\int |\mathrm{curl}\, \partial_t u|^2 dx + \int | - \Delta \partial_t u + \nabla \partial_t \pi|^2 dx \\
&= \int (-\partial_t u \cdot \nabla u - u \cdot \nabla \partial_t u)(-\Delta \partial_t u + \nabla \partial_t \pi) dx \\
&\leq (\|\nabla u\|_{L^3}\|\partial_t u\|_{L^6} + \|u\|_{L^\infty}\|\nabla \partial_t u\|_{L^2})\| - \Delta \partial_t u + \nabla \partial_t \pi\|_{L^2} \\
&\leq C\|u\|_{H^2}\|\partial_t u\|_{H^1}\| - \Delta \partial_t u + \nabla \partial_t \pi\|_{L^2} \\
&\leq \frac{1}{2}\int | - \Delta \partial_t u + \nabla \partial_t \pi|^2 dx + C\|u\|_{H^2}^2\|\partial_t u\|_{H^1}^2,
\end{aligned}$$

which yields

$$\int |\text{curl}\, \partial_t u|^2 dx + \int_{t_0}^{t} \|\partial_t u\|_{H^2}^2 d\tau \leq C(e+y)^{C_0\epsilon}. \tag{18}$$

On the other hand, it follows from (12), (11), (17) and (18) that

$$\begin{aligned}
\|u\|_{H^3} &\leq C(\|u\|_{L^2} + \|\Delta\omega\|_{L^2}) \\
&\leq C(1 + \|\partial_t\omega + u\cdot\nabla\omega - \omega\cdot\nabla u\|_{L^2}) \\
&\leq C(1 + \|\partial_t\omega\|_{L^2} + \|u\|_{L^\infty}\|\nabla\omega\|_{L^2} + \|\omega\|_{L^4}\|\nabla u\|_{L^4}) \\
&\leq C(1 + \|\partial_t\omega\|_{L^2} + \|u\|_{H^2}^2) \\
&\leq C(e+y)^{C_0\epsilon},
\end{aligned}$$

which gives

$$\|u\|_{L^\infty(0,T;H^3)} \leq C. \tag{19}$$

This completes the proof. □

## 3 Proof of Theorem 1.3

This section is denoted to the proof of Theorem 1.3. We only need to establish some a priori estimates.

First, using the formula $a \times (b \times c) = (a\cdot c)b - (a\cdot b)c$, and the fact that $|d| = 1$ implies $d\Delta d = -|\nabla d|^2$, we have the following equivalent equation

$$\frac{1}{2}d_t - \frac{1}{2}d \times d_t = \Delta d + d|\nabla d|^2. \tag{20}$$

Testing (20) by $d_t$ and using $d\cdot d_t = 0$, we see that

$$\frac{d}{dt}\int |\nabla d|^2 dx + \int |d_t|^2 dx = 0. \tag{21}$$

Testing (7) by $-\Delta d_t$ and using $|d| = 1$, we find that

$$\begin{aligned}
\frac{1}{2}\frac{d}{dt}\int |\Delta d|^2 dx + \int |\nabla d_t|^2 dx &= -\int (d|\nabla d|^2 + d \times \Delta d)\Delta d_t dx \\
= \int \nabla(d|\nabla d|^2 + d \times \Delta d)\cdot \nabla d_t dx
\end{aligned}$$

$$
\begin{aligned}
&\le C(\|\nabla d\|_{L^q}\|\nabla d\|^2_{L^{\frac{4q}{q-2}}} + \|\nabla d\|_{L^q}\|\Delta d\|_{L^{\frac{2q}{q-2}}} + \|\nabla\Delta d\|_{L^2})\|\nabla d_t\|_{L^2} \\
&\le C(\|\nabla d\|_{L^q}\|\Delta d\|_{L^{\frac{2q}{q-2}}} + \|\nabla\Delta d\|_{L^2})\|\nabla d_t\|_{L^2} \\
&\le C(\|\nabla d\|_{L^q}\|\Delta d\|_{L^2}^{1-\frac{3}{q}}\|d\|_{H^3}^{\frac{3}{q}} + \|d\|_{H^3})\|\nabla d_t\|_{L^2} \\
&\le \frac{1}{4}\|\nabla d_t\|^2_{L^2} + \epsilon\|d\|^2_{H^3} + C\|\nabla d\|_{L^q}^{\frac{2q}{q-3}}\|\Delta d\|^2_{L^2}
\end{aligned} \tag{22}
$$

for any $0 < \epsilon < 1$. Here we have used the Gagliardo-Nirenberg inequalities:

$$
\|\nabla d\|^2_{L^{\frac{4q}{q-2}}} \le C\|d\|_{L^\infty}\|\Delta d\|_{L^{\frac{2q}{q-2}}}, \tag{23}
$$

$$
\|\Delta d\|_{L^{\frac{2q}{q-2}}} \le C\|\Delta d\|_{L^2}^{1-\frac{3}{q}}\|d\|_{H^3}^{\frac{3}{q}}. \tag{24}
$$

Applying $\partial_i$ to (7), we get

$$
\partial_i d_t - \Delta\partial_i d = \partial_i(d|\nabla d|^2) + \partial_i d \times \Delta d + d \times \Delta\partial_i d.
$$

Testing the above equation by $\Delta\partial_i d$, summing over $i$, and using (23), (24), and $|d| = 1$, we derive

$$
\begin{aligned}
\|d\|_{H^3} &\le C(\|d\|_{L^2} + \|\nabla\Delta d\|_{L^2}) \\
&\le C + C\|\nabla d_t\|_{L^2} + C\|\nabla(d|\nabla d|^2)\|_{L^2} + C\|\partial_i d \times \Delta d\|_{L^2} \\
&\le C + C\|\nabla d_t\|_{L^2} + C\|\nabla d\|_{L^q}\|\nabla d\|^2_{L^{\frac{4q}{q-2}}} + C\|\nabla d\|_{L^q}\|\Delta d\|_{L^{\frac{2q}{q-2}}} \\
&\le C + C\|\nabla d_t\|_{L^2} + C\|\nabla d\|_{L^q}\|\Delta d\|_{L^{\frac{2q}{q-2}}} \\
&\le C + C\|\nabla d_t\|_{L^2} + C\|\nabla d\|_{L^q}\|\Delta d\|_{L^2}^{1-\frac{3}{q}}\|d\|_{H^3}^{\frac{3}{q}},
\end{aligned}
$$

which leads to

$$
\|d\|_{H^3} \le C + C\|\nabla d_t\|_{L^2} + C\|\nabla d\|_{L^q}^{\frac{q}{q-3}}\|\Delta d\|_{L^2}. \tag{25}
$$

Putting (25) into (22), taking $\epsilon$ small enough and using the Gronwall inequality, we deduce that

$$
\|d\|_{L^\infty(0,T;H^2)} + \|d\|_{L^2(0,T;H^3)} \le C, \tag{26}
$$

$$
\|d_t\|_{L^\infty(0,T;L^2)} + \|d_t\|_{L^2(0,T;H^1)} \le C. \tag{27}
$$

Applying $\partial_t$ to (7), testing by $-\Delta d_t$, and using $|d| = 1$, (26) and (27), we conclude that

$$\frac{1}{2}\frac{d}{dt}\int |\nabla d_t|^2 dx + \int |\Delta d_t|^2 dx = -\int [\partial_t(d|\nabla d|^2) + d_t \times \Delta d]\Delta d_t dx$$
$$\leq C(\|\nabla d\|_{L^6}^2 \|d_t\|_{L^6} + \|\nabla d\|_{L^\infty}\|\nabla d_t\|_{L^2} + \|d_t\|_{L^6}\|\Delta d\|_{L^3})\|\Delta d_t\|_{L^2}$$
$$\leq C(\|d_t\|_{H^1} + \|\nabla d\|_{L^\infty}\|\nabla d_t\|_{L^2} + \|\Delta d\|_{L^3}\|d_t\|_{H^1})\|\Delta d_t\|_{L^2}$$
$$\leq \frac{1}{4}\|\Delta d_t\|_{L^2}^2 + C\|d_t\|_{H^1}^2 + C(\|\nabla d\|_{L^\infty}^2 + \|\Delta d\|_{L^3}^2)\|d_t\|_{H^1}^2,$$

which gives

$$\|d_t\|_{L^\infty(0,T;H^1)} + \|d_t\|_{L^2(0,T;H^2)} \leq C, \tag{28}$$

from which and (25) it follows that

$$\|d\|_{L^\infty(0,T;H^3)} \leq C.$$

This completes the proof. □

**Acknowledgements** The research is partially supported by the NSFC (No. 11171154).

## References


1. H. Beirão da Veiga, A new regularity class for the Navier-Stokes equations in $\mathbb{R}^n$. Chin. Ann. Math. Ser. **B16**, 407–412 (1995)
2. G. Carbou, P. Fabrie, Regular solutions for Landau-Lifshitz equation in a bounded domain. Differ. Integr. Equ. **14**, 213–229 (2001)
3. P. Constantin, C. Fefferman, Direction of vorticity and the problem of global regularity for the Navier-Stokes equations. Indiana Univ. Math. J. **42**, 775–789 (1993)
4. L. Escauriaza, G.A. Serëgin, V. Sverak, $L_{3,\infty}$-solutions of Navier-Stokes equations and backward uniqueness. Uspekhi Mat. Nauk **58**, 3–44 (2003)
5. J. Fan, T. Ozawa, Logarithmically improved regularity criteria for Navier-Stokes and related equations. Math. Methods Appl. Sci. **32**, 2309–2318 (2009)
6. J. Fan, S. Jiang, G. Nakamura, Y. Zhou, Logarithmically improved regularity criteria for the Navier-Stokes and MHD equations. J. Math. Fluid Mech. **13**, 557–571 (2011)
7. Y. Giga, Solutions for semilinear parabolic equations in $L^p$ and regularity of weak solutions of the Navier-Stokes system. J. Differ. Equ. **62**, 186–212 (1986)
8. E. Hopf, Über die Anfangswertaufgabe für die hydrodynamischen Grundgleichungen. Math. Nachr. **4**, 213–231 (1951)
9. K. Kang, J. Kim, Regularity criteria of the magnetohydrodynamic equations in bounded domains or a half space. J. Differ. Equ. **253**(2), 764–794 (2012)
10. H. Kim, A blow-up criterion for the nonhomogeneous incompressible Navier-Stokes equations. SIAM J. Math. Anal. **37**, 1417–1434 (2006)

11. H. Kozono, Y. Shimada, Bilinear estimates in homogeneous Triebel-Lizorkin spaces and the Navier-Stokes equations. Math. Nachr. **276**, 63–74 (2004)
12. H. Kozono, T. Ogawa, Y. Taniuchi, The critical Sobolev inequalities in Besov spaces and regularity criterion to some semi-linear evolution equations. Math. Z. **242**, 251–278 (2002)
13. J. Leray, Sur le mouvement d'un liquide visqueux emplissant l'espace. Acta Math. **63**, 183–248 (1934)
14. T. Ogawa, Y. Taniuchi, A note on blow-up criterion to the 3D Euler equations in a bounded domain. J. Differ. Equ. **190**, 39–63 (2003)
15. J. Serrin, The initial value problem for the Navier-Stokes equations, in *Nonlinear Problems. Proceedings of Symposia*, Madison, WI (University of Wisconsin Press, Madison, 1963), pp. 69–98
16. M. Struwe, On a Serrin type regularity criterion for the Navier-Stokes equations in terms of the pressure. J. Math. Fluid Mech. **9**, 235–242 (2007)

# Local Regularity Results for the Instationary Navier-Stokes Equations Based on Besov Space Type Criteria

**Reinhard Farwig**

*To our colleague Prof. Yoshihiro Shibata on the occasion of his 60th birthday*

**Abstract** Consider a weak instationary solution $u$ of the Navier-Stokes equations in a domain $\Omega \subset \mathbb{R}^3$, i.e.,

$$u \in L^\infty\big(0,T;L^2(\Omega)\big) \cap L^2\big(0,T;W_0^{1,2}(\Omega)\big)$$

and $u$ solves the Navier-Stokes system in the sense of distributions. It is a famous open problem whether weak solutions are unique and smooth. A main step in the analysis of this problem is to show that the given weak solution is a strong one in the sense of J. Serrin, i.e., $u \in L^s\big(0,T;L^q(\Omega)\big)$ where $s > 2$, $q > 3$ and $\frac{2}{s} + \frac{3}{q} = 1$. In this review we report on recent results on this problem, considering first of all optimal initial values $u(0)$ to yield a local in time strong solution, then criteria to prove regularity locally on subintervals of $[0,T)$. Special emphasis is put on results for smooth bounded and also general unbounded domains. Most criteria are based on conditions of Besov space type.

**Keywords** General domains • Initial values • Navier-Stokes equations • Regularity • Serrin's class • Strong solutions • Uniqueness • Weak solutions

## 1 Introduction

The Navier-Stokes system is the most classical model to describe the flow of a viscous incompressible fluid. Despite of about 80 years of mathematical analysis, since the seminal paper of J. Leray [41] on the existence of global weak solutions in the whole space $\mathbb{R}^3$ and corresponding results of E. Hopf [34] for domains, basic

R. Farwig (✉)
Fachbereich Mathematik and International Research Training Group (IRTG 1529), Darmstadt-Tokyo, Technische Universität Darmstadt, Schlossgartenstr. 7, 64283 Darmstadt, Germany
e-mail: farwig@mathematik.tu-darmstadt.de



H. Amann et al. (eds.), *Recent Developments of Mathematical Fluid Mechanics*, Advances in Mathematical Fluid Mechanics, DOI 10.1007/978-3-0348-0939-9_11

questions on uniqueness and regularity of weak solutions are still open. These fundamental problems are also of importance for the general theory of partial differential equations and brought Clay Mathematics Institute in 2000 to classify the issue of regularity as one of the seven Millennium Prize Problem, see C. Fefferman [27].

Given a domain $\Omega \subset \mathbb{R}^3$ and a time interval $[0,T)$, $0 < T \leq \infty$, consider an external force $f : \Omega \times (0,T) \to \mathbb{R}^3$ and an initial value $u_0 : \Omega \to \mathbb{R}^3$. Then we are looking for a velocity field $u$ and an associated scalar pressure function $p$ such that

$$\begin{aligned} u_t - \Delta u + u \cdot \nabla u + \nabla p &= f, &\quad \operatorname{div} u &= 0 \text{ in } \Omega \times (0,T), \\ u(0) &= u_0, & u &= 0 \text{ on } \partial\Omega \times (0,T). \end{aligned} \tag{1}$$

For simplicity the coefficient of viscosity $\nu > 0$ has been set to $\nu = 1$. The nonlinear transport term $u \cdot \nabla u = \sum_{j=1}^3 u_j \partial_j u$ can also be written in the form

$$u \cdot \nabla u = \operatorname{div}(u \otimes u)$$

since $u$ is solenoidal; we recall that $u \otimes v = (u_i v_j)_{i,j=1}^3$ and $\operatorname{div}\big((F_{ij})_{i,j}\big) = \big(\sum_{i=1}^3 \partial_i F_{ij}\big)_{j=1}^3$ for a matrix field $F = (F_{ij})$.

In this article we use standard notation for Lebesgue, Sobolev and Bochner spaces, i.e. $\big(L^q(\Omega) = L^q, \|\cdot\|_q\big)$, $\big(W^{k,q}(\Omega) = W^{k,q}, \|\cdot\|_{W^{k,q}}\big)$, and $\big(L^s(0,T); L^q(\Omega) = L^s(L^q)), \|\cdot\|_{q,s;T} = \|\cdot\|_{L^s(L^q)}\big)$, $1 \leq s, q \leq \infty$, respectively. We do not differ between spaces of scalar-, vector- and matrix-valued functions. The index $\sigma$ will denote a subspace of solenoidal vector fields, the subscript 0 a subspace of functions with vanishing trace. Duality products of functions on $\Omega$ and $\Omega \times (0,T)$ will be denoted by $\langle \cdot,\cdot \rangle_\Omega = \langle \cdot,\cdot \rangle$ and $\langle \cdot,\cdot \rangle_{\Omega,T}$, respectively.

**Definition 1.1 (Definition of Weak and Strong Solutions)**

1. A *weak solution* $u$ (in the sense of Leray-Hopf) is a solution in the sense of distributions, i.e.

$$-\langle u, w_t \rangle_{\Omega,T} + \langle \nabla u, \nabla w \rangle_{\Omega,T} + \langle u \cdot \nabla u, w \rangle_{\Omega,T} = \langle f, w \rangle_{\Omega,T} + \langle u_0, w(0) \rangle \tag{2}$$

for all test functions $w \in C_0^\infty\big([0,T); C_{0,\sigma}^\infty(\Omega)\big)$, lying in the *Leray-Hopf class*

$$u \in \mathcal{LH}_T = L^\infty\big(0,T; L^2_\sigma(\Omega)\big) \cap L^2_{\mathrm{loc}}\big([0,T); H^1_0(\Omega)\big) \tag{3}$$

and satisfying the *strong energy inequality* (*SEI*)

$$\frac{1}{2}\|u(t)\|_2^2 + \int_{t_0}^{t} \|\nabla u\|_2^2 \, d\tau \leq \frac{1}{2}\|u(t_0)\|_2^2 + \int_{t_0}^{t} \langle f, u \rangle_\Omega \, d\tau \tag{4}$$

for almost all $t_0 \in [0,T)$ and for all $t \in [0,T)$.

If $u$ satisfies (4) for $t_0$ and all $t > t_0$ then we say that $u$ satisfies the *energy inequality* $(EI)_{t_0}$. To emphasize that $(EI)_0$ holds (only), $u$ is said to satisfy the *energy inequality* $(EI)$.

2. A weak solution $u \in \mathcal{LH}_T$ is called a *strong solution* (in the sense of Serrin) if there are exponents $s, q$ such that

$$u \in L^s\big(0, T; L^q(\Omega)\big), \quad s > 2,\ q > 3, \quad \frac{2}{s} + \frac{3}{q} = 1. \tag{5}$$

Under the assumption $\frac{2}{s} + \frac{3}{q} = 1$ the space $L^s\big(0, T; L^q(\Omega)\big)$ is called a *Serrin class*.

Let us recall several important results on weak and strong solutions.

*Remark 1.2*

1. Redefining a weak solution $u$ on a subset of $(0, T)$ of Lebesgue measure equal to 0 we may assume that

$$u : [0, T) \to L^2_\sigma(\Omega) \ \text{ is weakly continuous,} \tag{6}$$

or $u \in C^0_w\big([0, T); L^2_\sigma(\Omega)\big)$ for short. Condition (6) will be tacitly assumed in the following.

2. The existence of weak solutions is known for arbitrary domains $\Omega \subset \mathbb{R}^3$, initial values in $u_0 \in L^2_\sigma(\Omega)$ and forces $f \in L^1\big(0, T; L^2(\Omega)\big)$. More generally, $f$ may be assumed to be a functional of the form $f = \operatorname{div} F$, $F \in L^2\big(0, T; L^2(\Omega)\big)$ so that in (2)

$$\langle f, w\rangle_{\Omega,T} := -\langle F, \nabla w\rangle_{\Omega,T}.$$

Under this assumption a weak solution $u$ of (1) can be extended to $t = T$ and considered as a function $u \in C^0_w\big([0, T]; L^2_\sigma(\Omega)\big)$. Then the energy inequality (and the energy equality, if possible) does hold for all $t$ up to $t = T$.

Functionals of the form $f = \operatorname{div} F$ with $F \in L^2\big(0, T; L^2(\Omega)\big)$ are important since several parts of the theory require the nonlinear term to satisfy $u \cdot \nabla u = \operatorname{div}(u \otimes u)$ with $u \otimes u \in L^2\big(0, T; L^2(\Omega)\big)$.

3. Weak solutions can be constructed by several well-known methods, e.g., by Galerkin's method. However, in each case the weak solution satisfies the energy inequality $(EI)$ in (4). The reason for the inequality rather than an equality is the use of approximation techniques and the lower semi-continuity of the norm $\|\cdot\|_2$ with respect to weak convergence. When $\Omega$ is bounded, the weak solutions even satisfy the *strong energy inequality* $(SEI)$.

In the case of an exterior domain maximal regularity estimates yielding $L^s(L^q)$-estimates of the associated pressure $p$ are needed to construct a weak solution satisfying $(SEI)$, see [46]. In general smooth unbounded domains more special functions spaces are needed to construct (suitable) weak solutions satisfying the strong energy inequality. For details we refer to [14, 22] and Sect. 2.4.

4. It is an open problem whether each weak solution independent of the way it has been constructed satisfies (*EI*) or even (*SEI*).
5. One fundamental problem on weak solutions concerns their *uniqueness*. A classical theorem, the so-called *Serrin uniqueness theorem*, see [50], states that a weak solution $u$ lying in a Serrin class $L^s\big(0,T;L^q(\Omega)\big)$, see (5), is unique within the class of all Leray-Hopf type weak solutions. We note that this theorem exploits the fact that a Leray-Hopf type weak solution satisfies (*EI*). Uniqueness for a weak solution in the limit class $L^\infty\big(0,T;L^3(\Omega)\big)$ was proved by Kozono-Sohr [39] and in [25], see also Theorem 2.11 below.
6. The classical condition on weak solutions to satisfy the *energy equality* (*EE*), i.e.

$$\frac{1}{2}\|u(t)\|_2^2 + \int_0^t \|\nabla u\|_2^2\,\mathrm{d}\tau = \frac{1}{2}\|u_0\|_2^2 + \int_0^t \langle f, u\rangle_\Omega\,\mathrm{d}\tau \tag{7}$$

for all $t \in (0,T)$, is given

$$u \in L^4\big(0,T;L^4(\Omega)\big) \tag{8}$$

or equivalently $u \otimes u \in L^2\big(0,T;L^2(\Omega)\big)$. In this case, $u$ may be used as test function $w$ in (2) leading to the $L^1$-integrability of $(u\cdot\nabla u)u$ and a vanishing integral on $\Omega\times(0,T)$ due to the fact that $\langle u\cdot\nabla u, u\rangle_\Omega(t) = 0$ for a.a. $t \in (0,T)$. Note that in the $3D$ case—in contrast to the $2D$ case—a weak solution $u \in \mathcal{LH}_T$ is not an admissible test function in (2).
7. Assumptions different to (8) to guarantee (*EE*) were discussed by Shinbrot [51]: it suffices to assume that $u \in L^r\big(0,T;L^q(\Omega)\big)$ where $\frac{2}{r}+\frac{3}{q} \le 1+\frac{1}{q}$, $q \ge 4$. Actually, Shinbrot's condition together with the Leray-Hopf integrability $u \in L^\infty\big(0,T;L^2(\Omega)\big)$ implies by Hölder's inequality that $u \in L^4\big(0,T;L^4(\Omega)\big)$. A similar argument can be applied when $\frac{2}{r}+\frac{3}{q} \le 1+\frac{1}{r}$, $r \ge 4$ together with the fact that $u \in L^2\big(0,T;L^6(\Omega)\big)$.

   Cheskidov et al. [5] and together with Constantin [4] found much weaker conditions which concerning their scaling behavior are of the type $L^3\big(0,T;L^{9/2}(\Omega)\big)$. Note that $\frac{2}{3}+\frac{3}{9/2} = \frac{4}{3} > \frac{5}{4}$. Let $\Omega \subset \mathbb{R}^3$ be a bounded $C^2$-domain and $u$ a weak solution of (1). If $u \in L^3\big(0,T;\mathcal{D}(A_2^{5/12})\big)$, then $u$ satisfies (*EE*); see [5]. The even weaker condition $u \in L^3\big(0,T;B^{1/3}_{3,\infty}(\mathbb{R}^3)\big)$ is sufficient to get the same result for the whole space $\Omega = \mathbb{R}^3$, see [4]; here $B^{1/3}_{3,\infty}(\mathbb{R}^3)$ denotes a Besov space.

   An intermediate result was recently obtained in [13]: It suffices to use a condition on a $\frac{1}{2}$-derivative of $u$, namely $u \in L^3\big(0,T,\mathcal{D}(A^{1/4}_{18/7})\big)$. A similar condition is needed for flows in smooth unbounded domains; however, since the Stokes operator $A_{18/7}$ possibly cannot be defined for an unbounded domain it is replaced by the Stokes operator $\tilde{A}_{18/7}$ as in Sect. 2.4. Similar results were found in Lorentz spaces; for details we refer to [13].

8. The main open problem for weak solutions is the question of *regularity*, see [27]: is every weak solution $u$ (and an associated pressure $p$) of class $C^\infty$ in space and time provided that $u_0$ and $\partial\Omega$ are of class $C^\infty$? The classical result requires that $u$ lies in Serrin's class $L^s\big(0,T;L^q(\Omega)\big)$ as in (5); cf. [54, Chap. V, Theorems 1.8.1 and 1.8.2].

There are numerous results on *conditional regularity*, i.e., a posteriori conditions on a given weak solution $u$ to guarantee its regularity. Most of these criteria are of Serrin type controlling $\nabla u$, $\omega = \operatorname{rot} u$ or various components of $u$, $\nabla u$, $\omega$; other conditions work with the deformation tensor $\frac{1}{2}\big(\nabla u + (\nabla u)^T\big)$ or the pressure $p$. Rather than trying to summarize and describe these results the focus of this review is on a recent approach to use an optimal initial value condition on $u_0$ and on function values $u(t)$ of $u$ for all or almost all $t$. These and further related results, including also unbounded domains, can be found in the articles [7, 8, 11–14, 16, 20, 21, 23–26, 37, 47].

This article is organized as follows. In Sect. 2 we present the main results, postponing the proofs to Sect. 3. After fixing some notation and preliminaries in Sect. 2.1 the main Theorem 2.1 in Sect. 2.2 describing the optimal initial value condition is the basis for most of the following results and uses a Besov space characterization to be introduced in the beginning of Sect. 2.2. Next, Sect. 2.3 deals with further regularity criteria which either in some sense are beyond Serrin's condition or need the kinetic energy function. General unbounded domains will be considered in Sect. 2.4. Section 3 contains the proofs or sketches of them; the bounded domain case is described in Sects. 3.1 and 3.2, the case of general smooth domains in Sect. 3.3.

# 2 Main Results

## 2.1 Notation

First we introduce the Helmholtz projection on $L^q(\Omega)$,

$$P_q : L^q(\Omega) \to L^q_\sigma(\Omega) = \overline{C^\infty_{0,\sigma}(\Omega)}^{\|\cdot\|_q},$$

where $1 < q < \infty$ and $C^\infty_{0,\sigma}(\Omega) = \{u \in C^\infty_0(\Omega) : \operatorname{div} u = 0\}$. We recall that $P_q$ is a well-defined bounded projection for bounded and exterior $C^1$-domains and defines an algebraic and topological decomposition

$$L^q(\Omega) = L^q_\sigma(\Omega) \oplus G_q(\Omega)$$

with $L^q_\sigma(\Omega) = \mathcal{R}(P_q)$, the range of $P_q$, and $G_q(\Omega) = \{\nabla p \in L^q(\Omega) : p \in L^q_{\text{loc}}(\overline{\Omega})\} = \mathcal{N}(P_q)$, the kernel of $P_q$; for details see [29, 52].

The Stokes operator $A_q = -P_q\Delta$ on $L^q_\sigma(\Omega)$, $1 < q < \infty$, is defined by

$$\begin{aligned} \mathcal{D}(A_q) &= W^{2,q}(\Omega) \cap W^{1,q}_0(\Omega) \cap L^q_\sigma(\Omega), \\ A_q &: \mathcal{D}(A_q) \subset L^q_\sigma(\Omega) \to L^q_\sigma(\Omega), \quad u \mapsto A_q u = -P_q\,\Delta u. \end{aligned}$$

It is well-known, see e.g. [10, 30], that $-A_q$ generates a bounded analytic semigroup $\{e^{-tA_q}; t \geq 0\}$ for bounded and exterior domains of class $C^{1,1}$. Since $A_q$ coincides with $A_r$ on $\mathcal{D}(A_q) \cap \mathcal{D}(A_r)$, $1 < r, q < \infty$, we simply write $A$; by analogy, since $P_q u = P_r u$ for $u \in C^\infty_{0,\sigma}(\Omega)$, we simply write $P$. Note that in general non-smooth or general unbounded domains $P$ and $A$ may fail to exist, see [2, 43]. However, for $q = 2$ and any domain $\Omega \subset \mathbb{R}^3$, Hilbert space methods can be used to define $P_2$ and $A_2 = -P_2\Delta$ with the properties mentioned above.

The Stokes operator $A_q$, $1 < q < \infty$, being sectorial and generating a bounded analytic semigroup, admits fractional powers $A^\alpha_q$, $-1 \leq \alpha \leq 1$. First we consider the case of a bounded domain $\Omega \subset \mathbb{R}^3$ of class $C^{1,1}$. Then $A^\alpha_q$, $0 \leq \alpha \leq 1$, is an injective, closed and densely defined operator with domain $\mathcal{D}(A^\alpha_q) \subset L^q_\sigma(\Omega)$ and range $\mathcal{R}(A^\alpha_q) = L^q_\sigma(\Omega)$ such that $\mathcal{D}(A_q) \subset \mathcal{D}(A^\alpha_q) \subset \mathcal{D}(A^\beta_q) \subset L^q_\sigma(\Omega)$ for $0 \leq \beta \leq \alpha \leq 1$. For $-1 \leq \alpha < 0$ we define the bounded operators $A^\alpha_q = (A^{-\alpha}_q)^{-1} : L^q_\sigma(\Omega) \to \mathcal{R}(A^\alpha_q) = \mathcal{D}(A^{-\alpha}_q)$. Important properties are the embeddings (cf. [31])

$$\|v\|_q \leq c\|A^\alpha_r v\|_r, \quad v \in \mathcal{D}(A^\alpha_r), \quad 1 < r \leq q < \infty, \quad 2\alpha + \frac{3}{q} = \frac{3}{r} \tag{9}$$

$$\|\nabla v\|_q \sim \|A^{1/2}_q v\|_q, \quad v \in \mathcal{D}(A^{1/2}_q) = W^{1,q}_{0,\sigma}(\Omega), \quad 1 < q < \infty; \tag{10}$$

moreover, $\|\nabla v\|_2 = \|A^{1/2}_2 v\|_2$ for $v \in W^{1,2}_{0,\sigma}(\Omega)$. From semigroup theory and the fact that $\sigma(A_q) \subset (0, \infty)$ we know that there exist $c = c(q, \Omega) > 0$ and $\delta = \delta(q, \Omega) > 0$ such that

$$\|A^\alpha_q e^{-tA_q} v\|_q \leq ct^{-\alpha} e^{-\delta t}\|v\|_q, \quad v \in L^q_\sigma(\Omega), \quad 0 \leq \alpha \leq 1. \tag{11}$$

When $\Omega \subset \mathbb{R}^3$ is an exterior domain with $\partial\Omega \in C^{1,1}$, then inequalities (9)–(11) do still hold under certain restrictions, cf. [3, 33]: In (9) we need that $2\alpha + \frac{3}{q} = \frac{3}{r}$ and either

$$0 \leq \alpha \leq \frac{1}{2}, \quad 1 < r < 3 \quad \text{or} \quad 0 \leq \alpha \leq 1, \quad 1 < r < \frac{3}{2}. \tag{12}$$

Furthermore, the equivalence of norms in (10) is to be replaced by

$$\|A^{1/2}_q v\|_q \leq c\|\nabla v\|_q, \;\; 1 < q < \infty; \quad \|\nabla v\|_q \leq c\|A^{1/2}_q v\|_q, \;\; 1 < q < 3. \tag{13}$$

Since $0 \in \sigma(A) = [0, \infty)$, inequality (11) does hold only with $\delta = 0$.

Using the Helmholtz projection $P$ and the Stokes operator $A = -P\Delta$ we get rid of the pressure term $\nabla p$ in (1) and rewrite (1) as an abstract nonlinear evolution equation

$$u_t + Au + P(u \cdot \nabla u) = Pf \text{ in } (0,T), \quad u(0) = u_0 \tag{14}$$

in $L^q_\sigma(\Omega)$. Here we assume the condition $u_0 \in L^q_\sigma(\Omega)$ leading to $P_q u_0 = u_0$.

A weak solution to the (Navier-)Stokes system with external force $f = \operatorname{div} F$ can be considered as a mild solution of the related (non-)linear integral equation

$$u(t) = e^{-tA}u_0 + A^{1/2} \int_0^t e^{-(t-\tau)A}\big(A^{-1/2}P \operatorname{div}\big)(F - u \otimes u)(\tau)\, d\tau. \tag{15}$$

To explain the meaning of the formal operator $A^{-1/2}P$ div applied to matrix-valued $L^q(\Omega)$-functions $F$ we define for $\varphi \in \mathcal{D}(A_{q'}^{-1/2}) \subseteq L^{q'}_\sigma(\Omega)$

$$\langle A^{-1/2}P \operatorname{div} F, \varphi\rangle := -\langle F, \nabla A_{q'}^{-1/2}\varphi\rangle_\Omega \tag{16}$$

admitting the estimate $\big|\langle F, \nabla A_{q'}^{-1/2}\varphi\rangle\big| \leq \|F\|_q \|\nabla A_{q'}^{-1/2}\varphi\|_{q'} \leq c_{q'}\|\varphi\|_{q'}$. Hence

$$A^{-1/2}P \operatorname{div} \in \mathcal{L}\big(L^q(\Omega)\big), \quad \|A^{-1/2}P \operatorname{div}\|_{\mathcal{L}(L^q)} \leq c_{q'}(\Omega). \tag{17}$$

Note that (17) holds for all $1 < q < \infty$ when $\Omega$ is bounded, but for $q > \frac{3}{2}$ $(1 < q' < 3)$ only when $\Omega$ is an exterior domain.

For a further analysis of (15) we need the notion of maximal regularity. Consider the abstract inhomogeneous linear Cauchy problem

$$u_t + Au = f \quad \text{in } (0,T), \quad u(0) = u_0 \tag{18}$$

where $-A$ is the generator of an analytic semigroup $e^{-tA}$ on a Banach space $X$. Then $A$ is said to admit *maximal $L^p$-regularity* on $[0,T)$, $1 < p < \infty$, $0 < T \leq \infty$, if for suitable $u_0$ and all $f \in L^p(0,T;X)$ the solution of (18) given by the variation of constants formula

$$u(t) = e^{-tA}u_0 + \int_0^t e^{-(t-\tau)A}f(\tau)\, d\tau, \quad t \in [0,T), \tag{19}$$

is differentiable a.e., $u(t) \in \mathcal{D}(A)$ a.e. and $u_t$, $Au \in L^p(0,T;X)$. In this case the closed graph theorem yields a constant $C = C(p,T,X) > 0$ independent of $f$

such that

$$\|u_t; Au\|_{L^p(0,T;X)} \leq C\left(\int_0^T \|Ae^{-tA}u_0\|_X^p \, dt\right)^{1/p} + C\|f\|_{L^p(0,T;X)}. \tag{20}$$

It is well-known that maximal $L^p$-regularity for one $p \in (1,\infty)$ implies maximal $L^p$-regularity for all $p \in (1,\infty)$.

The Stokes operator $A = A_q$ on $X = L^q_\sigma(\Omega)$, $1 < q < \infty$, has maximal $L^p$-regularity on $[0,\infty)$ for bounded and exterior domains $\Omega \subset \mathbb{R}^3$ of class $C^{1,1}$; see e.g. [33, 42, 55]. In particular, for any $f \in L^s\big(0,\infty; L^q(\Omega)\big)$ the Cauchy problem (18) with $A_q = -P_q\Delta$, $u_0 = 0$ and $f$ replaced by $P_qf$ has a unique (strong) solution $u \in C^0_{\mathrm{loc}}\big([0,\infty); L^q_\sigma(\Omega)\big)$ such that $u_t, Au \in L^s\big(0,\infty; L^q(\Omega)\big)$, $1 < s < \infty$, and

$$\|u_t; Au\|_{q,s;T} \leq c\|f\|_{q,s;T}. \tag{21}$$

Moreover, if $\Omega$ is bounded and $1 < q < \infty$ or $\Omega$ is an exterior domain and $1 < q < \frac{3}{2}$, there exists an associated pressure $p$ such that

$$\|u_t; \nabla^2 u; \nabla p\|_{q,s;T} \leq c\|f\|_{q,s;T}. \tag{22}$$

For $u_0 \in \big(L^q_\sigma(\Omega), \mathcal{D}(A_q)\big)_{1-1/s,s} \subset L^q_\sigma(\Omega)$, a Besov space defined by real interpolation with norm

$$|||u_0|||_{1-1/s,s,q} := \|u_0\|_q + \left(\int_0^\infty \|Ae^{-tA}u_0\|_q^s \, d\tau\right)^{1/s},$$

the Cauchy problem (18) has a unique solution as above. However, in (21), (22) the term $|||u_0|||_{1-1/s,s,q}$ has to be added on the right-hand side. For details on the above real interpolation we refer to [56, Chaps. 1.13, 1.14]. The interval of integration, $(0,\infty)$, in the norm $|||\cdot|||_{1-1/s,s,q}$ may be replaced by $(0,\delta)$ for any $\delta > 0$, and, if $\Omega$ is bounded, the term $\|u_0\|_q$ may be omitted [56, Theorem 1.14.5]; cf. the discussion of $\|u_0\|_{\mathbb{B}^{-2/s}_{q,s}}$ in Sect. 2.2.

## 2.2 *Optimal Initial Values*

To describe an optimal condition on initial values $u_0 \in L^2_\sigma(\Omega)$ to allow for a local in time strong solution $u \in L^s\big(0,T; L^q(\Omega)\big)$ of Serrin type of the Navier-Stokes system (1) it is natural to require that the solution

$$E_0(t) := e^{-tA}u_0$$

of the corresponding linear Stokes system with vanishing external force has the property $E_0 \in L^s(0,T;L^q(\Omega))$. Actually, this condition which is well-known for the case of the whole space $\Omega = \mathbb{R}^3$ yields also a necessary and sufficient condition for smooth bounded and exterior domains, see [8, 11, 21, 24, 37], respectively. The integrability condition on $E_0$, say

$$\int_0^\infty \|e^{-\tau A}u_0\|_q^s \,\mathrm{d}\tau < \infty, \tag{23}$$

will be explained in terms of Besov spaces. Starting with the classical Besov space $B^{2/s}_{q',s'}(\Omega)$ for a domain $\Omega \subset \mathbb{R}^3$ (cf. [56, Chap. 4]) where $q'$, $s'$ are the conjugate exponents to $q$, $s$, respectively, and $\frac{2}{s} + \frac{3}{q} = 1$, solenoidal subspaces

$$\mathbb{B}^{2/s}_{q',s'}(\Omega) = B^{2/s}_{q',s'}(\Omega) \cap L^{q'}_\sigma(\Omega) = \left\{v \in B^{2/s}_{q',s'}(\Omega) : \operatorname{div} v = 0,\ N\cdot v|_{\partial\Omega} = 0\right\}$$

were defined in [1]. Actually, $\mathbb{B}^{2/s}_{q',s'}(\Omega)$ coincides with the real interpolation space $\left(L^{q'}_\sigma(\Omega), \mathcal{D}(A_{q'})\right)_{1/s,s'} \subset L^{q'}_\sigma(\Omega)$ [1, Proposition 3.4 (3.18)] yielding an optimal space of initial values $u_0$ such that $E_0(t) = e^{-tA}u_0$ satisfies $(E_0)_t$, $AE_0 \in L^{s'}\left(0,T;L^{q'}_\sigma(\Omega)\right)$, i.e., $E_0$ is a classical strong solution of the homogeneous Stokes problem with initial value $u_0$.

Here we do need the dual space

$$\mathbb{B}^{-2/s}_{q,s}(\Omega) := \left(\mathbb{B}^{2/s}_{q',s'}(\Omega)\right)^*. \tag{24}$$

By elementary properties of real interpolation and the duality theorem [56, Theorem 1.11.2]

$$\mathbb{B}^{-2/s}_{q,s}(\Omega) = \left(\left(\mathcal{D}(A_{q'}), L^{q'}_\sigma(\Omega)\right)_{1/s',s'}\right)^* = \left(\mathcal{D}(A_{q'})^*, L^q_\sigma(\Omega)\right)_{1/s',s}$$

since $\left(L^{q'}_\sigma(\Omega)\right)^* = L^q_\sigma(\Omega)$. Hence

$$\begin{aligned}\|u_0\|_{\mathbb{B}^{-2/s}_{q,s}} &\sim \|u_0\|_{\left(\mathcal{D}(A_{q'})^*, L^q_\sigma(\Omega)\right)_{1/s',s}} \sim \|A^{-1}u_0\|_{\left(L^q_\sigma(\Omega), \mathcal{D}(A_q)\right)_{1-1/s,s}} \\ &\sim \|A^{-1}u_0\|_q + \left(\int_0^\infty \|Ae^{-\tau A}(A^{-1}u_0)\|_q^s \,\mathrm{d}\tau\right)^{1/s}\end{aligned} \tag{25}$$

where the second equivalence of norms uses the identity $\langle A^{-1}u_0, A\varphi\rangle_\Omega = \langle u_0, \varphi\rangle_\Omega$ for $\varphi \in \mathcal{D}(A_{q'})$ and the equivalence $\|\varphi\|_{W^{2,q'}} \sim \|A_{q'}\varphi\|_{q'}$ (for bounded $\Omega$). The norm on the right-hand side of (25) is the norm of $A^{-1}u_0$ in $\mathbb{B}^{2/s}_{q,s}(\Omega)$. By [56, Theorem 1.14.5], the interval of integration $(0,\infty)$ may be replaced by any interval

$(0, \delta)$, $0 < \delta < \infty$, yielding an equivalent norm on $\mathbb{B}_{q,s}^{-2/s}(\Omega)$. Finally, for a bounded domain, the term $\|A^{-1}u_0\|_q$ in (25) may be omitted.

Given $\delta \in (0, \infty]$ we denote the space $\mathbb{B}_{q,s}^{-2/s}(\Omega)$ also by

$$\mathcal{B}_\delta^{q,s}(\Omega) = \left\{ u_0 : \|u_0\|_{\mathcal{B}_\delta^{q,s}} := \left( \int_0^\delta \|e^{-\tau A} u_0\|_q^s \, d\tau \right)^{1/s} < \infty \right\}. \tag{26}$$

Recall that $\mathcal{B}_\delta^{q,s}(\Omega) \subset \mathcal{D}(A_{q'})^*$ is a reflexive, separable Banach space and that all norms $\|\cdot\|_{\mathcal{B}_\delta^{q,s}(\Omega)}$, $\delta > 0$, are equivalent. For $\mathcal{B}_\infty^{q,s}(\Omega)$ we also write $\mathcal{B}^{q,s}(\Omega)$.

**Theorem 2.1** *Let $\Omega \subset \mathbb{R}^3$ be a bounded domain with boundary $\partial\Omega \in C^{1,1}$, let $0 < T \le \infty$, $2 < s < \infty$, $3 < q < \infty$ with $\frac{2}{s} + \frac{3}{q} = 1$ be given. We consider the Navier-Stokes system* (1) *with initial value $u_0 \in L_\sigma^2(\Omega)$ and an external force $f = \operatorname{div} F$ where $F \in L^2(0, T; L^2(\Omega)) \cap L^{s/2}(0, T; L^{q/2}(\Omega))$.*

(i) *There exists an absolute constant $\varepsilon_* = \varepsilon_*(q, \Omega) > 0$ with the following property: If*

$$\|u_0\|_{\mathcal{B}_T^{q,s}} + \|F\|_{q/2,s/2;T} \le \varepsilon_*, \tag{27}$$

*then there exists a unique strong solution $u \in L^s(0, T; L^q(\Omega))$ of* (1).

(ii) *The condition $u_0 \in \mathcal{B}_\infty^{q,s}(\Omega)$ is sufficient and necessary for the existence and uniqueness of a local in time strong solution $u \in L^s(0, T'; L^q(\Omega))$ of* (1) *for some $0 < T' \le T$.*

Let us recall further results and extensions of Theorem 2.1.

*Remark 2.2*

1. The condition $u_0 \in \mathcal{D}(A_2^{1/4})$ is due to Fujita and Kato [28] for a smooth bounded domain. This result bases on $L^2$-theory and can be generalized to arbitrary bounded and unbounded domains, see [54, Chap. V, Theorem 4.2.2]. Fabes et al. [6] as well as Miyakawa [45] proved that the condition $u_0 \in L^r(\Omega)$, $r > 3$, yields a local strong solution.
2. The condition $u_0 \in \mathcal{D}(A_2^{1/4})$ has the important property of being *scaling invariant*. Recall that with a solution $u$ of (1) also $u_\lambda(x, t) = \lambda u(\lambda x, \lambda^2 t)$, $\lambda > 0$, is a solution of (1) with the same Reynolds number (here $\mathrm{Re} = \frac{1}{\nu} = 1$). Since $\|u_\lambda\|_{L^s(L^q)} = \lambda^{1-(\frac{2}{s}+\frac{3}{q})}\|u\|_{L^s(L^q)}$ on their respective time intervals and domains, the norm $\|u_\lambda\|_{L^s(L^q)}$ is $\lambda$-independent if and only if $L^s(L^q)$ is a Serrin class; in this case, the space $L^s(L^q)$ (or its norm) is called scaling invariant.

   The corresponding condition for the initial value is related to $(u_0)_\lambda(x) = \lambda u_0(\lambda x)$. Scaling invariant initial value conditions are $u_0 \in \mathcal{D}(A^{1/4})$ with norm $\|A^{1/4}u_0\|_2$ and $u_0 \in L_\sigma^3(\Omega)$. The latter case was considered by Kato [35] and Giga [32].

The condition $u_0 \in L^3_\sigma(\Omega)$ can be weakened to assumptions in Lorentz spaces $L^{3,s}_\sigma(\Omega)$ when $q \le s < \infty$, see [53]. Here we mention the continuous embeddings

$$\mathcal{D}(A_2^{1/4}) \subset L^3_\sigma(\Omega) \subset L^{3,s}_\sigma(\Omega) \subset \mathcal{B}^{q,s}_\infty(\Omega) \tag{28}$$

where each space is scaling invariant; for the latter embedding which holds when $q \le s < \infty$ we refer to [1, (0.16)]. Replacing $\mathcal{B}^{q,s}_\infty(\Omega)$ by $\mathcal{B}^{q,s}_\delta(\Omega)$, the family of spaces $\mathcal{B}^{q,s}_\delta(\Omega)$ is scaling invariant in the sense that $\delta$ must be changed to $\lambda^{-2}\delta$ (and $\Omega$ to $\lambda^{-1}\Omega$).

3. For a smooth exterior domain $\Omega \subset \mathbb{R}^3$ similar results were obtained in [8, 37]. It is also shown that the assumption $F \in L^{s/2}(L^{q/2})$ (with $\frac{2}{s/2} + \frac{3}{q/2} = 2$, cf. Theorem 2.1) can be generalized to $F \in L^{s_*}(L^{q_*})$ with $\frac{2}{s_*} + \frac{3}{q_*} = 2$. Then the condition $\int_0^\infty \|e^{-\tau A}u_0\|_q^s \, d\tau < \infty$ is necessary and sufficient for the existence of a local strong solution $u \in L^s(L^q)$, $\frac{2}{s} + \frac{3}{q} = 1$.
4. For a general bounded or unbounded domain $\Omega \subset \mathbb{R}^3$—even with non-smooth boundary—only $L^2$-theory for the Stokes operator and the Helmholtz projection is available. In this case Theorem 2.1(i) holds with the exponents $q = 4$, $s = 8$, cf. [21, Sect. 4].
5. The largest space of initial values to yield solutions in scaling invariant function spaces was found by Koch and Tataru [36] for the whole space case $\mathbb{R}^3$. Assume that $u_0 \in BMO^{-1}$, i.e., $u_0$ can be written in the form $u_0 = \operatorname{div} f$ with some $f \in BMO$. Then there exists a local solution $u \in L^2_{\mathrm{loc}}(\mathbb{R}^3 \times [0,\infty))$ such that the scaling invariant norm

$$\sup_{x\in\mathbb{R}^3, R>0} \frac{1}{|B_R(x)|} \int_0^{R^2} \int_{B_R(x)} |u|^2 \, dy \, d\tau$$

is finite; here $B_R(x) \subset \mathbb{R}^3$ denotes the ball with center $x$ and radius $R$.

For simplicity let $F = 0$ in the following. It suggests itself to use Theorem 2.1(i) not only at $t_0 = 0$, but at all or almost all $t_0 \in [0,T)$ to show that a weak solution is a strong one. In view of (27) we need more information on the space $\mathcal{B}^{q,s}(\Omega)$ and on functions $u : [0,T) \to \mathcal{B}^{q,s}_\delta(\Omega)$. Although the constant $\varepsilon_*$ in (27) cannot be determined precisely, we will fix some $\varepsilon_* > 0$ so that Theorem 2.1(i) can be applied. For the following definition recall that for any nonzero $v \in L^2_\sigma(\Omega)$ the function $t \mapsto e^{-\tau A}v$ is nonzero for $t \ge 0$. Hence $V(t) := \|v\|_{\mathcal{B}^{q,s}_t}$ is a strictly increasing continuous function in $t > 0$ with range $(0, \|u\|_{\mathcal{B}^{q,s}_\infty})$.

**Definition 2.3** Let $\varepsilon_* > 0$ be fixed and let $0 \ne v \in L^2_\sigma(\Omega)$. Then

$$\delta(v) := \begin{cases} 0 & \text{if } v \notin \mathcal{B}^{q,s}(\Omega) \\ \delta \in (0,\infty) & \text{if } v \in \mathcal{B}^{q,s}(\Omega) \text{ and } \|v\|_{\mathcal{B}^{q,s}_\delta} = \varepsilon_* < \|v\|_{\mathcal{B}^{q,s}_\infty} \\ \infty & \text{if } v \in \mathcal{B}^{q,s}(\Omega) \text{ and } \|v\|_{\mathcal{B}^{q,s}_\infty} \le \varepsilon_* \end{cases}$$

With an abuse of notation we set $\|v\|_{\mathcal{B}^{q,s}} = \infty$ when $v \notin \mathcal{B}^{q,s}(\Omega)$.

**Lemma 2.4** *Let $u \in \mathcal{LH}_T$ be a Leray-Hopf type weak solution of* (1).

*(i) Let $\delta > 0$. Then the function $[0,T] \to [0,\infty]$, $t \mapsto \|u(t)\|_{\mathcal{B}^{q,s}_\delta}$ is lower semi-continuous. If additionally $u \in L^\infty\big([0,T); \mathcal{B}^{q,s}_\delta(\Omega)\big)$, then $u \in C^0_w\big([0,T]; \mathcal{B}^{q,s}_\delta(\Omega)\big)$ and*

$$\|u\|_{L^\infty(0,T;\mathcal{B}^{q,s}_\delta)} = \sup_{t\in[0,T]} \|u(t)\|_{\mathcal{B}^{q,s}_\delta}. \tag{29}$$

*(ii) The function $\delta(t) := \delta((u(t))$, see Definition* 2.3, *is upper semi-continuous in t.*

For the problem of local regularity we need the following terminology:

- $u$ is *left-sided $L^s(L^q)$-regular at $t_0 \in (0,T]$* if there exists $\varepsilon = \varepsilon(t_0) \in (0,t_0)$ such that $u \in L^s\big((t_0-\varepsilon, t_0); L^q(\Omega)\big)$
- $u$ is *right-sided $L^s(L^q)$-regular at $t_0 \in [0,T)$* if there exists $\varepsilon = \varepsilon(t_0) \in (0, T-t_0)$ such that $u \in L^s\big((t_0, t_0+\varepsilon); L^q(\Omega)\big)$
- $u$ is *$L^s(L^q)$-regular at $t_0 \in (0,T)$* if $u$ is left- as well as right-sided regular at $t_0$.

**Theorem 2.5** *Let $u \in \mathcal{LH}_T$ be a weak solution of* (1) *in a bounded smooth domain $\Omega \subset \mathbb{R}^3$, and let $2 < s < \infty$, $3 < q < \infty$, $\frac{2}{s} + \frac{3}{q} = 1$.*

*(i) Let $u$ satisfy $u(t) \in \mathcal{B}^{q,s}(\Omega)$ for a.a. $t \in [0,T)$. Given $t_1 \in (0,T)$ assume that for a.a. $t$ in a left-sided neighborhood of $t_1$*

$$\delta(t) \geq t_1 - t. \tag{30}$$

*Then $u$ is left-sided $L^s(L^q)$-regular at $t_1$.*

*(ii) Given $t_1 \in (0,T)$ assume that $u(t_1) \in \mathcal{B}^{q,s}(\Omega)$ and that $u$ satisfies $(EI)_{t_1}$, i.e., the strong energy inequality in $t_1$. Then $u$ is right-sided $L^s(L^q)$-regular at $t_1$.*

*(iii) Let $u$ satisfy $u(t) \in \mathcal{B}^{q,s}(\Omega)$ for all $t \in [0,T)$ and condition* (30) *at $t_1 \in (0,T)$. Then $u$ is regular at $t_1$. If condition* (30) *is satisfied at* every *$t_1 \in (0,T)$, then $u \in L^s_{\text{loc}}\big([0,T); L^q(\Omega)\big)$.*

*(iv) Let $u$ satisfy $u(t) \in \mathcal{B}^{q,s}(\Omega)$ for all $t \in [0,T)$. Assume that*

$$\lim_{t \nearrow t_1} \|u(t)\|_{\mathcal{B}^{q,s}_{t_1-t}} = 0. \tag{31}$$

*Then $u$ is $L^s(L^q)$-regular at $t_1$. If condition* (31) *is satisfied at* every *$t_1 \in (0,T)$, then $u \in L^s_{\text{loc}}\big([0,T); L^q(\Omega)\big)$.*

*Remark 2.6* Replacing the lower bound $\delta(t) \geq t_1 - t$ in (30) by the weaker condition $\delta(t) \geq \alpha(t_1 - t)$ for a fixed $\alpha \in (0,1)$, the arguments in the proof of Theorem 2.5 can be used iteratively to get a sequence of instants $(t_j)$, $t_j = t_1 - (1-\alpha)^j(t_1 - t_0)$, converging to $t_1$; hence $u \in L^s_{\text{loc}}\big(t_0, t_1; L^q(\Omega)\big)$. However, Theorem 2.1 yields no uniform bound of $\|u\|_{q,s;(t_0,t_j)}$ as $j \to \infty$ so that the statement $u \in L^s\big(t_0, t_1; L^q(\Omega)\big)$ cannot be guaranteed.

**Corollary 2.7** *Let $u$ be a weak solution of the Navier-Stokes system as in Theorem 2.5.*

*(i) Assume that $u \in L^s_{\text{loc}}\big([0,t_1); L^q_\sigma(\Omega)\big)$, but $u \notin L^s\big(0,t_1; L^q_\sigma(\Omega)\big)$. Then there exists an $\epsilon_* > 0$ such that for all $\alpha \in (\frac{1}{4}, \frac{1}{2}]$*

$$\|A^\alpha u(t)\|_2 > \epsilon_*(t_1 - t)^{1/4-\alpha} \quad \textit{for a.a. } t \in (0,t_1). \tag{32}$$

*(ii) Assume that $u$ satisfies $(EI)_{t_1}$ (this condition is e.g. satisfied when $t_1$ is a left-sided regular point of $u$), but is not right-sided regular at $t_1$. Then $u(t_1) \notin \mathcal{B}^{q,s}(\Omega)$ and for each $\delta > 0$ and $\alpha \in \big[\frac{1}{4}, \frac{1}{2}\big]$*

$$\|u(t)\|_3, \ \|A^\alpha u(t)\|_2, \ \|u(t)\|_{\mathcal{B}^{q,s}_\delta} \to \infty \ \textit{ as } \ t \searrow t_1.$$

## 2.3 Regularity Criteria Beyond Serrin's Condition and Energy Criteria

The regularity criteria of Theorem 2.5 have the disadvantage that the norm of $u(t)$ in $\mathcal{B}^{q,s}_{t_1-t}(\Omega)$ cannot be controlled directly. However, there are many applications of Theorem 2.5 yielding more concrete conditions.

**Corollary 2.8** *Let $u \in \mathcal{LH}_T$ be a weak solution of* (1) *on a bounded smooth domain $\Omega \subset \mathbb{R}^3$.*

(i) *If there exists $\delta \in (0,\infty]$ such that $u \in C^0\big([0,T); \mathcal{B}^{q,s}_\delta(\Omega)\big)$, then $u$ is a strong solution on $[0,T)$.*
(ii) *If there exists $\delta \in (0,\infty]$ such $\|u\|_{L^\infty(0,T;\mathcal{B}^{q,s}_\delta)} \le \varepsilon_*$, then $u$ is a strong solution on $[0,T)$.*
(iii) *In (i) and (ii) the space $\mathcal{B}^{q,s}_\delta(\Omega)$ can be replaced by any of the spaces $\mathcal{D}(A^{1/4})$, $L^3_\sigma(\Omega)$ and $L^{3,s}_\sigma(\Omega)$ ($s \ge q > 3$).*

The next criteria are based on Theorem 2.1, however, in a certain sense work beyond Serrin's condition at the expense of a further smallness assumption.

**Theorem 2.9** *Let $u \in \mathcal{LH}_T$ be a weak solution as in Corollary 2.8. Further let the exponents $q$, $r$, $s$ satisfy $2 < s < \infty$, $3 < q < \infty$, $\frac{2}{s} + \frac{3}{q} = 1$ and $1 \le r \le s$.*

*(i) Assume that*

$$\liminf_{\delta \to 0} \frac{1}{\delta^{1-r/s}} \int_{t_1-\delta}^{t_1} \|u(\tau)\|_q^r \, \mathrm{d}\tau = 0. \tag{33}$$

*Then $u$ is regular at $t_1$, i.e., $u \in L^s\big(t_1 - \varepsilon, t_1 + \varepsilon; L^q(\Omega)\big)$ for some $\varepsilon > 0$.*

*(ii) Assume that for* $0 \le t_0 < t_1 < T$ *and some* $t_1 < T' \le T$

$$\frac{1}{t_1 - t_0} \int_{t_0}^{t_1} (T' - \tau)^{r/s} \, \|u(\tau)\|_q^r \, d\tau \le \tilde{\varepsilon}_*. \tag{34}$$

*Then* $u$ *is regular at* $t_1$. *Here* $\tilde{\varepsilon}_* > 0$ *is a constant related to* $\varepsilon_*$ *in* (27).

A consequence of Theorem 2.9(i) is the well-known fact that a weak solution $u \in \mathcal{LH}_T$ is regular almost everywhere (even everywhere in $(0, T)$ except for a possible set $S \subset (0, T)$ of vanishing Hausdorff measure $\mathcal{H}^{1/2}(S) = 0$). Actually, since $u \in L^2\big(0, T; L^6(\Omega)\big)$, Lebesgue's differentiation theorem implies that $\frac{1}{\delta} \int_{t_1-\delta}^{t_1} \|u\|_6^2 \, d\tau \to \|u(t_1)\|_6^2$ $t_1$-a.e as $\delta \to 0$. Hence the term in (33) (with $q = 6$, $r = 2$, $s = 4$) vanishes $t_1$-a.e.

Finally, we describe a regularity criterion based on the *kinetic energy*

$$E_k(t) = \frac{1}{2} \|u(t)\|_2^2. \tag{35}$$

**Theorem 2.10** *Let* $u \in \mathcal{LH}_T$ *be a weak solution of* (1) *as in Corollary* 2.8. *Assume that at* $t_1 \in (0, T)$ *for* $\alpha \in (\frac{1}{2}, 1]$ *the left-sided* $\alpha$*-Hölder seminorm*

$$[E_k(t_1-)]_\alpha = \sup_{\delta > 0} \frac{|E_k(t_1 - \delta) - E_k(t_1)|}{\delta^\alpha} < \infty$$

*(with the supremum taken only for small* $\delta > 0$*) or that*

$$[E_k(t_1-)]_{1/2} \le \varepsilon_*. \tag{36}$$

*Then* $u$ *is regular at* $t_1$.

Note that for the Hölder exponent $\alpha = \frac{1}{2}$ we do need a smallness condition on the local left-sided Hölder seminorm. Actually, if $(t_0, t_1) \subset [0, T)$ is a maximal interval of regularity of a weak solution $u$, then due to (32)

$$\|\nabla u(\tau)\|_2 = \|A^{1/2} u(\tau)\|_2 \ge \varepsilon_* (t_1 - \tau)^{-1/4}, \quad 0 < \tau < t_1.$$

Hence the estimate

$$2c^2 \le \frac{1}{\delta^{1/2}} \int_{t_1-\delta}^{t_1} \|\nabla u(\tau)\|_2^2 \, d\tau \le \frac{1}{\delta^{1/2}} \big(E_k(t_1 - \delta) - E_k(t_1)\big)$$

for a.a. $\delta \in (t_1 - t_0, t_1)$ shows in this case that the condition (36) with an arbitrary (not sufficiently small) $\varepsilon_* > 0$ does not imply regularity. For the case including an external force we refer to [23].

For further regularity criteria beyond Serrin's condition see [16].

**Theorem 2.11** *Let $u \in \mathcal{LH}_T$ be a weak solution of the Navier-Stokes system* (1) *in a bounded smooth domain $\Omega \subset \mathbb{R}^3$ with $u_0 \in L^2_\sigma(\Omega)$.*

*(i) Assume that for some $2 < s < \infty$, $3 < q < \infty$, $\frac{2}{s} + \frac{3}{q} = 1$ there holds $u(t_0) \in \mathcal{B}^{q,s}(\Omega)$ for all $t_0 \in [0,T)$, and that $u$ satisfies the energy equality (EE). Then $u$ is uniquely determined by the initial value $u_0$ within the class of all weak solutions.*

*(ii) If $u \in L^4_{loc}\big([0,T); L^4(\Omega)\big)$ or*

$$u \in L^\infty_{loc}\big([0,T); L^{3,\infty}_\sigma(\Omega)\big), \tag{37}$$

*then $u$ even satisfies (EE).*

*Remark 2.12*

(i) Let $u$ be a weak solution of (1) as in Theorem 2.11. It is interesting to discuss uniqueness and regularity properties of $u$ satisfying the Serrin condition $u \in L^s_{\mathrm{loc}}([0,T); L^q(\Omega))$, $\frac{2}{s} + \frac{3}{q} = 1$ in the *limit case* $s = \infty, q = 3$. In this case, the arguments in the proof of Lemma 2.4(i) show that $u(t) \in L^3(\Omega)$ for each $t \in [0,T)$. Since $L^3(\Omega) \subset L^{3,\infty}(\Omega) \cap \mathcal{B}^{q,s}(\Omega)$, Theorem 2.11 yields the uniqueness property for $u$. On the other hand, from Corollary 2.8(iii) we see that the stronger assumption $u \in C\big([0,T); L^3_\sigma(\Omega)\big)$ is sufficient to get the regularity $u \in L^s_{\mathrm{loc}}\big([0,T); L^q(\Omega)\big)$ with Serrin exponents $s, q$.

(ii) Furthermore, for $u \in L^\infty_{\mathrm{loc}}\big([0,T); L^3_\sigma(\Omega)\big)$ we get the local right-side regularity property for each $t \in [0,T)$, see Theorem 2.5(ii). Hence Theorem 2.5 is a slightly weaker result than that in a series of papers on the celebrated $L^\infty\big(0,T; L^3(\Omega)\big)$-regularity result of Seregin et al. We refer to [49] for domains with a flat boundary, and to [44] where in domains with curved boundaries some additional condition on the pressure had to be assumed.

(iii) In the case of a general domain $\Omega \subset \mathbb{R}^3$, let it be bounded with rough boundary or unbounded, only an $L^2_\sigma(\Omega)$-theory of the Stokes operator is available. Here we get the following results, cf. [16, 21]: Let $u \in \mathcal{LH}_T$ be a weak solution satisfying (*SEI*). Assume that at $t_1 \in (0,T)$ one of the following scaling invariant conditions is satisfied: there exists $0 < \delta < t_1$ such that

$$\frac{1}{\delta}\int_{t_1-\delta}^{t_1} \|A^{1/4}u(\tau)\|_2 \, d\tau \le \varepsilon_*, \tag{38}$$

$$\frac{1}{\delta}\int_{t_1-\delta}^{t_1} \|u(\tau)\|_2 \, \|\nabla u(\tau)\|_2 \, d\tau \le \varepsilon_*, \tag{39}$$

$$\Big(\sup_{[t_1-\delta,t_1]} \|u(\cdot)\|_2^2\Big)\frac{1}{\delta}\int_{t_1-\delta}^{t_1} \|\nabla u(\tau)\|_2^2 \, d\tau \le \varepsilon_*, \tag{40}$$

where $\varepsilon_*$ in (38)–(40) is an absolute constant independent of the domain. Then $u$ is $L^8(L^4)$-regular at $t_1$.

## 2.4 The Navier-Stokes System in General Smooth Domains

**Definition 2.13** For $k \in \mathbb{N}_0$ and $\lambda \in (0, 1]$ a domain $\Omega \subset \mathbb{R}^n$ is called *uniform $C^{k,\lambda}$-domain*, if there are positive constants $\alpha, \beta, K$ such that for all $x_0 \in \partial\Omega$ there exist—after an orthogonal and an affine coordinate transform—a real-valued function $h$ of class $C^{k,\lambda}$ and a neighborhood $U_{\alpha,\beta,h}(x_0)$ of $x_0$ with the following properties: $h$ is defined on the closed ball $\overline{B'_\alpha(0)} \subseteq \mathbb{R}^{n-1}$ with $\|h\|_{C^{k,\lambda}} \le K$ and $h(0) = 0$ and, if $k \ge 1$, $h'(0) = 0$; moreover,

$$
\begin{aligned}
U_{\alpha,\beta,h}(x_0) &:= \{(y', y_n) \in \mathbb{R}^{n-1} \times \mathbb{R}: |y'| < \alpha, |h(y') - y_n| < \beta\},\\
U^-_{\alpha,\beta,h}(x_0) &:= \{(y', y_n) \in \mathbb{R}^{n-1} \times \mathbb{R}: |y'| < \alpha, h(y') - \beta < y_n < h(y')\}\\
&= \Omega \cap U_{\alpha,\beta,h}(x_0),\\
\partial\Omega \cap U_{\alpha,\beta,h}(x_0) &= \{(y', y_n) \in \mathbb{R}^{n-1} \times \mathbb{R}: h(y') = y_n\}.
\end{aligned}
$$

The triple $(\alpha, \beta, K)$ is called the *type* of $\Omega$ and will be denoted by $\tau(\Omega) = (\alpha, \beta, K)$. For a constant $C$ in some estimate we will write $C = C(\tau(\Omega))$ if it does depend only on $\alpha$, $\beta$ and $K$, but in no other way on $\Omega$. A uniform $C^k$-domain is defined in an obviously analogous way.

Note that bounded and exterior domains are included, as long as the boundary is smooth enough; in these cases the boundary $\partial\Omega$ is compact, implying that the constants $\alpha$, $\beta$, $K$ can be chosen uniformly.

Since the Helmholtz projection and the Stokes operator are not necessarily well-defined on a general smooth domain as described in Definition 2.13 we will introduce the spaces

$$
\tilde{L}^q(\Omega) := \begin{cases} L^q(\Omega) + L^2(\Omega), & \text{if } 1 \le q < 2,\\ L^q(\Omega) \cap L^2(\Omega), & \text{if } 2 \le q \le \infty. \end{cases} \tag{41}
$$

For bounded domains $\Omega$ it holds that $\tilde{L}^q(\Omega) = L^q(\Omega)$ with equivalent norms by Hölder's inequality. Note that functions in $\tilde{L}^q(\Omega)$ locally behave like $L^q$-functions, but globally like $L^2$-functions.

Moreover, we define for $1 < q < \infty$, $1 \le \rho \le \infty$ the Lorentz spaces

$$
\tilde{L}^{q,\rho}(\Omega) := \begin{cases} L^{q,\rho}(\Omega) + L^2(\Omega), & q < 2,\\ L^{q,\rho}(\Omega) \cap L^2(\Omega), & q > 2, \end{cases}
$$

letting the case $q = 2$ undefined.

For spaces of Sobolev-type we proceed analogously: For $k \in \mathbb{N}$ and $1 \le q \le \infty$ we let

$$
\tilde{W}^{k,q}(\Omega) := \begin{cases} W^{k,2}(\Omega) + W^{k,q}(\Omega), & 1 \le q < 2,\\ W^{k,2}(\Omega) \cap W^{k,q}(\Omega), & 2 \le q \le \infty. \end{cases}
$$

Similarly, we define the spaces $\tilde{W}_0^{1,q}(\Omega)$, $1 < q < 2$ and $2 \le q < \infty$, based on the classical Sobolev spaces $W_0^{1,q}(\Omega)$ and $W_0^{1,2}(\Omega)$.

The $\tilde{L}^q$- and $\tilde{W}^{k,q}(\Omega)$-spaces have the following properties; for a proof see [47, 48]:

- Let $1 \le q < \infty$. Then $(\tilde{L}^q(\Omega))^* = \tilde{L}^{q'}(\Omega)$
- Let $1 < q \le r \le \infty$. Then $\|u\|_{\tilde{L}^q} \le \|u\|_{\tilde{L}^r}$
- Let $1 \le r, p, q \le \infty$, $\frac{1}{r} = \frac{1}{p} + \frac{1}{q}$ and let $u \in \tilde{L}^p$, $v \in \tilde{L}^q$. Then $uv \in \tilde{L}^r$ and $\|uv\|_{\tilde{L}^r} \le \|u\|_{\tilde{L}^p} \|v\|_{\tilde{L}^q}$
- Let $1 < q, r < \infty$, $0 < \theta < 1$ and let $s$ be defined by $\frac{1}{s} = \frac{1-\theta}{q} + \frac{\theta}{r}$. Then in the sense of complex interpolation spaces

$$\left[\tilde{L}^q(\Omega), \tilde{L}^r(\Omega)\right]_\theta = \tilde{L}^s(\Omega)$$

- Let $1 \le r \ne q \le \infty$, $0 < \theta < 1$, $1 \le \rho \le \infty$, and define $1 < s \ne 2 < \infty$ by $\frac{1}{s} = \frac{1-\theta}{q} + \frac{\theta}{r}$. Then

$$\left(\tilde{L}^q(\Omega), \tilde{L}^r(\Omega)\right)_{\theta,\rho} = \tilde{L}^{s,\rho}(\Omega).$$

  For $s = 2$ and $\rho = 2$ we get

$$\left(\tilde{L}^q(\Omega), \tilde{L}^r(\Omega)\right)_{\theta,2} = L^2(\Omega).$$

- Let $m \in \mathbb{N}$, $1 \le q < \infty$ and $\Omega \subseteq \mathbb{R}^n$ be a uniform $C^2$-domain. Then

$$\tilde{W}^{m,q}(\Omega) \hookrightarrow \tilde{L}^r(\Omega)$$

  if either $q \le r \le \infty$ and $mq > n$, or $q \le r < \infty$ and $mq = n$, or $q \le r \le \frac{nq}{n-mq}$ and $mq < n$.

Concerning the Helmholtz projection on $\tilde{L}^q(\Omega)$ for a domain $\Omega \subseteq \mathbb{R}^n$ of uniform type $C^1$ we have the following result, see [17]. We define

$$\tilde{L}^q_\sigma(\Omega) := \begin{cases} L^q_\sigma(\Omega) + L^2_\sigma(\Omega), & 1 < q < 2 \\ L^q_\sigma(\Omega) \cap L^2_\sigma(\Omega), & 2 \le q < \infty \end{cases},$$

equipped with the norm of $\tilde{L}^q(\Omega)$, and gradient spaces by

$$\tilde{G}_q(\Omega) := \begin{cases} G_q(\Omega) + G_2(\Omega), & 1 < q < 2, \\ G_q(\Omega) \cap G_2(\Omega), & 2 \le q < \infty, \end{cases}$$

with norm $\|\cdot\|_{\tilde{G}_q(\Omega)} := \|\cdot\|_{\tilde{L}^q(\Omega)}$. Then the space $\tilde{L}^q(\Omega)$ admits the direct algebraic and topological decomposition

$$\tilde{L}^q(\Omega) = \tilde{L}^q_\sigma(\Omega) \oplus \tilde{G}_q(\Omega).$$

The corresponding projection $\tilde{P}_q$ from $\tilde{L}^q(\Omega)$ onto $\mathcal{R}(\tilde{P}_q) = \tilde{L}^q_\sigma(\Omega)$ and kernel $\mathcal{N}(\tilde{P}_q) = \tilde{G}_q(\Omega)$ has a norm bounded by a constant $c = c(q, \tau(\Omega))$. It satisfies the relation $(\tilde{P}_q)^* = \tilde{P}_{q'}$.

Using the Helmholtz projection $\tilde{P}_q$ we can define the Stokes operator $\tilde{A}_q$, $1 < q < \infty$, for a uniform $C^2$-domain $\Omega \subseteq \mathbb{R}^n$. Let

$$\mathcal{D}(\tilde{A}_q) := \begin{cases} \mathcal{D}_q + \mathcal{D}_2, & 1 < q < 2, \\ \mathcal{D}_q \cap \mathcal{D}_2, & 2 \le q < \infty, \end{cases}$$

where $\mathcal{D}_q := L^q_\sigma(\Omega) \cap W^{1,q}_0(\Omega) \cap W^{2,q}(\Omega)$. Then the Stokes operator $\tilde{A}_q : \mathcal{D}(\tilde{A}_q) \subseteq \tilde{L}^q_\sigma(\Omega) \to \tilde{L}^q_\sigma(\Omega)$ is defined by $\tilde{A}_q u := -\tilde{P}_q \Delta u$ and has the following properties, see [22]:

- $\tilde{A}_q$ is a densely defined closed operator in $\tilde{L}^q_\sigma(\Omega)$ satisfying $(\tilde{A}_q)^* = \tilde{A}_{q'}$.
- For all $\lambda \in \mathcal{S}_\theta := \{\lambda \in \mathbb{C} \setminus \{0\}; |\arg(\lambda)| < \theta\}$, $\frac{\pi}{2} < \theta < \pi$, the resolvent $(\lambda + \tilde{A}_q)^{-1} : \tilde{L}^q_\sigma(\Omega) \to \tilde{L}^q_\sigma(\Omega)$ is well-defined. Moreover, for $f \in \tilde{L}^q_\sigma(\Omega)$ the unique solution $u \in \tilde{L}^q_\sigma(\Omega)$ to $\lambda u + \tilde{A}_q u = f$ satisfies the estimate

  $$|\lambda| \|u\|_{\tilde{L}^q(\Omega)} + \|\nabla^2 u\|_{\tilde{L}^q(\Omega)} \le C \|f\|_{\tilde{L}^q(\Omega)}$$

  with a constant $C = C(q, \theta, \delta, \tau(\Omega))$, as long as $|\lambda| \ge \delta > 0$.
- $\tilde{A}_q$ generates an analytic semigroup $e^{-t\tilde{A}_q}$, $t \ge 0$, having the bound

  $$\|e^{-t\tilde{A}_q} f\|_{\tilde{L}^q(\Omega)} \le C e^{\delta t} \|f\|_{\tilde{L}^q(\Omega)}$$

  for all $f \in \tilde{L}^q_\sigma(\Omega)$ and $t \ge 0$ with a constant $C = C(q, \delta, \tau(\Omega))$, $\delta > 0$.
- Let $1 < r, q < \infty$, $0 < T < \infty$, and let a function $f \in L^r(0, T; \tilde{L}^q_\sigma(\Omega))$ as well as an initial value $u_0 \in \mathcal{D}(\tilde{A}_q)$ (for simplicity) be given. Then there exists a unique solution $u \in L^r(0, T; \mathcal{D}(\tilde{A}_q)) \cap W^{1,r}(0, T; \tilde{L}^q_\sigma(\Omega))$ of the Stokes system

  $$u_t + \tilde{A}_q u = f, \quad u(0) = u_0 \quad \text{in } \tilde{L}^q(\Omega).$$

  It satisfies the maximal regularity estimate

  $$\|u\|_{L^r(0,T;\mathcal{D}(\tilde{A}_q))} + \|u_t\|_{L^r(0,T;\tilde{L}^q)} \le C(\|u_0\|_{\mathcal{D}(\tilde{A}_q)} + \|f\|_{L^r(0,T;\tilde{L}^q)})$$

with a constant $C = C(q, r, T, \tau(\Omega))$, and can be represented by the variation of constants formula

$$u(t) = e^{-t\tilde{A}_q}u_0 + \int_0^t e^{-(t-\tau)\tilde{A}_q}f(\tau)d\tau, \quad 0 \le t \le T.$$

It is unknown whether the resolvent estimate holds uniformly in $\lambda \in \mathcal{S}_\theta$ as $|\lambda| \to 0$. Therefore, the semigroup may increase exponentially fast and the maximal regularity estimate is stated only for finite time intervals.

A further crucial property of the Stokes operator $1 + \tilde{A}_q$ is the fact that it admits bounded imaginary powers, see [40]. Hence complex interpolation methods can be used to describe domains of fractional powers $(1 + \tilde{A}_q)^\alpha$, $-1 \le \alpha \le 1$. To be more precise, for $0 \le \alpha \le 1$ let $\tilde{D}_q^\alpha = \tilde{D}_q^\alpha(\Omega) = \mathcal{D}((1 + \tilde{A}_q)^\alpha)$, the domain of the fractional power $(1 + \tilde{A}_q)^\alpha$, equipped with the norm $\|(1 + \tilde{A}_q)^\alpha \cdot \|_{\tilde{L}^q}$. If $-1 \le \alpha < 0$ define $\tilde{D}_q^\alpha$ as the completion of $\tilde{L}_\sigma^q(\Omega)$ in the norm $\|(1 + \tilde{A}_q)^\alpha \cdot \|_{\tilde{L}^q}$. These spaces are reflexive and satisfy the duality relation $(\tilde{D}_q^\alpha)^* = \tilde{D}_{q'}^{-\alpha}$. As special cases we get that $\tilde{D}_q^1 = \mathcal{D}(\tilde{A}_q)$, $\tilde{D}_q^{1/2} = \tilde{W}_0^{1,q} \cap \tilde{L}_\sigma^q(\Omega)$ (with norm $\|(1 + \tilde{A}_q)^{1/2} \cdot \|_{\tilde{L}^q}$ equivalent to $\| \cdot \|_{\tilde{W}^{1,q}(\Omega)}$), and $\tilde{D}_q^0 = \tilde{L}_\sigma^q(\Omega)$. Moreover, we obtain the interpolation result

$$\left[\tilde{D}_q^\alpha, \tilde{D}_q^\beta\right]_\theta = \tilde{D}_q^\gamma,$$

when $-1 \le \alpha \le \beta \le 1$ and $(1 - \theta)\alpha + \theta\beta = \gamma$, $\theta \in (0, 1)$.

This result is the basis to prove the following embedding estimate: Let $n \ge 3$, $0 \le \alpha \le 1$, $1 < q \le r < \infty$, and $\frac{1}{r} = \frac{1}{q} - \frac{2\alpha}{n}$. Then

$$\|u\|_{\tilde{L}^r(\Omega)} \le C\|(1 + \tilde{A}_q)^\alpha u\|_{\tilde{L}^q(\Omega)} \tag{42}$$

for all $u \in \tilde{D}_q^\alpha$ with a constant $C = C(\tau(\Omega), q, \alpha)$, see [48].

**Theorem 2.14** (**$\tilde{L}^r$-$\tilde{L}^q$-estimates, [48, Theorem 1]**) *Let $n \ge 3$, $1 < q \le r < \infty$, and $\alpha := \frac{n}{2}\left(\frac{1}{q} - \frac{1}{r}\right)$. Then for every $f \in \tilde{L}_\sigma^q(\Omega)$ and $t > 0$*

$$\|e^{-t\tilde{A}_r}f\|_{\tilde{L}^r(\Omega)} \le Ce^{\delta t}(1 + t)^\alpha t^{-\alpha}\|f\|_{\tilde{L}^q(\Omega)}, \tag{43}$$

$$\|\nabla e^{-t\tilde{A}_r}f\|_{\tilde{L}^r(\Omega)} \le Ce^{\delta t}(1 + t)^{\alpha+\frac{1}{2}}t^{-\alpha-\frac{1}{2}}\|f\|_{\tilde{L}^q(\Omega)} \tag{44}$$

*with a constant $C = C(\tau(\Omega), r, q, \delta) > 0$ independent of $t$ and with any $\delta > 0$.*

We note that the factor $(1 + t)^\alpha$ in (43), (44) is somehow natural in the context of $\tilde{L}^q$-spaces. Consider the exterior of a ball in $\mathbb{R}^3$ where we may choose $\delta = 0$ in (43). But an estimate of the form $\|e^{-t\tilde{A}_r}f\|_{\tilde{L}^r} \le Ct^{-\alpha}\|f\|_{\tilde{L}^q}$ cannot hold for all $t > 0$ with a constant $C$ independent of $t$. Indeed, otherwise, with $r = 6$, $q = 6/5$, $f \in C^\infty_{0,\sigma}(\Omega)$ we get that

$$\|e^{-t\tilde{A}_2} f\|_{L^2} \le \|e^{-t\tilde{A}_6} f\|_{\tilde{L}^6} \le Ct^{-1}\|f\|_{\tilde{L}^{6/5}} \le Ct^{-1}\|f\|_{L^2},$$

implying that the semigroup $e^{-t\tilde{A}_2} = e^{-tA_2}$ has some algebraic decay in $t$, which is known to be false.

Let $u \in \mathcal{LH}_T$ be a weak solution of the Navier-Stokes system (2) with initial value $u_0 \in L^2(\Omega)$ and—for simplicity—external force $f = 0$. Regularity results for this case (even including forces of the type $f_1 + \operatorname{div} f_2$) are obtained in the PhD thesis of F. Riechwald [47] and will be discussed in the following. The proofs are based on the theory of *very weak solutions*. For an introduction to this theory we refer to [1, 9, 15, 18]; a review can be found in [19].

Very weak solutions can be constructed most easily by duality arguments. Therefore, for finite $0 < T < \infty$ and $1 < s, q < \infty$, we introduce the function space

$$\mathcal{T}^{1,s',q'}(T,\Omega) := \{\phi \in L^{s'}(0,T;\tilde{D}^1_{q'}) \cap W^{1,s'}(0,T;\tilde{L}^{q'}(\Omega)) \colon \phi(T) = 0\}$$

equipped with the norm

$$\|\phi\|_{\mathcal{T}^{1,s',q'}} := \|\phi_t\|_{L^{s'}(0,T;\tilde{L}^{q'})} + \|\phi\|_{L^{s'}(0,T;\tilde{D}^1_{q'})}.$$

This reflexive Banach space is related to the backward Stokes problem

$$-\phi_t - \tilde{A}_{q'}\phi = v, \ \phi(T) = 0 \quad \text{in } \tilde{L}^{q'}(\Omega)$$

which admits for any $v \in L^{s'}\big(0,T;\tilde{L}^{q'}(\Omega)\big)$ a unique solution $\phi \in \mathcal{T}^{1,s',q'}(T;\Omega)$ satisfying the maximal regularity estimate

$$\|\phi\|_{\mathcal{T}^{1,s',q'}} \le C\|v\|_{\tilde{L}^{s'}(0,T;\tilde{L}^{q'})}$$

with a constant $C = C(q,s,T,\tau(\Omega))$. The dual space to $\mathcal{T}^{1,s',q'}(T;\Omega)$ will be denoted by $\mathcal{T}^{-1,s,q}(T,\Omega)$, its norm by $\|\cdot\|_{\mathcal{T}^{-1,s,q}}$.

**Definition 2.15** Let $\Omega \subseteq \mathbb{R}^n$ be a uniform $C^2$-domain, $0 < T < \infty$ and $2 < s < \infty$, $n < q < \infty$ and $2/s + n/q = 1$. For an initial value $u_0$ we define the functional $\mathcal{F}_{u_0}$ by

$$\langle \mathcal{F}_{u_0}, \phi\rangle = \langle u_0, \phi(0)\rangle_\Omega, \quad \phi \in \mathcal{T}^{1,s',q'}(T,\Omega),$$

and assume that $\mathcal{F}_{u_0} \in \mathcal{T}^{-1,s,q}(T,\Omega)$. Then we call $u \in L^s(0,T;\tilde{L}^q)$ a *very weak solution to the Navier-Stokes system* with data $\mathcal{F}$, if the conditions

$$\begin{aligned} -\langle u, \phi_t\rangle_{\Omega,T} - \langle u, \Delta\phi\rangle_{\Omega,T} - \langle u\otimes u, \nabla\phi\rangle_{T,\Omega} &= \langle \mathcal{F}_{u_0}, \phi\rangle, \\ \operatorname{div} u = 0,\quad u\cdot N &= 0 \text{ on } \partial\Omega \end{aligned}$$

hold for all $\phi \in \mathcal{T}^{1,s',q'}(T;\Omega)$.

We note that the conditions $\operatorname{div} u = 0$ in $\Omega$ and $u \cdot N = 0$ on $\partial\Omega$ on the normal component of $u$ can be formulated as a variational identity as well.

To show that the functional $\mathcal{F}_{u_0}$ in Definition 2.15 is meaningful let $\phi \in \mathcal{T}^{1,s',q'}(T,\Omega)$ and $v = -\phi_t + \tilde{A}_{q'}\phi$. Then from the variation of constants formula applied to $\phi$ at 0 and the maximal regularity result we get for $\langle \mathcal{F}_{u_0}, \phi\rangle = \langle u_0, \phi(0)\rangle_\Omega$ that

$$\begin{aligned} |\langle \mathcal{F}_{u_0}, \phi\rangle| &= \Big|\Big\langle u_0, \int_0^T e^{-(T-t)\tilde{A}_{q'}} v(T-t)dt\Big\rangle_\Omega\Big| \\ &= \Big|\int_0^T \big\langle e^{-t\tilde{A}_q} u_0, v(t)\big\rangle_\Omega dt\Big| \\ &\le \Big(\int_0^T \|e^{-t\tilde{A}_q}u_0\|^s_{\tilde{L}^q} dt\Big)^{1/s} \|v\|_{L^{s'}(0,T;\tilde{L}^{q'})} \\ &\le \Big(\int_0^T \|e^{-t\tilde{A}_q}u_0\|^s_{\tilde{L}^q} dt\Big)^{1/s} \|\phi\|_{\mathcal{T}^{1,s',q'}}. \end{aligned}$$

Hence the assumption $\int_0^T \|e^{-t\tilde{A}_q}u_0\|^s_{\tilde{L}^q} dt < \infty$ yields $\mathcal{F}_{u_0} \in \mathcal{T}^{-1,s,q}(T,\Omega)$.

In contrast to the case of a bounded domain the semigroup $e^{-t\tilde{A}_q}$ cannot shown to be exponentially decreasing. However, we may modify the norm $\big(\int_0^T \|e^{-t\tilde{A}_q}u_0\|^s_{\tilde{L}^q} dt\big)^{1/s}$ to equivalent norms by replacing $e^{-t\tilde{A}_q}$ by $e^{-t(1+\tilde{A}_q)}$ on $(0,T)$, $0 < T < \infty$, and then by integrating even from 0 to $\infty$. Hence, with the exponentially decreasing semigroup $e^{-t(1+\tilde{A}_q)}$, as in the bounded domain case, we are led to the reflexive and complete space of Besov type

$$\tilde{\mathcal{B}}^{q,s}_T(\Omega) = \Big\{u_0 : \|u_0\|_{\tilde{\mathcal{B}}^{q,s}_T} := \Big(\int_0^T \|e^{-\tau\tilde{A}}u_0\|^s_{\tilde{L}^q}\, d\tau\Big)^{1/s} < \infty\Big\}$$

which equals the real interpolation space $\big(\mathcal{D}(\tilde{A}_{q'})^*, \tilde{L}^q_\sigma(\Omega)\big)_{1/s',s}$ with an equivalent norm; cf. the discussion in Sect. 2.2.

Summarizing, we proved the estimate

$$|\langle \mathcal{F}_{u_0}, \phi\rangle| = |\langle u_0, \phi(0)\rangle_\Omega| \le \|u_0\|_{\tilde{\mathcal{B}}^{q,s}_T} \|\phi\|_{\mathcal{T}^{1,s',q'}}.$$

We note that when $n = 3$ (and $s \geq 3$) duality arguments, real interpolation and (43), (44) yield the embeddings

$$\tilde{D}_2^{1/4} \hookrightarrow \tilde{L}^3_\sigma(\Omega) \hookrightarrow \tilde{L}^{3,s}_\sigma(\Omega) \hookrightarrow \tilde{\mathcal{B}}^{q,s}_T(\Omega);$$

for details we refer to [47, Remark 6.2.2].

**Theorem 2.16 ([47, Theorems 6.4.2 and 6.4.4])** *Let $\Omega \subseteq \mathbb{R}^3$ be a uniform $C^2$-domain, $0 < T < \infty$, and let $2 < s < \infty$, $3 < q < \infty$, $\frac{2}{s} + \frac{3}{q} = 1$. There exists an $\varepsilon_* = \varepsilon_*(\tau(\Omega), q, T) > 0$ with the following property: If $u_0 \in \tilde{\mathcal{B}}^{q,s}_T(\Omega)$ and*

$$\|u_0\|_{\tilde{\mathcal{B}}^{q,s}_T} \leq \varepsilon_*,$$

*then there exists a unique very weak solution $u \in L^s(0, T; \tilde{L}^q(\Omega))$ to the Navier-Stokes system with initial data $u_0$.*

**Theorem 2.17** *Let $\Omega \subseteq \mathbb{R}^3$ be a uniform $C^2$-domain and exponents $s, q$ be given such that $\frac{16}{5} \leq s \leq 16$, $\frac{24}{7} \leq q \leq 8$, and $\frac{2}{s} + \frac{3}{q} = 1$. Moreover, let $u_0 \in L^2_\sigma(\Omega)$.*

*(i) There exists $\varepsilon_* = \varepsilon_*(\tau(\Omega), q, T) > 0$ with the following property: If $u_0$ satisfies*

$$\|u_0\|_{\tilde{\mathcal{B}}^{q,s}_T} \leq \varepsilon_* \quad \text{and} \quad \mathcal{F}_{u_0} \in \mathcal{T}^{-1,4,4}(T, \Omega),$$

*then there exists a unique weak solution $u \in \mathcal{LH}_T$ to the Navier-Stokes system with initial data $u_0$ satisfying the energy inequality (EI). Moreover, $u \in L^s(0, T; \tilde{L}^q_\sigma(\Omega))$. In particular, every $t \in (0, T)$ is a regular point of $u$.*

*(ii) Let $u$ be a weak solution in $\mathcal{LH}_T \cap L^s(0, T; \tilde{L}^q_\sigma(\Omega))$. Then $u_0 \in \tilde{\mathcal{B}}^{q,s}_T$.*

The condition $\mathcal{F}_{u_0} \in \mathcal{T}^{-1,4,4}(T, \Omega)$ is fulfilled if e.g. $u_0 \in L^2_\sigma(\Omega)$ satisfies

$$u_0 \in \tilde{L}^\gamma_\sigma(\Omega) \text{ for some } \frac{12}{5} < \gamma \leq 4 \quad \text{or} \quad \int_0^T \|e^{-\tau\tilde{A}_2} u_0\|^4_{\tilde{L}^4} d\tau < \infty.$$

Indeed, the first condition is reduced via (43) to the second condition. In the latter case, for $\phi$ and $v := -\phi_t + \tilde{A}_{q'}\phi$ we have

$$\begin{aligned}
\left|\langle u_0, \phi(0)\rangle_{\Omega,T}\right| &= \left|\int_0^T \left\langle u_0, e^{-\tau\tilde{A}_{q'}} v(\tau)\right\rangle_\Omega d\tau\right| = \left|\int_0^T \left\langle e^{-\tau\tilde{A}_2} u_0, v(\tau)\right\rangle_\Omega d\tau\right| \\
&\leq \left(\int_0^T \|e^{-\tau\tilde{A}_2} u_0\|^4_{\tilde{L}^4} d\tau\right)^{1/4} \|v\|_{L^{4/3}(0,T;\tilde{L}^{4/3})} \\
&\leq \left(\int_0^T \|e^{-\tau\tilde{A}_2} u_0\|^4_{\tilde{L}^4} d\tau\right)^{1/4} \|\phi\|_{\mathcal{T}^{1,4/3,4/3}}.
\end{aligned}$$

Hence $\mathcal{F}_{u_0} \in \mathcal{T}^{-1,4,4}(T, \Omega)$.

Theorem 2.17 the proof of which will be sketched in Sect. 3.3 yields an almost optimal characterization of initial values to allow for weak solutions in Serrin's class. Comparing with Theorem 2.1 in the bounded domain case the restrictions on $s, q$ and on $u_0$ in the sense that $\mathcal{F}_{u_0} \in \mathcal{T}^{-1,4,4}(T,\Omega)$ are needed for the identification of a weak with a very weak solution; this argument is more complicated in unbounded domains.

Next we will apply Theorem 2.17 not only at $t_0 = 0$, but at (almost all) $t_0 \in [0,T)$. In this context, the condition $\mathcal{F}_{u(t)} \in \mathcal{T}^{-1,4,4}(T-t,\Omega)$ is immediately satisfied for a.a. $t \in [0,T)$, since $u(t) \in L^2_\sigma(\Omega) \cap L^6(\Omega) \subset \tilde{L}^4_\sigma(\Omega)$ for a.a. $t$. Following Sect. 2.2 we define the controlling parameter $\delta(t)$ as in Definition 2.3, with the space $\mathcal{B}^{q,s}(\Omega)$ replaced by $\tilde{\mathcal{B}}^{q,s}(\Omega)$. From [14] we know that to each initial value $u_0 \in L^2_\sigma(\Omega)$ there does exist at least one weak solution satisfying the strong energy inequality (*SEI*). In the sequel we will always assume (*SEI*) to hold for the given weak solution. Of course, global, local and also one-sided regularity has to be defined in terms of the space $L^s(\tilde{L}^q)$.

Now we get the following version of Theorem 2.5.

**Theorem 2.18** *Let $u \in \mathcal{LH}_T$, $0 < T < \infty$, be a weak solution of* (1) *in a general $C^2$-domain $\Omega \subset \mathbb{R}^3$, and let $\frac{16}{5} \le s \le 16$, $\frac{24}{7} \le q \le 8$ satisfy $\frac{2}{s} + \frac{3}{q} = 1$.*

*(i) Let $u$ satisfy $u(t) \in \tilde{\mathcal{B}}^{q,s}_{T-t}(\Omega)$ for a.a. $t \in [0,T)$. Given $t_1 \in (0,T)$ assume that for a.a. $t$ in a left-sided neighborhood of $t_1$*

$$\delta(t) \ge t_1 - t. \tag{45}$$

*Then $u$ is left-sided $L^s(\tilde{L}^q)$-regular at $t_1$.*

*(ii) Given $t_1 \in (0,T)$ assume that $u(t_1) \in \tilde{\mathcal{B}}^{q,s}(\Omega)$ and that $u$ satisfies $(EI)_{t_1}$. Then $u$ is right-sided $L^s(\tilde{L}^q)$-regular at $t_1$.*

*(iii) Let $u$ satisfy $u(t) \in \tilde{\mathcal{B}}^{q,s}(\Omega)$ for all $t \in [0,T)$ and condition* (45) *at $t_1 \in (0,T)$. Then $u$ is regular at $t_1$. If condition* (45) *is satisfied at* every *$t_1 \in (0,T)$, then $u \in L^s_{\mathrm{loc}}\big([0,T);\tilde{L}^q(\Omega)\big)$.*

*(iv) Let $u$ satisfy $u(t) \in \tilde{\mathcal{B}}^{q,s}(\Omega)$ for all $t \in [0,T)$. Assume that at $t_1 \in (0,T)$*

$$\lim_{t \nearrow t_1} \|u(t)\|_{\tilde{\mathcal{B}}^{q,s}_{t_1-t}} = 0. \tag{46}$$

*Then $u$ is $L^s(\tilde{L}^q)$-regular at $t_1$. If condition* (46) *is satisfied at* every *$t_1 \in (0,T)$, then $u \in L^s_{\mathrm{loc}}\big([0,T);\tilde{L}^q(\Omega)\big)$.*

The proof of Theorem 2.18 almost follows word by word the proof of Theorem 2.5 and will be omitted. By analogy, we get a result similar to Theorem 2.9.

**Theorem 2.19** *Let $u \in \mathcal{LH}_T$ be a weak solution of* (1), *let $s, q$ be as in Theorem 2.18, and $1 \le r \le s$.*

(i) *Assume that*

$$\liminf_{\delta\to 0}\frac{1}{\delta^{1-r/s}}\int\limits_{t_1-\delta}^{t_1}\|u(\tau)\|_{\tilde{L}^q}^r\,\mathrm{d}\tau = 0.$$

*Then u is regular at* $t_1$.

(ii) *Assume that for* $0 \le t_0 < t_1 < T$ *and some* $t_1 < T' \le T$

$$\frac{1}{t_1 - t_0}\int\limits_{t_0}^{t_1}(T' - \tau)^{r/s}\,\|u(\tau)\|_{\tilde{L}^q}^r\,\mathrm{d}\tau \le \tilde{\varepsilon}_*.$$

*Then u is regular at* $t_1$.

Finally, we mention that also the criterion in Theorem 2.10 on the kinetic energy function holds for weak solutions in general uniform $C^2$-domains. For further results, even including non-vanishing external forces, we refer to [47, Chap. 8.2].

# 3 Proofs

## 3.1 Proofs of Results of Sect. 2.2

*Proof of Theorem 2.1* The proof is based on Banach's fixed theorem to obtain the existence of a locally unique solution of the nonlinear integral equation (15). Analyzing $E_0(t) = e^{-tA}u_0$ and

$$E_1(t) = \int\limits_0^t A^{1/2}e^{-(t-\tau)A}A^{-1/2}P\,\operatorname{div}F(\tau)\,\mathrm{d}\tau$$

separately, we are looking for a fixed point $\tilde{u} = \mathcal{F}\tilde{u}$ of the operator

$$\mathcal{F}\tilde{u}(t) = -\int\limits_0^t A^{1/2}e^{-(t-\tau)A}\big(A^{-1/2}P\,\operatorname{div}\big)(\tilde{u}+E)\otimes(\tilde{u}+E)(\tau)\,\mathrm{d}\tau$$

where $E = E_0 + E_1$. This fixed point will be found in a closed ball of the Banach space

$$X_T = \big\{v : [0,T) \to L^{q/2}_\sigma(\Omega) :$$
$$\big(A_{q/2}^{-1/2}v\big)_t,\ A_{q/2}^{1/2}v \in L^{s/2}\big(0,T;L^{q/2}_\sigma(\Omega)\big),\ A_{q/2}^{-1/2}v(0) = 0\big\}$$

equipped with the norm $\|v\|_{X_T} = \|(A_{q/2}^{-1/2}v)_t\|_{q/2,s/2;T} + \|A_{q/2}^{1/2}v\|_{q/2,s/2;T}$, using maximal regularity estimates. The crucial point is to identify the fixed point $\tilde{u}$, better to say $u = \tilde{u} + E$, with a weak solution of the Navier-Stokes system. For details we refer to [21]. □

*Proof of Lemma 2.4*

(i) We fix $t \in [0,T]$ and choose any sequence $(t_j) \subset [0,T]$ with $t_j \to t$ as $j \to \infty$. Since $(u(t_j)) \subset L^2_\sigma(\Omega)$ is bounded, we get from (9), (11) with $2\alpha + \frac{3}{q} = \frac{3}{2}$ that $\|e^{-\tau A}u(t_j)\|_q \le c\tau^{-\alpha}\|u(t_j)\|_2 \le C$ with a constant $C = C(\tau) > 0$. Moreover, the weak continuity of $u(\cdot)$ in $L^2_\sigma(\Omega)$ implies the convergence

$$\langle e^{-\tau A}u(t_j), \varphi\rangle = \langle u(t_j), e^{-\tau A}\varphi\rangle \to \langle u(t), e^{-\tau A}\varphi\rangle = \langle e^{-\tau A}u(t), \varphi\rangle$$

for all $\varphi$ in the dense subset $L^2_\sigma(\Omega) \cap L^{q'}_\sigma(\Omega)$ of $L^{q'}_\sigma(\Omega)$. Summarizing the last two arguments we conclude that $e^{-\tau A}u(t_j) \rightharpoonup e^{-\tau A}u(t)$ in $L^q_\sigma(\Omega)$; in particular, $\|e^{-\tau A}u(t)\|_q \le \liminf_j \|e^{-\tau A}u(t_j)\|_q$. Hence by Fatou's lemma

$$\int_0^\delta \|e^{-\tau A}u(t)\|_q^s \,d\tau \le \int_0^\delta \liminf_j \|e^{-\tau A}u(t_j)\|_q^s \,d\tau \le \liminf_j \int_0^\delta \|e^{-\tau A}u(t_j)\|_q^s \,d\tau,$$

i.e., $\|u(t)\|_{\mathcal{B}^{q,s}_\delta} \le \liminf_j \|u(t_j)\|_{\mathcal{B}^{q,s}_\delta}$.

Now assume that $u \in L^\infty(0,T;\mathcal{B}^{q,s}_\delta(\Omega))$. Let us consider $u$ as a function with values in the intersection space $\hat{\mathcal{B}}^{q,s}_\delta(\Omega) = L^2_\sigma(\Omega) \cap \mathcal{B}^{q,s}_\delta(\Omega)$. Obviously, $\hat{\mathcal{B}}^{q,s}_\delta(\Omega)$ is a reflexive, separable Banach space when equipped with the norm $\|\cdot\|_2 + \|\cdot\|_{\mathcal{B}^{q,s}_\delta}$. It is easy to see that there exists a Lebesgue null set $N \subset [0,T]$ such that $\|u\|_{L^\infty(0,T;\hat{\mathcal{B}}^{q,s}_\delta)} = \sup_{t\in[0,T]\setminus N} \|u(t)\|_{\hat{\mathcal{B}}^{q,s}_\delta}$. A similar result holds when the norm in $\hat{\mathcal{B}}^{q,s}_\delta$ is replaced by the norm in $\mathcal{B}^{q,s}_\delta$.

Fix any $t_0 \in [0,T]$ and choose a sequence $(t_j) \subset (0,T) \setminus N$ such that $t_j \to t_0$ as $j \to \infty$. Thus $(u(t_j))$ is a bounded sequence in $\hat{\mathcal{B}}^{q,s}_\delta(\Omega) \subset L^2_\sigma(\Omega)$. Due to (6) we get that $u(t_j) \rightharpoonup u(t_0) \in L^2_\sigma(\Omega)$. Moreover, by the reflexivity of $\hat{\mathcal{B}}^{q,s}_\delta(\Omega)$ we may conclude by standard arguments that $u(t_j) \rightharpoonup u(t_0)$ in $\mathcal{B}^{q,s}_\delta(\Omega)$. Hence $u(t_0) \in \mathcal{B}^{q,s}_\delta(\Omega)$ including all $t_0 \in N$ and $\|u(t_0)\|_{\mathcal{B}^{q,s}_\delta} \le \liminf_{j\to\infty} \|u(t_j)\|_{\mathcal{B}^{q,s}_\delta} \le \|u\|_{L^\infty(0,T;\hat{\mathcal{B}}^{q,s}_\delta)}$.

Similar ideas allow to prove that even $u \in C^0_w([0,T];\mathcal{B}^{q,s}_\delta(\Omega))$.

(ii) Assume that $\delta(t) \in [0,\infty)$. Then for $\beta > \delta(t)$ and any sequence $(t_j) \subset [0,T]$ with $t_j \to t$ as $j \to \infty$

$$\varepsilon_* < \int_0^\beta \|e^{-\tau A}u(t)\|_q^s \,d\tau \le \liminf_j \int_0^\beta \|e^{-\tau A}u(t_j)\|_q^s \,d\tau$$

which implies that $\int_0^\beta \|e^{-\tau A}u(t_j)\|_q^s \, d\tau > \varepsilon_*$ for large $j$. Hence $\delta(t_j) < \beta$, i.e., $\delta(\cdot)$ is upper semi-continuous at $t$. The case $\delta(t) = \infty$ is trivial. □

*Proof of Theorem 2.5*

(i) Let $t_1 \in (0,T)$. To show that $t_1$ is a left-sided regular point of $u$ we find due to the assumptions (30) and (*SEI*) a $t \in (0,t_1)$ such that $\|u(t)\|_{\mathcal{B}^{q,s}_{t_1-t}} \le \|u(t)\|_{\mathcal{B}^{q,s}_{\delta(t)}} \le \varepsilon_*$ and that $(EI)_t$ holds. Here $\varepsilon_* > 0$ is the constant from Theorem 2.1, see (27). By Serrin's uniqueness theorem and Theorem 2.1 we conclude that $u \in L^s\big(t_0, t_1; L^q(\Omega)\big)$. Hence $u$ is left-sided regular in $t_1$.

(ii) Since $u(t_1) \in \mathcal{B}^{q,s}_\infty(\Omega)$, there exists a strong solution $v \in L^s\big(t_1, t_1+\varepsilon; L^q(\Omega)\big)$, $\varepsilon > 0$, of (1) with initial value $v(t_1) = u(t_1)$. Moreover, by assumption, $(EI)_{t_1}$ holds for $u$. Hence Serrin's uniqueness theorem implies that $v = u$ in $[t_1, t_1+\varepsilon)$, and $u$ is right-sided regular in $t_1$.

(iii) To combine the results from (i) and (ii), in particular to apply (ii), it suffices to prove that $u$ satisfies $(EI)_{t_1}$ at $t_1$. Let $t_1 \in (0,T)$ be an instant where the validity of the energy inequality is not guaranteed by (*SEI*). By (i) $t_1$ is a left-sided regular point for $u$ and, consequently, $u \in L^s(t_0, t_1; L^q(\Omega))$ for some $0 < t_0 < t_1$. Therefore, $u$ satisfies (*EE*) for all initial times $t_0' \in (t_0, t_1)$, in particular

$$\frac{1}{2}\|u(t_1)\|_2^2 + \int_{t_0'}^{t_1} \|\nabla u\|_2^2 \, d\tau = \frac{1}{2}\|u(t_0')\|_2^2.$$

Thus $\lim_{t_0' \nearrow t_1} \|u(t_0')\|_2^2 = \|u(t_1)\|_2^2$. Moreover, by (*SEI*), there is a sequence $t_j \nearrow t_1$ such that

$$\frac{1}{2}\|u(t)\|_2^2 + \int_{t_j}^{t} \|\nabla u\|_2^2 \, d\tau \le \frac{1}{2}\|u(t_j)\|_2^2, \quad t_j \le t < T.$$

Passing to the limit $t_j \nearrow t_1$ we get that $u$ satisfies $(EI)_{t_1}$.

Finally, since $u(0) = u_0 \in \mathcal{B}^{q,s}_\infty(\Omega)$, we know that $u \in L^s\big(0, \varepsilon_0; L^q(\Omega)\big)$ for some $\varepsilon_0 > 0$. Now an elementary compactness argument proves that $u \in L^s_{\mathrm{loc}}\big([0,T); L^q(\Omega)\big)$.

(iv) Under the assumption (31) for $t_1$ we get for a.a. $t$ in a left-sided neighborhood of $t_1$ that $\|u(t)\|_{\mathcal{B}^{q,s}_{t_1-t}} < \varepsilon_*$. Hence $\delta(t) > t_1 - t$ for these $t$; moreover, we may assume $(EI)_t$. By the above arguments we conclude that $u$ lies in Serrin's class $L^s(L^q)$ on the interval $(t, t+\delta(t)) \supset (t, t_1)$, i.e., $u$ is $L^s(L^q)$-regular at $t_1$. □

*Proof of Corollary 2.7*

(i) Fix $\alpha \in (\frac{1}{4}, \frac{1}{2}]$ and let $q \geq q_0 \geq 2$ satisfy $2\alpha + \frac{3}{q} = \frac{3}{q_0}$. By Theorem 2.5(i) we get for all $t \in (0, t_1)$ where $(EI)_t$ is satisfied that

$$\begin{aligned} \epsilon_*^s &< \int_0^{t_1-t} \|e^{-\tau A} u(t)\|_q^s \, d\tau \leq c \int_0^{t_1-t} \|A^\alpha e^{-\tau A} u(t)\|_{q_0}^s \, d\tau \\ &\leq c \int_0^{t_1-t} \tau^{-\frac{s}{2}(\frac{3}{2}-\frac{3}{q_0})} \|A^\alpha u(t)\|_2^s \, d\tau \\ &= c(t_1 - t)^{-\frac{s}{4}+\alpha s} \|A^\alpha u(t)\|_2^s \end{aligned}$$

since $-\frac{s}{2}(\frac{3}{2} - \frac{3}{q_0}) = -\frac{s}{4} + \alpha s - 1 > -1$. Hence $\|A^\alpha u(t)\|_2 > \epsilon_*(t_1 - t)^{\frac{1}{4}-\alpha}$ for a.a. $t \in (0, t_1)$.

(ii) Since $u(t_1) \notin \mathcal{B}^{q,s}(\Omega)$, we have $\|u(t_1)\|_{\mathcal{B}^{q,s}_\delta} = \infty$ for each $\delta > 0$. The lower semi-continuity of the map $t \mapsto \|u(t)\|_{\mathcal{B}^{q,s}_\delta}$ implies that $\|u(t)\|_{\mathcal{B}^{q,s}_\delta} \to \infty$ as $t \searrow t_1$. Moreover, due to the embeddings (28) we get that $\|u(t)\|_{\mathcal{B}^{q,s}_\delta} \leq c\|u(t)\|_3 \leq c\|u(t)\|_{\mathcal{D}(A^{1/4})} \leq c\|u(t)\|_{\mathcal{D}(A^{1/2})}$. □

## 3.2 Proofs of Results of Sect. 2.3

*Proof of Corollary 2.8*

(i) For $0 < T' < T$ the function $u \in C^0([0, T']; \mathcal{B}^{q,s}_\sigma(\Omega))$ is uniformly continuous in $t \in [0, T']$ with values in $\mathcal{B}^{q,s}_\delta(\Omega)$. Given $\varepsilon_*$ from Theorem 2.1 the uniform continuity allows to find $\delta' \in (0, \delta]$ such that $\|u(t)\|_{\mathcal{B}^{q,s}_{\delta'}} \leq \varepsilon_*$ for all $t \in [0, T']$. Then a compactness argument on $[0, T']$ implies that $u \in L^s([0, T']; L^q(\Omega))$.

(ii) From Lemma 2.4 we know that $\|u(t)\|_{\mathcal{B}^{q,s}_\delta} \leq \|u\|_{L^\infty(0,T;\mathcal{B}^{q,s}_\delta)} \leq \varepsilon_*$ for all $t \in [0, T)$. Now a compactness argument as in (i) completes the proof.

(iii) The embeddings (28) and (i), (ii) immediately prove (iii). □

*Proof of Theorem 2.9*

(i) Assuming (i) we find $\delta > 0$ such that with $t_0 = t_1 - \delta$ and $T' = t_1 + \delta$

$$\frac{1}{t_1 - t_0} \int_{t_0}^{t_1} (T' - \tau)^{r/s} \|u(\tau)\|_q^r \, d\tau \leq \frac{2^{r/s}}{\delta^{1-r/s}} \int_{t_0}^{t_1} \|u(\tau)\|_q^r \, d\tau \leq \tilde{\varepsilon}_*,$$

i.e., (34) is satisfied. Thus it suffices to prove (ii).

(ii) It is sufficient to find $t \in (0, t_1)$ such that $\|u(t)\|_{\mathcal{B}^{q,s}_{T'-t}} \le \varepsilon_*$ and that $(EI)_t$ is satisfied. Indeed, by (34) there exists $t \in (t_0, t_1)$ such that

$$(T' - t)^{r/s} \|u(t)\|_q^r \le \tilde{\varepsilon}_*$$

or, equivalently, $(T' - t)\|u(t)\|_q^s \le \tilde{\varepsilon}_*^{s/r}$ and that $(EI)_t$ holds. Hence, employing the boundedness of the semigroup $e^{-\tau A}$ on $L^q_\sigma(\Omega)$,

$$\int_0^{T'-t} \|e^{-\tau A} u(t)\|_q^s \, d\tau \le C(T' - t)\|u(t)\|_q^s \le \varepsilon_*$$

with an appropriately chosen $\tilde{\varepsilon}_*$ in (34). □

*Proof of Theorem 2.10* Obviously the condition $[E_k(t_1-)]_\alpha < \infty$ for $\alpha \in (\frac{1}{2}, 1]$ implies that $[E_k(t_1-)]_{1/2} \le \varepsilon_*$ (with the supremum taken only for small $\delta > 0$). Hence it suffices to assume the second condition. With $r = 2$, $q = 6$ and $s = 4$ we get that

$$\begin{aligned} \int_{t_1-\delta}^{t_1} \|u(\tau)\|_q^r \, d\tau &\le c \int_{t_1-\delta}^{t_1} \|\nabla u(\tau)\|_2^2 \, d\tau \\ &\le c\big(E_k(t_1 - \delta) - E_k(t_1)\big) \\ &\le c\varepsilon_* \delta^{1/2} \end{aligned} \tag{47}$$

provided we choose only those $\delta > 0$ such that $(EI)_{t_1-\delta}$ holds. We conclude that (33) is satisfied and consequently that $u$ is regular at $t_1$. □

*Proof of Theorem 2.11*

(i) Assume that $u$ satisfies $(EE)$ and $u(t_0) \in \mathcal{B}^{q,s}(\Omega)$ for all $t_0 \in [0, T)$. Moreover, let $\tilde{u}$ be another weak solution satisfying $(SEI)$ for the same initial value $u_0 \in L^2_\sigma(\Omega)$. We obtain that $u \in L^s\big(0, \delta; L^q(\Omega)\big)$ with some $0 < \delta < T$. Then Serrin's uniqueness theorem implies that $u(t) = \tilde{u}(t)$ for $0 \le t < \delta$.

Let $[0, t_0)$, $0 < t_0 \le T$, be the largest half open interval such that $u(t) = \tilde{u}(t)$ is satisfied for each $t \in [0, t_0)$. If $t_0 < T$, then the weak $L^2$-continuity in time (6) implies that $u(t_0) = \tilde{u}(t_0)$.

Since $u$ satisfies $u(t_0) \in \mathcal{B}^{q,s}(\Omega)$ and $(EI)_{t_0}$ we conclude that $u \in L^s\big(t_0, t_0 + \delta; L^q(\Omega)\big)$. Moreover, since $\tilde{u}$ satisfies $(SEI)$, hence $(EI)_{t_j}$ for a sequence $(t_j)$ with $t_j \nearrow t_0$, and since $u = \tilde{u}$ satisfies $(EE)$ on $[0, t_0]$, we conclude that $\|\tilde{u}(t_j)\|_2 \to \|\tilde{u}(t_0)\|_2$. These arguments imply that $\tilde{u}$ satisfies $(EI)_{t_0}$. Consequently, by Serrin's uniqueness theorem, $u = \tilde{u}$ in $[0, t_0 + \delta)$. This is a contradiction to the construction of $t_0$.

(ii) For $u \in L^4(L^4)$ Remark 1.2(5) yields $(EE)$. Now assume that $u$ satisfies (37). By Hölder's inequality in Lorentz spaces [38, Lemma 2.1] and Sobolev's embedding $W_0^{1,2}(\Omega) \subset L^{6,2}(\Omega)$ [38, Lemma 2.2] we get that

$$\begin{aligned}\|uu\|_{L^2} &\le c\|uu\|_{L^{2,2}} \le c\|u\|_{L^{3,\infty}}\|u\|_{L^{6,2}}\\ &\le c\|u\|_{L^{3,\infty}}\|u\|_{W^{1,2}},\end{aligned}$$

where $c = c(\Omega) > 0$. Hence $u \in L^4(L^4)$. Now we proceed as above. □

## 3.3 *Proofs of Results of Sect. 2.4*

*Proof of Theorem 2.17 (Sketch)*

(i) Since $u_0 \in L^2_\sigma(\Omega)$, there exists a (global) weak solution $u$ satisfying the energy inequality $(EI)$. By Theorem 2.16 we also get the existence of a very weak solution $\tilde{u} \in L^s\big(0,T;\tilde{L}^q(\Omega)\big)$ with $\tilde{u}(0) = u_0$ which due to the assumption $\mathcal{F}_{u_0} \in \mathcal{T}^{-1,4,4}(T,\Omega)$ can be shown to have the additional property $\tilde{u} \otimes \tilde{u} \in L^2(0,T;L^2(\Omega))$. Hence $\tilde{u}$ is also a weak solution satisfying even the energy equality $(EE)$. Now Serrin's uniqueness theorem valid also in the $L^s(\tilde{L}^q)$-setting implies that $u = \tilde{u} \in L^s\big(0,T;\tilde{L}^q_\sigma(\Omega)\big)$. All $t \in (0,T)$ are thus regular points.

(ii) Since by assumption $u \in L^s\big(0,T;\tilde{L}^q(\Omega)\big)$ it suffices to show that $\tilde{u} = u - e^{-t\tilde{A}_q}u_0 \in L^s\big(0,T;\tilde{L}^q(\Omega)\big)$ as well. Here $\tilde{u}$ is a weak solution of the linear Stokes system with initial value 0, but external force $\mathcal{F} = -\operatorname{div} u \otimes u$. Hence it suffices to show that $\mathcal{F} \in \mathcal{T}^{-1,s,q}(T,\Omega)$. To this aim let $\phi \in \mathcal{T}^{1,s',q'}(T,\Omega)$ so that by Hölder's inequality in $\tilde{L}^q$-spaces

$$|\langle \mathcal{F}, \phi\rangle| = |(u \otimes u, \nabla\phi)_{T,\Omega}| \le \|u\|^2_{L^s(\tilde{L}^q)}\|\nabla\phi\|_{L^{(s/2)'}(\tilde{L}^{(q/2)'})}.$$

Concerning $\nabla\phi$ we use fractional powers of the operator $1 + \tilde{A}_{q'}$ and obtain for a.a. $0 \le t \le T$ that

$$\begin{aligned}\|\nabla\phi(t)\|_{\tilde{L}^{(q/2)'}} &\le C\|(1+\tilde{A}_{q'})^{1/2}\phi(t)\|_{\tilde{L}^{(q/2)'}}\\ &\le C\|(1+\tilde{A}_{q'})^{3/(2q)}(1+\tilde{A}_{q'})^{1/2}\phi(t)\|_{\tilde{L}^{q'}}\\ &\le C\|(1+\tilde{A}_{q'})^{1/s'}\phi(t)\|_{\tilde{L}^{q'}}.\end{aligned}$$

Finally, let $v := -\phi_t + \tilde{A}_{q'}\phi$ and use the variation of constants formula to write $\phi$ in terms of $v$. Hence, with a constant $C = C(T, \delta, s) > 0$,

$$\begin{aligned}
\|\nabla\phi\|^{(s/2)'}_{L^{(s/2)'}(\tilde{L}^{(q/2)'})} &\leq C\int_0^T \left(\int_0^t \|(1+\tilde{A}_{q'})^{1/s'} e^{-(t-\tau)\tilde{A}_{q'}} v(T-\tau)\|_{\tilde{L}^{q'}} d\tau\right)^{(s/2)'} dt \\
&\leq C\int_0^T \left(\int_0^t \frac{\|v(T-\tau)\|_{\tilde{L}^{q'}}}{(t-\tau)^{1/s'}} d\tau\right)^{(s/2)'} dt \\
&\leq C\left(\int_0^T \|v(T-\tau)\|^{s'}_{\tilde{L}^{q'}} dt\right)^{(s/2)'/s'} \\
&\leq C\|v\|^{(s/2)'}_{L^{s'}(\tilde{L}^{q'})},
\end{aligned}$$

where we also used the Hardy-Littlewood inequality. This proves that $\mathcal{F} \in \mathcal{T}^{-1,s,q}(T, \Omega)$. □

**Acknowledgements** This work was completed with the support of the International Research Training Group on Mathematical Fluid Mechanics Darmstadt-Tokyo (IRTG 1529).

## References

1. H. Amann, Nonhomogeneous Navier-Stokes equations with integrable low-regularity data, *in Nonlinear Problems in Mathematical Physics and Related Topics, II*. International Mathematical Series (Kluwer Academic/Plenum Publishing, New York, 2002), pp. 1–28
2. M.E. Bogovskiĭ, Decomposition of $L_p(\Omega, \mathbb{R}^n)$ into the direct sum of subspaces of solenoidal and potential vector fields. Soviet Math. Dokl. **33**, 161–165 (1986)
3. W. Borchers, T. Miyakawa, Algebraic $L^2$ decay for Navier-Stokes flows in exterior domains. Acta Math. **165**, 189–227 (1990)
4. A. Cheskidov, P. Constantin, S. Friedlander, R. Shvydkoy, Energy conservation and Onsager's conjecture for the Euler equations. Nonlinearity **21**, 1233–1252 (2008)
5. A. Cheskidov, S. Friedlander, R. Shvydkoy, On the energy equality for weak solutions of the 3D Navier-Stokes equations, in *Advances in Mathematical Fluid Mechanics* (Springer, Berlin 2010), pp. 171–175
6. E.B. Fabes, B.F. Jones, N.M. Rivière, The initial value problem from the Navier-Stokes equations with data in $L^p$. Arch. Ration. Mech. Anal. **45**, 222–240 (1972)
7. R. Farwig, C. Komo, Regularity of weak solutions to the Navier-Stokes equations in exterior domains. Nonlinear Differ. Equ. Appl. **17**, 303–321 (2010)
8. R. Farwig, C. Komo, Optimal initial value conditions for strong solutions of the Navier–Stokes equations in an exterior domain. Analysis (Munich) **33**, 101–119 (2013)
9. R. Farwig, J. Sauer, Very weak solutions of the stationary Stokes equations on unbounded domains of half space type. Math. Bohemica **140**, 81–109 (2015)
10. R. Farwig, H. Sohr, Generalized resolvent estimates for the Stokes system in bounded and unbounded domains. J. Math. Soc. Japan **46**, 607–643 (1994)
11. R. Farwig, H. Sohr, Optimal initial value conditions for the existence of local strong solutions of the Navier-Stokes equations. Math. Ann. **345**, 631–642 (2009)

12. R. Farwig, H. Sohr, On the existence of local strong solutions for the Navier-Stokes equations in completely general domains. Nonlinear Anal. **73**, 1459–1465 (2010)
13. R. Farwig, Y. Taniuchi, On the energy equality of Navier–Stokes equations in general unbounded domains. Arch. Math. **95**, 447–456 (2010)
14. R. Farwig, H. Kozono, H. Sohr, An $L^q$–approach to Stokes and Navier-Stokes equations in general domains. Acta Math. **195**, 21–53 (2005)
15. R. Farwig, G.P. Galdi, H. Sohr, A new class of weak solutions of the Navier-Stokes equations with nonhomogeneous data. J. Math. Fluid Mech. **8**, 423–444 (2006)
16. R. Farwig, H. Kozono, H. Sohr, Local in time regularity properties of the Navier-Stokes equations. Indiana Univ. Math. J. **56**, 2111–2131 (2007)
17. R. Farwig, H. Kozono, H. Sohr, On the Helmholtz decomposition in general unbounded domains. Arch. Math. **88**, 239–248 (2007)
18. R. Farwig, H. Kozono, H. Sohr, Very weak solutions of the Navier-Stokes equations in exterior domains with nonhomogeneous data. J. Math. Soc. Japan **59**, 127–150 (2007)
19. R. Farwig, H. Kozono, H. Sohr, Very weak, weak and strong solutions to the instationary Navier-Stokes system, in *Topics on Partial Differential Equations*, ed. by P. Kaplický, Š. Nečasová. Lecture Notes of the J. Nečas Center Mathematical Modeling, vol. 2 (MatfyzPress Publishing House/Charles University, Prague, 2007), pp. 1–54
20. R. Farwig, H. Kozono, H. Sohr, Energy-based regularity criteria for the Navier-Stokes equations. J. Math. Fluid Mech. **11**, 1–14 (2008)
21. R. Farwig, H. Sohr, W. Varnhorn, On optimal initial value conditions for local strong solutions of the Navier-Stokes equations. Ann. Univ. Ferrara **55**, 89–110 (2009)
22. R. Farwig, H. Kozono, H. Sohr, On the Stokes operator in general unbounded domains. Hokkaido Math. J. **38**, 111–136 (2009)
23. R. Farwig, H. Kozono, H. Sohr, Regularity of weak solutions for the Navier-Stokes equations via energy criteria, in *Advances in Mathematical Fluid Mechanics*, ed. by R. Rannacher, A. Sequeira (Springer, Berlin, 2010), pp. 215–227
24. R. Farwig, H. Sohr, W. Varnhorn, Necessary and sufficient conditions on local strong solvability of the Navier-Stokes systems. Appl. Anal. **90**, 47–58 (2011)
25. R. Farwig, H. Sohr, W. Varnhorn, Extensions of Serrin's uniqueness and regularity conditions for the Navier-Stokes equations. J. Math. Fluid Mech. **14**, 529–540 (2012)
26. R. Farwig, H. Sohr, W. Varnhorn, Besov space regularity conditions for weak solutions of the Navier-Stokes equations. J. Math. Fluid Mech. **16**, 307–320 (2014)
27. Ch. Fefferman, Existence and Smoothness of the Navier–Stokes Equations (2000). http://www.claymath.org/millennium/Navier-Stokes_Equations/navierstokes.pdf
28. H. Fujita, T. Kato, On the Navier-Stokes initial value problem. Arch. Ration. Mech. Anal. **16**, 269–315 (1964)
29. D. Fujiwara, H. Morimoto, An $L_r$-theorem of the Helmholtz decomposition of vector fields. J. Fac. Sci. Univ. Tokyo (1A) **24**, 685–700 (1977)
30. Y. Giga, Analyticity of the semigroup generated by the Stokes operator in $L_r$-spaces. Math. Z. **178**, 287–329 (1981)
31. Y. Giga, Domains of fractional powers of the Stokes operator in $L_r$-spaces. Arch. Ration. Mech. Anal. **89**, 251–265 (1985)
32. Y. Giga, Solution for semilinear parabolic equations in $L^p$ and regularity of weak solutions for the Navier-Stokes system. J. Differ. Equ. **61**, 186–212 (1986)
33. Y. Giga, H. Sohr, Abstract $L^q$-estimates for the Cauchy problem with applications to the Navier-Stokes equations in exterior domains. J. Funct. Anal. **102**, 72–94 (1991)
34. E. Hopf, Über die Anfangswertaufgabe für die hydrodynamischen Grundgleichungen. Math. Nachr. **4**, 213–231 (1950–1951)
35. T. Kato, Strong $L^p$-solutions of the Navier-Stokes equation in $\mathbb{R}^m$, with applications to weak solutions. Math. Z. **187**, 471–480 (1984)
36. H. Koch, D. Tataru, Well-posedness for the Navier-Stokes equations. Adv. Math. **157**, 22–35 (2001)

37. C. Komo, Necessary and sufficient conditions for local strong solvability of the Navier-Stokes equations in exterior domains. Technische Universität Darmstadt, FB Mathematik, J. Evol. Equ. **14**, 713–725 (2014)
38. H. Kozono, Uniqueness and regularity of weak solutions to the Navier-Stokes equations, in *Recent topics on Mathematical Theory of Viscous Incompressible fluid*, Tsukuba, 1996. Lecture Notes in Numerical and Applied Analysis, vol. 16 (Kinokuniya, Tokyo, 1998), pp. 161–208
39. H. Kozono, H. Sohr, Remark on uniqueness of weak solutions to the Navier-Stokes equations. Analysis **16**, 255–271 (1996)
40. P.C. Kunstmann, $H^\infty$-calculus for the Stokes operator on unbounded domains. Arch. Math. **91**, 178–186 (2008)
41. J. Leray, Sur le mouvement d'un liquide visqueux emplissant l'espace. Acta Math. **63**, 193–248 (1934)
42. P. Maremonti, V.A. Solonnikov, On nonstationary Stokes problem in exterior domains. Ann. Scuola Norm. Sup. Pisa Cl. Sci. (4) **24**, 395–449 (1997)
43. V.N. Maslennikova, M.E. Bogovskiĭ, Elliptic boundary value problems in unbounded domains with noncompact and nonsmooth boundaries. Rend. Sem. Mat. Fis. Milano **LVI**, 125–138 (1986)
44. A.S. Mikhailov, T.N. Shilkin, $L_{3,\infty}$-solutions to the 3D-Navier-Stokes system in a domain with a curved boundary. Zap. Nauchn. Semin. POMI **336**, 133–152 (2006) (Russian). J. Math. Sci. **143**, 2924–2935 (2007) (English)
45. T. Miyakawa, On the initial value problem for the Navier-Stokes equations in $L^p$-spaces. Hiroshima Math. J. **11**, 9–20 (1981)
46. T. Miyakawa, H. Sohr, On energy inequality, smoothness and large time behavior in $L^2$ for weak solutions of the Navier-Stokes equations in exterior domains. Math. Z. **199**, 455–478 (1988)
47. F. Riechwald, Very weak solutions to the Navier-Stokes equaitons in general unbounded domains. Ph.D. thesis, Technische Universität Darmstadt, Logos-Verlag, Berlin, 2011
48. F. Riechwald, Interpolation of sum and intersection spaces of $L^q$-type and applications to the Stokes problem in general unbounded domains. Ann. Univ. Ferrara 58, 167–181 (2012)
49. G. Seregin, On smoothness of $L_{3,\infty}$-solutions to the Navier-Stokes equations up to boundary. Math. Ann. **332**, 219–238 (2005)
50. J. Serrin, The initial value problem for the Navier-Stokes equations, in *Nonlinear Problems*, ed. by R.E. Langer (University of Wisconsin Press, Madison, WI, 1963)
51. M. Shinbrot, The energy equation for the Navier–Stokes system. SIAM J. Math. Anal. **5**, 948–954 (1974)
52. C.G. Simader, H. Sohr, A new approach to the Helmholtz decomposition and the Neumann problem in $L^q$-spaces for bounded and exterior domains, in *Advances in Mathematics for Applied Sciences*, vol. 11 (World Scientific, Singapore, 1992), pp. 1–35
53. H. Sohr, A regularity class for the Navier-Stokes equations in Lorentz spaces. J. Evol. Equ. **1**, 441–467 (2001)
54. H. Sohr, The Navier–Stokes equations, in *An Elementary Functional Analytic Approach*. Birkhäuser Advanced Texts (Birkhäuser, Basel, 2001)
55. V.A. Solonnikov, Estimates for solutions of nonstationary Navier-Stokes equations. J. Sov. Math. **8**, 467–529 (1977)
56. H. Triebel, *Interpolation Theory, Function Spaces, Differential Operators* (North-Holland, Amsterdam, 1978)

# On Global Well/Ill-Posedness of the Euler-Poisson System

**Eduard Feireisl**

**Abstract** We discuss the problem of well-posedness of the Euler-Poisson system arising, for example, in the theory of semi-conductors, models of plasma and gaseous stars in astrophysics. We introduce the concept of *dissipative* weak solution satisfying, in addition to the standard system of integral identities replacing the original system of partial differential equations, the balance of total energy, together with the associated *relative entropy inequality*. We show that strong solutions are unique in the class of dissipative solutions (weak-strong uniqueness). Finally, we use the method of convex integration to show that the Euler-Poisson system may admit even infinitely many weak dissipative solutions emanating from the same initial data.

**Keywords** Dissipative solution • Euler-Poisson system • Weak solution

## 1 Introduction

We consider the *Euler-Poisson system* of partial differential equations in the form

$$\partial_t n + \mathrm{div}_x \mathbf{J} = 0, \tag{1}$$

$$\partial_t \mathbf{J} + \mathrm{div}_x \left( \frac{\mathbf{J} \times \mathbf{J}}{n} \right) + \nabla_x (nT) = \pm n \mathbf{V}, \tag{2}$$

$$\frac{3}{2} \partial_t (nT) + \frac{3}{2} \mathrm{div}_x (T\mathbf{J}) - \Delta T + nT \mathrm{div}_x \left( \frac{\mathbf{J}}{n} \right) = 0, \tag{3}$$

$$\mathbf{V} = \nabla_x \Phi, \ -\Delta \Phi = n - 1. \tag{4}$$

In specific applications, $n$ is the density, $\mathbf{J}$ the flux, and $T$ the (absolute) temperature of charged particles, driven by the potential volume force proportional

E. Feireisl (✉)
Institute of Mathematics of the Academy of Sciences of the Czech Republic, Žitná 25, 115 67 Praha 1, Czech Republic
e-mail: feireisl@math.cas.cz



H. Amann et al. (eds.), *Recent Developments of Mathematical Fluid Mechanics*, Advances in Mathematical Fluid Mechanics, DOI 10.1007/978-3-0348-0939-9_12

to $n\nabla_x \Phi$, see Guo [10], Guo and Pausader [11], Juengel [13], among others. From the mathematical viewpoint, Eqs. (1), (2) represent a hyperbolic Euler system, with the density $n$ and the velocity $\mathbf{J}/n$, coupled with a parabolic "heat equation" (3), and the elliptic Poisson equation (4).

To avoid the technical problems caused by the presence of a kinematic boundary, we restrict ourselves to the spatially periodic boundary conditions, specifically all quantities are functions of the time $t \in (0, \tau)$ and the position $x$, belonging to the flat torus

$$\mathcal{T} = \left( [-1, 1]|_{\{-1,1\}} \right)^3 .$$

Accordingly, the problem is formally closed by prescribing the initial conditions

$$n(0, \cdot) = n_0, \ \mathbf{J}(0, \cdot) = \mathbf{J}_0, \ T(0, \cdot) = T_0. \tag{5}$$

For smooth and physically relevant initial data, meaning

$$n_0(x) \geq \underline{n} > 0, \ T_0(x) \geq \underline{T} > 0 \text{ for all } x \in \mathcal{T}, \tag{6}$$

the problem (1)–(5) admits a unique regular solution on a maximal existence interval $(0, \tau_{\max})$, see Alazard [1], Serre [15, 16]. On the other hand, the Euler system (1), (2) being hyperbolic, discontinuities in the form of shock waves are likely to develop in a finite time regardless the smoothness of the initial data, see Guo and Tahvildar-Zadeh [12]. However, as observed by Guo [10], the linearized Euler-Poisson system (1), (2), (4) (with $T = \text{const}$) coincides with the Klein-Gordon equation, where the dispersive effects due "plasma" oscillations prevents the formation of shocks in small irrotational solutions.

In view of these arguments, it is interesting to examine the problem of global existence in the class of weak solutions, satisfying, in addition, certain *admissibility criteria* that would guarantee well-posedness, that means, existence, uniqueness, and possibly stability for any physically relevant initial data.

Motivated by [8], we introduce a relative entropy (energy) functional associated to the system (1)–(5), together with a class of *dissipative* weak solutions. These are, roughly speaking, the weak solutions satisfying, in addition, the total energy balance. Then we show the weak-strong uniqueness property, namely, any dissipative solution coincides with the strong solution emanating from the same initial data as long as the latter exists. The strong solutions are unique within the class of weak solutions (cf. Berthelin and Vasseur [2], Dafermos [5], Germain [9] for related results).

The last part of the paper is devoted to the problem of well-posedness in the class of weak and/or dissipative solutions. Using an extension of the *variable coefficients* analogue of the results of DeLellis and Székelyhidi [6] developed in [4], we show that the Euler-Poisson system (1)–(5) admits infinitely many global-in-time weak solutions for any smooth initial data. Although one can still hope that some apparently non-physical solutions can be eliminated by imposing the total energy balance as an admissibility criterion (dissipative weak solutions), we identify a vast class of physically admissible initial data for which the problem possesses infinitely many dissipative weak solutions.

The paper is organized as follows. Section 2 contains some preliminary material including proper definitions of the weak and dissipative solutions. In Sect. 3, we show the weak-strong uniqueness property for the dissipative solutions. In Sect. 4, the existence of global-in-time weak solutions is established for any physically admissible smooth initial data. Some examples of ill-posedness within the class of dissipative weak solutions are discussed in Sect. 5. The various concepts of solutions and their basic properties are summarized in Sect. 6.

## 2 Preliminaries, Weak and Dissipative Solutions

We start rewriting (3) as an entropy balance. Specifically, dividing (3) on $T$ and using (1), we arrive at

$$\partial_t (nS(n,T)) + \mathrm{div}_x (S(n,T)\mathbf{J}) - \Delta \log(T) = |\nabla_x \log(T)|^2, \tag{7}$$

with the specific entropy

$$S(n,T) = \log\left(\frac{T^{3/2}}{n}\right).$$

### 2.1 Weak Solutions

We say that $[n, \mathbf{J}, T]$ is a *weak solution* to the problem (1)–(5) in $(0,\tau) \times \mathcal{T}$ if:

- $n > 0, T > 0$ a.a. in $[0,\tau) \times \mathcal{T}$, $\int_{\mathcal{T}} (n-1)\,\mathrm{d}x = 0$,

$$n \in L^\infty((0,\tau)\times\mathcal{T}) \cap C([0,\tau]; L^1(\mathcal{T})),$$
$$\mathbf{J} = L^\infty((0,\tau)\times\mathcal{T}; R^3) \cap C_{\mathrm{weak}}([0,\tau]; L^2(\mathcal{T}; R^3)),$$
$$T \in L^\infty((0,\tau)\times\mathcal{T}) \cap C([0,\tau]; L^1(\mathcal{T})),\ \nabla_x\vartheta \in L^2((0,\tau)\times\mathcal{T}; R^3);$$

- the equation of continuity (1) is replaced by a family of integral identities

$$\int_{\mathcal{T}} [n(s,\cdot)\varphi(s,\cdot) - n_0\varphi(0,\cdot)]\ \mathrm{d}x = \int_0^s \int_{\mathcal{T}} [n\partial_t\varphi + \mathbf{J}\cdot\nabla_x\varphi]\ \mathrm{d}x\,\mathrm{d}t \tag{8}$$

for any $s \in [0,\tau)$ and any test function $\varphi \in C^\infty([0,\tau]\times\mathcal{T})$;

- the momentum balance (2) is satisfied in the sense that

$$\int_{\mathcal{T}} [\mathbf{J}(s,\cdot)\cdot\varphi(s,\cdot) - \mathbf{J}_0\cdot\varphi(0,\cdot)]\ \mathrm{d}x \tag{9}$$

$$= \int_0^s \int_{\mathcal{T}} \left[ \mathbf{J}\cdot\partial_t\varphi + \frac{\mathbf{J}\otimes\mathbf{J}}{n} : \nabla_x\varphi + nT\mathrm{div}_x\varphi \pm n\nabla_x\Phi\cdot\varphi \right] \mathrm{d}x\,\mathrm{d}t$$

holds for any $s \in [0,\tau)$ and any test function $\varphi \in C^\infty([0,\tau]\times\mathcal{T};R^3)$;
- the heat equation (3) is replaced by a weak form of the entropy balance (7), specifically, the integral identity

$$\int_{\mathcal{T}} [nS(n,T)(s,\cdot)\varphi(s,\cdot) - n_0 S(n_0,T_0)\varphi(0,\cdot)]\ \mathrm{d}x = \int_0^s \int_{\mathcal{T}} |\nabla_x \log(T)|^2\,\mathrm{d}x\,\mathrm{d}t \tag{10}$$

$$+ \int_0^s \int_{\mathcal{T}} [nS(n,T)\partial_t\varphi + S(n,T)\mathbf{J}\cdot\nabla_x\varphi - \nabla_x\log(T)\cdot\nabla_x\varphi]\ \mathrm{d}x\,\mathrm{d}t$$

holds for any $s \in [0,\tau)$ and any test function $\varphi \in C^\infty([0,\tau]\times\mathcal{T})$;
- the potential $\Phi$ is the (unique) solution of the elliptic equation

$$-\Delta\Phi(s,\cdot) = n(s,\cdot) - 1 \text{ in } \mathcal{T},\ \int_{\mathcal{T}} \Phi(s,\cdot)\,\mathrm{d}x = 0 \text{ for all } s\in[0,\tau). \tag{11}$$

*Remark 2.1* The condition

$$\int_{\mathcal{T}} (n-1)\,\mathrm{d}x = 0$$

can be replaced by

$$\int_{\mathcal{T}} (n-\overline{n})\,\mathrm{d}x = 0 \text{ for a certain } \overline{n} > 0.$$

Accordingly, we have to take the potential $\Phi$ such that

$$-\Delta\Phi(s,\cdot) = n(s,\cdot) - \overline{n} \text{ in } \mathcal{T}.$$

*Remark 2.2* Apparently, replacing the heat equation (3) by the entropy balance (7) may not be an equivalent operation within the framework of weak solutions. On the other hand, however, the density $n$ as well as the temperature $T$ considered in the present paper will be regular enough for (7) to imply (3) and vice versa. The entropy formulation (10) is more convenient for introducing the concept of dissipative solution discussed below.

## 2.2 Relative Entropy (Energy), Dissipative Solutions

Similarly to [8], we introduce the ballistic free energy

$$H_\Theta(n,T) = n\left(\frac{3}{2}T - \Theta S(n,T)\right),$$

together with the relative entropy functional

$$\mathcal{E}\left(n,T,\mathbf{J}\middle|N,\Theta,\mathbf{V}\right) \tag{12}$$

$$= \int_{\mathcal{T}} \left[\frac{1}{2}n\left|\frac{\mathbf{J}}{n} - \frac{\mathbf{V}}{N}\right|^2 + \left(H_\Theta(n,T) - \frac{\partial H_\Theta(N,\Theta)}{\partial N}(n-N) - H_\Theta(N,\Theta)\right)\right] \mathrm{d}x.$$

*Remark 2.3* The relative entropy (12) coincides, modulo the multiplicative factor $\Theta$, with the relative entropy introduced in the context of hyperbolic conservation laws by Dafermos [5]. Thus, correctly speaking, the physical dimension of $\mathcal{E}$ is *energy* rather than *entropy*.

## 2.3 Relative Entropy Inequality

As we can check by direct manipulation, regular solutions of the system (1)–(5) satisfy the relative entropy inequality

$$\left[\mathcal{E}\left(n,T,\mathbf{J}\middle|N,\Theta,\mathbf{V}\right)\right]_{t=0}^{t=s} + \int_0^s \int_{\mathcal{T}} \Theta \frac{|\nabla_x T|^2}{T^2} \,\mathrm{d}x\,\mathrm{d}t \tag{13}$$

$$\leq \int_0^s \int_{\mathcal{T}} \frac{1}{N}(n\mathbf{V} - N\mathbf{J}) \cdot \partial_t\left(\frac{\mathbf{V}}{N}\right) \,\mathrm{d}x\,\mathrm{d}t$$

$$+ \int_0^s \int_{\mathcal{T}} \left[\frac{1}{nN}(n\mathbf{V} - N\mathbf{J}) \otimes \mathbf{J} : \nabla_x\left(\frac{\mathbf{V}}{N}\right) - nT\mathrm{div}_x\left(\frac{\mathbf{V}}{N}\right)\right] \mathrm{d}x\,\mathrm{d}t$$

$$- \int_0^s \int_{\mathcal{T}} \left[n\Big(S(n,T) - S(N,\Theta)\Big)\partial_t\Theta + \Big(S(n,T) - S(N,\Theta)\Big)\mathbf{J}\cdot\nabla_x\Theta\right] \mathrm{d}x\,\mathrm{d}t$$

$$+ \int_0^s \int_{\mathcal{T}} \left[\left(1 - \frac{n}{N}\right)\partial_t(N\Theta) - \frac{\mathbf{J}}{N}\cdot\nabla_x(N\Theta)\right] \mathrm{d}x + \int_0^s \int_{\mathcal{T}} \frac{\nabla_x T}{T}\cdot\nabla_x\Theta \,\mathrm{d}x\,\mathrm{d}t$$

$$\pm \int_0^s \int_{\mathcal{T}} \frac{1}{N}\nabla_x\Delta^{-1}[n-1]\,(n\mathbf{V} - N\mathbf{J}) \,\mathrm{d}x\,\mathrm{d}t$$

for a.a. $s \in [0, \tau)$ and any trio of smooth "test" functions

$$N, \ \Theta, \ \mathbf{V}, \ N > 0, \ \Theta > 0, \tag{14}$$

cf. [8].

## 2.4 Dissipative Solutions

We say that a trio $[n, \mathbf{J}, T]$ is a *dissipative solution* to the problem (1)–(5) in $(0, \tau) \times \mathcal{T}$ if:

- $n > 0, T > 0$ a.a. in $[0, \tau) \times \mathcal{T}$, $\int_{\mathcal{T}} (n - 1) \, dx = 0$,

$$n \in L^\infty((0, \tau) \times \mathcal{T}) \cap C([0, \tau]; L^1(\mathcal{T})),$$
$$\mathbf{J} = L^\infty((0, \tau) \times \mathcal{T}; R^3) \cap C_{\text{weak}}([0, \tau]; L^2(\mathcal{T}; R^3)),$$
$$T \in L^\infty((0, \tau) \times \mathcal{T}) \cap C([0, \tau]; L^1(\mathcal{T})), \ \nabla_x \vartheta \in L^2((0, \tau) \times \mathcal{T}; R^3);$$

- the relative entropy inequality (13) holds for any choice of smooth test functions $N, \Theta, \mathbf{V}$ satisfying (14).

# 3 Weak Strong Uniqueness

The concept of *dissipative solution* in the context of the incompressible Euler system was introduced by DiPerna and Lions (see [14]). It is interesting to note that the dissipative solutions apparently do not satisfy any system of differential equations but just the relative entropy inequality (10). However, the following *weak-strong uniqueness property* holds:

**Theorem 3.1** *Let* $[n, T, \mathbf{J}]$ *be a dissipative solution of the problem* (1)–(5) *in* $(0, \tau) \times \mathcal{T}$*, with the initial data* $[n_0, T_0, \mathbf{J}_0]$ *satisfying* (6). *Suppose that the problem* (1)–(5) *admits also a regular solution* $[\tilde{n}, \tilde{T}, \tilde{\mathbf{J}}]$,

$$\partial_t \tilde{n}, \ \partial_t \tilde{T}, \ \partial_t \tilde{\mathbf{J}}, \ \partial_x^m \tilde{n}, \ \partial_x^m \tilde{T}, \ \partial_x^m \tilde{\mathbf{J}} \in C([0, \tau) \times \mathcal{T}), \ m = 0, 1, 2,$$

*emanating form the same initial data* $[n_0, T_0, \mathbf{J}_0]$.

*Then*

$$n \equiv \tilde{n}, \ T \equiv \tilde{T}, \ \mathbf{J} \equiv \tilde{\mathbf{J}} \ \textit{in} \ [0, \tau) \times \mathcal{T}.$$

The proof of Theorem 3.1 is based on taking $N = \tilde{n}$, $\Theta = \tilde{T}$, $\mathbf{V} = \tilde{\mathbf{J}}$ as test functions in the relative entropy inequality (10) and "absorbing" the terms on the right-hand side of the resulting expression by means of the Gronwall argument. Since the dissipative solutions introduced in Sect. 2.4 are *bounded*, the proof of Theorem 3.1 is essentially the same as that of Theorem 6.1 [7, Sect. 6]. Note that the extra term

$$\pm \int_0^s \int_{\mathcal{T}} \frac{1}{N} \nabla_x \Delta^{-1} [n-1] \, (n\mathbf{V} - N\mathbf{J}) \, \mathrm{d}x \, \mathrm{d}t$$

in (10) can be handled without any additional difficulty.

Finally, note that the existence of *local-in-time* regular solutions to the problem (1)–(5) ranging in the standard energy Sobolev scale $W^{m,2}$ was established by Alazard [1], Serre [15], while the existence of possibly global-in-time dissipative solutions remains an outstanding open problem.

## 4 Existence of Weak Solutions for Physically Relevant Data

In the remaining part of the paper, we focus on the class of *weak* solutions to (1)–(5), and, in particular, on their relation to the dissipative solutions.

### 4.1 Global-in-Time Weak Solutions

We start with a rather striking result concerning the existence of global-in-time weak solutions in the sense specified in Sect. 2.1.

**Theorem 4.1** *Let $\tau > 0$ be given. Suppose that the initial data $[n_0, T_0, \mathbf{J}_0]$,*

$$n_0, T_0, \mathbf{J}_0 \in C^3(\mathcal{T}),$$

*satisfy* (6).

*Then the problem* (1)–(5) *possesses infinitely many weak solutions in* $[0,\tau) \times \mathcal{T}$. *In addition, the weak solutions* $[n, T, \mathbf{J}]$ *belong to the class*

$$n \in C^2([0,\tau) \times \mathcal{T}),$$

$$\partial_t T \in L^p(0,T; L^p(\mathcal{T})), \ \nabla_x^2 T \in L^p(0,\tau; L^p(\mathcal{T}; R^{3\times 3})) \ \textit{for any} \ 1 \le p < \infty,$$

$$\mathbf{J} \in C_{\mathrm{weak}}([0,\tau]; L^2(\mathcal{T}; R^3)) \cap L^\infty((0,\tau) \times \mathcal{T}; R^3), \ \mathrm{div}_x \mathbf{J} \in C^2([0,\tau) \times \mathcal{T}).$$

*Remark 4.2* It is easy to check that any weak solution enjoying the regularity properties specified in Theorem 4.1 satisfies the equation of continuity (1), the entropy balance equation (7) as well as the internal energy equation (3) a.a. in $[0,\tau) \times \mathcal{T}$. All possible singularities are therefore concentrated on the solenoidal component of the flux $\mathbf{J}$. On the other hand, the solutions are neither regular nor dissipative solutions of the problem in agreement with the conclusion of Theorem 3.1.

The remaining part of this section is devoted to the proof of Theorem 4.1.

### 4.1.1 Oscillatory Lemma

Similarly to [4], the weak solutions claimed in Theorem 4.1 are obtained by the method of convex integration, in particular, an extension to "variable coefficients" of the following result of De Lellis and Székelyhidi [6, Proposition 3], Chiodaroli [3, Sect. 6, formula (6.9)]:

**Lemma 4.3** *Let $[T_1, T_2]$, $T_1 < T_2$, be a time interval and $B \subset R^3$ a domain. Let $\tilde{n} \in (0, \infty)$, $\tilde{\mathbf{Z}} \in R^3$, $\tilde{\mathbb{U}} \in R^{3\times3}_{\mathrm{sym},0}$ be* constant *fields such that*

$$0 < \underline{n} < \tilde{n} < \overline{n}, |\tilde{\mathbf{Z}}| < \overline{Z}, \ |\tilde{\mathbb{U}}| < \overline{U}.$$

*Suppose that*

$$\mathbf{v} \in C_{\mathrm{weak}}([T_1, T_2]; L^2(B, R^3)) \cap C^1((T_1, T_2) \times \overline{B}; R^3)$$

*satisfies a linear system of equations*

$$\partial_t \mathbf{v} + \mathrm{div}_x \mathbb{U} = 0, \ \mathrm{div}_x \mathbf{v} = 0 \textit{ in } (T_1, T_2) \times B$$

*with some $\mathbb{U} \in C^1((T_1, T_2) \times \overline{B}; R^{3\times3}_{\mathrm{sym},0})$ such that*

$$\frac{3}{2}\lambda_{\max}\left[\frac{(\mathbf{v} + \tilde{\mathbf{Z}}) \otimes (\mathbf{v} + \tilde{\mathbf{Z}})}{\tilde{n}} - \frac{1}{3}\frac{|\mathbf{v} + \tilde{\mathbf{Z}}|^2}{\tilde{n}}\mathbb{I} - (\mathbb{U} + \tilde{\mathbb{U}})\right]$$
$$< e - \frac{1}{2}\frac{|\mathbf{v} + \tilde{\mathbf{Z}}|^2}{\tilde{n}} \textit{ in } (T_1, T_2) \times B$$

*for a certain function $e \in C([T_1; T_2] \times \overline{B})$.*

*Then there exist sequences $\{\mathbf{w}_n\}_{n=1}^{\infty} \subset C^{\infty}_c((T_1, T_2) \times B; R^3)$, $\{\mathbb{Y}_n\}_{n=1}^{\infty} \subset C^{\infty}_c((T_1, T_2) \times B; R^{3\times3}_{\mathrm{sym},0})$ such that $\mathbf{v}_n = \mathbf{v} + \mathbf{w}_n$, $\mathbb{U}_n = \mathbb{U} + \mathbb{Y}_n$ satisfy*

$$\partial_t \mathbf{v}_n + \mathrm{div}_x \mathbb{U}_n = 0, \ \mathrm{div}_x \mathbf{v}_n = 0 \textit{ in } (T_1, T_2) \times B$$

$$\frac{3}{2}\lambda_{\max}\left[\frac{(\mathbf{v}_n + \tilde{\mathbf{Z}}) \otimes (\mathbf{v}_n + \tilde{\mathbf{Z}})}{\tilde{n}} - \frac{1}{3}\frac{|\mathbf{v}_n + \tilde{\mathbf{Z}}|^2}{\tilde{n}}\mathbb{I} - (\mathbb{U}_n + \mathbb{U})\right]$$
$$< e - \frac{1}{2}\frac{|\mathbf{v}_n + \tilde{\mathbf{Z}}|^2}{\tilde{n}} \textit{ in } (T_1, T_2) \times B,$$

$$\mathbf{v}_n \to \mathbf{v} \in C_{\mathrm{weak}}([T_1, T_2]; L^2(B; R^3)),$$

*and*

$$\liminf_{n\to\infty} \int_{T_1}^{T_2} \int_B |\mathbf{v}_n - \mathbf{v}|^2 \, \mathrm{d}x \, \mathrm{d}t \tag{15}$$

$$\geq \Lambda \left(\underline{r}, \overline{n}, \overline{Z}, \overline{U}, \|e\|_{L^\infty((T_1,T_2)\times B)}\right) \int_{T_1}^{T_2} \int_B \left(e - \frac{1}{2}\frac{|\mathbf{v}+\tilde{\mathbf{Z}}|^2}{\tilde{n}}\right)^2 \mathrm{d}x \, \mathrm{d}t.$$

*Remark 4.4* The symbol $\lambda_{\max}[\mathbb{A}]$ denotes the maximal eigenvalue of a symmetric traceless matrix $\mathbb{A} \in R^{3\times 3}_{\mathrm{sym},0}$.

*Remark 4.5* Since $\tilde{\mathbf{v}}$, $\tilde{\mathbb{U}}$ are constant, we have

$$\partial_t(\mathbf{v} + \tilde{\mathbf{v}}) + \mathrm{div}_x(\mathbb{U} + \tilde{\mathbb{U}}) = 0.$$

Now, exactly as in [4, Sect. 3.4.1, Lemma 3.2], we can use the scale invariance of (15) to extend validity of Lemma 4.3 to *non-constant* fields $\tilde{n}$, $\tilde{\mathbf{Z}}$, $\tilde{\mathbb{U}}$:

**Lemma 4.6** *Let $[T_1, T_2]$, $T_1 < T_2$, be a time interval and $B \subset R^3$ a domain. Let*

$$\tilde{n} \in C^1([T_1,T_2]\times\overline{B}),\ \tilde{\mathbf{Z}} \in C^1([T_1,T_2]\times\overline{B}; R^3),\ \tilde{\mathbb{U}} \in C^1([T_1,T_2]\times\overline{B}; R^{3\times 3}_{\mathrm{sym},0})$$

*be given such that*

$$0 < \underline{n} < \tilde{n} < \overline{n}, |\tilde{\mathbf{Z}}| < \overline{Z},\ |\tilde{\mathbb{U}}| < \overline{U} \text{ in } (T_1,T_2)\times B.$$

*Suppose that*

$$\mathbf{v} \in C_{\mathrm{weak}}([T_1,T_2]; L^2(B,R^3)) \cap C^1((T_1,T_2)\times\overline{B}; R^3)$$

*satisfies a linear system of equations*

$$\partial_t\mathbf{v} + \mathrm{div}_x\mathbb{U} = 0,\ \mathrm{div}_x\mathbf{v} = 0 \text{ in } (T_1,T_2)\times B$$

*with some $\mathbb{U} \in C^1((T_1,T_2)\times\overline{B}; R^{3\times 3}_{\mathrm{sym},0})$ such that*

$$\frac{3}{2}\lambda_{\max}\left[\frac{(\mathbf{v}+\tilde{\mathbf{Z}})\otimes(\mathbf{v}+\tilde{\mathbf{Z}})}{\tilde{n}} - \frac{1}{3}\frac{|\mathbf{v}+\tilde{\mathbf{Z}}|^2}{\tilde{n}}\mathbb{I} - (\mathbb{U}+\tilde{\mathbb{U}})\right]$$
$$< e - \frac{1}{2}\frac{|\mathbf{v}+\tilde{\mathbf{Z}}|^2}{\tilde{n}} - \delta \text{ in } (T_1,T_2)\times B$$

*for a certain function $e \in C([T_1;T_2]\times\overline{B})$ and some $\delta > 0$.*

*Then there exist sequences* $\{\mathbf{w}_n\}_{n=1}^{\infty} \subset C_c^{\infty}((T_1,T_2)\times B;R^3)$, $\{\mathbb{Y}_n\}_{n=1}^{\infty} \subset C_c^{\infty}((T_1,T_2)\times B;R^{3\times 3}_{\mathrm{sym},0})$ *such that* $\mathbf{v}_n = \mathbf{v} + \mathbf{w}_n$, $\mathbb{U}_n = \mathbb{U} + \mathbb{Y}_n$ *satisfy*

$$\partial_t \mathbf{v}_n + \mathrm{div}_x \mathbb{U}_n = 0,\ \mathrm{div}_x \mathbf{v}_n = 0 \textit{ in } (T_1,T_2)\times B$$

$$\frac{3}{2}\lambda_{\max}\left[\frac{(\mathbf{v}_n+\tilde{\mathbf{Z}})\otimes(\mathbf{v}_n+\tilde{\mathbf{Z}})}{\tilde{n}} - \frac{1}{3}\frac{|\mathbf{v}_n+\tilde{\mathbf{Z}}|^2}{\tilde{n}}\mathbb{I} - (\mathbb{U}_n+\tilde{\mathbb{U}})\right]$$

$$< e - \frac{1}{2}\frac{|\mathbf{v}_n+\tilde{\mathbf{Z}}|^2}{\tilde{n}} \textit{ in } (T_1,T_2)\times B,$$

$$\mathbf{v}_n \to \mathbf{v} \in C_{\mathrm{weak}}([T_1,T_2];L^2(B;R^3)),$$

*and*

$$\liminf_{n\to\infty}\int_{T_1}^{T_2}\int_B |\mathbf{v}_n - \mathbf{v}|^2\ \mathrm{d}x\,\mathrm{d}t \tag{16}$$

$$\geq \Lambda\left(\underline{r},\overline{n},\overline{Z},\overline{U},\|e\|_{L^{\infty}((T_1,T_2)\times B)}\right)\int_{T_1}^{T_2}\int_B\left(e - \frac{1}{2}\frac{|\mathbf{v}+\tilde{\mathbf{Z}}|^2}{\tilde{n}}\right)^2\ \mathrm{d}x\,\mathrm{d}t.$$

*Remark 4.7* The present result may be viewed as a generalization of [4, Sect. 3.4.1, Lemma 3.2] to the case, where *both* $\mathbf{v}$ and $\mathbb{U}$ are perturbed by smooth fields $\tilde{Z}$, $\tilde{\mathbb{U}}$, respectively.

### 4.1.2 Reformulation of the Problem

We start by writing the field $\mathbf{J}$ is terms of its Helmholtz decomposition,

$$\mathbf{J} = \mathbf{v} + \nabla_x \Psi,\ \mathrm{div}_x \mathbf{v} = 0.$$

**Step 1:**
We observe that

$$\partial_t n = -\Delta \Psi,$$

in other words, the time evolution of $n$ is governed by the gradient component of $\mathbf{J}$. Consequently, we can fix a smooth trajectory $t \mapsto n(t,\cdot)$ in such a way that

$$n(0,\cdot) = n_0,\ \int_{\mathcal{T}}(n(t,\cdot)-1)\ \mathrm{d}x = 0,\ 0 < \underline{n} < n(t,\cdot) < \overline{n} \text{ for all } t \in [0,\tau],$$

and compute

$$\Psi(t,\cdot) = -\Delta^{-1}[\partial_t n], \ \int_{\mathcal{T}} \Psi(t,\cdot)\,\mathrm{d}x = 0.$$

Finally, we choose $\partial_t n(0,\cdot)$ in such a way that

$$\nabla_x \Psi(0,t) = \nabla_x \Psi_0, \ \mathbf{J}_0 = \mathbf{v}_0 + \nabla_x \Psi_0, \ \mathrm{div}_x \mathbf{v}_C = 0.$$

**Step 2:**
Given $n$, we compute the potential $\Phi$,

$$-\Delta\Phi(t,\cdot) = n(t,\cdot) - 1, \ \int_{\mathcal{T}} \Phi(t,\cdot)\,\mathrm{d}x = 0,$$

where, incidentally,

$$\partial_t \Phi = \Psi.$$

**Step 3:**
Having fixed $n$, $\Psi$, and $\Phi$, we may express the temperature $T = T[\mathbf{v}]$ using the internal energy balance (3). Exactly as in [4, Sect. 3.2], we deduce that for given $n$, $\Psi$, and

$$\mathbf{v} \in L^\infty((0,\tau)\times\mathcal{T};R^3), \ \mathbf{J} = \mathbf{v} + \nabla_x\Psi,$$

Eq. (3) admits a unique solution $T = T[\mathbf{v}]$ such that

$$T(t,x) > 0 \text{ for all } t \in [0,\tau], \ x \in \mathcal{T},$$

$$\partial_t T \in L^p(0,\tau;L^p(\mathcal{T})), \ \nabla_x^2 T \in L^p(0,\tau;L^p(\mathcal{T};R^{3\times3})) \text{ for any } 1 \le p < \infty, \tag{17}$$

where the bounds depend only on the data and on $\|\mathbf{v}\|_{L^\infty((0,\tau)\times\mathcal{T};R^3)}$. Moreover, it follows from the entropy balance (7) that

$$0 < \underline{T} \le T(t,x) < \overline{T} \text{ for all } t \in [0,\tau], \ x \in \mathcal{T}, \tag{18}$$

where the constants $\underline{T}$, $\overline{T}$ are *independent* of $\mathbf{v}$.
**Step 4:**
Finally, we rewrite the momentum equation as follows:

$$\partial_t \mathbf{v} + \mathrm{div}_x \left( \frac{(\mathbf{v} + \nabla_x\Psi)\otimes(\mathbf{v} + \nabla_x\Psi)}{n} \pm \nabla_x\Phi \otimes \nabla_x\Phi \right) \tag{19}$$

$$+\nabla_x \left( nT + \partial_t\Psi \mp \left( \Phi + \frac{1}{2}|\nabla_x\Phi|^2 \right) \right) = 0,$$

$$\mathrm{div}_x \mathbf{v} = 0, \ \mathbf{v}(0,\cdot) = \mathbf{v}_0, \tag{20}$$

where we have used the relation

$$n\nabla_x\Phi = \nabla_x\Phi + \frac{1}{2}\nabla_x|\nabla_x\Phi|^2 - \mathrm{div}_x\,(\nabla_x\Phi \otimes \nabla_x\Phi)\,.$$

### 4.1.3 Application of the Method of Convex Integration

We are ready to complete the proof of Theorem 4.1 following step by step the arguments of [4, Sect. 3.3]. To begin, we introduce the energy

$$\overline{e}[\mathbf{v}] = \chi - \frac{3}{2}nT[\mathbf{v}] - \frac{3}{2}\left(\partial_t\Psi \mp \left(\Phi + \frac{1}{6}|\nabla_x\Phi|^2\right)\right),\ \chi \in C^1[0,\tau], \tag{21}$$

together with the associated space of subsolutions:

$$X_0 = \Bigg\{\mathbf{v}\ \Bigg|\ \mathbf{v} \in L^\infty((0,\tau)\times\mathcal{T};R^3) \cap C^1((0,\tau)\times\mathcal{T};R^3) \cap C_{\mathrm{weak}}([0,\tau];L^2(\Omega;\mathcal{T})), \tag{22}$$

$$\partial_t\mathbf{v} + \mathrm{div}_x\mathbb{U} = 0,\ \mathrm{div}_x\mathbf{v} = 0 \text{ in } (0,\tau)\times\mathcal{T} \text{ for some } \mathbb{U} \in C^1((0,\tau)\times\mathcal{T};R^{3\times 3}_{\mathrm{sym},0}),$$

$$\mathbf{v}(0,\cdot) = \mathbf{v}(\tau,\cdot) = \mathbf{v}_0,$$

$$\inf_{t\in(\varepsilon,T),x\in\mathcal{T}}\Bigg\{\overline{e}[\mathbf{v}] - \frac{3}{2}\lambda_{\max}\Bigg[\frac{(\mathbf{v}+\nabla_x\Psi)\otimes(\mathbf{v}+\nabla_x\Psi)}{n}$$

$$-\left(\mathbb{U} \mp \nabla_x\Phi\otimes\nabla_x\Phi - \frac{1}{3}|\nabla_x\Phi|^2\mathbb{I}\right)\Bigg]\Bigg\} > 0$$

$$\text{for any } 0 < \varepsilon < T.$$

Next, as a consequence of (18), we can fix $\chi \in C^1[0,\tau]$ in (21) so that

$$\frac{3}{2}\lambda_{\max}\left[\frac{(\mathbf{v}_0+\nabla_x\Psi)\otimes(\mathbf{v}_0+\nabla_x\Psi)}{n} \mp \left(\nabla_x\Phi\otimes\nabla_x\Phi - \frac{1}{3}|\nabla_x\Phi|^2\mathbb{I}\right)\right]$$

$$< \overline{e}[\mathbf{v}_0] \text{ in } [0,\tau]\times\mathcal{T},$$

in particular, the constant function $\mathbf{v} \equiv \mathbf{v}_0$, together with $\mathbb{U} \equiv 0$, belong to $X_0$. Note that, as shown by De Lellis and Székelyhidi [6],

$$\frac{1}{2}|\mathbf{w}|^2 \le \frac{3}{2}\lambda_{\max}\left[\mathbf{w}\otimes\mathbf{w} - \mathbb{U}\right],\ \mathbf{w} \in R^3,\ \mathbb{U} \in R^{3\times 3}_{\mathrm{sym},0},$$

where the identity holds only if

$$\mathbb{U} = \mathbf{w} \otimes \mathbf{w} - \frac{1}{3}|\mathbf{w}|^2\mathbb{I}.$$

Now, exactly as in [4, Sect. 3.3], we define a *topological* space $X$ as a completion of $X_0$ with respect to the topology $C_{\text{weak}}([0,\tau]; L^2(\mathcal{T}; R^3))$ metrizable on bounded sets, noting that the space $X_0$ is non-empty as $\mathbf{v} = \mathbf{v}_0 \in X_0$.

We introduce a family of functionals

$$I_\varepsilon[\mathbf{v}] = \int_\varepsilon^\tau \int_{\mathcal{T}} \left( \frac{1}{2} \frac{|\mathbf{v} + \nabla_x \Psi|^2}{n} - \overline{e}[\mathbf{v}] \right) \, \mathrm{d}x \, \mathrm{d}t \text{ for } \mathbf{v} \in X, \ 0 < \varepsilon < \tau. \tag{23}$$

As a consequence of (17), (18), the functionals $I_\varepsilon$ are lower-semicontinuous in $X$.

Finally, we claim the following assertion that can be deduced from Lemma 4.6 by means of the arguments used in [4, Sect. 3.4]:

**Lemma 4.8** *Let* $\mathbf{v} \in X_0$ *such that*

$$I_\varepsilon[\mathbf{v}] < -\alpha < 0, \ 0 < \varepsilon < \tau/2.$$

*There there is* $\beta = \beta(\alpha) > 0$ *and a sequence* $\{\mathbf{v}_n\}_{n=1}^\infty \subset X_0$ *such that*

$$\mathbf{v}_n \to \mathbf{v} \text{ in } C_{\text{weak}}([0,\tau]; L^2(\mathcal{T}; R^3)), \ \liminf_{n\to\infty} I_\varepsilon[\mathbf{v}_n] \geq I_\varepsilon[\mathbf{v}] + \beta.$$

Following the arguments of [6] we obtain:

(1) cardinality of the space $X_0$ is infinite;
(2) the points of continuity of each $I_\varepsilon$ form a residual set in $X$;
(3) the set

$$\mathcal{C} = \bigcap_{m>1} \left\{ \mathbf{v} \in X \mid I_{1/m}[\mathbf{v}] \text{ is continuous} \right\},$$

being an intersection of a countable family of residual sets, is residual, in particular of infinite cardinality;

(4)

$$I_{1/m}[\mathbf{v}] = 0 \text{ for all } m > 1 \tag{24}$$

for each $\mathbf{v} \in \mathcal{C}$.

The relation (24) implies that

$$\frac{1}{2} \frac{|\mathbf{v} + \nabla_x \Psi|^2}{n} = \overline{e}[\mathbf{v}] = \chi - \frac{3}{2} n T[\mathbf{v}] - \frac{3}{2} \left( \partial_t \Psi \mp \left( \Phi + \frac{1}{6} |\nabla_x \Phi|^2 \right) \right)$$

for any $\mathbf{v} \in \mathcal{C}$,

$$\partial_t \mathbf{v} + \mathrm{div}_x \mathbb{U} = 0 \text{ in the sense of distributions in } (0, \tau) \times \mathcal{T},$$

where

$$\mathbb{U} = \frac{(\mathbf{v} + \nabla_x \Psi) \otimes (\mathbf{v} + \nabla_x \Psi)}{n} - \frac{1}{3} \frac{|\mathbf{v} + \nabla_x \Psi|^2}{n} \mathbb{I} \pm \left( \nabla_x \Phi \otimes \nabla_x \Phi - \frac{1}{3} |\nabla_x \Phi|^2 \mathbb{I} \right).$$

In other words, $\mathbf{v}$ satisfies (19), (20) in the sense of distributions. We have proved Theorem 4.1.

*Remark 4.9* The reader will have noticed that our construction of the weak solutions enables to *prescribe* the value of the density also for $t = \tau$.

## 5 Well-Posedness in the Class of Dissipative Weak Solutions

As we have seen in the previous part, although the problem (1)–(5) admits global-in-time weak solutions, it is not well-posed in this class. On the other hand, the dissipative solutions enjoy the property of weak-strong uniqueness, meaning they coincide with the unique (local) strong solution as long as the latter exists. We introduce an intermediate class of *dissipative weak* solutions, specifically, the weak solutions satisfying the relative entropy inequality (13).

As shown in [7], a weak solution is a dissipative solution as soon as it satisfies the total energy balance:

$$\int_{\mathcal{T}} \left[ \frac{1}{2} \frac{|\mathbf{J}|^2}{n} + \frac{3}{2} n T \mp \frac{1}{2} n \Phi \right] (s, \cdot) \, \mathrm{d}x = \int_{\mathcal{T}} \left[ \frac{1}{2} \frac{|\mathbf{J}_0|^2}{n_0} + \frac{3}{2} n_0 T_0 \mp \frac{1}{2} n_0 \Phi_0 \right] \mathrm{d}x \tag{25}$$

for a.a. $s \in (0, \tau)$, where we have use the identities

$$\int_{\mathcal{T}} \nabla_x \Phi \cdot \mathbf{J} \, \mathrm{d}x = - \int_{\mathcal{T}} \Phi \mathrm{div}_x \mathbf{J} \, \mathrm{d}x = \int_{\mathcal{T}} \Phi \partial_t n \, \mathrm{d}x$$
$$= - \int_{\mathcal{T}} \Phi \partial_t \Delta \Phi \, \mathrm{d}x = \frac{\mathrm{d}}{\mathrm{d}t} \frac{1}{2} \int_{\mathcal{T}} |\nabla_x \Phi|^2 \, \mathrm{d}x,$$

and

$$\int_{\mathcal{T}} |\nabla_x \Phi|^2 \, \mathrm{d}x = \int_{\mathcal{T}} n \Phi \, \mathrm{d}x.$$

Note that (25) combined with the weak formulation (8)–(11), is, in fact, stronger than the relative entropy inequality (13), where the latter still holds if (25) is replaced by an *inequality*:

$$\int_{\mathcal{T}} \left[ \frac{1}{2}\frac{|\mathbf{J}|^2}{n} + \frac{3}{2}nT \mp \frac{1}{2}n\Phi \right](s,\cdot)\,\mathrm{d}x = \int_{\mathcal{T}} \left[ \frac{1}{2}\frac{|\mathbf{J}_0|^2}{n_0} + \frac{3}{2}n_0T_0 \mp \frac{1}{2}n_0\Phi_0 \right]\mathrm{d}x$$

for a.a. $s \in (0,\tau)$.

A weak solution of the problem (1)–(5) satisfying the energy equality (25) will be termed *finite energy weak solution*. As a direct consequence of [7, Theorem 6.1] (cf. Theorem 3.1 above) we obtain:

**Theorem 5.1** *Let* $[n, T, \mathbf{J}]$ *be a finite energy weak solution of the problem* (1)–(5) *in* $(0,\tau) \times \mathcal{T}$, *with the initial data* $[n_0, T_0, \mathbf{J}_0]$ *satisfying* (6). *Suppose that the problem* (1)–(5) *admits also a regular solution* $[\tilde{n}, \tilde{T}, \tilde{\mathbf{J}}]$,

$$\partial_t\tilde{n},\ \partial_t\tilde{T},\ \partial_t\tilde{\mathbf{J}},\ \partial_x^m\tilde{n},\ \partial_x^m\tilde{T},\ \partial_x^m\tilde{\mathbf{J}} \in C([0,\tau)\times\mathcal{T}),\ m = 0,1,2,$$

*emanating form the same initial data* $[n_0, T_0, \mathbf{J}_0]$.

*Then*

$$n \equiv \tilde{n},\ T \equiv \tilde{T},\ \mathbf{J} \equiv \tilde{\mathbf{J}} \text{ in } [0,\tau)\times\mathcal{T}.$$

Thus, the stipulation of the total energy conservation (25) seems to eliminate the "non-physical" weak solutions obtained in Theorem 4.1. On the other hand, however, there might still be "irregular" initial data for which the problem (1)–(5) admits infinitely many finite energy weak solution. The precise statement reads:

**Theorem 5.2** *Let* $\tau > 0$ *and the initial data* $[n_0, T_0]$,

$$n_0, T_0 \in C^3(\mathcal{T}),$$

*satisfying* (6) *be given.*

*Then there exists a flux*

$$\mathbf{J}_0 \in L^\infty(\mathcal{T}; R^3)$$

*such that the problem* (1)–(5) *possesses infinitely many* finite energy *weak solutions in* $[0,\tau) \times \mathcal{T}$. *In addition, the weak solutions* $[n, T, \mathbf{J}]$ *belongs to the class specified in Theorem* 4.1.

The proof of Theorem 5.2 is essentially the same as that of [4, Theorem 4.2] and we leave it to the interested reader. Clearly, in accordance Theorem 3.1, the initial datum $\mathbf{J}_0$, the existence of which is claimed in Theorem 5.2, cannot be regular. On the other hand, the solutions obtained in Theorem 8 must have a "large" solenoidal part - to be compared with the result of Guo [10] on global existence of smooth solutions with *irrotational* initial data.

## 6 Conclusion

Unlike the standard Euler system, the Euler-Poisson system possesses global-in-time weak solutions for small irrotational initial data, see Guo [10]. On the other hand, we have shown that large data with a non-zero solenoidal component give rise to infinitely many weak solutions satisfying the global energy balance, see Theorem 5.2. In general, we have introduced several classes of solutions to the Euler-Poisson system (1)–(5), the properties of which can summarized as follows:

- **Strong (classical) solutions.** They are classical (differentiable) solutions of the system (1)–(5), emanating from regular initial data, that exist on a (possibly) short time interval for general (smooth) data $[n_0, \mathbf{J}_0, T_0]$. Global-in-time existence is to be expected for small irrotational data, see Guo [10].
- **Weak (distributional) solutions.** They satisfy (1)–(5) in the sense of distribution. Global-in-time weak solutions do exist for any initial data but they may be "unphysical" in the sense that they produce energy at the initial time.
- **Dissipative solutions.** The satisfy the relative entropy inequality and, consequently, they coincide with the (local) strong solution emanating from the same initial data as long as the latter exists. Global-in-time existence of dissipative solutions for general initial data is an open problem.
- **Finite energy weak solutions.** These are the weak solutions satisfying, in addition, the total energy balance. They are dissipative solutions so they coincide with a strong solutions as long as the latter exists. Global existence of finite energy weak solutions for general initial data is an open problem. On the other hand, there is a vast set of initial data (with an irregular flux $\mathbf{J}_0$), for which the problem (1)–(5) admits *infinitely many* global-in-time finite energy weak solutions.

**Acknowledgements** The research leading to these results has received funding from the European Research Council under the European Union's Seventh Framework Programme (FP7/2007-2013)/ ERC Grant Agreement 320078.

## References

1. T. Alazard, Low Mach number flows and combustion. SIAM J. Math. Anal. **38**(4), 1186–1213 (2006) (electronic)
2. F. Berthelin, A. Vasseur, From kinetic equations to multidimensional isentropic gas dynamics before shocks. SIAM J. Math. Anal. **36**, 1807–1835 (2005)
3. E. Chiodaroli, A counterexample to well-posedness of entropy solutions to the compressible Euler system. J. Hyperbolic Differ. Equ. **11**(3), 493–519 (2014)
4. E. Chiodaroli, E. Feireisl, O. Kreml, On the weak solutions to the equations of a compressible heat conducting gas. Ann. Inst. Henri Poincaré Anal Non Linéaire. **32**(1), 225–243 (2015)
5. C.M. Dafermos, The second law of thermodynamics and stability. Arch. Ration. Mech. Anal. **70**, 167–179 (1979)

6. C. De Lellis, L. Székelyhidi Jr., On admissibility criteria for weak solutions of the Euler equations. Arch. Ration. Mech. Anal. **195**(1), 225–260 (2010)
7. E. Feireisl, Relative entropies in thermodynamics of complete fluid systems. Discrete Cont. Dyn. Syst. Ser. A **32**, 3059–3080 (2012)
8. E. Feireisl, A. Novotný, Weak-strong uniqueness property for the full Navier-Stokes-Fourier system. Arch. Ration. Mech. Anal. **204**, 683–706 (2012)
9. P. Germain, Weak-strong uniqueness for the isentropic compressible Navier-Stokes system. J. Math. Fluid Mech. **13**(1), 137–146 (2011)
10. Y. Guo, Smooth irrotational flows in the large to the Euler-Poisson system in $R^{3+1}$. Commun. Math. Phys. **195**, 249–265 (1998)
11. Y. Guo, B. Pausader, Global smooth ion dynamics in the Euler-Poisson system. Commun. Math. Phys. **308**, 89–125 (2011)
12. Y. Guo, A.S. Tahvildar-Zadeh, Nonlinear partial differential equations, in *Contemporary Mathematics* (American Mathematical Society, Providence, RI, 1999), pp. 151–161
13. A. Jüengel, *Transport Equations for Semiconductors*. Lecture Notes in Physis, vol. 773 (Springer, Heidelberg, 2009)
14. P.-L. Lions, *Mathematical Topics in Fluid Dynamics: Vol. 1, Incompressible models* (Oxford Science Publication, Oxford, 1996)
15. D. Serre, Local existence for viscous system of conservation laws: $H^s$-data with $s > 1 + d/2$, in *Nonlinear Partial Differential Equations and Hyperbolic Wave Phenomena*, vol. 526. Contemporary Mathematics (American Mathematical Society, Providence, RI, 2010), pp. 339–358
16. D. Serre, The structure of dissipative viscous system of conservation laws. Phys. D **239**(15), 1381–1386 (2010)

# On the Motion of a Liquid-Filled Rigid Body Subject to a Time-Periodic Torque

**Giovanni P. Galdi, Giusy Mazzone, and Mahdi Mohebbi**

*Dedicated, with friendship and admiration, to Professor Yoshihiro Shibata, on the occasion of his 60th birthday*

**Abstract** In this paper we investigate the existence of time-periodic motions of a system constituted by a rigid body with an interior cavity completely filled with a viscous liquid, and subject to a time-periodic external torque acting on the rigid body. We then show that the system of equations governing the motion of the coupled system liquid-filled rigid body, has at least one corresponding time-periodic weak solution. Furthermore if the size of the torque is below a certain constant, the weak solution is in fact strong.

**Keywords** Navier-Stokes equations • Rigid body motions • Time-periodic solutions

## 1 Introduction

Problems involving the motion of a rigid body with a cavity filled with a viscous liquid are of fundamental interest in several applied research areas, such as dynamics of flight [10, 17], space technology [4, 14], and geophysical problems [16, 18]. Besides its important role in physical and engineering disciplines, the motion of such coupled systems generates a number of mathematical questions, which are intriguing and challenging, mostly due to the different properties of the components of the system: one dissipative (the liquid), the other conservative (the rigid body). However, despite their relevance, basic mathematical questions related to the above

G.P. Galdi (✉) • G. Mazzone
Department of Mechanical Engineering and Materials Science, University of Pittsburgh, 3700 O'Hara Street, Pittsburgh, PA 15261, USA
e-mail: galdi@pitt.edu; gim20@pitt.edu

M. Mohebbi
Department of Mechanical Engineering, SUNY Korea, 119 Songdo Moonhwa-ro, Yeonsu-gu, Incheon 406–840, Korea
e-mail: mahdi.mohebbi@sunykorea.ac.kr



H. Amann et al. (eds.), *Recent Developments of Mathematical Fluid Mechanics*, Advances in Mathematical Fluid Mechanics, DOI 10.1007/978-3-0348-0939-9_13

problem in its full generality appear to have received, to date, little or no attention; see, e.g., [8] and the reference cited therein.

Objective of this paper is to give a detailed analytical study of one of the above questions, when the coupled system body/liquid moves under the action of a time-periodic torque.

More specifically, let $\mathcal{B}$ be a rigid body containing in its interior a hollow cavity, $\mathcal{C}$, completely filled by a viscous liquid, $\mathcal{L}$. We shall assume that the resultant of the external forces applied to the coupled system $\mathcal{S} = \mathcal{B} \cup \mathcal{L}$ is identically zero, so that the center of mass, $G$, of $\mathcal{S}$ moves by uniform rectilinear motion with respect to an inertial frame $\mathcal{I}$. As a result, the relevant motion of $\mathcal{S}$ reduces to the one occurring around its center of mass, namely, with respect to a frame, $\mathcal{G}$, with the origin in $G$ and axes parallel to those of $\mathcal{I}$.

We shall next suppose that on $\mathcal{B}$ acts a torque, $\boldsymbol{m}$, varying in time in a time-periodic fashion:[1]

$$\boldsymbol{m} = f_i(t)\, \boldsymbol{e}_i(t)\,, \tag{1}$$

where $\{\boldsymbol{e}_i\}$ is a base in $\mathcal{G}$, while $f_i$, $i = 1, 2, 3$, are given $T$-periodic scalar functions of time $t$, that is, $f_i(t + T) = f_i(t)$, for all $t \in \mathbb{R}$.

In the wake of analogous classical problems formulated in absence of liquid, we propose to investigate whether, under the given assumptions, the coupled system $\mathcal{S}$ will execute a $T$-periodic motion in a *body-fixed* frame $\mathcal{S}$.

In order to handle the above question, it appears necessary to impose some restrictions on the functions $f_i$ in (1), as we shall show next.

To this end, we begin to observe that in the frame $\mathcal{S}$ the torque $\boldsymbol{m}$ becomes

$$\boldsymbol{M} = \boldsymbol{Q}^\top(t) \cdot \boldsymbol{m} = f_i(t)\, \boldsymbol{Q}^\top(t) \cdot \boldsymbol{e}_i\,, \tag{2}$$

where $\boldsymbol{Q} = \boldsymbol{Q}(t)$ is the (unknown) one-parameter family of elements of the special orthogonal group, $SO(3)$, associated with the change of frame $\mathcal{G} \to \mathcal{S}$, and $^\top$ denotes transpose. We then have

$$\frac{d\boldsymbol{Q}^\top}{dt} = \mathcal{W}(\boldsymbol{\Omega}) \cdot \boldsymbol{Q}^\top \;, \boldsymbol{Q}^\top(0) = \mathbf{1}\,, \quad \mathcal{W}(\boldsymbol{\Omega}) := \begin{bmatrix} 0 & \Omega_3 & -\Omega_2 \\ -\Omega_3 & 0 & \Omega_1 \\ \Omega_2 & -\Omega_1 & 0 \end{bmatrix}, \tag{3}$$

where $\boldsymbol{\Omega} = (\Omega_1, \Omega_2, \Omega_3)$ is the angular velocity of $\mathcal{B}$ in $\mathcal{S}$, and $\mathbf{1}$ is the identity matrix; see, e.g., [2, § 6.26]. Assuming the motion of $\mathcal{S}$ in the frame $\mathcal{S}$ to be $T$-periodic implies, in particular, that both $\boldsymbol{\Omega}$ and $\boldsymbol{M}$ have to be $T$-periodic as well. Therefore, from (3) and standard Floquet theory it follows that $\boldsymbol{Q}(t)$ has the following representation [19, Theorem 1]

$$\boldsymbol{Q}(t) = \boldsymbol{P}(t) \cdot e^{tS}\,, \quad t \in \mathbb{R}\,, \tag{4}$$

[1] We adopt summation convention over repeated indices.

where $\boldsymbol{S}$ is a real, skew-symmetric matrix, and

$$\boldsymbol{P}(t) \in SO(3)\,, \quad \boldsymbol{P}(t+T) = \boldsymbol{P}(t)\,, \text{ for all } t \in \mathbb{R}\,. \tag{5}$$

From the latter and from (2) we then deduce that for $\boldsymbol{M}$ to be a $T$-periodic function we must have

$$e^{TS} \cdot \boldsymbol{e}_i = \boldsymbol{e}_i\,, \quad i = 1, 2, 3\,,$$

namely, $\boldsymbol{e}_i$ must be parallel to the eigenvectors corresponding to the eigenvalue $\lambda = 0$ of $\boldsymbol{S}$. Since, in general, $\lambda = 0$ is simple, the existence of a $T$-periodic solution in the body-fixed frame $\mathcal{S}$ requires, in general, that $\boldsymbol{m}$ is directed along a *constant* direction. We thus have

$$\boldsymbol{m} = f(t)\,\boldsymbol{h}\,, \tag{6}$$

where $\boldsymbol{h}$ is a unit, time-independent vector in $\mathcal{G}$, and $f$ is a $T$-periodic function.

We next observe that, denoting by $\boldsymbol{k}_G$ the total angular momentum of $\mathscr{S}$ with respect to $G$, the balance of angular momentum in the frame $\mathcal{G}$ requires

$$\frac{d}{dt}\boldsymbol{k}_G = f(t)\,\boldsymbol{h}\,,$$

from which we at once deduce, in particular, that $|\boldsymbol{k}_G(t)|$ is $T$-periodic if and only if $f$ has a zero average over a period:

$$\int_0^T f(t)\,dt = 0\,. \tag{7}$$

However, $|\boldsymbol{k}_G(t)|$ is left invariant in the frame change $\mathcal{G} \to \mathcal{S}$, so that the searched $T$-periodicity of the motion of $\mathscr{S}$ with respect to the frame $\mathcal{S}$ requires that $f$ obeys (7). As a consequence of what just shown, we shall then require that the torque $\boldsymbol{m}$ acting on $\mathscr{B}$ satisfies (6)–(7).

Under these assumptions, the main goal of this paper consists in proving the existence of a motion of the coupled system $\mathscr{S}$ that is $T$-periodic with respect to the body-fixed frame $\mathcal{S}$. It is worth remarking that in $\mathcal{S}$ the direction of the torque becomes a function of time given by $\boldsymbol{H}(t) := \boldsymbol{Q}^\top(t) \cdot \boldsymbol{h}$, and since $\boldsymbol{Q}$ is not known, $\boldsymbol{H}$ becomes a further unknown. From the physical viewpoint, the latter circumstance means that, in order to perform such a periodic motion, the body has to find an "appropriate orientation" with respect to the direction of $\boldsymbol{m}$.

We thus show that, under the hypothesis that $f$ is $T$-periodic and square-summable over a period, the problem admits a corresponding (suitably defined) $T$-periodic *weak* solution. If, moreover, $f$ is essentially bounded with a sufficiently small norm, then the solution is *strong* and the relevant equations are satisfied almost everywhere in space-time.

We would like to complete this introductory section with two further observations. In the first place, it is natural to ask whether the motion of the coupled system $\mathcal{S}$ is time-periodic *also in the inertial frame* $\mathcal{G}$. The answer to this question is, in general, negative. In fact, denoting by $\boldsymbol{v}$ and $\boldsymbol{\omega}$ the velocity field of $\mathcal{L}$ and the angular velocity of $\mathcal{B}$ with respect to $\mathcal{G}$, we have

$$\boldsymbol{v}(\boldsymbol{x},t) = \boldsymbol{Q}(t)\cdot\boldsymbol{V}(\boldsymbol{Q}^\top(t)\cdot\boldsymbol{x},t)\,,\;\;\boldsymbol{\omega}(t) = \boldsymbol{Q}(t)\cdot\boldsymbol{\Omega}(t)\,. \tag{8}$$

Therefore, since both $\boldsymbol{V}$ and $\boldsymbol{\Omega}$ are time-periodic, in view of the representation (4)–(5) of $\boldsymbol{Q}$, we can only conclude that, in general, $\boldsymbol{v}$ and $\boldsymbol{\omega}$ are almost periodic functions of time. However, the component of $\boldsymbol{\omega}$ along the (fixed) direction $\boldsymbol{h}$ of the torque in $\mathcal{G}$ is $T$-periodic. Actually, by (4)–(5)

$$\boldsymbol{h} = \boldsymbol{P}(t)\cdot e^{tS}\cdot\boldsymbol{H}(t)\,,$$

and since $\boldsymbol{h} = \boldsymbol{H}(0)$, by the $T$-periodicity of $\boldsymbol{H}$ and $\boldsymbol{P}$ we deduce

$$\boldsymbol{h} = e^{TS}\cdot\boldsymbol{H}(0) = e^{TS}\cdot\boldsymbol{h}.$$

As a result, being $\boldsymbol{\Omega}(0) = \boldsymbol{\omega}(0)$ and using the $T$-periodicity of $\boldsymbol{\Omega}$ we conclude

$$\boldsymbol{\omega}(T)\cdot\boldsymbol{h} = e^{TS}\cdot\boldsymbol{\Omega}(T)\cdot\boldsymbol{h} = \boldsymbol{\omega}(0)\cdot e^{-TS}\cdot\boldsymbol{h} = \boldsymbol{\omega}(0)\cdot\boldsymbol{h}\,.$$

In the light of these considerations it is very likely that if the cavity $\mathcal{C}$ possesses suitable geometric symmetry, and the direction of $\boldsymbol{h}$ is constant and chosen along one of the principal axes of inertia of $\mathcal{S}$, the motion of $\mathcal{B}$ is periodic also in $\mathcal{G}$. We shall analyze this problem elsewhere.

Another interesting question concerns the *uniqueness* of the solutions. Also here the answer is, in general, negative. It is in fact sufficient to consider the case $f \equiv 0$ and observe that permanent rotations, where $\boldsymbol{\Omega}$ is constant and directed along any principal axis of inertia of $\mathcal{S}$ and $\boldsymbol{V} = \boldsymbol{0}$, are allowed time-periodic solution of arbitrary period. Therefore, the important issue here becomes the *stability* of time-periodic motions, aiming at selecting which, among the possibly many, is indeed observable and, as such, physically meaningful. This appears to be a very complicated task that will be the object of future work.

The paper is organized as follows. After giving in Sect. 2 the formulation of the problem with respect to the frame $\mathcal{S}$, in the subsequent Sect. 3 we furnish a corresponding weak formulation. The latter, essentially, requires that the relative velocity field, $\boldsymbol{V}$, of $\mathcal{L}$ in $\mathcal{S}$, satisfies the function properties of a solution à la Leray-Hopf (see [5]) while $\boldsymbol{\Omega}$ and $\boldsymbol{H}$ are requested to be only continuous functions of time. Successively, in Sect. 4, we show the existence of a weak solution (see Theorem 4.1). The proof is accomplished by an appropriate combination of the classical Galerkin method with a fixed point argument for triangulable manifolds based on the Lefschetz-Hopf theorem. Finally, in Sect. 5, we prove that if the size of $f$ is suitably restricted, the weak solution constructed in Theorem 4.1 is in fact strong, namely, $V$ possesses first time derivative and second spatial derivative that are square summable in space and time, while $\boldsymbol{\Omega}$ and $\boldsymbol{H}$ are suitably differentiable (Theorem 5.2).

## 2 Formulation of the Problem

We keep the notation presented in the introductory section. Assuming that the flow of the viscous liquid is described by the Navier-Stokes equations, we find that the governing equations of the liquid in the frame $\mathcal{S}$ attached to the body are given by Kopachevsky and Krein [11]

$$\left.\begin{array}{l}\dfrac{\partial \boldsymbol{V}}{\partial t}+\dot{\boldsymbol{\Omega}}\times \boldsymbol{y}+2\boldsymbol{\Omega}\times \boldsymbol{V}+\boldsymbol{V}\cdot \operatorname{grad}\boldsymbol{V}+\operatorname{grad}p-\nu\Delta \boldsymbol{V}=0\\ \operatorname{div}\boldsymbol{V}=0\end{array}\right\}\ \text{in } \mathcal{D}\times[0,T], \tag{9}$$

along with the boundary condition

$$\boldsymbol{V}(x,t)=\boldsymbol{0}\ \text{ for } (x,t)\in\partial\mathcal{D}\times[0,T]. \tag{10}$$

In the above equations, $\mathcal{D}$ is a bounded, simply connected domain of $\mathbb{R}^3$ representing the volume of $\mathcal{C}$, $\boldsymbol{V}$and $\boldsymbol{\Omega}$ are relative velocity field of $\mathcal{L}$ and angular velocity of $\mathcal{B}$, respectively, referred to $\mathcal{S}$, and $\nu>0$ is the coefficient of kinematic viscosity of $\mathcal{L}$. Finally,

$$p:=\frac{1}{\rho_{\mathcal{L}}}P(x,t)-\frac{1}{2}\left|\boldsymbol{\Omega}\times \boldsymbol{y}\right|^2$$

where $\rho_{\mathcal{L}}$ and $P$ are (constant) density and pressure field of $\mathcal{L}$.

As for the body $\mathcal{B}$, during its motion it has to obey the balance of angular momentum that, when referred to $\mathcal{S}$, takes the following form

$$\boldsymbol{J}\cdot\frac{d\boldsymbol{\Omega}}{dt}+\boldsymbol{\Omega}\times\boldsymbol{J}\cdot\boldsymbol{\Omega}=f(t)\boldsymbol{H}-\int_{\partial\mathcal{D}}\boldsymbol{y}\times\boldsymbol{T}(\boldsymbol{V},P)\cdot\boldsymbol{n}\,,\ \ t\in[0,T], \tag{11}$$

where $\boldsymbol{J}$ is the inertia tensor of $\mathcal{B}$, $\boldsymbol{T}(\boldsymbol{V},P)=-P\boldsymbol{1}+\frac{\rho_{\mathcal{L}}\nu}{2}(\operatorname{grad}\boldsymbol{V}+(\operatorname{grad}\boldsymbol{V})^\top)$ is the Cauchy stress tensor, and $\boldsymbol{n}$ is the unit outer normal to $\partial\mathcal{D}$. Furthermore, $\boldsymbol{H}:=\boldsymbol{Q}^\top(t)\cdot\boldsymbol{h}$, which, in view of (3), must satisfy the following (Poisson) equation

$$\frac{d\boldsymbol{H}}{dt}+\boldsymbol{\Omega}\times\boldsymbol{H}=\boldsymbol{0}. \tag{12}$$

It is now convenient to rewrite the system of equations (9)–(11) in a different but *equivalent* form that is also more significant from the physical viewpoint and more manageable from the mathematical one. To this end, we begin to notice that by a direct calculation one shows that (9)–(11) is equivalent to (9), (10) and the following one

$$\frac{d\boldsymbol{A}}{dt} + \boldsymbol{\Omega} \times \boldsymbol{A} = f(t)\,\boldsymbol{H}\,, \tag{13}$$

where $\boldsymbol{A}$ is the *angular momentum of the coupled system* $\mathcal{S}$, referred to $\mathcal{S}$, given by

$$\boldsymbol{A} := \rho_{\mathcal{L}} \int_{\mathcal{D}} \boldsymbol{y} \times \boldsymbol{V} + \boldsymbol{I} \cdot \boldsymbol{\Omega}\,, \tag{14}$$

with $\boldsymbol{I}$ inertia tensor of $\mathcal{S}$. Using (14) we can then eliminate $\boldsymbol{\Omega}$ and write the relevant equations only in terms of the unknowns $\boldsymbol{V}, p, \boldsymbol{A}$, and $\boldsymbol{H}$. Thus, observing that

$$\boldsymbol{\Omega} = \boldsymbol{I}^{-1} \cdot \boldsymbol{A} - \rho_{\mathcal{L}} \boldsymbol{I}^{-1} \cdot \left( \int_{\mathcal{D}} \boldsymbol{y} \times \boldsymbol{V}\, dV \right), \tag{15}$$

the system of equations (9), (10), (13), and (12) becomes (see also [11]):

$$\begin{aligned}
&\left.\begin{aligned}
&(\boldsymbol{1} - \boldsymbol{B}) \cdot \frac{\partial \boldsymbol{V}}{\partial t} + \left( \boldsymbol{I}^{-1} \cdot \frac{d\boldsymbol{A}}{dt} \right) \times \boldsymbol{y} + 2\left(\boldsymbol{I}^{-1} \cdot \boldsymbol{A}\right) \times \boldsymbol{V} \\
&\qquad -2\left[ \boldsymbol{I}^{-1} \cdot \left( \rho_{\mathcal{L}} \int_{\mathcal{D}} \boldsymbol{y} \times \boldsymbol{V}\, dV \right) \right] \times \boldsymbol{V} \\
&\qquad +\boldsymbol{V} \cdot \operatorname{grad} \boldsymbol{V} + \operatorname{grad} p - \nu \Delta \boldsymbol{V} = 0 \\
&\operatorname{div} \boldsymbol{V} = 0,
\end{aligned}\right\} \text{ in } \mathcal{D} \times [0, T]\,, \\
&\boldsymbol{V}(x,t) = \boldsymbol{0} \text{ on } \partial D \times [0,T]\,, \\
&\left.\begin{aligned}
&\frac{d\boldsymbol{A}}{dt} + (\boldsymbol{I}^{-1} \cdot \boldsymbol{A}) \times \boldsymbol{A} - \rho_{\mathcal{L}} \boldsymbol{I}^{-1} \cdot \left( \int_{\mathcal{D}} \boldsymbol{y} \times \boldsymbol{V}\, dV \right) \times \boldsymbol{A} = f(t) \boldsymbol{H} \\
&\frac{d\boldsymbol{H}}{dt} + (\boldsymbol{I}^{-1} \cdot \boldsymbol{A}) \times \boldsymbol{H} - \rho_{\mathcal{L}} \boldsymbol{I}^{-1} \cdot \left( \int_{\mathcal{D}} \boldsymbol{y} \times \boldsymbol{V}\, dV \right) \times \boldsymbol{H} = 0
\end{aligned}\right\} \text{ in } [0,T]\,,
\end{aligned} \tag{16}$$

where

$$\boldsymbol{B} : \ \boldsymbol{\psi} \in L^1(\mathcal{D}) \mapsto \boldsymbol{B} \cdot \boldsymbol{\psi} \in L^\infty(\mathcal{D}) \tag{17}$$

is such that

$$(\boldsymbol{B} \cdot \boldsymbol{\psi})(\boldsymbol{y}) = \rho_{\mathcal{L}} \boldsymbol{I}^{-1} \cdot \left( \int_{\mathcal{D}} \boldsymbol{y} \times \boldsymbol{\psi}\, dV \right) \times \boldsymbol{y}. \tag{18}$$

Our problem can be then formulated as follows. *Given a sufficiently smooth $T$-periodic function $f$, find a corresponding $T$-periodic solution to* (16)–(18).

For the resolution of this problem an important role is played by the properties of the operator $\boldsymbol{B}$. The latter have been studied and determined in [11, Chap. 1, Sect. 7.2.3], and are summarized in the following lemma.

**Lemma 2.1** *The operator* $\boldsymbol{B}$ *defined by* (17) *and* (18) *is non-negative and self-adjoint. Moreover,* $\mathbf{1} - \boldsymbol{B}$ *is a non-negative operator with a bounded inverse.*

## 3 Function Spaces and Weak Formulation

In this section we furnish a weak formulation of the time-periodic problem stated in Sect. 2. To this end, we introduce the main function spaces that we shall use throughout. For all the other (standard) notation used in this paper and not mentioned here, we refer to, e.g., [1]. Let

$$\mathcal{D}(\mathcal{D}) = \{\boldsymbol{\psi} \in C_0^\infty(\mathcal{D}) |\ \operatorname{div} \boldsymbol{\psi} = 0 \text{ in } \mathcal{D}\}.$$

We denote by $H(\mathcal{D})$ the closure of $\mathcal{D}(\mathcal{D})$ in the Lebesgue space $L^2(\mathcal{D})$ and by $H^k(\mathcal{D})$ the closure of $\mathcal{D}(\mathcal{D})$ in the Sobolev space $W^{k,2}(\mathcal{D})$.

As is well known, the Helmholtz-Weyl decomposition holds (see [6], Theorem III.1.1):

$$L^2(\mathcal{D}) = H(\mathcal{D}) \oplus G(\mathcal{D}),$$

where

$$G(\mathcal{D}) = \{\boldsymbol{v} \in L^2(\mathcal{D}) |\ \boldsymbol{v} = \operatorname{grad} p, \text{ for some } p \in L^1_{\text{loc}}(\mathcal{D})\}.$$

We indicate by $P$ the orthogonal projection of $L^2(\mathcal{D})$ onto $H(\mathcal{D})$. In the space $L^2(\mathcal{D})$ we shall use the customary notation for inner product and associate norm:

$$(\boldsymbol{u}, \boldsymbol{v}) := \int_{\mathcal{D}} \boldsymbol{u} \cdot \boldsymbol{v}\, dV, \qquad \|\boldsymbol{u}\|_2 := (\boldsymbol{u}, \boldsymbol{v})^{1/2},$$

defined for any $\boldsymbol{u}, \boldsymbol{v} \in L^2(\mathcal{D})$. The following functional

$$\langle \cdot, \cdot \rangle : (\boldsymbol{u}, \boldsymbol{v}) \in L^2(\mathcal{D}) \times L^2(\mathcal{D}) \mapsto \langle \boldsymbol{u}, \boldsymbol{u} \rangle := ((\mathbf{1} - \boldsymbol{B}) \cdot \boldsymbol{u}, \boldsymbol{v}), \tag{19}$$

defines a scalar product in $L^2(\mathcal{D})$ with the associate norm

$$\|\boldsymbol{u}\|_B := \langle \boldsymbol{u}, \boldsymbol{u} \rangle^{1/2} = ((\mathbf{1} - \boldsymbol{B}) \cdot \boldsymbol{u}, \boldsymbol{u})^{1/2},$$

which is equivalent to the norm $\|\cdot\|_2$. Indeed, since by Lemma 2.1, $\boldsymbol{B}$ is non-negative and $\mathbf{1} - \boldsymbol{B}$ admits a bounded inverse, we find

$$((\mathbf{1} - \boldsymbol{B}) \cdot \boldsymbol{u}, \boldsymbol{u}) = \|(\mathbf{1} - \boldsymbol{B}) \cdot \boldsymbol{u}\|_2^2 + (\boldsymbol{B} \cdot \boldsymbol{u}, \boldsymbol{u}) \geq C^2 \|\boldsymbol{u}\|_2^2$$

where, $C = C(\mathcal{D}) > 0$. Furthermore, again using the fact that $\boldsymbol{B}$ is non-negative, we deduce

$$((\mathbf{1} - \boldsymbol{B}) \cdot \boldsymbol{u}, \boldsymbol{u}) \leq (\boldsymbol{u}, \boldsymbol{u}),$$

so that

$$C\|\boldsymbol{u}\|_2 \leq \|\boldsymbol{u}\|_B \leq \|\boldsymbol{u}\|_2, \tag{20}$$

for every $\boldsymbol{u} \in L^2(\mathcal{D})$.

For $T > 0$, $q \in [1, \infty]$, and $k \in \mathbb{N}$ we set

$$\begin{aligned} L^q_T(\mathbb{R}) &\doteqdot \{u \in L^q_{\text{loc}}(\mathbb{R}) |\ u(t) = u(t+T), \text{ for a.a. } t \in \mathbb{R}\} \\ C^k_T(\mathbb{R}) &\doteqdot \{\xi \in C^k(\mathbb{R}) |\ \xi(t) = \xi(t+T), \text{ for all } t \in \mathbb{R}\}. \end{aligned}$$

Finally, given a Banach space $X$ with norm $\|\cdot\|_X$, we denote by $L^q(a, b; X)$, $q \in (1, \infty)$ [respectively, $C([a, b]; X)$], the class of functions $u : [a, b] \to X$, with $\|u\|_{L^q(a,b;X)} < \infty$ [respectively, $\max_{t \in [a,b]} \|u(t)\|_X < \infty$], where

$$\|u\|_{L^q(a,b;X)} := \begin{cases} \left(\int_a^b \|u(t)\|_X^q \, dt\right)^{1/q}, & \text{if } 1 \leq q < \infty, \\ \operatorname*{ess\ sup}_{t \in (a,b)} \|u(t)\|_X, & \text{if } q = \infty. \end{cases}$$

We are now in position to give the definition of $T$-periodic weak solution to the problem (16)–(18).

**Definition 3.1** A triple $(\boldsymbol{V}, \boldsymbol{A}, \boldsymbol{H})$ is a *T-periodic*, or simply *periodic weak solution* to the problem (16)–(18) if the following conditions hold.

1. $\boldsymbol{V} \in L^2(0, T; H^1(\mathcal{D})) \cap L^\infty(0, T; H(\mathcal{D})), \boldsymbol{A}, \boldsymbol{H} \in C_T$;
2. $(\boldsymbol{V}, \boldsymbol{A}, \boldsymbol{H})$ satisfies the following equations

$$\begin{aligned} &\int_0^T \left[ \langle \boldsymbol{V}, \boldsymbol{\psi} \rangle + \boldsymbol{A} \cdot \boldsymbol{I}^{-1} \cdot \left( \int_{\mathcal{D}} \boldsymbol{y} \times \boldsymbol{\psi} \, dV \right) \right] \frac{d\xi(t)}{dt} \, dt \\ &\quad = -\left[ -2\rho_{\mathcal{L}} \int_0^T \left( \int_{\mathcal{D}} \boldsymbol{y} \times \boldsymbol{V} \, dV \right) \cdot \boldsymbol{I}^{-1} \cdot \left( \int_{\mathcal{D}} \boldsymbol{V} \times \boldsymbol{\psi} \, dV \right) \xi(t) \, dt \right. \\ &\qquad + 2 \int_0^T \left( \int_{\mathcal{L}} \boldsymbol{A} \cdot \boldsymbol{I}^{-1} \cdot (\boldsymbol{V} \times \boldsymbol{\psi}) \, dV \right) \xi(t) \, dt \\ &\qquad \left. + \int_0^T (\boldsymbol{V} \cdot \operatorname{grad} \boldsymbol{\psi}, \boldsymbol{V}) \xi(t) \, dt - \nu \int_0^T (\operatorname{grad} \boldsymbol{V}, \operatorname{grad} \boldsymbol{\psi}) \xi(t) \, dt \right], \end{aligned} \tag{21}$$

for all $\boldsymbol{\psi} \in \mathcal{D}(\mathcal{D})$, $\xi \in C_T^\infty(\mathbb{R})$;

$$A(t) = A(0) - \int_0^t (I^{-1} \cdot A) \times A \, d\tau + \rho_{\mathcal{L}} I^{-1} \cdot \int_0^t \left( \int_{\mathcal{D}} y \times V \, dV \right) \times A \, d\tau + \int_0^t f(\tau) H \, d\tau, \qquad \text{for all } t \in [0, T]\,; \qquad (22)$$

$$H(t) = H(0) - \int_0^t (I^{-1} \cdot A) \times H \, d\tau + \rho_{\mathcal{L}} I^{-1} \cdot \int_0^t \left( \int_{\mathcal{D}} y \times V \, dV \right) \times H \, d\tau, \qquad \text{for all } t \in [0, T]. \qquad (23)$$

*Remark 3.2* A weak solution has, in fact, more regularity in time than that stated in the above definition. In fact, on the one hand, from (22) and (23) we deduce

$$A, \; H \in W^{1,r}(0, T)\,,$$

provided $f \in L^r_T(\mathbb{R})$, $r \in [1, \infty]$. On the other hand, proceeding as in [5, Lemma 2.2], and [7, Remark], one can show that $V(\cdot, t)$ is continuous in $[0, T]$ weakly in $L^2(\mathcal{D})$, and strongly in $H^{-1}(\mathcal{D})$. As a consequence, with the help of (21) it follows that $V$ is indeed periodic in time in the sense of the above topologies.

*Remark 3.3* If $(V, A, H)$ is a periodic weak solution to (16)–(18), then the corresponding angular velocity $\Omega$ is defined via (15) and belongs to $C_T$; see also Remark 3.2.

*Remark 3.4* If $(V, A, H)$ is a periodic weak solution to (16)–(18) and is sufficiently regular then, by a standard procedure one shows that there exists a scalar field $p = p(y, t)$ such that $V, p$, and $A$ satisfy $(16)_1$.

## 4 Existence of Periodic Weak Solutions

Objective of this section is to show the existence of a periodic weak solution to (16)–(18) under suitable assumptions on $f$. This will be achieved by combining the Faedo-Galerkin method with a fixed point argument. Specifically, we have the following.

**Theorem 4.1** *Let $f \in L^2_T(\mathbb{R})$ satisfy* (7), *and let $\mathcal{D}$ be a domain of $\mathbb{R}^3$. Then, there exists at least one periodic weak solution, $(V, A, H)$, to* (16)–(18).

*Proof* Let $\{\psi_n\}_{n\in\mathbb{N}}$ be a denumerable subset of $\mathcal{D}(\mathcal{D})$ whose linear hull is dense in $H^1(\mathcal{D})$, and let us normalize it as $\langle \psi_n, \psi_m \rangle = \delta_{nm}$. We look for "approximate solutions" of the type

$$V_n(y, t) := \sum_{k=1}^{n} c_{nk}(t) \psi_k(y), \;\; A_n := \sum_{i=1}^{3} \tilde{c}_{ni}(t) e_i, \;\; H_n := \sum_{j=1}^{3} \hat{c}_{nj}(t) e_k,$$

where $\{\boldsymbol{e}_i\}$ is a base of $\mathcal{S}$. The coefficients $c_{nk}$, $\tilde{c}_{ni}$ and $\hat{c}_{nj}$ are found by solving the following system of ordinary differential equations:

$$
\begin{aligned}
&\frac{d}{dt}\left[\langle \boldsymbol{V}_n, \boldsymbol{\psi}_r\rangle + \boldsymbol{A}_n \cdot \boldsymbol{I}^{-1} \cdot \left(\int_{\mathcal{D}} \boldsymbol{y} \times \boldsymbol{\psi}_r \, dV\right)\right] \\
&= -2\rho_{\mathcal{L}} \left(\int_{\mathcal{D}} \boldsymbol{y} \times \boldsymbol{V}_n \, dV\right) \cdot \boldsymbol{I}^{-1} \cdot \left(\int_{\mathcal{D}} \boldsymbol{V}_n \times \boldsymbol{\psi}_r \, dV\right) \\
&+ 2\int_{\mathcal{D}} \boldsymbol{A}_n \cdot \boldsymbol{I}^{-1} \cdot (\boldsymbol{V}_n \times \boldsymbol{\psi}_r) \, dV + (\boldsymbol{V}_n \cdot \operatorname{grad} \boldsymbol{\psi}_r, \boldsymbol{V}_n) - \nu(\operatorname{grad} \boldsymbol{V}_n, \operatorname{grad} \boldsymbol{\psi}_r), \\
&\frac{d\boldsymbol{A}_n}{dt} = -(\boldsymbol{I}^{-1} \cdot \boldsymbol{A}_n) \times \boldsymbol{A}_n + \rho_{\mathcal{L}} \boldsymbol{I}^{-1} \cdot \left(\int_{\mathcal{D}} \boldsymbol{y} \times \boldsymbol{V}_n \, dV\right) \times \boldsymbol{A}_n + f(t)\boldsymbol{H}_n, \\
&\frac{d\boldsymbol{H}_n}{dt} = -(\boldsymbol{I}^{-1} \cdot \boldsymbol{A}_n) \times \boldsymbol{H}_n + \rho_{\mathcal{L}} \boldsymbol{I}^{-1} \cdot \left(\int_{\mathcal{D}} \boldsymbol{y} \times \boldsymbol{V}_n \, dV\right) \times \boldsymbol{H}_n,
\end{aligned}
\tag{24}
$$

which, in terms of $c_{nk}$, $\tilde{c}_{ni}$ and $\hat{c}_{nj}$ reads as follows

$$
\begin{aligned}
&\frac{dc_{nr}}{dt} + \frac{d\tilde{c}_{ni}}{dt}\boldsymbol{e}_i \cdot \boldsymbol{I}^{-1} \cdot \left(\int_{\mathcal{D}} \boldsymbol{y} \times \boldsymbol{\psi}_r \, dV\right) \\
&= -2\rho_{\mathcal{L}} \left(\int_{\mathcal{D}} \boldsymbol{y} \times \boldsymbol{\psi}_k \, dV\right) \cdot \boldsymbol{I}^{-1} \cdot \left(\int_{\mathcal{D}} \boldsymbol{\psi}_m \times \boldsymbol{\psi}_r \, dV\right) c_{nk}c_{nm} \\
&\quad + 2\left[\int_{\mathcal{D}} \boldsymbol{e}_i \cdot \boldsymbol{I}^{-1} \cdot (\boldsymbol{\psi}_k \times \boldsymbol{\psi}_r) \, dV\right] \tilde{c}_{ni}c_{nk} \\
&\quad + (\boldsymbol{\psi}_k \cdot \operatorname{grad} \boldsymbol{\psi}_r, \boldsymbol{\psi}_m)c_{nk}c_{nm} - \nu(\operatorname{grad} \boldsymbol{\psi}_k, \operatorname{grad} \boldsymbol{\psi}_r)c_{nk}, \\
&\frac{d\tilde{c}_{ni}}{dt}\boldsymbol{e}_i = -\tilde{c}_{ni}\tilde{c}_{nj}(\boldsymbol{I}^{-1} \cdot \boldsymbol{e}_i) \times \boldsymbol{e}_j + c_{nk}\tilde{c}_{ni}\rho_{\mathcal{L}} \boldsymbol{I}^{-1} \cdot \left(\int_{\mathcal{D}} \boldsymbol{y} \times \boldsymbol{\psi}_k \, dV\right) \times \boldsymbol{e}_i \\
&\qquad\qquad + f(t)\hat{c}_{nj}\boldsymbol{e}_j, \\
&\frac{d\hat{c}_{nj}}{dt}\boldsymbol{e}_j = -\tilde{c}_{ni}\hat{c}_{nj}(\boldsymbol{I}^{-1} \cdot \boldsymbol{e}_i) \times \boldsymbol{e}_j + c_{nk}\hat{c}_{nj}\rho_{\mathcal{L}} \boldsymbol{I}^{-1} \cdot \left(\int_{\mathcal{D}} \boldsymbol{y} \times \boldsymbol{\psi}_k \, dV\right) \times \boldsymbol{e}_j.
\end{aligned}
\tag{25}
$$

By replacing the second equation in the first one, we get

$$
\begin{aligned}
\frac{dc_{nr}}{dt} &= \left[(\boldsymbol{I}^{-1} \cdot \boldsymbol{e}_i) \times \boldsymbol{e}_j\right] \cdot \boldsymbol{I}^{-1} \cdot \left(\int_{\mathcal{D}} \boldsymbol{y} \times \boldsymbol{\psi}_r \, dV\right) \tilde{c}_{ni}\tilde{c}_{nj} \\
&\quad - \rho_{\mathcal{L}} \left[\boldsymbol{I}^{-1} \cdot \left(\int_{\mathcal{D}} \boldsymbol{y} \times \boldsymbol{\psi}_k \, dV\right) \times \boldsymbol{e}_i\right] \cdot \boldsymbol{I}^{-1} \cdot \left(\int_{\mathcal{D}} \boldsymbol{y} \times \boldsymbol{\psi}_r \, dV\right) c_{nk}\tilde{c}_{ni} \\
&\quad - f(t)\boldsymbol{e}_j \cdot \boldsymbol{I}^{-1} \cdot \left(\int_{\mathcal{D}} \boldsymbol{y} \times \boldsymbol{\psi}_r \, dV\right) \hat{c}_{nj}
\end{aligned}
$$

$$
\begin{aligned}
&- 2\rho_{\mathcal{L}} \left( \int_{\mathcal{D}} \boldsymbol{y} \times \boldsymbol{\psi}_k \, dV \right) \cdot \boldsymbol{I}^{-1} \cdot \left( \int_{\mathcal{D}} \boldsymbol{\psi}_m \times \boldsymbol{\psi}_r \, dV \right) c_{nk} c_{nm} \\
&+ 2 \left[ \int_{\mathcal{D}} \boldsymbol{e}_i \cdot \boldsymbol{I}^{-1} \cdot (\boldsymbol{\psi}_k \times \boldsymbol{\psi}_r) \, dV \right] \tilde{c}_{ni} c_{nk} \\
&+ (\boldsymbol{\psi}_k \cdot \operatorname{grad} \boldsymbol{\psi}_r, \boldsymbol{\psi}_m) c_{nk} c_{nm} - \nu (\operatorname{grad} \boldsymbol{\psi}_k, \operatorname{grad} \boldsymbol{\psi}_r) c_{nk}.
\end{aligned}
$$

Setting

$$
\begin{aligned}
b_{rij} &:= \left[ (\boldsymbol{I}^{-1} \cdot \boldsymbol{e}_i) \times \boldsymbol{e}_j \right] \cdot \boldsymbol{I}^{-1} \cdot \left( \int_{\mathcal{D}} \boldsymbol{y} \times \boldsymbol{\psi}_r \, dV \right), \\
d_{rki} &:= -\rho_{\mathcal{L}} \left[ \boldsymbol{I}^{-1} \cdot \left( \int_{\mathcal{D}} \boldsymbol{y} \times \boldsymbol{\psi}_k \, dV \right) \times \boldsymbol{e}_i \right] \cdot \boldsymbol{I}^{-1} \cdot \left( \int_{\mathcal{D}} \boldsymbol{y} \times \boldsymbol{\psi}_r \, dV \right), \\
f_{rj}(t) &:= -f(t) \boldsymbol{e}_j \cdot \boldsymbol{I}^{-1} \cdot \left( \int_{\mathcal{D}} \boldsymbol{y} \times \boldsymbol{\psi}_r \, dV \right), \\
g_{rkm} &:= -2\rho_{\mathcal{L}} \left( \int_{\mathcal{D}} \boldsymbol{y} \times \boldsymbol{\psi}_k \, dV \right) \cdot \boldsymbol{I}^{-1} \cdot \left( \int_{\mathcal{D}} \boldsymbol{\psi}_m \times \boldsymbol{\psi}_r \, dV \right), \\
s_{rki} &:= 2 \left[ \int_{\mathcal{D}} \boldsymbol{e}_i \cdot \boldsymbol{I}^{-1} \cdot (\boldsymbol{\psi}_k \times \boldsymbol{\psi}_r) \, dV \right], \\
p_{rkm} &:= (\boldsymbol{\psi}_k \cdot \operatorname{grad} \boldsymbol{\psi}_r, \boldsymbol{\psi}_m), \qquad p_{rk} := -\nu (\operatorname{grad} \boldsymbol{\psi}_k, \operatorname{grad} \boldsymbol{\psi}_r) c_{nk},
\end{aligned}
$$

equation $(25)_1$ becomes

$$
\begin{aligned}
\frac{dc_{nr}}{dt} &= b_{rij} \tilde{c}_{ni} \tilde{c}_{nj} + d_{rki} c_{nk} \tilde{c}_{ni} + f_{rj}(t) \hat{c}_{nj} + g_{rkm} c_{nk} c_{nm} + s_{rik} \tilde{c}_{ni} c_{nk} \\
&\quad + p_{rkm} c_{nk} c_{nm} + p_{rk} c_{nk},
\end{aligned} \tag{26}
$$

where here and in the rest of the proof $i, j, \ell$ vary in the set $\{1, 2, 3\}$, whereas $r, k, m$ in the set $\{1, \ldots, n\}$. Concerning $(25)_2$ and $(25)_3$, taking the dot product of each side of both with $\boldsymbol{e}_\ell$, and putting

$$
u_{\ell ij} := - \left[ (\boldsymbol{I}^{-1} \cdot \boldsymbol{e}_i) \times \boldsymbol{e}_j \right] \cdot \boldsymbol{e}_\ell, \qquad w_{\ell kj} := \rho_{\mathcal{L}} \boldsymbol{e}_\ell \cdot \boldsymbol{I}^{-1} \cdot \left( \int_{\mathcal{D}} \boldsymbol{y} \times \boldsymbol{\psi}_k \, dV \right) \times \boldsymbol{e}_j,
$$

we deduce

$$
\frac{d\tilde{c}_{n\ell}}{dt} = u_{\ell ij} \tilde{c}_{ni} \tilde{c}_{nj} + w_{\ell ki} c_{nk} \tilde{c}_{ni} + f(t) \hat{c}_{n\ell}, \tag{27}
$$

and

$$\frac{d\hat{c}_{n\ell}}{dt} = u_{\ell ij}\tilde{c}_{ni}\hat{c}_{nj} + w_{\ell kj}c_{nk}\hat{c}_{nj}, \tag{28}$$

respectively. Following [15] we shall next prove that there exist initial data such that the system of ordinary differential equations (26)–(28) admits a corresponding solution $(c_{nr}, \tilde{c}_{n\ell}, \hat{c}_{n\ell})$ such that $c_{nr}(0) = c_{nr}(T)$, $\tilde{c}_{nr}(0) = \tilde{c}_{nr}(T)$ and $\hat{c}_{nr}(0) = \hat{c}_{nr}(T)$. To reach this goal, let

$$\boldsymbol{V}_{n,0} \in span\{\boldsymbol{\psi}_1, \ldots, \boldsymbol{\psi}_n\}\,,\ \boldsymbol{A}_{n,0} \in \mathbb{R}^3\,,\ \boldsymbol{H}_{n,0} \in S^2\,,$$

and set $c_{nr}(0) = c_{nr,0} := \langle \boldsymbol{V}_{n,0}, \boldsymbol{\psi}_r\rangle$, $\tilde{c}_{n\ell}(0) = \tilde{c}_{n\ell,0} := \boldsymbol{A}_{n,0} \cdot \boldsymbol{e}_\ell$, and $\hat{c}_{n\ell}(0) = \hat{c}_{n\ell,0} := \boldsymbol{H}_{n,0} \cdot \boldsymbol{e}_\ell$. Since $f, f_{r,j} \in C_T(\mathbb{R})$, by Picard theorem, there exists a unique solution, $(c_{nr}, \tilde{c}_{n\ell}, \hat{c}_{n\ell})$, to the Cauchy problem associated to (26)–(28) with $c_{nr}, \tilde{c}_{n\ell}, \hat{c}_{n\ell} \in C^1(0, T')$, $r = 1, \ldots, n$, $\ell = 1, 2, 3$, where $0 < T' \le T$. Multiplying both sides of (26) by $c_{nr}$, summing over $r = 1, \ldots, n$, and noticing that the terms corresponding to $d_{rki}, g_{rkm}, s_{rki}$ and $p_{rkm}$ vanish, we get

$$\frac{1}{2}\frac{d}{dt}\|\boldsymbol{V}_n\|_B^2 + \nu\|\text{grad}\,\boldsymbol{V}_n\|_2^2 = \rho_{\mathcal{L}}\left[(\boldsymbol{I}^{-1}\cdot\boldsymbol{A}_n)\times\boldsymbol{A}_n\right]\cdot\boldsymbol{I}^{-1}\cdot\int_{\mathcal{D}}\boldsymbol{y}\times\boldsymbol{V}_n\,dV$$
$$-f(t)\boldsymbol{H}_n\cdot\boldsymbol{I}^{-1}\cdot\int_{\mathcal{D}}\boldsymbol{y}\times\boldsymbol{V}_n\,dV. \tag{29}$$

Multiplying next both sides of (27) by $\tilde{c}_{n\ell}$, summing over $\ell = 1, 2, 3$, and taking into account that the terms corresponding to $u_{\ell ij}$ and $w_{\ell kj}$ vanish, it follows that

$$\frac{1}{2}\frac{d|\boldsymbol{A}_n|^2}{dt} = f(t)\boldsymbol{H}_n\cdot\boldsymbol{A}_n. \tag{30}$$

Finally, multiplying both sides of (28) by $\hat{c}_{n\ell}$ and summing over $\ell = 1, 2, 3$, similarly as above, we get

$$\frac{d|\boldsymbol{H}_n|^2}{dt} = 0\,, \tag{31}$$

which implies $|\boldsymbol{H}_n(t)| = |\boldsymbol{H}_{n,0}| = 1$. Thus, from (30) and (31) we infer

$$\frac{d|\boldsymbol{A}_n|}{dt} \le |f(t)| \tag{32}$$

and

$$|\boldsymbol{A}_n(t)| \le |\boldsymbol{A}_{n,0}| + \int_0^T |f(\tau)|\,d\tau, \qquad \text{for all } t \in [0, T'), n \in \mathbb{N}\,. \tag{33}$$

Moreover, by Poincaré and Cauchy–Schwarz inequalities, from (29) we deduce, on the one hand,

$$\frac{1}{2}\frac{d\|\boldsymbol{V}_n\|_B^2}{dt} + C_{1p}\|\mathrm{grad}\,\boldsymbol{V}_n\|_2^2 \le C_1|\boldsymbol{A}_n|^4 + C_2|f(t)|^2, \tag{34}$$

and, on the other hand, using one more time Poincaré inequality in conjunction with (20),

$$\frac{1}{2}\frac{d\|\boldsymbol{V}_n\|_B^2}{dt} + C_{2p}\|\boldsymbol{V}_n\|_B^2 \le C_1|\boldsymbol{A}_n|^4 + C_2|f(t)|^2, \tag{35}$$

where $C_{ip} = C_{ip}(\mathcal{D},\nu) > 0$, while $C_i = C_i(\mathcal{D},\mathcal{B},\nu) > 0$, $i = 1,2$. Using Gronwall's lemma in (35) furnishes,

$$\begin{aligned}\exp(C_3t)\|\boldsymbol{V}_n(t)\|_B^2 \le \|\boldsymbol{V}_{n,0}\|_B^2 + C_1\int_0^t \exp(C_3\tau)|\boldsymbol{A}_n(\tau)|^4\,d\tau \\ + C_2\int_0^t \exp(C_3\tau)|f(\tau)|^2\,d\tau\end{aligned} \tag{36}$$

which, by (33), implies

$$\begin{aligned}|c_{nr}(t)|^2 = \|\boldsymbol{V}_n(t)\|_B^2 &\le \exp(-C_3t)\|\boldsymbol{V}_{n,0}\|_B^2 \\ &\quad + C_1\exp(-C_3t)\int_0^t \exp(C_3\tau)|\boldsymbol{A}_n(\tau)|^4\,d\tau \\ &\quad + C_2\exp(-C_3t)\int_0^t \exp(C_3\tau)|f(\tau)|^2\,d\tau \\ &\le \exp(-C_3t)\|\boldsymbol{V}_{n,0}\|_B^2 + C_4\sup_{t\in[0,T]}|\boldsymbol{A}_n(t)|^4 + C_2\int_0^T |f(\tau)|^2\,d\tau,\end{aligned} \tag{37}$$

with $C_3$ and $C_4$ positive constants depending, at most, on $\mathcal{D}$, $\mathcal{B}$ and $\nu$. As a result, from (31), (33), and (37) we conclude $T' = T$. In order to build our periodic solution, we will use a suitable fixed point argument. To reach this goal, we multiply both sides of (27) by $\hat{c}_{n\ell}$ and sum over $\ell = 1,2,3$, then multiply (28) by $\tilde{c}_{n\ell}$ and sum over $\ell = 1,2,3$, and finally add the two resulting equations. We get

$$\frac{d(\boldsymbol{H}_n\cdot\boldsymbol{A}_n)}{dt} = f(t)|\boldsymbol{H}_n|^2 = f(t).$$

Since $f$ has zero average by assumption, it follows that

$$\boldsymbol{H}_{n,0}\cdot\boldsymbol{A}_{n,0} = \boldsymbol{H}_n(T)\cdot\boldsymbol{A}_n(T). \tag{38}$$

Taking the cross product of $(24)_2$ by $\boldsymbol{H}_n$ on the left, and then that of $(24)_3$ by $\boldsymbol{A}_n$ on the left, and summing the two equation so obtained, we deduce

$$\frac{d(\boldsymbol{H}_n \times \boldsymbol{A}_n)}{dt} = \boldsymbol{\Omega}_n \times (\boldsymbol{H}_n \times \boldsymbol{A}_n),$$

where $\boldsymbol{\Omega}_n$ is given by (15). It then follows that

$$\frac{1}{2}\frac{d|\boldsymbol{H}_n \times \boldsymbol{A}_n|^2}{dt} = \frac{1}{2}\left(\frac{d|\boldsymbol{A}_n|^2}{dt} - \frac{d(\boldsymbol{H}_n \cdot \boldsymbol{A}_n)^2}{dt}\right) = 0$$

where we have also used the fact that $|\boldsymbol{H}_n| = 1$. From the last displayed equation and (38), we conclude that

$$|\boldsymbol{A}_{n,0}| = |\boldsymbol{A}_n(T)|. \tag{39}$$

We next fix $R_1 > 0$ and take $|\boldsymbol{A}_{n,0}| \le R_1$. By (39), we obtain $|\boldsymbol{A}_n(T)| \le R_1$. Combining (37) and (33), we infer

$$\begin{aligned}\|\boldsymbol{V}_n(T)\|_B^2 &\le \exp(-C_3 T)\|\boldsymbol{V}_{n,0}\|_B^2 \\ &\quad + C_5\left(|\boldsymbol{A}_{n,0}|^4 + \|f\|_{L^1(0,T)}^4 + \|f\|_{L^2(0,T)}^2\right) \\ &\le \exp(-C_3 T)\|\boldsymbol{V}_{n,0}\|_B^2 + C_5 R_1^4 + C_6,\end{aligned} \tag{40}$$

where $C_5 = C_5(\mathcal{D}, \mathcal{B}, \nu) > 0$, and

$$C_6 := C_5\big(\|f\|_{L^1(0,T)}^4 + \|f\|_{L^2(0,T)}^2\big)\,.$$

Thus, choosing

$$R_2^2 \ge \frac{C_5 R_1^4 + C_6}{1 - \exp(-C_3 T)}, \tag{41}$$

from (40) we show that if $\|\boldsymbol{V}_{n,0}\|_B^2 \le R_2^2$, then $\|\boldsymbol{V}_n(T)\|_B^2 \le R_2^2$. Set $\mathbb{B} := \mathbb{B}_{R_2} \times \mathbb{B}_{R_1} \times S^2$, where $\mathbb{B}_{R_i}$ denotes the ball of radius $R_i$ in $\mathbb{R}^n$, $i = 1, 2$, while $S^2$ is the unit sphere in $\mathbb{R}^3$. Let $\Phi : \mathbb{B} \to \mathbb{B}$ be the map that takes any $(\boldsymbol{V}_{n,0}, \boldsymbol{A}_{n,0}, \boldsymbol{H}_{n,0}) \in \mathbb{B}$ to $(\boldsymbol{V}_n(T), \boldsymbol{A}_n(T), \boldsymbol{H}_n(T)) \in \mathbb{B}$, where $(\boldsymbol{V}_n(T), \boldsymbol{A}_n(T), \boldsymbol{H}_n(T))$ is the solution to (24) at time $T$. By a straightforward calculation (see e.g. [15]), one shows that $\Phi$ is continuous. Moreover, $\Phi$ is homotopic to the identity by the following homotopy

$$\mathcal{H} : \mathbb{B} \times [0, 1] \to \mathbb{B}$$

$$\mathcal{H}(\boldsymbol{V}_{n,0}, \boldsymbol{A}_{n,0}, \boldsymbol{H}_{n,0}, s) := (\boldsymbol{V}_n(sT), s\boldsymbol{A}_n(T) + (1 - s)\boldsymbol{A}_n(0), \boldsymbol{H}_n(sT)).$$

Notice that $\mathcal{H}$ is well defined, since by similar calculations which lead to the estimate (41), we show that $\boldsymbol{V}_n(sT) \in \mathbb{B}_{R_2}$ for all $s \in [0, 1]$.[2] Since the Euler characteristic of $\mathbb{B}$ is given by

$$\chi(\mathbb{B}) = \chi(\mathbb{B}_{R_2})\chi(\mathbb{B}_{R_1})\chi(S^2) = 1 \cdot 1 \cdot 2 \neq 0$$

(see [3]), by the Lefschetz-Hopf fixed-point theorem (see [3, Chap. IV, Sect. 23], ), $\Phi$ has at least one fixed point, from which it follows that there exist $(\boldsymbol{V}_{n,0}, \boldsymbol{A}_{n,0}, \boldsymbol{H}_{n,0})$ such that the solution to (24), $(\boldsymbol{V}_n, \boldsymbol{A}_n, \boldsymbol{H}_n)$, starting from $(\boldsymbol{V}_{n,0}, \boldsymbol{A}_{n,0}, \boldsymbol{H}_{n,0})$ satisfies

$$\boldsymbol{V}_n(\cdot, T) = \boldsymbol{V}_{n,0}(\cdot), \quad \boldsymbol{A}_n(T) = \boldsymbol{A}_{n,0}, \quad \boldsymbol{H}_n(T) = \boldsymbol{H}_{n,0}.$$

Now, multiplying $(24)_1$ by $\xi \in C_T^\infty(\mathbb{R})$ and integrating over $[0, T]$, we show that $\boldsymbol{V}_n$ satisfies the following

$$\begin{aligned}
&\int_0^T \left[ \langle \boldsymbol{V}_n, \boldsymbol{\psi}_r \rangle + \boldsymbol{A}_n \cdot \boldsymbol{I}^{-1} \cdot \left( \int_{\mathcal{D}} \boldsymbol{y} \times \boldsymbol{\psi}_r \, dV \right) \right] \frac{d\xi(t)}{dt} \, dt \\
&\quad = \left[ 2\rho_{\mathcal{L}} \int_0^T \left( \int_{\mathcal{D}} \boldsymbol{y} \times \boldsymbol{V}_n \, dV \right) \cdot \boldsymbol{I}^{-1} \cdot \left( \int_{\mathcal{D}} \boldsymbol{V}_n \times \boldsymbol{\psi}_r \, dV \right) \xi(t) \, dt \right. \\
&\qquad - 2 \int_0^T \left( \int_{\mathcal{D}} \boldsymbol{A}_n \cdot \boldsymbol{I}^{-1} \cdot (\boldsymbol{V}_n \times \boldsymbol{\psi}_r) \, dV \right) \xi(t) \, dt \\
&\qquad \left. - \int_0^T (\boldsymbol{V}_n \cdot \operatorname{grad} \boldsymbol{\psi}_r, \boldsymbol{V}_n) \xi(t) \, dt - \nu \int_0^T (\operatorname{grad} \boldsymbol{V}_n, \operatorname{grad} \boldsymbol{\psi}_r) \xi(t) \, dt \right],
\end{aligned} \tag{42}$$

for all $r = 1, \ldots, n$, and all $\xi \in C_T^\infty(\mathbb{R})$. Likewise, integrating $(24)_{2,3}$ over $[0, t]$, $t \in [0, T]$, we deduce that $\boldsymbol{A}_n$ and $\boldsymbol{H}_n$ satisfy

$$\begin{aligned}
\boldsymbol{A}_n(t) = \boldsymbol{A}_n(0) - \int_0^t (\boldsymbol{I}^{-1} \cdot \boldsymbol{A}_n) \times \boldsymbol{A}_n \, d\tau + \rho_{\mathcal{L}} \boldsymbol{I}^{-1} \cdot \int_0^t \left( \int_{\mathcal{D}} \boldsymbol{y} \times \boldsymbol{V}_n \, dV \right) \times \boldsymbol{A}_n \, d\tau \\
+ \int_0^t f(\tau) \boldsymbol{H}_n \, d\tau,
\end{aligned} \tag{43}$$

[2]The same argument does not work for $\boldsymbol{A}_n$, i.e. the solution map does not necessarily lie in $\mathbb{B}_{R_1}$ for all times $t \in [0, T]$. This is the reason for which we have used the linear homotopy for the $\boldsymbol{A}_n$ component.

and

$$\boldsymbol{H}_n(t) = \boldsymbol{H}_n(0) - \int_0^t (\boldsymbol{I}^{-1} \cdot \boldsymbol{A}_n) \times \boldsymbol{H}_n \, d\tau + \rho_{\mathcal{L}} \boldsymbol{I}^{-1} \cdot \int_0^t \left( \int_{\mathcal{D}} \boldsymbol{y} \times \boldsymbol{V}_n \, dV \right) \times \boldsymbol{H}_n \, d\tau, \tag{44}$$

respectively. Let us investigate the convergence properties of $(\boldsymbol{V}_n, \boldsymbol{A}_n, \boldsymbol{H}_n)$. By (26) and (37), together with Schwarz inequality, (33) and the fact that $|\boldsymbol{H}_n| = 1$, we can conclude that the sequence of functions $G_n^{(r)} := \langle \boldsymbol{V}_n(t), \boldsymbol{\psi}_r \rangle$ is uniformly bounded and equicontinuous. Then, by Ascoli-Arzelà theorem, we can extract a subsequence of $\{G_n^{(r)}\}_{n\in\mathbb{N}}$, which we will again denote by $\{G_n^{(r)}\}_{n\in\mathbb{N}}$, which is uniformly convergent to a continuous function $G^{(r)}(t)$. By a Cantor diagonalization argument, we can make that subsequence independent of $r$, so that $\{G_n^{(r)}\}_{n\in\mathbb{N}}$ converges to $G^r$, for all $r \in \mathbb{N}$, uniformly with respect to $t \in (0,T]$. Moreover, by (37) and by Banach-Alaoglu Theorem, we can also conclude that there exists $\boldsymbol{V}(t) \in H(\mathcal{L})$ such that

$$\lim_{n\to\infty} \langle \boldsymbol{V}_n(t) - \boldsymbol{V}(t), \boldsymbol{\psi}_r \rangle = 0 \text{ uniformly on } [0,T] \text{ and for all } r \in \mathbb{N}.$$

From the latter and (20), we can thus prove that $\{\boldsymbol{V}_n(t)\}_{n\in\mathbb{N}}$ converges weakly in $L^2(\mathcal{D})$, uniformly with respect to time, i.e.

$$\lim_{n\to\infty} \langle \boldsymbol{V}_n(t) - \boldsymbol{V}(t), \boldsymbol{w} \rangle = 0, \text{ uniformly in } t \in [0,T], \text{ for all } \boldsymbol{w} \in L^2(\mathcal{D}).$$

Concerning $\boldsymbol{A}_n$ and $\boldsymbol{H}_n$, since they are finite-dimensional functions, starting from (27) and (28), respectively, again by the Ascoli-Arzelà theorem and the Cantor diagonalization argument, we can find $\boldsymbol{A}, \boldsymbol{H} \in C_T$ such that

$$\lim_{n\to\infty} |(\boldsymbol{A}_n(t) - \boldsymbol{A}(t))| = 0, \quad \lim_{n\to\infty} |(\boldsymbol{H}_n(t) - \boldsymbol{H}(t))| = 0,$$

respectively, uniformly with respect to $t \in [0,T]$. From (37) and the weak compactness of $L^2(0,T;H(\mathcal{D}))$, we can conclude that $\boldsymbol{V} \in L^2(0,T;H^1(\mathcal{D})) \cap L^\infty(0,T;H(\mathcal{D}))$. Finally, by applying Friederichs inequality (see [6]) we conclude that

$$\boldsymbol{V}_n \to \boldsymbol{V} \quad \text{strongly in } L^2(0,T;H(\mathcal{D}))\,.$$

Summarizing, we have proved the existence of subsequences of $\{\boldsymbol{V}_n\}_{n\in\mathbb{N}}$, $\{\boldsymbol{A}_n\}_{n\in\mathbb{N}}$ and $\{\boldsymbol{H}_n\}_{n\in\mathbb{N}}$, and functions $(\boldsymbol{V}, \boldsymbol{A}, \boldsymbol{H})$ such that

$$
\begin{aligned}
\boldsymbol{V} \in L^2(0,T;H^1(\mathcal{D})) \cap L^\infty(0,T;H(\mathcal{D})),\ & \boldsymbol{A},\boldsymbol{H} \in C([0,T]),\\
\boldsymbol{V}_n \to \boldsymbol{V} \quad & \text{strongly in } L^2(0,T;H(\mathcal{D})),\\
\boldsymbol{V}_n \to \boldsymbol{V} \quad & \text{weakly in } L^2(0,T;H^1(\mathcal{D})),\\
\boldsymbol{V}_n(t) \to \boldsymbol{V}(t) \quad & \text{weakly in } L^2(\mathcal{D}),\ \text{uniformly in } t \in [0,T],\\
\boldsymbol{A}_n \to \boldsymbol{A} \quad & \text{uniformly in } t \in [0,T],\\
\boldsymbol{H}_n \to \boldsymbol{H} \quad & \text{uniformly in } t \in [0,T].
\end{aligned}
\tag{45}
$$

Since $\boldsymbol{A}_n(0) = \boldsymbol{A}_n(T)$ and $\boldsymbol{H}_n(0) = \boldsymbol{H}_n(T)$, from (45) we can conclude also that $\boldsymbol{A},\boldsymbol{H} \in C_T$. Finally, again employing (45) and taking into account the properties of $\{\boldsymbol{\psi}_n\}_{n\in\mathbb{N}}$, we can then pass to the limit in (42)–(44) and conclude that $(\boldsymbol{V},\boldsymbol{A},\boldsymbol{H})$ possesses all the properties of a periodic weak solution to (16)–(18). □

*Remark 4.2* For future reference, we wish to observe that the choice of the radius $R_1$ in the proof of the previous theorem is completely arbitrary, provided, of course, $R_2$ is taken appropriately according to (41). This follows from the fact that $|\boldsymbol{A}_n(t)|$ is $T$-periodic regardless of the choice of $R_1$; see (39).

## 5 Existence of Strong Periodic Solutions

In this section we will show existence of a strong solution, whenever the magnitude of the torque "sufficiently small". To this end, we need the following Gronwall-type Lemma.

**Lemma 5.1** *Let $y \in C^1((a,b)) \cap C([a,b])$, $a < b$, $y \geq 0$, satisfy the following inequality*

$$
y'(t) \leq -c_1 y(t) + c_2 y^\alpha(t) + c_3, \quad t \in (a,b) \tag{46}
$$

*where $c_1$, $c_2$, $c_3$ are positive constants and $\alpha > 1$. There exists $\delta > 0$ such that if*

$$
y(a) + \int_a^t y(s)\, ds \leq \delta, \qquad \text{for all } t \in [a,b], \tag{47}
$$

*then*

$$
y(t) < \frac{2c_3}{c_1}, \qquad \text{for all } t \in [a,b].
$$

*Proof* Let us argue by contradiction. Set $c := c_3/c_1$ and take $\delta < 2c$. By continuity, we have $y(t) < 2c$ in a right neighborhood of $t = a$. Denote then by $t_0 \in (a,b]$ the *first* point such that

$$
y(t_0) = 2\,c. \tag{48}
$$

From the inequality (46) we deduce

$$\frac{d\,(e^{c_1 t}y)}{dt} \le c_2 e^{c_1 t} y^{\alpha} + c_3 e^{c_1 t}\,,$$

which, once integrated between $a$ and $t_0$, by virtue of (47) and (48) furnishes

$$y(t_0) \le e^{-c_1(t_0-a)}y(a) + c_2(2c)^{\alpha-1}\int_a^{t_0} y(s)\,ds + c \le \delta\left(1 + c_2(2c)^{\alpha-1}\right) + c\,.$$

As a result, (48) is contradicted if we choose

$$\delta < \frac{c}{1 + c_2(2c)^{\alpha-1}}\,.$$

□

We are now in position to state and prove the main result of this section.

**Theorem 5.2** *Let $f \in L^{\infty}_{T}(\mathbb{R})$ satisfy* (7), *and assume $\mathcal{D}$ of class $C^2$. There is a positive constant $\kappa = \kappa(\mathcal{D}, \mathcal{B}, \mathcal{L}, T)$ such that if*

$$\|f\|_{L^{\infty}(0,T)} \le \kappa \tag{49}$$

*then there exists a periodic weak solution* $(\boldsymbol{V}, \boldsymbol{A}, \boldsymbol{H})$ *that, in addition, enjoys the following properties*

$$\begin{aligned}
&\boldsymbol{V} \in C([0,T], L^2(\mathcal{D})), \quad \frac{\partial \boldsymbol{V}}{\partial t} \in L^2(0,T,L^2(\mathcal{D})),\\
&\operatorname{grad} \boldsymbol{V} \in L^2(0,T;W^{1,2}(\mathcal{D})) \cap C([0,T],L^2(\mathcal{D})),\\
&\boldsymbol{A} \in W^{1,\infty}(0,T;\mathbb{R}^3)\,,\quad \boldsymbol{H} \in C([0,T];S^2) \cap W^{1,\infty}(0,T;\mathbb{R}^3).
\end{aligned}$$

*Finally, there is $p \in L^2(0,T,W^{1,2}(\mathcal{D}))$, such that* $(\boldsymbol{V}, \boldsymbol{A}, \boldsymbol{H}, p)$ *satisfies* (16) *almost everywhere in $\mathcal{D} \times [0,T]$.*

*Proof* The last statement is an immediate consequence of classical results about the existence of the pressure field for Navier-Stokes equations once the above properties of $\boldsymbol{V}$ and $\boldsymbol{A}$ have been established; see, e.g., [5]. To show the latter we will use the same Galerkin method employed in the proof of Theorem 4.1, the only difference being that this time we choose as orthonormal base of $H(\mathcal{D})$, again denoted by $\{\boldsymbol{\psi}_n\}_{n\in\mathbb{N}}$, the one constituted by the eigenfunctions of Stokes operator, with corresponding eigenvalues denoted by $\{\lambda_n\}_{n\in\mathbb{N}}$; see, e.g., [12, Chap. 2, Sect. 4]. As in the proof of Theorem 4.1, we can thus prove the existence of approximate periodic solutions $(\boldsymbol{V}_n, \boldsymbol{A}_n, \boldsymbol{H}_n)$ satisfying (37), (33) and $\boldsymbol{H}_n(t) \in S^2$ for all $t \in [0,T]$. We shall next show some further estimates that will lead to the improved regularity properties of the weak solution stated in the theorem. In the rest of the proof we shall denote by $C_i$, $i \in \mathbb{N}$, positive constants depending at most on $\mathcal{D}$, $\mathcal{B}$, $R_2$ and $T$, where $R_2$ is defined in (41). Moreover, we set $F := \|f\|_{L^\infty(0,T)}$. We begin by considering the magnitude of both sides of $(24)_2$ and then use Cauchy–Schwarz inequality. By taking into account (20), estimates (37), (33), that $|\boldsymbol{H}_n(t)| = 1$, and that $|\boldsymbol{A}_{n,0}| \leq R_1$, we thus obtain

$$\left|\frac{d\boldsymbol{A}_n}{dt}\right| \leq C_1\,(R_1^2 + R_1 + F^2 + F) := C_1\, D_{R_1,F}\,. \tag{50}$$

Similarly, from $(24)_3$, we show that

$$\left|\frac{d\boldsymbol{H}_n}{dt}\right| \leq C_2\,(R_1 + F + 1). \tag{51}$$

We next multiply $(25)_1$ by $\frac{dc_{nr}}{dt}$ and sum over $r = 1, \ldots, n$, to get

$$\begin{aligned}\left\|\frac{\partial \boldsymbol{V}_n}{\partial t}\right\|_B^2 + \frac{\nu}{2}\frac{d}{dt}\|\text{grad}\,\boldsymbol{V}_n\|_2^2 &= -\frac{d\boldsymbol{A}_n}{dt}\cdot \boldsymbol{I}^{-1}\cdot \int_{\mathcal{D}} \boldsymbol{y} \times \frac{\partial \boldsymbol{V}_n}{\partial t}\, dV \\ &- 2\boldsymbol{A}_n \cdot \boldsymbol{I}^{-1}\cdot \int_{\mathcal{D}} \boldsymbol{V}_n \times \frac{\partial \boldsymbol{V}_n}{\partial t}\, dV \\ &+ 2\rho_{\mathcal{L}} \left(\int_{\mathcal{D}} \boldsymbol{y} \times \boldsymbol{V}_n\, dV\right)\cdot \boldsymbol{I}^{-1}\cdot \left(\int_{\mathcal{D}} \boldsymbol{V}_n \times \frac{\partial \boldsymbol{V}_n}{\partial t}\, dV\right) \\ &- \int_{\mathcal{D}} (\boldsymbol{V}_n \cdot \text{grad}\,\boldsymbol{V}_n)\cdot \frac{\partial \boldsymbol{V}_n}{\partial t}\, dV.\end{aligned} \tag{52}$$

Let us estimate each term on the right-hand side of this equation. By (50), and Cauchy-Schwarz inequality we have

$$-\frac{d\boldsymbol{A}_n}{dt}\cdot \boldsymbol{I}^{-1}\cdot \int_{\mathcal{D}} \boldsymbol{y} \times \frac{\partial \boldsymbol{V}_n}{\partial t}\, dV \leq C_3 + D_{R_1,F}^2 \left|\frac{\partial \boldsymbol{V}_n}{\partial t}\right\|_2^2. \tag{53}$$

Again by Cauchy-Schwarz inequality and (33) we get

$$-2\boldsymbol{A}_n \cdot \boldsymbol{I}^{-1}\cdot \int_{\mathcal{D}} \boldsymbol{V}_n \times \frac{\partial \boldsymbol{V}_n}{\partial t}\, dV \leq C_4 + (R_1 + F)^2 \left\|\frac{\partial \boldsymbol{V}_n}{\partial t}\right\|_2^2 \tag{54}$$

By the same token, we show

$$2\rho_{\mathcal{L}} \left(\int_{\mathcal{D}} \boldsymbol{y} \times \boldsymbol{V}_n\, dV\right)\cdot \boldsymbol{I}^{-1}\cdot \left(\int_{\mathcal{D}} \boldsymbol{V}_n \times \frac{\partial \boldsymbol{V}_n}{\partial t}\, dV\right) \leq \frac{C_5}{2\varepsilon} + \varepsilon \left\|\frac{\partial \boldsymbol{V}_n}{\partial t}\right\|_2^2, \tag{55}$$

and

$$-\int_{\mathcal{D}} (\boldsymbol{V}_n \cdot \operatorname{grad} \boldsymbol{V}_n) \cdot \frac{\partial \boldsymbol{V}_n}{\partial t}\, dV \leq \frac{1}{2\varepsilon} \|\boldsymbol{V}_n \cdot \operatorname{grad} \boldsymbol{V}_n\|_2^2 + \varepsilon \left\| \frac{\partial \boldsymbol{V}_n}{\partial t} \right\|_2^2, \tag{56}$$

where $\varepsilon$ is an arbitrary positive number. As for the first term on the right-hand side of the previous inequality, it is well known that, for any $\varepsilon > 0$ [9, p. 649]

$$\|\boldsymbol{V}_n \cdot \operatorname{grad} \boldsymbol{V}_n\|_{L^2}^2 \leq K \,\|\operatorname{grad} \boldsymbol{V}_n\|_2^6 + \varepsilon \|P\Delta \boldsymbol{V}_n\|_2^2 ,$$

with $K = K(\mathcal{D}, \varepsilon) > 0$. Combining the latter with (56) we then conclude

$$\begin{aligned} -\int_{\mathcal{D}} (\boldsymbol{V}_n \cdot \operatorname{grad} \boldsymbol{V}_n) \cdot \frac{\partial \boldsymbol{V}_n}{\partial t}\, dV &\leq K_1 \|\operatorname{grad} \boldsymbol{V}_n\|_2^6 \\ &\quad + \varepsilon\Big( \|P\Delta \boldsymbol{V}_n\|_2^2 + \left\| \frac{\partial \boldsymbol{V}_n}{\partial t} \right\|_2^2 \Big), \end{aligned} \tag{57}$$

where $K_1 = K_1(\mathcal{D}, \varepsilon) > 0$. Taking into account (52)–(55), the last displayed equation, and (20), from (52) we deduce that

$$\begin{aligned} \big[C - D_{R_1,F}^2 - (R_1 + F)^2 - 3\varepsilon\big] \left\| \frac{\partial \boldsymbol{V}_n}{\partial t} \right\|_2^2 + \frac{\nu}{2} \frac{d}{dt} \|\operatorname{grad} \boldsymbol{V}_n\|_2^2 &\leq C_6 \\ + K_1 \|\operatorname{grad} \boldsymbol{V}_n\|_2^6 + \varepsilon \|P\Delta \boldsymbol{V}_n\|_2^2 . \end{aligned} \tag{58}$$

As a result, if we choose $\varepsilon < C/6$ and take $R_1 \equiv F$ (see Remark 4.2), from (58) it follows that there is a constant $\kappa_1 = \kappa_1(\mathcal{D}) > 0$ such that if

$$\|f\|_{L^\infty(0,T)} \leq \kappa_1 , \tag{59}$$

then

$$C \left\| \frac{\partial \boldsymbol{V}_n}{\partial t} \right\|_2^2 + \nu \frac{d}{dt} \|\operatorname{grad} \boldsymbol{V}_n\|_2^2 \leq C_7 + 2K_1 \|\operatorname{grad} \boldsymbol{V}_n\|_2^6 + 2\varepsilon \|P\Delta \boldsymbol{V}_n\|_2^2 . \tag{60}$$

In order to "absorb" the term $\varepsilon\|P\Delta \boldsymbol{V}_n\|_{L^2}^2$, let us multiply $(25)_1$ by $\lambda_r c_{nr}$ and sum over $r = 1, \ldots, n$. Proceeding in a way completely analogous to that leading to (53)–(55), and (57), we can show that, under an assumption similar to (59), it follows

$$\|P\Delta \boldsymbol{V}_n\|_{L^2}^2 \leq C_8 + K_2\Big( \left\| \frac{\partial \boldsymbol{V}_n}{\partial t} \right\|_2^2 + \|\operatorname{grad} \boldsymbol{V}_n\|_2^6 \Big), \tag{61}$$

where $K_2 = K_2(\mathcal{D}, \nu) > 0$ and $C_8$ depends also on $\nu$. Let us multiply both sides of (61) by $4\varepsilon$, and then add side by side the resulting inequality and inequality (60). If we take $\varepsilon$ sufficiently small we thus arrive at

$$K_3\left(\|P\Delta \boldsymbol{V}_n\|_2^2 + \left\|\frac{\partial \boldsymbol{V}_n}{\partial t}\right\|_{L^2}^2\right) + \frac{d}{dt}\|\text{grad}\, \boldsymbol{V}_n\|_2^2 \le C_9 + K_4\|\text{grad}\, \boldsymbol{V}_n\|_2^6 \tag{62}$$

where, $K_i = K_i(\mathcal{D}, \nu) > 0$, $i = 3, 4$, and $C_9$ depends also on $\nu$. At this point, we observe the following facts. In the first place, in view of classical estimates for the Stokes problem [6, Theorem IV.6.1], we get

$$\|\,\text{grad}\, \boldsymbol{V}_n\|_2 \le c\,\|P\Delta \boldsymbol{V}_n\|_2\,,$$

with $c = c(\mathcal{D}) > 0$. In the second place, recalling that $R_1 \equiv \|f\|_{L^\infty(0,T)}$ and that $f$ satisfies (59), from (41) we deduce that the radius $R_2$ may be chosen to become only a function of $\mathcal{D}$, $\mathcal{B}$, $\mathcal{L}$, and $T$. As a consequence, the constant $C_9$ in (62) depends only on the same quantities and $\rho_{\mathcal{L}}$. In view of all the above, from (62), we conclude that $y := \|\text{grad}\, \boldsymbol{V}_n\|_2^2$ satisfies the following differential inequality

$$y' \le -c_1 y + c_2 y^3 + c_3$$

where the constants $c_i$, $i = 1, 2$, depend, at most, on $\mathcal{D}$, $\mathcal{B}$, $\mathcal{L}$, and $T$. Our next objective is to show that $y = y(t)$ obeys the hypothesis (47) of Lemma 5.1 with a suitable choice of $a$ and $b$, provided $f$ satisfies a restriction of the type (49). To this end, we observe that integrating both sides of (34) between 0 and $T$ and using the $T$-periodicity of $\boldsymbol{V}_n$, along with (33) we show

$$\begin{aligned}\int_0^T \|\text{grad}\, \boldsymbol{V}_n(t)\|_{L^2}^2\, dt &\le k_1\left(|\boldsymbol{A}_{n,0}|^4 + \|f\|_{L^1(0,T)}^4 + \|f\|_{L^2(0,T)}^2\right)\\ &\le k_2\left(R_1^4 + \|f\|_{L^\infty(0,T)}^4 + \|f\|_{L^\infty(0,T)}^2\right)\\ &\le k_3\left(\|f\|_{L^\infty(0,T)}^4 + \|f\|_{L^\infty(0,T)}^2\right),\end{aligned} \tag{63}$$

where $k_i = k_i(\mathcal{D}, \mathcal{B}, \mathcal{L}, T) > 0$, $i = 1, 2, 3$, and in the last step we used the fact that $R_1 = \|f\|_{L^\infty(0,T)}$. From (63), the integral mean-value theorem, and the $T$-periodicity of $\boldsymbol{V}_n$ we deduce

$$\|\text{grad}\, \boldsymbol{V}_n(\bar{t})\|_{L^2}^2 + \int_{\bar{t}}^{\bar{t}+T} \|\text{grad}\, \boldsymbol{V}_n(t)\|_{L^2}^2\, dt \le 2k_3\left(\|f\|_{L^\infty(0,T)}^4 + \|f\|_{L^\infty(0,T)}^2\right), \tag{64}$$

for some $\bar{t} \in (0, T)$. From (64), Lemma 5.1 and again the $T$-periodicity of $\boldsymbol{V}_n$ we then derive that there is $\kappa_2 = \kappa_2(\mathcal{D}, \mathcal{B}, \mathcal{L}, T) > 0$ such that if

$$\|f\|_{L^\infty(0,T)} \leq \kappa_2 \tag{65}$$

it follows

$$\|\operatorname{grad} \boldsymbol{V}_n(t)\|_2 \leq 2\frac{c_3}{c_1} \quad \text{for all } t \in [0, T]. \tag{66}$$

Moreover, integrating (62) over a period, and taking into account (66) we also conclude

$$\int_0^T \|P\Delta \boldsymbol{V}_n(t)\|_2^2\, dt + \int_0^T \left\|\frac{\partial \boldsymbol{V}_n(t)}{\partial t}\right\|_2^2 dt \leq k_4. \tag{67}$$

where $k_4 = \kappa_4(\mathcal{D}, \mathcal{B}, \mathcal{L}, T) > 0$. Therefore, we conclude that setting $\kappa = \min\{\kappa_1, \kappa_2\}$, $\kappa_1$, $\kappa_2$ defined in (59) and (65), respectively, under the hypothesis (49) the approximating $T$-periodic solutions $(\boldsymbol{V}_n, \boldsymbol{A}_n, \boldsymbol{H}_n)$ constructed in the proof of Theorem 4.1 satisfy, in addition, the uniform bounds (50), (51), (66) and (67). As a consequence, the limiting fields $(\boldsymbol{V}, \boldsymbol{A}, \boldsymbol{H})$ defined through (45) satisfy all the properties stated in the theorem,[3] which is thus completely proved. □

**Acknowledgements** This work is partially supported by NSF grant DMS-1311983.

## References

1. R.A. Adams, J.J. Fournier, *Sobolev Spaces*. Pure and Applied Mathematics, 2nd edn. (Elsevier/Academic, Amsterdam, 2003)
2. V.I. Arnol'd, *Mathematical Methods of Classical Mechanics*. Graduate Texts in Mathematics, 2nd edn. (Springer, New York, 1989)
3. G.E. Bredon, *Topology and Geometry* (Springer, New York, 1993)
4. F.L. Chernousko, *Motion of a Rigid Body with Cavities Containing a Viscous Fluid*. (NASA Technical Translations, Moscow, 1972)
5. G.P. Galdi, *An Introduction to the Navier-Stokes Initial-Boundary Value Problem*. Fundamental Directions in Mathematical Fluid Mechanics (Birkhäuser, Basel, 2000), pp. 1–70
6. G.P. Galdi, *An Introduction to the Mathematical Theory of Navier-Stokes Equations: Steady-State Problems*. Springer Monographs in Mathematics, 2nd edn. (Springer, New York, 2011)
7. G.P. Galdi, A.L. Silvestre, Existence of time-periodic solutions to the Navier-Stokes equations around a moving body. Pac. J. Math. **223**, 251–267 (2006)
8. G.P. Galdi, G. Mazzone, P. Zunino, Inertial motions of a rigid body with a cavity filled with a viscous liquid. C. R. Méc. **341**, 760–765 (2013)

[3] The stated continuity property of grad $\boldsymbol{V}$ follows from classical interpolation results; see, e.g., [13, Théorème 2.1].

9. J.G. Heywood, The Navier-Stokes equations: on the existence, regularity and decay of solutions. Indiana Univ. Math. J. **29**, 639–681 (1980)
10. B.G. Karpov, Dynamics of Liquid-Filled Shell: Resonance and Effect of Viscosity. Ballistic Research Laboratories, Report no. 1279 (1965)
11. N.D. Kopachevsky, S.G. Krein, *Operator Approach to Linear Problems of Hydrodynamics, vol. 2: Nonself-Adjoint Problems for Viscous Fluids* (Birkhäuser Verlag, Basel/Boston/Berlin, 2000)
12. O.A. Ladyzhenskaya, *The Mathematical Theory of Viscous Incompressible Flow*, revised 2nd edn. (Gordon and Breach Science Publisher, New York, 1969)
13. J.L. Lions, Espaces intermédiaires entre espaces Hilbertiens et applications. Bull. Math. Soc. Sci. Math. Phys. R. P. Roumaine **2**, 419–432 (1958)
14. N.N. Moiseyev, V.V. Rumiantsev, *Dynamic Stability of Bodies Containing Fluids* (Springer, Berlin, 1968)
15. G. Prouse, Soluzioni periodiche dell'Equazione di Navier-Stokes. Atti Accad. Naz. Lincei Rend. Cl. Sci. Fis. Mat. Natur. **8**, 443–447 (1963)
16. P.H. Roberts, K. Stewartson, On the motion of a liquid in a spheroidal cavity of a precessing rigid body II. Proc. Camb. Philos. Soc. **61**, 279–288 (1965)
17. W.E. Scott, The Free Flight Stability of Liquid-Filled Shell, Part 1a. Ballistic Research Laboratories, Report no. 120 (1960)
18. K. Stewartson, P.H. Roberts, On the motion of a liquid in a spheroidal cavity of a precessing rigid body. J. Fluid Mech. **17**, 1–20 (1963)
19. F.S. Van Vleck, A note on the relation between periodic and orthogonal fundamental solutions of linear systems II. Am. Math. Mon. **71**, 774–776 (1964)

# Seeking a Proof of Xie's Inequality: On the Conjecture That $\mu_m \to \infty$

**John G. Heywood**

*Dedicated to Yoshihiro Shabata on his 60th birthday*

**Abstract** I pursue an argument of Wenzheng Xie, as furthered in several of my papers, to prove a particular point-wise bound for solutions of the three-dimensional steady Stokes problem. If proven, it will provide the basis for an existence and regularity theory for the non-stationary Navier-Stokes equations, free of assumptions about the regularity of the boundary of the flow region. It will be valid for flow in an arbitrary open set.

In his doctoral thesis, Xie proved an analogous bound for solutions of the Poisson problem for the Laplacian, considering it as a model problem. His proof carries over to the Stokes problem except at one point where the maximum principle is invoked. Subsequently, I've proposed a variant of Xie's argument that circumvents the maximum principle, but requires instead a proof that a certain sequence of functions introduced in Xie's argument tends to become singular. I've expressed this as a further conjecture, which is studied here for both the Stokes problem and for the Poisson problem, the latter being considered as a model problem.

**Keywords** Estimates for the stokes and poisson problems • Navier-stokes equations • Non-smooth boundaries

## 1 Introduction

Let $\Omega$ be an arbitrary open subset of $R^3$. Let $\mathbf{D}(\Omega) = \{\varphi \in \mathbf{C}_0^\infty(\Omega) : \nabla \cdot \varphi = 0\}$. Let $\mathbf{J}(\Omega)$ and $\mathbf{J}_0(\Omega)$ be the completions of $\mathbf{D}(\Omega)$ in the $\mathbf{L}^2$-norm $\|\cdot\|$ and the Dirichlet norm $\|\nabla\cdot\|$, respectively. Then, given $\mathbf{u} \in \mathbf{J}_0(\Omega)$, there is at most one $\mathbf{f} \in \mathbf{J}(\Omega)$ such that $(\nabla\mathbf{u}, \nabla\varphi) = -(\mathbf{f}, \varphi)$, for all $\varphi \in \mathbf{D}(\Omega)$. If such a function $\mathbf{f}$ exists, it is denoted by $\tilde{\Delta}\mathbf{u}$, thus defining an operator $\tilde{\Delta}$, referred to as the "Stokes operator", from a subspace of $\mathbf{J}_0(\Omega)$ to $\mathbf{J}(\Omega)$. Under these assumptions

J.G. Heywood (✉)
Department of Mathematics, University of British Columbia, Vancouver, BC, Canada
e-mail: heywood@math.ubc.ca



H. Amann et al. (eds.), *Recent Developments of Mathematical Fluid Mechanics*, Advances in Mathematical Fluid Mechanics, DOI 10.1007/978-3-0348-0939-9_14

it is conjectured that

$$\sup_{\Omega} |\mathbf{u}|^2 \leq \frac{1}{3\pi} \|\nabla \mathbf{u}\| \left\| \tilde{\Delta} \mathbf{u} \right\| . \tag{1}$$

The constant has been proven to be optimal, by Wenzheng Xie in [6]. It cannot be improved, even for particular choices of the domain.

In attempting to prove (1), Xie [4, 5] succeeded in proving

$$\sup_{\Omega} |u|^2 \leq \frac{1}{2\pi} \|\nabla u\| \|\Delta u\| \tag{2}$$

for functions $u$ that belong to the completion $\hat{H}_0^1(\Omega)$ of $C_0^\infty(\Omega)$ in the Dirichlet norm $\|\nabla \cdot\|$, and have a square-integrable Laplacian $\Delta u$, defined here as a function $f \in L^2(\Omega)$ such that $(\nabla u, \nabla \varphi) = -(f, \varphi)$, for all $\varphi \in C_0^\infty(\Omega)$. Again, the domain $\Omega$ is taken to be an arbitrary open subset of $R^3$, and the constant is optimal. The inequality is valid for either scalar or vector-valued functions, the vector-valued case being a simple corollary of the scalar case. In our paper [3] with Xie, we have based an existence and regularity theory for the vector Burgers equation on the inequality (2). It is valid for solutions defined in arbitrary three-dimensional open sets. Once the inequality (1) is proven, this theory for the Burgers equation will serve as a model for a similar existence and regularity theory for the Navier-Stokes equations.

In [6], Xie showed that the argument he used to prove (2) in [4, 5] carries over to a proof of (1), except for his use of the maximum principle in proving that for every point $y$ in a smoothly bounded domain $\Omega$,

$$\left\| G_\mu(\cdot, y) \right\|_{L^2(\Omega)} \leq \left\| g_\mu(\cdot, y) \right\|_{L^2(R^3)}, \tag{3}$$

where $G_\mu$ is the Green's function for the Helmholtz operator $-\Delta + \mu$, and $g_\mu$ is the corresponding fundamental singularity. Although he conjectured an analogous inequality for the spectral Stokes operator $-\tilde{\Delta} + \mu$, it has yet to be proven. However, we have proven in [1], even for the spectral Stokes operator, that the ratio of the two sides of (3) tends to 1 as $\mu \to \infty$. This will suffice in place of (3) for the proof of either (1) or (2), provided the relevant values of $\mu$ tend to infinity. Xie's argument in [4, 5] involves a sequence $\{\overline{u}_m\}$ of finite sums of eigenfunctions. And his argument in [6] for the Stokes operator involves an analogous sequence $\{\overline{\mathbf{u}}_m\}$, defined here in Lemma 2.1 below. The values $\overline{\mu}_m$ of $\mu$ that need to be considered provide a measure of the regularity of the functions $\overline{\mathbf{u}}_m$, which must be shown to become increasingly singular. Specifically, what remains to be shown in this approach to the problem is that

$$\overline{\mu}_m \equiv \left\| \tilde{\Delta} \overline{\mathbf{u}}_m \right\|^2 / \|\nabla \overline{\mathbf{u}}_m\|^2 \to \infty, \quad \text{as } m \to \infty. \tag{4}$$

In this paper we take several new steps towards a proof of (4), furthering our confidence that it will ultimately be proven, and along with it (1).

In Sect. 2, we will review the main steps already taken in attempting to generalize the argument for (2), given in [4] and [5], to a proof of (1). These begin with the results of Xie's paper [6], which provide a complete proof of (1) modulo Xie's conjecture that an analogue of (3) holds for the Stokes equations. Then we state our result from [1], which provides a means of circumventing Xie's conjecture if our conjecture (4), that $\overline{\mu}_m \to \infty$, can be proven.

In Sect. 3, we seek information about $\overline{\mu}_m$ by showing that it is a root of an equation $f_m(\mu, y, \mathbf{e}) = 0$, numbered as (21) below. For a fixed point $y$ and unit vector $\mathbf{e}$, the function $f_m(\mu, y, \mathbf{e})$ is an expression involving the first $m$ eigenvalues of the Stokes operator along with the projections $\mathbf{e} \cdot \boldsymbol{\varphi}_n(y)$ of the associated eigenfunctions $\boldsymbol{\varphi}_n$. The functions $f_m(\mu, y, \mathbf{e})$ are shown to converge to a limit $f_\infty(\mu, y, \mathbf{e})$ as $m \to \infty$. If it can be shown that $f_\infty(\mu, y, \mathbf{e}) < 0$ for all $\mu \geq 0$, then the desired conclusion (4) follows easily. Thus we are led to a further conjecture, that $f_\infty(\mu, y, \mathbf{e}) < 0$ for all $\mu \geq 0$. These results along with this new conjecture are analogous to ones we have given in [2] for the Poisson problem. The conjectures, for both the Stokes and Poisson problems, are formulated below as (27) and (28), respectively. Even though (2) has been proven using (3), the conjecture (28), like (27), remains unproven, and seems an appropriate model problem to consider in trying to prove (27). The remainder of this paper concerns our attempt to prove (28).

The conjecture (28) concerns a function $f_\infty(\mu, y)$ that is expressed in terms of the eigenvalues and eigenfunctions of the Laplacian. In Sect. 4 we show that $f_\infty(\mu, y)$ can also be expressed in terms of certain integrals of Green's functions, and an identity is proven for one of these integrals that greatly simplifies its evaluation.

In Sect. 5, we begin by using the results of Sect. 4 to prove that the inequality (28) holds if $\Omega$ is a ball and $y$ is its center. This is shown by a precise calculation, resulting in a remarkable identity, (46) below. This result is then extended to centrally located open subregions of points $y$ within nearly spherical domains $\Omega$, using the maximum principle. It follows for such domains and points, that $\overline{\mu}_m \to \infty$ for the sequence of functions considered in Xie's proof of (2). We have every expectation that these results are generic, for all smoothly bounded domains, and all points within them, and for the Stokes operator as well as for the Laplacian. If confirmed, the conjecture (4) and the inequality (1) will be proven, for all domains.

In [4, 5], Xie showed that if $\Omega = R^3$, then the functions $u(x) = \left(1 - e^{-\alpha|x|}\right)/\alpha\,|x|$ satisfy $\sup_\Omega |u|^2 = (1/2\pi)\,\|\nabla u\|\,\|\Delta u\|$ for every $\alpha > 0$. As $\alpha \to \infty$, the functions $u$ become increasingly singular at the origin, in that $\mu \equiv \|\Delta u\|^2 / \|\nabla u\|^2 \to \infty$, while becoming very small and nearly constant away from the origin. Thus, if $\Omega$ is a domain with boundary, and the origin is interior to $\Omega$, the functions $u$ can be truncated near the boundary with arbitrarily small changes in $\mu$, provided $\alpha$ is large enough. This was the argument Xie used to prove that the constant $1/2\pi$ in (2) cannot be improved, even for particular choices of $\Omega$. Xie also showed, in [7], that this $\alpha$-dependent family of functions $u$ is unique (up to multiples or shifts in position) in possessing the property of equalizing the two sides of (2). That suggests that any function that nearly equalizes the two sides of (2) must be close to one of these $\alpha$-dependent functions. If the domain has a boundary, that will only be possible for large $\alpha$ (to minimize the effects of truncation) and thus large $\mu$. It is this reasoning that led us to conjecture (4), and hope for a proof by contradiction. Of course, the existence of this $\alpha$-dependent family of functions also shows that one must expect the proof of (4) to be quite subtle. Indeed, if $\Omega = R^3$, it provides functions that equalize the two sides of (2) for which one has, not $\mu \to \infty$, but even $\mu \to 0$, by letting $\alpha \to 0$. Everything said in this paragraph carries over to the Stokes equations, using results provided by Xie in [6] and [8]. In particular, in [8], an $\alpha$-dependent family of functions $u$ that equalizes the two sides of (1) was found and proven to be unique, up to shifts in position, magnitude, and a choice of direction.

## 2 Background Results from [6] and [1]

We begin by reviewing the results of Xie's paper [6]. He began by making two reductions of the problem. First, it was shown that if the inequality (1) holds for all smoothly bounded domains, with a fixed constant independent of the domain, then it holds for all domains (meaning for arbitrary open sets). This was proved by considering an expanding sequence of smoothly bounded domains.

The second reduction concerns eigenfunction expansions in smoothly bounded domains. The orthonormal eigenfunctions $\{\boldsymbol{\varphi}_n\}$ and associated eigenvalues $\{\lambda_n\}$ of the Stokes operator satisfy

$$\tilde{\Delta}\boldsymbol{\varphi}_n = -\lambda_n \boldsymbol{\varphi}_n\,, \qquad \boldsymbol{\varphi}_n|_{\partial\Omega} = 0\,, \tag{5}$$

and linear combinations of the eigenfunctions are dense in $\mathbf{J}(\Omega)$ and $\mathbf{J}_0(\Omega)$. Xie showed that if (1) holds for every finite linear combination

$$\mathbf{u}_m = \sum_{n=1}^{m} c_n \boldsymbol{\varphi}_n$$

of the eigenfunctions, then it holds for all $\mathbf{u} \in \mathbf{J}_0(\Omega)$ with $\tilde{\Delta}\mathbf{u} \in \mathbf{J}(\Omega)$. The proofs of these reductions are elegant, but irrelevant to what follows.

Now let $\mathbf{e}$ denote an arbitrary three-dimensional unit vector. Clearly

$$\sup_{\Omega} |\mathbf{u}|^2 = \sup_{y\in\Omega} \sup_{\mathbf{e}} (\mathbf{e}\cdot\mathbf{u}(y))^2\,. \tag{6}$$

In attempting to prove (1), Xie went on to consider the ratio of its left to right sides, beginning with the following lemma.

**Lemma 2.1** *For any fixed choice of $y$, $\mathbf{e}$ and $m$, let $\bar{\mathbf{u}}_m = \sum_{n=1}^{m} \bar{c}_n \boldsymbol{\varphi}_n$ maximize*

$$R(y, \mathbf{e}, m; c_1, \cdots, c_m) \equiv \frac{(\mathbf{e}\cdot\mathbf{u}_m(y))^2}{\|\nabla\mathbf{u}_m\|\,\left\|\tilde{\Delta}\mathbf{u}_m\right\|}\,. \tag{7}$$

*among all functions of the form* $\mathbf{u}_m = \sum_{n=1}^{m} c_n \boldsymbol{\varphi}_n$ *Then*

$$\frac{\left(\mathbf{e}\cdot\overline{\mathbf{u}}_m(y)\right)^2}{\|\nabla\overline{\mathbf{u}}_m\|\left\|\tilde{\Delta}\overline{\mathbf{u}}_m\right\|} = 4\sqrt{\overline{\mu}_m}\sum_{n=1}^{m}\left(\frac{\mathbf{e}\cdot\boldsymbol{\varphi}_n(y)}{\lambda_n+\overline{\mu}_m}\right)^2 \tag{8}$$

*where*

$$\overline{\mu}_m = \frac{\left\|\tilde{\Delta}\overline{\mathbf{u}}_m\right\|^2}{\|\nabla\overline{\mathbf{u}}_m\|^2}. \tag{9}$$

To prove this, note first that $R$ is smooth and homogeneous as a function of the variables $(c_1,\cdots,c_m) \in R^3 \setminus \{0\}$. Being homogeneous, it is constant along rays from the origin, and thus assumes all of its values on any sphere about the origin. Therefore it has a maximum and attains it at a point $(\overline{c}_1,\cdots,\overline{c}_m)$ where $R$ is smooth. At such a point $\partial R/\partial c_n = 0$, for $n = 1,\cdots m$. With a bit of work these equations can be reassembled to obtain (8).

In the next step, the terms summed on the right of (8) were identified as the squares of the Fourier coefficients of a Green's function. It was proven that:

**Lemma 2.2** *Let* $\mathbf{G}(x,y;\mu,\mathbf{e})$ *be the Green's function for the spectral Stokes problem*

$$(\Delta-\mu)\,\mathbf{G} + \nabla P = \delta(x-y)\,\mathbf{e}, \quad \nabla\cdot\mathbf{G} = 0, \quad \mathbf{G}|_{\partial\Omega} = 0. \tag{10}$$

*Then*

$$\int_\Omega |\mathbf{G}(x,y;\mu,\mathbf{e})|^2\,dx = \sum_{n=1}^{\infty}\left(\frac{\mathbf{e}\cdot\boldsymbol{\varphi}_n(y)}{\lambda_n+\mu}\right)^2. \tag{11}$$

Next, the rather complicated fundamental singularity $\mathbf{g}(x,y;\mu,\mathbf{e})$ for the problem (10) was found explicitly and its square was integrated.

**Lemma 2.3** *The fundamental singularity* $\mathbf{g}(x,y;\mu,\mathbf{e})$ *for the problem (10) satisfies*

$$\int_{R^3} |\mathbf{g}(x,y;\mu,\mathbf{e})|^2\,dx = \frac{1}{12\pi\sqrt{\mu}}. \tag{12}$$

At this point in the corresponding argument for the Poisson problem, Xie proved (3) using the maximum principle. Along with analogues of the preceding lemmas, that completed the proof of (2). To circumvent this use of the maximum principle, which is not available for the Stokes equations, we proved the following lemma in [1].

**Lemma 2.4** *There holds*

$$\frac{\int_{\Omega} |\mathbf{G}(x,y;\mu,\mathbf{e})|^2\,dx}{\int_{R^3} |\mathbf{g}(x,y;\mu,\mathbf{e})|^2\,dx} \to 1, \quad \textit{as } \mu \to \infty. \tag{13}$$

We mention that the analogue of this for the scalar Helmholtz operator $-\Delta + \mu$ is easily proven, because the fundamental singularity decays exponentially. But in the case of the Stokes equations, $\mathbf{g}(x,y;\mu,\mathbf{e})$ has terms that decay only like $1/r^3$, considerably complicating the proof of (13).

Combining Lemmas 2.1–2.4, we have

$$\begin{aligned}\frac{(\mathbf{e}\cdot\overline{\mathbf{u}}_m(y))^2}{\|\nabla\overline{\mathbf{u}}_m\|\,\left\|\tilde{\Delta}\overline{\mathbf{u}}_m\right\|} &= 4\sqrt{\overline{\mu}_m}\sum_{n=1}^{m}\left(\frac{\mathbf{e}\cdot\boldsymbol{\varphi}_n(y)}{\lambda_n+\overline{\mu}_m}\right)^2 \\ &\le 4\sqrt{\overline{\mu}_m}\int_{\Omega}|\mathbf{G}(x,y;\overline{\mu}_m,\mathbf{e})|^2\,dx \\ &\to \frac{1}{3\pi},\end{aligned} \tag{14}$$

provided $\overline{\mu}_m \to \infty$ as $m \to \infty$, which is needed to apply Lemma 2.4 in the last step. Since the left side of ( 14) increases with $m$, (14) implies the desired inequality (1). Thus it only remains to show that $\overline{\mu}_m \to \infty$, as $m \to \infty$.

## 3 An Equation $f_m(\mu, y, \mathbf{e}) = 0$ Satisfied by $\mu = \overline{\mu}_m$

Due to the use of (3) rather than (13) in Xie's proof of (2), the values of $\overline{\mu}_m$ became irrelevant, and were even eliminated by cancellation. In trying to get some information about them, we have discovered an equation that they satisfy. We will now provide the full details of this for the Stokes operator, paralleling an analogous argument we have given for the Laplacian in [2].

To determine $\overline{\mu}_m$, we reconsider the maximization of

$$\frac{(\mathbf{e}\cdot\mathbf{u}_m(y))^2}{\|\nabla\mathbf{u}_m\|\,\left\|\tilde{\Delta}\mathbf{u}_m\right\|} \equiv \frac{\left(\sum_{n=1}^{m} c_n\mathbf{e}\cdot\boldsymbol{\varphi}_n(y)\right)^2}{\left(\sum_{n=1}^{m}\lambda_n c_n^2\right)^{1/2}\left(\sum_{n=1}^{m}\lambda_n^2 c_n^2\right)^{1/2}}, \tag{15}$$

for fixed $y$, $\mathbf{e}$ and $m$, using the method of Lagrange multipliers. Since the function to be maximized is a homogeneous function of the variables $c_1, \cdots, c_m$, it is constant on lines through the origin in the space of these variables. Thus, corresponding to any maximizing point, there will be another, on the same line through the origin, satisfying $\mathbf{e} \cdot \mathbf{u}_m(y) = 1$. Thus, we may revisit Lemma 2.1 above and make this an additional condition to be satisfied by $\overline{\mathbf{u}}_m$. Henceforth we will assume that Lemma 2.1 has been so modified. Hence the maximizing choice of $\overline{\mathbf{u}}_m$, with coefficients $(\overline{c}_1, \cdots, \overline{c}_m)$, must minimize the denominator, or the square of the denominator,

$$f(c_1, \cdots, c_m) \equiv \left(\sum_{n=1}^{m} \lambda_n c_n^2\right)\left(\sum_{n=1}^{m} \lambda_n^2 c_n^2\right) \tag{16}$$

subject to the constraint

$$g(c_1, \cdots, c_m) \equiv \sum_{n=1}^{m} c_n \mathbf{e} \cdot \boldsymbol{\varphi}_n(y) = 1. \tag{17}$$

Therefore there exists a Lagrange multiplier $\lambda$ such that

$$\overline{c}_n \left[\lambda_n \sum_{k=1}^{m} \lambda_k^2 \overline{c}_k^2 + \lambda_n^2 \sum_{k=1}^{m} \lambda_k \overline{c}_k^2\right] = \lambda \mathbf{e} \cdot \boldsymbol{\varphi}_n(y), \quad \text{for } n = 1, \cdots, m. \tag{18}$$

Multiplying this by $\overline{c}_n$ and summing over $n$, from 1 to $m$, one obtains

$$2f(\overline{c}_1, \cdots, \overline{c}_m) = \sum_{n=1}^{m} \lambda \overline{c}_n \mathbf{e} \cdot \boldsymbol{\varphi}_n(y) = \lambda.$$

To eliminate $\lambda$, multiply this by $\mathbf{e} \cdot \boldsymbol{\varphi}_n(y)$, getting

$$2\mathbf{e} \cdot \boldsymbol{\varphi}_n(y) f(\overline{c}_1, \cdots, \overline{c}_m) = \lambda \mathbf{e} \cdot \boldsymbol{\varphi}_n(y),$$

and substitute for the right side the left side of (18) to obtain

$$2\mathbf{e} \cdot \boldsymbol{\varphi}_n(y) f(\overline{c}_1, \cdots, \overline{c}_m) = \overline{c}_n \left[\lambda_n \sum_{k=1}^{m} \lambda_k^2 \overline{c}_k^2 + \lambda_n^2 \sum_{k=1}^{m} \lambda_k \overline{c}_k^2\right].$$

Solving this for $\overline{c}_n$ gives

$$\overline{c}_n = \frac{2\mathbf{e} \cdot \boldsymbol{\varphi}_n(y) \left(\sum_{k=1}^{m} \lambda_k \overline{c}_k^2\right)\left(\sum_{k=1}^{m} \lambda_k^2 \overline{c}_k^2\right)}{\lambda_n \sum_{k=1}^{m} \lambda_k^2 \overline{c}_k^2 + \lambda_n^2 \sum_{k=1}^{m} \lambda_k \overline{c}_k^2}. \tag{19}$$

Letting

$$A_m = \|\nabla \overline{\mathbf{u}}_m\|^2 \quad \text{and} \quad B_m = \left\|\tilde{\Delta} \overline{\mathbf{u}}_m\right\|^2,$$

(19) can be rewritten as

$$\overline{c}_n = \frac{2\mathbf{e} \cdot \boldsymbol{\varphi}_n(y) A_m B_m}{B_m \lambda_n + A_m \lambda_n^2}. \tag{20}$$

**Lemma 3.1** *The ratio $\overline{\mu}_m = B_m/A_m$ is a root of the equation*

$$f_m(\mu, y, \mathbf{e}) \equiv \sum_{n=1}^{m} \frac{(\mathbf{e} \cdot \boldsymbol{\varphi}_n(y))^2}{(\mu + \lambda_n)^2} \left[\frac{\mu}{\lambda_n} - 1\right] = 0. \tag{21}$$

*Proof* Square (20), multiply by $\lambda_n$ and sum to get

$$\begin{aligned} A_m &= 4A_m^2 B_m^2 \sum_{n=1}^{m} \frac{(\mathbf{e} \cdot \boldsymbol{\varphi}_n(y))^2 \lambda_n}{(B_m \lambda_n + A_m \lambda_n^2)^2} \\ &= 4A_m B_m \sum_{n=1}^{m} \frac{(\mathbf{e} \cdot \boldsymbol{\varphi}_n(y))^2}{(\overline{\mu}_m + \lambda_n)^2} \left(\frac{\overline{\mu}_m}{\lambda_n}\right). \end{aligned} \tag{22}$$

Similarly, square (20), multiply by $\lambda_n^2$ and sum to get

$$\begin{aligned} B_m &= 4A_m^2 B_m^2 \sum_{n=1}^{m} \frac{(\mathbf{e} \cdot \boldsymbol{\varphi}_n(y))^2 \lambda_n^2}{(B_m \lambda_n + A_m \lambda_n^2)^2} \\ &= 4B_m^2 \sum_{n=1}^{m} \frac{(\mathbf{e} \cdot \boldsymbol{\varphi}_n(y))^2}{(\overline{\mu}_m + \lambda_n)^2}. \end{aligned} \tag{23}$$

Multiplying (23) by $A_m/B_m$ gives

$$A_m = 4A_m B_m \sum_{n=1}^{m} \frac{(\mathbf{e} \cdot \boldsymbol{\varphi}_n(y))^2}{(\overline{\mu}_m + \lambda_n)^2}. \tag{24}$$

Subtracting (24) from (22) and dividing by $4A_m B_m$ gives the desired result,

$$f_m(\overline{\mu}_m, y, \mathbf{e}) \equiv \sum_{n=1}^{m} \frac{(\mathbf{e} \cdot \boldsymbol{\varphi}_n(y))^2}{(\overline{\mu}_m + \lambda_n)^2} \left[\frac{\overline{\mu}_m}{\lambda_n} - 1\right] = 0. \tag{25}$$

The next lemma follows immediately from the definition of $f_m(\mu, y, \mathbf{e})$.

**Lemma 3.2** *Each function $f_m(\mu, y, \mathbf{e})$ is negative on the interval $0 \leq \mu < \lambda_1$ and positive on the interval $\lambda_m < \mu < \infty$. Hence $\lambda_1 \leq \overline{\mu}_m \leq \lambda_m$.*

The next lemma follows immediately from the convergence of the series on the right of (11).

**Lemma 3.3** *The infinite series*

$$f_\infty(\mu, y, \mathbf{e}) \equiv \sum_{n=1}^{\infty} \frac{(\mathbf{e}\cdot\varphi_n(y))^2}{(\mu+\lambda_n)^2}\left[\frac{\mu}{\lambda_n}-1\right] \tag{26}$$

*is uniformly and absolutely convergent for* $\mu$ *in any bounded subinterval of* $[0,\infty)$.

From these lemmas we reach the following conclusion.

**Proposition 3.1** *Consider a fixed point* $y$ *of* $\Omega$, *and a fixed unit vector* $\mathbf{e}$. *Suppose* $f_\infty(\mu, y, \mathbf{e}) < 0$ *for all* $\mu \geq 0$. *Then the inequality* $|\mathbf{e}\cdot\mathbf{u}(y)|^2 \leq \frac{1}{3\pi}\|\nabla\mathbf{u}\|\left\|\tilde{\Delta}\mathbf{u}\right\|$ *holds for every* $\mathbf{u} \in \mathbf{J}_0(\Omega)$ *that has a square-integrable Stokes operator* $\tilde{\Delta}\mathbf{u}$. *This implies (1) if the hypothesis holds for every* $y \in \Omega$, *and every unit vector* $\mathbf{e}$.

*Proof* Since the sequence of functions $f_m(\mu, y, \mathbf{e})$ is uniformly convergent to the negative function $f_\infty(\mu, y, \mathbf{e})$ on every bounded subinterval of $[0,\infty)$, the roots of $f_m(\mu, y, \mathbf{e})$ are pushed out further and further to the right as $m$ increases. Thus $\overline{\mu}_m \to \infty$ as $m \to \infty$, and this is all that remained to be shown at the end of the previous section. The final conclusion (1) is reached remembering (6).

It follows from the preceding results that if the following conjecture is proven, then the inequality (1) will be proven.

*Conjecture 1* For every smoothly bounded three-dimensional domain $\Omega$, and for every $y \in \Omega$, and every unit vector $\mathbf{e}$, we suspect that

$$f_\infty(\mu, y, \mathbf{e}) \equiv \sum_{n=1}^{\infty} \frac{(\mathbf{e}\cdot\varphi_n(y))^2}{(\mu+\lambda_n)^2}\left[\frac{\mu}{\lambda_n}-1\right] < 0, \qquad \text{for all } \mu \in [0,\infty). \tag{27}$$

In trying to prove (27), we have looked again to the corresponding problem for the Laplacian. Let $\{\phi_n\}$ be the $L^2$-orthonormal eigenfunctions of the Laplacian $\Delta$, satisfying $-\Delta\phi_n = \lambda_n\phi_n$, $\phi_n|_{\partial\Omega} = 0$. We will attempt to prove the following conjecture as a model problem. If proven, it will provide a proof of (2) that is independent of (3).

*Conjecture 2* For every smoothly bounded three-dimensional domain $\Omega$, and for every $y \in \Omega$, we suspect that

$$f_\infty(\mu, y) \equiv \sum_{n=1}^{\infty} \frac{\phi_n^2(y)}{(\mu+\lambda_n)^2}\left[\frac{\mu}{\lambda_n}-1\right] < 0, \qquad \text{for all } \mu \in [0,\infty). \tag{28}$$

## 4 Green's Function Formulation of $f_\infty(\mu, y) < 0$

The inequality (28) can be expressed in terms of Green's functions. For a smoothly bounded domain $\Omega \subset R^3$, and arbitrary point of singularity $y \in \Omega$, the Green's functions for the Laplacian $-\Delta$ and for the Helmholtz operator $-\Delta + \mu$ are

$$G_0(x, y) = \frac{1}{4\pi r} + w_0(x, y) \quad \text{and} \quad G_\mu(x, y) = \frac{e^{-\sqrt{\mu} r}}{4\pi r} + w_\mu(x, y) \tag{29}$$

respectively, where $r \equiv |x - y|$ and the functions $w_0(x, y)$ and $w_\mu(x, y)$ satisfy

$$\begin{cases} \Delta w_0(\cdot, y) = 0 \\ w_0(x, y)\big|_{x \in \partial\Omega} = -\dfrac{1}{4\pi r} \end{cases} \quad \text{and} \quad \begin{cases} \Delta w_\mu(\cdot, y) = \mu w_\mu(\cdot, y) \\ w_\mu(x, y)\big|_{x \in \partial\Omega} = -\dfrac{e^{-\sqrt{\mu} r}}{4\pi r} \end{cases}. \tag{30}$$

**Lemma 4.1** *For all $y \in \Omega$ and $\mu \in [0, \infty)$,*

$$f_\infty(\mu, y) = \int_\Omega G_\mu(x, y) G_0(x, y)\, dx - 2\left\|G_\mu(\cdot, y)\right\|^2. \tag{31}$$

*Proof* Since $\phi_n$ satisfies

$$(-\Delta + \mu)\phi_n = (\lambda_n + \mu)\phi_n, \qquad \phi_n|_{\partial\Omega} = 0,$$

it can be represented using the Green's function for the Helmholtz operator as

$$\phi_n(y) = \int_\Omega G_\mu(x, y)(\lambda_n + \mu)\phi_n(x)\, dx.$$

It follows that the $n^{th}$ Fourier coefficient of $G_\mu(\cdot, y)$ is

$$\frac{\phi_n(y)}{\mu + \lambda_n} = \int_\Omega G_\mu(x, y)\phi_n(x)\, dx. \tag{32}$$

Hence, simply rearranging (28), we have

$$\begin{aligned} f_\infty(\mu, y) &= \sum_{n=1}^\infty \frac{\mu\phi_n^2(y)}{\lambda_n(\mu + \lambda_n)^2} - \sum_{n=1}^\infty \frac{\phi_n^2(y)}{(\mu + \lambda_n)^2} \\ &= \sum_{n=1}^\infty \left[\frac{\phi_n^2(y)}{\lambda_n(\mu + \lambda_n)} - \frac{\phi_n^2(y)}{(\mu + \lambda_n)^2}\right] - \sum_{n=1}^\infty \frac{\phi_n^2(y)}{(\mu + \lambda_n)^2} \\ &= \sum_{n=1}^\infty \frac{\phi_n^2(y)}{\lambda_n(\mu + \lambda_n)} - 2\sum_{n=1}^\infty \frac{\phi_n^2(y)}{(\mu + \lambda_n)^2} \\ &= \int_\Omega G_\mu(x, y) G_0(x, y)\, dx - 2\left\|G_\mu(\cdot, y)\right\|^2. \end{aligned}$$

**Lemma 4.2** *For all* $x, y \in \Omega$, *there holds*

$$\mu \int_{\Omega} G_{\mu}(z, y)\, G_0(z, x)\, dz = G_0(x, y) - G_{\mu}(x, y). \tag{33}$$

*Proof* This identity is a representation of $G_0(x, y) - G_{\mu}(x, y)$, considered as a function of $x$ with $y$ fixed, in terms of the Green's function $G_0$ for the Laplacian. For fixed $y$ and $x \neq y$, it follows from (29) and (30) that the function $G_0(x, y) - G_{\mu}(x, y)$ satisfies

$$-\triangle \left(G_0(x, y) - G_{\mu}(x, y)\right) = \mu G_{\mu}(x, y) \tag{34}$$

and vanishes on $\partial\Omega$. Thus we need only check that it is sufficiently regular to justify the representation (33). In fact, the singularity of $G_0 - G_{\mu}$ at $x = y$ is not very bad since the difference between the fundamental singularities, $u(r) \equiv 1/4\pi r - e^{-\sqrt{\mu} r}/4\pi r$, satisfies

$$\lim_{r \to 0} u(r) = \sqrt{\mu}/4\pi, \tag{35}$$

and is therefore continuous. Furthermore, $u \in H^2_{loc}\left(R^3\right)$, since it can be approximated by a sequence of functions

$$u_n(r) \equiv \begin{cases} u(1/n) + \dfrac{u'(1/n)}{2n}\left(n^2 r^2 - 1\right), & \text{for } r \leq 1/n \\ u(r), \quad \text{for } r \geq 1/n, \end{cases}$$

that belong to $C^1\left(R^3\right)$, are piecewise $C^2$ in $R^3$, and converge to $u(r)$ both pointwise and in the norm of $H^2\left(R^3\right)$. Thus, considering $y$ as fixed, the functions $g^n_{\mu}(x, y) \equiv u_n(r) + \left(w_0(x, y) - w_{\mu}(x, y)\right)$ admit the representation

$$g^n_{\mu}(x, y) = \int_{\Omega} -\triangle\, g^n_{\mu}(z, y)\; G_0(z, x)\; dz,$$

which converges to (33) as $n \to \infty$.

**Lemma 4.3** *For all* $y \in \Omega$, *there holds*

$$\int_{\Omega} G_{\mu}(x, y)\, G_0(x, y)\, dx = \frac{1}{4\pi\sqrt{\mu}} + \frac{w_0(y, y) - w_{\mu}(y, y)}{\mu}. \tag{36}$$

*Proof* One obtains (36) from (33) as follows. First, replace $G_0(x, y)$ and $G_{\mu}(x, y)$ on the right of (33) by their component parts from (29), (30). Then take the limit $x \to y$, remembering (35). Finally, relabel the integration variable $z$ as $x$.

Although not needed here, there is also a representation

$$\mu \int_{\Omega} G_0(z,y)\, G_\mu(z,x)\, dz = G_0(x,y) - G_\mu(x,y) \tag{37}$$

analogous to (33), but in terms of the Green's function $G_\mu$. It can be proved directly by noting that $(-\Delta + \mu)\big(G_0(x,y) - G_\mu(x,y)\big) = \mu G_0(x,y)$, for $x \neq y$, or it can be simply obtained as a corollary of (33) and the symmetry of $G_0(x,y)$ and $G_\mu(x,y)$.

## 5 Conditions That Ensure $f_\infty(\mu, y) < 0$ for All $\mu \geq 0$

We will consider two cases. First, we will take $\Omega$ to be a sphere and $y$ to be its center. In this case we can evaluate $f_\infty(\mu, y)$ exactly, through an explicit computation using (31) and (36). Then, appealing to the maximum principle, we will generalize the argument to a somewhat wider class of domains and points, including some non-spherical domains and some open regions of points $y$.

### 5.1 *The Case of y Being the Center of a Spherical Domain*

Consider the spherical domain $\Omega = \{x \in R^3 : |x| < R\}$ and the point $y = 0$. Then we can find $G_0(x,0)$ and $G_\mu(x,0)$ explicitly. Indeed,

$$w_0(x,0) = -\frac{1}{4\pi R} \quad \text{and} \quad w_\mu(x,0) = -\left(\frac{e^{-\sqrt{\mu}R}}{e^{\sqrt{\mu}R} - e^{-\sqrt{\mu}R}}\right)\frac{e^{\sqrt{\mu}r} - e^{-\sqrt{\mu}r}}{4\pi r}. \tag{38}$$

Setting

$$a = \left(\frac{e^{-\sqrt{\mu}R}}{e^{\sqrt{\mu}R} - e^{-\sqrt{\mu}R}}\right), \tag{39}$$

observe that

$$w_\mu(0,0) = \lim_{x\to 0} w_\mu(x,0) = -a \lim_{r\to 0}\frac{e^{\sqrt{\mu}r} - e^{-\sqrt{\mu}r}}{4\pi r} = -a\frac{\sqrt{\mu}}{2\pi}. \tag{40}$$

Combining (38) and (40) with (36), we have

$$\int_{\Omega} G_{\mu}(x,0)\, G_0(x,0)\ dx = \frac{1}{4\pi\sqrt{\mu}} - \frac{1}{4\pi\mu R} + a\frac{1}{2\pi\sqrt{\mu}}. \tag{41}$$

To compute $2\left\|G_{\mu}(\cdot,0)\right\|^2$, we note that

$$G_{\mu}(x,0) = \frac{e^{-\sqrt{\mu}r}}{4\pi r} - a\frac{e^{\sqrt{\mu}r} - e^{-\sqrt{\mu}r}}{4\pi r},$$

and thus

$$\begin{aligned}
&2\left\|G_{\mu}(\cdot,0)\right\|^2\\
&= 2\int_0^R \left(\frac{e^{-2\sqrt{\mu}r}}{16\pi^2 r^2} - 2a\frac{1-e^{-2\sqrt{\mu}r}}{16\pi^2 r^2} + a^2\frac{e^{2\sqrt{\mu}r} - 2 + e^{-2\sqrt{\mu}r}}{16\pi^2 r^2}\right) 4\pi r^2 dr\\
&= \frac{1}{2\pi}\int_0^R \left(e^{-2\sqrt{\mu}r} - 2a\left(1-e^{-2\sqrt{\mu}r}\right) + a^2\left(e^{2\sqrt{\mu}r} - 2 - e^{-2\sqrt{\mu}r}\right)\right) dr\\
&= \frac{1}{2\pi}\left[-\frac{e^{-2\sqrt{\mu}r}}{2\sqrt{\mu}} - 2a\left(r + \frac{e^{-2\sqrt{\mu}r}}{2\sqrt{\mu}}\right) + a^2\left(\frac{e^{2\sqrt{\mu}r}}{2\sqrt{\mu}} - 2r - \frac{e^{-2\sqrt{\mu}r}}{2\sqrt{\mu}}\right)\right]_0^R\\
&= \frac{1-e^{-2\sqrt{\mu}R}}{4\pi\sqrt{\mu}} + \frac{a}{2\pi}\frac{1-e^{-2\sqrt{\mu}R}}{\sqrt{\mu}} - \frac{aR + a^2R}{\pi} + \frac{a^2}{4\pi}\frac{e^{2\sqrt{\mu}R} - e^{-2\sqrt{\mu}R}}{\sqrt{\mu}}.
\end{aligned} \tag{42}$$

Substituting (41) and (42) into (31) gives

$$\begin{aligned}
f_{\infty}(\mu,0) &= \frac{1}{4\pi\sqrt{\mu}} - \frac{1}{4\pi\mu R} + a\frac{1}{2\pi\sqrt{\mu}} - \frac{1-e^{-2\sqrt{\mu}R}}{4\pi\sqrt{\mu}} - \frac{a}{2\pi}\frac{1-e^{-2\sqrt{\mu}R}}{\sqrt{\mu}}\\
&\quad + \frac{aR + a^2R}{\pi} - \frac{a^2}{4\pi}\frac{e^{2\sqrt{\mu}R} - e^{-2\sqrt{\mu}R}}{\sqrt{\mu}}\\
&= -\frac{1}{4\pi R\mu} + \frac{1+2a}{4\pi}\frac{e^{-2\sqrt{\mu}R}}{\sqrt{\mu}} + \frac{aR + a^2R}{\pi} - \frac{a^2}{4\pi}\frac{e^{2\sqrt{\mu}R} - e^{-2\sqrt{\mu}R}}{\sqrt{\mu}}.
\end{aligned} \tag{43}$$

Multiplying by $4\pi\sqrt{\mu}$, setting $\eta = \sqrt{\mu}R$, and noting that

$$a \equiv \frac{e^{-\eta}}{e^{\eta} - e^{-\eta}} = \frac{1 + e^{-2\eta}}{e^{2\eta} - e^{-2\eta}} \quad \text{and} \quad 1 + a = \frac{e^{\eta}}{e^{\eta} - e^{-\eta}} \tag{44}$$

we obtain

$$
\begin{aligned}
4\pi\sqrt{\mu}f_\infty(\mu,0) &= -\frac{1}{\eta} + (1+2a)\,e^{-2\eta} + 4a\,(1+a)\,\eta - a^2\left(e^{2\eta} - e^{-2\eta}\right) \\
&= -\frac{1}{\eta} + (1+2a)\,e^{-2\eta} + 4a\,(1+a)\,\eta - a\left(1 + e^{-2\eta}\right) \\
&= -\frac{1}{\eta} + (1+a)\,e^{-2\eta} + 4a\,(1+a)\,\eta - a \\
&= -\frac{1}{\eta} + \frac{e^{-\eta}}{e^{\eta} - e^{-\eta}} + \frac{4}{e^{\eta} - e^{-\eta}}\frac{\eta}{e^{\eta} - e^{-\eta}} - \frac{e^{-\eta}}{e^{\eta} - e^{-\eta}} \\
&= -\frac{1}{\eta} + \frac{4\eta}{\left(e^{\eta} - e^{-\eta}\right)^2} = -\frac{1}{\eta} + \frac{\eta}{\sinh^2\eta}\,.
\end{aligned}
\tag{45}
$$

We conclude that $f_\infty(\mu,0)$ is negative for all $\mu \geq 0$, since $\sinh\eta > \eta$ for all $\eta > 0$, and because we know from its definition in (28) that $f_\infty(\mu,0) < 0$ for $0 \leq \mu \leq \lambda_1$. Thus we have proven the following proposition.

**Proposition 5.1** *Let* $\Omega = \{x \in R^3 : |x| < R\}$ *and* $y = 0$*. Then the function* $f_\infty(\mu,0)$ *is negative for all* $\mu \geq 0$*. Therefore,* $\mu_m \to \infty$ *as* $m \to \infty$ *in Xie's proof that* $|u(0)|^2 \leq \frac{1}{2\pi}\|\nabla u\|\,\|\Delta u\|$*. Furthermore, we have proven a rather remarkable identity for the eigenfunctions and eigenvalues of the Laplacian in a spherical domain. The values of the eigenfunctions at the center of a sphere satisfy*

$$
4\pi\sum_{n=1}^{\infty}\frac{\phi_n^2(0)}{(\mu+\lambda_n)^2}\left[\frac{\mu}{\lambda_n} - 1\right] = -\frac{1}{\mu R} + \frac{R}{\sinh^2\sqrt{\mu}R}, \quad \textit{for all } \mu \in [0,\infty)\,. \tag{46}
$$

Taking the limit of the right side as $\mu \to 0$, one easily finds that

$$
\lim_{\mu\to 0} 4\pi f_\infty(\mu,0) = -R/3\,.
$$

The graph of $f_\infty(\mu,0)$ has the appearance we expected. Taking a fixed value of $R$, say $R = 1$, the graph of $y = 4\pi f_\infty(\mu,0)$ is:

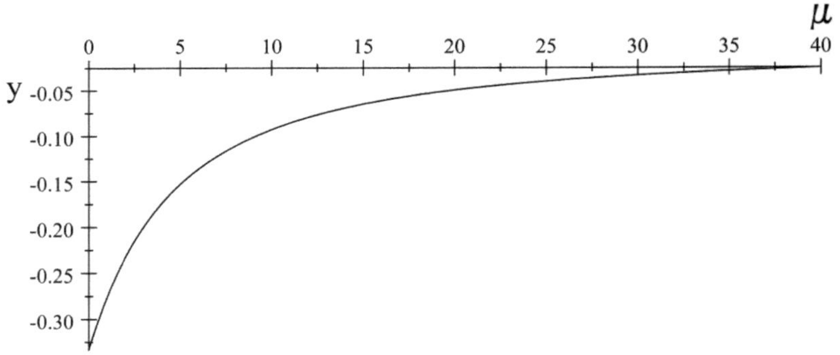

## 5.2 Some General Considerations for Arbitrary y in Arbitrary Domains

Consider now an arbitrary bounded three-dimensional domain $\Omega$, and an arbitrary point $y \in \Omega$. Without loss of generality, we may assume that $y = 0$. Then choose $\rho$ and $R$ such that the concentric spherical domains $\Omega_\rho = \{x \in R^3 : |x| < \rho\}$ and $\Omega_R = \{x \in R^3 : |x| < R\}$ satisfy $\Omega_\rho \subset \Omega \subset \Omega_R$. We will need not only the Green's functions $G_0(x, y)$ and $G_\mu(x, y)$ for the domain $\Omega$, but also the corresponding Green's functions for $\Omega_\rho$ and $\Omega_R$, which will be denoted by $G_{\rho,0}$ and $G_{\rho,\mu}$ and by $G_{R,0}$ and $G_{R,\mu}$, respectively. It follows from the maximum principle that these Green's functions are everywhere positive and that in their common regions of definition they satisfy $G_{\rho,0}(x, y) \le G_0(x, y) \le G_{R,0}(x, y)$ and $G_{\rho,\mu}(x, y) \le G_\mu(x, y) \le G_{R,\mu}(x, y)$. Therefore (31) implies

$$f_\infty(\mu, 0) \le \int_{\Omega_R} G_{R,\mu}(x, 0)\, G_{R,0}(x, 0)\, dx - 2\left\|G_{\rho,\mu}(\cdot, 0)\right\|^2_{\Omega_\rho}. \tag{47}$$

Letting

$$a = \left(\frac{e^{-\sqrt{\mu}R}}{e^{\sqrt{\mu}R} - e^{-\sqrt{\mu}R}}\right) \quad \text{and} \quad b = \left(\frac{e^{-\sqrt{\mu}\rho}}{e^{\sqrt{\mu}\rho} - e^{-\sqrt{\mu}\rho}}\right) \tag{48}$$

it follows from (47), (41), and the analogue of (42) with $R$ replaced by $\rho$, that

$$\begin{aligned} &f_\infty(\mu, 0) \\ &\le \frac{1}{4\pi\sqrt{\mu}} - \frac{1}{4\pi R\mu} + a\frac{\sqrt{\mu}}{2\pi\mu} - \frac{1 - e^{-2\sqrt{\mu}\rho}}{4\pi\sqrt{\mu}} - \frac{b}{2\pi}\frac{1 - e^{-2\sqrt{\mu}\rho}}{\sqrt{\mu}} \\ &\quad + \frac{b\rho + b^2\rho}{\pi} - \frac{b^2}{4\pi}\frac{e^{2\sqrt{\mu}\rho} - e^{-2\sqrt{\mu}\rho}}{\sqrt{\mu}} \\ &= -\frac{1}{4\pi R\mu} + \frac{1 + 2b\,e^{-2\sqrt{\mu}\rho}}{4\pi}\frac{}{\sqrt{\mu}} + \frac{b\rho + b^2\rho}{\pi} + \frac{a - b}{2\pi\sqrt{\mu}} - \frac{b^2}{4\pi}\frac{e^{2\sqrt{\mu}\rho} - e^{-2\sqrt{\mu}\rho}}{\sqrt{\mu}}. \end{aligned} \tag{49}$$

## 5.3 The Case of Concentric Spheres About y Satisfying $\Omega_\rho \subset \Omega \subset \Omega_R$, with $R = 2\rho$

Let $R > 0$ be arbitrary, and let $\rho = R/2$. As above, $a$ and $b$ are defined by (48). Setting $\eta = \sqrt{\mu}R$, we have again (44), and similarly

$$b \equiv \frac{e^{-\eta/2}}{e^{\eta/2} - e^{-\eta/2}} = \frac{1 + e^{-\eta}}{e^{\eta} - e^{-\eta}} \quad \text{and} \quad 1 + b = \frac{e^{\eta/2}}{e^{\eta/2} - e^{-\eta/2}}. \tag{50}$$

Furthermore

$$a = b\frac{e^{-\eta/2}}{e^{\eta/2}+e^{-\eta/2}} \quad \text{and}$$

$$a - b = b\left(\frac{e^{-\eta/2}}{e^{\eta/2}+e^{-\eta/2}} - \frac{e^{\eta/2}+e^{-\eta/2}}{e^{\eta/2}+e^{-\eta/2}}\right) = \frac{-1}{e^{\eta}-e^{-\eta}}\,. \tag{51}$$

Therefore, multiplying (49) by $4\pi\sqrt{\mu}$, we obtain

$$\begin{aligned}
&4\pi\sqrt{\mu} f_\infty(\mu,0)\\
&\leq -\frac{1}{\eta} + (1+2b)\,e^{-\eta} + 4b\,(1+b)\,\eta/2 + 2\,(a-b) - b^2\,(e^{\eta}-e^{-\eta})\\
&= -\frac{1}{\eta} + (1+2b)\,e^{-\eta} + 4b\,(1+b)\,\eta/2 + 2\,(a-b) - b\,(1+e^{-\eta})\\
&= -\frac{1}{\eta} + (1+b)\,e^{-\eta} + 4b\,(1+b)\,\eta/2 + 2\,(a-b) - b\\
&= -\frac{1}{\eta} + \frac{e^{-\eta/2}}{e^{\eta/2}-e^{-\eta/2}} + \frac{4e^{-\eta/2}}{e^{\eta/2}-e^{-\eta/2}}\frac{e^{\eta/2}\eta/2}{e^{\eta/2}-e^{-\eta/2}}\\
&\quad +2\,(a-b) - \frac{e^{-\eta/2}}{e^{\eta/2}-e^{-\eta/2}}\\
&= -\frac{1}{\eta} + \frac{2\eta}{\left(e^{\eta/2}-e^{-\eta/2}\right)^2} + \frac{-2}{e^{\eta}-e^{-\eta}} \equiv g(\eta)\ .
\end{aligned} \tag{52}$$

To prove that $f_\infty(\mu,0) < 0$ for all $\mu > 0$, we will show that $g(\eta) < 0$ for all $\eta > 0$. To prove this, we need only show that

$$h(\eta) \equiv \eta\,(e^{\eta}-e^{-\eta})\,g(\eta) = -\,(e^{\eta}-e^{-\eta}) + \frac{2\eta^2\,(e^{\eta}-e^{-\eta})}{\left(e^{\eta/2}-e^{-\eta/2}\right)^2} - 2\eta$$

is negative for all $\eta > 0$. And indeed it is, since

$$\lim_{\eta\to 0}\left(-\,(e^{\eta}-e^{-\eta}) + \frac{2\eta^2\,(e^{\eta}-e^{-\eta})}{\left(e^{\eta/2}-e^{-\eta/2}\right)^2} - 2\eta\right) = 0$$

and

$$\begin{aligned}
&\frac{d}{d\eta}\left(-\,(e^{\eta}-e^{-\eta}) + \frac{2\eta^2\,(e^{\eta}-e^{-\eta})}{\left(e^{\eta/2}-e^{-\eta/2}\right)^2} - 2\eta\right)\\
&\qquad = -\frac{e^{-\eta}}{(e^{\eta}-1)^2}\left(-e^{2\eta}+2\eta e^{\eta}+1\right)^2 < 0\,, \quad \text{for all } \eta > 0\,.
\end{aligned}$$

Although the estimate (52) tends to zero as $\eta = \sqrt{\mu}R \to 0$, we know from the definition of $f_\infty(\mu,0)$ that $f_\infty(\mu,0) < 0$ for $0 \leq \mu \leq \lambda_1$. It follows that $f_\infty(\mu,0) < 0$ for all $\mu \geq 0$. This concludes the proof of the following proposition.

**Proposition 5.2** *Let $y$ be a point of $\Omega$ such that, for some $R > 0$, the ball of radius $R$ centered at $y$ contains $\Omega$, while the ball of radius $R/2$ centered at $y$ is contained in $\Omega$. Then the function $f_\infty(\mu, y)$ is negative for all $\mu \geq 0$, and therefore $\mu_m \to \infty$ as $m \to \infty$ in Xie's proof that $|u(y)|^2 \leq \frac{1}{2\pi} \|\nabla u\| \|\Delta u\|$.*

It is clear from this proposition that there will be many domains, of various geometrical shapes, for which there will be interior open subregions of points $y$ such that $f_\infty(\mu, y)$ is negative for all $\mu \geq 0$. The following corollary is an easily proven example of this.

**Corollary 5.1** *If $\Omega$ is a sphere of radius $R$, then, for all points $y$ within a distance $R/3$ of its center, there holds $f_\infty(\mu, y) < 0$ for all $\mu \geq 0$.*

*Remark* The argument used in proving Proposition 5.2, utilizing estimation by the maximum principle between spherical domains, does not extend beyond the ratio $\rho = R/2$. Experimenting numerically with $R = 1$ and $\rho = 1/2.001$, it only implies that $\lim_{\mu \to 0} f_\infty(\mu, 0) \leq 1.3256 \times 10^{-5}$, while we know that $f_\infty(\mu, 0) < 0$ for $0 \leq \mu \leq \lambda_1$, for any domain.

## References


1. J.G. Heywood, An alternative to Xie's conjecture concerning the Stokes problem in nonsmooth domains. Ann. Univ. Ferrara VII **46**, 267–284 (2000)
2. J.G. Heywood, On a conjecture concerning the Stokes problem in nonsmooth domains, in *Mathematical Fluid Mechanics; Recent Results and Open Questions*, ed. by J. Neustupa, P. Penel. Advances in Mathematical Fluid Mechanics, vol. 2 (Birkhäuser, Basel, 2001), pp. 195–206
3. J.G. Heywood, W. Xie, Smooth solutions of the vector Burgers equation in non-smooth domains. Diff. Integr. Equ. **10**, 961–974 (1997)
4. W. Xie, A sharp inequality for Poisson's equation in arbitrary domains and its applications to Burgers' equation. Ph.D. Thesis, University of British Columbia, March 1991
5. W. Xie, A sharp pointwise bound for functions with $L^2$-Laplacians and zero boundary values on arbitrary three-dimensional domains. Indiana Univ. Math. J. **40**, 1185–1192 (1991)
6. W. Xie, On a three-norm inequality for the Stokes operator in nonsmooth domains, in *The Navier-Stokes Equations II: Theory and Numerical Methods*. Lecture Notes in Mathematics, vol. 1530 (Springer, Berlin, 1992), pp. 310–315
7. W. Xie, Integral representations and $L^\infty$ bounds for solutions of the Helmholts equation on arbitrary open sets in $R^2$ and $R^3$. Diff. Integr. Equ. **8**, 689–698 (1995)
8. W. Xie, Sharp Sobolev interpolation inequalities for the Stokes operator. Diff. Integr. Equ. **10**, 393–399 (1997)

# Bounded Analyticity of the Stokes Semigroup on Spaces of Bounded Functions

**Matthias Hieber and Paolo Maremonti**

*Dedicated to Prof. Y. Shibata on the occasion of his 60th-Birthday*

**Abstract** Let $\Omega \subset \mathbb{R}^n$, $n \geq 3$, be an exterior domain with smooth boundary. It is shown that the Stokes semigroup on $L^\infty_\sigma(\Omega)$ is a bounded analytic semigroup on this space.

**Keywords** Bounded analytic semigroups • $L^\infty$-estimates • Maximum modulus theorem • Stokes semigroup

**Mathematics Subject Classification (2010).** 35Q35, 35K90, 76D07

## 1 Introduction

Let $\Omega \subset \mathbb{R}^n$, $n \geq 3$, be an exterior domain with boundary of class $C^m$ with $m \geq 3$ and $m > n/2$. We consider the Stokes equation on $\Omega$ subject to Dirichlet boundary conditions, i.e.

$$\begin{cases} \partial_t u - \Delta u + \nabla \pi_u = 0 & \text{in } J \times \Omega, \\ \operatorname{div} u = 0 & \text{in } J \times \Omega, \\ u = 0 & \text{on } J \times \partial\Omega, \\ u(0) = u_0 & \text{in } \Omega, \end{cases} \tag{1}$$

M. Hieber (✉)
Fachbereich Mathematik, Schlossgartenstr. 7, 64289 Darmstadt, Germany

University of Pittsburgh, 607 Benedum Engineering Hall, Pittsburgh, PA 15261, USA
e-mail: hieber@mathematik.tu-darmstadt.de

P. Maremonti
Dipartimento di Mathematica, Via Vivaldi 43, 81100 Caserta, Italy
e-mail: paolo.maremonti@unina2.it



H. Amann et al. (eds.), *Recent Developments of Mathematical Fluid Mechanics*, Advances in Mathematical Fluid Mechanics, DOI 10.1007/978-3-0348-0939-9_15

with *non decaying* initial data $u_0$. Here, $J := (0, \infty)$ and $u : J \times \Omega \to \mathbb{R}^n$ denotes the velocity field and $\pi_u : J \times \Omega \to \mathbb{R}$ the pressure of the associated fluid. The situation of initial data $u_0 \in L^p_\sigma(\Omega)$ for $1 < p < \infty$ was considered by very many authors and it is well known that in this case the Stokes equation admits a unique solution $u(t) = T(t)u_0$, where $T$ denotes the Stokes semigroup on $L^p_\sigma(\Omega)$ with generator $A = P\Delta$. Here $P$ denotes the Helmholtz projection $L^p(\Omega) \to L^p_\sigma(\Omega)$. It was shown by Borchers and Sohr [5] that the Stokes semigroup on $L^p_\sigma(\Omega)$ is a bounded analytic semigroup of angle $\pi/2$. The latter property has many consequences, for example the construction of global, strong solution to the Navier-Stokes equations for initial data being small in $L^p_\sigma(\Omega)$ for $p \geq n$ and also for weak solutions; see e.g [10, 11]. In this article we are interested in the situation where $u_0 \in L^\infty_\sigma(\Omega)$. Here $L^\infty_\sigma(\Omega)$ is defined as

$$L^\infty_\sigma(\Omega) := \{f \in L^\infty(\Omega) : \int_\Omega f \cdot \nabla\varphi dx = 0 \text{ for all } \varphi \in \widehat{W}^{1,1}(\Omega)\},$$

where $\widehat{W}^{1,1}(\Omega) = \{\varphi \in W^{1,1}_{\ell oc}(\Omega) \text{ with } \nabla\varphi \in L^1(\Omega)\}$. Note that the Helmholtz projection introduced as above is no longer a bounded operator if $p = \infty$ and that also the Stokes operator needs to be defined in a different way, in this case. It was proved recently by Abe and Giga in [1, 2] and by Abe et al. [3] by different approaches that the Stokes operator associated to the above equation generates an analytic semigroup $T$ on $L^\infty_\sigma(\Omega)$ of angle $\pi/2$. In addition, Maremonti showed in [14] that $T$ is a bounded semigroup on $L^\infty_\sigma(\Omega)$, i.e. there exists a constant $M > 0$ such that $\|T(t)\|_{\mathcal{L}(L^\infty_\sigma(\Omega))} \leq M$ for all $t > 0$. It is the aim of this paper to show that, similarly to the situation of $1 < p < \infty$, the semigroup $T$ is a bounded analytic semigroup on $L^\infty_\sigma(\Omega)$. Our result may be thus viewed as a certain extension of the result by Borchers and Sohr for $p \in (1, \infty)$ to the case of $p = \infty$. By saying that $T$ is bounded analytic semigroup, we mean that $T$ admits a bounded analytic extension to the sector $\Sigma_\theta := \{z \in \mathbb{C}\backslash\{0\} : |\arg z| < \theta\}$ for some $\theta \in (0, \pi/2)$. For more information on bounded analytic semigroups, we refer e.g. to [4]. Note, however, that a bounded semigroup which is analytic is not necessarily a bounded analytic semigroup. For example, let $X = \mathbb{C}$ and $T(t) = e^{it}$ for $t > 0$. We remark that the corresponding result for the half space $\mathbb{R}^n_+$ was already proved in [8]. Note, however, that the usual localization procedure known for the case of $L^p$ with $p \in (1, \infty)$ does not work for the case $p = \infty$. Thus a different approach is needed. Roughly speaking, our approach may be described as follows: we decompose the solution $(u, q)$ of Eq. (1) as $(u, \pi_u) = (z, 0) + (v, \pi_v)$, where $(z, 0)$ is the solution of the Stokes problem on $\mathbb{R}^n$ with initial data $\tilde{u}_0$ and $(v, \pi_v)$ is the solution of the Stokes problem on $(0, T) \times \Omega$ with inhomogeneous boundary data $\gamma z$. By $\gamma z$ we mean the restriction of the continuous function $z$ on $\partial\Omega$. The strategy of our proof relies on the real characterization of bounded analytic semigroups, i.e. we show that

$$\sup_{t>0} \|tAT(t)\|_{\mathcal{L}(L^\infty_\sigma(\Omega))} < \infty,$$

where $A$ denotes the generator of $T$. We show in particular that there exists a constant $M > 0$ such that $u$ given by $u(t) = T(t)u_0$ satisfies

$$\|tu'(t)\|_{L^\infty_\sigma(\Omega)} \leq M\|u_0\|_{L^\infty_\sigma(\Omega)}, \quad t > 0. \tag{2}$$

Our first main result reads as follows.

**Theorem 1.1** *Let $\Omega \subset \mathbb{R}^n$, $n \geq 3$, be an exterior domain with boundary of class $C^m$, where $m \geq \max\{3, n/2\}$ and let $T$ be the Stokes semigroup on $L^\infty_\sigma(\Omega)$. Then $T$ is a bounded analytic semigroup on $L^\infty_\sigma(\Omega)$.*

*Remark 1.2* The above theorem implies that there exist $\varepsilon > 0$ and $M > 0$ such that

$$\|\lambda R(\lambda, A)\|_{\mathcal{L}(L^\infty_\sigma(\Omega))} \leq M, \quad \lambda \in \Sigma_{\pi/2+\varepsilon}, \tag{3}$$

where $A$ denotes the generator of the Stokes semigroup on $L^\infty_\sigma(\Omega)$ defined above. Here $R(\lambda, A) = (\lambda - A)^{-1}$ for all $\lambda \in \varrho(A)$ and $\varrho(A)$ denoting the resolvent set of $A$. As already mentioned above, the above estimates (3) generalizes to a certain extend the corresponding estimates for $p \in (1, \infty)$ due to Borchers and Sohr [5] to the situation of $L^\infty_\sigma(\Omega)$.

Let us now turn our attention to uniform $L^\infty$-estimates for $\nabla T(t)$ or $\nabla^2 T(t)$ with respect to $t > 0$. Our second main result then reads as follows.

**Theorem 1.3** *Let $T$ be the Stokes semigroup on $L^\infty_\sigma(\Omega)$. Then there exists a constant $M > 0$ such that*

$$\frac{t^{1/2}}{t^{\frac{1}{2}}+1}\|\nabla T(t)\|_{\mathcal{L}(L^\infty_\sigma(\Omega))} + \frac{t}{t+1}\|\nabla^2 T(t)\|_{\mathcal{L}(L^\infty_\sigma(\Omega))} \leq M, \, t > 0. \tag{4}$$

*Remark 1.4* Let us point out that the above estimate (4) is sharp in the following sense: it is impossible to obtain estimates of the form

$$t^\alpha \|\nabla T(t)\|_{\mathcal{L}(L^\infty(\Omega)))} \leq M, \, t > 0 \quad \text{or} \tag{5}$$

$$t^\beta \|\nabla^2 T(t)\|_{\mathcal{L}(L^\infty(\Omega))} \leq M, \, t > 0, \tag{6}$$

for some $\alpha, \beta > 0$. In order to see this, we follow the arguments employed in [14]. Note first that following e.g. [12], the solution $u(t) = T(t)u_0$ of the Stokes problem (1) satisfies

$$\|\nabla u(t)\|_{L^q(\Omega)} \leq Mt^{-\frac{n}{2p}}\|u_0\|_{L^p_\sigma(\Omega)}, \, q \geq p \geq n, \quad t > 0. \tag{7}$$

Moreover, in [12], it is proved that estimate (7) is sharp, in the sense that for all $\varepsilon > 0$ and $M > 0$, there exists a $u_0$ and $t > 1$ such that

$$\|\nabla u(t)\|_{L^q(\Omega)} > Mt^{-\frac{n}{2p}-\varepsilon}\|u_0\|_{L^p(\Omega)}. \tag{8}$$

If (5) would be true for some for some $\alpha > 0$ and all

$$u_0 \in \{\widetilde{u}_0 \in C(\overline{\Omega}) \text{ and } |\widetilde{u}_0(x)| \to 0 \text{ for } |x| \to \infty\} \subset L^\infty_\sigma(\Omega),$$

then, for all $r \in [n, \infty)$, by the semigroup property of $T$

$$\|\nabla u(t)\|_{L^\infty(\Omega)} \leq Ct^{-\alpha}\|u(\tfrac{t}{2})\|_{L^\infty(\Omega)} \leq Ct^{-\alpha-\frac{n}{2r}}\|u_0\|_r,\ t>0,\ \text{for all } u_0 \in L^r_\sigma(\Omega).$$

An application of the Riesz-Thorin theorem yields

$$\|\nabla u(t)\|_{L^q(\Omega)} \leq Mt^{-\alpha\theta-\frac{n}{2p}}, \frac{1}{q} = \frac{1}{p}\frac{r-p}{r-n}, \theta = \frac{r}{p}\frac{p-n}{r-n}, t > 0, \tag{9}$$

for all $p \in [n, r]$. This contradicts the fact that $t^{-\frac{n}{2p}}$ is the sharp decay rate for $t > 1$ for $\|\nabla T(t)\|_{\mathcal{L}(L^p(\Omega), L^q(\Omega))}$. Similarly, also the estimate for $\|\nabla^2 T\|_{\mathcal{L}(L^\infty(\Omega))}$ is sharp. Indeed, it follows e.g. from Lemma 8.1 of [12] that

$$\|\nabla^2 u(t)\|_{L^{\frac{n}{2}}(\Omega)} \leq Mt^{-1}\|u_0\|_{L^{\frac{n}{2}}_\sigma(\Omega)}, \quad t > 0. \tag{10}$$

Again, if (6) would be true for some $\beta > 0$, then, the same arguments give

$$\|\nabla^2 u(t)\|_{L^q(\Omega)} \leq Mt^{-\beta\theta-\frac{n}{2p}}\|u_0\|_{L^p_\sigma(\Omega)},\ \frac{1}{q} = \frac{2}{p}\frac{(r-p)}{2r-n},\ \theta = \frac{r}{p}\frac{2p-n}{2r-n},\ t > 0, \tag{11}$$

for all $p \in [\frac{n}{2}, r]$. By interpolation we would obtain

$$\|\nabla u(t)\|_q \leq \|\nabla^2 u(t)\|_q^{\frac{1}{2}}\|u(t)\|_q^{\frac{1}{2}} \leq Mt^{\frac{n}{4q}-\frac{n}{2p}-\theta\frac{\beta}{2}}\|u_0\|_{L^p_\sigma(\Omega)}, \quad t > 0,$$

which for $p \in (\frac{nr(1+\beta)}{n+2\beta r}, r)$, however, contradicts the optimal decay rate of $\nabla u(t)$ given in (7).

The plan of this paper is as follows. We begin in Sect. 2 by collecting existence, uniqueness and regularity results for the Stokes equation. Subsequentially, in Sect. 3, we present the proofs of our main results, Theorems 1.1 and 1.3. For further information regarding maximum modulus estimates for bounded domains and $L^\infty$ estimates for the Stokes semigroup in the half space $\mathbb{R}^n_+$ we refer to [6] and [8, 16].

## 2 Results on the Stokes Equations: Existence, Uniqueness, Regularity

We start this section with the following result due to Abe and Giga, saying in particular that, for $u_0 \in L^\infty_\sigma(\Omega)$ the solution of the Stokes equation (1) is governed by an analytic semigroup on $L^\infty_\sigma(\Omega)$.

**Theorem 2.1 ([2, Theorem 1.5])** *Let $u_0 \in L^\infty_\sigma(\Omega)$. Then the Stokes equation* (1) *admits a unique solution $(u, \nabla q) \in C^{2,1}(\overline{\Omega}\times(0,T])\times C(\overline{\Omega}\times(0,T]))$ for some $T > 0$ and there exists a constant $C > 0$, independent of $u_0$, such that*

$$\sup_{0\le t\le T} \|N(u,q)\|_\infty(t) \le C\|u_0\|_\infty,$$

*where*

$$N(u,q)(x,t) = |u(t,x)| + t|u_t(t,x)| + t^{1/2}|\nabla u(t,x)| + t|\nabla^2 u(t,x)| + t|\nabla q(t,x)|,$$

*for $x \in \Omega$ and $t \in (0,T]$.*

For the rest of this chapter, we assume that the initial data $u_0$ is smooth in the following sense. More precisely, set

$$\mathcal{C}(\Omega) := \{u \in C(\Omega) \cap L^\infty_\sigma(\Omega) : u = 0 \text{ on } \partial\Omega\},$$

and

$$\mathcal{C}(\overline{\Omega}) := \mathcal{C}(\Omega) \cap C(\overline{\Omega}).$$

We then assume in the following that $u_0 \in \mathcal{C}(\overline{\Omega}) \cap C^1(\overline{\Omega})$. In order to prove Theorem 1.1 we decompose $(u, q)$ as

$$u = z + v, \qquad q_u = \pi_z + \pi_v,$$

where $z$ is the solution of the following Stokes problem on $\mathbb{R}^n$

$$\begin{cases} \partial_t z - \Delta z = 0 & \text{in } J \times \mathbb{R}^n, \\ \operatorname{div} z = 0 & \text{in } J \times \mathbb{R}^n, \\ z(0) = \tilde{u_0} & \text{in } \mathbb{R}^n, \end{cases} \tag{12}$$

and where $\tilde{u_0}$ is defined as the trivial extension by zero of $u_0$ to $\mathbb{R}^n$. Moreover, $(v, \pi_v)$ is defined as the solution of the problem

$$\begin{cases} \partial_t v - \Delta v + \nabla\pi_v = 0 & \text{in } J \times \Omega, \\ \operatorname{div} v = 0 & \text{in } J \times \Omega, \\ v = -\gamma z & \text{on } J \times \partial\Omega, \\ v(0) = 0 & \text{in } \Omega, \end{cases} \tag{13}$$

where $\gamma z$ denotes the trace of $z$, where $z$ is the solution of (12). We look for a solution $(v, \pi_v)$ of the form

$$v = w + F, \qquad \pi_v = \pi_w + P,$$

where the pair $(w, \pi_w)$ solves the equation

$$\begin{cases} \partial_t w - \Delta w + \nabla \pi_w = -F_t + G & \text{in } J \times \Omega\,, \\ \operatorname{div} w = 0 & \text{in } J \times \Omega\,, \\ w = 0 & \text{on } J \times \partial\Omega\,, \\ w(0) = 0 & \text{in } \Omega\,, \end{cases} \tag{14}$$

and where $F$ is a suitable extension of $\gamma z$ to $J \times \Omega$ defined precisely as well as $G$ in the following lemma.

**Lemma 2.2** *Let $z$ be the unique solution of* (12), *$\gamma z(t)$ be its trace on $\partial\Omega$ for all $t \in J$. Let $p \in (1, \infty)$ and $k \in \mathbb{N}$. Then there exists an extension $F \in C(\overline{\Omega}) \cap C^k(\Omega)$ of $\gamma z$ on $(0, T) \times \Omega$ having compact support in $\overline{\Omega}$ satisfying div $F = 0$ for all $t > 0$, as well as the following properties:*

(i) $\|D_t^k F(t)\|_\infty \le Ct^{-k}\|u_0\|_\infty, t > 0, k \ge 0,$
(ii) $\|D_t^k F(t)\|_\infty \le Ct^{-k+\frac{1}{2}}\|\nabla u_0\|_\infty, t > 0, k \ge 0,$
(iii) $\Delta F = \nabla P + G$ *on* $J \times \Omega$ *for some* $G \in C^2(\Omega)$ *having compact support in* $\Omega$,
(iv) $\lim_{t\to 0}\|F(t)\|_\infty = \lim_{t \to 0} t\|F_t(t)\|_\infty = 0,$
(v) $\|D_t^k G(t)\|_\infty \le Ct^{-k}, t > 0, k \ge 0,$
(vi) $t\|\nabla F_t(t)\|_p + t\|D_t^3 F(t)\|_p + \|\nabla G(t)\|_p \le Ct^{-\frac{1}{2}}\|u_0\|_\infty,\ t > 0,$

*for some $C > 0$ independent of $u_0$.*

For a proof of the above properties we refer to [14, Corollary 1 and 2]. We now consider regularity properties of the inhomogeneous Stokes equation (14). To this end, for $\mu > 0$, $q \in [1, \infty]$ and $(t, x, L, u_0) \in \mathbb{R}^+ \times \mathbb{R}^n \times \mathbb{R}^+ \times L^\infty(\mathbb{R}^n)$ we set

$$M_q(t, x, L, u_0) := \|u_0\|_{L^q(B(x,L))} t^{-\frac{n}{2q}} + \|u_0\|_\infty t^{\mu/2}(L + t^{\frac{1}{2}})^{-\mu}.$$

Then, by Maremonti [14, Theorem 5.4], the following result concerning the inhomogenous equation (14) holds true.

**Proposition 2.3** *Let $F$ and $G$ as above, $s \in (1, 2)$ and $p \in (1, \infty)$. Given $T > 0$, there exist a unique $w \in C([0, T]; L^p_\sigma(\Omega))$ and a unique $\nabla \pi_w \in L^s(0, T; L^p(\Omega))$ with $w' \in L^s(0, T; L^p(\Omega))$ solving Eq.* (14). *Moreover, for $p \in (\frac{n}{n-2}, \infty)$ and $q \in (\frac{n}{2}, \infty]$ there exists a constant $C > 0$ such that*

$$\frac{t}{t+1}\|w'(t)\|_p + \sum_{|\alpha|=0}^{2} \left(\frac{t}{t+1}\right)^{\frac{|\alpha|}{2}} \|D^\alpha w(t)\|_p + \frac{t}{t+1}\|\nabla \pi_w(t)\|_p \le C \sup_{\xi \in \partial\Omega} M_q(t, \xi, L, u_0), t > 0, L > 0.$$

*Moreover, $tw', w \in C([0, T); L^p(\Omega))$ and* $\lim_{t\to 0} t\|w'(t)\|_p = \lim_{t \to 0}\|w(t)\|_p = 0.$

We remark that Theorem 5.4. of [14] does not state explicitly the $L^p$-continuity of $tw'$ and the related limit property; however both follow directly from the step 2 in the proof of Theorem 5.4 of [14]. Next, for $T > 0$ we set $\widetilde{J} = (\frac{t}{2}, \infty)$ for $t \geq T$. Consider then the initial boundary value problem

$$\begin{cases} \partial_t \widetilde{w} - \Delta \widetilde{w} + \nabla \pi_{\widetilde{w}} = F_{tt} + G_t & \text{in } \widetilde{J} \times \Omega, \\ \operatorname{div} \widetilde{w} = 0 & \text{in } \widetilde{J} \times \Omega, \\ \widetilde{w} = 0 & \text{on } \widetilde{J} \times \partial\Omega, \\ \widetilde{w}(\frac{t}{2}) = 0 & \text{in } \Omega, \end{cases} \tag{15}$$

where $F$ and $G$ are the same as in (14). Then the following holds.

**Lemma 2.4** *Assume that $F$ and $G$ are given as in Lemma 2.2 and let $r, p \in (1, \infty)$. Then, given $T > 0$, there exist a unique $\widetilde{w} \in C(\widetilde{J}; L^p_\sigma(\Omega))$ and a unique $\nabla \pi_{\widetilde{w}} \in L^r(\widetilde{J}; L^p(\Omega))$ with $\widetilde{w}' \in L^r(\widetilde{J}; L^p(\Omega))$ solving Eq. (15). Moreover, for $p \in (\frac{n}{n-2}, \infty)$ there exists $C > 0$ such that*

$$\frac{s}{s+1} \|\widetilde{w}'(s)\|_p + \sum_{|\alpha|=0}^{2} \left(\frac{s}{s+1}\right)^{\frac{|\alpha|}{2}} \|D^\alpha \widetilde{w}(s)\|_p + \frac{s}{s+1} \|\nabla \pi_{\widetilde{w}}(s)\|_p$$
$$\leq C s^{-1} \|u_0\|_\infty, \quad s > t/2.$$

We only sketch a proof following the lines of [14]. In fact, taking into account Lemma 2.2 (i) and (v) and following the arguments given in the third step of the proof of Theorem 5.4 in [14], we obtain the estimate

$$\|\widetilde{w}(s)\|_p + s\|\widetilde{w}_s(s)\|_p \leq C s^{-1} \|u_0\|_\infty, \ s > \frac{t}{2}, \tag{16}$$

for some $C$ independent of $u_0$. The $L^p$-theory of the steady Stokes problem yields

$$\|D^2 \widetilde{w}(s)\|_p + \|\nabla \pi_{\widetilde{w}}\|_p \leq C(\|\widetilde{w}_s(s)\|_p + \|\widetilde{w}(s)\|_p), \quad s > \frac{t}{2}. \tag{17}$$

and as a consequence of estimate (16)

$$\|D^2 \widetilde{w}(s)\|_p + \|\nabla \pi_{\widetilde{w}}\|_p \leq C(s^{-2} + s^{-1}) \|u_0\|_\infty, \quad s > \frac{t}{2}. \tag{18}$$

Combining the Gagliardo-Nirenberg inequality with (16)–(18) we obtain

$$\|\nabla \widetilde{w}(s)\|_p \leq C \|D^2 \widetilde{w}(s)\|_p^{\frac{1}{2}} \|\widetilde{w}(s)\|_p^{\frac{1}{2}} \leq C(s^{-\frac{3}{2}} + s^{-1}) \|u_0\|_\infty, \ s > \frac{t}{2}, \tag{19}$$

which finally proves the lemma.

Next, we state a result which was proved in [14], Theorem 6.1. To this end, for $\varrho > d := \operatorname{diam} \mathbb{R}^n \setminus \Omega$ set $\Omega_\varrho := \Omega \cap B(0, \varrho)$.

**Proposition 2.5** *Let $u_0 \in \mathcal{C}(\overline{\Omega}) \cap C^1(\overline{\Omega})$ and let $(u, \pi_u)$ be the solution to equation (1). Then, for $p \in (\frac{n}{n-2}, \infty), q \in (\frac{n}{2}, \infty]$ and $\varrho > d$ there exists a constant $C > 0$ such that*

$$\frac{t}{t+1}\|u_t(t)\|_{L^p(\Omega_\varrho)} + \sum_{|\alpha|=0}^{2}(\frac{t}{t+1})^{\frac{|\alpha|}{2}}\|D^\alpha u(t)\|_{L^p(\Omega_\varrho)} \leq C \sup_{y\in B(0,\varrho)} M_q(t,y,L,u_0), \quad t>0, L>0. \tag{20}$$

*Moreover,* $\lim_{t\to 0}\|u(t)-u_0\|_\infty = 0$ *and for $\eta \in (0, \frac{1}{2})$ there exists a constant $C > 0$ such that*

$$(\frac{t}{t+1})^{\frac{1}{2}+\eta}|\pi_u(t,x)| \leq C \sup_{y\in B(0,\varrho))} M_q(t,y,L,u_0)(1+|x|)^{2-n}, (t,x) \in (0,T)\times\overline{\Omega}, L>0. \tag{21}$$

*The constant $C$ is independent of $t, L$ and $u_0$. In addition,*

$$(u, \pi_u) \in C^{2,\lambda}(\overline{\Omega}) \times C^{1,\lambda}(\overline{\Omega}) \text{ and } u_t, \nabla^2 u \in C^{0,\frac{\lambda}{2}}((0,T)\times\overline{\Omega}).$$

We finally recall from Maremonti [13] Theorem 3.2 a results concerning the Stokes initial boundary value problem

$$\begin{cases} \partial_t u - \Delta u + \nabla\pi_u = 0 & \text{in } J\times\Omega, \\ \operatorname{div} u = 0 & \text{in } J\times\Omega, \\ u = 0 & \text{on } J\times\partial\Omega, \\ (u(0),\varphi(0)) = (\psi_0,\varphi) & \varphi\in\mathcal{C}_0(\Omega) \end{cases} \tag{22}$$

for $\psi_0 \in C_0^1(\Omega)$. Denote by $J^{1,p}(\Omega)$ the completion of $\mathcal{C}_0(\Omega)$ with respect to the $W^{1,p}$-norm.

**Proposition 2.6** *Let $\psi_0 \in C_0^1(\Omega)$, $\eta > 0$ and $q \in (1,\infty)$. Then Eq. (22) admits a unique solution $(\psi, \pi_\psi)$ satisfying*

$$\psi \in C([0,T); L^q_\sigma(\Omega)) \cap L^q(\eta,T; W^{2,q}(\Omega)\cap J^{1,q}(\Omega)) \text{ and }$$
$$\nabla\pi_\psi, \psi_t \in L^q(\eta,T;L^q(\Omega)).$$

*Moreover, for $q \in (1,\infty]$ there exists a constant $C > 0$, independent of $u_0$, such that*

$$\|\psi(t)\|_q \leq C\|\psi_0\|_1 t^{-\mu_1}, t>0, \quad \mu_1 = \frac{n}{2}(1-\frac{1}{q}),$$
$$\|\psi_t(t)\|_q \leq C\|\psi_0\|_1 t^{-\mu_2}, t>0, \quad \mu_2 = 1+\mu_1$$

*In addition,* $\lim_{t\to 0}(\psi(t),\varphi) = (\psi_0,\varphi)$ *for any $\varphi \in \mathcal{C}_0(\Omega)$.*

For the duality arguments in the following chapter we also use the following lemma which can be found in [14].

**Lemma 2.7** *Let be* $v \in C([0,T); L^p_\sigma(\Omega))$, $\psi \in C([0,T); L^{p'}_\sigma(\Omega))$ *and* $\psi_0 \in C_0(\Omega)$. *Assume that* $\lim_{t\to 0}(\psi(t), \varphi) = (\psi_0, \varphi)$ *for all* $\varphi \in C(\Omega)$. *Then,*

$$\lim_{\varepsilon\to 0}(v(t-\varepsilon), \psi(\varepsilon)) = (v(t), \psi_0), \quad t \in (0,T).$$

## 3 Proof of the Main Results

We subdivide the proof of Theorem 1.1 into three steps.

*Step 1: The case of smooth initial data, i.e.* $u_0 \in C(\overline{\Omega}) \cap C^1(\overline{\Omega})$.
Taking into account Theorem 2.1 we only have to prove the estimate (2) for arbitrary $t > T$. To this end, note first that the solution $z$ of the Stokes problem on $\mathbb{R}^n$ satisfies

$$\|z'(t)\|_\infty \le \frac{C}{t}\|u_0\|_\infty, \quad t > 0. \tag{23}$$

for some $C > 0$ since $z(t) = e^{t\Delta}\tilde{u}_0$. Secondly, Lemma 2.2 implies that

$$\|F'(t)\|_\infty \le \frac{C}{t}\|u_0\|_\infty, \quad t > 0. \tag{24}$$

It thus remains to prove

$$\|w'(t)\|_\infty \le Ct^{-1}\|u_0\|_\infty \text{ for all } t > T.$$

To this end, we multiply the first equation of (14) by $t$ and differentiate then with respect to $t$. We then obtain the equation

$$\begin{cases} (tw)_{tt} - \Delta(tw)_t + (t\nabla\pi_w)_t = -(tF_t - tG)_t + w_t & \text{in } J \times \Omega, \\ \operatorname{div} w = 0 & \text{in } J \times \Omega, \\ w = 0 & \text{on } J \times \partial\Omega, \\ w(0) = 0 & \text{in } \Omega. \end{cases} \tag{25}$$

We continue by considering the solution $(\psi, \pi_\psi)$ to problem (22) with initial data $\psi_0 \in C^1_0(\Omega)$. Its existence is ensured by Proposition 2.6. We then multiply the first equation of (25) by $\psi(t-\tau)$ for $\tau \in (0,t)$ and $t > T$ and integrate by parts on $(0,t) \times \Omega$. By virtue of Lemma 2.7 and taking into account that

$\lim_{s\to 0}\|w(s)\|_p = \lim_{s\to 0} s\|w'(s)\|_p = 0$, the following integral relation holds:

$$(tw_t(t),\psi_0) = -\int_0^t (\frac{\partial}{\partial\tau}(\tau F_\tau(\tau)),\psi(t-\tau))d\tau + \int_0^t (\frac{\partial}{\partial\tau}(\tau G(\tau)),\psi(t-\tau))d\tau$$
$$+\int_0^t (w_\tau(\tau),\psi(t-\tau))d\tau - (w(t),\psi_0).$$

Lemma 2.2 (ii) implies that $\lim_{t\to 0} t(F_t(t),\psi(t-\tau)) = 0$ and Proposition 2.3 yields $\lim_{t\to 0}\|w(t)\|_p = 0$. Thus, integrating by parts yields

$$|(tw_t(t),\psi_0)| \le \int_0^{t-T/2} |\tau(F_\tau(\tau),\psi_\tau(t-\tau))|d\tau + \int_0^{t-T/2} |(w(\tau),\psi_\tau(t-\tau))|d\tau$$
$$+|(t-\frac{T}{2})(F_t((t-T/2),\psi(T/2))| + |(w(t-T/2),\psi(T/2))| + |(w(t),\psi_0)|$$
$$+\int_0^t |(\frac{\partial}{\partial\tau}(\tau G(\tau)),\psi(t-\tau))|d\tau + \int_{t-T/2}^t |(\frac{\partial}{\partial\tau}(\tau F_\tau(\tau)),\psi(t-\tau))|d\tau$$
$$+\int_{t-T/2}^t |(w_\tau(\tau),\psi(t-\tau))|d\tau =: \sum_{i=1}^8 I_i(t).$$

Let us start with estimating the term $I_1+I_2$. By Lemma 2.2 (i) and Proposition 2.3

$$\|\tau F_\tau(\tau)\|_\infty + \|w(\tau)\|_p \le C\|u_0\|_\infty, \quad t>0.$$

Further, since $F$ has compact support, Hölder's inequality with $r\in(1,\frac{n}{2})$ and $r'$ its conjugate exponent as well as Proposition 2.6 yield

$$I_1(t)+I_2(t) \le C\|u_0\|_\infty \int_0^{t-T/2} \|\psi_\tau(t-\tau)\|_{r'}d\tau \le C\|u_0\|_\infty\|\psi_0\|_1, \quad t>T,$$

for some $C\equiv C(T)>0$. Next, estimating the terms $I_3$ and $I_4$ similarly as above, we obtain

$$I_3(t)+I_4(t) \le C\|u_0\|_\infty\|\psi(T/2)\|_{r'} \le C\|u_0\|_\infty T^{-\frac{n}{2r}}\|\psi_0\|_1 \quad t>T,$$

for some $C\equiv C(T)>0$. Concerning $I_5$, note that Proposition 2.3 together with Sobolev's embedding ensures that $\|w(t)\|_\infty \le C(t^{-1}+1)\|\psi_0\|_\infty$. Hence,

$$I_5(t) \le \|w(t)\|_\infty\|\psi_0\|_1 \le C(t^{-1}+1)\|u_0\|_\infty\|\psi_0\|_1, \quad t>0.$$

Next, splitting the integral at $T/2$ we obtain by Hölder's inequality

$$I_6(t) = \int_0^{t-T/2} |(\frac{\partial}{\partial\tau}(\tau G(\tau)), \psi(t-\tau))| d\tau + \int_{t-T/2}^{t} |(\frac{\partial}{\partial\tau}(\tau G(\tau)), \psi(t-\tau))| d\tau$$

$$\leq \int_0^{t-T/2} \|\frac{\partial}{\partial\tau}(\tau G(\tau))\|_1 \|\psi(t-\tau)\|_\infty d\tau +$$

$$\int_{t-T/2}^{t} \|\frac{\partial}{\partial\tau}(\tau G(\tau))\|_p \|\psi(t-\tau)\|_{p'} d\tau.$$

Choosing $p > \frac{n}{2}$ and employing Lemma 2.2 (v) we obtain

$$I_6(t) \leq C\|u_0\|_\infty \|\psi_0\|_1, \quad t > T,$$

for some $C \equiv C(T) > 0$. Finally, we turn our attention to $I_7$ and $I_8$. Since $F$ has compact support, choosing $p > n/2$ and using Hölder's inequality as well as Proposition 2.3 we obtain

$$I_7(t) + I_8(t) \leq C \int_{t-T/2}^{t} \|\frac{\partial}{\partial\tau}(\tau F_\tau(\tau))\|_p \|\psi(t-\tau)\|_{p'} d\tau +$$

$$\int_{t-T/2}^{t} \|w_\tau(\tau)\|_p \|\psi(t-\tau)\|_{p'} d\tau$$

$$\leq C(T^{-1}+1)T^{1-\frac{n}{2p}} \|u_0\|_\infty \|\psi_0\|_1, \quad t > T.$$

Summing up, we deduce that given $T > 0$ there exists a constant $C(T) > 0$ such that

$$t\|w_t(t)\|_\infty \leq C(T)\|u_0\|_\infty, \quad t > T, \tag{26}$$

with $C(T)$ independent of $u_0$. Finally, combining the estimates (23), (24) with estimate (26), we see that

$$\|u_t\|_\infty = \|z_t + F_t + w_t\|_\infty \leq \frac{C(T)}{t}\|u_0\|_\infty, \quad t > T. \tag{27}$$

*Step 2: The case of $u_0 \in L^\infty_\sigma(\Omega)$*
We start the second step with an approximation result due to Abe and Giga, see [2], Lemma 5.1.

**Lemma 3.1** *Let $\Omega$ be an exterior domain with Lipschitz boundary. Then, for $u \in L^\infty_\sigma(\Omega)$, there exists a sequence $(u_m) \subset C^1(\overline{\Omega}) \cap C(\overline{\Omega})$ and a constant $C > 0$, independent of $u$, such that*

$$\|u_m\|_\infty \leq C\|u\|_\infty, \quad m \in \mathbb{N},$$

*and* $u_m \to u$ *a.e. in* $\Omega$. *In particular,*

$$\|u_m - u\|_{L^p(\Omega_\varrho)} = 0$$

*for* $p \in [1, \infty)$ *and* $\rho > 0$.

Assume now that $u_0 \in L^\infty_\sigma(\Omega)$ and let $(u_0^m) \subset \mathcal{C}(\overline{\Omega}) \cap C^1(\overline{\Omega})$ be a sequence converging to $u_0$ a.e. in $\Omega$ according to Lemma 3.1. By Proposition 2.5 there exists a sequence of solutions $(u^m, \pi_{u^m})$ to Eq. (1) enjoying the estimates (27) and (21). The estimate (20) implies further that for all $q \in (\frac{n}{2}, \infty)$ and all $\varrho > d =$ diam $\mathbb{R}^n \backslash \Omega$ ,

$$\frac{t}{t+1}\|u_t^k(t) - u_t^m(t)\|_{L^p(\Omega_\varrho)} + \sum_{|\alpha|=0}^{2} (\frac{t}{t+1})^{\frac{|\alpha|}{2}} \|D^\alpha u^k(t) - D^\alpha u^m(t)\|_{L^p(\Omega_\varrho)}$$
$$\leq C \sup_{B(0,\varrho)} M_q(t, y, L, u_0^k - u_0^m), \ t > 0, L > 0.$$

Moreover, by (21), for each $\eta \in (0, \frac{1}{2})$

$$|\pi^k(t,x) - \pi^m(t,x)| \leq C(t^{-\frac{1}{2}-\eta} + 1) \sup_{\widetilde{B(O,\varrho)}} M_q(t, y, L, u_0^k - u_0^m)(1 + |x|)^{2-n},$$

with $C$ independent of $k, m \in \mathbb{N}$ and $L$. Following [14] we then see that

(i) for all $[\eta, T_0] \subset (0, T)$ and all compact $\overline{\Omega_\varrho} \subset \overline{\Omega}$, $(u^m)$ is a Cauchy sequence in $C([\eta, T_0] \times \overline{\Omega_\varrho})$,
(ii) for all $t > 0$ and all $\rho > 0$, $(u^m)$ is a Cauchy sequence in $W^{2,p}(\Omega_\varrho)$ and $(u_t^m)$ is a Cauchy sequence in $L^p(\Omega_\varrho)$,
(iii) $(\pi^m)$ is a Cauchy sequence in $C((0, T) \times \Omega)$.

Taking the limit with respect to the above family of seminorms yields that the limit $(u, \pi_u)$ satisfies Eq. (1). Moreover, $u_t$ is a Hölder continuous function in space and in time. Thus, the estimate (2) follows from the estimate (27) proved for smooth data. Indeed, since the sequence $(u_t^m)$ converges in $L^p(\Omega_\varrho)$ for all $t > 0$, we find a subsequence $(u_t^{m_k})$ converging to $u_t(t, x)$ for a.e. $(t, x) \in \{t\} \times \Omega$. Hence, by estimate (27)

$$|u_t(t,x)| \leq |u_t(t,x) - u_t^{m_k}(t,x)| + |u_t^{m_k}(t,x)|$$
$$\leq |u_t(t,x) - u_t^{m_k}(t,x)| + Ct^{-1}\|u_0\|_\infty.$$

Thus, we deduce (2) by letting $m_k \to \infty$ for such $(t, x)$. Finally, since $u_t$ is Hölder continuous in $x$, we may extend the latter estimate pointwise.

*Step 3: Bounded analyticity of T*
For $u_0 \in L^\infty_\sigma(\Omega)$ let $T(t)u_0 := u(t)$, where $(u, \pi)$ is the solution of the Stokes equation (1). Then $T$ is the semigroup on $L^\infty_\sigma(\Omega)$ and we denote its generator by

$A$. Taking into account Steps 1 and 2, it follows from Corollary 3.7.12 of [4] that $\{z \in \mathbb{C} : Re z > 0\} \subset \varrho(A)$ and that

$$\sup_{Re\lambda>0} \|\lambda R(\lambda, A)\|_{\mathcal{L}(L^\infty_\sigma(\Omega))} < \infty.$$

Thus, there exists $\delta > 0$ and $M > 0$ such that

$$\|\lambda R(\lambda, A)\| \leq M, \quad \lambda \in \Sigma_{\pi/2+\delta}. \tag{28}$$

Thus $A$ generates a bounded analytic semigroup on $L^\infty_\sigma(\Omega)$. The proof of Theorem 1.1 is complete.
In order to prove Theorem 1.3 consider the time derivative of $(w, \pi_w)$, which by (14) then is a solution of

$$\begin{cases} \partial_{ss} w - \Delta w_s + \nabla \pi_{w_s} = F_{ss} + G_s & \text{in } \widetilde{J} \times \Omega, \\ \operatorname{div} w_s = 0 & \text{in } \widetilde{J} \times \Omega, \\ w_s = 0 & \text{on } \widetilde{J} \times \partial\Omega, \\ w_s(\frac{t}{2}) = w_t(\frac{t}{2}) & \text{on } \{\frac{t}{2}\} \times \Omega. \end{cases} \tag{29}$$

We are looking for a solution of this problem in the form

$$w_s(s) = \widetilde{w}(s) + \widehat{w}(s), \quad \pi_w(s) = \pi_{\widetilde{w}}(s) + \pi_{\widehat{w}}(s) \quad s > \frac{t}{2}, \tag{30}$$

where the pair $(\widetilde{w}, \pi_{\widetilde{w}})$ is the solution of problem (15) and $(\widehat{w}, \pi_{\widehat{w}})$ is the solution of the following problem

$$\begin{cases} \partial_s \widehat{w} - \Delta \widehat{w} + \nabla \pi_{\widehat{w}} = 0 & \text{in } \widetilde{J} \times \Omega, \\ \operatorname{div} \widehat{w} = 0 & \text{in } \widetilde{J} \times \Omega, \\ \widehat{w} = 0 & \text{on } \widetilde{J} \times \partial\Omega, \\ \widehat{w}(\frac{t}{2}) = w_t(\frac{t}{2}) & \text{on } \{t\} \times \Omega. \end{cases} \tag{31}$$

The $L^p$-theory for the Stokes problem (see e.g. [5, 10, 12, 15]) combined with estimate (26) for the initial data $u_t(\frac{t}{2})$ yields for $p > n$

$$(s - \frac{t}{2})^{\gamma_1} \|\nabla \widehat{w}(s)\|_p \leq C \|w_t(\frac{t}{2})\|_p \leq C t^{-1} \|u_0\|_\infty, \quad s > \frac{t}{2}, \tag{32}$$

for some $C > 0$ and $\gamma_1 = \begin{cases} \frac{1}{2} & \text{if } s \in (\frac{t}{2}, \frac{t}{2} + 1), \\ \frac{n}{2p} & \text{if } s \geq \frac{t}{2} + 1. \end{cases}$ By (30), (19) and (32) we obtain

$$\|\nabla w_s(s)\|_p \leq c(s - \frac{t}{2})^{-\gamma_1} t^{-1} \|u_0\|_\infty, \; s > \frac{t}{2}. \tag{33}$$

Considering problem (14) as a stationary Stokes boundary value problem with data $F_t + G - w_t$, we obtain via well known estimates in exterior domains for the Stokes problem (see [9])

$$\|D^3 w(s)\|_p + \|D^2 \pi_w\|_p \le C\left(\|\nabla w_s(s)\|_p + \|\nabla F_s(s)\|_p + \|\nabla G(s)\|_p + \|w(s)\|_p\right)$$

provided $s > \frac{t}{2}$. Now, combining Lemma 2.2 (vi) with Lemma 2.4 for $\|w(s)\|_p$ and with (33) for $\|\nabla w(s)\|_p$, respectively, yields

$$\|D^3 w(s)\|_p \le C[(s - \frac{t}{2})^{-\gamma_1} t^{-1} + s^{-\frac{3}{2}} + s^{-1} + 1]\|u_0\|_\infty, \; s > \frac{t}{2}. \tag{34}$$

By virtue of the Gagliardo-Nirenberg inequality for non-homogeneous boundary values, see [7], and assuming $p > n$, we get

$$\|D^2 w(s)\|_\infty \le C\|D^3 w(s)\|_p^a \|w(s)\|_p^{1-a}, \; a = \frac{n + 2p}{3p}.$$

Hence, employing estimates (34) and Lemma 2.4 for $\|w(s)\|_p$ and also assuming $s = t$ we see that

$$\|D^2 w(t)\|_\infty \le C(t^{-\delta} + 1)\|u_0\|_\infty, \; t \ge T$$

for some $\delta > 0$. Since the solution of the Stokes problem (1) is given by $u = z + F + w$ and by virtue of estimate Lemma 2.2 (vi), we deduce that

$$\|D^2 u(t)\|_\infty \le C(t^{-\delta} + 1)\|u_0\|_\infty, \; t \ge T.$$

Taking into account Theorem 2.1 by Abe and Giga, this completes the proof of Theorem 1.3 for the second derivatives. We finally note that the case of first derivatives follows by means of the interpolation.

## References

1. K. Abe, Y. Giga, Analyticity of the Stokes semigroup in spaces of bounded functions. Acta Math. **211**, 1–46 (2013)
2. K. Abe, Y. Giga, The $L^\infty$-Stokes semigroup in exterior domains. J. Evol. Equ. **14**, 1–28 (2014)
3. K. Abe, Y. Giga, M. Hieber, Stokes resolvent estimates in spaces of bounded functions. Ann. Sci. Éc. Norm. Supér. **48**, 537–559 (2015)
4. W. Arendt, C. Batty, M. Hieber, F. Neubrander, in *Vector-Valued Laplace Transforms and Cauchy Problems*. Birkhauser Monographs in Mathematics, vol. 96, 2nd edn. (Birkhäuser, Basel, 2012)
5. W. Borchers, H. Sohr, On the semigroup of the Stokes operator for exterior domains in $L^q$-spaces. Math. Z. **196**, 415–425 (1987)

6. T. Chang, H. Choe, Maximum modulus estimates for the solution of the Stokes equation. J. Differ. Equ. **254**, 2682–2704 (2013)
7. F. Crispo, P. Maremonti, An interpolation inequality in exterior domains. Rend. Semin. Mat. Univ. Padova **112**, 11–39 (2004)
8. W. Desch, M. Hieber, J. Prüss, $L^p$-theory of the Stokes equation in a half space. J. Evol. Equ. **1**, 115–142 (2001)
9. G.P. Galdi, *An Introduction to the Mathematical Theory of the Navier-Stokes Equations*, 2nd edn. (Springer, New York, 2011)
10. Y. Giga, H. Sohr, Abstract $L^p$-estimates for the Cauchy problem with applications to the Navier-Stokes equations in exterior domains. J. Funct. Anal. **102**, 72–94 (1991)
11. T. Kato, Strong $L^p$-solutions of the Navier-Stokes equations in $\mathbb{R}^m$, with applications to weak solutions. Math. Z. **187**, 471–480 (1984)
12. P. Maremonti, V.A. Solonnikov, On nonstationary Stokes problem in exterior domains. Ann. Sc. Norm. Sup. Pisa **24**, 395–449 (1997)
13. P. Maremonti, A remark on the Stokes problem with initial data in $L^1$. J. Math. Fluid Mech.**13**, 469–480 (2011)
14. P. Maremonti, On the Stokes problem in exterior domains: the maximum modulus theorem. AIMS J. A **34**, 2135–2171 (2014)
15. Y. Shibata, R. Shimada, On a generalized resolvent estimate for the Stokes system with Robin boundary conditions. J. Math. Soc. Jpn. **59**, 469–519 (2007)
16. V.A. Solonnikov, On nonstationary Stokes problem and Navier-Stokes problem in a half-space with initial data nondecreasing at infinity. J. Math. Sci. **114**, 1726–1740 (2003)

# On the Weak Solution of the Fluid-Structure Interaction Problem for Shear-Dependent Fluids

**Anna Hundertmark, Mária Lukáčová-Medviďová, and Šárka Nečasová**

**Abstract** In this paper the coupled fluid-structure interaction problem for incompressible non-Newtonian shear-dependent fluid flow in two-dimensional time-dependent domain is studied. One part of the domain boundary consists of an elastic wall. Its temporal evolution is governed by the generalized string equation with action of the fluid forces by means of the Neumann type boundary condition. The aim of this work is to present the limiting process for the auxiliary $(\kappa, \varepsilon, k)$-problem. The weak solution of this auxiliary problem has been studied in our recent work (Hundertmark-Zaušková, Lukáčová-Medviďová, Nečasová, On the existence of weak solution to the coupled fluid-structure interaction problem for non-Newtonian shear-dependent fluid, J. Math. Soc. Japan (in press)).

**Keywords** Existence of weak solution • Fluid-structure interaction • Hemodynamics • Non-Newtonian fluids • Shear-thickening fluids • Shear-thinning fluids

## 1 Problem Definition

The problem of a fluid interaction with a moving or deformable structure is important in many applications like biomechanics, hydroelasticity, aeroelasticity, sedimentation, modeling of blood flow, etc. We consider a two-dimensional fluid motion governed by the momentum and the continuity equation

$$\rho\partial_t \boldsymbol{v} + \rho\,(\boldsymbol{v}\cdot\nabla)\,\boldsymbol{v} - \mathrm{div}\boldsymbol{\tau} + \nabla\pi = 0, \quad \mathrm{div}\,\boldsymbol{v} = 0, \tag{1}$$

with $\rho$ denoting the constant density of fluid, $\boldsymbol{v} = (v_1, v_2)$ the velocity vector, $\pi$ the pressure and $\boldsymbol{\tau}$ the shear stress tensor.

A. Hundertmark • M. Lukáčová-Medviďová
Institute for Mathematics, Johannes Gutenberg University, Staudingerweg 9, Mainz, Germany
e-mail: hundertm@uni-mainz.de; lukacova@mathematik.uni-mainz.de

Š. Nečasová (✉)
Academy of Sciences of Czech Republic, Žitná 25, Praha, Czech Republic
e-mail: matus@math.cas.cz

 
H. Amann et al. (eds.), *Recent Developments of Mathematical Fluid Mechanics*, Advances in Mathematical Fluid Mechanics, DOI 10.1007/978-3-0348-0939-9_16

Let us first specify the shear-dependent fluids that will be considered in this paper. We assume that

$$\boldsymbol{\tau} = \boldsymbol{\tau}(e(\boldsymbol{v})) = 2\mu(|e(\boldsymbol{v})|)e(\boldsymbol{v}), \quad \text{where } e(\boldsymbol{v}) = \frac{1}{2}(\nabla \boldsymbol{v} + \nabla \boldsymbol{v}^T)$$

is the symmetric deformation tensor. Moreover we assume that there exists a potential $\mathcal{U} \in C^2(\mathbb{R}^{2\times 2})$ of the stress tensor $\boldsymbol{\tau}$, such that for some $1 < p < \infty$, $C_1, C_2 > 0$ we have for all $\eta, \xi \in \mathbb{R}^{2\times 2}_{sym}$ and $i, j, k, l \in \{1, 2\}$, cf. [11]

$$\frac{\partial \mathcal{U}(\eta)}{\partial \eta_{ij}} = \boldsymbol{\tau}_{ij}(\eta), \qquad \mathcal{U}(\mathbf{0}) = \frac{\partial \mathcal{U}(\mathbf{0})}{\partial \eta_{ij}} = 0, \tag{2}$$

$$\frac{\partial^2 \mathcal{U}(\eta)}{\partial \eta_{mn} \partial \eta_{rs}} \xi_{mn} \xi_{rs} \geq C_1 \, (1 + |\eta|)^{p-2} |\xi|^2, \tag{3}$$

$$\left| \frac{\partial^2 \mathcal{U}(\eta)}{\partial \eta_{ij} \partial \eta_{kl}} \right| \leq C_2 (1 + |\eta|)^{p-2}. \tag{4}$$

One particular example satisfying the above properties is a stress tensor, which contains shear-dependent viscosity obeying the power-law model, cf. [8, 11, 12, 16]

$$\mu(|e(\boldsymbol{v})|) = \mu(1 + |e(\boldsymbol{v})|^2)^{\frac{p-2}{2}} \quad p > 1. \tag{5}$$

For $p < 2$ the viscosity is a decreasing function of the shear rate, i.e., shear-thinning. For $p > 2$ we have shear-thickening property and this model is an analogy of the so-called Ladyzhenskaja's fluid; for $p = 3$ it yields the Smagorinskij model of turbulence. In numerical simulations presented in our recent papers [8, 10] the shear-thinning model of Carreau has been used in order to model blood flow in compliant vessels. For the simplicity of presentation we will consider here only the case of shear-thickening fluids, i.e. $p \geq 2$. The generalization for shear-thinning fluids may be done in an analogous way as here, using an appropriate techniques for shear-thinning fluids, see results of Diening, Růžička and Wolf [5, 15].
The two-dimensional deformable computational domain

$$\Omega(\eta(t)) \equiv \{(x_1, x_2);\ 0 < x_1 < L,\ 0 < x_2 < R_0(x_1) + \eta(x_1, t)\},\ 0 < t < T$$

is given by a reference radius function $R_0(x_1)$ and the unknown free boundary function $\eta(x_1, t)$ describing the domain deformation. The fluid and the geometry of the computational domain are coupled through the following Dirichlet boundary condition on the deformable part of the boundary $\Gamma_w(t)$

$$\boldsymbol{v}(x_1, R_0(x_1) + \eta(x_1, t), t) = \left(0, \frac{\partial \eta(x_1, t)}{\partial t}\right), \tag{6}$$

where $\Gamma_w(t) = \{(x_1, x_2);\ x_2 = R_0(x_1) + \eta(x_1, t),\ x_1 \in (0, L)\}$. The normal component of the fluid stress tensor $\mathbf{T}_f \boldsymbol{n}$ and the outside pressure $P_w$ provide the forcing terms for the boundary displacement $\eta$, that is modeled by the generalized string equation:

$$\frac{\partial^2 \eta}{\partial t^2} - a\frac{\partial^2 \eta}{\partial x_1^2} + b\eta + c\frac{\partial^5 \eta}{\partial t \partial x_1^4} - a\frac{\partial^2 R_0}{\partial x_1^2} = \frac{-g}{\tilde{E}\rho}\left[\tilde{\mathbf{T}}_f + \tilde{P_w}\mathbf{I}\right]\tilde{\boldsymbol{n}} \cdot \boldsymbol{e}_2 \text{ on } \Gamma_w^0. \tag{7}$$

Here $\Gamma_w^0 := \Gamma_w(t)|_{t=0}$, $[(\tilde{\mathbf{T}}_f + \tilde{P_w}\mathbf{I})\tilde{\boldsymbol{n}}](\tilde{x}) = [(\mathbf{T}_f + P_w\mathbf{I})\boldsymbol{n}](x)$, $x \in \Gamma_w(t)$, $\tilde{x} \in \Gamma_w^0$ and $\mathbf{T}_f = \boldsymbol{\tau} - \pi\mathbf{I}$. Moreover, $\boldsymbol{n}, \tilde{\boldsymbol{n}}$ denote the unit outward normals on $\Gamma_w(t)$, $\Gamma_w^0$, respectively and $\boldsymbol{n}|\boldsymbol{n}| = (-\partial_{x_1}(R_0 + \eta), 1)^T$. The coefficient $g = \frac{(R_0+\eta)\sqrt{1+(\partial_{x_1}(R_0+\eta))^2}}{R_0\sqrt{1+(\partial_{x_1}R_0)^2}}$ arises from the transformation from the Eulerian frame of the fluid forces into the Lagrangian formulation of the string.

Equation (7) is equipped with the following boundary and initial conditions

$$\eta(0,t) = \eta_{x_1}(0,t) = \eta(L,t) = \eta_{x_1}(L,t) = \eta(x_1,0) = \frac{\partial \eta}{\partial t}(x_1,0) = 0. \tag{8}$$

Positive coefficients $\tilde{E}$, $a$, $b$, $c$ appearing in (7) are given as follows [8],

$$\tilde{E} = \rho_w \hbar, \qquad a = \frac{|\sigma_z|}{\left(1 + \left(\frac{\partial R_0}{\partial x_1}\right)^2\right)^2}, \quad b = \frac{\mathcal{E}}{(R_0 + \eta)R_0}, \qquad c > 0,$$

where $\mathcal{E}$ is the Young modulus, $\hbar$ the wall thickness, $\rho_w$ the density of the vessel wall tissue, the coefficient $c = \gamma/(\rho_w \hbar)$, $\gamma$ positive constant. $|\sigma_z| = G\kappa$ is the longitudinal stress, $\kappa = 1$ is the Timoshenko's shear correction factor and $G$ is the shear modulus, equal to $G = \mathcal{E}/2(1+\sigma)$ with $\sigma = 1/2$ for incompressible materials. Note that the coefficients $a$, $b$ are non-constant, however, according to the assumption (16) below they are upper- and down-bounded. In what follows, we linearize the term $b = \frac{\mathcal{E}}{(R_0+\eta)R_0}$ by $\frac{\mathcal{E}}{\rho_w R_0^2}$ and for the sake of simplicity we work with constant coefficients $a$, $b$, $c$.

Equation (7) can be transformed as follows.

$$E\rho\left[\frac{\partial^2 \eta}{\partial t^2} - a\frac{\partial^2 \eta}{\partial x_1^2} + b\eta + c\frac{\partial^5 \eta}{\partial t \partial x_1^4} - a\frac{\partial^2 R_0}{\partial x_1^2}\right](x_1, t) =$$
$$\left[-\mathbf{T}_f \boldsymbol{n}|\boldsymbol{n}| \cdot \boldsymbol{e}_2 - P_w\right](x_1, R_0(x_1) + \eta(x_1,t), t), \tag{9}$$

$x_1 \in (0, L)$. Here $E = \tilde{E}\sqrt{1 + (\partial_{x_1} R_0)^2}$. We assume that $E$ is bounded.

We complete the system (1) with the following boundary and initial conditions: on the inflow part of the boundary, which we denote $\Gamma_{in}$, we set

$$v_2(0,x_2,t) = 0, \ \left(2\mu(|e(\boldsymbol{v})|)\frac{\partial v_1}{\partial x_1} - \pi + P_{in} - \frac{\rho}{2}|v_1|^2\right)(0,x_2,t) = 0 \quad (10)$$

for any $0 < x_2 < R_0(0)$, $0 < t < T$ and for a given function $P_{in} = P_{in}(x_2,t)$. On the opposite, outflow part of the boundary $\Gamma_{out}$, we set

$$v_2(L,x_2,t) = 0, \ \left(2\mu(|e(\boldsymbol{v})|)\frac{\partial v_1}{\partial x_1} - \pi + P_{out} - \frac{\rho}{2}|v_1|^2\right)(L,x_2,t) = 0 \quad (11)$$

for any $0 < x_2 < R_0(L)$, $0 < t < T$ and for a given function $P_{out} = P_{out}(x_2,t)$. Note that we require that the so-called kinematic pressure is prescribed on the inflow and outflow boundary. This implies that the fluxes of kinetic energy on inflow and outflow boundary will disappear in the weak formulation. Finally, on the remaining part of the boundary, $\Gamma_c$, we set the flow symmetry condition

$$v_2(x_1,0,t) = 0\,, \qquad \mu(|e(\boldsymbol{v})|)\frac{\partial v_1}{\partial x_2}(x_1,0,t) = 0 \quad (12)$$

for any $0 < x_1 < L$, $0 < t < T$. The initial conditions read

$$\boldsymbol{v}(x_1,x_2,0) = \boldsymbol{0} \quad \text{for any } 0 < x_1 < L,\ 0 < x_2 < R_0(x_1). \quad (13)$$

The problem defined in (1)–(13) is a generalization of the problem for Newtonian fluid previously studied by Filo and Hundertmark in [6, 17]. Here the original generalized string model of Quarteroni [13], Quarteroni and Formaggia [14] with a regularization term $\eta_{txx}$ has been used. The iterative process with respect to the domain deformation, cf. item 3 below and Sect. 4, has been be completed only for the $(\kappa,\varepsilon)$-approximation of the original problem and the convergence with respect to domain deformation was an open problem. In the present paper, similarly as in [4], we use a modified model for the structure equation having a viscoelastic term $\eta_{txxxx}$. For this model we show global existence in time of weak solution of unsteady, fully coupled fluid-structure problem. The existence result holds until a contact of the elastic boundary with a fixed boundary part. The question of existence of weak solution of fully coupled fluid-structure interaction problem with the original Quarteroni's generalized string model for generalized Newtonian fluids is still an open problem.

The main result of this paper is formulated in Theorem 1.2. For the existence proof a suitable approximation of the problem (1)–(13), see Sect. 2, is constructed.

1. *$\varepsilon$-approximation* (22): the space of solenoidal functions on a moving domain is approximated by the artificial compressibility approach,

2. *κ-approximation* (20), (21): the boundary conditions (6)–(7) has been splitted and the deformable boundary becomes semi-pervious for $\kappa < \infty$,
3. *h-approximation*: the domain deformation is assume to be given by a sufficiently smooth function $\delta(x_1, t)$; the weak formulation on a deformable domain $\Omega(\delta(t)) =: \Omega(h(t))$ is transformed to a reference domain $D = (0, L) \times (0, 1)$ using the known radius $h := R_0 + \delta$, see (23).

Letting $\varepsilon \to 0$, $\kappa \to \infty$ and finally the fixed point procedure for the domain deformation complete the proof. In [9] the above fluid-structure interaction problem has been studied and the existence of weak solution for fixed parameters $\kappa, \varepsilon$ and given deformation $\delta$, such that $h = R_0 + \delta$, i.e., to the $(\kappa, \varepsilon, h)$-approximation of the problem (1)–(13) has been proven in details. In this work we only present the limiting processes for $\varepsilon \to 0$, $\kappa \to \infty$ and the fixed point procedure with respect to the domain deformation regarding the geometric nonlinearity of our problem.

## 1.1 Weak Formulation

In this section our aim is to present the weak formulation of the problem (1)–(13). Assuming that $\eta$ is enough regular (see below) and taking into account the results from [4] we can define the functional spaces that gives sense to the trace of velocity from $W^{1,p}(\Omega(\eta(t)))$ and thus to define the weak solution of the problem. We assume that $R_0 \in C_0^2(0, L)$.

**Definition 1.1 (Weak Formulation)** We say that $(\boldsymbol{v}, \eta)$ is a weak solution of (1)–(13) on $[0, T)$ if the following conditions hold

- $\boldsymbol{v} \in L^p(0, T; W^{1,p}(\Omega(\eta(t)))) \cap L^\infty(0, T; L^2(\Omega(\eta(t))))$,
- $\eta \in W^{1,\infty}(0, T; L^2(0, L)) \cap H^1(0, T; H_0^2(0, L))$,
- $\operatorname{div} \boldsymbol{v} = 0$ a.e. on $\Omega(\eta(t))$,
- $\boldsymbol{v}\big|_{\Gamma_w(t)} = (0, \eta_t)$ for a.e. $x \in \Gamma_w(t)$, $t \in (0, T)$, $v_2\big|_{\Gamma_{in} \cup \Gamma_{out} \cup \Gamma_c} = 0$,

$$
\begin{aligned}
&\int_0^T \int_{\Omega(\eta(t))} \Big\{ -\rho \boldsymbol{v} \cdot \frac{\partial \boldsymbol{\varphi}}{\partial t} + 2\mu(|e(\boldsymbol{v})|) e(\boldsymbol{v}) e(\boldsymbol{\varphi}) + \rho \sum_{i,j=1}^{2} v_i \frac{\partial v_j}{\partial x_i} \boldsymbol{\varphi}_j \Big\} \, dx \, dt \\
&+ \int_0^T \int_0^{R_0(L)} \Big( P_{out} - \frac{\rho}{2} |v_1|^2 \Big) \varphi_1(L, x_2, t) \, dx_2 \, dt \\
&- \int_0^T \int_0^{R_0(0)} \Big( P_{in} - \frac{\rho}{2} |v_1|^2 \Big) \varphi_1(0, x_2, t) \, dx_2 \, dt \\
&+ \int_0^T \int_0^L P_w \varphi_2(x_1, R_0(x_1) + \eta(x_1, t), t) - a \frac{\partial^2 R_0}{\partial x_1^2} \xi \, dx_1 \, dt \\
&+ \int_0^T \int_0^L -\frac{\partial \eta}{\partial t} \frac{\partial \xi}{\partial t} + c \frac{\partial^3 \eta}{\partial x_1^2 \partial t} \frac{\partial^2 \xi}{\partial x_1^2} + a \frac{\partial \eta}{\partial x_1} \frac{\partial \xi}{\partial x_1} + b \eta \, \xi \, dx_1 \, dt = 0
\end{aligned}
\tag{14}
$$

for every test functions

$$
\begin{aligned}
&\boldsymbol{\varphi}(x_1, x_2, t) \in H^1(0, T; W^{1,p}(\Omega(\eta(t)))) \text{ such that} \\
&\operatorname{div} \boldsymbol{\varphi} = 0 \text{ a.e on } \Omega(\eta(t)), \\
&\varphi_2\big|_{\Gamma_w(t)} \in H^1(0, T; H_0^2(\Gamma_w(t))), \ \ \varphi_2\big|_{\Gamma_{in}\cup\Gamma_{out}\cup\Gamma_c} = \varphi_1\big|_{\Gamma_w(t)} = 0 \quad \text{and} \\
&\xi(x_1, t) = E\rho\, \varphi_2(x_1, R_0(x_1) + \eta(x_1, t), t).
\end{aligned} \tag{15}
$$

**Theorem 1.2 (Main Result: Existence of a Weak Solution)** *Let $p \geq 2$. Assume that the boundary data fulfill $P_{in} \in L^{p'}(0, T; L^2(0, R_0(0)))$, $P_{out} \in L^{p'}(0, T; L^2(0, R_0(L)))$, $P_w \in L^{p'}(0, T; L^2(0, L))$, $\frac{1}{p} + \frac{1}{p'} = 1$. Furthermore, assume that the properties (2)–(4) for the viscous stress tensor hold. Then for $T \leq T^*$, $T^*$ depending on the data $R_0$, $P_{in}$, $P_{out}$, $P_w$, $K$, $\alpha$, cf. (16), and $\alpha \leq \min\{R_{min}, \frac{1}{R_{min}+R_{max}}\}$, $R_{min} \leq R_0 \leq R_{max}$ there exists a weak solution $(\boldsymbol{v}, \eta)$ of the problem (1)–(13) such that*

*(i)* $\boldsymbol{v} \in L^p(0, T; W^{1,p}(\Omega(\eta(t)))) \cap L^\infty(0, T; L^2(\Omega(\eta(t))))$, $\eta \in W^{1,\infty}(0, T; L^2(0, L)) \cap H^1(0, T; H_0^2(0, L))$,

*(ii)* $\boldsymbol{v}\big|_{\Gamma_w(t)} = (0, \eta_t)$ *for a.e.* $x \in \Gamma_w(t)$, $t \in (0, T)$, $v_2\big|_{\Gamma_{in}\cup\Gamma_{out}\cup\Gamma_c} = 0$,

*(iii)* $\boldsymbol{v}$ *satisfies the condition* $\operatorname{div} \boldsymbol{v} = 0$ *a.e on* $\Omega(\eta(t))$ *and (14) holds.*

*Remark 1.3* Let us point out that $T^*$ denotes a time when the elastic boundary reaches the bottom boundary. If $T^* = \infty$ we have an existence of the global weak solution, otherwise the existence of the weak solution holds until the elastic boundary reaches the bottom boundary, see Sect. 4 for further details.

## 2 Auxiliary Problem: $(\kappa, \varepsilon, h)$-Approximation

In what follows we will formulate a suitable approximation of the original problem (1)–(13).

First of all we approximate the deformable boundary $\Gamma_w$ by a given function $h = R_0 + \delta$, $\delta \in H^1(0, T; H_0^2(0, L)) \cap W^{1,\infty}(0, T; L^2(0, L))$, $R_0(x_1) \in C^2[0, L]$ satisfying for all $x_1 \in [0, L]$

$$
0 < \alpha \leq h(x_1, t) \leq \alpha^{-1}, \quad \left|\frac{\partial h(x_1, t)}{\partial x_1}\right| + \int_0^T \left|\frac{\partial h(x_1, t)}{\partial t}\right|^2 dt \leq K < \infty \tag{16}
$$

$h(0, t) = R_0(0)$, $h(L, t) = R_0(L)$.

We look for a solution $(\boldsymbol{v}, \pi, \eta)$ of the following problem

$$
\rho \frac{\partial \boldsymbol{v}}{\partial t} + \rho(\boldsymbol{v} \cdot \nabla)\boldsymbol{v} = \operatorname{div} \boldsymbol{\tau} - \nabla \pi \quad \text{in} \quad \Omega(h(t)), \tag{17}
$$

and for all $x_1 \in (0, L)$, see (9), $0 < t < T$

$$- E\rho \left[ \frac{\partial^2 \eta}{\partial t^2} - a\frac{\partial^2 \eta}{\partial x_1^2} + b\eta + c\frac{\partial^5 \eta}{\partial t \partial x_1^4} - a\frac{\partial^2 R_0}{\partial x_1^2} \right] (x_1, t) =$$

$$\left[ \mu(|e(\boldsymbol{v})|) \left\{ -\left( \frac{\partial v_2}{\partial x_1} + \frac{\partial v_1}{\partial x_2} \right) \frac{\partial h}{\partial x_1} + 2\frac{\partial v_2}{\partial x_2} \right\} - \pi + P_w \right] (\bar{x}, t), \tag{18}$$

$$\boldsymbol{v}(\bar{x}, t) = \left( 0, \frac{\partial \eta}{\partial t}(x_1, t) \right), \tag{19}$$

$\bar{x} = (x_1, h(x_1, t))$.

Furthermore, in the analysis of problem (1)–(13) the boundary condition (6)–(7), cf. (18)–(19), is splitted in the following way, see [6]

$$\left[ \mu(|e(\boldsymbol{v})|) \left\{ -\left( \frac{\partial v_2}{\partial x_1} + \frac{\partial v_1}{\partial x_2} \right) \frac{\partial h}{\partial x_1} + 2\frac{\partial v_2}{\partial x_2} \right\} - \pi + P_w \right] (\bar{x}, t)$$
$$-\frac{\rho}{2} v_2 \Big( v_2(\bar{x}, t) - \frac{\partial h}{\partial t}(x_1, t) \Big) = \rho\kappa \Big[ \frac{\partial \eta}{\partial t}(x_1, t) - v_2(\bar{x}, t) \Big] \tag{20}$$

and

$$-E\Big[ \frac{\partial^2 \eta}{\partial t^2} - a\frac{\partial^2 \eta}{\partial x_1^2} + b\eta + c\frac{\partial^5 \eta}{\partial t \partial x_1^4} - a\frac{\partial^2 R_0}{\partial x_1^2} \Big](x_1, t) = \kappa \Big[ \frac{\partial \eta}{\partial t}(x_1, t) - v_2(\bar{x}, t) \Big]$$
$$\text{with } \kappa \gg 1. \tag{21}$$

We will show later, that the approximation with $\kappa$ is reasonable. One of the possible physical interpretations for introducing finite $\kappa$ comes from the mathematical modeling of semi-pervious boundary, where this type of boundary condition occurs. In our case, the boundary $\Gamma_w$ seems to be partly permeable for finite $\kappa$, but letting $\kappa \to \infty$ it becomes impervious. In fact, we prove the existence of solution if $\kappa \to \infty$ and thus we get the original boundary condition (18)–(19).

Furthermore, we overcome the difficulties with solenoidal spaces by means of the artificial compressibility. We approximate the continuity equation similarly as in [6] with

$$\varepsilon \left( \frac{\partial \pi_\varepsilon}{\partial t} - \Delta \pi_\varepsilon \right) + \operatorname{div} \boldsymbol{v}_\varepsilon = 0 \quad \text{in } \Omega(h(t)),\ t \in (0, T)$$

$$\frac{\partial \pi_\varepsilon}{\partial \boldsymbol{n}} = 0, \text{ on } \partial\Omega(h(t)),\ t \in (0, T), \quad \pi_\varepsilon(0) = 0 \text{ in } \Omega(h(0)) \quad \varepsilon > 0. \tag{22}$$

By letting $\varepsilon \to 0$ we show that $\boldsymbol{v}_\varepsilon \to \boldsymbol{v}$, where $\boldsymbol{v}$ is the weak solution of (1). For fixed $\varepsilon$, due to the lack of solenoidal property for velocity, we have the additional term in momentum equation $\frac{\rho}{2} v_i \operatorname{div} \boldsymbol{v}$, which we include into the convective term, see (27).

Our approximated problem is defined on a moving domain depending on function $h = R_0 + \delta$. Now we will reformulate it to a fixed rectangular domain. Set

$$\begin{aligned} \boldsymbol{u}(y_1, y_2, t) &\stackrel{\text{def}}{=} \boldsymbol{v}(y_1, h(y_1, t)y_2, t) \\ q(y_1, y_2, t) &\stackrel{\text{def}}{=} \rho^{-1}\pi(y_1, h(y_1, t)y_2, t) \\ \sigma(y_1, t) &\stackrel{\text{def}}{=} \frac{\partial \eta}{\partial t}(y_1, t) \end{aligned} \tag{23}$$

for $y \in D = \{(y_1, y_2);\ 0 < y_1 < L,\ 0 < y_2 < 1\}, 0 < t < T$.
We define the following space

$$\begin{aligned} \boldsymbol{V} &\equiv \left\{\boldsymbol{w} \in W^{1,p}(D) :\ w_1 = 0 \ \text{ on } S_w,\ w_2 = 0 \ \text{ on } S_{in} \cup S_{out} \cup S_c\right\}, \\ S_w &= \{(y_1, 1) : 0 < y_1 < L\}, \qquad S_{in} = \{(0, y_2) : 0 < y_2 < 1\}, \\ S_{out} &= \{(L, y_2) : 0 < y_2 < 1\}, \qquad S_c = \{(y_1, 0) : 0 < y_1 < L\}. \end{aligned} \tag{24}$$

Let us introduce the following notations

$$\text{div}_h \boldsymbol{u} \stackrel{\text{def}}{=} \frac{\partial u_1}{\partial y_1} - \frac{y_2}{h}\frac{\partial h}{\partial y_1}\frac{\partial u_1}{\partial y_2} + \frac{1}{h}\frac{\partial u_2}{\partial y_2},$$

$$\begin{aligned} a_1(q, \phi) = \int_D \Bigg\{ &\left[h\left(\frac{\partial q}{\partial y_1} - \frac{y_2}{h}\frac{\partial h}{\partial y_1}\frac{\partial q}{\partial y_2}\right)\right]\frac{\partial \phi}{\partial y_1} \\ &+ \left[\frac{1}{h}\frac{\partial q}{\partial y_2} - y_2\frac{\partial h}{\partial y_1}\left(\frac{\partial q}{\partial y_1} - \frac{y_2}{h}\frac{\partial h}{\partial y_1}\frac{\partial q}{\partial y_2}\right)\right]\frac{\partial \phi}{\partial y_2}\Bigg\}\, dy, \end{aligned} \tag{25}$$

viscous term

$$((\boldsymbol{u}, \boldsymbol{\psi})) = \int_D h\boldsymbol{\tau}_{ij}(\hat{e}(\boldsymbol{u}))\hat{e}_{ij}(\boldsymbol{\psi})dy, \tag{26}$$

$$\boldsymbol{\tau}_{ij}(\hat{e}(\boldsymbol{u})) = 2\rho^{-1}\mu(|\hat{e}(\boldsymbol{u})|)\hat{e}_{ij}(\boldsymbol{u}), \qquad \hat{e}_{ij}(\boldsymbol{u}) = \frac{1}{2}(\hat{\partial}_i(u_j) + \hat{\partial}_j(u_i)),$$

$$\hat{\partial}_1 = \left(\frac{\partial}{\partial y_1} - \frac{y_2}{h}\frac{\partial h}{\partial y_1}\frac{\partial}{\partial y_2}\right), \qquad \hat{\partial}_2 = \frac{1}{h}\frac{\partial}{\partial y_2},$$

convective term

$$\begin{aligned} b(\boldsymbol{u}, z, \boldsymbol{\psi}) = &\int_D \left(hu_1\left(\frac{\partial z}{\partial y_1} - \frac{y_2}{h}\frac{\partial h}{\partial y_1}\frac{\partial z}{\partial y_2}\right) + u_2\frac{\partial z}{\partial y_2}\right) \cdot \boldsymbol{\psi} + \frac{h}{2} z \cdot \boldsymbol{\psi}\ \text{div}_h \boldsymbol{u}\, dy \\ &-\frac{1}{2}\int_0^1 R_0 u_1 z_1 \psi_1\, (L, y_2)\, dy_2 + \frac{1}{2}\int_0^1 R_0 u_1 z_1 \psi_1\, (0, y_2)\, dy_2 \\ &-\frac{1}{2}\int_0^L u_2 z_2 \psi_2\, (y_1, 1)\, dy_1. \end{aligned} \tag{27}$$

*Remark 2.1* Note, that the transformed stress tensor $\boldsymbol{\tau}_{ij} = 2\rho^{-1}\mu(|\hat{e}(\boldsymbol{u})|)\hat{e}_{ij}(\boldsymbol{u})$ from (26) with $\mu(|\hat{e}(\boldsymbol{u})|)$ defined in (5) also satisfies (2)–(4).

*Remark 2.2* Since the terms defined in (25)–(27) are dependent on the domain deformation $h$, it will be sometimes useful to denote this explicitly, e.g., $b(\boldsymbol{u}, z, \boldsymbol{\psi}) = b_h(\boldsymbol{u}, z, \boldsymbol{\psi})$, $\hat{e}(\boldsymbol{u}) = \hat{e}_h(\boldsymbol{u})$.

**Definition 2.3 (Weak Solution of $(\kappa, \varepsilon, h)$-Approximate Problem)** Let $\boldsymbol{u} \in L^p(0,T;\boldsymbol{V}) \cap L^\infty(0,T;L^2(D))$, $q \in L^2(0,T;H^1(D)) \cap L^\infty(0,T;L^2(D))$ and $\sigma \in L^\infty(0,T;L^2(0,L)) \cap L^2(0,T;H_0^2(0,L))$. A triple $\boldsymbol{w} = (\boldsymbol{u}, q, \sigma)$ is called a weak solution of the regularized problem (1)–(13) if the following equation holds

$$
\begin{aligned}
&-\int_0^T \left\langle \frac{\partial (h\boldsymbol{u})}{\partial t}, \boldsymbol{\psi} \right\rangle dt = \\
&\int_0^T \int_D \left( -\frac{\partial h}{\partial t}\frac{\partial (y_2 \boldsymbol{u})}{\partial y_2} \cdot \boldsymbol{\psi} + b(\boldsymbol{u}, \boldsymbol{u}, \boldsymbol{\psi}) - h\, q \operatorname{div}_h \boldsymbol{\psi} \right) dy + ((\boldsymbol{u}, \boldsymbol{\psi}))\, dt \\
&+ \int_0^T \int_0^1 h(L,t) q_{out} \psi_1\,(L, y_2, t) - h(0,t) q_{in} \psi_1\,(0, y_2, t)\, dy_2 dt \\
&+ \int_0^T \int_0^L \left( q_w + \frac{1}{2} u_2 \frac{\partial h}{\partial t} + \kappa\,(u_2 - \sigma) \right) \psi_2\,(y_1, 1, t)\, dy_1 dt \\
&+\varepsilon \int_0^T \left\langle \frac{\partial (hq)}{\partial t}, \phi \right\rangle dt \qquad (28) \\
&+ \int_0^T \int_D \left( -\varepsilon \frac{\partial h}{\partial t}\frac{\partial (y_2 q)}{\partial y_2} \phi + \varepsilon a_1(q, \phi) + h \operatorname{div}_h \boldsymbol{u}\, \phi \right) dy\, dt \\
&+\frac{\varepsilon}{2} \int_0^T \int_0^L \frac{\partial h}{\partial t}(y_1, t) q \phi(y_1, 1, t)\, dy_1 dt + \\
&+ \int_0^T \int_0^L \left( \frac{\partial \sigma}{\partial t}\xi + c \frac{\partial^2 \sigma}{\partial y_1^2}\frac{\partial^2 \xi}{\partial y_1^2} + a \frac{\partial}{\partial y_1} \int_0^t \sigma(y_1, s) ds \frac{\partial \xi}{\partial y_1} \right. \\
&\qquad \left. + b \int_0^t \sigma(y_1, s) ds\, \xi - a \frac{\partial^2 R_0}{\partial y_1^2}\xi + \frac{\kappa}{E}\,(\sigma - u_2)\,\xi \right) (y_1, t)\, dy_1 dt
\end{aligned}
$$

for every $(\boldsymbol{\psi}, \phi, \xi) \in H_0^1(0,T;\boldsymbol{V}) \times L^2(0,T;H^1(D)) \times L^2(0,T;H_0^2(0,L))$.

Here we remind $E = \tilde{E}\sqrt{1 + (\partial_{y_1} R_0)^2}$. For simplicity and without lost of generality we assume in what follows that $E$, $a$, $b$, $c$ are constant, cf. (16).

## 2.1 Existence of $(\kappa, \varepsilon, h)$-Approximate Weak Solution

For the proof of weak solution to the auxiliary problem defined in Definition 2.3 following properties of viscous and convective forms are useful[1]; for their proofs see [9, Sect. 3.1].

**Lemma 2.4 (Coercivity of the Viscous Form)** *The viscous form defined in (26) satisfies for any* $2 \leq p < \infty$ *the following estimates: there exists* $\tilde{\delta} = \tilde{\delta}(K, \alpha) > 0$ *such that*

$$1)\ ((\boldsymbol{u}, \boldsymbol{u})) \geq \tilde{\delta} \|\boldsymbol{u}\|_{1,p}^{p} + \tilde{\delta} \|\boldsymbol{u}\|_{1,2}^{2},$$

$$2)\ ((\boldsymbol{u}^1, \boldsymbol{u}^1 - \boldsymbol{u}^2)) - ((\boldsymbol{u}^2, \boldsymbol{u}^1 - \boldsymbol{u}^2)),$$

$$\geq \tilde{\delta} \int_D |\hat{e}(\boldsymbol{u}^1) - \hat{e}(\boldsymbol{u}^2)|^2 + |\hat{e}(\boldsymbol{u}^1) - \hat{e}(\boldsymbol{u}^2)|^p,$$

$$3)\ ((\boldsymbol{u}^1, \boldsymbol{u}^1 - \boldsymbol{u}^2)) - ((\boldsymbol{u}^2, \boldsymbol{u}^1 - \boldsymbol{u}^2)) \geq 0.$$

**Lemma 2.5 (Boundedness of the Viscous Form)** *Let* $\boldsymbol{u}$, $\boldsymbol{v} \in V$, *then for* $2 \leq p < \infty$ *it holds*

$$((\boldsymbol{u}, \boldsymbol{v})) \leq C \|\boldsymbol{u}\|_{1,p}^{p-1} \|\boldsymbol{v}\|_{1,p} + C_0 \|\boldsymbol{u}\|_{1,p} \|\boldsymbol{v}\|_{1,p}, \quad C_0 > 0. \tag{29}$$

**Lemma 2.6 (Nonlinear Convective Term $b(\boldsymbol{u}, \boldsymbol{z}, \boldsymbol{\psi})$)** *For the trilinear form* $b(\boldsymbol{u}, \boldsymbol{z}, \boldsymbol{\psi})$, *defined in (27) the following properties hold*

$$b(\boldsymbol{u}, \boldsymbol{z}, \boldsymbol{\psi}) = \frac{1}{2} B(\boldsymbol{u}, \boldsymbol{z}, \boldsymbol{\psi}) - \frac{1}{2} B(\boldsymbol{u}, \boldsymbol{\psi}, \boldsymbol{z}), \tag{30}$$

$$\text{where} \quad B(\boldsymbol{u}, \boldsymbol{z}, \boldsymbol{\psi}) \equiv \int_D \left( h u_1 \left( \frac{\partial \boldsymbol{z}}{\partial y_1} - \frac{y_2}{h} \frac{\partial h}{\partial y_1} \frac{\partial \boldsymbol{z}}{\partial y_2} \right) + u_2 \frac{\partial \boldsymbol{z}}{\partial y_2} \right) \cdot \boldsymbol{\psi}\, dy.$$

*Moreover for* $p \geq 2$ *we have*

$$|B(\boldsymbol{u}, \boldsymbol{z}, \boldsymbol{\psi})| \leq c \|\boldsymbol{u}\|_{1,p} \|\boldsymbol{z}\|_{1,p} \|\boldsymbol{\psi}\|_{1,p}.$$

In our recent work [9] the following result has been proved.

**Theorem 2.7 (Existence of $(\kappa, \varepsilon, h)$-Approximate Weak Solution)** *Let* $\varepsilon$, $\kappa$, *be fixed. Assume (2)–(4), a given function* $h$, *such that (16) holds,* $q_{in}, q_{out} \in L^{p'}(0, T; L^2(0, 1))$, $q_w \in L^{p'}(0, T; L^2(0, L))$. *Then there exists a weak solution of*

[1] We use here notations $\|\cdot\|_p := \|\cdot\|_{L^p(D)}$, $\|\cdot\|_{1,p} := \|\cdot\|_{W^{1,p}(D)}$.

*the* $(\kappa, \varepsilon, h)$*-approximated problem transformed to the fixed domain, in the sense of integral identity (28). Moreover,*

$$\frac{\partial(hu)}{\partial t} \in \begin{cases} L^{p'}(0,T;V^*) \; for\, 2 < p < \infty, \\ L^{p'}(0,T;V^*) \oplus L^{4/3}((0,T)\times D), \\ \qquad\qquad\qquad\qquad for\, p = 2, \end{cases} \qquad \frac{\partial(hq)}{\partial t} \in L^2(0,T;H^{-1}(D)),$$

*such that*

$$\int_0^T \left\langle \frac{\partial(hu)}{\partial t}, \psi \right\rangle dt = -\int_0^T \int_D hu \cdot \frac{\partial \psi}{\partial t} dy\, dt.$$

*Proof* See [9, Sects. 3 and 4]. □

## 3 Problem with $\varepsilon = 0$, $\kappa = \infty$

The weak solution from Theorem 2.7 depends on the parameters $\varepsilon$, $\kappa$ and $h$. Passing to the limit with $\varepsilon \to 0$, $\kappa \to \infty$ we will obtain the weak solution of the original problem (1)–(13) defined on $\Omega(h)$. By this procedure we will prove the existence of the weak solution for the domain deformation $h$. We realize the limiting process by passing to the limit in both parameters at once, taking $\kappa = \varepsilon^{-1}$ and letting $\kappa \to \infty$.

In what follows we point out the dependence of weak solution on parameters $\varepsilon$, $\kappa$: $\boldsymbol{u}_\kappa$, $q_\kappa$, $\sigma_\kappa$. In this section, we omit the notation of the dependence on $h$.

In analogy to the estimate [9, Sect. 4.1, Estimate (4.7)] we obtain the following a priori estimate by testing (28) with $(\boldsymbol{u}_\kappa, q_\kappa, \sigma_\kappa)$, using the lemmas from Sect. 2.1 and after analogous manipulations as in [9]. Note, that the right hand side is independent on $\varepsilon, \kappa$.

$$\begin{aligned} &\max_{0\le t\le T} \int_D h(t)\left(|\boldsymbol{u}_\kappa|^2 + \varepsilon |q_\kappa|^2\right)(t) dy + \frac{E}{2}\int_0^L |\sigma_\kappa(t)|^2 dy_1 \\ &+ \int_0^T \int_D \tilde{\delta} |\nabla \boldsymbol{u}_\kappa|^p + \frac{2\alpha\varepsilon}{2+K^2} |\nabla q_\kappa|^2 dy + E\int_0^L c\left|\frac{\partial^2 \sigma_\kappa}{\partial y_1^2}\right|^2 dy_1\, dt \\ &+ \int_0^L \frac{aE}{2}\left|\int_0^t \frac{\partial \sigma_\kappa(s)}{\partial y_1} ds\right|^2 + \frac{bE}{2}\left|\int_0^t \sigma_\kappa(s) s\right|^2 dy_1 \\ &+ \int_0^T\int_0^L 2\kappa\, |\sigma_\kappa - u_{2\kappa}|^2\, dy_1\, dt \le \tilde{M} \int_0^T \mathcal{P}^{p'} + c_1 \left\| \frac{\partial^2 R_0}{\partial y_1^2} \right\|^2_{L^2(0,L)} dt\,. \end{aligned} \tag{31}$$

Here $c_1 = c_1(p, E, a, c)$, $\tilde{M} = \tilde{M}(p, K, \alpha)$ and $\mathcal{P} := \|q_{in}\|_{L^2(0,1)} + \|q_{out}\|_{L^2(0,1)} + \|q_w\|_{L^2(0,L)}$.

### 3.1 Limiting Process $\kappa = \varepsilon^{-1} \to \infty$

First of all let us investigate the solenoidal property of the weak solution in the limiting case, i.e., for $\kappa = \infty$. The estimate (31) implies the weak convergence of

$$(\boldsymbol{u}_\kappa, \sqrt{\varepsilon} q_\kappa, \sigma_\kappa) \rightharpoonup (\boldsymbol{u}, \tilde{q}, \sigma) \quad (32)$$
$$\text{in } L^p(0,T;\boldsymbol{V}) \times L^2(0,T;H^1(D)) \times L^2(0,T;H^2(0,L))$$

as $\kappa \to \infty$. Moreover, after inserting test functions $(\boldsymbol{0}, \phi, 0)$ into (28) for sufficiently smooth $\phi$ we obtain

$$\int_0^T \int_D h\phi \operatorname{div} \boldsymbol{u}_\kappa \leq \quad (33)$$
$$\sqrt{\varepsilon} C \|\sqrt{\varepsilon} q_\kappa\|_{L^2(0,T;H^1(D))} (\|\phi\|_{L^2(0,T;H^1(D))} + \|\partial_t \phi\|_{L^2((0,T)\times D)}),$$

Using the boundedness of $\sqrt{\varepsilon} q_\kappa$ in $L^2(0,T;H^1(D))$ and letting $\varepsilon = \kappa^{-1} \to 0$ in (33) we get

$$\operatorname{div}_h \boldsymbol{u} = 0 \qquad a.e. \text{ on } (0,T) \times D.$$

This fact allows us to confine later the space of test functions to the solenoidal space, i.e. $\operatorname{div}_h \boldsymbol{\psi} = 0$ a.e. on $D$.

In the limiting process for $\kappa \to \infty$ we cannot use the Lions-Aubin lemma in order to obtain strong convergences in appropriate spaces for $(\boldsymbol{u}_\kappa, \sigma_\kappa) \to (\boldsymbol{u}, \sigma)$, since the estimates of the time derivatives $\partial_t \boldsymbol{u}_\kappa$, $\partial_t \sigma_\kappa$, depend on $\kappa$, see [9, Sect. 4.1]. In fact, we have to use another argument to obtain the strong convergence. We follow the lines of [6, Sect. 8] and use the equicontinuity in time as in Alt, Luckhaus cf. [1, Lemma 1.9]. It can be shown that

$$\int_0^{T-\tau} \int_D |(h\boldsymbol{u}_\kappa)(t+\tau) - (h\boldsymbol{u}_\kappa)(t)|^2 + \varepsilon |(hq_\kappa)(t+\tau) - (hq_\kappa)(t)|^2 dydt$$
$$+ \int_0^{T-\tau} \int_0^L |(h\sigma_\kappa)(t+\tau) - (h\sigma_\kappa)(t)|^2 dy_1 dt \leq C(K,\alpha)\tau, \quad (34)$$

where $C$ is a positive constant independent on $\tau, \kappa, \varepsilon$. The proof of (34) can be found in [9, Sect. 5.1] and we omit its presentation here.

The estimate (34) and the compactness argument from [1, Lemma 1.9] imply the following strong convergences for $\kappa \to \infty$

$$\boldsymbol{u}_\kappa \to \boldsymbol{u} \text{ in } L^1((0,T) \times D), \qquad \sigma_\kappa \to \sigma \text{ in } L^1((0,T) \times (0,L)).$$

Using the standard interpolations of spaces $L^r(Q_T)$ and $L^s(S_T)$, $Q_T = (0,T) \times D$, $S_T = (0,T) \times (0,L)$ and boundedness of $\boldsymbol{u}, \sigma$ in $L^4(Q_T)$, $L^6(S_T)$, respectively, we obtain

$$\boldsymbol{u}_\kappa \to \boldsymbol{u} \text{ in } L^r((0,T) \times D), \qquad \sigma_\kappa \to \sigma \text{ in } L^s((0,T) \times (0,L)),$$

where $1 \le r < 4$, $1 \le s < 6$ for $\kappa \to \infty$.

Now let us consider test functions $\boldsymbol{\psi} \in L^p(0,T;X)$, $\boldsymbol{\psi}(T) = 0$, $\xi = E\boldsymbol{\psi}_2|_{S_w}$, where

$$X = \{\boldsymbol{\psi} \in V_{div}\ ;\ \boldsymbol{\psi}_2|_{S_w} \in H_0^2(0,L)\}, \tag{35}$$
$$V_{div} := \{f \in \boldsymbol{V},\ \mathrm{div}_h f = 0 \ a.e. \text{ on } D\},\ \text{cf. (24)}$$

in (28). With this choice of test functions the boundary terms with $\kappa$ are canceled.

Now, we can pass to the limit in $\kappa \to \infty$ in (28), where $\kappa = \varepsilon^{-1}$. We use the weak convergences of $\boldsymbol{u}_\kappa$ in $L^p(0,T;V_{div})$, $\sqrt{\varepsilon}q_\kappa$ in $L^2(0,T;H^1(D))$, $\sigma_\kappa$ in $L^2(0,T;H^2(0,L))$, see (32) and strong convergence of $h\boldsymbol{u}_\kappa$ in $L^r((0,T) \times D)$, $0 \le r < 4$. The convergence of the viscous term follows from the monotonicity of the viscous operator and the Minty-Browder theorem, see also [9, Sect. 4.1]. Analogous arguments in order to obtain convergence in the viscous term when $k \to \infty$ will be presented in Sect. 4.

The convergence of the convective term for $\boldsymbol{\psi} \in H^1(0,T;X)$ can be obtained for all $p > 2$ in following way. For case $p = 2$ see [6, Sect. 8]. In order to obtain $\int_0^T b(\boldsymbol{u}_\kappa, \boldsymbol{u}_\kappa, \boldsymbol{\psi}) \to \int_0^T b(\boldsymbol{u}, \boldsymbol{u}, \boldsymbol{\psi})$ one needs to show that $\int_0^T |B(\boldsymbol{u}_\kappa - \boldsymbol{u}, \boldsymbol{u}_\kappa, \boldsymbol{\psi})| \to 0$, $\int_0^T |B(\boldsymbol{u}, \boldsymbol{u} - \boldsymbol{u}_\kappa, \boldsymbol{\psi})| \to 0$. Indeed, using the Hölder inequality and imbedding $L^{\frac{2p}{p-2}}(D) \hookrightarrow W^{1,p}(D)$ we have

$$\int_0^T |B(\boldsymbol{u}_\kappa - \boldsymbol{u}, \boldsymbol{u}_\kappa, \boldsymbol{\psi})| \le C(K,\alpha) \int_0^T \|\boldsymbol{u}_\kappa - \boldsymbol{u}\|_2 \|\boldsymbol{u}_\kappa\|_{1,p} \|\boldsymbol{\psi}\|_{\frac{2p}{p-2}} \tag{36}$$
$$\le C(K,\alpha) \|\boldsymbol{\psi}\|_{H^1(0,T;W^{1,p}(D))} \|\boldsymbol{u}_\kappa - \boldsymbol{u}\|_{L^2((0,T)\times D)} \|\boldsymbol{u}_\kappa\|_{L^p(0,T;W^{1,p}(D))}.$$

Thus $\int_0^T |B(\boldsymbol{u}_\kappa - \boldsymbol{u}, \boldsymbol{u}_\kappa, \boldsymbol{\psi})| \to 0$. Further $\int_0^T |B(\boldsymbol{u}, \boldsymbol{u} - \boldsymbol{u}_\kappa, \boldsymbol{\psi})| \to 0$ due to the weak convergence of $\boldsymbol{u}_\kappa$ in $L^p(0,T;V_{div})$.

The convergence of the terms containing $\sqrt{\varepsilon}q_\kappa$ can be realized by the weak convergence in the corresponding spaces. The term $\int_0^T \int_D hq_\kappa \mathrm{div}_h \boldsymbol{\psi}$ is canceled due to the solenoidal test functions.

Finally, after the limiting process $\kappa \to \infty$ in (28) using above considerations for all $\boldsymbol{\psi} \in H_0^1(0,T;X)$ and $\xi = E\boldsymbol{\psi}_2|_{S_w}$ we arrive at

$$\int_0^T \int_D \left\{ h\boldsymbol{u} \cdot \frac{\partial \boldsymbol{\psi}}{\partial t} + \frac{\partial h}{\partial t} \frac{\partial (y_2 \boldsymbol{u})}{\partial y_2} \cdot \boldsymbol{\psi} \right\} dy = \tag{37}$$
$$\int_0^T \left\{ ((\boldsymbol{u}, \boldsymbol{\psi}))_h + b_h(\boldsymbol{u}, \boldsymbol{u}, \boldsymbol{\psi}) \right.$$

$$+ \int_0^1 h(L)q_{out}(y_2,t)\psi_1(L,y_2,t) - h(0)q_{in}(y_2,t)\psi_1(0,y_2,t)\,dy_2$$
$$+ \int_0^L \left(q_w + \frac{1}{2}\frac{\partial h}{\partial t}u_2\right)\psi_2(y_1,1,t)\,dy_1$$
$$+ \int_0^L -\sigma\frac{\partial\xi}{\partial t} + c\frac{\partial^2\sigma}{\partial y_1^2}\frac{\partial^2\xi}{\partial y_1^2} + a\frac{\partial}{\partial y_1}\int_0^t \sigma(y_1,s)ds\frac{\partial\xi}{\partial y_1}$$
$$- a\frac{\partial^2 R_0}{\partial y_1^2}\xi + b\int_0^t \sigma(y_1,s)ds\,\xi(y_1,t)\,dy_1\Big\}\,dt.$$

In order to investigate the meaning of the left hand side of the above equality we define the ALE-type time derivative $\bar{\partial}_t$

$$\bar{\partial}_t(h\boldsymbol{u}) := \frac{\partial(h\boldsymbol{u})}{\partial t} - \frac{\partial h}{\partial t}\frac{1}{h}\frac{\partial(y_2 h\boldsymbol{u})}{\partial y_2}. \tag{38}$$

Note that $\bar{\partial}_t(h\boldsymbol{u}) = h\partial_t^y\boldsymbol{u}$, where $\partial_t^y := \left(\frac{\partial}{\partial t} - \frac{\partial h}{\partial t}\frac{y_2}{h}\frac{\partial}{\partial y_2}\right)$ denotes in fact the time derivative transformed to the rectangle domain $D$, i.e., in coordinates $(y_1, y_2)$.

The right hand side of (37) is bounded for every $\boldsymbol{\psi} \in \mathcal{M}$,

$$\mathcal{M} = \{\omega \in L^p(0,T;X) \text{ for } p > 2; \\ \omega \in L^p(0,T;X) \cap L^4((0,T)\times D) \text{ for } p = 2\}. \tag{39}$$

Thus it can be identified with some functional $\chi \in \mathcal{M}^*$. Then using integration by parts with respect to $y_2$ on the left hand side, backward transformation from $D$ to the moving domain $\Omega(h(t))$ and the separation of variables it can be shown that $\chi = \bar{\partial}_t(h\boldsymbol{u}) \in L^{p'}(0,T;X^*)$, see [9, Appendix A] for more details. Thus we can replace

$$\int_0^T\int_D\left\{h\boldsymbol{u}\cdot\frac{\partial\boldsymbol{\psi}}{\partial t} + \frac{\partial h}{\partial t}\frac{\partial(y_2\boldsymbol{u})}{\partial y_2}\cdot\boldsymbol{\psi}\right\}dy\,dt = -\int_0^T\langle\bar{\partial}_t(h\boldsymbol{u}),\boldsymbol{\psi}\rangle_X.$$

Finally, we transform (37) from the rectangle $D$ to the moving domain $\Omega(h(t))$ and obtain the existence of a weak solution to our original problem (1)–(13) with the Dirichlet boundary condition $\partial_t\eta = v_2|_{\Gamma_w(h(t))}$ for a prescribed domain deformation $h$.

**Theorem 3.1 (Existence of Weak Solution for $\varepsilon = 0$, $\kappa = \infty$)** *Assume that $h \in H^1(0,T;H_0^2(0,L)) \cap W^{1,\infty}(0,T;L^2(0,L))$ satisfies (16). Let the boundary data fulfill $q_{in}, q_{out} \in L^{p'}(0,T;L^2(0,1))$, $q_w \in L^{p'}(0,T;L^2(0,L))$. Furthermore, assume that* *the properties (2)–(4) for the viscous stress tensor hold. Then there exists a weak solution* $(\boldsymbol{v}, \eta)$ *of the problem (1)–(13), such that*

(i) $(\boldsymbol{u}, \eta) \in [L^p(0,T;\boldsymbol{V}) \times H^1(0,T;H_0^2(0,L))] \cap [L^\infty(0,T;L^2(D)) \times W^{1,\infty}(0,T;L^2(0,L))]$, *where* $\boldsymbol{u}$ *is defined in (23),*

(ii) *the time derivative* $\bar{\partial}_t(h\boldsymbol{u}) \in L^{p'}(0,T;X^*)$ *for* $p > 2$ *and* $\bar{\partial}_t(h\boldsymbol{u}) \in L^{p'}(0,T;X^*) \oplus L^{4/3}((0,T)\times D)$ *for* $p = 2$,

$$\int_0^T \int_D \left\{ h\boldsymbol{u} \cdot \frac{\partial \boldsymbol{\psi}}{\partial t} + \frac{\partial h}{\partial t} \frac{\partial (y_2 \boldsymbol{u})}{\partial y_2} \cdot \boldsymbol{\psi} \right\} dy\, dt = -\int_0^T \left\langle \bar{\partial}_t(h\boldsymbol{u}), \boldsymbol{\psi} \right\rangle dt,$$

*where* $\bar{\partial}_t(h\boldsymbol{u}) = \frac{\partial(h\boldsymbol{u})}{\partial t} - \frac{1}{h}\frac{\partial h}{\partial t}\frac{\partial(y_2 h\boldsymbol{u})}{\partial y_2} = h\partial_t^y \boldsymbol{u}$, *for every test function* $\boldsymbol{\psi} \in \mathcal{M} \cap H_0^1(0,T;X)$,

(iii) $\boldsymbol{v}$ *satisfies the condition* $div\, \boldsymbol{v} = 0$ *a.e on* $\Omega(h(t))$, $v_2(x_1, h(x_1,t), t) = \partial_t \eta(x_1, t)$ *for a.e.* $x_1 \in (0,L)$, $t \in (0,T)$

*and the following integral identity holds*

$$\begin{aligned}
&\int_0^T \int_{\Omega(h(t))} \left\{ -\rho \boldsymbol{v} \cdot \frac{\partial \boldsymbol{\varphi}}{\partial t} + 2\mu(|e(\boldsymbol{v})|) e(\boldsymbol{v}) e(\boldsymbol{\varphi}) + \rho \sum_{i,j=1}^{2} v_i \frac{\partial v_j}{\partial x_i} \boldsymbol{\varphi}_j \right\} dx\, dt \\
&+ \int_0^T \int_0^{R_0(L)} \left( P_{out} - \frac{\rho}{2} |v_1|^2 \right) \varphi_1(L, x_2, t)\, dx_2\, dt \\
&- \int_0^T \int_0^{R_0(0)} \left( P_{in} - \frac{\rho}{2} |v_1|^2 \right) \varphi_1(0, x_2, t)\, dx_2\, dt \\
&+ \int_0^T \int_0^L \left( P_w - \frac{\rho}{2} v_2 \left( v_2 - \frac{\partial h}{\partial t} \right) \right) \varphi_2(x_1, h(x_1,t), t)\, dx_1\, dt \\
&+ \int_0^T \int_0^L -\frac{\partial \eta}{\partial t} \frac{\partial \xi}{\partial t} + c \frac{\partial^3 \eta}{\partial x_1^2 \partial t} \frac{\partial^2 \xi}{\partial x_1^2} + a \frac{\partial \eta}{\partial x_1} \frac{\partial \xi}{\partial x_1} dx_1\, dt \\
&+ \int_0^T \int_0^L -a \frac{\partial^2 R_0}{\partial x_1^2} \xi + b\eta\, \xi\, dx_1\, dt \;=\; 0
\end{aligned}$$

*for every test functions*

$$\begin{aligned}
&\boldsymbol{\varphi}(x_1, x_2, t) = \boldsymbol{\psi}\left( x_1, \frac{x_2}{h(x_1,t)}, t \right) \quad \text{such that} \\
&\boldsymbol{\psi} \in H_0^1(0,T;\boldsymbol{V}), \quad \boldsymbol{\psi}_2\big|_{S_w} \in H_0^1(0,T;H_0^2(0,L)), \\
&div\, \boldsymbol{\varphi} = 0 \quad a.e.\ on\ \Omega(h(t)), \\
&and \quad \xi(x_1,t) = E\rho\, \varphi_2(x_1, h(x_1,t), t).
\end{aligned}$$

Note that the structure equation is fulfilled in a slightly modified sense,

$$E\rho\left[\frac{\partial^2\eta}{\partial t^2} - a\frac{\partial^2\eta}{\partial x_1^2} + b\eta + c\frac{\partial^5\eta}{\partial t\partial x_1^4} - a\frac{\partial^2 R_0}{\partial x_1^2}\right](x_1,t) =$$
$$\left[-(\mathbf{T}_f + P_w\mathbf{I})\boldsymbol{n}|\boldsymbol{n}|\cdot\boldsymbol{e}_2 + \frac{\rho}{2}\partial_t\eta(\partial_t\eta - \partial_t h)\right](x_1, h(x_1,t), t)$$

a.e. on $(0,T)\times(0,L)$, compare (9).

## 4 Fixed Point Procedure

Until now we have proved the existence of weak solution of the original problem in a domain given by a known deformation function $\delta$, i.e. $h = R_0 + \delta$, $\delta \in H^1(0,T;H_0^2(0,L)) \cap W^{1,\infty}(0,T;L^2(0,L))$, $R_0(x_1) \in C^2[0,L]$. In this section we show the existence of the weak solution of (14), which implies, that the domain deforms according to the function $\eta(x_1,t)$, i.e. $h = R_0 + \eta$. This will be realized with the use of the Schauder fixed point theorem. First, applying the compactness argument based on the equicontinuity in time we obtain that bounded sequence $(\boldsymbol{v}^{(k)}, \eta^{(k)})$ defined on $\Omega(\delta^{(k)})$ for some sequence $\delta^{(k)} \to \delta$ converges to the limit $(\boldsymbol{v},\eta)$ defined on $\Omega(\delta)$. Consequently, the Schauder fixed point argument implies, that the weak solution $\eta$ is associated with the time dependent domain $\Omega(\eta)$. Finally we obtain the main result: existence of weak solution for a fully coupled fluid structure interaction problem (1)–(13).

Let us denote the space $Y = H^1(0,T;L^2(0,L))$. For each test function $\boldsymbol{\psi} \in L^p(0,T;X)$, $\boldsymbol{\psi}(T) = 0$, $\xi = E\boldsymbol{\psi}_2\big|_{S_w}$, recalling (35), and for any $h = R_0 + \delta \in Y$, such that (16) holds, we construct solutions $\boldsymbol{u}$, $\eta$ of the following problem defined on the reference domain $D$, ($\sigma = \partial_t\eta$),

$$\begin{aligned}
&-\int_0^T \left\langle\bar{\partial}_t(h\boldsymbol{u}), \boldsymbol{\psi}\right\rangle dt \qquad (40)\\
&= \int_0^T \Big\{ ((\boldsymbol{u},\boldsymbol{\psi}))_h + b_h(\boldsymbol{u},\boldsymbol{u},\boldsymbol{\psi})\\
&\quad + \int_0^1 h(L)q_{out}(y_2,t)\psi_1(L,y_2,t) - h(0)q_{in}(y_2,t)\psi_1(0,y_2,t)\,dy_2\\
&\quad + \int_0^L \left(q_w + \frac{1}{2}\frac{\partial h}{\partial t}\sigma\right)\psi_2(y_1,1,t)\,dy_1\\
&\quad + \left\langle\partial_t\sigma,\xi\right\rangle + \int_0^L c\frac{\partial^2\sigma}{\partial y_1^2}\frac{\partial^2\xi}{\partial y_1^2} + a\frac{\partial}{\partial y_1}\int_0^t \sigma(y_1,s)ds\frac{\partial\xi}{\partial y_1}\\
&\qquad - a\frac{\partial^2 R_0}{\partial y_1^2}\xi + b\int_0^t \sigma(y_1,s)ds\,\xi(y_1,t)\,dy_1 \Big\}\,dt\,.
\end{aligned}$$

Further, let the ball $B_{\alpha,K}$ be defined by

$$B_{\alpha,K} = \Big\{\delta \in Y;\ \|\delta\|_Y \le C(\alpha,K),\ 0 < \alpha \le R_0(y_1) + \delta(y_1,t) \le \alpha^{-1},$$
$$\left|\frac{\partial \delta(y_1,t)}{\partial y_1}\right| \le K,\ \delta(y_1,0) = 0,\ \forall y_1 \in [0,L],\ \forall t \in [0,T],$$
$$\int_0^T \left|\frac{\partial \delta(y_1,t)}{\partial t}\right|^2 dt \le K,\ \forall y_1 \in [0,L]\Big\},$$

where $C(\alpha,K)$ is a suitable constant large enough with respect to $K, \alpha$ and the data.

By choosing $\delta \in B_{\alpha,K}$ the following energy estimate holds for all $2 \le p < \infty$ uniformly in $\delta$,

$$\begin{aligned} &\|\boldsymbol{u}\|^2_{L^\infty(0,T;L^2(D))} + \|\boldsymbol{u}\|^p_{L^p(0,T;W^{1,p}(D))} \\ &\quad + \|\eta_t\|^2_{L^\infty(0,T;L^2(0,L))} + \|\eta_t\|^2_{L^2(0,T;H^2(0,L))} + \|\eta\|^2_{L^\infty(0,T;H^1(0,L))} \\ &\quad \le c(T,p,K,\alpha)\left(\|\mathcal{P}\|^{p'}_{L^{p'}(0,T)} + \|R_0\|^2_{C^2[0,L]}\right). \end{aligned} \tag{41}$$

This estimate is obtained by multiplying (40) by $\boldsymbol{\psi} = \boldsymbol{u}$ and $\xi = Eu_2|_{S_w} = E\eta_t$, cf. (31).

Now, let us define the following mapping by (40),

$$\mathcal{F} : B_{\alpha,K} \to Y;$$
$$\mathcal{F}(\delta) = \eta, \quad (\delta = h - R_0).$$

Our aim is to apply the Schauder fixed point theorem and prove that the mapping $\mathcal{F}$ has at least one fixed point. This implies the existence of the weak solution to our problem (14).

First we check that $\mathcal{F}(B_{\alpha,K}) \subset B_{\alpha,K}$. Note that our a priori estimate (41) yields $\|\eta_{y_1}\|_{C([0,T]\times[0,L])} \le K$, $\|\eta_t\|_{L^2(0,T;C[0,L])} \le K$ and $\|\eta\|_Y \le C(\alpha,K)$ for given data $P_{in}$, $P_{out}$, $P_w$, $R_0$, given $K, \alpha$; $\alpha < R_{min} := \min_{y_1\in[0,L]} R_0(y_1)$ and for sufficiently small time $\tilde{T}$. Moreover, since $H^1(0,T;H^2(0,L)) \hookrightarrow C(0,T;C^1[0,L])$ and $\eta(y_1,0) = 0$, there exist a maximal time $T_{max}$, such that

($i$) $\|\eta\|_\infty := \|\eta\|_{C([0,T_{max}]\times[0,L])} \le R_{min} - \alpha$. This yields that $\min_{t\in(0,T_{max})}\min_{y_1\in(0,L)} (R_0 + \eta) \ge R_{min} - \|\eta\|_\infty \ge \alpha$. Thus we can avoid a contact of the deforming wall with the solid bottom.

($ii$) Further, we require that the domain deformation is bounded from above, $\|R_0 + \eta\|_\infty \le \alpha^{-1}$.

Having ($i$), the condition ($ii$) is satisfied if $R_{min} - \alpha \le \alpha^{-1} - R_{max}$. Thus, for instance if $\alpha^{-1} \ge R_{min} + R_{max}$. Consequently, $\mathcal{F}(B_{\alpha,K}) \subset B_{\alpha,K}$ as far as

$t \leq T^* := \min\{T_{max}, \tilde{T}\}$ for given data $P_{in}$, $P_{out}$, $P_w$, $R_0$, $K$ and $\alpha$ such that $\alpha \leq \min\{R_{min}, \frac{1}{R_{min}+R_{max}}\}$.

Secondly, we verify the relative compactness of the mapping $\mathcal{F}$ in $Y$. Let us consider a sequence $\{\delta^{(k)}\}_{k=1}^{\infty}$ in $B_{\alpha,K}$. Denote by $\boldsymbol{u}^{(k)}$ and $\eta^{(k)} \equiv \mathcal{F}(\delta^{(k)})$ the weak solution of (40) for $h = h^{(k)} := R_0 + \delta^{(k)}$. Note, that due to the apriori estimate (41) we have weak convergences of $\eta^{(k)}$, $\boldsymbol{u}^{(k)}$ in the corresponding spaces. In Sect. 4.1, cf. Lemma 4.1, we show the equicontinuity in time, which implies the strong convergences of $\eta^{(k)}$ in $Y$ as well as the strong convergence $\boldsymbol{u}^{(k)}$ in $L^2((0,T) \times D)$.

Finally, in Sect. 4.2 we check the continuity of the mapping $\mathcal{F}$ with respect to the strong topology in $Y$.

## 4.1 Relative Compactness of the Fixed Point Mapping $\mathcal{F}$

In this section we verify the relative compactness of the mapping $\mathcal{F}$ using the integral equicontinuity in time and the Riesz-Fréchet-Kolmogorov compactness argument. We prove Lemma 4.1, which provides the integral equicontinuity of $\eta^{(k)}$ and additionally of $\boldsymbol{u}^{(k)}$, that holds independently on $k$. To this end we need to find suitable divergence free test functions in order to control difference of velocity at different time instances. In order to obtain such test functions we follow a construction presented in [4], see also the reference [16] therein.

We introduce, in analogy to [4, Lemma 3], the following extensions of the domain and the weak solution. Let $B_M$ be a box domain

$$B_M \equiv (0, L) \times (0, M) \in \mathbb{R}^2 \tag{42}$$

for some $M > \alpha^{-1}$ specified later. Let us consider a sequence $\{\delta^{(k)}\}_{k=1}^{\infty}$ in $B_{\alpha,K}$ and $h^{(k)} := R_0 + \delta^{(k)}$. We define an extension of solution $\boldsymbol{u}^{(k)}(y,t) = \boldsymbol{v}^{(k)}(x,t)$ of (40) where $h = h^{(k)}$ into $B_M$,

$$\bar{\boldsymbol{v}}^{(k)} = \begin{cases} \boldsymbol{v}^{(k)} & \text{in } \Omega(h^{(k)}(t)) \\ (0, \eta_t^{(k)}) & \text{in } B_M \backslash \Omega(h^{(k)}(t)). \end{cases} \tag{43}$$

Further, for $\gamma > 1$ and any function $\boldsymbol{f}(x_1, x_2)$ we define $\boldsymbol{f}_\gamma$ as follows

$$\boldsymbol{f}_\gamma(x_1, x_2) = (\gamma f_1(x_1, \gamma x_2), f_2(x_1, \gamma x_2)).$$

Note that if $\boldsymbol{f}$ is divergence free, then $\boldsymbol{f}_\gamma$ is divergence free, too.

In what follows we will use the following property, which is valid for any $\boldsymbol{f} \in H^1(B_M)$, for $\gamma = 1 + \frac{C\sqrt{\tau}}{\alpha}$ and $M \geq 2\alpha^{-1}$, see [9, Lemma 9.2]

$$|\boldsymbol{f}_\gamma - \boldsymbol{f}| \leq |\boldsymbol{f}(x_1, \gamma x_2, t) - \boldsymbol{f}(x_1, x_2, t)| + (\gamma - 1)|\boldsymbol{f}(x_1, \gamma x_2, t)|. \tag{44}$$

**Lemma 4.1** *For the weak solution* $(\boldsymbol{v}^{(k)}, \eta_t^{(k)}) = (\boldsymbol{u}^{(k)}, \sigma^{(k)})$ *of the problem (40), where* $h = h^{(k)}$, *it holds*

$$\int_0^{T-\tau}\int_{B_M} \chi_t^{(k)} |\bar{\boldsymbol{v}}^{(k)}(t+\tau) - \bar{\boldsymbol{v}}^{(k)}(t)|^2 + \int_0^{T-\tau}\int_0^L |\eta_t^{(k)}(t+\tau) - \eta_t^{(k)}(t)|^2 \leq C(\tau^{1/p} + \tau^{1/2}). \tag{45}$$

*Here* $\chi_t^{(k)}$ *denotes the characteristic function of* $\Omega(h^{(k)}(t))$. *The constant* $C = C(K, \alpha)$ *does not depend on k.*

*Proof* We recall that $h^{(k)} = R_0 + \delta^{(k)}$, but for the sake of simplicity we omit the superscript $(k)$ in this proof and we denote $h := R_0 + \delta^{(k)}$, $\bar{\boldsymbol{v}} := \bar{\boldsymbol{v}}^{(k)}$, $\eta := \eta^{(k)}$. To prove the statement of this lemma, we will use following two properties.

1. For each $\boldsymbol{\psi} \in H^1(0, T; X)$, cf. (35), $\boldsymbol{\psi}(T) = 0$ it holds

$$-\int_0^{\tilde{\tau}} \langle \bar{\partial}_t(h\boldsymbol{u}), \boldsymbol{\psi} \rangle dt \tag{46}$$

$$= \int_0^{\tilde{\tau}}\int_D h\boldsymbol{u}\frac{\partial \boldsymbol{\psi}}{\partial t} + \frac{\partial h}{\partial t}\frac{\partial(y_2 \boldsymbol{u})}{\partial y_2}\boldsymbol{\psi}\, dydt - \int_D h\boldsymbol{u}(\tilde{\tau}, y)\boldsymbol{\psi}(\tilde{\tau}, y)dy.$$

For classical time derivative, this property is clear. For our distributive derivative $\bar{\partial}$ it can be proven using test function $\boldsymbol{\psi} = \zeta(y,t)\varphi_\epsilon(t)$, where $\zeta \in H^1(0, T; X)$, $\varphi_\epsilon(t) = \max\{0, \min\{1, \frac{\tilde{\tau}+\epsilon-t}{\epsilon}\}\}$ and passing $\epsilon \to 0$, cf.[17].

2. By inserting any time independent test function $\boldsymbol{\psi} = \boldsymbol{\psi}(y)$ into (46) and subtracting (46) for $\tilde{\tau} = t + \tau$, and $\tilde{\tau} = t$ we obtain

$$-\int_t^{t+\tau} \langle \partial_t \boldsymbol{v}, \boldsymbol{\varphi}(x,t) \rangle_{X_\Omega} ds \tag{47}$$

$$= \int_t^{t+\tau}\int_D \frac{\partial h}{\partial t}\frac{\partial(y_2 \boldsymbol{u})}{\partial y_2}\boldsymbol{\psi}(y) dyds - \int_D [h\boldsymbol{u}(t+\tau) - h\boldsymbol{u}(t)]\boldsymbol{\psi}(y)dy.$$

Here the integral on the left hand side has been transformed into $\Omega(h(t))$, $\boldsymbol{\psi} = \boldsymbol{\psi}(y) = \boldsymbol{\varphi}(x)$, $y \in D$, $x \in \Omega(h(t))$ and the space $X_\Omega$ is defined as

$$X_\Omega = \{\boldsymbol{\varphi} \in W^{1,p}(\Omega);\ \operatorname{div}\varphi = 0 \text{ a.e. on } \Omega,$$
$$\varphi_2|_{\Gamma_w} \in H_0^2(0,L), \varphi_1|_{\Gamma_w} = \varphi_2|_{\Gamma_{in}\cup\Gamma_{out}\cup\Gamma_c} = 0\}, \quad \Omega = \Omega(h(t)).$$

Now, let us integrate (47) over $\int_0^{T-\tau} dt$. The first term on the right hand side (integrated over $\int_0^{T-\tau}$) can be bounded with $C\tau$ independently on $k$ for test functions (52) specified later . The second term on the right hand of (47) can be rewritten due to the transformation to the $\Omega(h(t))$

$$\int_0^{T-\tau} \int_{\Omega(h(t+\tau))} \boldsymbol{v}(x_{t+\tau}, t+\tau)\boldsymbol{\varphi}(x_{t+\tau})dx - \int_{\Omega(h(t))} \boldsymbol{v}(x_t, t)\boldsymbol{\varphi}(x_t)\, dx\, dt. \tag{48}$$

Note, that the space coordinate $x_t \equiv x(t) \in \Omega(h(t))$ depends on time, hence the test functions $\boldsymbol{\varphi}$ implicitly depend on time, which is pointed out above.

Using the previously defined extensions of the solution $\bar{\boldsymbol{v}}$ and some further manipulations we can rewrite (48) as follows

$$\begin{aligned}
&\int_0^{T-\tau} \int_{\Omega(h(t))} \bar{\boldsymbol{v}}(x_{t+\tau}, t+\tau)\boldsymbol{\varphi}(x_{t+\tau}) - \boldsymbol{v}(x_t, t)\boldsymbol{\varphi}(x_t)dx \\
&+ \int_{B_M} (\chi_{t+\tau} - \chi_t)\bar{\boldsymbol{v}}(x_{t+\tau}, t+\tau)\boldsymbol{\varphi}(x_{t+\tau})\, dx\, dt = \\
&\int_0^{T-\tau}\int_{\Omega(h(t))} \underbrace{[\bar{\boldsymbol{v}}\,(x_{t+\tau}, t+\tau) - \boldsymbol{v}(x_t, t)]\boldsymbol{\varphi}(x_t)}_{\text{(I)}} + \underbrace{[\boldsymbol{\varphi}(x_{t+\tau}) - \boldsymbol{\varphi}(x_t)]\bar{\boldsymbol{v}}(x_{t+\tau}, t+\tau)}_{\text{(II)}} \\
&+ \int_{B_M} \underbrace{(\chi_{t+\tau} - \chi_t)\bar{\boldsymbol{v}}(x_{t+\tau}, t+\tau)\boldsymbol{\varphi}(x_{t+\tau})}_{\text{(III)}}\, dx\, dt.
\end{aligned} \tag{49}$$

Here $\chi_t$, $\chi_{t+\tau}$ are the characteristic functions of $\Omega(h(t))$, $\Omega(h(t+\tau))$, respectively. In what follows we estimate the term (II) for any test function $\boldsymbol{\varphi} \in L^p(0, T; X_\Omega)$. Further, we concentrate on the terms (I), (III) using specific test functions.

From the imbeddings in one dimension we have $\delta \in C^{0,1/2}([0, T]; H^1(0, L))$, cf. (58), thus

$$\|\delta(t+\tau) - \delta(t)\|_{L^\infty((0,T)\times(0,L))} \leq C\sqrt{\tau}. \tag{50}$$

Using (50) we can estimate the term (II):

$$\begin{aligned}
\text{(II)} &\leq \int_0^{T-\tau} \left( \int_{\Omega(h(t))} |\boldsymbol{\varphi}(x_{t+\tau}) - \boldsymbol{\varphi}(x_t)|^2 dx \right)^{1/2} \|\bar{\boldsymbol{v}}\|_{L^2(\Omega(h(t))}\| dt \\
&= \int_0^{T-\tau} \left( \int_{\Omega(h(t))} \left| \int_{x_2(t)}^{x_2(t+\tau)} \partial_s \boldsymbol{\varphi}(x_1, s) ds \right|^2 dx \right)^{1/2} \|\bar{\boldsymbol{v}}\|_{L^2(\Omega(h(t))}\| dt \\
&\leq \int_0^{T-\tau} \left( \int_{B_M} |\nabla \boldsymbol{\varphi}|^2 dx |x_2(t+\tau) - x_2(t)|^2 \right)^{1/2} \|\bar{\boldsymbol{v}}\|_{L^2(\Omega(h(t))}\| dt \\
&\leq \|\boldsymbol{\varphi}\|_{L^2(0,T;H^1(B_M))} \|\delta(t+\tau) - \delta(t)\|_{L^\infty((0,T)\times(0,L))} \|\bar{\boldsymbol{v}}\|_{L^2((0,T)\times B_M)} \\
&\leq C\sqrt{\tau}.
\end{aligned} \tag{51}$$

Now we specify proper test functions, that will be used in what follows. For $x_t = x(t) \in \Omega(h(t))$, $\gamma > 1$ and fixed $t$, $\tau$ we set

$$\varphi(x_t) = \bar{v}_\gamma(x_{t+\tau}, t+\tau) - \bar{v}_\gamma(x_t, t), \qquad (52)$$
$$\xi(x_1) = E(\partial_t \eta(x_1, t+\tau) - \partial_t \eta(x_1, t)).$$

Note that since $v$ is divergence-free, the test function $\varphi$ is also divergence-free.[2] Moreover, taking into account (50), for $\gamma \geq 1 + \frac{C\sqrt{\tau}}{\alpha}$ and $x_2 \in \Gamma_w(t)$ the coordinate $\gamma x_2$ exceeds the moving domain $\Omega(h(s))$, since we have $\gamma(R_C + \delta(s)) \geq R_0 + \delta(s) + \|\delta(t+\tau) - \delta(t)\|_\infty$, $s = t, t+\tau$. According to the construction, such a test function fulfill the boundary condition

$$E\varphi(x_1, R_0(x_1) + \delta(x_1, t)) = E(0, \partial_t \eta(x_1, t+\tau) - \partial_t \eta(x_1, t)) \equiv (0, \xi(x_1)).$$

Let us estimate now the term (III). Since $\partial_t \eta$ is bounded in $L^\infty(0, T; L^2(0, L))$ independently on $k$, we have

$$\int_{B_M} |\chi_{t+\tau} - \chi_t|^2 = \int_0^L |\delta(t+\tau) - \delta(t)|^2 = \int_0^L \left| \int_t^{t+\tau} \partial_t \delta(s) ds \right|^2 \leq C\tau. \qquad (53)$$

Thus, the term (III) can be bounded for $\varphi$ from (52) as follows.

$$\text{(III)} \leq \int_0^{T-\tau} \|\chi_{t+\tau} - \chi_t\|_{L^2(B_M)} \|\bar{v}\|_{L^4(B_M)} \|\varphi\|_{L^4(B_M)} dt \leq C\sqrt{\tau}. \qquad (54)$$

For the test functions from (52) the term (I) equals

$$\begin{aligned} \text{(I)} &= \int_0^{T-\tau} \int_{\Omega(h(t))} [\bar{v}(t+\tau) - \bar{v}(t)][\bar{v}_\gamma(t+\tau) - \bar{v}_\gamma(t)] dx dt \\ &= \int_0^{T-\tau} \int_{\Omega(h(t))} \underbrace{[\bar{v}(t+\tau) - \bar{v}(t)]^2}_{\text{(Ia)}} + \qquad (55) \\ &\quad \underbrace{[\bar{v}(t+\tau) - \bar{v}(t)].\Big([\bar{v}_\gamma(t+\tau) - \bar{v}(t+\tau)] - [\bar{v}_\gamma(t) - \bar{v}(t)]\Big)}_{\text{(Ib)}} \, dx dt \end{aligned}$$

For the simplicity we used shorter notations here, e.g., $\bar{v}(t+\tau) := \bar{v}(x_{t+\tau}, t+\tau)$. The term (Ia) appears on the left hand side of the assertion of this lemma; the term

[2] Since $\varphi(x_{t+\tau}) = \bar{v}_\gamma(x_{t+2\tau}, t+2\tau) - \bar{v}_\gamma(x_{t+\tau}, t+\tau)$, we have to integrate over $\int_C^{T-2\tau} dt$ in the estimate of the term (II), or we define $\varphi(x_{t+\tau}) = 0$ if $t + \tau > T$.

(Ib) need to be estimated from above. We illustrate the estimate of some chosen terms of (Ib) as follows. Estimates of other terms are analogous.

In the sequel we take $\gamma = 1 + \frac{C\sqrt{\tau}}{\alpha}$ and $M \geq 2\alpha^{-1}$. For these parameters we have according to lemma (44),

$$\int_0^{T-\tau} \int_{\Omega(h(t))} \bar{v}(t+\tau)[\bar{v}_\gamma(t) - \bar{v}(t)]dxdt \leq$$

$$C_\alpha\sqrt{\tau} \int_0^{T-\tau} \|\bar{v}(t+\tau)\|_{L^2(B_M)} \|\bar{v}(t)\|_{H^1(B_M)} dt \leq C_\alpha\sqrt{\tau}.$$

To complete the proof, the remaining terms coming from the fluid equations, i.e., the convective term, the viscous term, boundary terms and the equation for $\eta$ have to be estimated. We illustrate here only the calculations for the nonlinear viscous term and omit tedious but standard calculations for other terms, previously performed also in [6].

After subtracting the weak formulation (40) for $\int_0^{t+\tau} ds - \int_0^t ds$, inserting test functions constructed above (independent on $s$) into (40) and integrating over $\int_0^{T-\tau} dt$ we obtain from the viscous term

$$\int_0^{T-\tau} \int_t^{t+\tau} \int_{\Omega(h(s))} \tau_{ij}(e[\mathbf{v}(s)]).e[\bar{\mathbf{v}}_\gamma(t+\tau) - \bar{\mathbf{v}}_\gamma(t)]dx\,ds\,dt. \tag{56}$$

For the simplicity, we set $\omega := \bar{\mathbf{v}}_\gamma(t+\tau)$ or $\omega := \bar{\mathbf{v}}_\gamma(t)$ in (56). Using the fact, that $|\tau_{ij}(e(\mathbf{v}))| \leq C_5(1 + |e(\mathbf{v})|)^{p-1}$, which can be derived from (2), (4), cf. [11, Lemma 1.19], (56) can be bounded as follows,

$$\begin{aligned}
&\leq \int_0^{T-\tau} \int_t^{t+\tau} \int_{\Omega(h(s))} C_5(1 + |e[\mathbf{v}(s)]|)^{p-1} e[\omega]dx\,ds\,dt \\
&\leq C(K,\alpha) \int_0^{T-\tau} \int_t^{t+\tau} \|1 + \nabla \mathbf{v}(s)\|^{p-1}_{L^p(\Omega(h(s)))} \|\nabla\omega\|_{L^p(\Omega(h(s)))} ds\,dt \\
&\leq C(K,\alpha) \int_0^{T-\tau} \left( \int_t^{t+\tau} \|1 + \nabla \mathbf{v}(s)\|^p_{L^p(\Omega(h(s)))} ds \right)^{\frac{p-1}{p}} \|\nabla\omega\|_{L^p(B_M)} \tau^{\frac{1}{p}} dt \\
&\leq C(K,\alpha)\tau^{\frac{1}{p}} \left( \int_0^T \|1 + \nabla \mathbf{v}(s)\|^p_{L^p(\Omega(h(s)))} ds \right)^{\frac{p-1}{p}} \int_0^{T-\tau} \|\nabla\omega\|_{L^p(B_M)} dt \\
&\leq C(K,\alpha)\tau^{\frac{1}{p}} \|1 + \nabla \mathbf{v}\|^{p-1}_{L^p(0,T;L^p(\Omega(h)))} \|\nabla\omega\|_{L^1(0,T;L^p(B_M))} \leq C(K,\alpha)\tau^{\frac{1}{p}}.
\end{aligned}$$

The estimates of remaining terms on the right hand side can be obtained using the so-called Steklov averages analogously as in e.g., [9, Sect. 5] or [6, Sect. 8] and we leave them to the valued reader. The proof of the lemma is now completed. □

Due to the (53) it is also easy to obtain from (45) that

$$\int_0^{T-\tau}\int_{B_M}|\chi_{t+\tau}^{(k)}\bar{\boldsymbol{v}}^{(k)}(t+\tau)-\chi_t^{(k)}\bar{\boldsymbol{v}}^{(k)}(t)|^2+\int_0^{T-\tau}\int_0^L |\eta_t^{(k)}(t+\tau)-\eta_t^{(k)}(t)|^2$$
$$\leq C(\tau^{1/p}+\tau^{1/2}). \tag{57}$$

This result implies that $\chi_t^{(k)}\bar{\boldsymbol{v}}^{(k)}(t)$, and consequently $\bar{\boldsymbol{v}}^{(k)}(t)$ is relatively compact in $L^2((0,T)\times B_M)$.

Consequently, the Riesz-Fréchet-Kolmogorov compactness argument [2, Theorem IV.26] based on (57) implies the relative compactness of $\partial_t\eta^{(k)}, \bar{\boldsymbol{v}}^{(k)}$ in $L^2(0,T;L^2(0,L))$, $L^2(0,T;L^2(B_M))$, respectively. Additionally, the standard interpolations give us the compactness of $\bar{\boldsymbol{v}}^{(k)}$ in $L^r((0,T)\times B_M)$, $1\leq r<4$ and $\partial_t\eta^{(k)}$ in $L^s((0,T)\times(0,L))$, $1\leq s<6$.

## 4.2 Continuity of the Mapping $\mathcal{F}$

As already shown above $\eta^{(k)}$ converges strongly to some $\eta$ in $Y$ as $k\to\infty$. In this section we show by limiting process for $k\to\infty$ in (40) that for any convergent subsequence $\delta^{(k)}\in B_{\alpha,K}$, $\delta^{(k)}\to\delta$ in $Y$ we have

$$\mathcal{F}(\delta^{(k)})=\eta^{(k)}\to\mathcal{F}(\delta)\text{ and that }\eta\equiv\mathcal{F}(\delta).$$

First, we know that $\eta^{(k)}\to\eta$ in $H^1(0,T;L^2(0,L))$. Due to the boundedness of $\eta$ from apriori estimate (41) and the imbeddings in one dimension we have even stronger result—the uniform convergence of $\partial_{y_1}\eta^{(k)}$ in $C([0,T]\times[0,L])$. Indeed,

$$L^\infty(0,T;H^2(0,L))\cap W^{1,\infty}(0,T;L^2(0,L))$$
$$\hookrightarrow C^{0,1-\beta}(0,T;H^{2\beta}(0,L)) \tag{58}$$

for $0<\beta<1$. From the continuous imbedding of $H^{2\beta}(0,L)$ into $H^{2\beta-\epsilon}(0,L)$ and the Arzelá-Ascoli Lemma we conclude that a subsequence of $\eta^{(k)}$ converges strongly in $C([0,T];H^s(0,L))$, $0<s<2$. Since for $s>3/2$ we also have continuous imbedding $H^s(0,L)\hookrightarrow C^1[0,L]$, we can conclude, that $\eta^{(k)}\to\eta$ strongly in $C(0,T;C^1[0,L])$.

Let us summarize available convergences

$$\begin{aligned}
\boldsymbol{u}^{(k)} &\rightharpoonup \boldsymbol{u} && \text{weakly in } L^p(0,T;W^{1,p}(D)),\\
\bar{\boldsymbol{v}}^{(k)} &\to \bar{\boldsymbol{v}} && \text{strongly in } L^r((0,T)\times B_M),\ 1\leq r<4,\\
\boldsymbol{u}^{(k)} &\to \boldsymbol{u} && \text{strongly in } L^r((0,T)\times D),\ 1\leq r<4,\\
\eta^{(k)} &\rightharpoonup \eta && \text{weakly in } H^1(0,T;H^2(0,L)),
\end{aligned} \tag{59}$$

$$\eta^{(k)} \rightharpoonup^* \eta \quad \text{weakly* in } L^\infty(0,T;L^2(0,L))$$
$$\eta^{(k)} \to \eta \quad \text{uniformly in } C(0,T;C^1[0,L]),$$
$$\partial_t \eta^{(k)} \to \partial_t \eta \quad \text{strongly in } L^s((0,T)\times(0,L)),\ 1 \le s < 6.$$
$$\bar{\partial}_t (h\boldsymbol{u})^{(k)} \rightharpoonup \chi \quad \text{weakly in } \begin{cases} L^{p'}(0,T;V^*) \text{ for } 2 < p < \infty \\ L^{p'}(0,T;V^*) \oplus L^{4/3}((0,T)\times D) \text{ for } p = 2. \end{cases}$$

The last statement of (59) follows from (41) and from the boundedness of the functional $\bar{\partial}_t (h\boldsymbol{u})^{(k)}$. Let us present the estimation of nonlinear terms on the right hand side. Indeed, from Lemma 2.5 it follows $\int_0^T ((\boldsymbol{u}^{(k)}, \boldsymbol{\psi})) \le C(K,\alpha)\|\boldsymbol{\psi}\|_{L^p(0,T;W^{1,p}(D))}$. The non-linear convective term can be estimated using Lemma 2.6, the Hölder inequality and the imbedding of $W^{1,p}(D)$ into $L^{\frac{2p}{p-2}}(D)$ for $p > 2$ as follows

$$\int_0^T B_{h^{(k)}}(\boldsymbol{u}^{(k)}, \boldsymbol{u}^{(k)}, \boldsymbol{\psi}) \le C(K,\alpha) \int_0^T \|\boldsymbol{u}^{(k)}\|_{1,p} \|\boldsymbol{u}^{(k)}\|_2 \|\boldsymbol{\psi}\|_{\frac{2p}{p-2}}$$
$$\le C(K,\alpha) \|\boldsymbol{u}^{(k)}\|_{L^\infty(0,T;L^2(D))} \|\boldsymbol{u}^{(k)}\|_{L^p(0,T;W^{1,p}(D))} \|\boldsymbol{\psi}\|_{L^{p'}(0,T;W^{1,p}(D))},$$

which is bounded due to (41) for all $\boldsymbol{\psi} \in L^p(0,T;\boldsymbol{V})$. Analogously the term $\int_0^T B_{h^{(k)}}(\boldsymbol{u}^{(k)}, \boldsymbol{\psi}, \boldsymbol{u}^{(k)})$ is bounded, which leads to[3]

$$\int_0^T b_{h^{(k)}}(\boldsymbol{u}^{(k)}, \boldsymbol{u}^{(k)}, \boldsymbol{\psi}) \le C(K,\alpha)\|\boldsymbol{\psi}\|_{L^p(0,T;W^{1,p}(D))} \text{ for } p \in (2,\infty). \tag{60}$$

Further estimates of the remaining terms on the right hand side conclude the proof of $(59)_8$.

In what follows we have to verify, that $\mathcal{F}(\delta^{(k)}) \to \mathcal{F}(\delta)$ and that the limit $\eta$ from (59) is the weak solution associated with $\delta$, and thus $\mathcal{F}(\delta) = \eta$.

### 4.2.1 Limiting Process $k \to \infty$

Now let us consider (40) with $\boldsymbol{u}^{(k)}$ instead of $\boldsymbol{u}$, $h^{(k)}$ instead of $h$ and $\sigma^{(k)} = \partial_t \eta^{(k)}$ instead of $\sigma$.

First of all we have to realize, that due to the solenoidal property, which depends on $h^{(k)}$, the test functions are also implicitly dependent on $k$. This fact presents a difficulty when we pass with $k \to \infty$. Nevertheless we can construct sufficiently smooth test functions $\tilde{\boldsymbol{\psi}}(y,t) = \tilde{\boldsymbol{\varphi}}(x,t)$, which are independent on $k$ and divergence free in $\Omega(h)$ (i.e. $\mathrm{div}_h \tilde{\boldsymbol{\psi}} = 0$). They are also well defined on infinitely many

[3] For $p = 2$ this estimate is valid for $\boldsymbol{\psi} \in L^p(0,T;\boldsymbol{V}) \cap L^4((0,T)\times D)$, cf. [6].

approximate domains $\Omega(h^{(k)})$ and dense in the space of admissible test functions $L^p(0,T;X)$, cf. (35). Such a test functions $\tilde{\varphi}$ can be constructed on $(0,T) \times B_M$ as algebraic sum, see [4, Remark 3]

$$\tilde{\varphi} = \varphi_0 + \varphi_1,$$

where $\varphi_0$ is a smooth function with compact support in $\Omega(h)$, $\operatorname{div}\varphi_0 = 0$ on $\Omega(h)$ and $\varphi_0$ is extended by 0 to $(0,T) \times B_M$. Further, having $\xi \in H^1(0,T;H_0^2(0,L))$ we define $\varphi_1 \stackrel{\text{def}}{=} (0, \xi(x_1)/E)$ on $B_M \backslash B_\alpha$, $B_\alpha = (0,L) \times (0,\alpha) \in \mathbb{R}^2$, the constant $E$ comes from (15). Note that $\operatorname{div}\varphi_1 = 0$ on $B_M \backslash B_\alpha$. Moreover, $\varphi_1$ such that $\int_{\partial B_\alpha} \varphi_1 \cdot n = \int_0^\alpha \varphi_1^1(L,x_2,t) - \varphi_1^1(0,x_2,t)dx_2 + \int_0^L \frac{\xi}{E}(x_1,t)dx_1 = 0$ can be extended into $B_\alpha$ by a divergence-free extension, whereas remaining boundary conditions on $\Gamma_{in}$, $\Gamma_{out}$, $\Gamma_c$ are preserved, see e.g., [7, p. 144]. Note, that due to the uniform convergence of $\eta^{(k)}$ we have that $\sup \varphi_0 \subset \Omega(h^{(N)})$ for sufficiently large $N$ and the function $\varphi_0$ is well defined on each $\Omega(h^{(N)})$ for such $N$. Moreover $\varphi_1$ is defined on $\Omega(h^{(k)})$ for each $k$. For more details on this construction we refer a reader to [3, Sect. 7, pp. 35–36], compare [4].

Having $\tilde{\psi}(y,t) = \tilde{\psi}(x_1, \frac{x_2}{h(x_1,t)}, t) = \tilde{\varphi}(x,t)$, $x \in \Omega(h)$, $y \in D$, let us construct the set of admissible test functions $\psi^{(k)}$ by transformation of $\tilde{\varphi}$ from $\Omega(h^{(k)})$ into $D$,

$$\psi^{(k)}(y_1,y_2,t) := \tilde{\psi}(x_1, \frac{x_2}{h^{(k)}(x_1,t)}, t) = \tilde{\varphi}(x_1,x_2,t), \qquad x \in \Omega(h^{(k)}),\ y \in D. \tag{61}$$

The test functions (61) have the following property

$$\psi^{(k)} : D \to \mathbb{R}^2;\ \operatorname{div}_{h^{(k)}} \psi^{(k)} = 0, \quad E\psi_2^{(k)}(y_1,1,t) = \xi(y_1,t), \quad \text{and}$$

$$\left.\begin{array}{l} \psi^{(k)} \to \tilde{\psi}, \\ \hat{e}_{h^{(k)}}(\psi^{(k)}) \to \hat{e}(\tilde{\psi}) \end{array}\right\} \quad \text{uniformly on } (0,T) \times D.$$

This property follows from the special construction of $\tilde{\varphi}$, the property (63) below and the uniform convergence of $h^{(k)}$ and $\partial_{y_1} h^{(k)}$ that follows from (58). The test functions (61) satisfy the boundary conditions on $S_{in}$, $S_{out}$, $S_c$, cf. (24) as well.

Thus it is enough to consider test functions $\psi = \tilde{\psi}$, which are independent on $k$ and smooth enough. The limiting process in the test functions follows afterwards using the uniform convergence $\psi^{(k)}$ and $\hat{e}(\psi^{(k)})$.

In the following lines we will present the limiting process for $k \to \infty$ in chosen nonlinear terms. Let us first consider the convective term and show

$$\int_0^T \left(b_{h^{(k)}}(\boldsymbol{u}^{(k)}, \boldsymbol{u}^{(k)}, \psi) - b_h(\boldsymbol{u}, \boldsymbol{u}, \psi)\right) dt \to 0.$$

Recalling (30), the following terms appear in the above expression

$$\int_0^T B_h(\boldsymbol{u}, \boldsymbol{u}^{(k)} - \boldsymbol{u}, \boldsymbol{\psi}) + B_{h^{(k)}}(\boldsymbol{u}^{(k)} - \boldsymbol{u}, \boldsymbol{u}^{(k)}, \boldsymbol{\psi}) + B_{(h^{(k)}-h)}(\boldsymbol{u}, \boldsymbol{u}^{(k)}, \boldsymbol{\psi})dt.$$

To show the convergence of above integrals, we restrict ourselves only to the terms containing $\partial y_1 h^{(k)}$, convergence of terms with $h^{(k)}$ is analogous. Let us consider

$$\int_0^T \int_D \left( \frac{\partial \boldsymbol{u}^{(k)}}{\partial y_2} - \frac{\partial \boldsymbol{u}}{\partial y_2} \right) \cdot \boldsymbol{\psi} u_1 \frac{\partial h}{\partial y_1} + \frac{\partial \boldsymbol{u}^{(k)}}{\partial y_2} \cdot \boldsymbol{\psi} \left( u_1^{(k)} - u_1 \right) \frac{\partial h^{(k)}}{\partial y_1}$$
$$\frac{\partial \boldsymbol{u}^{(k)}}{\partial y_2} \cdot \boldsymbol{\psi} \left( \frac{\partial h^{(k)}}{\partial y_1} - \frac{\partial h}{\partial y_1} \right) u_1^{(k)} dy\, dt.$$

The convergence of the first term is obvious due to the weak convergence of $\boldsymbol{u}^{(k)}$ in $L^p(0,T;W^{1,p}(D))$. The strong convergence of $\boldsymbol{u}^{(k)}$ in $L^{p'}((0,T) \times (D))$ and the uniform convergence of $\partial_{y_1} h^{(k)}$ imply the convergence in the remaining two terms.

Now we denote $\hat{e}^{(k)} := \hat{e}_{h^{(k)}}$, $\hat{e} := \hat{e}_h$, cf. (26) and Remark 2.2. The limiting process in the viscous term will be realized as follows.

$$\int_0^T ((\boldsymbol{u}^{(k)}, \boldsymbol{\psi}))_{h^{(k)}} - ((\boldsymbol{u}, \boldsymbol{\psi}))_h dt \tag{62}$$
$$= \int_0^T \int_D h \left[ \tau_{ij}(\hat{e}^{(k)}(\boldsymbol{u}^{(k)}))\hat{e}_{ij}^{(k)}(\boldsymbol{\psi}) - \tau_{ij}(\hat{e}(\boldsymbol{u}))\hat{e}_{ij}(\boldsymbol{\psi}) \right]$$
$$+ \left[ h^{(k)} - h \right] \tau_{ij}(\hat{e}^{(k)}(\boldsymbol{u}^{(k)}))\hat{e}_{ij}^{(k)}(\boldsymbol{\psi})\, dy\, dt$$
$$= \underbrace{\int_0^T \int_D h\, \tau_{ij}(\hat{e}^{(k)}(\boldsymbol{u}^{(k)})) \left[ \hat{e}_{ij}^{(k)}(\boldsymbol{\psi}) - \hat{e}_{ij}(\boldsymbol{\psi}) \right] dy\, dt}_{\text{(I)}}$$
$$+ \underbrace{\int_0^T \int_D h \left[ \tau_{ij}(\hat{e}^{(k)}(\boldsymbol{u}^{(k)})) - \tau_{ij}(\hat{e}(\boldsymbol{u})) \right] \hat{e}_{ij}(\boldsymbol{\psi})\, dy\, dt}_{\text{(II)}}$$
$$+ \underbrace{\int_0^T \int_D \left[ h^{(k)} - h \right] \tau_{ij}(\hat{e}^{(k)}(\boldsymbol{u}^{(k)}))\hat{e}_{ij}^{(k)}(\boldsymbol{\psi})\, dy\, dt}_{\text{(III)}}.$$

It is easy to see that the term (III) goes to zero. Using the fact that

$$\hat{e}_h(\boldsymbol{u}) = \nabla \boldsymbol{u} F(h) + (\nabla \boldsymbol{u} F(h))^T \in \mathbb{R}^{2\times 2}_{sym}; \quad F(h) = \frac{1}{2} \begin{bmatrix} 1 & 0 \\ \frac{-y_2}{h}\frac{\partial h}{\partial y_1} & \frac{1}{h} \end{bmatrix} \tag{63}$$

and due to the uniform convergence of $h^{(k)}$ in $C(0,T;C^1[0,L])$ the convergence in all components of $F$ is obvious and we obtain that (I) $\to 0$.

In order to show the convergence in the term (II), we will use the Minty- Trick. Let us denote for better readability $\xi^k := \hat{e}_{h^{(k)}}(\boldsymbol{u}^{(k)})$ $\xi := \hat{e}_h(\boldsymbol{u})$ and $\phi := \hat{e}(\boldsymbol{\psi})$. Define the operator $\mathcal{A}$, $\mathcal{A} : L^p((0,T)\times D) \to L^{p'}((0,T)\times D)$,

$$\langle \mathcal{A}(\xi^k), \phi \rangle := \int_0^T \int_D h\, \tau_{ij}(\xi^k)\phi\, dy\, dt.$$

Lemma 2.4 implies the monotonicity of operator $\mathcal{A}$, $\langle \mathcal{A}(\xi^k) - \mathcal{A}(\xi), \xi^k - \xi \rangle \geq 0$. Further, from assumptions (16) on $h^{(k)}$ and Lemma 2.5 we obtain for $\boldsymbol{u}, \boldsymbol{\psi} \in L^p(0,T;W^{1,p}(D))$ and for any $k$

$$\left|\langle A(\xi^k), \phi \rangle\right| \leq c(K,\alpha)\|\phi\|_{L^p(0,T;L^p(D))}.$$

Thus $\mathcal{A}$ is bounded in $L^{p'}((0,T)\times D)$ and consequently $\mathcal{A}(\xi^k) \rightharpoonup f$. Moreover, from the weak convergence of $\nabla \boldsymbol{u}^{(k)}$ and the uniform convergence of $h^{(k)}$ in $C(0,T;C^1[0,L])$ we obtain the weak convergence $\xi^k \rightharpoonup \xi$.

Now we prove that $\lim_{k\to\infty}\langle A(\xi^k), \xi^k \rangle = \langle f, \xi \rangle$, i.e., we show that $\lim_{k\to\infty}\langle A(\xi^k), \xi^k - \xi \rangle = 0$. To this end, we limit in the rest terms of the weak formulation (40) with $\boldsymbol{u}^{(k)}$, $h^{(k)}$, $\sigma^{(k)}$ instead of $\boldsymbol{u}$, $h$, $\sigma$ using test functions $\boldsymbol{\psi} = \boldsymbol{u}^{(k)} - \boldsymbol{u}$. In what follows we present the limiting process in the nonlinear convective term. Recalling (30) we can write

$$\int_0^T b_{h^{(k)}}(\boldsymbol{u}^{(k)}, \boldsymbol{u}^{(k)}, \boldsymbol{u}^{(k)} - \boldsymbol{u})\, dt = \frac{1}{2}\int_0^T B_{h^{(k)}}(\boldsymbol{u}^{(k)}, \boldsymbol{u}^{(k)}, \boldsymbol{u}^{(k)} - \boldsymbol{u}) - B_{h^{(k)}}(\boldsymbol{u}^{(k)}, \boldsymbol{u}^{(k)} - \boldsymbol{u}, \boldsymbol{u}^{(k)})\, dt. \tag{64}$$

We can estimate the first term on the right hand side using the Hölder and the interpolation inequality $\|\varphi\|_4 \leq c\|\varphi\|_{1,2}^{1/2}\|\varphi\|_2^{1/2}$, cf. [9, Lemma 3.1]

$$\int_0^T B_{h^{(k)}}(\boldsymbol{u}^{(k)}, \boldsymbol{u}^{(k)}, \boldsymbol{u}^{(k)} - \boldsymbol{u})dt \leq$$

$$C(K,\alpha)\|\boldsymbol{u}^{(k)} - \boldsymbol{u}\|_{L^2(L^2)}\left(\|\boldsymbol{u}^{(k)}\|^{\frac{3}{2}}_{L^2(W^{1,2})} + \|\boldsymbol{u}^{(k)}\|_{L^2(W^{1,2})}\|\boldsymbol{u}\|^{\frac{1}{2}}_{L^2(W^{1,2})}\right),$$

here $L^2(W^{1,2}) := L^2(0,T;W^{1,2}(D))$. Thus, the strong convergence of $\boldsymbol{u}^{(k)}$ in $L^2((0,T)\times D)$ implies, that the first term on the right hand side of (64) tends to 0. Further, we can rewrite the second term as

$$B_{h^{(k)}}(\boldsymbol{u}^{(k)}, \boldsymbol{u}^{(k)} - \boldsymbol{u}, \boldsymbol{u}^{(k)}) = B_h(\boldsymbol{u}, \boldsymbol{u}^{(k)} - \boldsymbol{u}, \boldsymbol{u}) + B_{(h^{(k)}-h)}(\boldsymbol{u}, \boldsymbol{u}^{(k)} - \boldsymbol{u}, \boldsymbol{u})$$
$$+ B_{h^{(k)}}(\boldsymbol{u}^{(k)} - \boldsymbol{u}, \boldsymbol{u}^{(k)} - \boldsymbol{u}, \boldsymbol{u}^{(k)}) + B_{h^{(k)}}(\boldsymbol{u}, \boldsymbol{u}^{(k)} - \boldsymbol{u}, \boldsymbol{u}^{(k)} - \boldsymbol{u}).$$

Due to the weak convergence of $\nabla \boldsymbol{u}^{(k)}$, uniform convergence of $h^{(k)}$ and the strong convergence of $\boldsymbol{u}^{(k)}$, cf. (59) we obtain also for the second term $\int_0^T B_{h^{(k)}}(\boldsymbol{u}^{(k)}, \boldsymbol{u}^{(k)} - \boldsymbol{u}, \boldsymbol{u}^{(k)})dt \to 0$. This concludes the proof of convergence in the convective term (64). The limiting process in the remaining terms of (40) is obvious and we omit it here.

Consequently, we have obtained $\lim_{k\to\infty} \langle A(\xi^k), \xi^k \rangle = \langle f, \xi \rangle$ and the Minty-Trick argument implies that $f = \mathcal{A}(\xi)$, i.e.

$$\mathcal{A}(\xi^k) \rightharpoonup \mathcal{A}(\xi) \quad \text{and thus} \quad \langle \mathcal{A}(\xi^k), \phi \rangle \to \langle \mathcal{A}(\xi), \phi \rangle$$

for any $\phi \in L^p((0,T) \times D)$ as $k \to \infty$.

This concludes the limiting process in (40) and the Sect. 4.2. We have found out that $\mathcal{F}(\delta^{(k)}) \to \mathcal{F}(\delta)$ as $k \to \infty$ and that $\eta$ is the weak solution of (40) associated with the limit $\delta$, ($h = R_0 + \delta$), thus $\mathcal{F}(\delta) = \eta$.

Finally, using the continuity of the mapping $\mathcal{F}$, its relative compactness in $Y$ and the property $\mathcal{F}(B_{\alpha,K}) \subset B_{\alpha,K}$ we deduce from the Schauder fixed point theorem, that there exists at least one fixed point of the mapping $\mathcal{F}$ defined by the weak formulation (40), $\mathcal{F}(\eta) = \eta$. Thus, we obtain the existence of at least one weak solution (14) of the original unsteady fluid-structure interaction problem (1)–(13). The proof of the Theorem 1.2 is now completed. □

**Remark on the Global Existence Result** Let us point out that we have obtained the existence of weak solution until some time $T^*$. We remind that this time is obtained in order to achieve the fixed point of the mapping $\mathcal{F}$ and to avoid the contact of the elastic boundary $\Gamma_w(t)$ with the fixed boundary for given data $P_{in}, P_{out}, P_w, R_0$ and $\alpha$, $K$. Similarly as in [4, Grandmont et al.], we can prolongate the solution in time and even obtain the global existence until the contact with the solid bottom.

Indeed, we can construct a non-decreasing sequence of times $\{T^* = T_1^*, \ldots, T_{m-1}^*, T_m^*, \ldots\}$, such that for given $\alpha$, $K$, $\alpha \leq \min\{R_{min}, \frac{1}{R_{min}+R_{max}}\}$, starting from initial data in time $T_{m-1}^*$, we have the existence of weak solution for some time $T_{m-1}^* + T := T_m^*$. We distinguish between two situations. Either $\sup T_m^* = \infty$, which means, that the contact with the solid bottom never happens and we obtain global existence. Otherwise $\sup T_m^* := T^{**} < \infty$ for given $\alpha$. In this case we can decrease $\alpha$. If the time interval of the existence cannot be prolongated for chosen $\alpha$, we have to decrease $\alpha$ again. This is repeated until $\alpha$ reaches 0. The later represents the contact with the solid boundary at some time $T^{**} + \bar{T}$, where $\bar{T} \geq 0$.

**Acknowledgements** The present research has been financed by the DFG project ZA 613/1-1, the Nečas Centrum for Mathematical Modelling LC06052 (financed by MSMT) and the Grant of the Czech Republic, No. P201/11/1304. It has also been partially supported by the 6th EU-Framework Programme under the Contract No. DEASE: MEST-CT-2005-021122 and the DST-DAAD project based personnel exchange program with Indian Institute of Science, Bangalore. We would like to thank Ján Filo (Comenius University, Bratislava) for fruitful discussions on the topic.

## References

1. H.W. Alt, S. Luckhaus, Quasilinear elliptic-parabolic differential equations. Math. Z. **183**, 311–341 (1983)
2. H. Brezis, *Analyse Fonctionelle- Théorie et Applications* (Masson, Paris, 1983)
3. S. Čanić, B. Muha, Existence of a weak solution to a nonlinear fluid-structure interaction problem modeling the flow of an incompressible, viscous fluid in a cylinder with deformable walls. Arch. Ration. Mech. Anal. **207**(4), 919–968 (2013)
4. A. Chambolle, B. Desjardin, M.J. Esteban, C. Grandmont, Existence of weak solutions for unsteady fluid-plate interaction problem. J. Math. Fluid. Mech. **4**(3), 368–404 (2005)
5. L. Diening, M. Růžička, J. Wolf, Existence of weak solutions for unsteady motions of generalized Newtonian fluids. Ann. Sc. Norm. Super. Pisa Cl. Sci. **9**(1), 1–46 (2010)
6. J. Filo, A. Zaušková, 2D Navier-Stokes equations in a time dependent domain with Neumann type boundary conditions. J. Math. Fluid Mech. **12**(1), 1–46 (2010)
7. G.P. Galdi, *An Introduction to the Theory of Navier-Stokes Equations I* (Springer, New York, 1994)
8. A. Hundertmark-Zaušková, M. Lukáčová-Medviďová, Numerical study of shear-dependent non-Newtonian fluids in compliant vessels. Comput. Math. Appl. **60**, 572–590 (2010)
9. A. Hundertmark-Zaušková, M. Lukáčová-Medviďová, Š. Nečasová On the existence of weak solution to the coupled fluid-structure interaction problem for non-Newtonian shear-dependent fluid. J. Math. Soc. Japan (in press)
10. M. Lukáčová-Medviďová, G. Rusnáková, A. Hundertmark-Zaušková, Kinematic splitting algorithm for fluid-structure interaction in hemodynamics. Comput. Methods Appl. Mech. Eng. **265**, 83–106 (2013)
11. J. Málek, J. Nečas, M. Rokyta, M. Růžička, *Weak and Measure-Valued Solutions to Evolutionary PDEs* (Chapman and Hall, London, 1996)
12. J. Málek, K.R. Rajagopal, Mathematical issues concerning the Navier-Stokes equations and some of its generalizations, in *Handbook of Differential Equations*, ed. by C.M. Dafermos, E. Feireisl (North-Holland Publishing Company, Amsterdam, 2005)
13. A. Quarteroni, Mathematical and numerical simulation of the cardiovascular system, in *Proceedings of the ICM*, Beijing, vol. 3, 2002, pp. 839–850
14. A. Quarteroni, L. Formaggia, Computational models in the human body, in *Handbook of Numerical Analysis*, vol. XII, ed. by P.G. Ciarlet, Guest Editor N. Ayache (Elsevier/North Holland, Amsterdam, 2004)
15. J. Wolf, Existence of weak solution to the equations of non-stationary motion of non-Newtonian fluids with shear rate dependent viscosity. J. Math. Fluid Mech. **9**(1), 104–138 (2007)
16. K.K. Yeleswarapu, Evaluation of continuum models for characterizing the constitutive behavior of blood. Ph.D. Thesis, University of Pittsburgh, Pittsburgh, 1996
17. A. Zaušková, 2D Navier–Stokes equations in a time dependent domain. Ph.D. Thesis, Comenius University, Bratislava, 2007

# Stability of Time Periodic Solutions for the Rotating Navier-Stokes Equations

**Tsukasa Iwabuchi, Alex Mahalov, and Ryo Takada**

*Dedicated to Professor Yoshihiro Shibata on his sixtieth birthday*

**Abstract** We consider the stability problem of time periodic solutions for the rotating Navier-Stokes equations. For the non-rotating case, it is known that time periodic solutions to the original Navier-Stokes equations are asymptotically stable under the smallness assumptions both on the time periodic solutions and on the initial disturbances. We shall treat the high-rotating cases, and prove the asymptotic stability of large time periodic solutions for large initial perturbations.

**Keywords** Asymptotic stability • The rotating Navier-Stokes equations • Time periodic solutions

## 1 Introduction

Let us consider the time periodic problems for the rotating Navier-Stokes equations, describing the motion of viscous incompressible fluids in the rotational framework:

$$\begin{cases} \dfrac{\partial v}{\partial t} - \Delta v + \Omega e_3 \times v + (v \cdot \nabla) v + \nabla q = f & t \in \mathbb{R}, x \in \mathbb{R}^3, \\ \operatorname{div} v = 0 & t \in \mathbb{R}, x \in \mathbb{R}^3, \end{cases} \tag{1}$$

T. Iwabuchi
Department of Mathematics, Osaka City University and OCAMI, Sumiyoshi-ku, Osaka 558-8585, Japan
e-mail: iwabuchi@sci.osaka-cu.ac.jp

A. Mahalov
School of Mathematical and Statistical Science, Arizona State University, PO Box 871804, Tempe, AZ 85287, USA
e-mail: mahalov@asc.edu

R. Takada (✉)
Mathematical Institute, Tohoku University, Sendai 980–8578, Japan
e-mail: takada@math.tohoku.ac.jp



H. Amann et al. (eds.), *Recent Developments of Mathematical Fluid Mechanics*, Advances in Mathematical Fluid Mechanics, DOI 10.1007/978-3-0348-0939-9_17

where $v = v(t,x) = (v_1(t,x), v_2(t,x), v_3(t,x))$ and $q = q(t,x)$ denote the unknown velocity field and the unknown pressure of the fluid at the point $(t,x) \in \mathbb{R} \times \mathbb{R}^3$, respectively, while $f = f(t,x) = (f_1(t,x), f_2(t,x), f_3(t,x))$ denotes the given time periodic external force. Here $\Omega \in \mathbb{R}$ represents the speed of rotation around the vertical unit vector $e_3 = (0,0,1)$, which is called the Coriolis parameter. In [12], Koh et al. showed the existence and uniqueness of periodic in time solutions to (1) in the class

$$v \in BC(\mathbb{R}; \dot{W}^{s,p}(\mathbb{R}^3))^3 \quad \text{with} \quad \sup_{t\in\mathbb{R}} \|v(t)\|_{\dot{W}^{s,p}} \leqslant C|\Omega|^{\frac{3}{2}(\frac{1}{3}+\frac{s}{3})-\frac{1}{p}} \tag{2}$$

for $0 < s < 3/5$, $1/3 + s/9 \leqslant 1/p < \min\{1/3 + s/3, 1 - s\}$ and some constant $C = C(s,p)$.

The purpose of this paper is to show the stability of time periodic solutions to (1) in the class (2). If $v(0,x)$ is initially perturbed by $a$, the perturbed flow $w(t,x)$ is governed by the following initial value problem for the rotating Navier-Stokes equations:

$$\begin{cases} \dfrac{\partial w}{\partial t} - \Delta w + \Omega e_3 \times w + (w\cdot\nabla)w + \nabla\pi = f & t > 0, x \in \mathbb{R}^3, \\ \operatorname{div} w = 0 & t > 0, x \in \mathbb{R}^3, \\ w(0,x) = v(0,x) + a(x) & x \in \mathbb{R}^3. \end{cases} \tag{3}$$

In this paper, we shall prove that for every time periodic solution $v \in L^\infty(\mathbb{R}; \dot{W}^{s,p}(\mathbb{R}^3))^3$ and every initial disturbance $a \in \dot{H}^s(\mathbb{R}^3)^3$ with $1/2 < s < 3/5$ and $1/3 + s/9 \leqslant 1/p < 1 - s$, there exists a unique global mild solution $w$ to (3) such that the global space-time integral

$$\int_0^\infty \|w(t) - v(t)\|_{\dot{W}^{s,p}}^\theta \, dt \quad \text{for some } \theta = \theta(s,p) \text{ with } 2 < \theta < \infty$$

converges to zero with algebraic decay rates as $|\Omega| \to \infty$, provided that the speed of rotation is sufficiently high. Note that in the high-rotating case we do not impose the smallness assumptions either on time periodic solutions or on initial disturbance. Introducing a new unknown vector $u := w - v$, we can reduce the original equations (3) to the following ones with the initial disturbance $a$:

$$\begin{cases} \dfrac{\partial u}{\partial t} - \Delta u + \Omega e_3 \times u & \\ \qquad + (u\cdot\nabla)v + (v\cdot\nabla)u + (u\cdot\nabla)u + \nabla p = 0 & t > 0, x \in \mathbb{R}^3, \\ \operatorname{div} u = 0 & t > 0, x \in \mathbb{R}^3, \\ u(0,x) = a(x) & x \in \mathbb{R}^3. \end{cases} \tag{4}$$

Hence our problem on the stability of time periodic solutions $v$ can be reduced to an investigation into the unique existence of the global solution $u$ to (4) and its asymptotic behavior.

Let us review the known results on the existence of time periodic solutions to (1) and its stability. In the non-rotating case $\Omega = 0$, (1) corresponds to the original Navier-Stokes equations and there are a number of literatures concerning our problem. We shall only mention the results on time periodic solutions in unbounded domains. Maremonti [17, 18] considered the whole space $\mathbb{R}^3$ and the half space $\mathbb{R}^3_+$, respectively, and proved the unique existence of time periodic solution and its asymptotic stability under small perturbations. Kozono and Nakao [14] introduced the notion of mild solutions to the time periodic problems for (1), and proved the existence and uniqueness of time periodic mild solutions in the whole space $\mathbb{R}^d$, the half space $\mathbb{R}^d_+$ with $d \geqslant 3$ and exterior domains in $\mathbb{R}^d$ with $d \geqslant 4$. The asymptotic stability of the periodic solutions constructed in [14] was proved by Taniuchi [19]. Yamazaki [20, 21] treated the 3-dimensional exterior domains and generalized the results of Kozono and Nakao [14], Taniuchi [19] for Morrey spaces, respectively. We remark that all of the stability results are obtained under the assumption that both time periodic solutions and initial perturbations are small enough.

In the rotating case $\Omega \in \mathbb{R}$, following the idea in [14] several authors obtained the existence results of time periodic mild solutions to (1). Konieczny and Yoneda [13] proved the existence of stationary mild solutions to (1). Kozono et al. [16] showed the uniform solvabilities of (1) and (4) and generalized the results of Kozono and Nakao [14], Taniuchi[19] for all $\Omega \in \mathbb{R}$. In the high-rotating case $|\Omega| \gg 1$, as is already known in [2–6], the Coriolis force $\Omega e_3 \times u$ exhibits a dispersion phenomenon which is closely related to dispersive estimates for the operator

$$e^{\pm i\Omega t \frac{D_3}{|D|}} f(x) := \int_{\mathbb{R}^3} e^{ix\cdot\xi \pm i\Omega t \frac{\xi_3}{|\xi|}} \hat{f}(\xi) d\xi$$

see Lemma 2.2 in Sect. 2. Based on such estimates, the authors in [9, 12] proved the unique existence of time periodic solutions to (1) for large external forces provided that the speed of rotation is sufficiently large.

Now let us state our main result. In order to solve (4), we consider the following integral equations:

$$\begin{aligned} u(t) = T_\Omega(t)a &- \int_0^t T_\Omega(t-\tau)\mathbb{P}\nabla \cdot [v(\tau) \otimes u(\tau)]\, d\tau \\ &- \int_0^t T_\Omega(t-\tau)\mathbb{P}\nabla \cdot [u(\tau) \otimes v(\tau)]\, d\tau \qquad (5) \\ &- \int_0^t T_\Omega(t-\tau)\mathbb{P}\nabla \cdot [u(\tau) \otimes u(\tau)]\, d\tau, \end{aligned}$$

where $\mathbb{P} = (\delta_{jk}+R_jR_k)_{1\leqslant j,k\leqslant 3}$ denotes the Helmholtz projection onto the divergence-free vector fields and $T_\Omega(\cdot)$ denotes the semigroup corresponding to the linear problem of (4), which is given explicitly by

$$T_\Omega(t)g = \mathscr{F}^{-1}\left[\cos\left(\Omega\frac{\xi_3}{|\xi|}t\right)e^{-|\xi|^2t}I\hat{g}(\xi) + \sin\left(\Omega\frac{\xi_3}{|\xi|}t\right)e^{-|\xi|^2t}R(\xi)\hat{g}(\xi)\right]$$

for $t \geqslant 0$ and the divergence-free vector field $g$. Here $\{R_j\}_{j=1}^3$ denote the Riesz transforms, $I$ is the identity matrix in $\mathbb{R}^3$ and $R(\xi)$ is the skew-symmetric matrix related to the symbol of the Riesz transforms, which is defined by

$$R(\xi) := \frac{1}{|\xi|}\begin{pmatrix} 0 & \xi_3 & -\xi_2 \\ -\xi_3 & 0 & \xi_1 \\ \xi_2 & -\xi_1 & 0 \end{pmatrix}, \quad \xi \in \mathbb{R}^3 \setminus \{0\}.$$

For the derivation of the explicit form of the semigroup $T_\Omega(t)$, we refer to Babin et al. [1–3] and Hieber and Shibata [8]. We call that $u$ is a mild solution to (4) if $u$ satisfies (5) in some appropriate function space.

Our theorem on the stability of time periodic solutions now reads:

**Theorem 1.1** *Let $s, p$ and $\theta$ satisfy*

$$\frac{1}{2} < s < \frac{3}{5}, \quad \frac{1}{3} + \frac{s}{9} \leqslant \frac{1}{p} < 1 - s, \quad \frac{3}{4} - \frac{3}{2p} \leqslant \frac{1}{\theta} < \frac{5}{8} - \frac{3}{2p} + \frac{s}{4}. \tag{6}$$

*Then, there exist positive constants $\delta_1 = \delta_1(s,p)$ and $\delta_2 = \delta_2(s,p,\theta)$ such that for every $\Omega \in \mathbb{R} \setminus \{0\}$, every time periodic solution $v \in L^\infty(\mathbb{R}; \dot{W}^{s,p}(\mathbb{R}^3))$ and every initial disturbance $a \in \dot{H}^s(\mathbb{R}^3)$ with* $\operatorname{div} a = 0$ *satisfying*

$$\|v\|_{L^\infty(\mathbb{R};\dot{W}^{s,p})} \leqslant \delta_1|\Omega|^{\frac{3}{2}\left\{\left(\frac{1}{3}+\frac{s}{3}\right)-\frac{1}{p}\right\}}, \quad \|a\|_{\dot{H}^s} \leqslant \delta_2|\Omega|^{\frac{1}{2}\left(s-\frac{1}{2}\right)}, \tag{7}$$

*(4) possesses a unique mild solution $u$ in the class $C([0,\infty); \dot{H}^s(\mathbb{R}^3)) \cap L^\theta(0,\infty; \dot{W}^{s,p}(\mathbb{R}^3))$ satisfying* $\operatorname{div} u = 0$. *Furthermore, there exists a positive constant $C = C(p,\theta)$ such that*

$$\|u\|_{L^\theta(0,\infty;\dot{W}^{s,p})} \leqslant C|\Omega|^{-\left\{\frac{1}{\theta}-\left(\frac{3}{4}-\frac{3}{2p}\right)\right\}}\|a\|_{\dot{H}^s} \tag{8}$$

*for all $\Omega \in \mathbb{R} \setminus \{0\}$.*

*Remark 1.2* Theorem 1.1 states that the time periodic solution $v$ to (1) is asymptotically stable as $t \to \infty$ in the sense that the global in time $L^\theta(0,\infty; \dot{W}^{s,p}(\mathbb{R}^3))$-norm $\|w - v\|_{L^\theta(0,\infty;\dot{W}^{s,p})}$ of the difference between the perturbed flow $w$ to (3) and $v$ is finite. Furthermore, in the case $1/\theta > 3/4 - 3/(2p)$ and $|\Omega| \gg 1$, (8) yields that the perturbed flow $w$ behaves like the time periodic flow $v$ in the sense that

$$\|w - v\|_{L^\theta(0,\infty;\dot{W}^{s,p})} = O(|\Omega|^{-\left\{\frac{1}{\theta}-\left(\frac{3}{4}-\frac{3}{2p}\right)\right\}}) \quad \text{as } |\Omega| \to \infty.$$

*Remark 1.3* In the non-rotating case $\Omega = 0$, the stability of time periodic solutions to the original Navier-Stokes equations was proved under the smallness assumptions both on the time periodic solutions and on the initial disturbances (see [15, 19, 20]). On the other hand, since the powers of $|\Omega|$ in (7) are positive, we can obtain the unique global solution to (4) for large time periodic solutions $v$ and initial perturbations $a$ provided $\Omega$ satisfies

$$|\Omega| \geqslant \max\left\{\left(\delta_1^{-1}\|v\|_{L^\infty(\mathbb{R};\dot{W}^{s,p})}\right)^{\frac{2}{3\{(\frac{1}{3}+\frac{s}{3})-\frac{1}{p}\}}}, \left(\delta_2^{-1}\|a\|_{\dot{H}^s}\right)^{\frac{2}{s-\frac{1}{2}}}\right\}.$$

*Remark 1.4* The size conditions (7) to time periodic solutions and initial disturbances are closely related to the scaling invariance to (4). Indeed, if $(u, v, p)$ solves (4) with the parameter $\Omega$ then so does $(u_\lambda, v_\lambda, p_\lambda)$ with $\Omega_\lambda$ for all $\lambda > 0$, where

$$u_\lambda(t,x) := \lambda u(\lambda^2 t, \lambda x), \quad v_\lambda(t,x) := \lambda v(\lambda^2 t, \lambda x),$$
$$p_\lambda(t,x) := \lambda^2 p(\lambda^2 t, \lambda x), \quad \Omega_\lambda := \lambda^2 \Omega.$$

The size conditions (7) are invariant under the above scalings. This property could be regarded as a counterpart of the scaling invariance to the original Navier-Stokes equations such as $\|u_\lambda(0,\cdot)\|_{\dot{H}^{\frac{1}{2}}(\mathbb{R}^3)} = \|u(0,\cdot)\|_{\dot{H}^{\frac{1}{2}}(\mathbb{R}^3)}$ for all $\lambda > 0$.

In the periodic case $\mathbb{T}^3$, it seems to be difficult to obtain the characterization (7) for the size condition on initial data and time periodic solutions because of the presence of resonant domains and the lack of dispersion effect. Concerning the existence theorem and the global regularity results on $\mathbb{T}^3$, we refer to Babin et al. [1–3], and also Flandoli and Mahalov [7] for the stochastic case.

*Remark 1.5* In the non-rotating case $\Omega = 0$, the solutions to (4) constructed in [15, 19, 20] satisfy the integral equation (5) in the weak form. This difficulty is caused by the estimate for the convection terms $(u \cdot \nabla)v + (v \cdot \nabla)u$: Since a time periodic flow does not satisfy the time global estimates like $\sup_{t>0} t^{1/2}\|\nabla v(t)\|_{L^3(\mathbb{R}^3)} < \infty$, the standard Kato iteration scheme [11] does not work well. On the other hand, our solution in Theorem 1.1 satisfies (5) in the strong sense. This is due to the use of dispersive effects of the Coriolis force and the space-time integral estimates.

This paper is organized as follows. In Sect. 2, we recall the dispersive estimates due to the Coriolis force and prepare the linear estimates for the semigroup $T_\Omega(t)$. In Sect. 3, we prove the bilinear estimates associated with the convection terms. In Sect. 4, we give the proof of Theorem 1.1.

Throughout this paper, we denote by $C$ the constants which may differ from line to line. In particular, $C = C(\cdot, \dots, \cdot)$ will denote the constant which depends only on the quantities appearing in parentheses.

## 2 Linear Estimates

Let $\mathscr{S}(\mathbb{R}^3)$ be the Schwartz class, and let $\mathscr{S}'(\mathbb{R}^3)$ be the space of tempered distributions. Let $\{\varphi_j\}_{j\in\mathbb{Z}} \subset \mathscr{S}(\mathbb{R}^3)$ be a smooth partition of unity in the frequency space satisfying the following properties:

$$
\begin{aligned}
0 \leqslant \varphi_0(\xi) \leqslant 1 \quad &\text{for all } \xi \in \mathbb{R}^3, \\
\operatorname{supp} \varphi_0 \subset \{\xi \in \mathbb{R}^3 \mid 1/2 \leqslant |\xi| \leqslant 2\}&, \\
\varphi_j(\xi) = \varphi_0\left(\frac{\xi}{2^j}\right) \quad &\text{for all } j \in \mathbb{Z}
\end{aligned}
$$

and

$$
\sum_{j\in\mathbb{Z}} \varphi_j(\xi) = 1 \quad \text{for all } \xi \in \mathbb{R}^3 \setminus \{0\}.
$$

For $j \in \mathbb{Z}$ and $f \in \mathscr{S}'(\mathbb{R}^3)$, we define the frequency localized projection operator $\Delta_j$ by

$$
\widehat{\Delta_j f}(\xi) := \varphi_j(\xi)\hat{f}(\xi).
$$

Then, we recall the definition of the homogeneous Besov space $\dot{B}^s_{p,q}(\mathbb{R}^3)$.

**Definition 2.1** For $s \in \mathbb{R}$ and $1 \leqslant p, q \leqslant \infty$, the homogeneous Besov space $\dot{B}^s_{p,q}(\mathbb{R}^3)$ is defined to be the set of all tempered distributions $f \in \mathscr{S}'(\mathbb{R}^3)$ such that

$$
\|f\|_{\dot{B}^s_{p,q}} := \left\| \{2^{sj} \|\Delta_j f\|_{L^p}\}_{j\in\mathbb{Z}} \right\|_{\ell^q} < \infty.
$$

Next, we define the operators $e^{\pm it\frac{D_3}{|D|}}$ by the Fourier integral

$$
e^{\pm it\frac{D_3}{|D|}} f(x) := \int_{\mathbb{R}^3} e^{ix\cdot\xi \pm it\frac{\xi_3}{|\xi|}} \hat{f}(\xi) d\xi, \quad t \in \mathbb{R},\ x \in \mathbb{R}^3.
$$

Then, we can rewrite the semigroup $T_\Omega(t)$ as

$$\begin{aligned} T_\Omega(t)f &= \mathscr{F}^{-1}\left[\cos\left(\Omega\frac{\xi_3}{|\xi|}t\right)e^{-|\xi|^2t}I\hat{f}(\xi) + \sin\left(\Omega\frac{\xi_3}{|\xi|}t\right)e^{-|\xi|^2t}R(\xi)\hat{f}(\xi)\right] \\ &= \frac{1}{2}e^{i\Omega t\frac{D_3}{|D|}}e^{t\Delta}(I+\mathcal{R})f + \frac{1}{2}e^{-i\Omega t\frac{D_3}{|D|}}e^{t\Delta}(I-\mathcal{R})f \end{aligned} \tag{9}$$

for $t \geqslant 0$ and $\Omega \in \mathbb{R}$, where $e^{t\Delta}$ denotes the standard heat semigroup and $\mathcal{R}$ denotes the matrix of singular integral operators defined by

$$\mathcal{R} := \begin{pmatrix} 0 & R_3 & -R_2 \\ -R_3 & 0 & R_1 \\ R_2 & -R_1 & 0 \end{pmatrix}.$$

The operators $e^{\pm i\Omega t\frac{D_3}{|D|}}$ represent the oscillation parts of $T_\Omega(t)$. Let us first recall the dispersive estimates for $e^{\pm i\Omega t\frac{D_3}{|D|}}$.

**Lemma 2.2 ([12, Lemma 2.2])** *For $2 \leqslant p \leqslant \infty$, there exists a positive constant $C = C(p)$ such that*

$$\left\| e^{\pm it\frac{D_3}{|D|}} f \right\|_{\dot{B}^s_{p,q}} \leqslant C(1+|t|)^{-(1-\frac{2}{p})} \|f\|_{\dot{B}^{s+3(1-\frac{2}{p})}_{p',q}}$$

*for all $t \in \mathbb{R}$, $s \in \mathbb{R}$, $1 \leqslant q \leqslant \infty$, and $f \in \dot{B}^{s+3(1-\frac{2}{p})}_{p',q}(\mathbb{R}^3)$, where $1/p + 1/p' = 1$.*

By the explicit formula (9) of the semigroup $T_\Omega(t)$, Lemma 2.2 and the smoothing properties of the heat semigroup yield the following $L^q$-$L^p$ estimates for $T_\Omega(t)$.

**Lemma 2.3 ([12, Lemma 3.2])** *Let $\alpha \in (\mathbb{N} \cup \{0\})^3$, and let $p$ and $q$ satisfy $2 \leqslant p < \infty$ and $1 < q \leqslant p'$, where $1/p + 1/p' = 1$. Then there exists a positive constant $C = C(\alpha, p, q)$ such that*

$$\left\| \partial_x^\alpha T_\Omega(t)f \right\|_{L^p} \leqslant C(1+|\Omega|t)^{-(1-\frac{2}{p})} t^{-\frac{3}{2}(\frac{1}{q}-\frac{1}{p})-\frac{|\alpha|}{2}} \|f\|_{L^q}$$

*for all $t > 0$, $\Omega \in \mathbb{R}$ and $f \in L^q(\mathbb{R}^3)^3$.*

Using the $TT^*$ method and Lemma 2.3, we have the following homogeneous and inhomogeneous space-time estimates for our semigroup $T_\Omega(t)$.

**Lemma 2.4 ([12, Lemma 3.3])** *Let $2 < p < 6$, and let $\theta$ satisfy*

$$\frac{3}{4} - \frac{3}{2p} \leqslant \frac{1}{\theta} < \min\left\{\frac{1}{2}, \frac{5}{4} - \frac{5}{2p}\right\}.$$

*Then there exists a positive constant* $C = C(p, \theta)$ *such that*

$$\|T_\Omega(\cdot)f\|_{L^\theta(0,\infty:L^p)} \leqslant C|\Omega|^{-\left\{\frac{1}{\theta}-(\frac{3}{4}-\frac{3}{2p})\right\}}\|f\|_{L^2} \tag{10}$$

*for all* $\Omega \in \mathbb{R}\setminus\{0\}$ *and* $f \in L^2(\mathbb{R}^3)^3$. *In particular, in the critical case* $1/\theta = 3/4 - 3/2p$, (10) *holds for all* $\Omega \in \mathbb{R}$.

**Lemma 2.5 ([12, Lemma 3.4])** *Let* $p, q$ *and* $\theta$ *satisfy*

$$2 < p < 3, \quad 1 - \frac{1}{p} \leqslant \frac{1}{q} < \frac{1}{p} + \frac{1}{3}$$

*and*

$$\max\left\{0, \frac{1}{2} - \frac{3}{2}\left(\frac{1}{q} - \frac{1}{p}\right) - \left(1 - \frac{2}{p}\right)\right\} < \frac{1}{\theta} \leqslant \frac{1}{2} - \frac{3}{2}\left(\frac{1}{q} - \frac{1}{p}\right),$$

*respectively. Then there exists a positive constant* $C = C(p, q, \theta)$ *such that*

$$\left\|\int_0^t T_\Omega(t-\tau)\mathbb{P}\nabla f(\tau)d\tau\right\|_{L^\theta(0,\infty;L^p)} \leqslant C|\Omega|^{-\left\{\frac{1}{2}-\frac{3}{2}(\frac{1}{q}-\frac{1}{p})-\frac{1}{\theta}\right\}}\|f\|_{L^{\frac{\theta}{2}}(0,\infty;L^q)} \tag{11}$$

*for all* $\Omega \in \mathbb{R}\setminus\{0\}$ *and* $f \in L^{\frac{\theta}{2}}(0,\infty; L^q(\mathbb{R}^3))$. *In particular, in the case* $1/\theta = 1/2 - 3(1/q - 1/p)/2$, (11) *holds for all* $\Omega \in \mathbb{R}$.

In order to estimate the convection terms $(u\cdot\nabla)v + (v\cdot\nabla)u$, we first prepare the following inhomogeneous space-time estimate, which is a variant of Lemma 2.5.

**Lemma 2.6** *Let* $p$ *and* $q$ *satisfy* $2 < p < 3$ *and*

$$1 - \frac{1}{p} \leqslant \frac{1}{q} < \frac{1}{p} + \frac{1}{3}, \quad \frac{7}{3p} - \frac{1}{3} < \frac{1}{q} < \frac{1}{p} + \frac{1}{3}.$$

*Then there exists a positive constant* $C = C(p, q)$ *such that*

$$\left\|\int_0^t T_\Omega(t-\tau)\mathbb{P}\nabla f(\tau)d\tau\right\|_{L^\theta(0,\infty;L^p)} \leqslant C|\Omega|^{-\frac{3}{2}\left\{\left(\frac{1}{3}+\frac{1}{p}\right)-\frac{1}{q}\right\}}\|f\|_{L^\theta(0,\infty;L^q)}$$

*for all* $\Omega \in \mathbb{R}\setminus\{0\}$, $1 \leqslant \theta \leqslant \infty$ *and* $f \in L^\theta(0,\infty; L^q(\mathbb{R}^3))$.

*Proof* Since $\mathbb{P}$ is bounded in $L^q(\mathbb{R}^3)$, it follows from Lemma 2.3 that

$$\left\|\int_0^t T_\Omega(t-\tau)\mathbb{P}\nabla f(\tau)d\tau\right\|_{L^p} \leqslant C\int_0^t k_\Omega(t-\tau)\|f(\tau)\|_{L^q}d\tau,$$

where

$$k_\Omega(t) := \{1 + |\Omega|t\}^{-(1-\frac{2}{p})} t^{-\frac{3}{2}(\frac{1}{q}-\frac{1}{p})-\frac{1}{2}}.$$

Hence by the Hausdorff-Young inequality we have

$$\begin{aligned}\left\| \int_0^t T_\Omega(t-\tau)\mathbb{P}\nabla f(\tau)d\tau \right\|_{L^\theta(0,\infty;L^p)} &\leqslant C \left\| \int_0^t k_\Omega(t-\tau)\|f(\tau)\|_{L^q} d\tau \right\|_{L^\theta(0,\infty)} \\ &\leqslant C\, \|k_\Omega\|_{L^1(0,\infty)}\, \|f\|_{L^\theta(0,\infty;L^q)}.\end{aligned}$$

Here, by the change of variable $t \to t/|\Omega|$, we see that

$$\begin{aligned}\|k_\Omega\|_{L^1(0,\infty)} &= \int_0^\infty \frac{1}{(1+|\Omega|t)^{1-\frac{2}{p}}} \frac{1}{t^{\frac{3}{2}(\frac{1}{q}-\frac{1}{p})+\frac{1}{2}}} dt \\ &= |\Omega|^{-\frac{3}{2}\{(\frac{1}{3}+\frac{1}{p})-\frac{1}{q}\}} \int_0^\infty \frac{1}{(1+t)^{1-\frac{2}{p}}} \frac{1}{t^{\frac{3}{2}(\frac{1}{q}-\frac{1}{p})+\frac{1}{2}}} dt. \qquad (12)\end{aligned}$$

By the assumptions on $p$ and $q$, the last integral in (12) converges. This completes the proof of Lemma 2.6. □

## 3 Nonlinear Estimates

In this section, we shall prove several bilinear estimates. First, we recall the following bilinear estimates in the Sobolev spaces of fractional order.

**Lemma 3.1 ([10, Lemma 2.7])** *Let $s, p$ and $q$ satisfy*

$$0 \leqslant s < 3, \quad \frac{s}{3} < \frac{1}{p} < \frac{1}{2} + \frac{s}{6}, \quad \frac{1}{q} = \frac{2}{p} - \frac{s}{3}.$$

*Then there exists a positive constant $C = C(s, p, q)$ such that*

$$\|fg\|_{\dot{W}^{s,q}} \leqslant C \|f\|_{\dot{W}^{s,p}} \|g\|_{\dot{W}^{s,p}}$$

*for all $f, g \in \dot{W}^{s,p}(\mathbb{R}^3)$.*

Combining Lemmas 2.5 and 3.1, we have the following bilinear estimates for the nonlinear term $(u \cdot \nabla)u$.

**Lemma 3.2 ([12, Lemma 4.2])** *Let $s, p$ and $\theta$ satisfy*

$$0 < s < \frac{3}{2}, \quad \frac{1}{3} + \frac{s}{9} \leqslant \frac{1}{p} < \min\left\{\frac{1}{2}, \frac{1}{3} + \frac{s}{3}\right\}$$

*and*

$$\max\left\{0, \frac{1}{2p} + \frac{s}{2} - \frac{1}{2}\right\} < \frac{1}{\theta} \leqslant \frac{1}{2} - \frac{3}{2p} + \frac{s}{2}.$$

*Then there exists a positive constant $C = C(s, p, \theta)$ such that*

$$\begin{aligned} &\left\| \int_0^t T_\Omega(t-\tau)\mathbb{P}\nabla \cdot [u(\tau) \otimes v(\tau)]\, d\tau \right\|_{L^\theta(0,\infty;\dot{W}^{s,p})} \\ &\quad \leqslant C|\Omega|^{-\left\{(\frac{1}{2}-\frac{3}{2p}+\frac{s}{2})-\frac{1}{\theta}\right\}} \|u\|_{L^\theta(0,\infty;\dot{W}^{s,p})} \|v\|_{L^\theta(0,\infty;\dot{W}^{s,p})} \end{aligned} \tag{13}$$

*for all $\Omega \in \mathbb{R} \setminus \{0\}$ and $u, v \in L^\theta(0,\infty; \dot{W}^{s,p}(\mathbb{R}^3))^3$. In particular, in the case $1/\theta = 1/2 - 3/2p + s/2$, (13) holds for all $\Omega \in \mathbb{R}$.*

Similarly to Lemma 3.2, we shall show the inhomogeneous bilinear estimates for the convection terms $(u \cdot \nabla)v + (v \cdot \nabla)u$.

**Lemma 3.3** *Let $s$ and $p$ satisfy*

$$0 < s < \frac{3}{5}, \quad \frac{1}{3} + \frac{s}{9} \leqslant \frac{1}{p} < \min\left\{\frac{1}{3} + \frac{s}{3}, 1 - s\right\}.$$

*Then there exists a positive constant $C = C(s, p)$ such that*

$$\begin{aligned} &\left\| \int_0^t T_\Omega(t-\tau)\mathbb{P}\nabla \cdot [u(\tau) \otimes v(\tau)]\, d\tau \right\|_{L^\theta(0,\infty;\dot{W}^{s,p})} \\ &\quad \leqslant C|\Omega|^{-\frac{3}{2}\left\{(\frac{1}{3}+\frac{s}{3})-\frac{1}{p}\right\}} \|u\|_{L^\theta(0,\infty;\dot{W}^{s,p})} \|v\|_{L^\infty(0,\infty;\dot{W}^{s,p})} \end{aligned}$$

*for all $1 \leqslant \theta \leqslant \infty$, $u \in L^\theta(0,\infty; \dot{W}^{s,p}(\mathbb{R}^3))^3$, $v \in L^\infty(0,\infty; \dot{W}^{s,p}(\mathbb{R}^3))^3$ and $\Omega \in \mathbb{R} \setminus \{0\}$.*

*Proof* Set $1/q := 2/p - s/3$. Then, the exponents $(p, q)$ and $(s, p, q)$ satisfy the assumptions of Lemmas 2.6 and 3.1, respectively. Therefore, it follows from Lemmas 2.6 and 3.1 that

$$\begin{aligned} &\left\| \int_0^t T_\Omega(t-\tau)\mathbb{P}\nabla \cdot [u(\tau) \otimes v(\tau)]\, d\tau \right\|_{L^\theta(0,\infty;\dot{W}^{s,p})} \\ &\leqslant C|\Omega|^{-\frac{3}{2}\left\{(\frac{1}{3}+\frac{1}{p})-(\frac{2}{p}-\frac{s}{3})\right\}} \|u \otimes v\|_{L^\theta(0,\infty;\dot{W}^{s,q})} \\ &\leqslant C|\Omega|^{-\frac{3}{2}\left\{(\frac{1}{3}+\frac{s}{3})-\frac{1}{p}\right\}} \left\| \|u(\tau)\|_{\dot{W}^{s,p}} \|v(\tau)\|_{\dot{W}^{s,p}} \right\|_{L^\theta(0,\infty)} \\ &\leqslant C|\Omega|^{-\frac{3}{2}\left\{(\frac{1}{3}+\frac{s}{3})-\frac{1}{p}\right\}} \|u(\tau)\|_{L^\theta(0,\infty;\dot{W}^{s,p})} \|v(\tau)\|_{L^\infty(0,\infty;\dot{W}^{s,p})}. \end{aligned}$$

Hence we obtain Lemma 3.3. □

## 4 Proof of the Main Result

In this section, we shall give the proof of Theorem 1.1. The strategy of the proof is based on the idea in [10].

*Proof of Theorem 1.1* Let the exponents $s, p$ and $\theta$ satisfy (6). Then these exponents satisfy the assumptions of Lemmas 2.4, 3.2 and 3.3. Indeed, for $1/2 < s < 3/5$, it is easy to see that

$$\frac{1}{6} < \frac{1}{3} + \frac{s}{9} < \min\left\{\frac{1}{3} + \frac{s}{3}, 1 - s\right\} = 1 - s < \min\left\{\frac{1}{2}, \frac{1}{3} + \frac{s}{3}\right\} = \frac{1}{2}.$$

Furthermore, for such $s$ and $p$, we have

$$0 = \max\left\{0, \frac{1}{2p} + \frac{s}{2} - \frac{1}{2}\right\} < \frac{3}{4} - \frac{3}{2p} < \frac{5}{8} - \frac{3}{2p} + \frac{s}{4} < \begin{cases} \min\left\{\frac{1}{2}, \frac{5}{4} - \frac{5}{2p}\right\} = \frac{5}{4} - \frac{5}{2p}, \\ \frac{1}{2} - \frac{3}{2p} + \frac{s}{2}. \end{cases}$$

Let $\Omega \in \mathbb{R}\backslash\{0\}$, and suppose that $u_0 \in \dot{H}^s(\mathbb{R}^3)^3$ satisfies $\operatorname{div} u_0 = 0$. By Lemma 2.4, there exists a positive constant $C_0 = C_0(p, \theta)$ such that

$$\|T_\Omega(\cdot)a\|_{L^\theta(0,\infty;\dot{W}^{s,p})} \leqslant C_0|\Omega|^{-\left\{\frac{1}{\theta} - \left(\frac{3}{4} - \frac{3}{2p}\right)\right\}} \|a\|_{\dot{H}^s}. \tag{14}$$

Then, we define the map $\Phi$ and the solution space $(X, d_X)$ with the metric $d_X$ by

$$\begin{aligned}\Phi(u)(t) := T_\Omega(t)a &- \int_0^t T_\Omega(t-\tau)\mathbb{P}\nabla\cdot[v(\tau)\otimes u(\tau)]\,d\tau \\ &- \int_0^t T_\Omega(t-\tau)\mathbb{P}\nabla\cdot[u(\tau)\otimes v(\tau)]\,d\tau \\ &- \int_0^t T_\Omega(t-\tau)\mathbb{P}\nabla\cdot[u(\tau)\otimes u(\tau)]\,d\tau\end{aligned}$$

and

$$\begin{aligned}X &:= \Big\{u \in L^\theta(0,\infty;\dot{W}^{s,p}(\mathbb{R}^3))^3 \;\Big|\; \\ &\qquad \|u\|_{L^\theta(0,\infty;\dot{W}^{s,p})} \leqslant 2C_0|\Omega|^{-\left\{\frac{1}{\theta} - \left(\frac{3}{4} - \frac{3}{2p}\right)\right\}}\|a\|_{\dot{H}^s}\Big\}, \\ d_X(u, w) &:= \|u - w\|_{L^\theta(0,\infty;\dot{W}^{s,p})},\end{aligned}$$

respectively. From (14), Lemmas 3.2 and 3.3, there exist positive constants $C_1 = C_1(s,p)$ and $C_2 = C_2(s,p,\theta)$ such that

$$
\begin{aligned}
&\|\Phi(u)\|_{L^\theta(0,\infty;\dot W^{s,p})}\\
&\leqslant C_0|\Omega|^{-\left\{\frac{1}{\theta}-(\frac{3}{4}-\frac{3}{2p})\right\}}\|a\|_{\dot H^s}\\
&\quad + C_1|\Omega|^{-\frac{3}{2}\left\{(\frac{1}{3}+\frac{s}{3})-\frac{1}{p}\right\}}\|v\|_{L^\infty(\mathbb{R};\dot W^{s,p})}\|u\|_{L^\theta(0,\infty;\dot W^{s,p})}\\
&\quad + C_2|\Omega|^{-\left\{(\frac{1}{2}-\frac{3}{2p}+\frac{s}{2})-\frac{1}{\theta}\right\}}\|u\|^2_{L^\theta(0,\infty;\dot W^{s,p})}\\
&\leqslant C_0|\Omega|^{-\left\{\frac{1}{\theta}-(\frac{3}{4}-\frac{3}{2p})\right\}}\|a\|_{\dot H^s}\\
&\quad + 2C_0C_1|\Omega|^{-\frac{3}{2}\left\{(\frac{1}{3}+\frac{s}{3})-\frac{1}{p}\right\}-\left\{\frac{1}{\theta}-(\frac{3}{4}-\frac{3}{2p})\right\}}\|v\|_{L^\infty(\mathbb{R};\dot W^{s,p})}\|a\|_{\dot H^s}\\
&\quad + 4C_0^2C_2|\Omega|^{-\left\{(\frac{1}{2}-\frac{3}{2p}+\frac{s}{2})-\frac{1}{\theta}\right\}-2\left\{\frac{1}{\theta}-(\frac{3}{4}-\frac{3}{2p})\right\}}\|a\|^2_{\dot H^s}\\
&= C_0|\Omega|^{-\left\{\frac{1}{\theta}-(\frac{3}{4}-\frac{3}{2p})\right\}}\|a\|_{\dot H^s}\\
&\quad\times\left\{1+2C_1|\Omega|^{-\frac{3}{2}\left\{(\frac{1}{3}+\frac{s}{3})-\frac{1}{p}\right\}}\|v\|_{L^\infty(\mathbb{R};\dot W^{s,p})}+4C_0C_2|\Omega|^{-\frac{1}{2}(s-\frac{1}{2})}\|a\|_{\dot H^s}\right\}
\end{aligned}
\tag{15}
$$

for all $u \in X$. Moreover, it follows from Lemmas 3.2 and 3.3 that there exist positive constants $C_3 = C_3(s,p)$ and $C_4 = C_4(s,p,\theta)$ such that

$$
\begin{aligned}
&\|\Phi(u)-\Phi(w)\|_{L^\theta(0,\infty;\dot W^{s,p})}\\
&\leqslant \left\|\int_0^t T_\Omega(t-\tau)\mathbb{P}\nabla\cdot[v(\tau)\otimes(u(\tau)-w(\tau))]\,d\tau\right\|_{L^\theta(0,\infty;\dot W^{s,p})}\\
&\quad+\left\|\int_0^t T_\Omega(t-\tau)\mathbb{P}\nabla\cdot[(u(\tau)-w(\tau))\otimes v(\tau)]\,d\tau\right\|_{L^\theta(0,\infty;\dot W^{s,p})}\\
&\quad+\left\|\int_0^t T_\Omega(t-\tau)\mathbb{P}\nabla\cdot[(u(\tau)-w(\tau))\otimes u(\tau)]\,d\tau\right\|_{L^\theta(0,\infty;\dot W^{s,p})}\\
&\quad+\left\|\int_0^t T_\Omega(t-\tau)\mathbb{P}\nabla\cdot[w(\tau)\otimes(u(\tau)-w(\tau))]\,d\tau\right\|_{L^\theta(0,\infty;\dot W^{s,p})}\\
&\leqslant C_3|\Omega|^{-\frac{3}{2}\left\{(\frac{1}{3}+\frac{s}{3})-\frac{1}{p}\right\}}\|v\|_{L^\infty(\mathbb{R};\dot W^{s,p})}\|u-w\|_{L^\theta(0,\infty;\dot W^{s,p})}\\
&\quad+C_4|\Omega|^{-\left\{(\frac{1}{2}-\frac{3}{2p}+\frac{s}{2})-\frac{1}{\theta}\right\}}\left(\|u\|_{L^\theta(0,\infty;\dot W^{s,p})}+\|w\|_{L^\theta(0,\infty;\dot W^{s,p})}\right)\\
&\quad\times\|u-w\|_{L^\theta(0,\infty;\dot W^{s,p})}
\end{aligned}
$$

$$
\begin{aligned}
&\leqslant C_3|\Omega|^{-\frac{3}{2}\left\{\left(\frac{1}{3}+\frac{s}{3}\right)-\frac{1}{p}\right\}}\|v\|_{L^\infty(\mathbb{R};\dot{W}^{s,p})}\|u-w\|_{L^\theta(0,\infty;\dot{W}^{s,p})}\\
&\quad + 4C_0C_4|\Omega|^{-\left\{\left(\frac{1}{2}-\frac{3}{2p}+\frac{s}{2}\right)-\frac{1}{\theta}\right\}-\left\{\frac{1}{\theta}-\left(\frac{3}{4}-\frac{3}{2p}\right)\right\}}\|a\|_{\dot{H}^s}\|u-w\|_{L^\theta(0,\infty;\dot{W}^{s,p})}\\
&=\Big\{C_3|\Omega|^{-\frac{3}{2}\left\{\left(\frac{1}{3}+\frac{s}{3}\right)-\frac{1}{p}\right\}}\|v\|_{L^\infty(\mathbb{R};\dot{W}^{s,p})}\\
&\quad + 4C_0C_4|\Omega|^{-\frac{1}{2}\left(s-\frac{1}{2}\right)}\|a\|_{\dot{H}^s}\Big\}\|u-w\|_{L^\theta(0,\infty;\dot{W}^{s,p})}
\end{aligned}
\tag{16}
$$

for all $u, w \in X$. Hence if the time periodic solution $v$ and the initial disturbance $a$ satisfy

$$
\|v\|_{L^\infty(\mathbb{R};\dot{W}^{s,p})} \leqslant \min\left\{\frac{1}{4C_1}, \frac{1}{4C_3}\right\}|\Omega|^{\frac{3}{2}\left\{\left(\frac{1}{3}+\frac{s}{3}\right)-\frac{1}{p}\right\}}
$$

and

$$
\|a\|_{\dot{H}^s} \leqslant \min\left\{\frac{1}{8C_0C_2}, \frac{1}{16C_0C_4}\right\}|\Omega|^{\frac{1}{2}\left(s-\frac{1}{2}\right)},
$$

it follows from (15) and (16) that

$$
\begin{aligned}
\|\Phi(u)\|_{L^\theta(0,\infty;\dot{W}^{s,p})} &\leqslant 2C_0|\Omega|^{-\left\{\frac{1}{\theta}-\left(\frac{3}{4}-\frac{3}{2p}\right)\right\}}\|a\|_{\dot{H}^s},\\
\|\Phi(u)-\Phi(w)\|_{L^\theta(0,\infty;\dot{W}^{s,p})} &\leqslant \frac{1}{2}\|u-w\|_{L^\theta(0,\infty;\dot{W}^{s,p})}
\end{aligned}
$$

provided that $u, w \in X$. Therefore, by the contraction mapping principle, there exists a unique solution $u \in X$ satisfying (5) for all $t > 0$.

It remains to show that the mild solution $u \in X$ to (4) satisfies $u \in C([0,\infty);\dot{H}^s(\mathbb{R}^3))^3$. Take $1/q := 2/p - s/3$. It is easy to see that $1 < q < 2$ and the exponents $(s, p, q)$ satisfy the assumption of Lemma 3.1. Then, by Lemmas 2.4 and 3.1 and the Hölder inequality, we have

$$
\begin{aligned}
\|u(t)\|_{\dot{H}^s} \leqslant \|T_\Omega(t)a\|_{\dot{H}^s} &+ \left\|\int_0^t T_\Omega(t-\tau)\mathbb{P}\nabla\cdot[v(\tau)\otimes u(\tau)]\,d\tau\right\|_{\dot{H}^s}\\
&+ \left\|\int_0^t T_\Omega(t-\tau)\mathbb{P}\nabla\cdot[u(\tau)\otimes v(\tau)]\,d\tau\right\|_{\dot{H}^s}\\
&+ \left\|\int_0^t T_\Omega(t-\tau)\mathbb{P}\nabla\cdot[u(\tau)\otimes u(\tau)]\,d\tau\right\|_{\dot{H}^s}
\end{aligned}
$$

$$
\begin{aligned}
&\leqslant C\|u_0\|_{\dot{H}^s} + C\int_0^t (t-\tau)^{-\frac{3}{2}(\frac{1}{q}-\frac{1}{2})-\frac{1}{2}} \|v(\tau)\otimes u(\tau)\|_{\dot{W}^{s,q}} d\tau \\
&\quad + C\int_0^t (t-\tau)^{-\frac{3}{2}(\frac{1}{q}-\frac{1}{2})-\frac{1}{2}} \|u(\tau)\otimes v(\tau)\|_{\dot{W}^{s,q}} d\tau \\
&\quad + C\int_0^t (t-\tau)^{-\frac{3}{2}(\frac{1}{q}-\frac{1}{2})-\frac{1}{2}} \|u(\tau)\otimes u(\tau)\|_{\dot{W}^{s,q}} d\tau \\
&\leqslant C\|u_0\|_{\dot{H}^s} + C\int_0^t (t-\tau)^{-\frac{3}{p}+\frac{s}{2}+\frac{1}{4}} \|v(\tau)\|_{\dot{W}^{s,p}} \|u(\tau)\|_{\dot{W}^{s,p}} d\tau \\
&\quad + C\int_0^t (t-\tau)^{-\frac{3}{p}+\frac{s}{2}+\frac{1}{4}} \|u(\tau)\|^2_{\dot{W}^{s,p}} d\tau \\
&\leqslant C\|u_0\|_{\dot{H}^s} \\
&\quad + \left(\int_0^t (t-\tau)^{(-\frac{3}{p}+\frac{s}{2}+\frac{1}{4})\theta'} d\tau\right)^{\frac{1}{\theta'}} \|v\|_{L^\infty(\mathbb{R};\dot{W}^{s,p})} \|u\|_{L^\theta(0,\infty;\dot{W}^{s,p})} \\
&\quad + \left(\int_0^t (t-\tau)^{(-\frac{3}{p}+\frac{s}{2}+\frac{1}{4})(\frac{\theta}{2})'} d\tau\right)^{\frac{1}{(\frac{\theta}{2})'}} \|u\|^2_{L^\theta(0,\infty;\dot{W}^{s,p})} \qquad (17)
\end{aligned}
$$

for $t \geqslant 0$. Since $1/\theta < 5/8 - 3/2p + s/4 < 5/4 - 3/p + s/2$, the time integrals on the right hand side of (17) converge and

$$
\begin{aligned}
\left(\int_0^t (t-\tau)^{(-\frac{3}{p}+\frac{s}{2}+\frac{1}{4})\theta'} d\tau\right)^{\frac{1}{\theta'}} &= Ct^{(\frac{5}{4}-\frac{3}{p}+\frac{s}{2})-\frac{1}{\theta}}, \\
\left(\int_0^t (t-\tau)^{(-\frac{3}{p}+\frac{s}{2}+\frac{1}{4})(\frac{\theta}{2})'} d\tau\right)^{\frac{1}{(\frac{\theta}{2})'}} &= Ct^{2\left\{(\frac{5}{8}-\frac{3}{2p}+\frac{s}{4})-\frac{1}{\theta}\right\}}.
\end{aligned}
$$

This shows that $u(t) \in \dot{H}^s(\mathbb{R}^3)^3$ for $t \geqslant 0$. Similarly, we see that $u \in C([0,\infty);\dot{H}^s(\mathbb{R}^3))^3$. This completes the proof of Theorem 1.1. □

**Acknowledgements** T. Iwabuchi is partially supported by JSPS Grant-in-Aid for Young Scientists (B) #25800069.

A. Mahalov is partially supported by NSF DMS grant 1419593.

R. Takada was partially supported by JSPS Grant-in-Aid for Research Activity Start-up #25887005.

## References


1. A. Babin, A. Mahalov, B. Nicolaenko, Long-time averaged Euler and Navier-Stokes equations for rotating fluids, in *Advanced Series in Nonlinear Dynamics*, vol. 7 (World Scientific Publishing, River Edge, NJ, 1995), pp. 145–157
2. A. Babin, A. Mahalov, B. Nicolaenko, Regularity and integrability of 3D Euler and Navier-Stokes equations for rotating fluids. Asymptot. Anal. **15**, 103–150 (1997)
3. A. Babin, A. Mahalov, B. Nicolaenko, Global regularity of 3D rotating Navier-Stokes equations for resonant domains. Indiana Univ. Math. J. **48**, 1133–1176 (1999)
4. A. Babin, A. Mahalov, B. Nicolaenko, 3D Navier-Stokes and Euler equations with initial data characterized by uniformly large vorticity. Indiana Univ. Math. J. **50**, 1–35 (2001)
5. J.-Y. Chemin, B. Desjardins, I. Gallagher, E. Grenier, Anisotropy and dispersion in rotating fluids, in *Nonlinear Partial Differential Equations and Their Applications*. Collège de France Seminar, Paris, 1997/1998, vol. XIV. Studies in Mathematics and Its Applications, vol. 31, (North-Holland, Amsterdam, 2002), pp. 171–192
6. J.-Y. Chemin, B. Desjardins, I. Gallagher, E. Grenier, *Mathematical Geophysics*. Oxford Lecture Series in Mathematics and Its Applications, vol. 32 (The Clarendon Press/Oxford University Press, Oxford, 2006)
7. F. Flandoli, A. Mahalov, Stochastic three-dimensional rotating Navier-Stokes equations: averaging, convergence and regularity. Arch. Ration. Mech. Anal. **205**, 195–237 (2012)
8. M. Hieber, Y. Shibata, The Fujita-Kato approach to the Navier-Stokes equations in the rotational framework. Math. Z. **265**, 481–491 (2010)
9. T. Iwabuchi, R. Takada, Time periodic solutions to the Navier-Stokes equations in the rotational framework. J. Evol. Equ. **12**, 985–1000 (2012)
10. T. Iwabuchi, R. Takada, Global solutions for the Navier-Stokes equations in the rotational framework. Math. Ann. **357**, 727–741 (2013)
11. T. Kato, Strong $L^p$-solutions of the Navier-Stokes equation in $\boldsymbol{R}^m$, with applications to weak solutions, Math. Z. **187**, 471–480 (1984)
12. Y. Koh, S. Lee, R. Takada, Dispersive estimates for the Navier-Stokes equations in the rotational framework. Adv. Differ. Equ. **19**, 857–878 (2014)
13. P. Konieczny, T. Yoneda, On dispersive effect of the Coriolis force for the stationary Navier-Stokes equations. J. Differ. Equ. **250**, 3859–3873 (2011)
14. H. Kozono, M. Nakao, Periodic solutions of the Navier-Stokes equations in unbounded domains. Tohoku Math. J. (2) **48**, 33–50 (1996)
15. H. Kozono, T. Ogawa, On stability of Navier-Stokes flows in exterior domains. Arch. Ration. Mech. Anal. **128**, 1–31 (1994)
16. H. Kozono, Y. Mashiko, R. Takada, Existence of periodic solutions and their asymptotic stability to the Navier-Stokes equations with the Coriolis force. J. Evol. Equ. **14**, 565–601 (2014)
17. P. Maremonti, Existence and stability of time-periodic solutions to the Navier-Stokes equations in the whole space. Nonlinearity **4**, 503–529 (1991)
18. P. Maremonti, Some theorems of existence for solutions of the Navier-Stokes equations with slip boundary conditions in half-space. Ricerche mat. **40**, 81–135 (1991)
19. Y. Taniuchi, On stability of periodic solutions of the Navier-Stokes equations in unbounded domains. Hokkaido Math. J. **28**, 147–173 (1999)
20. M. Yamazaki, The Navier-Stokes equations in the weak-$L^n$ space with time-dependent external force. Math. Ann. **317**, 635–675 (2000)
21. M. Yamazaki, Solutions in the Morrey spaces of the Navier-Stokes equation with time-dependent external force. Funkcial. Ekvac. **43**, 419–460 (2000)

# Weighted $L^p - L^q$ Estimates of Stokes Semigroup in Half-Space and Its Application to the Navier-Stokes Equations

**Takayuki Kobayashi and Takayuki Kubo**

*Dedicated to Professor Yoshihiro Shibata on the occasion of his sixtieth birthday*

**Abstract** We consider the Navier-Stokes equations in half-space and in $L^p$ space with Muckenhoupt weight and show the $L^p - L^q$ estimates of Stokes semigroup with $\langle x' \rangle^{s_1} \langle x_n \rangle^{s_n}$ type weight. Finally as the application of the weighted $L^p - L^q$ estimates, we shall obtain the weighted asymptotic behavior of the solution to the Navier-Stokes equations.

**Keywords** Half-space • Navier-Stokes equations • Weighted $L^p - L^q$ estimates

## 1 Introduction

Let $n \geq 2$. In this paper, we shall consider the Navier-Stokes equations in half-space $\mathbb{R}^n_+$ defined by $\mathbb{R}^n_+ = \{x = (x', x_n) \in \mathbb{R}^n \mid x_n > 0\}$ and in weighted $L^p$ space. The Navier-Stokes equations are given by

$$\begin{cases} \partial_t u - \Delta u + (u \cdot \nabla)u + \nabla \pi = 0, \quad \nabla \cdot u = 0 & \text{in } \mathbb{R}^n_+ \times (0, \infty), \\ u(x,t) = 0 & \text{on } \partial\mathbb{R}^n_+ \times (0, \infty), \\ u(x,0) = u_0(x) & \text{in } \mathbb{R}^n_+, \\ \lim_{|x| \to \infty} u(x,t) = 0. & \end{cases} \tag{NS}$$

T. Kobayashi
Division of Mathematical Science, Department of Systems Innovation, Graduate School of Engineering Science, Osaka University Machikaneyamacho 1-3, Toyonakashi 560-8531, Japan
e-mail: kobayashi@sigmath.es.osaka-u.ac.jp

T. Kubo (✉)
Division of Mathematics, Faculty of Pure and Applied Sciences, University of Tsukuba, Ibaraki 305-8571, Japan
e-mail: tkubo@math.tsukuba.ac.jp



H. Amann et al. (eds.), *Recent Developments of Mathematical Fluid Mechanics*, Advances in Mathematical Fluid Mechanics, DOI 10.1007/978-3-0348-0939-9_18

Here $u(x,t) = (u_1(x,t), \dots, u_n(x,t))$ and $\pi(x,t)$ denote unknown velocity field and scalar pressure and $u_0(x)$ is given vector function.

The Navier-Stokes equations (NS) have been already studied by many mathematicians. In particular, $L^p$ theory has been developed. Indeed, it is well known for several bounded and unbounded domains $\Omega$ that if the initial data $u_0$ is in the space of $L^n_\sigma(\Omega)$ of $L^n$ solenoidal vector fields and if $\|u_0\|_{L^n(\Omega)}$ is sufficiently small, then (NS) admits a unique strong solution $u$ satisfying

$$t^{\frac{n}{2}(\frac{1}{n}-\frac{1}{p})}u \in BC([0,\infty); L^p(\Omega)) \qquad (n \le p \le \infty), \tag{1}$$

$$t^{\frac{1}{2}+\frac{n}{2}(\frac{1}{n}-\frac{1}{p})}\nabla u \in BC([0,\infty); L^p(\Omega)) \qquad (n \le p < \infty). \tag{2}$$

One of the standard argument to prove the unique existence of solution satisfying (1) and (2) is to prove the following $L^p - L^q$ estimates of Stokes semigroup $e^{-tA}$ : for $f \in L^p_\sigma(\Omega)$,

$$\|e^{-tA}f\|_{L^q(\Omega)} \le Ct^{-\frac{n}{2}(\frac{1}{p}-\frac{1}{q})}\|f\|_{L^p(\Omega)}, \tag{3}$$

$$\|\nabla e^{-tA}f\|_{L^q(\Omega)} \le Ct^{-\frac{n}{2}(\frac{1}{p}-\frac{1}{q})-\frac{1}{2}}\|f\|_{L^p(\Omega)}. \tag{4}$$

In fact, in the whole space case, Kato [16] derived (3) and (4) for $1 < p \le q < \infty$ by using Young's inequality with the concrete solution formula to linearized problem of (NS) and considered the corresponding integral equation to (NS). He could prove the unique existence of the strong solution to (NS) with (3) and (4) by the fixed point theorem (for others domains, see e.g. [1, 4–6, 12, 13, 18–20, 23]).

On the other hand, the solvability of the Navier-Stokes equations and decay rate of the solution in the weighted $L^p$ spaces has been also considered by many mathematicians. There is abundant literature for the Cauchy problem related to the weighted space and the existence results are almost optimal. However, there is less literature for the half-space problem related to the weighted space and much improvement is required. Here focussing on the half-space problem, we describe the known results concerning on the decay estimate for Stokes problem.

Bae and Choe[3] and Fujigaki and Miyakawa [11] derived more rapid $L^p$ estimate of the Stokes flow with initial data in the weighted $L^p$ space:

$$\|e^{-tA}a\|_{L^q(\mathbb{R}^n_+)} \le Ct^{-\frac{1}{2}-\frac{n}{2}(\frac{1}{p}-\frac{1}{q})}\|x_n a\|_{L^p(\mathbb{R}^n_+)}.$$

for $1 < p \le q < \infty$, $1 \le p < q < \infty$. Bae [2] extended their results to a rapid $L^1$ estimate and a rapid $L^\infty$ estimate with an initial data with the special condition $\int_{\mathbb{R}} a(x)dx_i = 0$ for some $i = 1, \dots, n-1$ and in this case,

$$\|e^{-tA}a\|_{L^p(\mathbb{R}^n_+)} \le Ct^{-\frac{1}{2}}\|x_n a\|_{L^p(\mathbb{R}^n_+)}, \qquad p = 1, \infty.$$

Recently, Kobayashi and Kubo [17] proved the following different type $L^p - L^q$ estimates with $\omega^s(x) := \langle x\rangle^{sp}$ for $0 \le s < (n-1)(1-\frac{1}{p})$:

$$\|\langle x\rangle^s e^{-tA}a\|_{L^q} \le Ct^{-\frac{n}{2}(\frac{1}{p}-\frac{1}{q})}\|\langle x\rangle^s a\|_{L^p} + Ct^{-\frac{n}{2}(\frac{1}{r}-\frac{1}{q})+\frac{s}{2}}\|a\|_{L^r},$$

$$\|\langle x\rangle^s \nabla e^{-tA}a\|_{L^q} \le Ct^{-\frac{n}{2}(\frac{1}{p}-\frac{1}{q})-\frac{1}{2}}\|\langle x\rangle^s a\|_{L^p} + Ct^{-\frac{n}{2}(\frac{1}{r}-\frac{1}{q})+\frac{s}{2}-\frac{1}{2}}\|a\|_{L^r}$$

for $1 < p, r \le q < \infty$ and $t \ge 2$. For the other weights, Jin [14] proved the following weighted $L^1 - L^q$ estimates:

$$\||x'|^s e^{-tA}a\|_{L^q} \le Ct^{-\frac{1}{2}-\frac{n}{2}(1-\frac{1}{q})}\||x'|^s x_n a\|_{L^1} + Ct^{-\frac{n}{2}(1-\frac{1}{q})+\frac{s-1}{2}}\|x_n a\|_{L^1}$$

for $0 \le s < (n-1)(1-\frac{1}{q})$ and $1 < q < \infty$ and

$$\|x_n^s e^{-tA}a\|_{L^q} \le Ct^{-\frac{1}{2}-\frac{n}{2}(1-\frac{1}{q})}\|x_n^{s+1}a\|_{L^1} + Ct^{-\frac{n}{2}(1-\frac{1}{q})+\frac{s-1}{2}}\|x_n a\|_{L^1}$$

for $0 \le s < n(1-\frac{1}{q})$ and $1 < q < \infty$. Moreover in [15], he proved the $L^1 - L^\infty$ estimate corresponding to [14] under some conditions.

Goal of this paper is to show the different type $L^p - L^q$ estimates of Stokes semigroup with the weight function of $\langle x'\rangle^{s_1}\langle x_n\rangle^{s_n}$ type as follows:

$$\|\langle x'\rangle^{s_1'}\langle x_n\rangle^{s_n'}e^{-tA}a\|_{L^q} \le Ct^{-\frac{n}{2}(\frac{1}{p}-\frac{1}{q})}(1+t)^{-\frac{(s_1+s_n)-(s_1'+s_n')}{2}}\|\langle x'\rangle^{s_1}\langle x_n\rangle^{s_n}a\|_{L^p}.$$

Moreover we consider its application to the Navier-Stokes equations and we obtain the weighted asymptotic behavior of the solution to the Navier-Stokes equation as $t \to \infty$.

## 2 Main Results and Notation

In this paper, we shall consider the Navier-Stokes equations (NS) in half-space and in the weighted $L^p$ space. The half-space $\mathbb{R}^n_+$ is defined by $\mathbb{R}^n_+ = \{x = (x', x_n) \in \mathbb{R}^n \mid x_n > 0\}$. The weighted $L^p$ space with Muckenhoupt weight $w \in \mathcal{A}_p$ is defined by

$$L^p_w(\mathbb{R}^n_+) = \left\{u \in L^1_{loc}(\overline{\mathbb{R}^n_+}) \mid \|u\|_{L^p_w(\mathbb{R}^n_+)} = \|uw^{1/p}\|_{L^p(\mathbb{R}^n_+)}\right\} \tag{5}$$

for $1 < p < \infty$ (see [8] for the definition of Muckenhoupt class $\mathcal{A}_p$).

In order to consider the Navier-Stokes equations by Kato's argument [16], we need the $L^p - L^q$ estimates of Stokes semigroup in the weighted $L^p$ space. In [9, 10], Fröhlich proved that the Stokes operator $A_{\mathbb{R}^n_+}$ is defined by $A_{\mathbb{R}^n_+} = -P_{\mathbb{R}^n_+}\Delta$ with

Helmholtz decomposition $P_{\mathbb{R}^n_+}$ and it generates an analytic semigroup $e^{-tA_{\mathbb{R}^n_+}}$. In this paper, we fix $w(x) = w^{sp}(x) = \langle x' \rangle^{s_1 p} \langle x_n \rangle^{s_n p}$ as the weight function and assume

$$-\frac{n-1}{p} < s_1 < (n-1)\left(1 - \frac{1}{p}\right), \quad -\frac{1}{p} < s_n < 1 - \frac{1}{p}. \tag{C$_p$}$$

We notice that $w(x)$ belongs to the Muckenhoupt class $\mathcal{A}_p$ $(1 < p < \infty)$ when the exponent $s = (s_1, s_n)$ of the weight function satisfies $(\mathrm{C_p})$.

The main result is the following $L^p - L^q$ estimates of Stokes semigroup with weight $w(x)$.

**Theorem 2.1 (Weighted $L^p - L^q$ Estimates)** *Let $n \geq 2$ and $1 < p \leq q < \infty$. Let $s = (s_1, s_n)$ and $s' = (s_1', s_n')$ satisfy*

$$-\frac{n-1}{q} < s_1' \leq s_1 < (n-1)\left(1 - \frac{1}{p}\right), \quad -\frac{1}{q} < s_n' \leq s_n < 1 - \frac{1}{p} \tag{C$_{p,q}$}$$

*and let $w(x) = w^{sp}(x)$.*

(i) *For $a \in L^p_w(\mathbb{R}^n_+)$, the following estimate holds: for $t > 0$,*

$$\|w^{s'} e^{-tA} a\|_{L^q(\mathbb{R}^n_+)} \leq C t^{-\frac{n}{2}(\frac{1}{p} - \frac{1}{q})} (1+t)^{-\frac{(s_1+s_n)-(s_1'+s_n')}{2}} \|w^s a\|_{L^p(\mathbb{R}^n_+)}. \tag{6}$$

(ii) *For $a \in L^p_w(\mathbb{R}^n_+)$, the following estimate holds: for $t > 0$,*

$$\|w^{s'} \nabla e^{-tA} a\|_{L^q} \leq C t^{-\frac{n}{2}(\frac{1}{p} - \frac{1}{q}) - \frac{1}{2}} (1+t)^{-\frac{(s_1+s_n)-(s_1'+s_n')}{2}} \|w^s a\|_{L^p(\mathbb{R}^n_+)}. \tag{7}$$

We next consider the application of weighted $L^p - L^q$ estimates to Navier-Stokes equations (NS). By applying the Helmholtz projection $P_{\mathbb{R}^n_+}$ to (NS), we can rewrite (NS) as follows:

$$\partial_t u + A_{\mathbb{R}^n_+} + P_{\mathbb{R}^n_+}[(u \cdot \nabla) u(t)] = 0, \quad u(0) = u_0.$$

By Duhamel's principle, we obtain the integral equations:

$$u(t) = e^{-tA_{\mathbb{R}^n_+}} u_0 - \int_0^t e^{-(t-\tau)A_{\mathbb{R}^n_+}} P_{\mathbb{R}^n_+}[(u \cdot \nabla) u](\tau) d\tau. \tag{IE}$$

By using the weighted Theorem 2.1, we can obtain the following theorem:

**Theorem 2.2** *Let $n \geq 2$ and let $s = (s_1, s_n)$ satisfy $0 \leq s_1 < (n-1)(1 - \frac{1}{n})$ and $0 \leq s_n < 1 - \frac{1}{n}$. Then there exists $\delta > 0$ such that if $a \in L^n_{w,\sigma}$ satisfied $\|w^s u_0\|_{L^n} < \delta$,* *then* (IE) *admits a unique solution* $u(t) \in BC([0,\infty); L^n_{w,\sigma})$ *satisfying*

$$\|w^{s'} u(t)\|_{L^q} = O\left(t^{-\frac{1}{2}+\frac{n}{2q}-\frac{(s_1+s_n)-(s_1'+s_n')}{2}}\right), \tag{8}$$

$$\|w^{s'} \nabla u(t)\|_{L^q} = O\left(t^{-1+\frac{n}{2q}-\frac{(s_1+s_n)-(s_1'+s_n')}{2}}\right) \tag{9}$$

*for* $n \le q < \infty, -(n-1)/q < s_1' \le s_1$ *and* $-1/q < s_n' \le s_n$ *as* $t \to \infty$.

## 3 Preliminaries

In this section, we shall introduce some definition and some lemmas which play important role for our proof. We first introduce the lemma concerning the weight function in Muckenhoupt class $\mathcal{A}_p$. The weight $w \in \mathcal{A}_p$ has the important property that regular singular integral operators are continuous on $L^p_w(\mathbb{R}^n)$ into itself (see [22] for detail).

**Lemma 3.1** *Let* $1 < p < \infty$, $w \in \mathcal{A}_p$ *and let* $T$ *be a regular singular integral operator. Then* $T$ *is bounded on* $L^p_w(\mathbb{R}^n)$. *More precisely, there is a positive constant* $C$ *such that for all* $f \in L^p_w(\mathbb{R}^n)$, *we have*

$$\|Tf\|_{L^p_w} \le C\|f\|_{L^p_w}.$$

*Remark 3.2* By Lemma 3.1, the Riesz transform $R_jf$ and the partial Riesz transform $S_jf$ defined by

$$R_jf := \mathcal{F}_\xi^{-1}\left[\frac{i\xi_j}{|\xi|}\mathcal{F}_x[f](\xi)\right], \qquad j = 1,\dots,n, \tag{10}$$

$$S_jf := \mathcal{F}_{\xi'}^{-1}\left[\frac{i\xi_j}{|\xi'|}\mathcal{F}_{x'}[f](\xi', x_n)\right], \qquad j = 1,\dots,n-1, \tag{11}$$

are continuous on $L^p_w(\mathbb{R}^n)$ and $L^p_w(\mathbb{R}^n_+)$ into itself respectively. Here $\mathcal{F}_x$ and $\mathcal{F}_{x'}$ denote the Fourier transform with respect to $x$ and the partial Fourier transform with respect to $x'$.

We next introduce the Helmholtz decomposition of the weighted space $L^p_w(\mathbb{R}^n_+)$ $(1 < p < \infty)$. The following Helmholtz decomposition is proved by Fröhlich[9]: $L^p_w(\mathbb{R}^n_+) = L^p_{w,\sigma}(\mathbb{R}^n_+) \oplus \nabla \dot{H}^{1,p}_w(\mathbb{R}^n_+)$, where

$$L^p_{w,\sigma}(\mathbb{R}^n_+) = \overline{\{u \in C_0^\infty \mid \nabla \cdot u = 0 \quad \text{in } \mathbb{R}^n_+\}}^{\|\cdot\|_{L^p_w(\mathbb{R}^n_+)}},$$

$$\dot{H}^{1,p}_w(\mathbb{R}^n_+) = \{\pi \in L^1_{\mathrm{loc}}(\overline{\mathbb{R}^n_+}) \mid \nabla\pi \in L^p_w(\mathbb{R}^n_+)\}.$$

Their result implies that the Helmholtz projection $P_{\mathbb{R}^n_+}$ from $L^p_w(\mathbb{R}^n_+)$ to $L^p_{w,\sigma}(\mathbb{R}^n_+)$ is bounded. By the Helmholtz projection, the Stokes operator $A_{\mathbb{R}^n_+}$ in $L^p_{w,\sigma}(\mathbb{R}^n_+)$ is defined by $A_{\mathbb{R}^n_+} = -P_{\mathbb{R}^n_+}\Delta$ with domain $D(A_{\mathbb{R}^n_+}) = \{u \in W^{2,p}_w \cap L^p_{w,\sigma}(\mathbb{R}^n_+) \mid u = 0 \text{ on } \partial\mathbb{R}^n_+\}$. Moreover in [10], he proved that the Stokes operator generates an analytic semigroup $e^{-tA_{\mathbb{R}^n_+}}$.

For simplicity, we often use the abbreviations $A$ for $A_{\mathbb{R}^n_+}$ and $A_{\mathbb{R}^n}$ and $P$ for $P_{\mathbb{R}^n_+}$ and $P_{\mathbb{R}^n}$ if there is no confusion.

## 4 Proof of Theorem 2.1

In this section, we consider the weighted $L^p - L^q$ estimates of Stokes semigroup in half-space. First step is the following lemma which says the weighted $L^p - L^q$ estimates for the whole space.

**Lemma 4.1** *Let $n \geq 2$, $1 < p \leq q < \infty$ and let $s = (s_1, s_n)$ and $s' = (s'_1, s'_n)$ satisfy the condition* ($\mathrm{C_{p,q}}$). *Then the following estimate holds.*

$$\begin{aligned} &\|w^{s'}\partial_t^k \nabla^\alpha e^{-tA} Pf\|_{L^q(\mathbb{R}^n)} \\ &\leq Ct^{-\frac{n}{2}(\frac{1}{p}-\frac{1}{q})-\frac{|\alpha|}{2}-k}(1+t)^{-\frac{(s_1+s_n)-(s'_1+s'_n)}{2}}\|w^s f\|_{L^p(\mathbb{R}^n)} \end{aligned} \tag{12}$$

*for $k \in \mathbb{N}_0 = \mathbb{N} \cup \{0\}$, $\alpha \in \mathbb{N}^n_0$ and $t > 0$.*

*Proof* We recall the following estimate proved by Murata [21].

$$\int_{\mathbb{R}^n} E(x-y)\langle y\rangle^{-r} dr \leq M(1+t)^{-\frac{r}{2}} \quad (0 \leq r < n). \tag{13}$$

where $E_n(x) = E_n(t,x) = (4\pi t)^{-\frac{n}{2}} e^{-\frac{|x|^2}{4t}}$. The following estimates are obtained from (13):

$$\int_{\mathbb{R}} E_1(x_n - y_n)\langle y_n\rangle^{-r} dy_n \leq M(1+t)^{-\frac{r}{2}} \quad (0 \leq r < 1), \tag{14}$$

$$\int_{\mathbb{R}^{n-1}} E_{n-1}(x'-y')\langle y'\rangle^{-r} dy' \leq M(1+t)^{-\frac{r}{2}} \quad (0 \leq r < n-1). \tag{15}$$

We shall prove (12) by using these estimates. Since we can prove the case $|\alpha| \neq 0$ or $k > 0$ in the same way as $|\alpha| = k = 0$, we shall prove only the case where $|\alpha| = 0,\ k = 0$. We first prove the case where $0 \leq s'_1 \leq s_1 < (n-1)(1-\frac{1}{p})$ and $0 \leq s'_n \leq s_n \leq 1 - \frac{1}{p}$. Taking the fact that the Stokes semigroup in $\mathbb{R}^n$ is described by using heat kernel $e^{-tA}f = E * f$ for $f \in L^q_{w,\sigma}(\mathbb{R}^n)$ into account, it is sufficient to estimate $E * f$. Since we see

$$
\begin{aligned}
&|(E * f)(x)| \\
&= \left| \int_{\mathbb{R}^n} |w^s(y) f(y)|^{1-\frac{p}{q}} \left( w^{-\frac{sqr}{q-r}}(y) E_n^r(x-y) \right)^{\frac{1}{r}-\frac{1}{q}} \left( (w^s(y) f(y))^p E_n^r(x-y) \right)^{\frac{1}{q}} dy \right| \\
&\le \|w^s f\|_{L^p}^{1-\frac{p}{q}} \left( \int_{\mathbb{R}^n} w(y)^{-\frac{sqr}{q-r}} E_n^r(x-y) dy \right)^{\frac{1}{r}-\frac{1}{q}} \left( \int_{\mathbb{R}^n} |w^s(y) f(y)|^p E_n^r(x-y) dy \right)^{\frac{1}{q}}
\end{aligned}
$$

for $1 + \frac{1}{q} = \frac{1}{p} + \frac{1}{r}$, we obtain

$$
\begin{aligned}
&\|w^{s'}(E * f)\|_{L^q}^q \\
&\le \|w^s f\|_{L^p}^{q-p} \\
&\cdot \int_{\mathbb{R}^n} \left( \int_{\mathbb{R}^n} (w^s(y) w^{s'}(x))^{-\frac{qr}{q-r}} E_n^r(x-y) dy \right)^{\frac{q}{r}-1} \left( \int_{\mathbb{R}^n} |w^s f(y)|^p E_n^r(x-y) dy \right) dx \\
&= \|wf\|_{L^p}^q \sup_x \left( \int_{\mathbb{R}^n} (w(y) w'(x))^{-\frac{qr}{q-r}} E_n^r(x-y) dy \right)^{\frac{q}{r}-1} \int E_n^r(x) dx.
\end{aligned}
$$

We notice that

$$
\begin{aligned}
&\int_{\mathbb{R}^n} (w^s(y) w^{s'}(x))^{-\frac{qr}{q-r}} E_n^r(x-y) dy \\
&= \int_{\mathbb{R}^{n-1}} (\langle y' \rangle^{s_1} \langle x' \rangle^{s_1'})^{-\frac{qr}{q-r}} E_{n-1}^r(x'-y') dy' \int_{\mathbb{R}} (\langle y_n \rangle^{s_n} \langle x_n \rangle^{s_n'})^{-\frac{qr}{q-r}} E_1^r(x_n - y_n) dy_n.
\end{aligned}
$$

Here we know that $\langle z \rangle^a E_n(z) \le C t^{-\frac{n}{2}} (1+t)^{\frac{a}{2}} e^{-\frac{|z|^2}{8t}}$ for $a \ge C$ and

$$
\begin{aligned}
\int_{\mathbb{R}^n} \langle y \rangle^{-a} E_n^r(x-y) dy &\le \int_{\mathbb{R}^n} \langle y \rangle^{-a} E_n(x-y) E_n^{r-1}(x-y) dy' \\
&\le M t^{-\frac{n}{2}(r-1)} (1+t)^{-\frac{a}{2}}
\end{aligned}
$$

for $0 \le a < \frac{n}{2}$ by (14). Since $\langle x \rangle^a \le C(\langle x-y \rangle^a + \langle y \rangle^a)$ for $a > 0$, by (14) and (15) we have

$$
\begin{aligned}
&\int_{\mathbb{R}^{n-1}} (\langle y' \rangle^{s_1} \langle x' \rangle^{s_1'})^{-\frac{qr}{q-r}} E_{n-1}^r(x'-y') dy' \\
&\le C \int_{\mathbb{R}^{n-1}} \langle y' \rangle^{-\frac{s_1 qr}{q-r}} \left( \langle x'-y' \rangle^{\frac{s_1' qr}{q-r}} + \langle y' \rangle^{\frac{s_1' qr}{q-r}} \right) E_{n-1}^r(x'-y') dy' \\
&\le C t^{\frac{s_1' qr}{2(q-r)}} \int_{\mathbb{R}^{n-1}} \langle y' \rangle^{-\frac{s_1 qr}{q-r}} \tilde{E}_{n-1}^r(x'-y') dy'
\end{aligned}
$$

$$+ C \int_{\mathbb{R}^{n-1}} \langle y' \rangle^{-\frac{(s_1 - s_1')qr}{q-r}} E_{n-1}^r (x' - y') dy'$$

$$\le Ct^{-\frac{n-1}{2}(r-1) + \frac{s_1' qr}{2(q-r)} - \frac{s_1 qr}{2(q-r)}} + Ct^{-\frac{n-1}{2}(r-1) - \frac{(s_1 - s_1')qr}{2(q-r)}}$$

$$= Ct^{-\frac{n-1}{2}(r-1) - \frac{(s_1 - s_1')qr}{2(q-r)}},$$

where $\tilde{E}_n(x) = (4\pi t)^{-\frac{n}{2}} e^{-\frac{|x|^2}{8t}}$. Therefore we see

$$\int_{\mathbb{R}^n} (w^s(y) w^{s'}(x))^{-\frac{qr}{q-r}} E_n^r (x - y) dy$$

$$\le Ct^{-\frac{n-1}{2}(r-1) - \frac{(s_1 - s_1')qr}{2(q-r)}} Ct^{-\frac{1}{2}(r-1) - \frac{(s_n - s_n')qr}{2(q-r)}}$$

$$= Ct^{-\frac{n}{2}(r-1) - \frac{(s_1 - s_1')qr}{2(q-r)} - \frac{(s_n - s_n')qr}{2(q-r)}},$$

which implies

$$\|w^{s'}(E * f)\|_{L^q} \le Ct^{-\frac{n}{2}(r-1)(\frac{1}{r} - \frac{1}{q}) - \frac{(s_1 + s_n) - (s_1' - s_n')}{2}} t^{-\frac{n}{2}(\frac{r}{q} - \frac{1}{q})} \|w^s f\|_{L^p}$$

$$= Ct^{-\frac{n}{2}(1 - \frac{1}{r}) - \frac{(s_1 + s_n) - (s_1' + s_n')}{2}} \|w^s f\|_{L^p}$$

$$= Ct^{-\frac{n}{2}(\frac{1}{p} - \frac{1}{q}) - \frac{(s_1 + s_n) - (s_1' + s_n')}{2}} \|w^s f\|_{L^p}.$$

Therefore we obtain the estimate for $0 \le s_1' \le s_1 < (n-1)(1 - \frac{1}{p})$ and $0 \le s_n' \le s_n < 1 - \frac{1}{p}$. Since we can obtain the other cases in a similar way, we may omit the proof for the other cases. Therefore we obtain the desired result. □

By using Lemma 4.1, we shall prove Theorem 2.1.

*Proof of Theorem 2.1* In half-space $\mathbb{R}^n$, we have the solution formula obtained by Ukai [23]. Let $R_j$ and $S_j$ be the Riesz transform and the partial Riesz transform defined by (10) and (11). And let $rf = f|_{\mathbb{R}^n_+}$ and let $\gamma f|_{\partial\mathbb{R}^n_+}$ and $e_0$ be zero extension operator from $\mathbb{R}^n_+$ to $\mathbb{R}^n$ with value 0. Finally, let $E(t)$ be the solution operator for the heat equations in $\mathbb{R}^n_+$, which is derived from $E_0(t)$ by odd extension from $\mathbb{R}^n_+$ to $\mathbb{R}^n$. Then the solution $(u(t), \pi(t))$ of the non-stationary Stokes equations in $\mathbb{R}^n_+$ is

$$u(t) = WE(t)Vu_0, \qquad \pi(t) = -Dr\left(\partial_n E(t) V_1 u_0\right),$$

where

$$W = \begin{pmatrix} I & -SU \\ 0 & U \end{pmatrix}, \qquad V = \begin{pmatrix} V_1 \\ V_2 \end{pmatrix}$$

with

$$S = {}^t(S_1, \dots, S_{n-1}), \qquad R' = {}^t(R_1, \dots, R_{n-1}),$$
$$U = \gamma R' \cdot S(R' \cdot S + R_n)e_0, \quad V_1 a = a' + Sa^n, \qquad V_2 a = -S \cdot a' + a^n.$$

and $D$ is the Poisson operator for the Dirichlet problem of the Laplace equation in $\mathbb{R}^n_+$. By Remark 3.2, we can reduce to the whole space case, so that we can obtain the following estimate for the half-space. □

## 5 Proof of Theorem 2.2

In this section, we shall show the proof of Theorem 2.2. Our method is based on the contraction mapping principle with Theorem 2.1. Since this method is well-known, we shall describe the outline of the proof (see [7] for example).

As the underlying space, we set

$$\mathcal{I}_\varepsilon = \Big\{u \in BC([0,\infty) : L^n_{w,\sigma}(\mathbb{R}^n_+)) \mid \\ \lim_{t\to 0+} \{[u(\cdot) - a]_{n.0,t} + [u]_{p,\mu(p),t} + [\nabla u]_{n,\frac{1}{2},t}\} = 0, |||u|||_t < \varepsilon\Big\},$$

where $p \in (n, \infty)$ and $s \in [0, n-1)$ are fixed numbers, $\mu(p) = \frac{1}{2} - \frac{n}{2p}$, $\varepsilon > 0$ is a small number determined late and

$$[u]_{p,\ell,t} = \sup_{0<\tau<t} \left(\tau^\ell \|w^s u(\cdot,\tau)\|_{L^q}\right),$$
$$|||u|||_t = [u]_{n,0,t} + [u]_{p,\mu(p),t} + [\nabla u]_{n,\frac{1}{2},t}.$$

If we set

$$H(u,v)(t) = \int_0^t e^{-(t-\tau)A} P[(u\cdot\nabla)v](\tau)d\tau,$$
$$\Phi(u)(t) = e^{-tA}u_0 - H(u,u)(t),$$

then (IE) is written in the form $u(t) = \Phi(u)(t)$. Our first goal is to show that $\Phi$ is a contraction mapping from $\mathcal{I}_\varepsilon$ into itself with suitable choice of $\varepsilon$.

For this purpose, we shall begin to prove the estimate for nonlinear term $H(u,v)(t)$ which we can prove easily by using Theorem 2.1.

**Lemma 5.1** *There hold the following estimates:*

$$[H(u,v)]_{p,\mu(p),t} \le C[u]_{p,\mu(p),t}[\nabla v]_{n,\frac{1}{2},t} \qquad n \le p \le \infty,$$

$$[H(u,v)]_{p,\ell,t} \le C[u]_{p,\ell,t}[\nabla v]_{n,\frac{1}{2},t} \qquad 0 \le \ell < \frac{1}{2},$$

$$[\nabla H(u,v)]_{n,\frac{1}{2},t} \le C[u]_{p,\mu(p),t}[\nabla v]_{n,\frac{1}{2},t} \qquad n < p < \infty.$$

By Theorem 2.1 and Lemma 5.1, we see that there exist constants $C$ and $D$ such that

$$|||\Phi|||_t \le \left|\left|\left|e^{-tA}u_0\right|\right|\right|_t + |||H(u,u)|||_t \le C\|w^s u_0\|_{L^n} + D|||u|||_t^2$$

If $u \in \mathcal{I}_\varepsilon$, then it follows

$$|||\Phi|||_t \le C\|w^s u_0\|_{L^n} + D\varepsilon^2.$$

Therefore if we choose $\varepsilon$ so small that $D\varepsilon < \frac{1}{2}$ and if the initial data $u_0 \in L^n_{w,\sigma}(\mathbb{R}^n_+)$ satisfies the condition $C\|w^s u_0\|_{L^n} < \frac{\varepsilon}{2}$, then we see $|||\Phi(u)|||_t \le \varepsilon$.

In order to prove that $\Phi(u)$ is the map to $\mathcal{I}_\varepsilon$ into itself, we have to prove

$$\lim_{t\to 0+}\left([e^{-tA}u_0 - u_0]_{n.0,t} + [e^{-tA}u_0]_{p,\mu(p),t} + [\nabla e^{-tA}u_0]_{n,\frac{1}{2},t}\right) = 0. \tag{16}$$

Equation (16) is easily proved by Theorem 2.1 and density property that $C^\infty_{0,\sigma}(\mathbb{R}^n_+)$ is dense in $L^p_{w,\sigma}(\mathbb{R}^n_+)$. Summing up, we can show that $\Phi$ is map from $\mathcal{I}_\varepsilon$ into itself.

In order to show $\Phi$ has a fixed point, we have to prove that $\Phi$ is a contraction mapping. For any $u, v \in \mathcal{I}_\varepsilon$, we have

$$|||\Phi(u) - \Phi(v)|||_t \le D|||u - v|||_t(|||u|||_t + |||v|||_t) \le 2D\varepsilon|||u - v|||_t.$$

If necessary, choosing $\varepsilon$ so small that $2D\varepsilon < \frac{1}{2}$, we see that $\Phi$ is a contraction mapping on $\mathcal{I}_\varepsilon$. Therefore we can obtain the unique solution $u \in BC([0,\infty); L^n_{w,\sigma}(\mathbb{R}^n_+))$ to (IE).

Next we shall prove the asymptotic behavior as follows:

$$\|w^{s'}u(t)\|_{L^q} = O\left(t^{-\frac{1}{2}+\frac{n}{2q}-\frac{(s_1+s_n)-(s_1'+s_n')}{2}}\right), \tag{17}$$

$$\|w^{s'}\nabla u(t)\|_{L^q} = O\left(t^{-1+\frac{n}{2q}-\frac{(s_1+s_n)-(s_1'+s_n')}{2}}\right) \tag{18}$$

for $n \le q < \infty$, $-\frac{n-1}{q} < s_1' \le s_1$ and $-\frac{1}{q} < s_n' \le s_n$ as $t \to \infty$.

We first prove (17) for $s_1' = s_n' = 0$. $u(t)$ is described by $u(t) = e^{-tA}u_0 - H(u,u) = u_1(t) + u_2(t)$ and we can prove the following estimate for $u_1(t)$ by Theorem 2.1:

$$\|u_1(t)\|_{L^q} \le Ct^{-\frac{1}{2}+\frac{n}{2q}}(1+t)^{-\frac{s_1+s_n}{2}}\|w^s u_0\|_{L^n}$$

for $n < q < \infty$. We estimate $u_2(t)$. To this end, we set

$$M = \sup_{0<t<\infty} \left( t^{\frac{1}{2}-\frac{n}{2q}} (1+t)^{\frac{s_1+s_n}{2}} \|u(t)\|_{L^q} \right) \tag{19}$$

and we notice that $u \in \mathcal{I}_\varepsilon$ implies

$$\sup_{0<t<\infty} \left( t^{\frac{1}{2}} \|w^s \nabla u(t)\|_{L^n} \right) \le \varepsilon. \tag{20}$$

We have

$$\begin{aligned}
\|u_2(t)\|_{L^q} &\le \int_0^t (t-\tau)^{-\frac{1}{2}} (1+t-\tau)^{-\frac{s_1+s_n}{2}} \|u(\tau)\|_{L^q} \|w^s \nabla u(\tau)\|_{L^n} d\tau \\
&\le C\varepsilon M \int_0^t (t-\tau)^{-\frac{1}{2}} (1+t-\tau)^{-\frac{s_1+s_n}{2}} \tau^{-\frac{1}{2}+\frac{n}{2q}} (1+\tau)^{-\frac{s_1-s_n}{2}} \tau^{-\frac{1}{2}} d\tau \\
&\le C\varepsilon M t^{-\frac{1}{2}+\frac{n}{2q}-\frac{s_1+s_n}{2}}.
\end{aligned}$$

Therefore we obtain

$$t^{\frac{1}{2}-\frac{n}{2q}} (1+t)^{\frac{s_1+s_n}{2}} \|u(t)\|_{L^q} \le C\|w^s u_0\|_{L^n} + C\varepsilon M.$$

Here choosing $\varepsilon$ smaller, if necessary, we see

$$\sup_{0<t<\infty} \left( t^{\frac{1}{2}-\frac{n}{2q}} (1+t)^{\frac{s_1+s_n}{2}} \|u(t)\|_{L^q} \right) \le C\|w^s u_0\|_{L^n}.$$

Next we consider the case $s'_1 \neq 0$ or $s'_n \neq 0$. Since we can obtain the estimate for $u_1(t)$:

$$\|w^{s'} u_1(t)\|_{L^q} \le Ct^{-\frac{1}{2}+\frac{n}{2q}} (1+t)^{-\frac{(s_1+s_n)-(s'_1+s'_n)}{2}} \|w^s u_0\|_{L^n}$$

by Theorem 2.1, we shall consider the term $u_2(t)$. By (19) and (20), we see

$$\begin{aligned}
&\|w^{s'} u_2(t)\|_{L^q} \\
&\le C \int_0^t (t-\tau)^{-\frac{1}{2}} (1+t-\tau)^{-\frac{(s_1+s_n)-(s'_1+s'_n)}{2}} \|u(\tau)\|_{L^q} \|w^s \nabla u(\tau)\|_{L^n} d\tau \\
&\le C\varepsilon M \int_0^t (t-\tau)^{-\frac{1}{2}} (1+t-\tau)^{-\frac{(s_1+s_n)-(s'_1+s'_n)}{2}} \tau^{-1+\frac{n}{2q}} (1+\tau)^{-\frac{s_1+s_n}{2}} d\tau \\
&\le C\varepsilon M t^{-\frac{1}{2}+\frac{n}{2q}} (1+t)^{-\frac{(s_1+s_n)-(s'_1+s'_n)}{2}}.
\end{aligned}$$

Choosing $\varepsilon$ smaller, if necessary, we obtain

$$\sup_{0<t<\infty}\left(t^{\frac{1}{2}-\frac{n}{2q}}(1+t)^{\frac{(s_1+s_n)-(s_1'+s_n')}{2}}\|w^{s'}u(t)\|_{L^q}\right)\le C\|w^s u_0\|_{L^n}.$$

Finally we consider the asymptotic behavior for $\|w^{s'}\nabla u(t)\|_{L^q}$. We have

$$\|w^{s'}\nabla u_1(t)\|_{L^q}\le Ct^{-1+\frac{n}{2q}}(1+t)^{-\frac{(s_1+s_n)-(s_1'+s_n')}{2}}\|w^s u_0\|_{L^n}$$

by Theorem 2.1 and

$$\begin{aligned}
&\|w^{s'}\nabla u_2(t)\|_{L^q}\\
&\le C\int_0^t (t-\tau)^{-1-\frac{n}{2}(\frac{1}{r}-\frac{1}{q})}(1+t-\tau)^{-\frac{(s_1+s_n)-(s_1'+s_n')}{2}}\|u(\tau)\|_{L^r}\|w^s\nabla u(\tau)\|_{L^n}d\tau\\
&\le C\varepsilon M\int_0^t (t-\tau)^{-1-\frac{n}{2}(\frac{1}{r}-\frac{1}{q})}(1+t-\tau)^{-\frac{(s_1+s_n)-(s_1'+s_n')}{2}}\tau^{-1+\frac{n}{2q}}(1+\tau)^{-\frac{s_1+s_n}{2}}d\tau\\
&\le Ct^{-1+\frac{n}{2q}}(1+t)^{-\frac{(s_1+s_n)-(s_1'+s_n')}{2}}\varepsilon M
\end{aligned}$$

by (19) and (20) for $q<r$. Therefore we obtain

$$\sup_{0<t<\infty}\left(t^{1-\frac{n}{2q}}(1+t)^{\frac{(s_1+s_n)-(s_1'+s_n')}{2}}\|w^{s'}\nabla u(t)\|_{L^q}\right)\le C\|w^s u_0\|_{L^n}.$$

Summing up, we obtain the desired results.

**Acknowledgements** This research was partly supported by JSPS Grant-in-Aid for Scientific Research (C) 225-40202.

## References

1. T. Abe, Y. Shibata, On a resolvent estimate of the Stokes equation on an infinite layer, Part 2 $\lambda=0$ case. J. Math. Fluid Mech. **5**, 245–274 (2003)
2. H.O. Bae, Analyticity and asymptotics for the Stokes Solutions in a weighted space. J. Math. Anal. Appl. **269**, 149–171 (2002)
3. H.O. Bae, H.J. Choe, Decay rate for the incompressible flows in half space. Math. Z. **238**, 799–816 (2001) .
4. W. Borchers, T. Miyakawa, $L^2$ decay for the Navier-Stokes flow in halfspaces. Math. Ann. **282**, 139–155 (1988)
5. W. Dan, Y. Shibata, On the $L_q$ - $L_r$ estimates of the Stokes semigroup in a two dimensional exterior domain. J. Math. Soc. Jpn. **51**, 181–207 (1999)
6. W. Dan, T. Kobayashi, Y. Shibata, On the local energy decay approach to some fluid flow in exterior domain, in *Recent Topics on Mathematical Theory of Viscous Incompressible Fluid.*

Lecture Notes in Numerical and Applied Mathematics, vol.16 (Kinokuniya, Tokyo, 1998), pp. 1–51
7. Y. Enomoto, Y. Shibata, On the rate of decay of the Ossen semigroup in exterior domains and its application to Navier-Stokes equation. J. Math. Fluid Mech. **7**, 339–367 (2005)
8. R. Farwig, H. Sohr, Weighted $L^q$-theory for the Stokes resolvent in exterior domains. J. Math. Soc. Jpn. **49**(2), 251–288 (1997)
9. A. Fröhlich, Helmholtz decomposition of weighted $L^q$ spaces for Muckenhoupt weights. Ann. Univ. Ferrara-Sez. VII -Sc. Mat. **XLVI**, 11–19 (2000)
10. A. Fröhlich, The Stokes operator in weighted $L^q$-spaces I: weighted estimates for the Stokes Resolvent Problem in a half-space. J. Math. Fluid Mech. **5**, 166–199 (2003)
11. Y. Fujigaki, T. Miyakawa, Asymptotic profiles of nonstationary incompressible Navier-Stokes flows in the half-space. Methods Appl. Anal. **8**(1), 121–157 (2001)
12. Y. Giga, T. Miyakawa, Solutions in $L^r$ of the Navier-Stokes initial value problem. Arch. Ration. Mech. Anal. **89**, 267–281 (1985)
13. H. Iwashita, $L_q$-$L_r$ estimates for solutions of the nonstationary Stokes equations in an exterior domain and the Navier-Stokes initial value problems in $L^q$ spaces. Math. Ann. **285**, 265–288 (1989)
14. B.J. Jin, Weighted $L^q - L^1$ estimate of the Stokes flow in the half-space. Nonlinear Anal. **72**, 1031–1043 (2010)
15. B.J. Jin, Spatial and temporal decay estimate of the Stokes flow of weighted $L^1$ initial data in the half-space. Nonlinear Anal. **73**, 1394–1407 (2010)
16. T. Kato, Strong $L^p$-solutions of the Navier-Stokes equation in $\mathbb{R}^m$, with applications to weak solutions. Math. Z. **187**, 471–480 (1984)
17. T. Kobayashi, T. Kubo, Weighted $L^p$-theory for the Stokes resolvent in some unbounded domains. Tsukuba J. Math. **37**(2), 179–205 (2013)
18. T. Kubo, The Stokes and Navier-Stokes equations in an aperture domain. J. Math. Soc. Jpn. **59**(3), 837–859 (2007)
19. T. Kubo, Y. Shibata, On the Stokes and Navier-Stokes equations in a perturbed half-space. Adv. Differ. Equ. **10**(6), 695–720 (2005)
20. M. McCracken, The resolvent problem for the Stokes equation on halfspaces in $L^p$. SIAM J. Math. Anal. **12**, 201–228 (1981)
21. M. Murata, Large time asymptotics for fundamental solutions of diffusion equations. Tohoku Math. J. **37**, 151–195 (1985)
22. E.M. Stein, *Singular Integrals and Differentiability Properties of Functions* (Princeton University Press, Princeton, 1970)
23. S. Ukai, A solution formula for the stokes equation in $\mathbb{R}^n_+$. Commun. Pure Appl. Math. **40**, 611–621 (1987)

# On Vorticity Formulation for Viscous Incompressible Flows in $\mathbf{R}^3_+$

**Humiya Kosaka and Yasunori Maekawa**

*Dedicated to Professor Yoshihiro Shibata on his sixtieth birthday*

**Abstract** In this paper we study the vorticity equations for viscous incompressible flows in the half space under the no-slip boundary condition on the velocity field. In particular, the boundary condition for the vorticity field is presented explicitly, and the solution formula for the linearized problem is obtained.

**Keywords** Biot-Savart law • Navier-Stokes equations • Vorticity equations

## 1 Introduction

We consider the Navier-Stokes equations for viscous incompressible flows in the three-dimensional half space:

$$(\mathrm{NS}) \begin{cases} \partial_t u - \nu\Delta u + u\cdot\nabla u + \nabla p = 0 \ \text{ in } \ (0,T)\times\mathbf{R}^3_+, \\ \nabla\cdot u = 0 \ \text{ in } \ [0,T)\times\mathbf{R}^3_+, \\ u = 0 \ \text{ on } [0,T)\times\partial\mathbf{R}^3_+, \\ u|_{t=0} = a \ \text{ in } \ \mathbf{R}^3_+. \end{cases}$$

Here $u = (u_1,u_2,u_3)^\top$ and $p$ are the unknown velocity field and pressure field, respectively, while $a = (a_1,a_2,a_3)^\top$ is a given initial velocity field, and $\nu$ is a positive constant which represents the kinematic viscosity of the fluid. We use the standard notations of derivatives; $\partial_t = \partial/\partial t$, $\Delta = \sum_{j=1}^3 \partial^2/\partial^2_{x_j}$, and $\nabla = (\partial/\partial_{x_1}, \partial/\partial_{x_2}, \partial/\partial_{x_3})$.

H. Kosaka
Hyogo Himeji East High School, Honmachi 68-70, Himeji 670-0012, Hyogo, Japan
e-mail: sunnyday19881003@gmail.com

Y. Maekawa (✉)
Mathematical Institute, Tohoku University, 6-3 Aoba, Aramaki, Aoba, Sendai 980-8578, Japan
e-mail: maekawa@m.tohoku.ac.jp



H. Amann et al. (eds.), *Recent Developments of Mathematical Fluid Mechanics*, Advances in Mathematical Fluid Mechanics, DOI 10.1007/978-3-0348-0939-9_19

There is a huge literature on the time-local solvability of the initial-boundary value problem for the Navier-Stokes equations in a domain $\Omega$ of $\mathbf{R}^n$, $n \geq 2$. If the initial data possesses a suitable regularity a typical approach to solve the Navier-Stokes equations is to use the semigroup theory for linear operators, which converts the differential equations to the associated integral equations. The solution to the integral equation is called the mild solution. This approach was established by Kato and Fujita [1, 8] for initial data in $H^{1/2}(\Omega)$, and by Giga-Miyakawa [5] for initial data in $L^p(\Omega)$, $p \geq n$ based on the analysis of the Stokes operator by Giga [3, 4]. The $L^p$ theory for the Navier-Stokes equations was also developed by Solonnikov [17] when $n = 3$, and by Weissler [19] and Ukai [18] when $\Omega = \mathbf{R}^n_+$.

In the fluid dynamics the vorticity field, denoted by $\omega$ in this paper, is known to play an important role. The vorticity field is defined as the curl of the velocity field, that is,

$$\omega = \nabla \times u = (\partial_2 u_3 - \partial_3 u_2, \partial_3 u_1 - \partial_1 u_3, \partial_1 u_2 - \partial_2 u_1)^\top .$$

In the three-dimensional Euclidian space $\mathbf{R}^3$ the time-local solvability of the vorticity equations was obtained by Giga-Miyakawa [6] when the initial vorticity belongs to a scaling invariant Morrey space which includes $L^{3/2}(\mathbf{R}^3)$. We note that the vorticity field is a scalar field for the two-dimensional flow, and the detailed analysis has been developed by now; see Majda-Bertozzi [11], Giga-Giga-Saal [7]. However, when the fluid domain has a nontrivial boundary the study of the vorticity equations (under the no-slip boundary condition on the velocity) is still less developed. The main obstacle is the boundary condition on the vorticity field. In fact, as far as the author knows, there is little literature which explicitly describes the boundary condition on the vorticity field. Indeed, in the presence of nontrivial boundary the vorticity equations have been formulated in terms of a variational form; see, e.g., Quartapelle [12], Ruas [14], Rautmann [13], and references therein. However, in this variational formulation the boundary condition itself is rather implicit, and also, the analysis is essentially restricted to the $L^2$ framework. Hence there still remains difficulty in studying the behavior of the vorticity near the boundary in details. Recently, the $L^p$ approach to the vorticity equations was provided by the author [9] for two-dimensional flows in the half plane $\mathbf{R}^2_+$. There, the boundary condition on the vorticity field is explicitly derived from the Biot-Savart law. This approach is useful to capture the anisotropic nature of the vorticity near the boundary, and it is applied to the analysis of the inviscid limit problem for the Navier-Stokes equations in $\mathbf{R}^2_+$ by Maekawa [10].

The first goal of this paper is to derive the boundary condition on the vorticity field in $\mathbf{R}^3_+$ by applying the argument in [9]. The resulting vorticity formulation, which is shown to be equivalent with (NS), is described as follows.

$$(\mathrm{V}) \begin{cases} \partial_t \omega - \nu \Delta \omega + \nabla \times (u \cdot \nabla u) = 0 & \text{in } (0,T) \times \mathbf{R}^3_+, \\ u = \nabla \times \Psi(\omega) & \text{in } (0,T) \times \mathbf{R}^3_+, \\ \nu K \omega = -N(\nabla \times (u \cdot \nabla u)) & \text{on } (0,T) \times \partial \mathbf{R}^3_+, \\ \omega|_{t=0} = b & \text{in } \quad \mathbf{R}^3_+. \end{cases}$$

Here $b = \nabla \times a$ and $K$ is a linear operator defined by (5), which consists of the Neumann derivative, the Dirichlet-Neumann map, and the Neumann-Dirichlet map, while $N(\nabla \times (u \cdot \nabla u))$ is a nonlinear term defined by (6). The relation $u = \nabla \times \Psi(\omega)$ represents the Biot-Savart law, whose definition will be stated in (1).

Our second goal is to analyze the linearized problem for (V), and then it is applied to show the time-local solvability of (V) as follows.

**Theorem 1.1** *Assume that $a \in \dot{W}^{1,3/2}_{0,\sigma}(\mathbf{R}^3_+)$. Then there is $T > 0$ such that there exists a unique mild solution $\omega \in C\left([0,T)\,;(L^{3/2}(\mathbf{R}^3_+))^3\right)$ to (V) satisfying*

$$\sup_{0<t<T}\left(t^{\frac{1}{10}}\|\omega(t)\|_{L^{\frac{5}{3}}(\mathbf{R}^3_+)} + t^{\frac{3}{5}}\|\nabla\omega(t)\|_{L^{\frac{5}{3}}(\mathbf{R}^3_+)}\right) < \infty,$$

$$\lim_{T'\to 0}\sup_{0<t<T'}\left(t^{\frac{1}{10}}\|\omega(t)\|_{L^{\frac{5}{3}}(\mathbf{R}^3_+)} + t^{\frac{3}{5}}\|\nabla\omega(t)\|_{L^{\frac{5}{3}}(\mathbf{R}^3_+)}\right) = 0.$$

Here $\dot{W}^{1,p}_{0,\sigma}(\mathbf{R}^3_+)$, $1 < p < \infty$, is the completion with respect to the homogeneous norm $\|\nabla u\|_{L^p(\mathbf{R}^3_+)}$ of the space of all smooth, divergence-free vector fields with compact support in $\mathbf{R}^3_+$.

This paper is organized as follows. In Sect. 2 we derive the vorticity formulation for the inhomogeneous Stokes equations in $\mathbf{R}^3_+$. The vorticity formulation for the nonlinear problem (NS) is easily obtained from this result. Section 3 is devoted to the analysis for the linearized vorticity equations. We first establish the solution formula by using the Fourier-Laplace transform in Sect. 3.1 and then prove the $L^p$-$L^q$ estimates in Sect. 3.2. These $L^p$-$L^q$ estimates are applied to the nonlinear vorticity equations, which will be presented in Sect. 4.

## 2 Vorticity Formulation for the Stokes Problem

In this section we consider the Stokes equations

$$(\mathrm{S})\begin{cases}\partial_t u - \nu\Delta u + \nabla p = f & \text{in } (0,T)\times\mathbf{R}^3_+,\\ \nabla\cdot u = 0 & \text{in } [0,T)\times\mathbf{R}^3_+,\\ u = 0 & \text{on } [0,T)\times\partial\mathbf{R}^3_+,\\ u|_{t=0} = a & \text{in } \mathbf{R}^3_+.\end{cases}$$

Here $f = (f_1, f_2, f_3)^\top$ is a given external force. The aim of this section is to establish the vorticity formulation of (S). By taking the curl in the first and the last equations of (S), we obtain the equations for the vorticity field $\omega = \nabla \times u$ as follows.

$$(\mathrm{LV}')\begin{cases}\partial_t\omega - \nu\Delta\omega = \nabla\times f & \text{in } (0,T)\times\mathbf{R}^3_+,\\ \omega|_{t=0} = \nabla\times a & \text{in } \mathbf{R}^3_+.\end{cases}$$

The problem here is, of course, the boundary conditions on $\omega$, which must be compatible with the no-slip boundary condition on $u$ in (S). Roughly speaking, the boundary conditions on $\omega$ have to be imposed so that the following requirements are justified:

(1) the velocity field given by the Biot-Savart law solves (S),
(2) the solution $\omega$ satisfies the divergence-free condition, that is, $\nabla \cdot \omega = 0$ in $(0, T) \times \mathbf{R}^3_+$.

The condition (2) above is a natural requirement in view of the relation $\omega = \nabla \times u$. In Sect. 2.1 we derive the boundary conditions on $\omega$ based on the requirements (1) and (2), and in Sect. 2.2 we show that the resulting boundary conditions indeed give the equivalent formulation to the Stokes equations (S).

## *2.1 Boundary Condition on Vorticity Fields*

The aim of this section is to derive the boundary condition on the vorticity field $\omega = \nabla \times u$, where $u$ is the solution to the Stokes problem (S). To this end we first recall the Biot-Savart law in $\mathbf{R}^3_+$. For a given function $f \in C_0^\infty(\mathbf{R}^3_+)$ we denote by $(-\Delta_D)^{-1}f$ the solution to the Dirichlet problem: $-\Delta h = f$ in $\mathbf{R}^3_+$, $h = 0$ on $\partial \mathbf{R}^3_+$. Similarly, we denote by $(-\Delta_N)^{-1}f$ the solution to the Neumann problem: $-\Delta h = f$ in $\mathbf{R}^3_+$, $\partial_3 h = 0$ on $\partial \mathbf{R}^3_+$.

**Definition 2.1 (Biot-Savart Law)** For a given $f = (f', f_3)^\top \in (C_0^\infty(\overline{\mathbf{R}^3_+}))^2 \times C_0^\infty(\overline{\mathbf{R}^3_+})$ the velocity field $v$ recovered from $f$ via the Biot-Savart law is defined by

$$v = \nabla \times \Psi(f), \qquad \Psi(f) = \begin{pmatrix} \Psi'(f') \\ \Psi_3(f_3) \end{pmatrix} = \begin{pmatrix} (-\Delta_D)^{-1} f' \\ (-\Delta_N)^{-1} f_3 \end{pmatrix}. \tag{1}$$

*Remark 2.2* Note that the definition of $v$ in (1) yields $\nabla \cdot v = 0$ in $\mathbf{R}^3_+$ and $v_3 = 0$ on $\partial \mathbf{R}^3_+$. Moreover, if $f$ is of the form $\nabla \times h$ with $h \in (C_0^\infty(\overline{\mathbf{R}^3_+}))^3$ satisfying $\nabla \cdot h = 0$ in $\mathbf{R}^3_+$ and $h = 0$ on $\partial \mathbf{R}^3_+$, then the direct calculation using the integration by parts gives $v = h$. In particular, if $u$ is the smooth solution to (S) which decays fast enough at spatial infinity then one can verify the relation $u = \nabla \times \Psi(\omega)$, where $\omega = \nabla \times u$.

Next we introduce the Dirichlet-Neumann map and the Neumann-Dirichlet map, which play an essential role in the boundary conditions on vorticity fields. We denote by $\gamma$ the trace operator to the boundary $\partial \mathbf{R}^3_+$ for functions on $\overline{\mathbf{R}^3_+}$, which satisfies $\gamma f = f|_{x_3=0}$ if $f$ is continuous on $\overline{\mathbf{R}^3_+}$.

**Definition 2.3 (Dirichlet-Neumann Map and Neumann-Dirichlet Map)** Let $f \in C_0^\infty(\mathbf{R}^2)$. We denote by $E_D f$ and $E_N f$, respectively, the solution to the Dirichlet problem: $\Delta E_D f = 0$ in $\mathbf{R}^3_+$, $E_D f = f$ on $\partial \mathbf{R}^3_+$, and the solution to the Neumann problem: $\Delta E_N f = 0$ in $\mathbf{R}^3_+$, $-\partial_3 E_N f = f$ on $\partial\mathbf{R}^3_+$. Then the Dirichlet-Neumann map $\Lambda_{DN}$ and the Neumann-Dirichlet map $\Lambda_{ND}$ are respectively defined by

$$\Lambda_{DN} f = -\gamma\partial_3 E_D f, \qquad \Lambda_{ND} f = \gamma E_N f. \tag{2}$$

Let $f \in C_0^\infty(\overline{\mathbf{R}^3_+})$. Then by the definition of (2) we have

$$\begin{aligned}\gamma\partial_3(-\Delta_D)^{-1}\Delta f &= \gamma\partial_3(-\Delta_D)^{-1}\Delta_D(f - E_D\gamma f)\\ &= \gamma\partial_3(f - E_D\gamma f) = -\gamma\partial_3 f - \Lambda_{DN}\gamma f\end{aligned} \tag{3}$$

and for $j = 1, 2$,

$$\begin{aligned}\gamma\partial_j(-\Delta_N)^{-1}\Delta f &= \gamma\partial_j(-\Delta_N)^{-1}\Delta_N(f + E_N\gamma\partial_3 f)\\ &= -\gamma\partial_j f - \partial_j\Lambda_{ND}\gamma\partial_3 f.\end{aligned} \tag{4}$$

The relations (3) and (4) will be used in the proof of the next proposition. set $\nabla' = (\partial_1, \partial_2)$ and $\omega' = (\omega_1, \omega_2)$.

**Proposition 2.4** *Assume that $f \in (C_0^\infty(\overline{\mathbf{R}^3_+}))^3$ and that $a \in (C_0^\infty(\overline{\mathbf{R}^3_+}))^3$ satisfies $\nabla \cdot a = 0$ in $\mathbf{R}^3_+$ and $a = 0$ on $\partial\mathbf{R}^3_+$. Let $u$ be the smooth solution to* (S) *which decays fast enough at spatial infinity. Then the vorticity field $\omega = \nabla \times u$ is subject to the boundary condition*

$$\nu K\omega = N(\nabla \times f) \qquad \text{on} \quad (0,T) \times \partial\mathbf{R}^3_+, \tag{LBC}$$

*where $K\omega$ and $N(h)$ are respectively defined by*

$$K\omega = \begin{pmatrix} -\gamma\partial_3\omega_1 - \Lambda_{DN}\gamma\omega_1 - \partial_1\Lambda_{ND}\gamma\nabla'\cdot\omega' + \gamma\partial_1\omega_3 \\ -\gamma\partial_3\omega_2 - \Lambda_{DN}\gamma\omega_2 - \partial_2\Lambda_{ND}\gamma\nabla'\cdot\omega' + \gamma\partial_2\omega_3 \\ -\gamma\partial_3\omega_3 - \gamma\nabla'\cdot\omega' \end{pmatrix}, \tag{5}$$

$$N(h) = \begin{pmatrix} -\gamma\partial_3(-\Delta_D)^{-1}h_1 + \gamma\partial_1(-\Delta_N)^{-1}h_3 \\ -\gamma\partial_3(-\Delta_D)^{-1}h_2 + \gamma\partial_2(-\Delta_N)^{-1}h_3 \\ 0 \end{pmatrix}. \tag{6}$$

*Remark 2.5* Our derivation of the boundary condition below is based the Biot-Savart law, and hence, the vorticity fields need to be sufficiently localized so that the Biot-Savart law makes sense.

*Proof of Proposition 2.4* Set $v = \nabla \times \Psi(\omega)$. Then by using (LV') we have the following equivalence on $(0, T) \times \partial\mathbf{R}^3_+$:

$$\begin{aligned}\gamma\partial_t v_2 = 0 &\Leftrightarrow \gamma\partial_3\Psi_1(\partial_t\omega_1) - \gamma\partial_1\Psi_3(\partial_t\omega_3) = 0 \\ &\Leftrightarrow \gamma\partial_3(-\Delta_D)^{-1}(\nu\Delta\omega_1) - \gamma\partial_1(-\Delta_N)^{-1}(\nu\Delta\omega_3) \\ &\qquad = -\gamma\partial_3(-\Delta_D)^{-1}(\nabla \times f)_1 + \gamma\partial_1(-\Delta_N)^{-1}(\nabla \times f)_3.\end{aligned}$$

Hence (3) and (4) yield

$$\begin{aligned}\gamma\partial_t v_2 = 0 \Leftrightarrow\ & \nu\{-\gamma\partial_3\omega_1 - \Lambda_{DN}\gamma\omega_1 + \partial_1\Lambda_{ND}\gamma\partial_3\omega_3 + \gamma\partial_1\omega_3\} \\ &= -\gamma\partial_3(-\Delta_D)^{-1}(\nabla \times f)_1 + \gamma\partial_1(-\Delta_N)^{-1}(\nabla \times f)_3.\end{aligned} \tag{7}$$

In the same manner, we have

$$\begin{aligned}\gamma\partial_t v_1 = 0 \Leftrightarrow\ & \nu\{-\gamma\partial_3\omega_2 - \Lambda_{DN}\gamma\omega_2 + \partial_2\Lambda_{ND}\gamma\partial_3\omega_3 + \gamma\partial_2\omega_3\} \\ &= -\gamma\partial_3(-\Delta_D)^{-1}(\nabla \times f)_2 + \gamma\partial_2(-\Delta_N)^{-1}(\nabla \times f)_3.\end{aligned} \tag{8}$$

When $\omega$ satisfies the divergence-free condition the term $\Lambda_{ND}\gamma\partial_3\omega_3$ in (7) and (8) can be replaced by $-\Lambda_{ND}\gamma\nabla' \cdot \omega'$, while the trace of the equality $\nabla \cdot \omega = 0$ gives

$$-\gamma\partial_3\omega_3 - \gamma\nabla' \cdot \omega' = 0. \tag{9}$$

As stated in Remark 2.2, $\nabla \times \Psi(\omega) = u$ holds if $\omega = \nabla \times u$ and $u$ satisfies $\nabla \cdot u = 0$ and $\gamma u = 0$. Hence, under the assumption of Proposition 2.4 we have $v = u$, and thus, $\gamma\partial_t v = 0$ is valid. Then (7)–(9) give the condition (LBC). The proof is complete.

## 2.2 Equivalence Between the Stokes Problem and Its Vorticity Formulation

In the present section we show that the solution to the linearized vorticity equations

$$(\mathrm{LV}) \begin{cases} \partial_t\omega - \nu\Delta\omega = \nabla \times f & \text{in } (0, T) \times \mathbf{R}^3_+, \\ \nu K\omega = N(\nabla \times f) & \text{on } (0, T) \times \partial\mathbf{R}^3_+, \\ \omega|_{t=0} = \nabla \times a & \text{in } \mathbf{R}^3_+. \end{cases}$$

gives the solution to the Stokes equations ($S$) with the aid of the Biot-Savart law. Thus, combining with Proposition 2.4, we conclude that (LV) is an equivalent formulation to ($S$). The main result of this section is stated as follows.

**Proposition 2.6** *Assume that* $f \in (C_0^\infty(\overline{\mathbf{R}^3_+}))^3$ *and that* $a \in (C_0^\infty(\overline{\mathbf{R}^3_+}))^3$ *satisfies* $\nabla \cdot a = 0$ *in* $\mathbf{R}^3_+$ *and* $a = 0$ *on* $\partial\mathbf{R}^3_+$. *Let* $\omega$ *be the smooth solution to* (LV) *which decays fast enough at spatial infinity. Then* $u = \nabla \times \Psi(\omega)$ *is the smooth solution to* (S) *with a suitable pressure* $p$.

*Proof* Note that $u|_{t=0} = \nabla \times \Psi(\nabla \times a) = a$ by Remark 2.2. Firstly we show that $\nabla \cdot \omega = 0$ in $[0, T) \times \mathbf{R}^3_+$. To see this let us take the divergence in the first and the last equations of (LV). Then the function $\psi = \nabla \cdot \omega$ satisfies

$$\begin{cases} \partial_t \psi - \nu \Delta \psi = 0 & \text{in } (0,T) \times \mathbf{R}^3_+, \\ \psi = 0 & \text{on } (0,T) \times \partial\mathbf{R}^3_+, \\ \psi|_{t=0} = 0 & \text{in } \mathbf{R}^3_+. \end{cases} \tag{10}$$

Here we have used the boundary condition (9). Thus, by the uniqueness of the solution to the linear heat equations, $\psi$ must be identically zero, i.e., $\nabla \cdot \omega = 0$ in $[0, T) \times \mathbf{R}^3_+$. By the definition of $K$ we then have

$$\begin{aligned} &\nu\{-\gamma\partial_3\omega_1 - \Lambda_{DN}\gamma\omega_1 + \partial_1\Lambda_{ND}\gamma\partial_3\omega_3 + \gamma\partial_1\omega_3\} \\ &\qquad = -\gamma\partial_3(-\Delta_D)^{-1}(\nabla \times f)_1 + \gamma\partial_1(-\Delta_N)^{-1}(\nabla \times f)_3, \end{aligned} \tag{11}$$

and

$$\begin{aligned} &\nu\{-\gamma\partial_3\omega_2 - \Lambda_{DN}\gamma\omega_2 + \partial_2\Lambda_{ND}\gamma\partial_3\omega_3 + \gamma\partial_2\omega_3\} \\ &\qquad = -\gamma\partial_3(-\Delta_D)^{-1}(\nabla \times f)_2 + \gamma\partial_2(-\Delta_N)^{-1}(\nabla \times f)_3. \end{aligned} \tag{12}$$

Now let us recall that the derivation of (7) and (8) requires only (LV'), (3), and (4). Thus we see $\gamma\partial_t u = \gamma\partial_t \nabla \times \Psi(\omega) = 0$ in $(0, T)$, i.e., $\gamma u = \gamma a = 0$ in $(0, T)$. Next we observe that $h := \nabla \cdot \Psi(\omega)$ satisfies

$$\gamma h = \gamma\partial_1\Psi_1(\omega_1) + \gamma\partial_2\Psi_2(\omega_2) + \gamma\partial_3\Psi_3(\omega_3) = 0 \tag{13}$$

by the definition of $\Psi$ in (1). On the other hand, $h$ satisfies $\Delta h(t) = \nabla \cdot \Delta\Psi(\omega(t)) = -\nabla \cdot \omega(t) = 0$ in $\mathbf{R}^3_+$ for each $t \in (0, T)$, which implies $h = 0$ by (13). Then we have

$$\nabla \times u = \nabla \times \nabla \times \Psi(\omega) = -\Delta\Psi + \nabla\nabla \cdot \Psi(\omega) = \omega + \nabla h = \omega.$$

Collecting these, we have shown that $u = \nabla \times \Psi(\omega)$ satisfies

$$\nabla \cdot u = 0, \quad \nabla \times u = \omega, \quad \gamma u = 0, \quad u|_{t=0} = a. \tag{14}$$

Hence, in view of (S) it suffices to show that $u$ satisfies the first equation of (S). Let $v \in C_0^\infty((0,T) \times C_{0,\sigma}^\infty(\mathbf{R}_+^3))$. Then $v$ is written as $v = \nabla \times \Psi(w)$ with $w = \nabla \times v$ by Remark 2.2, and we have from the integration by parts,

$$\begin{aligned}
\int_0^T \langle \partial_t u - \nu\Delta u, v\rangle_{L^2(\mathbf{R}_+^3)} dt &= \int_0^T \langle \partial_t u + \nu\nabla \times \omega, \nabla \times \Psi(w)\rangle_{L^2(\mathbf{R}_+^3)} dt \\
&= \int_0^T \langle \partial_t \nabla \times u + \nu\nabla \times \nabla \times \omega, \Psi(w)\rangle_{L^2(\mathbf{R}_+^3)} dt \\
&\quad + \nu \int_0^T \langle \mathbf{e_3} \times \nabla \times \omega, \Psi(w)\rangle_{L^2(\partial\mathbf{R}_+^3)} dt \\
&= \int_0^T \langle \partial_t \omega - \nu\Delta\omega, \Psi(w)\rangle_{L^2(\mathbf{R}_+^3)} dt \\
&= \int_0^T \langle \nabla \times f, \Psi(w)\rangle_{L^2(\mathbf{R}_+^3)} dt = \int_0^T \langle f, v\rangle_{L^2(\mathbf{R}_+^3)} dt.
\end{aligned}$$

Here $\mathbf{e_3} = (0,0,1)^\top$. Thus, there is a scalar function $p \in L^2(0,T; W_{loc}^{1,1}(\mathbf{R}_+^3))$ such that $\nabla p = f - \partial_t u + \nu\Delta u$ in $(0,T) \times \mathbf{R}_+^3$; see [2, 15]. The pressure $p$ is smooth in $(0,T) \times \mathbf{R}_+^3$, for $u = \nabla \times \Psi(\omega)$ and $f$ are assumed to be smooth. The proof is complete.

## 3 Analysis of Linearized Vorticity Equations

The aim of this section is to derive the solution formula of the linearized vorticity equations and to establish the $L^p - L^q$ estimates of solutions for the linearized problem. We note that, for the Stokes equations in $\mathbf{R}_+^n$, the solution formula was obtained by Solonnikov [16, 17] and Ukai [18].

### *3.1 Solution Formula of Linearized Vorticity Equations*

The aim of this section is to derive the solution formula of the linearized vorticity equations by using the Fourier-Laplace transform

$$\mathcal{J}[f](s,\xi',x_3) := \int_0^\infty e^{-st} \int_{\mathbf{R}^2} e^{-ix'\cdot\xi'} f(t,x',x_3)dx'dt. \tag{15}$$

We will often write $\hat{f}$ for $\mathcal{J}[f]$. Let as consider the linear problem

$$(\mathrm{L}) \begin{cases} \partial_t \omega - \nu\Delta\omega = f \text{ in } (0,T)\times\mathbf{R}^3_+, \\ \nu K\omega = g \text{ on } [0,T)\times\partial\mathbf{R}^3_+, \\ \omega|_{t=0} = b \text{ in } \mathbf{R}^3_+, \end{cases} \tag{16}$$

for given smooth functions $f, g$ and $b$. Note that (LV) is a special case of (L). We start from the next lemma.

**Lemma 3.1** *Let $K$ be the linear operator defined by (5). Then we have*

$$\mathcal{J}[K\omega] = \begin{pmatrix} -\gamma\partial_3\hat{\omega}_1 - |\xi'|\gamma\hat{\omega}_1 + \frac{\xi_1}{|\xi'|}(\xi'\cdot\gamma\hat{\omega}') + i\xi_1\gamma\hat{\omega}_3 \\ -\gamma\partial_3\hat{\omega}_2 - |\xi'|\gamma\hat{\omega}_2 + \frac{\xi_2}{|\xi'|}(\xi'\cdot\gamma\hat{\omega}') + i\xi_2\gamma\hat{\omega}_3 \\ -\partial_3\gamma\hat{\omega}_3 - i(\xi'\cdot\gamma\hat{\omega}') \end{pmatrix}, \tag{17}$$

*where $\hat{\omega} := \mathcal{J}[\omega](s,\xi',x_3)$.*

*Proof* It suffices to calculate $\mathcal{J}[\Lambda_{DN}\gamma\omega_j]$ and $\mathcal{J}[\partial_j\Lambda_{ND}\gamma\nabla'\omega']$. It is well known that the partial Fourier-Laplace transforms of $E_D h$ and $E_N h$ are given by

$$\mathcal{J}[E_D h](s,\xi',x_3) = e^{-x_3|\xi'|}\hat{h}(s,\xi'), \quad \mathcal{J}[E_N h](s,\xi',x_3) = \frac{e^{-x_3|\xi'|}}{|\xi'|}\hat{h}(s,\xi').$$

Thus, we have

$$\mathcal{J}[\Lambda_{DN}\gamma\omega_j] = -\gamma\partial_3 e^{x_3|\xi'|}\gamma\hat{\omega}_j = |\xi'|\gamma\hat{\omega}_j,$$

and

$$\mathcal{J}[\partial_j\Lambda_{ND}\gamma\nabla'\cdot\omega'] = \frac{i\xi_j}{|\xi'|}\mathcal{J}[\gamma\nabla'\cdot\omega'] = -\frac{\xi_j}{|\xi'|}(\xi'\cdot\gamma\hat{\omega}_j).$$

The proof is complete.

Let $f, g \in (C_0^\infty([0,\infty)\times\overline{\mathbf{R}^3_+}))^3$ and $b = 0$. Then by performing the Fourier-Laplace transform the equations (L) is converted to the boundary value problem for the system of ordinary differential equations

$$\begin{cases} s\hat{\omega} + \nu|\xi'|^2\hat{\omega} - \nu\partial_3^2\hat{\omega} = \hat{f}, \; x_3 > 0, \\ \nu\mathcal{J}[K\omega] = \hat{g}, \; x_3 = 0, \end{cases} \tag{18}$$

where $\operatorname{Re} s > 0$ and $\xi' \in \mathbf{R}^2$. Set

$$\lambda = \sqrt{\frac{s}{\nu} + |\xi'|^2}.$$

Then the general solution of the first equations of (18) is given by

$$\hat{\omega} = ce^{-x_3\lambda} + \int_0^{x_3} \frac{e^{-(x_3-y_3)\lambda}}{2\lambda\nu}\hat{f}dy_3 + \int_{x_3}^{\infty} \frac{e^{-(y_3-x_3)\lambda}}{2\lambda\nu}\hat{f}dy_3, \tag{19}$$

where $c$ is independent of $x_3$. Computing $\gamma\hat{\omega}$ and $\gamma\partial_3\hat{\omega}$, we have

$$\gamma\hat{\omega} = c + \frac{1}{2\lambda\nu}V_\lambda(\hat{f}), \quad \gamma\partial_3\hat{\omega} = -\lambda c + \frac{2}{\lambda}V_\lambda(\hat{f}),$$

where $V_\lambda(\hat{f}) := \int_0^\infty e^{-y_3\lambda}\hat{f}dy_3$. Hence by the boundary condition in (18) and by applying Lemma 3.1

$$Lc - \frac{1}{2\lambda\nu}RV_\lambda(\hat{f}) = \frac{1}{\nu}\hat{g}, \tag{20}$$

where

$$L = \begin{pmatrix} \lambda - \frac{\xi_2^2}{|\xi'|} & \frac{\xi_1\xi_2}{|\xi'|} & i\xi_1 \\ \frac{\xi_1\xi_2}{|\xi'|} & \lambda - \frac{\xi_1^2}{|\xi'|} & i\xi_2 \\ -i\xi_1 & -i\xi_2 & \lambda \end{pmatrix}, \quad R = \begin{pmatrix} \lambda + \frac{\xi_2^2}{|\xi'|} & -\frac{\xi_1\xi_2}{|\xi'|} & -i\xi_1 \\ -\frac{\xi_1\xi_2}{|\xi'|} & \lambda + \frac{\xi_1^2}{|\xi'|} & -i\xi_2 \\ i\xi_1 & i\xi_2 & \lambda \end{pmatrix}. \tag{21}$$

Since the determinant of $L$ is given by $(\lambda - |\xi'|)^2(\lambda + |\xi'|)$ the matrix $L$ is invertible when $\mathrm{Re}\, s > 0$ and $\xi' \in \mathbf{R}^2$, and $L^{-1} = L^{-1}(s, \xi')$ is given by

$$L^{-1}(s,\xi') = \frac{1}{\lambda^2 - |\xi'|^2}\begin{pmatrix} \lambda + \frac{\xi_2^2}{|\xi'|} & -\frac{\xi_1\xi_2}{|\xi'|} & -i\xi_1 \\ -\frac{\xi_1\xi_2}{|\xi'|} & \lambda + \frac{\xi_1^2}{|\xi'|} & -i\xi_2 \\ i\xi_1 & i\xi_2 & \lambda \end{pmatrix} = \frac{\nu}{s}R(s,\xi'). \tag{22}$$

Thus we have

$$c = \frac{1}{2\lambda\nu}\frac{\nu}{s}R^2V_\lambda(\hat{f}) + \frac{1}{s}R\hat{g}. \tag{23}$$

For later use we introduce the matrix $M$ defined by

$$M := \begin{pmatrix} 2|\xi'|^2 + 2\lambda\frac{\xi_2^2}{|\xi'|} & -2\lambda\frac{\xi_1\xi_2}{|\xi'|} & -2i\lambda\xi_1 \\ -2\lambda\frac{\xi_1\xi_2}{|\xi'|} & 2|\xi'|^2 + 2\lambda\frac{\xi_1^2}{|\xi'|} & -2i\lambda\xi_2 \\ 2i\lambda\xi_1 & 2i\lambda\xi_2 & 2|\xi'|^2 \end{pmatrix}. \tag{24}$$

Then $R$ and $M$ have the relations

$$R = \frac{1}{2\lambda}M + \frac{s}{\nu\lambda}I, \quad R^2 = M + \frac{s}{\nu}I, \tag{25}$$

where $I$ is the identity matrix. Substituting (23) into (19) and using (25), we obtain

$$\hat{\omega} = \left\{ \frac{1}{2\lambda s} M V_\lambda(\hat{f}) + \frac{1}{2\lambda\nu} V_\lambda(\hat{f}) + \frac{1}{2\lambda s} M\hat{g} + \frac{1}{\lambda\nu}\hat{g} \right\} e^{-x_3\lambda}$$
$$+ \int_0^{x_3} \frac{e^{-(x_3-y_3)\lambda}}{2\lambda\nu} \hat{f} dy_3 + \int_{x_3}^{\infty} \frac{e^{-(y_3-x_3)\lambda}}{2\lambda\nu} \hat{f} dy_3. \tag{26}$$

The solution formula is then derived by taking the inverse Fourier-Laplace transform in (26). We will freely use the following relations in the Fourier-Laplace transform:

$$e^{-\lambda x_3} \sim -2\nu \frac{\partial G_\nu(x,t)}{\partial x_3}, \quad \frac{e^{-\lambda x_3}}{\lambda} \sim 2\nu G_\nu(x,t), \quad \frac{1}{s}\hat{f} \sim \int_0^t f(\tau)d\tau, \tag{27}$$

where

$$G_\nu(t,x) := (4\pi\nu t)^{-\frac{3}{2}} \exp\left(-\frac{|x|^2}{4\nu t}\right)$$

is the three-dimensional Gaussian. For given $f, h \in (C_0^\infty(\overline{\mathbf{R}^3_+}))^3$ and $g \in C_0^\infty(\mathbf{R}^2)$ we set

$$f * h = \int_{\mathbf{R}^3_+} f(x-y)h(y)dy,$$

$$f \star h = \int_{\mathbf{R}^3_+} f(x-y^*)h(y)dy, \quad y^* = (y_1, y_2, -y_3),$$

$$f \star (g\mathcal{H}^2_{\partial\mathbf{R}^3_+}) = \int_{\mathbf{R}^2} f(x'-y', x_3)g(y')dy'.$$

Using (27), we see

$$\mathcal{J}^{-1}\left[\frac{1}{2\lambda s} M V_\lambda(\hat{f}) e^{-x_3\lambda}\right] = \int_0^\infty \mathcal{J}^{-1}\left[\frac{e^{-(x_3+y_3)\lambda}}{2\lambda s} M\hat{f}\right] dy_3$$
$$= \nu \int_0^\infty \int_0^t \mathcal{M} \int_0^\tau \int_{\mathbf{R}^2} G_\nu(\tau - r, x'-y', x_3+y_3) f(r,y) dy' dr d\tau dy_3$$
$$= \nu \int_0^t \mathcal{M} \int_0^\tau G_\nu(\tau - r) \star f(r) dr d\tau$$

Here $\mathcal{M}$ is a second order differential operator whose symbol is given by $M$, that is,

$$\mathcal{M} = \begin{pmatrix} -2\Delta' + 2\partial_3\partial_2^2(-\Delta')^{-\frac{1}{2}} & -2\partial_3\partial_1\partial_2(-\Delta')^{-\frac{1}{2}} & 2\partial_3\partial_1 \\ -2\partial_3\partial_1\partial_2(-\Delta')^{-\frac{1}{2}} & -2\Delta' + 2\partial_3\partial_1^2(-\Delta')^{-\frac{1}{2}} & 2\partial_3\partial_2 \\ -2\partial_3\partial_1 & -2\partial_3\partial_2 & -2\Delta' \end{pmatrix}. \tag{28}$$

In the same way we have the following equalities:

$$\mathcal{J}^{-1}\left[\frac{1}{2\lambda\nu}V_\lambda(\hat{f})e^{-x_3\lambda}\right] = \int_0^t G_\nu(t-\tau)\star f(\tau)d\tau,$$

$$\mathcal{J}^{-1}\left[\frac{1}{2\lambda s}M\hat{g}e^{-x_3\lambda}\right] = \nu\int_0^t \mathcal{M}\int_0^\tau G_\nu(\tau-r)\star\left(g(r)\mathcal{H}^2_{\partial\mathbf{R}^3_+}\right)drd\tau,$$

$$\mathcal{J}^{-1}\left[\frac{1}{\lambda\nu}\hat{g}e^{-x_3\lambda}\right] = 2\int_0^t G_\nu(t-\tau)\star\left(g(\tau)\mathcal{H}^2_{\partial\mathbf{R}^3_+}\right)d\tau.$$

Collecting these above, the inverse Fourier-Laplace transform for the first term of the right-hand side of (26) is given by

$$\mathcal{J}^{-1}\left[\left\{\frac{1}{2\lambda s}MV_\lambda(\hat{f}) + \frac{1}{2\lambda\nu}V_\lambda(\hat{f}) + \frac{1}{2\lambda s}M\hat{g} + \frac{1}{\lambda\nu}\hat{g}\right\}e^{-x_3\lambda}\right]$$
$$= \nu\int_0^t \mathcal{M}\int_0^\tau G_\nu(\tau-r)\star f(r)drd\tau + \int_0^t G_\nu(t-\tau)\star f(\tau)d\tau$$
$$+\nu\int_0^t \mathcal{M}\int_0^\tau G_\nu(\tau-r)\star\left(g(r)\mathcal{H}^2_{\partial\mathbf{R}^3_+}\right)drd\tau + 2\int_0^t G_\nu(t-\tau)\star\left(g(\tau)\mathcal{H}^2_{\partial\mathbf{R}^3_+}\right)d\tau.$$

As for the second and the third terms of the right-hand side of (26), we have

$$\mathcal{J}^{-1}\left[\int_0^{x_3}\frac{e^{-(x_3-y_3)\lambda}}{2\lambda\nu}\hat{f}dy_3 + \int_{x_3}^\infty\frac{e^{-(y_3-x_3)\lambda}}{2\lambda\nu}\hat{f}dy_3\right]$$
$$= \int_0^t\int_0^{x_3}\int_{\mathbf{R}^2} G_\nu(t-\tau, x'-y', x_3-y_3)f(\tau,y)dy'dy_3d\tau$$
$$+\int_0^t\int_{x_3}^\infty\int_{\mathbf{R}^2} G_\nu(t-\tau, x'-y', y_3-x_3)f(\tau,y)dy'dy_3d\tau$$
$$= \int_0^t\int_{\mathbf{R}^3_+} G_\nu(t-\tau, x-y)f(\tau,y)dyd\tau = \int_0^t G_\nu(t-\tau)*f(\tau)d\tau.$$

where we have used $G_\nu(t,x) = G_\nu(t,x^*)$ by the symmetry of the Gaussian. Recalling that the kernel of the heat semigroup $\{e^{\nu t\Delta_N}\}_{t\geq 0}$ under the homogeneous Neumann boundary condition in $\mathbf{R}^3_+$ is given by $G_\nu(t,x-y) + G_\nu(t,x-y^*)$, we have arrived at the following representation of $\omega$:

$$\omega(t) = \int_0^t e^{\nu(t-\tau)\Delta_N}\left(f(\tau) + g(r)\mathcal{H}^2_{\partial\mathbf{R}^3_+}\right)d\tau$$
$$+ \nu\int_0^t \mathcal{M}\int_0^\tau G_\nu(\tau - r)\star\left(f(r) + g(r)\mathcal{H}^2_{\partial\mathbf{R}^3_+}\right)drd\tau. \tag{29}$$

It is straightforward to see the function

$$\phi(t) = e^{\nu t\Delta_N}b + \int_0^t \mathcal{M}G_\nu(\tau)\star bd\tau \tag{30}$$

satisfies (L) with $f = g = 0$. Hence, the solution of (L) is expressed as

$$\omega(t) = e^{\nu t\Delta_N}b + \int_0^t \mathcal{M}G_\nu(\tau)\star bd\tau + \int_0^t e^{\nu(t-\tau)\Delta_N}\left(f(\tau) + g(r)\mathcal{H}^2_{\partial\mathbf{R}^3_+}\right)d\tau$$
$$+ \nu\int_0^t \mathcal{M}\int_0^\tau G_\nu(\tau - r)\star\left(f(r) + g(r)\mathcal{H}^2_{\partial\mathbf{R}^3_+}\right)drd\tau. \tag{31}$$

Next we denote by $(-\Delta_{\mathbf{R}^3})^{-1}f$ the solution to the Poisson equation: $-\Delta h = f$ in $\mathbf{R}^3$. Then we have

$$(-\Delta_{\mathbf{R}^3})^{-1}f = E *_{\mathbf{R}^3} f, \qquad E(x) = \frac{1}{2\pi|x|},$$

where $*_{\mathbf{R}^3}$ stands for the convolution in $\mathbf{R}^3$. Since the Gaussian $G_\nu$ satisfies the heat equation $\partial_t G_\nu = \nu\Delta G_\nu$ in $\mathbf{R}_+ \times \mathbf{R}^3$ we can write

$$G_\nu(t) = -\nu^{-1}(-\Delta_{\mathbf{R}^3})^{-1}\partial_t G_\nu(t) = -\nu^{-1}E *_{\mathbf{R}^3} \partial_t G_\nu(t).$$

Thus the integration by parts yields

$$\nu\int_0^\tau G_\nu(\tau - r)\star\left(f(r) + g(r)\mathcal{H}^2_{\partial\mathbf{R}^3_+}\right)dr$$
$$= -\int_0^\tau \left(E *_{\mathbf{R}^3} \partial_\tau G(\tau - r)\right)\star\left(f(r) + g(r)\mathcal{H}^2_{\partial\mathbf{R}^3_+}\right)dr$$
$$= -\partial_\tau\left\{\int_0^\tau \left(E *_{\mathbf{R}^3} G_\nu(\tau - r)\right)\star\left(f(r) + g(r)\mathcal{H}^2_{\partial\mathbf{R}^3_+}\right)dr\right\}$$
$$+ E\star\left(f(\tau) + g(\tau)\mathcal{H}^2_{\partial\mathbf{R}^3_+}\right).$$

Set

$$\Gamma_\nu(t) = -\mathcal{M}E *_{\mathbf{R}^3} G_\nu(t), \qquad \Gamma_\nu(0) \star f = \lim_{t\to 0} \Gamma_\nu(t) \star f = -\mathcal{M}E \star f. \tag{32}$$

Then the last term of (31) is written as

$$\begin{aligned} &\nu \int_0^t \mathcal{M} \int_0^\tau G_\nu(\tau - r) \star \left(f(r) + g(r)\mathcal{H}^2_{\partial\mathbf{R}^3_+}\right) dr d\tau \\ &= \int_0^t \Gamma_\nu(t-\tau) \star \left(f(\tau) + g(\tau)\mathcal{H}^2_{\partial\mathbf{R}^3_+}\right) d\tau - \int_0^t \Gamma_\nu(0) \star \left(f(\tau) + g(\tau)\mathcal{H}^2_{\partial\mathbf{R}^3_+}\right) d\tau. \end{aligned} \tag{33}$$

Similarly, we have

$$\int_0^t \mathcal{M}G_\nu(\tau) \star b d\tau = \Gamma_\nu(t) \star b - \Gamma_\nu(0) \star b. \tag{34}$$

Combining (33) and (34) with (31), we have arrived at the following

**Proposition 3.2** *Let $b \in (C_0^\infty(\mathbf{R}^3_+))^3$ and $f, g \in (C_0^\infty([0,\infty) \times \overline{\mathbf{R}^3_+}))^3$. Then the integral equation of* (L) *is given by*

$$\begin{aligned} \omega(t) &= e^{\nu t \Delta_N} b + \Gamma_\nu(t) \star b - \Gamma_\nu(0) \star b + \int_0^t e^{\nu(t-\tau)\Delta_N} \left(f(\tau) + g(\tau)\mathcal{H}^2_{\partial\mathbf{R}^3_+}\right) d\tau \\ &+ \int_0^t \Gamma_\nu(t-\tau) \star \left(f(\tau) + g(\tau)\mathcal{H}^2_{\partial\mathbf{R}^3_+}\right) d\tau - \int_0^t \Gamma_\nu(0) \star \left(f(\tau) + g(\tau)\mathcal{H}^2_{\partial\mathbf{R}^3_+}\right) d\tau. \end{aligned} \tag{35}$$

By the definition of (32) the operator $\Gamma_\nu(0)\star$ defines a singular integral operator. In particular, the smoothing effect is lost near the boundary. The next cancellation property is important in solving the nonlinear vorticity equations.

**Lemma 3.3** *Let $f \in (C_0^\infty(\overline{\mathbf{R}^3_+}))^3$ and let $g = N(f)$, where $N$ is defined by* (6). *Then it follows that*

$$\Gamma_\nu(0) \star \left(f + g\mathcal{H}^2_{\partial\mathbf{R}^3_+}\right) = 0 \qquad \text{in } \mathbf{R}^3_+. \tag{36}$$

*Proof* First we note that $\Delta E(x - y^*) = 0$ for $x_3,\ y_3 > 0$. Then, by the integration by parts we have

$$\begin{aligned} &\Gamma_\nu(0) \star \left(\left(-\partial_3(-\Delta_D)^{-1} f'\right)\mathcal{H}^2_{\partial\mathbf{R}^3_+}\right) \\ &= \int_{\mathbf{R}^3_+} \nabla_y \Gamma_\nu(0, x-y^*) \cdot \nabla_y(-\Delta_D)^{-1} f' dy + \int_{\mathbf{R}^3_+} \Gamma_\nu(0, x-y^*)\Delta(-\Delta_D)^{-1} f' dy \end{aligned}$$

$$
\begin{aligned}
&= -\int_{\mathbf{R}^3_+} \Delta_y \Gamma_\nu(0, x-y^*)(-\Delta_D)^{-1} f' dy - \int_{\mathbf{R}^3_+} \Gamma_\nu(0, x-y^*) f' dy \\
&= -\int_{\mathbf{R}^3_+} \Gamma_\nu(0, x-y^*) f' dy.
\end{aligned} \tag{37}
$$

Similarly, we have

$$
\begin{aligned}
E \star \Big(\big(-\partial_3^2(-\Delta_N)^{-1} f_3\big)\mathcal{H}^2_{\partial\mathbf{R}^3_+}\Big) &= -\int_{\mathbf{R}^3_+} E(x-y^*)\partial_3 f_3 dy' \\
&= \int_{\mathbf{R}^3_+} \partial_3 E(x-y^*) f_3 dy + E \star (f_3 \mathcal{H}^2_{\partial\mathbf{R}^3_+}).
\end{aligned} \tag{38}
$$

Next, by the calculation (38) and the concrete expression of $\mathcal{M} = (m_{ij})_{1\le i,j\le 3}$, we have again from the integration by parts,

$$
\begin{aligned}
&\text{the first component of the vector of } \Gamma_\nu(0) \star \{(\nabla'(-\Delta_N)^{-1} f_3)\mathcal{H}^2_{\partial\mathbf{R}^3_+}\} \\
&= m_{11} E \star \{(\partial_1(-\Delta_N)^{-1} f_3)\mathcal{H}^2_{\partial\mathbf{R}^3_+}\} + m_{12} E \star \{(\partial_1(-\Delta_N)^{-1} f_2)\mathcal{H}^2_{\partial\mathbf{R}^3_+}\} \\
&= (m_{11}\partial_1 + m_{12}\partial_2) E \star \{((-\Delta_N)^{-1} f_3)\mathcal{H}^2_{\partial\mathbf{R}^3_+}\} \\
&= -2\partial_1 E \star \{(\Delta'(-\Delta_N)^{-1} f_3)\mathcal{H}^2_{\partial\mathbf{R}^3_+}\}, \\
&= -2\partial_1 E \star \{-f_3\mathcal{H}^2_{\partial\mathbf{R}^3_+} - (\partial_3^2(-\Delta_N)^{-1} f_3)\mathcal{H}^2_{\partial\mathbf{R}^3_+}\} \\
&= -2\partial_1 E \star \left(-f_3\mathcal{H}^2_{\partial\mathbf{R}^3_+} + f_3\mathcal{H}^2_{\partial\mathbf{R}^3_+}\right) - 2\int_{\mathbf{R}^3_+} \partial_1\partial_3 E(x-y^*) f_3 dy \\
&= -2\partial_1\partial_3 E \star f_3.
\end{aligned} \tag{39}
$$

Applying the same argument to the other components, we have

$$
\Gamma_\nu(0) \star \{(\nabla'(-\Delta_N)^{-1} f_3)\mathcal{H}^2_{\partial\mathbf{R}^3_+}\} = \begin{pmatrix} -2\partial_1\partial_3 E \star f_3 \\ -2\partial_2\partial_3 E \star f_3 \\ 2\Delta' E \star f_3 \end{pmatrix}. \tag{40}
$$

Combining (37) with (40), we obtain

$$
\begin{aligned}
&\Gamma_\nu(0) \star \left(f + g\mathcal{H}^2_{\partial\mathbf{R}^3_+}\right) \\
&= \Gamma_\nu(0) \star \left[\begin{pmatrix} f_1 \\ f_2 \\ f_3 \end{pmatrix} + \begin{pmatrix} -\partial_3(-\Delta_D)^{-1} f_1 + \partial_1(-\Delta_N)^{-1} f_3 \\ -\partial_3(-\Delta_D)^{-1} f_2 + \partial_2(-\Delta_N)^{-1} f_3 \\ 0 \end{pmatrix} \mathcal{H}^2_{\partial\mathbf{R}^3_+}\right]
\end{aligned}
$$

$$= \Gamma_\nu(0) \star \begin{pmatrix} 0 \\ 0 \\ f_3 \end{pmatrix} + \begin{pmatrix} -2\partial_1\partial_3 E \star f_3 \\ -2\partial_2\partial_3 E \star f_3 \\ 2\Delta' E \star f_3 \end{pmatrix}$$

$$= \begin{pmatrix} m_{13} E \star f_3 \\ m_{23} E \star f_3 \\ m_{33} E \star f_3 \end{pmatrix} + \begin{pmatrix} -2\partial_1\partial_3 E \star f_3 \\ -2\partial_2\partial_3 E \star f_3 \\ 2\Delta' E \star f_3 \end{pmatrix} = 0\,.$$

The proof is complete.

**Corollary 3.4** *Assume that* $a \in \dot{W}^{1,p}_{0,\sigma}(\mathbf{R}^3_+)$, $1 < p < \infty$, *and* $b = \nabla \times a$. *Then*

$$\Gamma_\nu(0) \star b = 0 \qquad \text{in } \mathbf{R}^3_+. \tag{41}$$

*Proof* Since $a \in \dot{W}^{1,p}_{0,\sigma}(\mathbf{R}^3_+)$, there exists a function sequence $\{f_n\}_{n\in\mathbf{N}} \subset C^\infty_{0,\sigma}(\mathbf{R}^3_+)$ such that $\lim_{n\to\infty} \|\nabla(a - f_n)\|_{L^p(\mathbf{R}^3_+)} = 0$. By $\nabla \cdot f_n = 0$ and $\gamma f_n = 0$ we have the Biot-Savart law: $f_n = \nabla \times \Psi(h_n)$ with $h_n = \nabla \times f_n$; see Remark 2.2. Hence $N(h_n) = \gamma f_n = 0$ holds. Then, by Lemma 3.3 we obtain

$$\Gamma_\nu(0) \star h_n = \Gamma_\nu(0) \star \left(h_n + N(h_n)\mathcal{H}^2_{\partial\mathbf{R}^3_+}\right) = 0. \tag{42}$$

Thus we see

$$\begin{aligned} \|\Gamma_\nu(0) \star b\|_{L^p(\mathbf{R}^3_+)} &= \|\Gamma_\nu(0) \star (b - h_n)\|_{L^p(\mathbf{R}^3_+)} \le C\|b - h_n\|_{L^p(\mathbf{R}^3_+)} \\ &= \|\nabla \times (a - f_n)\|_{L^p(\mathbf{R}^3_+)} \xrightarrow{n\to\infty} 0 \end{aligned} \tag{43}$$

The proof is complete.

Proposition 3.2, Lemma 3.3, and Corollary 3.4 yield

**Proposition 3.5** *Assume that* $a \in \dot{W}^{1,p}_{0,\sigma}(\mathbf{R}^3_+)$, $1 < p < \infty$, *and* $b = \nabla \times a$. *Let* $f \in (C^\infty_0([0,\infty) \times \overline{\mathbf{R}^3_+}))^3$. *Then the integral equation of* (LV) *is given by*

$$\begin{aligned} \omega(t) &= \left(e^{\nu t\Delta_N} + \Gamma_\nu(t)\star\right) b \\ &\quad + \int_0^t \left(e^{\nu(t-s)\Delta_N} + \Gamma_\nu(t-s)\star\right) \left\{\nabla \times f + N(\nabla \times f)\mathcal{H}^2_{\partial\mathbf{R}^3_+}\right\} ds. \end{aligned} \tag{44}$$

## 3.2 Estimates of Solution to Linearized Vorticity Equations

In this section we analyze the solution formula derived in Sect. 3.1 and establish the $L^p$-$L^q$ estimates for the evolution operator of the linearized vorticity equation. The main result of this section is the following

**Proposition 3.6** *Let $f, g \in (C_0^\infty(\overline{\mathbf{R}^3_+}))^3$ and $1 < q \le p < \infty$. Then we have*

$$\|\nabla^\alpha e^{\nu t \Delta_N} f\|_{L^p(\mathbf{R}^3_+)} + \|\nabla^\alpha \Gamma_\nu(t) \star f\|_{L^p(\mathbf{R}^3_+)} \le C(\nu t)^{-\frac{3}{2}(\frac{1}{q}-\frac{1}{p})-\frac{|\alpha|}{2}} \|f\|_{L^q(\mathbf{R}^3_+)}, \tag{45}$$

$$\|\nabla^\alpha e^{\nu t \Delta_N}(g\mathcal{H}^2_{\partial\mathbf{R}^3_+})\|_{L^p(\mathbf{R}^3_+)} + \|\nabla^\alpha \Gamma_\nu(t) \star (g\mathcal{H}^2_{\partial\mathbf{R}^3_+})\|_{L^p(\mathbf{R}^3_+)}$$
$$\le C(\nu t)^{-\frac{1}{2}-\frac{1}{q}+\frac{3}{2p}-\frac{|\alpha|}{2}} \|g\|_{L^q(\mathbf{R}^2)}. \tag{46}$$

*Here $\nabla^\alpha = \partial_1^{\alpha_1} \cdots \partial_3^{\alpha_3}$ for a given multi-index $\alpha$.*

*Proof* Since the estimates for the heat semigroup $e^{\nu t \Delta_N}$ are well known we focus on the estimates for $\Gamma_\nu(t)\star$. First we observe that $\Gamma_\nu(t) \star f = -(\mathcal{M}E * G_\nu(t)) \star f = -\mathcal{M}E * (G_\nu(t) \star f)$ and that the operator $\mathcal{M}E*$ is a singular integral operator. Then we have

$$\begin{aligned}
\|\nabla^\alpha \Gamma_\nu(t) \star f\|_{L^p(\mathbf{R}^3_+)} &= \|\mathcal{M}E * \nabla^\alpha (G_\nu(t) \star f)\|_{L^p(\mathbf{R}^3_+)} \\
&\le C\|\nabla^\alpha (G_\nu(t) \star f)\|_{L^p(\mathbf{R}^3_+)} \\
&\le C\|\nabla^\alpha G_\nu(t)\|_{L^r(\mathbf{R}^3_+)} \|f\|_{L^q(\mathbf{R}^3_+)} \\
&\le C(\nu t)^{-\frac{3}{2}\left(\frac{1}{q}-\frac{1}{p}\right)-\frac{|\alpha|}{2}} \|f\|_{L^q(\mathbf{R}^3_+)},
\end{aligned} \tag{47}$$

where $1/p = 1/r + 1/q - 1$. Hence (45) is proved. Next, by applying the Young inequality for the convolution in the $(x_1, x_2)$ variables, we have

$$\begin{aligned}
\|\nabla^\alpha \Gamma_\nu(t) \star (g\mathcal{H}^2_{\partial\mathbf{R}^3_+})\|_{L^p(\mathbf{R}^3_+)} &\le C\|\nabla^\alpha G_\nu(t) \star (g\mathcal{H}^2_{\partial\mathbf{R}^3_+})\|_{L^p(\mathbf{R}^3_+)} \\
&\le C\left\{\int_0^\infty \|\nabla_x^\alpha G_\nu(t,\cdot,x_3)\|^p_{L^r(\mathbf{R}^2)} dx_3\right\}^{\frac{1}{p}} \|g\|_{L^q(\mathbf{R}^2)},
\end{aligned}$$

where $1/p = 1/r + 1/q - 1$. Then (46) follows from

$$\left\{\int_0^\infty \|\nabla_x^\alpha G_\nu(t,\cdot,x_3)\|^p_{L^r(\mathbf{R}^2)} dx_3\right\}^{\frac{1}{p}} \le C(\nu t)^{-\frac{1}{2}-\frac{1}{q}+\frac{3}{2p}-\frac{|\alpha|}{2}}.$$

The proof is complete.

In view of (44) we also need

**Proposition 3.7** *Let $f \in (C_0^\infty(\overline{\mathbf{R}^3_+}))^3$ and let $N(f)$ be the function defined by* (6)*. Assume that $1 < q < 3$ and $r = 2q/(3-q)$. Then it follows that*

$$\|N(f)\|_{L^r(\mathbf{R}^2)} \le C\|f\|_{L^q(\mathbf{R}^3_+)}. \tag{48}$$

*Proof* Let $q'$ be the Hölder conjugate of $q$. Then the Hölder inequality yields for $h \in L^q(\mathbf{R}^3_+)$,

$$\begin{aligned}
|\gamma\partial_i E * h\,(x')| &\le C\int_{\mathbf{R}^2}\int_0^\infty \frac{1}{|x'-y'|^2+y_3^2}|h(y)|dy_3dy' \\
&\le C\int_{\mathbf{R}^2}\left(\int_0^\infty \frac{dy_3}{(|x'-y'|^2+y_3^2)^{q'}}\right)^{\frac{1}{q'}} \|h(y')\|_{L^q(\mathbf{R}_+)}dy' \\
&= C\int_{\mathbf{R}^2}\frac{1}{|x'-y'|^{1+\frac{1}{q}}}\|h(y')\|_{L^q(\mathbf{R}_+)}dy'.
\end{aligned}$$

Thus, the Hardy-Littlewood-Sobolev inequality leads to

$$\|\gamma\partial_i E * h\|_{L^r(\mathbf{R}^2)} \le C\|h\|_{L^q(\mathbf{R}^3_+)}. \tag{49}$$

Now (48) directly follows from (49) by the definition of $N(f)$. The proof is complete.

Combining Proposition 3.6 with Proposition 3.7, we have

**Corollary 3.8** *Let $1 < q < 3$ and $2q/(3-q) \le p < \infty$. Then it follows that*

$$\begin{aligned}
&\|\nabla^\alpha e^{\nu t\Delta_N}(N(f)\mathcal{H}^2_{\partial\mathbf{R}^3_+})\|_{L^p(\mathbf{R}^3_+)} + \|\nabla^\alpha\Gamma_\nu(t)\star(N(f)\mathcal{H}^2_{\partial\mathbf{R}^3_+})\|_{L^p(\mathbf{R}^3_+)} \\
&\qquad\le C(\nu t)^{-\frac{3}{2}(\frac{1}{q}-\frac{1}{p})-\frac{|\alpha|}{2}}\|f\|_{L^q(\mathbf{R}^3_+)}.
\end{aligned} \tag{50}$$

## 4 Solvability of Nonlinear Vorticity Equations

In this section we will solve the nonlinear vorticity equations for the initial data in $L^{3/2}(\mathbf{R}^3_+)$, which is a natural scaling invariant space. To this end let as introduce the Banach space $X_T$ and $Y_T$, $T > 0$, as follows.

$$X_T = \{f \in C([0,T);(L^{\frac{3}{2}}(\mathbf{R}^3_+))^3) \mid \|f\|_{X_T} = \sup_{0<t<T}\|f(t)\|_{L^{\frac{3}{2}}(\mathbf{R}^3_+)} < \infty\},$$

$$Y_T = \{f \in L^1((0,T);(L^{\frac{5}{3}}(\mathbf{R}^3_+))^3) \mid$$

$$\|f\|_{Y_T} = \sup_{0<t<T} t^{\frac{1}{10}} \|f(t)\|_{L^{\frac{5}{3}}(\mathbf{R}^3_+)} + \sup_{0<t<T} t^{\frac{3}{5}} \|\nabla f(t)\|_{L^{\frac{5}{3}}(\mathbf{R}^3_+)} < \infty\}. \tag{51}$$

## 4.1 Estimates of Bilinear Forms

Let $J(f) = \nabla \times \Psi(f)$ be the Biot-Savart law defined by (1). To deal with the nonlinear term in (V) it is convenient to introduce the following bilinear forms.

$$B(f,h) = -(J(f) \cdot \nabla)J(h), \tag{52}$$

$$\mathbb{B}(f,h)(t) = \int_0^t \left(e^{\nu(t-s)\Delta_N} + \Gamma_\nu(t-s)\star\right) \left\{\nabla \times B(f,h) + N(\nabla \times B(f,h))\mathcal{H}^2_{\partial \mathbf{R}^3_+}\right\} ds. \tag{53}$$

The next proposition is the key for the estimate of the bilinear forms.

**Proposition 4.1** *Let $f, h \in (W^{1,5/3}(\mathbf{R}^3_+))^3$. Then there is a positive constant $C$ such that*

$$\|\nabla \times B(f,h)\|_{L^{\frac{15}{13}}(\mathbf{R}^3_+)} \le C\left(\|f\|^{\frac{1}{2}}_{L^{\frac{5}{3}}(\mathbf{R}^3_+)} \|\nabla f\|^{\frac{1}{2}}_{L^{\frac{5}{3}}(\mathbf{R}^3_+)} \|h\|^{\frac{1}{2}}_{L^{\frac{5}{3}}(\mathbf{R}^3_+)} \|\nabla h\|^{\frac{1}{2}}_{L^{\frac{5}{3}}(\mathbf{R}^3_+)} + \|f\|_{L^{\frac{5}{3}}(\mathbf{R}^3_+)} \|\nabla h\|_{L^{\frac{5}{3}}(\mathbf{R}^3_+)}\right). \tag{54}$$

*Proof* By using the definition of $B(f,h)$ we see

$$\|\nabla \times B(f,h)\|_{L^{\frac{15}{13}}(\mathbf{R}^3_+)} \le \left(\|\nabla J(f) \cdot \nabla J(h)\|_{L^{\frac{15}{13}}(\mathbf{R}^3_+)} + \|J(f) \cdot \nabla\nabla \times J(h)\|_{L^{\frac{15}{13}}(\mathbf{R}^3_+)}\right). \tag{55}$$

The Calderón-Zygmund inequality and Gagliardo-Nirenberg inequality imply

$$\|\nabla J(f)\|_{L^{\frac{30}{13}}(\mathbf{R}^3_+)} \le \|\nabla^2 \Psi(f)\|_{L^{\frac{30}{13}}(\mathbf{R}^3_+)} \le C\|f\|^{\frac{1}{2}}_{L^{\frac{5}{3}}(\mathbf{R}^3_+)} \|\nabla f\|^{\frac{1}{2}}_{L^{\frac{5}{3}}(\mathbf{R}^3_+)}$$

Thus the first term of the right-hand side of (55) is estimated as

$$\|\nabla J(f) \cdot \nabla J(h)\|_{L^{\frac{15}{13}}(\mathbf{R}^3_+)} \le C\|f\|^{\frac{1}{2}}_{L^{\frac{5}{3}}(\mathbf{R}^3_+)} \|\nabla f\|^{\frac{1}{2}}_{L^{\frac{5}{3}}(\mathbf{R}^3_+)} \|h\|^{\frac{1}{2}}_{L^{\frac{5}{3}}(\mathbf{R}^3_+)} \|\nabla h\|^{\frac{1}{2}}_{L^{\frac{5}{3}}(\mathbf{R}^3_+)}. \tag{56}$$

To estimate the second term of the right-hand side of (55) we first observe the identity

$$\nabla \times J(h) = \nabla \times \nabla \times \Psi(h) = h + \nabla(-\Delta_D)^{-1}\nabla \cdot h.$$

Here we have used the definition of $\Psi$ in (1) and the equalities

$$\nabla' \cdot (-\Delta_D)^{-1}h' = (-\Delta_D)^{-1}\nabla' \cdot h', \quad \partial_3(-\Delta_N)^{-1}h_3 = (-\Delta_D)^{-1}\partial_3 h_3.$$

Thus we have form the Calderón-Zygmund inequality, $\|\nabla\nabla \times J(h)\|_{L^{\frac{5}{3}}(\mathbf{R}^3_+)} \leq C\|\nabla h\|_{L^{\frac{5}{3}}(\mathbf{R}^3_+)}$. Hence, the Hölder inequality and the Hardy-Littlewood-Sobolev inequality imply

$$\begin{aligned}\|J(f) \cdot \nabla\nabla \times J(h)\|_{L^{\frac{15}{13}}(\mathbf{R}^3_+)} &\leq \|J(f)\|_{L^{\frac{15}{4}}(\mathbf{R}^3_+)}\|\nabla\nabla \times J(h)\|_{L^{\frac{5}{3}}(\mathbf{R}^3_+)} \\ &\leq C\|f\|_{L^{\frac{5}{3}}(\mathbf{R}^3_+)}\|\nabla h\|_{L^{\frac{5}{3}}(\mathbf{R}^3_+)}.\end{aligned} \tag{57}$$

Then (54) follows from (55)–(57). The proof is complete.

Proposition 4.1 leads to the estimate of $\mathbb{B}(f, h)$ as follows.

**Corollary 4.2** *Let $f, h \in Y_T$. Then there is a positive constant $C_{1,\nu}$ depending only on $\nu$ such that*

$$\|\mathbb{B}(f, h)\|_{X_T} + \|\mathbb{B}(f, h)\|_{Y_T} \leq C_{1,\nu}\|f\|_{Y_T}\|h\|_{Y_T}. \tag{58}$$

*Proof* We note that Proposition 4.1 leads to

$$\|\nabla \times B(f, h)\|_{L^{\frac{15}{13}}(\mathbf{R}^3_+)} \leq Cs^{-\frac{7}{10}}\|f\|_{Y_T}\|h\|_{Y_T}, \quad 0 < s < T.$$

Then by using Proposition 3.6 and Corollary 3.8 we have

$$\begin{aligned}&\sup_{0<t<T} t^{\frac{1}{10}}\|\mathbb{B}(f, h)(t)\|_{L^{\frac{5}{3}}(\mathbf{R}^3_+)} \\ &\leq C\sup_{0<t<T} t^{\frac{1}{10}}\int_0^t (\nu(t-s))^{-\frac{3}{2}(\frac{13}{15}-\frac{3}{5})}\|\nabla \times B(f, h)(s)\|_{L^{\frac{15}{13}}(\mathbf{R}^3_+)}ds \\ &\leq C\nu^{-\frac{2}{5}}\|f\|_{Y_T}\|h\|_{Y_T}\sup_{0<t<T} t^{\frac{1}{10}}\int_0^t (t-s)^{-\frac{2}{5}}s^{-\frac{7}{10}}ds \leq C_{1,\nu}\|f\|_{Y_T}\|h\|_{Y_T}.\end{aligned}$$

In the same way we obtain

$$\sup_{0<t<T} \|\mathbb{B}(f,h)(t)\|_{L^{\frac{3}{2}}(\mathbf{R}^3_+)} \le C_\nu \|f\|_{Y_T} \|h\|_{Y_T} \sup_{0<t<T} \int_0^t (t-s)^{-\frac{3}{10}} s^{-\frac{7}{10}} ds$$
$$\le C_{1,\nu} \|f\|_{Y_T} \|h\|_{Y_T},$$

and

$$\sup_{0<t<T} t^{\frac{3}{5}} \|\nabla \mathbb{B}(f,h)\|_{L^{\frac{5}{3}}(\mathbf{R}^3_+)} \le C_\nu \|f\|_{Y_T} \|h\|_{Y_T} \sup_{0<t<T} t^{\frac{3}{5}} \int_0^t (t-s)^{-\frac{9}{10}} s^{-\frac{7}{10}} ds$$
$$\le C_{1,\nu} \|f\|_{Y_T} \|h\|_{Y_T}.$$

The proof is complete.

## 4.2 *Proof of Theorem 1.1*

In this section we apply the estimates established in the previous section and give a proof of Theorem 1.1. Recalling Proposition 3.5, we set

$$\Phi(\omega)(t) = \big(e^{\nu t \Delta_N} + \Gamma_\nu(t) \star \big) b + \mathbb{B}(\omega, \omega)(t), \tag{59}$$

where $b \in (L^{3/2}(\mathbf{R}^3_+))^3$ and $\mathbb{B}(f,h)$ is the bilinear form defined by (53). As usual, the mild solution to (V) is obtained as a fixed point of the map $\Phi$, which will be constructed by the standard iteration procedure. We set an iteration sequence $\{\omega_k\}_{k=0}^\infty$ as

$$\omega_0 = \Phi(0), \qquad \omega_k = \Phi(\omega_{k-1}), \qquad k \in \mathbb{N}. \tag{60}$$

Let us start from the next lemma.

**Lemma 4.3** *There exists a positive constant $\varepsilon_0$ depending only on $\nu$ such that if $\|\omega_0\|_{Y_T} < \varepsilon_0/2$ then $\|\omega_k\|_{Y_T} < \varepsilon_0$ for all $k \in \mathbb{N} \cup \{0\}$, and $\{\omega_k\}_{k=0}^\infty$ is a Cauchy sequence in $Y_T$.*

*Proof* Let $C_{1,\nu} > 0$ be the constant in Corollary 4.2. Set $\varepsilon_0 = 1/(4C_{1,\nu})$, which depends only on $\nu > 0$. The assertion is proved by the induction on $k$. The case $k = 0$ is trivial because of the assumption of the lemma. Assuming that $\|\omega_k\|_{Y_T} < \varepsilon_0$, we apply (58) to estimate the norm of $\omega_{k+1} = \Phi(\omega_k)$ in $Y_T$ and get

$$\|\omega_{k+1}\|_{Y_T} \le \|\omega_0\|_{Y_T} + \|\mathbb{B}(\omega_k,\omega_k)\|_{Y_T} \le \frac{\varepsilon_0}{2} + C_{1,\nu}\|\omega_k\|_{Y_T}^2 \le \frac{\varepsilon_0}{2} + C_{1,\nu}\varepsilon_0^2 < \varepsilon_0,$$

due to the choice of $\varepsilon_0$, as desired. To show the last statement in Lemma 4.3 we observe that

$$\begin{aligned}\|\omega_{m+1}-\omega_m\|_{Y_T} &= \|\mathbb{B}(\omega_m,\omega_m)-\mathbb{B}(\omega_{m-1},\omega_{m-1})\|_{Y_T}\\ &\le \|\mathbb{B}(\omega_m-\omega_{m-1},\omega_m)\|_{Y_T}+\|\mathbb{B}(\omega_{m-1},\omega_m-\omega_{m-1})\|_{Y_T}\\ &\le 2C_{1,\nu}\varepsilon_0\|\omega_m-\omega_{m-1}\|_{Y_T}\le \frac{1}{2}\|\omega_m-\omega_{m-1}\|_{Y_T}.\end{aligned}$$

Here we have used (58), $\|\omega_k\|_{Y_T}<\varepsilon$, and the definition of $\varepsilon_0$. Thus we obtain

$$\|\omega_k-\omega_l\|_{Y_T}\le \sum_{m=l}^{k-1}\|\omega_{m+1}-\omega_m\|_{Y_T}\le \sum_{m=l}^{k-1}\left(\frac{1}{2}\right)^m\|\omega_1-\omega_0\|_{Y_T}\xrightarrow[k,l\to\infty]{} 0.$$

The proof is complete.

Next lemma is used to verify the assumption in Lemma 4.3.

**Lemma 4.4** *For any $b\in(L^{3/2}(\mathbf{R}^3_+))^3$ it follows that $\lim_{T\to 0}\|\omega_0\|_{Y_T}=0$.*

*Proof* For any $\varepsilon>0$ there is $b_\varepsilon\in(C_0^\infty(\mathbf{R}^3_+))^3$ such that $\|b-b_\varepsilon\|_{L^{\frac{3}{2}}(\mathbf{R}^3_+)}<\varepsilon$. By the equality $\omega_0(t)=\big(e^{\nu t\Delta_N}+\Gamma_\nu(t)\star\big)(b-b_\varepsilon)+\big(e^{\nu t\Delta_N}+\Gamma_\nu(t)\star\big)b_\varepsilon$ and Corollary 4.2 we have

$$\|\omega_0\|_{Y_T}\le C_\nu\|b-b_\varepsilon\|_{L^{\frac{3}{2}}(\mathbf{R}^3_+)}+C_{1,\nu}T^{\frac{1}{10}}\|b_\varepsilon\|_{L^{\frac{5}{3}}(\mathbf{R}^3_+)}. \tag{61}$$

Thus , if $T$ is small enough then the right-hand side of (61) is bounded from above by $(C_\nu+1)\varepsilon$. The proof is complete.

Lemmas 4.3–4.4 yield

**Theorem 4.5** *Let $b\in(L^{3/2}(\mathbf{R}^3_+))^3$. Then there exists $T>0$ such that there is a unique $\omega\in Y_T$ satisfying $\omega=\Phi(\omega)$ in $Y_T$ and $\|\omega\|_{Y_T}\le\varepsilon_0$. Here $\varepsilon_0$ is the number in Lemma 4.3.*

The proof of Theorem 4.5 is straightforward and omitted here. Theorem 4.5 and the next theorem complete the proof of Theorem 1.1.

**Theorem 4.6** *The function $\omega\in Y_T$ in Theorem 4.5 belongs to $X_T$ and satisfies $\lim_{T'\to 0}\|\omega\|_{Y_{T'}}=0$. Moreover, if $b$ is of the form $b=\nabla\times a$ with $a\in\dot W^{1,3/2}_{0,\sigma}(\mathbf{R}^3_+)$ then $\lim_{t\to 0}\omega(t)=b$ in $(L^{3/2}(\mathbf{R}^3_+))^3$.*

*Proof* By the construction $\omega$ satisfies $\omega=\omega_0+\mathbb{B}(\omega,\omega)$. From Proposition 3.6 and Corollary 4.2 we have

$$\|\omega\|_{X_T}\le\|\omega_0\|_{X_T}+\|\mathbb{B}(\omega,\omega)\|_{X_T}\le C\|b\|_{L^{\frac{3}{2}}(\mathbf{R}^3_+)}+C_{1,\nu}\|\omega\|^2_{Y_T}<\infty.$$

It is easy to see that $\omega(t)$ is continuous with respect to $t$ in $(L^{3/2}(\mathbf{R}^3_+))^3$, and the details are omitted here. To show $\lim_{T'\to 0} \|\omega\|_{Y_{T'}} = 0$ we observe from Corollary 4.2 that

$$\begin{aligned}\|\omega\|_{Y_{T'}} \le \|\omega_0\|_{Y_{T'}} + \|\mathbb{B}(\omega,\omega)\|_{Y_{T'}} &\le \|\omega_0\|_{Y_{T'}} + C_{1,\nu}\|\omega\|^2_{Y_{T'}} \\ &\le \|\omega_0\|_{Y_{T'}} + \frac{1}{2}\|\omega\|_{Y_{T'}},\end{aligned}$$

by the choice of $\varepsilon_0$ in the proof of Lemma 4.3. Hence we have $\|\omega\|_{Y_{T'}} \le 2\|\omega_0\|_{Y_{T'}} \to 0$ as $T' \to 0$ by Lemma 4.4. Next we consider the behavior of $\omega$ at $t \to 0$ in $(L^{3/2}(\mathbf{R}^3_+))^3$. From the estimate $\|\mathbb{B}(\omega,\omega)(t)\|_{L^{3/2}(\mathbf{R}^3_+)} \le C_{1,\nu}\|\omega\|^2_{Y_t} \to 0$, $t \to 0$, it suffices to deal with the term $\omega_0(t) = (e^{\nu t\Delta_N} + \Gamma_\nu(t)\star)b$. It is clear that $e^{\nu t\Delta_N}b \to b$ as $t \to 0$ in $(L^{3/2}(\mathbf{R}^3_+))^3$. As for the term $\Gamma_\nu(t) \star b$, we observe that

$$\Gamma_\nu(t) \star b - \Gamma_\nu(0) \star b = \int_0^t \mathcal{M}G_\nu(\tau) \star b d\tau \qquad \text{in } \mathbf{R}^3_+,$$

and hence, if $b \in (C_0^\infty(\mathbf{R}^3_+))^3$ then we have from the definition of $\mathcal{M}$ in (28),

$$\|\Gamma_\nu(t) \star b - \Gamma_\nu(0) \star b\|_{L^{\frac{3}{2}}(\mathbf{R}^3_+)} \le C\int_0^t \tau^{-\frac{1}{2}}\|\nabla' b\|_{L^{\frac{3}{2}}(\mathbf{R}^3_+)} d\tau \le Ct^{\frac{1}{2}}\|\nabla' b\|_{L^{\frac{3}{2}}(\mathbf{R}^3_+)},$$

which tends to 0 as $t \to 0$. On the other hand, Proposition 3.6 implies $\sup_{t>0}\|\Gamma_\nu(t) \star b\|_{L^{3/2}(\mathbf{R}^3_+)} \le C\|b\|_{L^{3/2}(\mathbf{R}^3_+)}$, while $\|\Gamma_\nu(0) \star b\|_{L^{3/2}(\mathbf{R}^3_+)} \le C\|b\|_{L^{3/2}(\mathbf{R}^3_+)}$ holds since $\Gamma_\nu(0)\star = -\mathcal{M}E\star$ is a singular integral operator. Hence the density argument yields $\Gamma_\nu(t) \star b \to \Gamma_\nu(0) \star b$ as $t \to 0$ in $(L^{3/2}(\mathbf{R}^3_+))^3$ for all $b \in (L^{3/2}(\mathbf{R}^3_+))^3$. Thus we have arrived at

$$\lim_{t\to 0}\omega(t) = \lim_{t\to 0}\omega_0(t) = b + \Gamma_\nu(0) \star b \qquad \text{in } (L^{\frac{3}{2}}(\mathbf{R}^3_+))^3, \qquad b \in (L^{\frac{3}{2}}(\mathbf{R}^3_+))^3. \tag{62}$$

If $a \in \dot{W}^{1,3/2}_{0,\sigma}(\mathbf{R}^3_+)$ and $b = \nabla \times a$ then $\Gamma_\nu(0) \star b = 0$ by Corollary 3.4, which shows the last statement of Theorem 4.6. The proof is complete.

## References


1. H. Fujita, T. Kato, On the Navier-Stokes initial value problem I. Arch. Ration. Mech. Anal. **16**, 269–315 (1964)
2. G.P. Galdi, *An Introduction to the Mathematical Theory of the Navier-Stokes Equations*, vol. I (Springer, New York, 1994)

3. Y. Giga, Analyticity of the semigroup generated by the Stokes operator in $L_r$ spaces. Math. Z. **178**, 297–329 (1981)
4. Y. Giga, Domains of fractional powers of the Stokes operator in $L_r$ spaces. Arch. Ration. Mech. Anal. **89**, 251–265 (1985)
5. Y. Giga, T. Miyakawa, Solutions in $L^r$ of the Navier-Stokes initial value problem. Arch. Ration. Mech. Anal. **89**, 267–281 (1985)
6. Y. Giga, T. Miyakawa, Navier-Stokes flow in $\mathbf{R}^3$ with measures as initial vorticity and Morrey spaces. Commun. Partial Diff. Equ. **14**, 577–618 (1989)
7. M.-H. Giga, Y. Giga, J. Saal, *Nonlinear Partial Differential Equations: Asymptotic Behavior of Solutions and Self-similar Solutions* (Birkhäuser, Boston, Basel, Berlin, 2010)
8. T. Kato, H. Fujita, On the nonstationary Navier-Stokes system. Rend. Sem. Mat. Univ. Padova **32**, 243–260 (1962)
9. Y. Maekawa, Solution formula for the vorticity equations in the half plane with application to high vorticity creation at zero viscosity limit. Adv. Diff. Equ. **18**, 101–146 (2013)
10. Y. Maekawa, On the inviscid limit problem of the vorticity equations for viscous incompressible flows in the half plane. Commun. Pure Appl. Math. **67**, 1045–1128 (2014)
11. A.J. Majda, A.L. Bertozzi, *Vorticity and Incompressible Flows* (Cambridge University Press, Cambridge, 2002)
12. L. Quartapelle, *Numerical Solution of the Incompressible Navier-Stokes Equations* (Birkäuser, Basel, 1993)
13. R. Rautmann, A direct approach to vorticity transport & diffusion, in *Kyoto Conference on the Navier-Stokes Equations and Their Applications*, vol. B1 (RIMS Kokyuroku Bessatsu, Kyoto, 2007), pp. 305–329
14. V. Ruas, A new formulation of the three-dimensional velocity-vorticity system in viscous incompressible flow. Z. Angew. Math. Mech. **79**(1), 29–36 (1999)
15. H. Sohr, *The Navier-Stokes Equations. An Elementary Functional Analytic Approach* (Birkhäuser, Basel, 2001)
16. V.A. Solonnikov, Estimates of the solutions of a nonstationary linearized system of Navier-Stokes equations. Am. Math. Soc. Transl. **75**(2), 1–116 (1968)
17. V.A. Solonnikov, Estimates for solutions of nonstationary Navier-Stokes equations. J. Sov. Math. **8** 467–529 (1977)
18. S. Ukai, A solution formula for the Stokes equation in $\mathbf{R}^n_+$. Commun. Pure Appl. Math. **11**, 611–621 (1987)
19. F.B. Weissler, The Navier-Stokes initial value problem in $L^p$. Arch. Ration. Mech. Anal. **74**, 219–230 (1980)

# A Weak Solution to the Navier–Stokes System with Navier's Boundary Condition in a Time-Varying Domain

**Jiří Neustupa and Patrick Penel**

*Dedicated to Professor Yoshihiro Shibata on the occasion of his 60th birthday.*

**Abstract** We assume that $\Omega^t$ (for $t \in [0, T]$) is a time-varying domain in $\mathbb{R}^3$. Particularly, $\Omega^t$ can be a region around colliding bodies. Under certain conditions on $\Omega^t$ and the way it varies, we prove the weak solvability of the Navier–Stokes system with Navier's slip boundary condition in $Q_{(0,T)} := \{(\mathbf{x}, t);\ 0 < t < T,\ \mathbf{x} \in \Omega^t\}$.

**Keywords** Navier–Stokes equations • Slip boundary condition • Weak solutions

## 1 Introduction and Formulation of the Problem

The global (in time) weak solvability of the Navier–Stokes system in a fixed domain with the no-slip boundary conditions has already been known for a long time due to the classical results of J. Leray (1934), E. Hopf (1951) and others. The existence of a weak solution to the Navier–Stokes system with the no-slip Dirichlet boundary condition in a time-varying domain $\Omega^t$ with a prescribed form at each time $t \in [0, T]$ is also known, see [4] (a smooth time-varying domain) and [7] (an arbitrary time-varying domain, parts of the boundary may strike). Analogous results concerning the Navier–Stokes equations for compressible fluid in a smooth time-varying domain can be found in [2] (the no-slip boundary condition) and [3] (Navier's slip boundary condition).

If the considered time-varying domain includes the region around a family of moving and possibly also colliding bodies then one has to overcome a series of new difficulties, e.g. some constants in the imbedding inequalities or in the estimates

---

J. Neustupa (✉)
Institute of Mathematics, Czech Academy of Sciences, Žitná 25, 115 67 Prague 1, Czech Republic
e-mail: neustupa@math.cas.cz

P. Penel
Université du Sud-Toulon-Var, BP 20132, 83957 La Garde, France
e-mail: patrick.penel@gmail.com

 
H. Amann et al. (eds.), *Recent Developments of Mathematical Fluid Mechanics*, Advances in Mathematical Fluid Mechanics, DOI 10.1007/978-3-0348-0939-9_20

of traces depend on $t$ and they blow up if $t$ approaches the instant of the collision. Starovoitov [10] derived necessary conditions for the existence of a weak solution of the Navier–Stokes equation in a time-variable domain $\Omega^t$ that is an exterior of several solid bodies moving in the fluid, considering the no-slip Dirichlet boundary condition. The results of Starovoitov [10] show that if the bodies have $C^2$ boundaries then they may strike only with the speed equal to zero, otherwise the weak solution does not exist.

Motivated by this state, we study the flow of a viscous incompressible fluid in a time-variable domain $\Omega^t$ under the assumptions that $\Omega^t$ at each time $t \in [0, T]$ is a priori known and the velocity of the fluid satisfies Navier's slip boundary condition on the boundary $\partial\Omega^t$. The motion of $\Omega^t$ admits collisions of parts of the boundary. We prove the existence of a weak solution. We show that unlike the case with Dirichlet's boundary condition, "smooth" parts of the boundary may strike with a non-zero speed. A simplified situation is treated in our previous article [9]. We are aware of the fact that the complete model of the interaction between the fluid and the solid part requires a more complex description (the response of the solid part to forces generated by the fluid, the information what part of the kinetic energy is absorbed by the bodies at the instants of collisions). However, even in the presented setting, due to the choice of the slip boundary condition and the admittance of collisions of bodies in the fluid, possibly with a non-zero speed, the results are original.

## 1.1 *Classical Formulation of the Initial-Boundary Value Problem*

Let $B_R(\mathbf{0})$ be a ball in $\mathbb{R}^3$. We assume that $\Omega^t \subset \overline{\Omega^t} \subset B_R(\mathbf{0})$ is a time-varying domain filled by a moving fluid, while $S^t := \overline{B_R(\mathbf{0})} \smallsetminus \Omega^t$ represents a solid environment (e.g. a finite family of bodies moving in the fluid in a tank whose shape can also generally vary in dependence on time). We assume that the motion of $S^t$ is known: the velocity at point $\mathbf{x} \in S^t$ at time $t$ is $\mathbf{V}(\mathbf{x}, t)$.

For $I \subset [0, T]$, we denote $Q_I := \{(\mathbf{x}, t);\ t \in I,\ \mathbf{x} \in \Omega^t\}$, $\mathcal{S}_I := \{(\mathbf{x}, t);\ t \in I,\ \mathbf{x} \in S^t\}$ and $\Gamma_I := \{(\mathbf{x}, t);\ t \in I,\ \mathbf{x} \in \Gamma^t := \partial\Omega^t\}$. We study the initial-boundary value problem

$$\partial_t \mathbf{v} + \mathbf{v} \cdot \nabla \mathbf{v} + \nabla p = \nu \Delta \mathbf{v} + \mathbf{f} \qquad \text{a.e. in } Q_{(0,T)}, \tag{1}$$

$$\div \mathbf{v} = 0 \qquad \text{a.e. in } Q_{(0,T)}, \tag{2}$$

$$\mathbf{v} \cdot \mathbf{n} = \mathbf{V} \cdot \mathbf{n} \qquad \text{a.e. in } \Gamma_{(0,T)}, \tag{3}$$

$$[\mathbb{T}_{\mathrm{d}}(\mathbf{v}) \cdot \mathbf{n}]_\tau + \gamma(\mathbf{v} - \mathbf{V}) = \mathbf{0} \qquad \text{a.e. in } \Gamma_{(0,T)}, \tag{4}$$

$$\mathbf{v} = \mathbf{v}_0 \qquad \text{a.e. in } \Omega^0 \times \{0\}, \tag{5}$$

Symbols $\mathbf{v}$, $p$, $\mathbf{f}$, $\nu$ and $\gamma$ successively denote the velocity, the pressure, the specific external body force, the kinematic coefficient of viscosity and the coefficient of friction between the fluid and the boundary. (Both $\nu$ and $\gamma$ are supposed to be positive constants.) The outer normal vector field on the boundary $\Gamma^t$ of $\Omega^t$ is denoted by $\mathbf{n}$. Condition (3) expresses the impermeability of $\Gamma^t$. Condition (4) was proposed by H. Navier in 1824. It expresses the requirement that the tangential component of the stress acting on the boundary should be proportional to the velocity of the fluid relative to the material boundary. Here, $\mathbb{T}_{\mathrm{d}}(\mathbf{v})$ denotes the dynamic stress tensor associated with the flow $\mathbf{v}$. It has the form $\mathbb{T}_{\mathrm{d}}(\mathbf{v}) = 2\nu\,(\nabla\mathbf{v})_s$ (where $(\nabla\mathbf{v})_s$ is the symmetrized gradient of $\mathbf{v}$) in incompressible Newtonian fluids. The subscript $\tau$ denotes the component, tangential to $\Gamma^t$. The operators div, $\nabla$ and $\Delta$ always act only in the spatial variables.

The problem (1)–(5) is treated in Sects. 2–6. In Sect. 7, we deal with an example when $\Omega^t$ is a special region around two moving bodies, striking at the time instant $t^c \in (0, T)$.

## 1.2 Notation of Norms and Function Spaces

Vector functions and spaces of vector functions are denoted by boldface letters.

- $\|\,.\,\|_{2;\,\Omega^t}$ is the norm in $\mathbf{L}^2(\Omega^t)$. The meaning of $\|\,.\,\|_{q;\,\Omega^t}$ or $\|\,.\,\|_{2;\,\Gamma^t}$ is similar.
- $\mathbf{C}^{\infty}_{0,\sigma}(\Omega^t)$ is the linear space of infinitely differentiable divergence-free vector-functions in $\Omega^t$ with a compact support in $\Omega^t$. The closure of $\mathbf{C}^{\infty}_{0,\sigma}(\Omega^t)$ in $\mathbf{L}^q(\Omega^t)$ (for $1 < q < \infty$) is denoted $\mathbf{L}^q_{\sigma}(\Omega^t)$.
- $\mathbf{W}^{1,2}_{\sigma}(\Omega^t) := \mathbf{W}^{1,2}(\Omega^t) \cap \mathbf{L}^2_{\sigma}(\Omega^t)$ (with the norm $\|\,.\,\|_{1,2;\,\Omega^t}$ as in $\mathbf{W}^{1,2}(\Omega^t)$).

## 1.3 Assumptions on Solid Part $S^t$ and Domain $\Omega^t$

We denote by $\mathbf{Y}(t;\,\vartheta,\mathbf{x})$ the position (at time $t$) of the point of $S^t$ whose position at time $\vartheta$ was $\mathbf{x}$. We assume that

(a1) function $\mathbf{Y}$ is continuous in the set $\{(t,\vartheta,\mathbf{x}) \in \mathbb{R}^5;\ 0 \le t,\ \vartheta \le T,\ \mathbf{x} \in S^{\vartheta}\}$ and the corresponding velocity $\mathbf{V}(\mathbf{x},t) := (\mathrm{d}/\mathrm{d}t)\,\mathbf{Y}(t;\,\vartheta,\mathbf{x})\big|_{\vartheta=t}$ has continuous and bounded derivatives in $S_{[0,T]\smallsetminus\mathcal{T}^c}$, where $\mathcal{T}^c$ is a finite family of "critical" time instants $0 < t^c_1 < \ldots < t^c_M < T$ when different parts of $S^t$ may touch (strike) themselves.

We assume that $\Omega^t$ is a Lipschitzian domain for $t \in [0,T] \smallsetminus \mathcal{T}^c$ and

(a2) $\displaystyle\int_{\Gamma^t} \boldsymbol{\phi}\cdot\nabla\mathbf{n}\cdot\boldsymbol{\phi}\;\mathrm{d}\Gamma \;\le\; c_1\,\|\boldsymbol{\phi}\|_{2;\,\Omega^t}\big(\|\boldsymbol{\phi}\|_{2;\,\Omega^t} + \|\nabla\boldsymbol{\phi}\|_{2;\,\Omega}\big)$ for $t \in [0,T] \smallsetminus \mathcal{T}^c$, all $\boldsymbol{\phi} \in \mathbf{W}^{1,2}_{\sigma}(\Omega^t)$ and some $c_1$ independent of $t$.

Furthermore, we assume that there exists a divergence-free vector function **a** in $Q_{[0,T]}$ with properties (a3)–(a7). These are in fact assumptions on domain $\Omega^t$ and the way it varies in dependence on $t$, because the existence of an appropriate function **a** satisfying conditions (a3)–(a7) is closely connected with the shape of $\Omega^t$ and its time-variability. The construction of function **a** is the main point in Sect. 7, where we present the concrete example. Conditions (a3)–(a7) are:

(a3) $\mathbf{a} \cdot \mathbf{n} = \mathbf{V} \cdot \mathbf{n}$ a.e. on $\Gamma_{(0,T)}$,

(a4) **a** and $\nabla \mathbf{a}$ are continuous in $Q_{[0,T] \smallsetminus \mathcal{T}^c}$ and the function $\theta_1(t) := \|\mathbf{a}(\,.\,,t)\|_{1,2;\,\Omega^t}$ is in $L^2(0,T)$,

(a5) $\theta_2(t) := \|\mathbf{a}(.\,,t) - \mathbf{V}(.\,,t)\|_{2;\,\Gamma^t} \in L^2(0,T)$,

(a6) there exist functions $\theta_3 \in L^1(0,T)$, $\theta_4 \in L^2(0,T)$ and $\theta_5 \in L^1(0,T)$, continuous in $[0,T] \smallsetminus \mathcal{T}^c$, such that for $t \in [0,T] \smallsetminus \mathcal{T}^c$ and $\boldsymbol{\phi} \in \mathbf{W}^{1,2}_\sigma(\Omega^t)$ we have

$$\left| \int_{\Omega^t} \big[ \partial_t \mathbf{a}(\,.\,,t) + \mathbf{a}(\,.\,,t) \cdot \nabla \mathbf{a}(\,.\,,t) \big] \cdot \boldsymbol{\phi} \; \mathrm{d}\mathbf{x} \right| \le \theta_3(t)\, \|\boldsymbol{\phi}\|_{2;\,\Omega^t} + \theta_4(t)\, \|\nabla \boldsymbol{\phi}\|_{2;\,\Omega^t}, \tag{6}$$

$$\left| \int_{\Omega^t} \boldsymbol{\phi} \cdot \nabla \boldsymbol{\phi} \cdot \mathbf{a}(\,.\,,t) \; \mathrm{d}\mathbf{x} \right| \le \frac{\nu}{10}\, \|\nabla \boldsymbol{\phi}\|^2_{2;\,\Omega^t} + \frac{\gamma}{4}\, \|\boldsymbol{\phi}\|^2_{2;\,\Gamma^t} + \theta_5(t)\, \|\boldsymbol{\phi}\|^2_{2;\,\Omega^t}, \tag{7}$$

(a7) the initial-value problem

$$\frac{\mathrm{d}}{\mathrm{d}t} \mathbf{X}(t;\, \vartheta, \mathbf{x}) \;=\; \mathbf{a}\big(\mathbf{X}(t;\, \vartheta, \mathbf{x}), t\big), \qquad \mathbf{X}(\vartheta;\, \vartheta, \mathbf{x}) = \mathbf{x} \tag{8}$$

has a solution $\mathbf{X}(t;\, \vartheta, \mathbf{x})$, defined for $t \in [0,T]$, $\vartheta \in [0,T]$ and $\mathbf{x} \in \Omega^\vartheta$, such that for $t$, $\vartheta \in [0,T] \smallsetminus \mathcal{T}^c$, the mapping $\mathbf{x} \mapsto \mathbf{X}(t;\, \vartheta, \mathbf{x})$ is a $C^1$-diffeomorphism of $\Omega^\vartheta$ onto $\Omega^t$.

Note that while mapping **Y** acts in the solid part, mapping **X** acts in the fluid part. They "meet" on set $\Gamma_{[0,T]}$. Function **a** enables us, except others, to transform the inhomogeneous boundary condition (3) into the homogeneous one. The conditions (a3) and (a7) are used in the weak formulation of the problem (1)–(5), the other conditions are needed in the derivation of appropriate energy estimates.

# 2 Weak Formulation of the Initial-Boundary Value Problem (1)–(5) and the Main Theorem

## *2.1 A Formal Derivation of the Weak Formulation*

Assume that $\boldsymbol{\phi}$ is an infinitely differentiable divergence-free vector-function in $B_R(\mathbf{0}) \times [0,T]$, satisfying the condition $\boldsymbol{\phi} \cdot \mathbf{n} = 0$ a.e. on $\Gamma_{(0,T)}$. Assume that **v** is a "sufficiently smooth" solution of (1)–(5) of the form $\mathbf{v} = \mathbf{a} + \mathbf{u}$ where $\mathbf{a}$ is the function with the properties named in Sect. 2. Conditions (3) and (a3) imply that $\mathbf{u} \cdot \mathbf{n} = 0$ a.e. in $\Gamma_{(0,T)}$. Let us multiply Eq. (1) by $\boldsymbol{\phi}$ and integrate in $Q_{(0,T)}$. The integral of $\{\partial_t \mathbf{u} + (\mathbf{a} \cdot \nabla)\mathbf{u}\} \cdot \boldsymbol{\phi}$ can be expressed as follows:

$$\begin{aligned}
&\int_0^T \int_{\Omega^t} \{\partial_t \mathbf{u}(\mathbf{x},t) + \mathbf{a}(\mathbf{x},t) \cdot \nabla \mathbf{u}(\mathbf{x},t)\} \cdot \boldsymbol{\phi}(\mathbf{x},t)\, \mathrm{d}\mathbf{x}\, \mathrm{d}t \\
&\quad = \int_0^T \int_{\Omega^0} \frac{\mathrm{d}}{\mathrm{d}t} \mathbf{u}\big(\mathbf{X}(t;\,0,\mathbf{x}_0),t\big) \cdot \boldsymbol{\phi}\big(\mathbf{X}(t;\,0,\mathbf{x}_0),t\big)\, \mathrm{d}\mathbf{x}_0\, \mathrm{d}t \\
&\quad = -\int_{\Omega^0} \mathbf{u}_0(\mathbf{x}_0) \cdot \boldsymbol{\phi}(\mathbf{x}_0,0)\, \mathrm{d}\mathbf{x}_0 \\
&\qquad - \int_0^T \int_{\Omega^0} \mathbf{u}\big(\mathbf{X}(t;\,0,\mathbf{x}_0),t\big) \cdot \frac{\mathrm{d}}{\mathrm{d}t} \boldsymbol{\phi}\big(\mathbf{X}(t;\,0,\mathbf{x}_0),t\big)\, \mathrm{d}\mathbf{x}_0\, \mathrm{d}t \\
&\quad = -\int_{\Omega^0} \mathbf{u}_0(\mathbf{x}_0) \cdot \boldsymbol{\phi}(\mathbf{x}_0,0)\, \mathrm{d}\mathbf{x}_0 \\
&\qquad - \int_0^T \int_{\Omega^t} \mathbf{u}(\mathbf{x},t) \cdot \big\{\partial_t \boldsymbol{\phi}(\mathbf{x},t) + \mathbf{a}(\mathbf{x},t) \cdot \nabla \boldsymbol{\phi}(\mathbf{x},t)\big\}\, \mathrm{d}\mathbf{x}\, \mathrm{d}t,
\end{aligned}$$

where $\mathbf{u}_0 = \mathbf{v}_0 - \mathbf{a}(.\,,0)$. The integral of $\mathbf{u} \cdot \nabla \mathbf{v} \cdot \boldsymbol{\phi}$ in $\Omega^t$ can be transformed to the integral of $\mathbf{u} \cdot \nabla \mathbf{a} \cdot \boldsymbol{\phi}$ minus the integral of $\mathbf{u} \cdot \nabla \boldsymbol{\phi} \cdot \mathbf{u}$ by means of the integration by parts. Further, denoting the components of the vectors $\mathbf{a}$, $\mathbf{v}$, $\mathbf{u}$ and $\boldsymbol{\phi}$ by the same slanted letters with indices, we have

$$\begin{aligned}
\int_{\Omega^t} \nu \Delta \mathbf{v} \cdot \boldsymbol{\phi}\, \mathrm{d}\mathbf{x} &= \int_{\Omega^t} \nu\,(\partial_j^2 v_i)\phi_i\, \mathrm{d}\mathbf{x} \\
&= \int_{\Gamma^t} \nu\,(\partial_j v_i)\, n_j\, \phi_i\, \mathrm{d}\Gamma - \int_{\Omega^t} \nu\,(\partial_j v_i)(\partial_j \phi_i)\, \mathrm{d}\mathbf{x} \\
&= \int_{\Gamma^t} \nu\,(\partial_j v_i + \partial_i v_j)\, n_j\, \phi_i\, \mathrm{d}\mathbf{x} - \int_{\Gamma^t} \nu\,(\partial_i v_j)\, n_j\, \phi_i\, \mathrm{d}\mathbf{x} - \int_{\Omega^t} \nu\,(\partial_j v_i)(\partial_j \phi_i)\, \mathrm{d}\mathbf{x} \\
&= -\int_{\Gamma^t} \gamma\,(\mathbf{v} - \mathbf{V}) \cdot \boldsymbol{\phi}\, \mathrm{d}\Gamma - \int_{\Omega^t} 2\nu\,(\nabla \mathbf{v})_s : \nabla \boldsymbol{\phi}\, \mathrm{d}\mathbf{x},
\end{aligned} \tag{9}$$

where the subscript $s$ denotes the symmetric part. We use the identities $\nu\,(\partial_j v_i + \partial_i v_j)\, n_j\, \phi_i = [\mathbb{T}_\mathrm{d}(\mathbf{v}) \cdot \mathbf{n}] \cdot \boldsymbol{\phi} = -\gamma\,(\mathbf{v} - \mathbf{V}) \cdot \boldsymbol{\phi}$, following from the boundary condition (4). Thus, writing everywhere $\mathbf{a} + \mathbf{u}$ instead of $\mathbf{v}$, we obtain the integral equation

$$\int_0^T \int_{\Omega^t} \{-(\partial_t \boldsymbol{\phi} + \mathbf{a}\cdot\nabla\boldsymbol{\phi})\cdot\mathbf{u} - \mathbf{u}\cdot\nabla\boldsymbol{\phi}\cdot\mathbf{a} + \mathbf{u}\cdot\nabla\mathbf{u}\cdot\boldsymbol{\phi} + 2\nu\,[\nabla(\mathbf{a}+\mathbf{u})]_s : \nabla\boldsymbol{\phi}\}\,\mathrm{d}\mathbf{x}\,\mathrm{d}t$$

$$+ \int_0^T \int_{\Gamma^t} \gamma\,(\mathbf{a}+\mathbf{u}-\mathbf{V})\cdot\boldsymbol{\phi}\,\mathrm{d}\Gamma\,\mathrm{d}t$$

$$= \int_0^T \int_{\Omega^t} \mathbf{g}\cdot\boldsymbol{\phi}\,\mathrm{d}\mathbf{x}\,\mathrm{d}t + \int_{\Omega^0} \mathbf{u}_0\cdot\boldsymbol{\phi}(.\,,0)\,\mathrm{d}\mathbf{x} \tag{10}$$

where $\mathbf{g} := \mathbf{f} - \partial_t\mathbf{a} - \mathbf{a}\cdot\nabla\mathbf{a}$. We arrive at the definition:

**Definition 2.1** Let $\mathbf{u}_0 \in \mathbf{L}^2_\sigma(\Omega^0)$ and $\mathbf{f} \in L^2(0,T;\, \mathbf{L}^2(\Omega^t))$. Put $\mathbf{g} = \mathbf{f} - \partial_t\mathbf{a} - \mathbf{a}\cdot\nabla\mathbf{a}$. We call the function $\mathbf{v} \equiv \mathbf{a} + \mathbf{u}$ a *weak solution* of the problem (1)–(5) if $\mathbf{u} \in L^2(0,T;\, \mathbf{W}^{1,2}_\sigma(\Omega^t)) \cap L^\infty(0,T;\, \mathbf{L}^2_\sigma(\Omega^t))$, the trace of $\mathbf{u}$ on $\Gamma_{(0,T)}$ is in $L^2(0,T;\, \mathbf{L}^2(\Gamma^t))$ and $\mathbf{u}$ satisfies (10) for all infinitely differentiable divergence-free vector-functions $\boldsymbol{\phi}$ in $B_R(\mathbf{0})\times[0,T]$, satisfying the condition $\boldsymbol{\phi}\cdot\mathbf{n} = 0$ a.e. on $\Gamma_{(0,T)}$.

Our main theorem, whose proof is given in Sects. 4–6, says:

**Theorem 2.2** *Suppose that domain $\Omega^t$ and function* $\mathbf{a}$ *satisfy the assumptions (a1)–(a7). Then the weak solution of the problem* (1)–(5), *introduced in Definition* 2.1, *exists.*

This result, with a brief description of the approach, has already been announced in our note [8].

## 3 An a Priori Energy-Type Estimate of a Solution of the Problem (1)–(5)

In this section, we present a formal derivation of the energy-type inequality, assuming that $(\mathbf{v},p)$ a "sufficiently smooth" solution of (1)–(5). We are going to derive the inequality

$$\|\mathbf{u}(.\,,t)\|^2_{2;\,\Omega^t} + \nu\int_0^t \|\nabla\mathbf{u}(.\,,s)\|^2_{2;\,\Omega^s}\,\mathrm{d}s + \gamma\int_0^t \|\mathbf{u}(.\,,s)\|^2_{2;\,\Gamma^s}\,\mathrm{d}s$$

$$\leq \|\mathbf{u}_0\|^2_{2;\,\Omega^0} + \int_0^t \omega_1(s)\,\|\mathbf{u}(.\,,s)\|^2_{2;\,\Omega^s}\,\mathrm{d}s + \omega_2(t) \tag{11}$$

where $\omega_1$ and $\omega_2$ are certain functions, integrable in $(0,T)$.

Multiplying (1) (with $\mathbf{v} = \mathbf{a} + \mathbf{u}$) by $\mathbf{u}$ and integrating in $\Omega^t$, we obtain

$$\int_{\Omega^t} \{(\partial_t\mathbf{u} + \mathbf{a}\cdot\nabla\mathbf{u} + \mathbf{u}\cdot\nabla\mathbf{a})\cdot\mathbf{u} + 2\nu\,[\nabla(\mathbf{a}+\mathbf{u})]_s : \nabla\mathbf{u}\}\,\mathrm{d}\mathbf{x}$$

$$+\gamma\int_{\Gamma^t}(\mathbf{a}+\mathbf{u}-\mathbf{V})\cdot\mathbf{u}\,\mathrm{d}\Gamma = \int_{\Omega^t}[\mathbf{f} - \partial_t\mathbf{a} - \mathbf{a}\cdot\nabla\mathbf{a}]\cdot\mathbf{u}\,\mathrm{d}\mathbf{x}. \tag{12}$$

Using the transformation $\mathbf{x} \mapsto \mathbf{y} = \mathbf{X}(t+h;\, t, \mathbf{x})$ of $\Omega^t \smallsetminus \mathfrak{s}^t$ onto $\Omega^{t+h} \smallsetminus \mathfrak{s}^{t+h}$ (see condition (a7)), we can rewrite the integral of $(\partial_t \mathbf{u} + \mathbf{a} \cdot \nabla \mathbf{u}) \cdot \mathbf{u}$ as follows:

$$\int_{\Omega^t} \big[\partial_t \mathbf{u} + \mathbf{a} \cdot \nabla \mathbf{u}\big] \cdot \mathbf{u}\, \mathrm{d}\mathbf{x} = \bigg[\int_{\Omega^t} \frac{\mathrm{d}}{\mathrm{d}\vartheta} \frac{1}{2} \big|\mathbf{u}\big(\mathbf{X}(\vartheta;\, t, \mathbf{x}), \vartheta\big)\big|^2\, \mathrm{d}\mathbf{x}\bigg]_{\vartheta = t}$$

$$= \lim_{h \to 0} \frac{1}{2h} \bigg[\int_{\Omega^t} \Big(\big|\mathbf{u}\big(\mathbf{X}(t+h;\, t, \mathbf{x}), t+h\big)\big|^2 - \big|\mathbf{u}\big(\mathbf{X}(t;\, t, \mathbf{x}), t\big)\big|^2\Big)\, \mathrm{d}\mathbf{x}\bigg]$$

$$= \lim_{h \to 0} \frac{1}{2h} \bigg[\int_{\Omega^{t+h}} |\mathbf{u}(\mathbf{y}, t+h)|^2\, \mathrm{d}\mathbf{y} - \int_{\Omega^t} |\mathbf{u}(\mathbf{x}, t)|^2\, \mathrm{d}\mathbf{x}\bigg] = \frac{\mathrm{d}}{\mathrm{d}t} \frac{1}{2} \int_{\Omega^t} |\mathbf{u}|^2\, \mathrm{d}\mathbf{x}.$$

Further, we successively rewrite and estimate the integrals in (12):

- Due to assumption (a5), we have $\gamma \int_{\Gamma^t} (\mathbf{a} + \mathbf{u} - \mathbf{V}) \cdot \mathbf{u}\, \mathrm{d}\Gamma \ge \frac{3\gamma}{4} \|\mathbf{u}\|^2_{2;\, \Gamma^t} - \gamma\, \theta_2^2(t)$.
- The integral of $(\nabla \mathbf{a})_s : \nabla \mathbf{u}$ can be estimated from below by $-\frac{1}{10} \|\nabla \mathbf{u}\|^2_{2;\, \Omega^t} - \frac{5}{2} \theta_1^2(t)$.
-

$$2\nu \int_{\Omega^t} (\nabla \mathbf{u})_s : \nabla \mathbf{u}\, \mathrm{d}\mathbf{x} = \nu \int_{\Omega^t} |\nabla \mathbf{u}|^2\, \mathrm{d}\mathbf{x} + \nu \int_{\Gamma^t} (\partial_i u_j)\, u_i\, n_j\, \mathrm{d}\Gamma$$

$$= \nu\, \|\nabla \mathbf{u}\|^2_{2;\, \Omega^t} + \nu \int_{\Gamma^t} \partial_i (u_j n_j)\, u_i\, \mathrm{d}\Gamma - \nu \int_{\Gamma^t} u_j\, (\partial_i n_j)\, u_i\, \mathrm{d}\Gamma.$$

The second integral on the right hand side is equal to zero. The third integral can be estimated by means of assumption (a2). We obtain

$$2\nu \int_{\Omega^t} (\nabla \mathbf{u})_s : \nabla \mathbf{u}\, \mathrm{d}\mathbf{x} \ge \frac{9\nu}{10} \|\nabla \mathbf{u}\|^2_{2;\, \Omega^t} - \Big[\nu c_1 + \frac{5\nu}{2} c_1^2\Big] \|\mathbf{u}\|^2_{2;\, \Omega^t}. \tag{13}$$

- The modulus of the integral of the product $(\partial_t \mathbf{a} + \mathbf{a} \cdot \nabla \mathbf{a}) \cdot \mathbf{u}$ in (12) can be estimated by inequality (6) in assumption (a6):

$$\bigg|\int_{\Omega^t} (\partial_t \mathbf{a} + \mathbf{a} \cdot \nabla \mathbf{a}) \cdot \mathbf{u}\, \mathrm{d}\mathbf{x}\bigg| \le \theta_3(t)\, \|\mathbf{u}\|^2_{2;\, \Omega^t} + \frac{1}{4}\, \theta_3(t) + \frac{\nu}{10} \|\nabla \mathbf{u}\|^2_{2;\, \Omega^t} + \frac{5}{2\nu}\, \theta_4^2(t).$$

- The integral of $\mathbf{u} \cdot \nabla \mathbf{a} \cdot \mathbf{u}$ can be estimated by means of inequality (7) in assumption (a6).

Substituting all these estimates to (12) and integrating with respect to time from 0 to $t$, we obtain inequality (11) with

$$\omega_1(t) = 2\nu c_1 + 5\nu c_1^2 + \theta_3(t) + 2\theta_5(t) + \tfrac{1}{2},$$

$$\omega_2(t) = 2\gamma\, \theta_2^2(t) + \tfrac{1}{4} \theta_3(t) + 5\nu\, \theta_1^2(t) + \tfrac{5}{2} \nu\, \theta_4^2(t) + \tfrac{1}{2} \|\mathbf{f}\|^2_{2;\, \Omega^t}.$$

# 4 The Time Discretization and Stationary Boundary-Value Problems

In this section, we begin the proof of Theorem 2.2 by the definition and studies of stationary problems which follows from (10) by the time discretization.

## 4.1 A Partition of the Interval $[0, T]$

As the functions $\theta_1^2$, $\theta_3$, $\theta_4^2$ and $\theta_5$ are integrable in $(0, T)$ and continuous in $[0, T] \smallsetminus \mathcal{T}^c$, there exists a bound $\Theta > 0$ such that to each $N \in \mathbb{N}$ there exists a partition $P_N : 0 = t_0 < t_1 < \ldots < t_N = T$ of the interval $[0, T]$ with the properties

(a) $\|P_N\| := \max_{n=1,\ldots,N} d_n < 2T/N$ (where $d_n := t_n - t_{n-1}$),
(b) $\{t_1;\, t_2;\, \ldots;\, t_N\} \cap \mathcal{T}^c = \emptyset$,

$$\text{(c)} \quad \sum_{n=1}^{N} \bigl[\theta_1^2(t_n) + \theta_2^2(t_n) + \theta_3(t_n) + \theta_4^2(t_n) + \theta_5(t_n)\bigr]\, d_n \;\leq\; \Theta, \tag{14}$$

$$\text{(d)} \quad \lim_{N\to\infty} \max_{n=1,\ldots,N} \bigl[\theta_1^2(t_n) + \theta_2^2(t_n) + \theta_3(t_n) + \theta_4^2(t_n) + \theta_5(t_n)\bigr]\, d_n \;=0. \tag{15}$$

We further consider number $N \in \mathbb{N}$ to be fixed in this section.

## 4.2 Notation

We denote $\Omega_n := \Omega^{t_n}$, $S_n := S^{t_n}$ and $\Gamma_n := \Gamma^{t_n} = \partial\Omega^{t_n}$ for $n = 0, 1, \ldots, N$.

Due to technical reasons, we extend function $\mathbf{f}(\,.\,, t)$ by zero to $B_R(\mathbf{0}) \smallsetminus \Omega^t$.

For $\mathbf{x} \in \Omega_n$, we denote by $[\nabla\mathbf{a}]_n(\mathbf{x})$ (respectively $\mathbf{f}_n(\mathbf{x})$) the mean value of $\nabla\mathbf{a}(\mathbf{x},\, .)$ (respectively $\mathbf{f}(\mathbf{x},\, .)$ on the time interval $(t_{n-1}, t_n)$. We put $\mathbf{g}_n(\mathbf{x}) := \mathbf{f}_n(\mathbf{x}) - [\partial_t\mathbf{a}(\mathbf{x}, t_n) + \mathbf{a}(\mathbf{x}, t_n)\cdot\nabla\mathbf{a}(\mathbf{x}, t_n)]$. Finally, for $\mathbf{x} \in \Gamma_n$, we denote by $\mathbf{A}_n(\mathbf{x})$ (respectively by $\mathbf{V}_n(\mathbf{x})$) the mean value of $\mathbf{a}(\mathbf{Y}(.\,; t_n, \mathbf{x}),\, .)$ (respectively $\mathbf{V}(\mathbf{Y}(.\,; t_n, \mathbf{x}),\, .)$) on $(t_{n-1}, t_n)$.

## 4.3 Stationary Boundary Value Problems: The Weak Formulation and Existence of a Solution

We put $\mathbf{U}_0 := \mathbf{u}_0$ and we denote by $\mathbf{U}_n$ approximate values of the unknown function $\mathbf{u}$ on the time levels $t_n$ ($n = 1, 2, \ldots, N$). On the $n$-th time level, we assume that

$\mathbf{U}_{n-1}$ is already a known function from $\mathbf{L}^2_\sigma(\Omega_n)$ and *we look for* $\mathbf{U}_n \in \mathbf{W}^{1,2}_\sigma(\Omega_n)$ *such that*

$$\int_{\Omega_n} \big\{\big[\mathbf{U}_n - \mathbf{U}_{n-1} \circ \mathbf{X}(t_{n-1};\, t_n,\, .)\big] \cdot \boldsymbol{\Phi} - d_n\, \mathbf{U}_n \cdot \nabla\boldsymbol{\Phi} \cdot \mathbf{a}(.\,, t_n)$$

$$+ d_n\, \mathbf{U}_n \cdot \nabla\mathbf{U}_n \cdot \boldsymbol{\Phi}\big\}\; \mathrm{d}\mathbf{x} + \int_{\Omega_n} 2d_n\nu\, \big\{[\nabla\mathbf{a}]_n + \nabla\mathbf{U}_n\big\}_s : \nabla\boldsymbol{\Phi}\; \mathrm{d}\mathbf{x}$$

$$+ \int_{\Gamma_n} d_n\gamma\, (\mathbf{A}_n + \mathbf{U}_n - \mathbf{V}_n) \cdot \boldsymbol{\Phi}\; \mathrm{d}\Gamma \;=\; \int_{\Omega_n} d_n\, \mathbf{g}_n \cdot \boldsymbol{\Phi}\; \mathrm{d}\mathbf{x} \tag{16}$$

*for all* $\boldsymbol{\Phi} \in \mathbf{W}^{1,2}_\sigma(\Omega_n)$. Here, $\mathbf{U}_{n-1} \circ \mathbf{X}(t_{n-1};\, t_n,\, .)$ denotes the function $\mathbf{x} \mapsto \mathbf{U}_{n-1}\big(\mathbf{X}(t_{n-1};\, t_n, \mathbf{x})\big)$. The difference $\mathbf{U}_n(\mathbf{x}) - \mathbf{U}_{n-1}\big(\mathbf{X}(t_{n-1};\, t_n, \mathbf{x})\big)$ (for $\mathbf{x} \in \Omega_n$) approximates the time derivative $(\mathrm{d}/\mathrm{d}t)\mathbf{u}\big(\mathbf{X}(t;\, t_n, \mathbf{x}), t\big)$ at the time $t = t_n$, multiplied by $d_n$.

**Lemma 4.1** *Let $N$ be so large that the maximum over $n = 1, \ldots, N$ in* (15) *is less than one and $n \in \{1; \ldots; N\}$. Then Eq.* (16) *has a solution* $\mathbf{U}_n$ *in* $\mathbf{W}^{1,2}_\sigma(\Omega_n)$.

*Principle of the Proof* We write, for simplicity, only $\mathbf{U}$ and $d$ instead of $\mathbf{U}_n$ and $d_n$. Since $t_n \notin \mathcal{T}^c$, domain $\Omega_n$ is Lipschitzian. Hence $\mathbf{W}^{1,2}_\sigma(\Omega_n) \hookrightarrow \mathbf{L}^q(\Omega_n)$ for $1 \le q \le 6$ and there exists a continuous operator of traces from $\mathbf{W}^{1,2}_\sigma(\Omega_n)$ into $\mathbf{L}^2(\Gamma_n)$. One can verify for given $\mathbf{U} \in \mathbf{W}^{1,2}_\sigma(\Omega_n)$ that

$$\int_{\Omega_n} \big\{\mathbf{U} \cdot \boldsymbol{\Phi} - d\,\mathbf{U} \cdot \nabla\boldsymbol{\Phi} \cdot \mathbf{a}(.\,, t_n) + d\,\mathbf{U} \cdot \nabla\mathbf{U} \cdot \boldsymbol{\Phi} + 2d\nu\,(\nabla\mathbf{U})_s : \nabla\boldsymbol{\Phi}\big\}\; \mathrm{d}\mathbf{x}$$

$$+ \int_{\Gamma_n} d\gamma\, \mathbf{U} \cdot \boldsymbol{\Phi}\; \mathrm{d}\Gamma$$

is a bounded linear functional in dependence on $\boldsymbol{\Phi} \in \mathbf{W}^{1,2}_\sigma(\Omega_n)$. Thus, it can be written in the form $\big\langle \mathcal{A}(\mathbf{U}), \boldsymbol{\Phi}\big\rangle_{\Omega_n}$, where $\mathcal{A}(\mathbf{U})$ belongs to the dual space $\mathbf{W}^{-1,2}_\sigma(\Omega_n)$ to $\mathbf{W}^{1,2}_\sigma(\Omega_n)$ and $\langle\, .\,,\, .\,\rangle_{\Omega_n}$ denotes the duality between $\mathbf{W}^{-1,2}_\sigma(\Omega_n)$ and $\mathbf{W}^{1,2}_\sigma(\Omega_n)$. Similarly, the difference

$$\int_{\Omega_n} \big\{\mathbf{U}_{n-1} \circ \mathbf{X}(t_{n-1};\, t_n,\, .) \cdot \boldsymbol{\Phi} - 2d\nu\, [\nabla\mathbf{a}]_n : \nabla\boldsymbol{\Phi}\big\}\; \mathrm{d}\mathbf{x}$$

$$- \int_{\Gamma_n} d_n\gamma\, (\mathbf{A}_n - \mathbf{V}_n) \cdot \boldsymbol{\Phi}\; \mathrm{d}\Gamma + d \int_{\Omega_n} \mathbf{g}_n \cdot \boldsymbol{\Phi}\; \mathrm{d}\mathbf{x}$$

can be expressed as $\langle \mathcal{F}, \boldsymbol{\Phi}\rangle_{\Omega_n}$ where $\mathcal{F} \in \mathbf{W}^{-1,2}_\sigma(\Omega_n)$. We can now write Eq. (16) (for the unknown $\mathbf{U}$) in the equivalent form as an operator equation in the space $\mathbf{W}^{-1,2}_\sigma(\Omega_n)$: $\mathcal{A}(\mathbf{U}) = \mathcal{F}$. Operator $\mathcal{A}$ is bounded and demicontinuous from $\mathbf{W}^{1,2}_\sigma(\Omega_n)$ to $\mathbf{W}^{-1,2}_\sigma(\Omega_n)$. Using inequalities (7) and (15), one can verify that operator $\mathcal{A}$ is coercive if $d$ is so small (i.e. $N$ is so large) that $\max_{n=1,\ldots,N}$ in (15) is less than one. Moreover, one can also verify that operator $\mathcal{A}$ satisfies all the assumptions of the Leray-Lions theorem (see e.g. [5]) for these $d$. Due to this theorem, the equation $\mathcal{A}(\mathbf{U}) = \mathcal{F}$ has a solution $\mathbf{U} \in \mathbf{W}^{1,2}_{\sigma}(\Omega_n)$. □

## 4.4 Estimates of Solutions of the Weak Problem (16)

We assume that the assumptions of Lemma 4.1 hold. We derive a discrete version of the energy inequality (11). Using $\mathbf{\Phi} = \mathbf{U}_n$ in (16), we obtain:

$$\begin{aligned}
&\frac{1}{2}\,\|\mathbf{U}_n\|^2_{2;\,\Omega_n} + \frac{1}{2}\int_{\Omega_n} \big|\mathbf{U}_n - \mathbf{U}_{n-1}\circ \mathbf{X}(t_{n-1};\, t_n,\, .)\big|^2\, \mathrm{d}\mathbf{x} \\
&\quad + 2d_n\nu \int_{\Omega_n} (\nabla\mathbf{U}_n)_s : \nabla\mathbf{U}_n\, \mathrm{d}\mathbf{x} + \int_{\Gamma_n} d_n\gamma\, |\mathbf{U}_n|^2\, \mathrm{d}\Gamma \\
&\leq \frac{1}{2}\,\|\mathbf{U}_{n-1}\|^2_{2;\,\Omega_{n-1}} + \left| d_n \int_{\Omega_n} \mathbf{g}_n\cdot\mathbf{U}_n\, \mathrm{d}\mathbf{x}\right| + \left| d_n \int_{\Omega_n} \mathbf{U}_n\cdot\nabla\mathbf{U}_n\cdot\mathbf{a}(.\,,t_n)\, \mathrm{d}\mathbf{x}\right| \\
&\quad + \left| 2d_n\nu \int_{\Omega_n} \big([\nabla\mathbf{a}]_n\big)_s : \nabla\mathbf{U}_n\, \mathrm{d}\mathbf{x}\right| + \left| \int_{\Gamma_n} d_n\gamma\, (\mathbf{A}_n - \mathbf{V}_n)\cdot\mathbf{U}_n\, \mathrm{d}\Gamma\right|. \qquad (17)
\end{aligned}$$

By analogy with (13), we have

$$2\nu \int_{\Omega_n} (\nabla\mathbf{U}_n)_s : \nabla\mathbf{U}_n\, \mathrm{d}\mathbf{x} \;\geq\; \frac{9\nu}{10}\,\|\nabla\mathbf{U}_n\|^2_{2;\,\Omega_n} - \nu\left[c_1 + \frac{5}{2}c_1^2\right]\|\mathbf{U}_n\|^2_{2;\,\Omega_n}. \qquad (18)$$

The integral of $[\partial_t\mathbf{a}(.\,,t_n)+\mathbf{a}(.\,,t_n)\cdot\nabla\mathbf{a}(.\,,t_n)]\cdot\mathbf{U}_n$ (the part of $\mathbf{g}_n\cdot\mathbf{U}_n$) can be estimated by means of inequality (6):

$$\begin{aligned}
&\left|\int_{\Omega_n} \big[\partial_t\mathbf{a}(.\,,t_n) + \mathbf{a}(.\,,t_n)\cdot\nabla\mathbf{a}(.\,,t_n)\big]\cdot\mathbf{U}_n\, \mathrm{d}\mathbf{x}\right| \\
&\quad\leq \theta_3(t_n)\,\|\mathbf{U}_n\|_{2;\,\Omega_n} + \theta_4(t_n)\,\|\nabla\mathbf{U}_n\|_{2;\,\Omega_n} \\
&\quad\leq \frac{\nu}{10}\,\|\nabla\mathbf{U}_n\|^2_{2;\,\Omega_n} + \frac{5}{2\nu}\,\theta_4^2(t_n) + \frac{1}{2}\,\theta_3(t_n)\,\big[\|\mathbf{U}_n\|^2_{2;\,\Omega_n} + 1\big].
\end{aligned}$$

The integral of $\mathbf{U}_n\cdot\nabla\mathbf{U}_n\cdot\mathbf{a}(.\,,t_n)$ can be estimated by means of inequality (7). The integral of $\big([\nabla\mathbf{a}]_n\big)_s : \nabla\mathbf{U}_n$ can be estimated as follows:

$$\left| 2\nu \int_{\Omega_n} \big([\nabla\mathbf{a}]_n\big)_s : \nabla\mathbf{U}_n\, \mathrm{d}\mathbf{x}\right| \;\leq\; \frac{\nu}{10}\,\|\nabla\mathbf{U}_n\|^2_{2;\,\Omega_n} + 10\nu\,\overline{\theta^2_{1n}},$$

where $\overline{\theta^2_{1n}}$ denotes the mean value of $\theta_1^2$ on $(t_{n-1}, t_n)$. (Recall that $\theta_1(t) := \|\mathbf{a}(.\,,t)\|_{1,2;\,\Omega^t}$.) Finally, the integral of $(\mathbf{A}_n - \mathbf{V}_n)\cdot\mathbf{U}_n$ on $\Gamma_n$ can be estimated by means of assumption (a5):

$$\begin{aligned}
&\left|\gamma \int_{\Gamma_n} \big(\mathbf{A}_n(\mathbf{x}_n) - \mathbf{V}_n(\mathbf{x}_n)\big) \cdot \mathbf{U}_n(\mathbf{x}_n)\, \mathrm{d}\Gamma(\mathbf{x}_n)\right| \\
&= \left|\frac{\gamma}{d_n} \int_{t_{n-1}}^{t_n} \int_{\Gamma_n} \big[\mathbf{a}\big(\mathbf{Y}(t;\, t_n, \mathbf{x}_n), t\big) - \mathbf{V}\big(\mathbf{Y}(t;\, t_n, \mathbf{x}_n), t\big)\big] \cdot \mathbf{U}_n(\mathbf{x}_n)\, \mathrm{d}\Gamma(\mathbf{x}_n)\, \mathrm{d}t\right| \\
&\leq \frac{\gamma}{4}\, \|\mathbf{U}_n\|_{2;\,\Gamma_n}^2 + \frac{\gamma}{d_n} \int_{t_{n-1}}^{t_n} \int_{\Gamma_n} \big|\mathbf{a}\big(\mathbf{Y}(t;\, t_n, \mathbf{x}_n), t\big) - \mathbf{V}\big(\mathbf{Y}(t;\, t_n, \mathbf{x}_n), t\big)\big|^2\, \mathrm{d}\Gamma(\mathbf{x}_n)\, \mathrm{d}t \\
&= \frac{\gamma}{4}\, \|\mathbf{U}_n\|_{2;\,\Gamma_n}^2 + \frac{\gamma}{d_n} \int_{t_{n-1}}^{t_n} \int_{\Gamma^t} \big|\mathbf{a}(\mathbf{x}, t) - \mathbf{V}(\mathbf{x}, t)\big|^2\, \mathrm{d}\Gamma(\mathbf{x})\, \mathrm{d}t \;=\; \frac{\gamma}{4}\, \|\mathbf{U}_n\|_{2;\,\Gamma_n}^2 + \gamma\, \overline{\theta_{2n}^2}.
\end{aligned}$$

Substituting these estimates to (17), summing for $n = 1, \ldots, j$ (where $1 \leq j \leq N$, $j \in \mathbb{N}$), and multiplying by two, we obtain the inequality

$$\begin{aligned}
&\|\mathbf{U}_j\|_{2;\,\Omega_j}^2 + \sum_{n=1}^{j} \big\|\mathbf{U}_n - \mathbf{U}_{n-1} \circ \mathbf{X}(t_{n-1};\, t_n,\, .)\big\|_{2;\,\Omega_n}^2 + \nu \sum_{n=1}^{j} d_n\, \|\nabla \mathbf{U}_n\|_{2;\,\Omega_n}^2 \\
&+ \gamma \sum_{n=1}^{j} d_n\, \|\mathbf{U}_n\|_{2;\,\Gamma_n}^2 \;\leq\; \|\mathbf{U}_0\|_{2;\,\Omega_0}^2 + \sum_{n=1}^{j} \omega_{1n}\, \|\mathbf{U}_n\|_{2;\,\Omega_n}^2 - \sum_{n=1}^{j} \omega_{2n}, \qquad (19)
\end{aligned}$$

where

$$\begin{aligned}
\omega_{1n} &= 2d_n \Big(c_1 \nu + \frac{5}{2}\, c_1^2 \nu + 1\Big) + 2d_n\, \theta_3(t_n) + 2d_n\, \theta_5(t_n), \\
\omega_{2n} &= \frac{5}{\nu}\, d_n\, \theta_4^2(t_n) + \int_{t_{n-1}}^{t_n} \|\mathbf{f}(.\,, t)\|_{2;\,\Omega^t}^2\, \mathrm{d}t + 20 d_n \nu\, \overline{\theta_{1n}^2} + 2\gamma d_n\, \overline{\theta_{2n}^2}.
\end{aligned}$$

Put $\lambda_N(s) := \omega_{1n}$ for $t_{n-1} < s \leq t_n$. It follows from the definition of $\overline{\theta_{1n}^2}$, $\overline{\theta_{2n}^2}$ and from (14) that

$$\int_0^T \lambda_N(s)\, \mathrm{d}s \;\leq\; 2\Big(c_1 \nu + \frac{5}{2} c_1^2 \nu + 1\Big) T + \int_0^T \theta_3(s)\, \mathrm{d}s + 2\Theta \;:=\; c_2, \qquad (20)$$

$$\sum_{n=1}^{N} \omega_{2n} \;\leq\; \frac{5}{\nu}\, \Theta + \int_0^T \big[\|\mathbf{f}(.\,, s)\|_{2;\,\Omega^s}^2 + 20\nu\, \theta_1^2(s) + 2\gamma\, \theta_2^2(s)\big]\, \mathrm{d}s \;:=\; c_3. \qquad (21)$$

Both $c_2$ and $c_3$ are independent of $N$. Moreover, due to (15), $\omega_{1n} \to 0$ as $N \to +\infty$ uniformly with respect to $n \in \{1;\, \ldots;\, N\}$.

# 5 Non-stationary Approximations and Their Weak Convergence

For $t_{n-1} < t \le t_n$ (where $n = 1, \ldots, N$), we define

$$\mathbf{u}^N(\,.\,,t) := \begin{cases} \mathbf{U}_n \text{ in } \Omega_n, \\ \mathbf{0} \quad \text{in } B_R(\mathbf{0}) \smallsetminus \Omega_n, \end{cases} \qquad \mathbb{U}^N(\,.\,,t) := \begin{cases} \nabla\mathbf{U}_n \text{ in } \Omega_n, \\ \mathbb{O} \qquad \text{in } B_R(\mathbf{0}) \smallsetminus \Omega_n, \end{cases}$$

$$\mathbf{u}^N_*(\,.\,,t) := \mathbf{u}^N\big(\mathbf{X}(t_n;\, t,\, .\,), t\big).$$

The values of $\mathbf{u}^N_*(\,.\,,t)$ at points on $\Gamma^t$ are considered to be the traces of function $\mathbf{u}^N_*(\,.\,,t)$, restricted to $\Omega^t$.

## 5.1 Estimates of the Sequences $\{\mathbf{u}^N\}$, $\{\mathbb{U}^N\}$ and $\{\mathbf{u}^N_*\}$

Inequalities (19) and (21) imply that if $N$ is so large that $\omega_{1n} < \frac{1}{2}$ for all $n \in \{1; \ldots; N\}$ then

$$\frac{1}{2}\,\|\mathbf{u}^N(.\,,t)\|^2_{2;\,B_R(\mathbf{0})} + \nu \int_0^t \|\mathbb{U}^N(.\,,s)\|^2_{2;\,B_R(\mathbf{0})}\,\mathrm{d}s + \gamma \int_0^t \|\mathbf{u}^N_*(.\,,s)\|^2_{2;\,\Gamma^s}\,\mathrm{d}s$$
$$\le\; \|\mathbf{u}_0\|^2_{2;\,\Omega^0} + \int_0^t \lambda_N(s)\,\|\mathbf{u}^N(.\,,s)\|^2_{2;\,B_R(\mathbf{0})}\,\mathrm{d}s + c_3. \tag{22}$$

Applying Gronwall's lemma and estimate (20), we get

$$\|\mathbf{u}^N(.\,,t)\|^2_{2;\,B_R(\mathbf{0})} \;\le\; const.\big(\|\mathbf{u}_0\|^2_{2;\,\Omega^0} + c_3\big) \;:=\; c_4 \qquad (\text{for } 0 \le t \le T), \tag{23}$$

$$\nu \int_0^T \|\mathbb{U}^N(.\,,s)\|^2_{2;\,B_R(\mathbf{0})}\,\mathrm{d}s + \gamma \int_0^T \|\mathbf{u}^N_*(.\,,s)\|^2_{2;\,\Gamma^s}\,\mathrm{d}s \tag{24}$$
$$\le\; c_2 c_4 + c_3 + 2c_4 \;:=\; c_5. \tag{25}$$

Constants $c_4$ and $c_5$ are independent of $N$. Inequalities (23) and (25) imply that

$$\|\mathbf{U}_n\|_{2;\,\Omega_n} \;\le\; c_4 \qquad (n = 1, \ldots, N), \tag{26}$$

$$\nu \sum_{n=1}^{N} d_n\,\|\nabla\mathbf{U}_n\|^2_{2;\,\Omega_n} + \gamma \sum_{n=1}^{N} d_n\,\|\mathbf{U}_n\|^2_{2;\,\Gamma_n} \;\le\; c_5. \tag{27}$$

## 5.2 Weak Convergence of Subsequences

Estimates (23) and (25) imply that there exist subsequences of $\{\mathbf{u}^N\}$, $\{\mathbb{U}^N\}$ and $\{\mathbf{u}_*^N\}$ (we denote them again by $\{\mathbf{u}^N\}$, $\{\mathbb{U}^N\}$ and $\{\mathbf{u}_*^N\}$) and functions $\mathbf{u} \in L^\infty(0,T;\, \mathbf{L}^2(B_R(\mathbf{0})))$, $\mathbb{U} \in L^2(0,T;\, L^2(B_R(\mathbf{0}))^9)$ and $\mathbf{u}_* \in \mathbf{L}^2(\Gamma_{(0,T)})$ such that

$$\mathbf{u}^N \rightharpoonup \mathbf{u} \qquad \text{weakly-}* \text{ in } L^\infty(0,T;\, \mathbf{L}^2(B_R(\mathbf{0}))) \qquad \text{for } N \to \infty, \tag{28}$$

$$\mathbb{U}^N \rightharpoonup \mathbb{U} \qquad \text{weakly in } L^2(0,T;\, L^2(B_R(\mathbf{0}))^9) \qquad \text{for } N \to \infty, \tag{29}$$

$$\mathbf{u}_*^N \rightharpoonup \mathbf{u}_* \qquad \text{weakly in } \mathbf{L}^2(\Gamma_{(0,T)}) \qquad \text{for } N \to \infty. \tag{30}$$

**Lemma 5.1** *(a)* $\mathbb{U} = \nabla\mathbf{u}$ *in the sense of distributions in* $Q_{(0,T)}$,
*(b)* $\mathbf{u} \in L^2(0,T;\, \mathbf{W}_\sigma^{1,2}(\Omega^t))$,
*(c)* $\mathbf{u}_* = \mathrm{tr}(\mathbf{u})$ *on* $\Gamma_{(0,T)}$ *(where* $\mathrm{tr}(\mathbf{u})$ *is the trace of the function* $\mathbf{u}|_{Q_{(0,T)}}$ *on* $\Gamma_{(0,T)}$*).*

*Principle of the Proof*

(a) Let $\mathbb{F} \in C_0^\infty(Q_{(0,T)})^9$. Let us extend $\mathbb{F}$ by $\mathbb{O}$ to $[B_R(\mathbf{0}) \times (0,T)] \smallsetminus Q_{(0,T)}$. The support of $\mathbb{F}$ belongs to $\cup_{n=1}^N \Omega_n \times [t_{n-1}, t_n)$ for all sufficiently large $N$, hence $\mathbb{U}^N = \nabla\mathbf{u}^N$ on $\operatorname{supp}\mathbb{F}$. Statement (a) now follows from the identities

$$\begin{aligned}\int_0^T\int_{\Omega^t} \mathbb{F} : \mathbb{U}\, \mathrm{d}\mathbf{x}\, \mathrm{d}t &= \lim_{N\to+\infty} \int_0^T\int_{B_R(\mathbf{0})} \mathbb{F} : \nabla\mathbf{u}^N\, \mathrm{d}\mathbf{x}\, \mathrm{d}t \\ &= -\lim_{N\to+\infty} \int_0^T\int_{B_R(\mathbf{0})} \mathrm{Div}\,\mathbb{F}\cdot\mathbf{u}^N\, \mathrm{d}\mathbf{x}\, \mathrm{d}t = -\int_0^T\int_{\Omega^t} \mathrm{Div}\,\mathbb{F}\cdot\mathbf{u}\, \mathrm{d}\mathbf{x}\, \mathrm{d}t.\end{aligned}$$

(b) Since $\mathbf{u} \in L^\infty(0,T;\, \mathbf{L}^2(\Omega^t))$, $\mathbb{U} \in L^2(0,T;\, L^2(\Omega^t)^9)$ and $\nabla\mathbf{u} = \mathbb{U}$ a.e. in $Q_{(0,T)}$, we deduce that $\mathbf{u} \in L^2(0,T;\, \mathbf{W}^{1,2}(\Omega^t))$. If $p \in L^2(0,T;\, W^{1,2}(B_R(\mathbf{0})))$ then $\int_{Q_{(0,T)}} \mathbf{u}\cdot\nabla p\ \mathrm{d}\mathbf{x}\, \mathrm{d}t$ equals the same integral in $B_R(\mathbf{0}) \times (0,T)$, and it further equals the limit for $N \to \infty$ of the integrals $\int_0^T \int_{B_R(\mathbf{0})} \mathbf{u}^N\cdot\nabla p\, \mathrm{d}\mathbf{x}\, \mathrm{d}t$. Due to the definition of function $\mathbf{u}^N$, all these integrals are equal to zero. Consequently, $\mathbf{u}(\,.\,,t) \in \mathbf{L}_\sigma^2(\Omega^t)$ for a.a. $t \in (0,T)$.

(c) The last statement of Lemma 5.1 can be proven so that we show that the weak limit of $\mathbf{u}_*^N$ in $\mathbf{L}^2(B_R(\mathbf{0}) \times (0,T))$ is a.e. equal to function $\mathbf{u}$. □

### 5.3 Substitution of the Approximations to Integral Equation (10)

The approximations $\mathbf{u}^N$ (represented by $\mathbf{u}_*^N$ on $\Gamma^t$) naturally satisfy the integral equation (10) with certain error $\mathcal{E}^N(\boldsymbol{\phi})$. Thus, if we denote

$$\mathcal{I}_1(\mathbf{u}^N, \mathbf{u}_*^N, \boldsymbol{\phi}) := \int_0^T \int_{\Gamma^t} \gamma(\mathbf{a} + \mathbf{u}_*^N - \mathbf{V}) \cdot \boldsymbol{\phi} \, \mathrm{d}\Gamma \, \mathrm{d}t$$

$$- \int_0^T \int_{\Omega^t} \big[(\partial_t \boldsymbol{\phi} + \mathbf{a} \cdot \nabla \boldsymbol{\phi}) \cdot \mathbf{u}^N + \mathbf{u}^N \cdot \nabla \boldsymbol{\phi} \cdot \mathbf{a} - 2\nu [\nabla(\mathbf{a} + \mathbf{u}^N)]_s : \nabla \boldsymbol{\phi}\big] \, \mathrm{d}\mathbf{x} \, \mathrm{d}t,$$

$$\mathcal{I}_2(\mathbf{u}^N, \boldsymbol{\phi}) := \int_0^T \int_{\Omega^t} \mathbf{u}^N \cdot \mathbb{U}^N \cdot \boldsymbol{\phi} \, \mathrm{d}\mathbf{x} \, \mathrm{d}t$$

then (10) yields

$$\mathcal{I}_1(\mathbf{u}^N, \mathbf{u}_*^N, \boldsymbol{\phi}) + \mathcal{I}_2(\mathbf{u}^N, \boldsymbol{\phi}) = \int_0^T \int_{\Omega^t} \mathbf{g} \cdot \boldsymbol{\phi} \, \mathrm{d}\mathbf{x} \, \mathrm{d}t + \int_{\Omega^0} \mathbf{u}_0 \cdot \boldsymbol{\phi}(.\,,0) \, \mathrm{d}\mathbf{x} + \mathcal{E}^N(\boldsymbol{\phi}). \tag{31}$$

Using the types of convergence named in (28)–(30) and statements (a) and (c) of Lemma 5.1, we can deduce that $\mathcal{I}_1(\mathbf{u}^N, \mathbf{u}_*^N, \boldsymbol{\phi}) \to \mathcal{I}_1(\mathbf{u}, \mathbf{u}^*, \boldsymbol{\phi})$ as $N \to +\infty$. In order to show that $\mathbf{u}$ is a weak solution of the problem (1)–(5), it remains to verify that

$$\lim_{N \to +\infty} \mathcal{I}_2(\mathbf{u}^N, \boldsymbol{\phi}) = \int_0^T \int_{\Omega^t} \mathbf{u} \cdot \nabla \mathbf{u} \cdot \boldsymbol{\phi} \, \mathrm{d}\mathbf{x} \, \mathrm{d}t, \tag{32}$$

$$\lim_{N \to +\infty} \mathcal{E}^N(\boldsymbol{\phi}) = 0. \tag{33}$$

The validity of (32) is the crucial part of the proof. It is treated in the next section. The proof of (33) uses the structure of functions $\mathbf{u}^N$, $\mathbf{u}_*^N$, the smoothness of the test function $\boldsymbol{\phi}$, and especially the fact that functions $\mathbf{U}_n$ (appearing in the definition of $\mathbf{u}^N$) satisfy Eq. (16). Although it is technical and laborious, it is more or less standard and we therefore omit it in this text.

## 6 The Limit Process in the Nonlinear Term $\mathcal{I}_2(\mathbf{u}^N, \boldsymbol{\phi})$

The existence of the limit on the left hand side of (32) follows from Eq. (31), from (33) and from the existence of the limit of $\mathcal{I}_1(\mathbf{u}^N, \mathbf{u}_*^N, \boldsymbol{\phi})$ as $N \to +\infty$. Thus, it is sufficient to check the value of the limit in (32) only for an arbitrary subsequence of $\{\mathbf{u}^N, \mathbb{U}^N\}$. Further in this section, we assume that $\boldsymbol{\phi}$ is a fixed test function with the properties named in Definition 2.1. In Sects. 6.1–6.3, we explain that it is sufficient to prove (32) with certain modifications $\boldsymbol{\phi}^*$, $\boldsymbol{\phi}^{**}$ and $\boldsymbol{\phi}_j^{**}$ (for $j = 1, \dots, J$) of function $\boldsymbol{\phi}$.

## 6.1 Definition of Function $\boldsymbol{\phi}^*$

Recall that $\mathcal{T}^c = \{t_1^c; \dots; t_M^c\} \subset (0, T)$ is the family of critical time instants when the parts of the boundary of $\Omega^t$ may strike themselves. Let $\epsilon_1 > 0$ be given. Then, due to (23) and (25), there exists $\kappa > 0$ so small that

$$\left| \sum_{m=1}^{M} \int_{t_m^c-\kappa}^{t_m^c+\kappa} \int_{\Omega^t} \mathbf{u}^N \cdot \mathbb{U}^N \cdot \boldsymbol{\phi} \, \mathrm{d}\mathbf{x} \, \mathrm{d}t \right| \le C \sum_{m=1}^{M} \int_{t_m^c-\kappa}^{t_m^c+\kappa} \|\mathbb{U}^N(.\,,t)\|_{2;\,\Omega^t}^2 \, \mathrm{d}t < \epsilon_1 \tag{34}$$

for all $N \in \mathbb{N}$ sufficiently large. (The constant on the right hand side depends on $c_4$, $c_5$, $\nu$ and the maximum of $|\boldsymbol{\phi}|$.) Let $\eta$ be an infinitely differentiable cut-off function of variable $t$ defined on the interval $[0, T]$, with values in $[0, 1]$, such that $\eta(t) := 1$ if $\mathrm{dist}(t; \mathcal{T}^c) \ge \kappa$ and $\eta(t) := 0$ if $\mathrm{dist}(t; \mathcal{T}^c) \le \frac{1}{2}\kappa$. The function $\boldsymbol{\phi}^*(\mathbf{x}, t) := \eta(t)\, \boldsymbol{\phi}(\mathbf{x}, t)$ equals zero for $t \in [0, T]$ such that $\mathrm{dist}(t; \mathcal{T}^c) \le \frac{1}{2}\kappa$ and

$$\left| \int_0^T \int_{\Omega^t} \mathbf{u}^N \cdot \mathbb{U}^N \cdot (\boldsymbol{\phi} - \boldsymbol{\phi}^*) \, \mathrm{d}\mathbf{x} \, \mathrm{d}t \right| < \epsilon_1 .$$

As $\epsilon_1$ can be chosen arbitrarily small, it is sufficient to prove (32) with function $\boldsymbol{\phi}^*$ instead of $\boldsymbol{\phi}$.

## 6.2 Definition of Function $\boldsymbol{\phi}^{**}$

Since domain $\Omega^t$ is Lipschitzian for each $t \in [0, T] \setminus \mathcal{T}^c$, it has a cone property (see [1, p. 66]x) and $\mathbf{W}_\sigma^{1,2}(\Omega^t) \hookrightarrow \mathbf{L}^6(\Omega^t)$. Moreover, if we restrict ourselves to times $t \in I(\kappa)$, where

$$I(\kappa) := \left\{ t \in [0, T];\ \mathrm{dist}(t; \mathcal{T}^c) > \tfrac{1}{2}\kappa \right\},$$

then the cone parameters in the definition of the cone property of domain $\Omega^t$ can be chosen to be independent of $t$. Hence the constant in the corresponding imbedding inequality also becomes independent of $t$, see [1, p. 103]. Consequently,

$$\|\mathbf{u}^N(.\,,t)\|_{6;\,B_R(\mathbf{0})} \le \mathit{const.} \left( \|\mathbf{u}^N(.\,,t)\|_{2;\,B_R(\mathbf{0})} + \|\mathbb{U}^N(.\,,t)\|_{2;\,B_R(\mathbf{0})} \right)$$

for all $t \in I(\kappa)$. From this information and from (26), we deduce that the product $\mathbf{u}^N \cdot \mathbb{U}^N$ belongs to the space $L^2\big(I(\kappa); \mathbf{L}^1(B_R(\mathbf{0}))\big) \cap L^1\big(I(\kappa); \mathbf{L}^{3/2}(B_R(\mathbf{0}))\big)$. By interpolation, we obtain the inclusion $\mathbf{u}^N \cdot \mathbb{U}^N \in L^r\big(I(\kappa); \mathbf{L}^s(B_R(\mathbf{0}))\big)$ for $r \geq 1$, $s \geq 1$ such that $2/r + 3/s = 4$. Particularly, $\mathbf{u}^N \cdot \mathbb{U}^N \in \mathbf{L}^{5/4}\big(B_R(\mathbf{0}) \times I(\kappa)\big)$.

Function $\boldsymbol{\phi}^*$ can be approximated by infinitely differentiable divergence-free vector-functions with a compact support in $Q_{[0,T)}$ with an arbitrary accuracy in the norm of the space $\mathbf{L}^5\big(B_R(\mathbf{0}) \times I(\kappa)\big)$. Hence, given $\epsilon_2 > 0$, there exists a vector-function $\boldsymbol{\phi}^{**}$ that satisfies

$$\left| \int_0^T \int_{\Omega^t} \mathbf{u}^N \cdot \mathbb{U}^N \cdot \boldsymbol{\phi}^* \, \mathrm{d}\mathbf{x} - \int_0^T \int_{\Omega^t} \mathbf{u}^N \cdot \mathbb{U}^N \cdot \boldsymbol{\phi}^{**} \, \mathrm{d}\mathbf{x} \right| < \epsilon_2$$

for all $N \in \mathbb{N}$ large enough. Since $\epsilon_2$ can be chosen to be arbitrarily small, we can prove (32) only with the function $\boldsymbol{\phi}^{**}$ instead of $\boldsymbol{\phi}$ (respectively instead of $\boldsymbol{\phi}^*$).

## 6.3 *Partition of Function $\boldsymbol{\phi}^{**}$*

Let $J \in \mathbb{N}$. We denote $\tau_j = jT/J$ (for $j = 0, \dots, J$). There exist $J + 1$ infinitely differentiable functions $\varphi_0, \dots, \varphi_J$ on $[0, T]$ with their values in the interval $[0, 1]$ such that $\operatorname{supp} \varphi_0 \subset I_0 := [\tau_0, \tau_1)$, $\operatorname{supp} \varphi_j \subset I_j := (\tau_{j-1}, \tau_{j+1})$ (for $j = 1, \dots, J-1$), $\operatorname{supp} \varphi_J \subset I_J := (\tau_{J-1}, \tau_J]$ and $\sum_{j=0}^{J} \varphi_j(t) = 1$ for $0 \leq t \leq T$. Now we put $\boldsymbol{\phi}_j^{**} := \varphi_j \boldsymbol{\phi}^{**}$ (for $j = 0, 1, \dots, J$). The functions $\boldsymbol{\phi}_j^{**}$ are divergence-free, they have compact supports in $Q_{I_j}$ and $\sum_{j=0}^{J} \boldsymbol{\phi}_j^{**} = \boldsymbol{\phi}^{**}$ in $Q_{[0,T]}$.

Denote by $G_j$ the set $\big\{\mathbf{x} \in \mathbb{R}^3;\ \exists\, t \in I_j : (\mathbf{x}, t) \in \operatorname{supp} \varphi_j^{**}\big\}$. If $J$ is large enough then the distance between $G_j$ and $\Gamma^t$ is greater than one half of the distance between $\operatorname{supp} \boldsymbol{\phi}^{**}$ and $\Gamma_{[0,T]}$ for all $t \in I_j$. Thus, there exists a bounded open set $O_j$ in $B_R(\mathbf{0})$ with the boundary of the class $C^{1,1}$ such that $G_j \subset O_j \subset \overline{O_j} \subset \Omega^t$ for all $t \in I_j$ (see Fig. 1). So, we conclude that in order to prove (32), it is sufficient to show

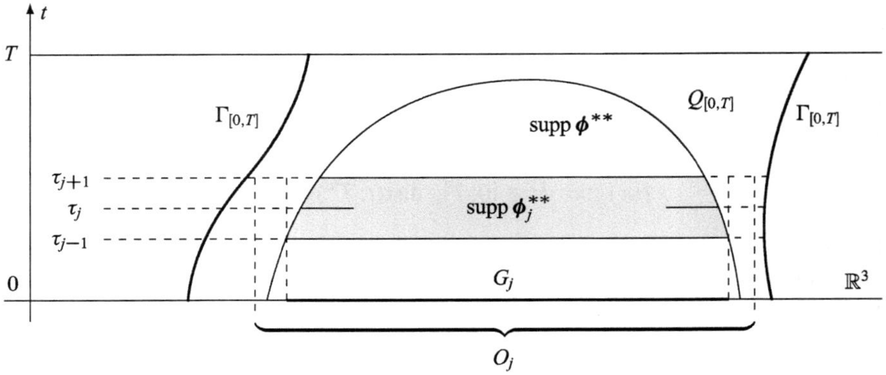


**Fig. 1** Sets $Q_{[0,T]}$, $\operatorname{supp} \boldsymbol{\phi}^{**}$, $\operatorname{supp} \boldsymbol{\phi}_j^{**}$, $G_j$ and $O_j$

that

$$\lim_{N\to+\infty} \int_{I_j}\int_{O_j} \mathbf{u}^N \cdot \nabla \mathbf{u}^N \cdot \boldsymbol{\phi}_j^{**} \, \mathrm{d}\mathbf{x}\, \mathrm{d}t = \int_{I_j}\int_{O_j} \mathbf{u} \cdot \nabla \mathbf{u} \cdot \boldsymbol{\phi}_j^{**} \, \mathrm{d}\mathbf{x}\, \mathrm{d}t \tag{35}$$

for each $j = 0, 1, \dots, J$. (Since $G_j \times I_j \subset \cup_{n=1}^{N} \Omega_n \times (t_{n-1}, t_n]$ for sufficiently large $N$, we can write $\nabla \mathbf{u}^N$ instead of $\mathbb{U}^N$ in (35).) We further assume that $j$ is fixed.

## 6.4 *The Local Helmholtz Decomposition of Function* $\mathbf{u}^N$

We denote by $P_\sigma^j$ the Helmholtz projection in $\mathbf{L}^2(O_j)$. Denote $\mathbf{w}_j^N := P_\sigma^j \mathbf{u}^N$. The function $(I - P_\sigma^j)\mathbf{u}^N$ has the form $\nabla \varphi_j^N$ for an appropriate function $\varphi_j^N$. Thus, the left hand side of (35) can be written in the form

$$\lim_{N\to+\infty} \int_{I_j}\int_{O_j} \big[\mathbf{w}_j^N \cdot \nabla \mathbf{w}_j^N + \mathbf{w}_j^N \cdot \nabla^2 \varphi_j^N + \nabla \varphi_j^N \cdot \nabla \mathbf{w}_j^N$$
$$+ \nabla \varphi_j^N \cdot \nabla^2 \varphi_j^N\big] \cdot \boldsymbol{\phi}_j^{**} \, \mathrm{d}\mathbf{x}\, \mathrm{d}t. \tag{36}$$

Since $\nabla \varphi_j^N \cdot \nabla^2 \varphi_j^N = \nabla\big(\frac{1}{2}|\nabla \varphi_j^N|^2\big)$ and $\boldsymbol{\phi}_j^{**}(\,.\,,t) \in \mathbf{L}_\sigma^2(O_j)$, the integral of $\nabla \varphi_j^N \cdot \nabla^2 \varphi_j^N \cdot \boldsymbol{\phi}_j^{**}$ equals zero.

Due to (28) and (29), the coincidence of $\mathbb{U}^N$ with $\nabla \mathbf{u}^N$ in $O_j \times I_j$ and the boundedness of operator $P_\sigma^j$ in $\mathbf{L}^2(O_j)$ and in $\mathbf{W}^{1,2}(O_j)$, we have

$$\mathbf{w}_j^N \rightharpoonup \mathbf{w}_j := P_\sigma^j \mathbf{u}, \quad \text{and} \quad \nabla \varphi_j^N \rightharpoonup \nabla \varphi_j := (I - P_\sigma^j)\mathbf{u} \quad \text{for } N \to +\infty \tag{37}$$

weakly in $L^2(I_j;\, \mathbf{W}^{1,2}(O_j))$ and weakly-$*$ in $L^\infty(I_j;\, \mathbf{L}_\sigma^2(O_j))$.

## 6.5 *Strong Convergence of a Subsequence of* $\{\mathbf{w}_j^N\}$

We are going to show that there exists a subsequence of $\{\mathbf{w}_j^N\}$ that tends to $\mathbf{w}_j$ strongly in $L^2(I_j;\, \mathbf{L}_\sigma^2(O_j))$ as $N \to +\infty$. We use the next lemma, see [6, Theorem 5.2].

**Lemma 6.1** *Let* $0 < \alpha < \frac{1}{2}$ *and let* $H_0$, $H$ *and* $H_1$ *be Hilbert spaces such that* $H_0 \hookrightarrow\hookrightarrow H \hookrightarrow H_1$. *Let* $\mathcal{H}^\alpha(\mathbb{R};\, H_0, H_1)$ *denote the Banach space* $\big\{w \in L^2(\mathbb{R};\, H_0);\ |\vartheta|^\alpha\, \widehat{w}(\vartheta) \in L^2(\mathbb{R};\, H_1)\big\}$ *with the norm*

$$|||w|||_{\alpha;\mathbb{R}} := \Big(\|w\|^2_{L^2(\mathbb{R};\, H_0)} + \|\, |\vartheta|^\alpha\, \widehat{w}(\vartheta)\|^2_{L^2(\mathbb{R};\, H_1)}\Big)^{1/2}.$$

*(Here $\widehat{w}(\vartheta)$ is the Fourier transform of $w(t)$.) Let $\mathcal{H}^\alpha(a,b;\, H_0,\, H_1)$ further denote the Banach space of restrictions of functions from $\mathcal{H}^\alpha(\mathbb{R};\, H_0,\, H_1)$ onto the interval $(a,b)$, with the norm $|||w|||_{\alpha;\,(a,b)} := \inf |||z|||_{\alpha;\,\mathbb{R}}$, where the infimum is taken over all $z \in \mathcal{H}^\alpha(\mathbb{R};\, H_0,\, H_1)$ such that $z = w$ a.e. in $(a,b)$. Then $\mathcal{H}^\alpha(0,T;\, H_0,\, H_1) \hookrightarrow\hookrightarrow L^2(a,b;\, H)$.*

We apply Lemma 6.1 with $(a,b) = I_j$, $H_0 = \mathbf{W}^{1,2}_\sigma(O_j)$, $H = \mathbf{L}^2_\sigma(O_j)$ and $H_1 = \mathbf{W}^{-1,2}_{0,\sigma}(O_j)$. (Here $\mathbf{W}^{-1,2}_{0,\sigma}(O_j)$ denotes the dual to $\mathbf{W}^{1,2}_{0,\sigma}(O_j)$, where $\mathbf{W}^{1,2}_{0,\sigma}(O_j) := \mathbf{W}^{1,2}_0(O_j) \cap \mathbf{L}^2_\sigma(O_j)$. The norm in $\mathbf{W}^{-1,2}_{0,\sigma}(O_j)$ is denoted by $\|\,.\,\|_{-1,2;\,O_j}$.)

We claim that $\{\mathbf{w}_j^N\}$ is bounded in the space $\mathcal{H}^\alpha(I_j;\, H_0,\, H_1)$. The boundedness of $\{\mathbf{w}_j^N\}$ in $L^2(I_j;\, H_0)$ follows from (23), (25), from the equality $\mathbb{U}^N = \nabla \mathbf{u}^N$ in $O_j \times I_j$ and from the boundedness of operator $P^j_\sigma$ in $\mathbf{L}^2(O_j)$ and in $\mathbf{W}^{1,2}(O_j)$. Thus, we only need to verify that the sequence $\{|\vartheta|^\alpha\, \widehat{\mathbf{w}}_j^N\}$ is bounded in the space $L^2(I_j;\, H_1)$, i.e. in $L^2(I_j;\, \mathbf{W}^{-1,2}_{0,\sigma}(O_j))$. Let $\mathbf{z}_j^N$ be the extension by zero of $\mathbf{w}_j^N$ from the time interval $I_j$ onto $\mathbb{R}$. Recall the notation from Sects. 4.1 and 4.2 and the definitions of functions $\mathbf{u}^N$ (see Sect. 5) and $\mathbf{w}^N$ ($:= P^j_\sigma \mathbf{u}^N$ in $O_j$). Let $\Lambda_j^N$ be the set of the indices $n \in \{1;\ldots;N\}$ such that $(t_{n-1}, t_n) \cap I_j \neq \emptyset$. Set $\Lambda_j^N$ has the form $\Lambda_j^N = \{l;\, l+1;\ldots;q\}$ where $1 \le l \le q \le N$. If $N$ is large enough then $O_j \subset \Omega_n$ for all $n \in \Lambda_j^N$. Thus, we have

$$
\begin{aligned}
\widehat{\mathbf{z}}_j^N(\vartheta) &= \int_{-\infty}^{+\infty} \mathrm{e}^{-2\pi\,\mathrm{i}\,t\vartheta}\, \mathbf{w}_j^N(t)\, \mathrm{d}t = \sum_{n\in\Lambda_j^N} \int_{t_{n-1}}^{t_n} \mathrm{e}^{-2\pi\,\mathrm{i}\,t\vartheta}\, P^j_\sigma \mathbf{U}_n\, \mathrm{d}t \\
&= \sum_{n=l}^{q} \frac{1}{2\pi\,\mathrm{i}\,\vartheta} \left[\mathrm{e}^{-2\pi\,\mathrm{i}\,t_{n-1}\vartheta} - \mathrm{e}^{-2\pi\,\mathrm{i}\,t_n\vartheta}\right] P^j_\sigma \mathbf{U}_n \\
&= \frac{1}{2\pi\,\mathrm{i}\,\vartheta} \left[\mathrm{e}^{-2\pi\,\mathrm{i}\,t_{l-1}}\, P^j_\sigma \mathbf{U}_l - \mathrm{e}^{-2\pi\,\mathrm{i}\,t_q}\, P^j_\sigma \mathbf{U}_q\right] \\
&\quad + \frac{1}{2\pi\,\mathrm{i}\,\vartheta} \sum_{n=l+1}^{q} \mathrm{e}^{-2\pi\,\mathrm{i}\,t_{n-1}\vartheta}\, [P^j_\sigma \mathbf{U}_n - P^j_\sigma \mathbf{U}_{n-1}].
\end{aligned} \tag{38}
$$

If $|\vartheta| \le 1$ then, using (38) and (26), we can estimate the norm of $|\vartheta|^\alpha\, \widehat{\mathbf{z}}_j^N(\vartheta)$ in $\mathbf{W}^{-1,2}_{0,\sigma}(O_j)$:

$$
\big\|\, |\vartheta|^\alpha\, \widehat{\mathbf{z}}_j^N(\vartheta) \big\|_{-1,2;\,O_j} \le C(O_j)\, |\vartheta|^\alpha \sum_{n=l}^{q} d_n\, \|\mathbf{U}_n\|_{2:\,O_j} \le C(O_j)\, |\vartheta|^\alpha. \tag{39}
$$

If $|\vartheta| > 1$ then we must proceed more subtly:

$$
\begin{aligned}
\big\|\, |\vartheta|^\alpha\, \widehat{\mathbf{z}}_j^N(\vartheta) \big\|_{-1,2;\,O_j} &\le \frac{|\vartheta|^{\alpha-1}}{2\pi} \left(\|P^j_\sigma \mathbf{U}_l\|_{-1,2;\,O_j} + \|P^j_\sigma \mathbf{U}_q\|_{-1,2;\,O_j}\right) \\
&\quad + \frac{|\vartheta|^{\alpha-1}}{2\pi} \sum_{n=l+1}^{q} \|P^j_\sigma \mathbf{U}_n - P^j_\sigma \mathbf{U}_{n-1}\|_{-1,2;\,O_j}
\end{aligned}
$$

$$\leq C(O_j)\,|\vartheta|^{\alpha-1}\left(\|\mathbf{U}_l\|_{2:\,O_j}+\|\mathbf{U}_q\|_{2:\,\Omega_{(j)}}\right)$$
$$+\frac{|\vartheta|^{\alpha-1}}{2\pi}\sum_{n=l+1}^{q}\sup_{\psi_n}\frac{1}{\|\boldsymbol{\psi}_n\|_{1,2;\,O_j}}\left|\int_{O_j}(\mathbf{U}_n-\mathbf{U}_{n-1})\cdot\boldsymbol{\psi}_n\,\mathrm{d}\mathbf{x}\right|, \qquad (40)$$

where the supremum is taken over all $\boldsymbol{\psi}_n \in \mathbf{W}^{1,2}_{0,\sigma}(O_j)$ such that $\|\boldsymbol{\psi}_n\|_{1,2;\,O_j} > 0$. The sum in (40) can be split to two parts and estimated from above by $\mathcal{Z}_1 + \mathcal{Z}_2$ where

$$\mathcal{Z}_1 = \sum_{n=l+1}^{q}\sup_{\psi_n}\frac{1}{\|\boldsymbol{\psi}_n\|_{1,2;\,O_j}}\left|\int_{O_j}\left[\mathbf{U}_n(\mathbf{x})-\mathbf{U}_{n-1}\big(\mathbf{X}(t_{n-1};\,t_n,\mathbf{x})\big)\right]\cdot\boldsymbol{\psi}_n(\mathbf{x})\,\mathrm{d}\mathbf{x}\right|,$$

$$\mathcal{Z}_2 = \sum_{n=l+1}^{q}\sup_{\psi_n}\frac{1}{\|\boldsymbol{\psi}_n\|_{1,2;\,O_j}}\left|\int_{O_j}\left[\mathbf{U}_{n-1}(\mathbf{x})-\mathbf{U}_{n-1}\big(\mathbf{X}(t_{n-1};\,t_n,\mathbf{x})\big)\right]\cdot\boldsymbol{\psi}_n(\mathbf{x})\,\mathrm{d}\mathbf{x}\right|.$$

Function $\boldsymbol{\psi}_n$, extended by zero to $\Omega_n \smallsetminus O_j$, belongs to $\mathbf{W}^{1,2}_{\sigma}(\Omega_n)$. Hence the integral of $\left[\mathbf{U}_n(\mathbf{x})-\mathbf{U}_{n-1}\big(\mathbf{X}(t_{n-1};\,t_n,\mathbf{x})\big)\right]\cdot\boldsymbol{\psi}_n(\mathbf{x})$ in $O_j$ equals the integral of the same function in $\Omega_n$ and it can be therefore expressed by means of (16). The surface integral on $\Gamma_n$ equals zero because the function $\boldsymbol{\psi}_n$ is zero on $\Gamma_n$. All other terms can be estimated by means of (26), (27), standard inequalities based on Sobolev's imbedding theorem (applied in $O_j$) and Hölder's inequality. Thus, we show that $\mathcal{Z}_1 \leq C(O_j)$. In order to estimate $\mathcal{Z}_2$, we use the identities

$$\mathbf{U}_{n-1}(\mathbf{x})-\mathbf{U}_{n-1}\big(\mathbf{X}(t_{n-1};\,t_n,\mathbf{x})\big) = \int_{t_{n-1}}^{t_n}\frac{\mathrm{d}}{\mathrm{d}\xi}\,\mathbf{U}_{n-1}\big(\mathbf{X}(\xi;\,t_n,\mathbf{x})\big)\,\mathrm{d}\xi$$
$$= \int_{t_{n-1}}^{t_n}\mathbf{a}\big(\mathbf{X}(\xi;\,t_n,\mathbf{x}),\xi\big)\cdot\nabla\mathbf{U}_{n-1}\big(\mathbf{X}(\xi;\,t_n,\mathbf{x})\big)\,\mathrm{d}\xi.$$

Then $\mathcal{Z}_2$ can be estimated by means of inequality (27) so that we also obtain $\mathcal{Z}_2 \leq C(O_j)$. Substituting the estimates of $\mathcal{Z}_1$ and $\mathcal{Z}_2$ to (40), we get

$$\left\|\,|\vartheta|^{\alpha}\,\widehat{\mathbf{z}}^{N}_{j}(\vartheta)\right\|_{-1,2;\,O_j} \leq C\Big(O_j, c_4, c_5, \textstyle\int_0^T \theta_1(t)\,\mathrm{d}t\Big)\,|\vartheta|^{\alpha-1}. \qquad (41)$$

The constant on the right hand side is independent of $N$. Recall that inequality (41) holds for $|\vartheta| > 1$. Since the exponent $\alpha$ satisfies $0 < \alpha < \frac{1}{2}$, the right hand side of (41) is integrable on $(-\infty,-1)\cup(1,+\infty)$ with power 2. This, together with (39), implies that the sequence $\{|\vartheta|^{\alpha}\,\widehat{\mathbf{z}}^{N}_{j}(\vartheta)\}$ is bounded in $L^2(\mathbb{R};\,\mathbf{W}^{-1,2}_{0,\sigma}(O_j))$. Thus, the sequence $\{\mathbf{w}^N_j\}$ is bounded in $\mathcal{H}^{\alpha}\big(I_j;\,\mathbf{W}^{1,2}_{\sigma}(O_j),\,\mathbf{W}^{-1,2}_{0,\sigma}(O_j)\big)$. This space is reflexive, hence there exists a subsequence (we denote it again by $\{\mathbf{w}^N_j\}$) that converges weakly in $\mathcal{H}^{\alpha}\big(I_j;\,\mathbf{W}^{1,2}_{\sigma}(O_j),\,\mathbf{W}^{-1,2}_{0,\sigma}(O_j)\big)$. Due to (37), the limit is $\mathbf{w}_j$. Applying Lemma 6.1, we obtain: $\mathbf{w}^N_j \longrightarrow \mathbf{w}_j = P^j_{\sigma}\mathbf{u}$ strongly in $L^2\big(I_j;\,\mathbf{L}^2(O_j)\big)$.

### 6.6 Completion of the Proof of Theorem 2.2

This strong convergence, together with the weak convergence (37), enables us to pass to the limit in the first three terms in (36). The procedure is standard (see e.g. [6] or [11]), therefore we omit the details. Using also the identity $\int_{O_j} (\nabla \varphi_j \cdot \nabla) \nabla \varphi_j \cdot \boldsymbol{\phi}_j^{**} \, \mathrm{d}\mathbf{x} = 0$, we verify the validity of (35), and consequently also the validity of (32). This confirms that $\mathbf{u}$ is a weak solution of the problem (1)–(5).

## 7 Example: The Flow Around Two Striking Bodies with Conic or $C^{1+\alpha}$ Axially Symmetric Front Surfaces

### 7.1 The Geometrical Configuration

We assume that two compact solid bodies $\mathcal{B}_1$ and $\mathcal{B}_2$ with Lipschitzian surfaces move in a fluid in a fixed bounded tank $D$ in the time interval $[0, T]$, and they strike and rebound at the time instant $t^c \in (0, T)$. We denote by $S_1^t$ (respectively $S_2^t$) the compact region in $D$, occupied by the first (respectively second) body at time $t$. Thus, the time-variable domain $\Omega^t$, where we consider the motion of the fluid, has the form $\Omega^t = D \smallsetminus (S_1^t \cup S_2^t)$ and set $\mathcal{T}^c$ of critical times in $(0, T)$ is the one point set $\mathcal{T}^c = \{t^c\}$. Furthermore, we assume that there exists $\tau > 0$ such that for $0 < |t - t^c| < \tau$

(a8) the distance $\delta^t$ between the bodies equals zero at the time instant $t^c$, it is positive and has a bounded second derivative $\ddot{\delta}^t$ for $0 < |t - t^c| < \tau$, the first derivative satisfies $\dot{\delta}^t \le 0$ in $(t^c - \tau, t^c)$ and $\dot{\delta}^t \ge 0$ in $(t^c, t^c + \tau)$,

(a9) body $\mathcal{B}_1$ moves with the velocity $\mathbf{V}_1^t = (0, 0, \frac{1}{2}\dot{\delta}^t)$ and body $\mathcal{B}_2$ moves with the velocity $\mathbf{V}_2^t = (0, 0, -\frac{1}{2}\dot{\delta}^t)$, where the velocities are expressed in a local Cartesian coordinate system $y_1, y_2, y_3$ which may generally depend on time,

(a10) there exists $r > 0$ such that for $\mathbf{y}' \in B_r(\mathbf{0}')$ (where $\mathbf{y}' := (y_1, y_2)$ and $\mathbf{0}' = (0, 0)$), the front surfaces of bodies $\mathcal{B}_1$ and $\mathcal{B}_2$ coincide with the graphs of the two functions

$$y_3 = \tfrac{1}{2}\delta^t + \kappa_1 |\mathbf{y}'|^{\beta_1}, \qquad y_3 = -\tfrac{1}{2}\delta^t - \kappa_2 |\mathbf{y}'|^{\beta_2},$$

where $\kappa_1 > 0$, $\kappa_2 \ge 0$, $1 \le \beta_1 \le \beta_2 \le 2$. We denote by $\Gamma_1^t$ (respectively $\Gamma_2^t$) the graph of the first (respectively the second) function.

The described configuration of bodies $\mathcal{B}_1$ and $\mathcal{B}_2$ close to the instant and place of the collision is sketched on Fig. 2.

We denote by $\Omega_c^t$ the critical sub-domain of $\Omega^t$, where the collision occurs: $\Omega_c^t := \{\mathbf{y} = (\mathbf{y}', y_3) \in \mathbb{R}^3;\ |\mathbf{y}'| < r \text{ and } -\frac{1}{2}\delta^t - \kappa_2 |\mathbf{y}'|^{\beta_2} < y_3 < \frac{1}{2}\delta^t + \kappa_1 |\mathbf{y}'|^{\beta_1}\}$.

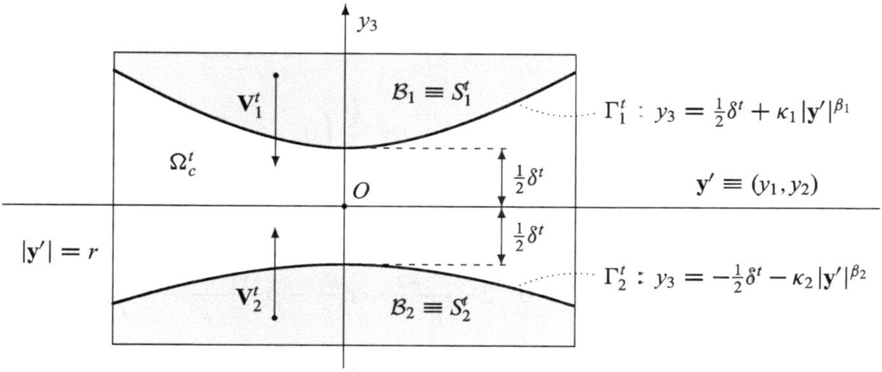


**Fig. 2** The shapes of bodies $\mathcal{B}_1$ and $\mathcal{B}_2$ near the point of the collision at times $t$ close to the instant $t^c$ of the collision

The transforming equations between $x_1, x_2, x_3$ and $y_1, y_2, y_3$ have the form $y_i = A_{ij}^t x_j + U_i^t$ $(i,j = 1,2,3)$, where $A^t = (A_{ij}^t)$ is a $3 \times 3$ unitary matrix. In accordance with the general assumptions on the motion of the solid part $\mathcal{S}^t$ from subsection 1.3, we may assume that all the entries $A_{ij}^t$ and $U_i^t$ are functions from $C^1\big([0,t^c) \cup (t^c,T]\big)$, continuous on $[0,T]$.

The integral of $\boldsymbol{\phi} \cdot \nabla\mathbf{n} \cdot \boldsymbol{\phi}$ on $\Gamma^t$ in condition (a2) can be split to the integral on $\Gamma_1^t \cup \Gamma_2^t$ (where $\nabla\mathbf{n}$ is negative semi-definite) and the integral on $\Gamma^t \smallsetminus (\Gamma_1^t \cup \Gamma_2^t)$ (where $\nabla\mathbf{n}$ is bounded and one can use the continuity of the operator of traces from $\mathbf{W}_\sigma^{1,2}(\Omega^t)$ into $\mathbf{L}^2\big(\Gamma^t \smallsetminus (\Gamma_1^t \cup \Gamma_2^t)\big)$ with a constant in the corresponding inequality independent of $t$). Thus, condition (a2) holds in the concrete situation, considered in this section.

In order to apply Theorem 2.2, we need to construct a divergence-free function **a**, satisfying conditions (a3)–(a7) formulated in subsection 1.4. **We confine ourselves to the definition of function a only in the "critical" domain $\Omega_c^t$ for $t \in (t^c - \tau, t^c) \cup (t^c, t^c + \tau)$.** We consider an appropriate extension of **a** to the set $Q_{[0,T]\smallsetminus\{t^c\}}$ to be only a matter of standard techniques and we do not describe it here in order not to extend the paper. In accordance with this philosophy, we sketch the verification of conditions (a3)–(a7) only for the part of function **a**, defined in $\Omega_c^t$ at times $t$ satisfying $0 < |t - t^c| < \tau$. Moreover, due to the aforementioned properties of the transformation between the coordinate systems $y_1, y_2, y_3$ and $x_1, x_2, x_3$, we may verify all the conditions in the (generally moving) frame $y_1, y_2, y_3$.

## 7.2 *Definition of Function* a *in* $\Omega_c^t$

It is further advantageous to work in the cylindrical coordinates $\rho, \varphi, y_3$, where $\rho = |\mathbf{y}'| = (y_1^2 + y_2^2)^{1/2}$. Furthermore, we denote $g_1 \equiv g_1(\rho, t) := \frac{1}{2}\delta^t + \kappa_1 \rho^{\beta_1}$,

$g_2 \equiv g_2(\rho,t) := \frac{1}{2}\delta^t + \kappa_2\rho^{\beta_2}$ and $g := g_1 + g_2$. The primes always denote the derivatives with respect to $\rho$. (Thus, e.g. $g_2' = \partial_\rho g_2$.) We use the vectorial potential

$$\mathbf{w}(\rho,\varphi,y_3,t) \equiv (w_\rho, w_\varphi, w_3) := \frac{\dot{\delta}^t}{4}\left(0,\ \rho\frac{y_3+g_2}{g},\ 0\right)$$

in the closure of $\Omega_c^t$ and we define $\mathbf{a} \equiv (a_1,a_2,a_3)$ by the formula

$$\mathbf{a} := \mathbf{curl}\,\mathbf{w} = \frac{\dot{\delta}^t}{4}\left(-\frac{\rho}{g},\ 0,\ 2\frac{y_3+g_2}{g} + \rho\frac{g_2'}{g} - \rho\frac{(y_3+g_2)\,g'}{g^2}\right). \tag{42}$$

## 7.3 Conditions (a3) and (a5)

Conditions (a3) and (a5) can be verified only on surface $\Gamma_1^t$; the situation on $\Gamma_2^t$ is the same. The outer normal vector on surface $\Gamma_1^t$ is $\mathbf{n} = \left(-g_1',\ 0,\ 1\right)/\sqrt{(g_1')^2+1}$. Substituting $y_3 = g_1$ to $\mathbf{a}$ and calculating the product $\mathbf{a}\cdot\mathbf{n}$, we obtain

$$\mathbf{a}\big|_{\Gamma_1^t}\cdot\mathbf{n} = \frac{\dot{\delta}^t}{4}\left(-\frac{\rho}{g},\ 0,\ 2-\frac{\rho g_1'}{g}\right)\cdot\frac{(-g_1',\ 0,\ 1)}{\sqrt{(g_1')^2+1}} = \left(0,\ 0,\ \tfrac{1}{2}\dot{\delta}^t\right)\cdot\mathbf{n} = \mathbf{V}_1^t\cdot\mathbf{n}.$$

Thus, condition (a3) holds. Similarly, using the explicit form of $\mathbf{a}$ and the assumptions on $\delta^t$, we verify condition (a5).

## 7.4 Conditions (a4) and (a7)

The non-zero cylindrical components of $\nabla\mathbf{a}$ are

$$\partial_\rho a_\rho = \frac{\dot{\delta}^t}{4}\left[\frac{1}{g} - \frac{\rho g'}{g^2}\right], \qquad \partial_3 a_3 = \frac{\dot{\delta}^t}{4}\left[\frac{2}{g} - \frac{\rho g'}{g^2}\right],$$

$$\partial_\rho a_3 = \frac{\dot{\delta}^t}{4}\,(y_3+g_2)\left[\frac{3g_2'+\rho g_2''}{(y_3+g_2)\,g} - \frac{3g'+\rho g''}{g^2} - \frac{2\rho g_2' g'}{(y_3+g_2)\,g^2} + \frac{2\rho(g')^2}{g^3}\right].$$

Using these formulas, one can verify that $\mathbf{a}$ and $\nabla\mathbf{a}$ are continuous in $\{(\mathbf{y},t)\in\mathbb{R}^4;\ |t-t^c|<\tau,\ \mathbf{y}\in\Omega_c^t\}\smallsetminus\{(\mathbf{0},t^c)\}$ and the norm $\|\mathbf{a}(\,.\,,t)\|_{1,2;\,\Omega_c^t}$ is integrable with an arbitrary positive power in $(t^c-\tau, t^c+\tau)$. These properties, together with an appropriate extension of $\mathbf{a}$ to $Q_{[0,T]\smallsetminus\{t^c\}}$, imply the validity of conditions (a4) and (a7). Moreover, $\mathbf{a}$ can be extended to $Q_{[0,T]\smallsetminus\{t^c\}}$ so that $\|\mathbf{a}(\,.\,,t)\|_{1,2;\,\Omega^t}$ is integrable with an arbitrary positive power in $(0,T)$ as well.

## 7.5 Condition (a6)

Of inequalities (6) and (7) in condition (a6), we focus on (7). We shall see that inequality (7) induces certain restriction on the size of $\dot{\delta}^t$ in the neighbourhood of the critical time instant $t^c$ of the collision in the case when exponents $\beta_1$ and $\beta_2$ are both equal to two. Thus, let $\boldsymbol{\phi} \in \mathbf{W}^{1,2}_{\sigma}(\Omega^t)$. In accordance with the appointment made in Sect. 7.1, we confine ourselves to the critical sub-domain $\Omega^t_c$ and times $t \in (t^c - \tau, t^c + \tau) \smallsetminus \{t^c\}$. We have

$$\left| \int_{\Omega^t_c} \boldsymbol{\phi} \cdot \nabla \boldsymbol{\phi} \cdot \mathbf{a}(.\,,t) \, d\mathbf{y} \right| \leq \|\nabla \boldsymbol{\phi}\|_{2;\,\Omega^t_c} \left( \int_{\Omega^t_c} |\mathbf{a}(.\,,t)|^2 \, |\boldsymbol{\phi}|^2 \, d\mathbf{y} \right)^{1/2}. \tag{43}$$

One can verify that the decisive contribution to the integral on the right hand side comes from the component of $a_\rho$. Hence, considering only this component, we have

$$\begin{aligned}
\int_{\Omega^t_c} |a_\rho|^2 \, |\boldsymbol{\phi}|^2 \, d\mathbf{y} &= \left(\frac{\dot{\delta}^t}{4}\right)^2 \int_0^r \rho \, d\rho \int_0^{2\pi} d\varphi \int_{-g_2}^{g_1} \frac{\rho^2}{g^2} \left[\phi_\rho^2 + \phi_\varphi^2 + \phi_3^2\right] dy_3 \\
&\leq \left(\frac{\dot{\delta}^t}{4}\right)^2 \int_0^r \frac{\rho^3 \, d\rho}{g^2} \int_0^{2\pi} d\varphi \int_{-g_2}^{g_1} \sum_{j \in \{\rho;\,\varphi;\,3\}} \left[\phi_j(\rho,\varphi,g_1) + \int_{g_1}^{y_3} \partial_3 \phi_j(\rho,\varphi,\xi) \, d\xi\right]^2 dy_3 \\
&\leq \frac{(\dot{\delta}^t)^2}{8} \sum_{j \in \{\rho;\,\varphi;\,3\}} \int_0^r \frac{\rho^3 \, d\rho}{g^2} \int_0^{2\pi} d\varphi \left[g\, \phi_j^2(\rho,\varphi,g_1) + g^2 \int_{-g_2}^{g_1} \left[\partial_3 \phi_j(\rho,\varphi,\xi)\right]^2 d\xi\right] \\
&\leq \frac{(\dot{\delta}^t)^2}{8} \sum_{j \in \{\rho;\,\varphi;\,3\}} \int_0^r \rho^3 \, d\rho \int_0^{2\pi} d\varphi \left[\frac{\phi_j^2(\rho,\varphi,g_1)}{g} + \int_{-g_2}^{g_1} \left[\partial_3 \phi_j(\rho,\varphi,\xi)\right]^2 d\xi\right].
\end{aligned}$$

We can assume without loss of generality that $r \leq 1$. Then the last expression is less than or equal to

$$\begin{aligned}
&C\,(\dot{\delta}^t)^2 \, r^2 \left[ r^{-\beta_1} \int_0^r \int_0^{2\pi} d\varphi \, |\boldsymbol{\phi}(\rho,\varphi,g_1)|^2 \, \rho \, d\rho + \int_0^r \int_{-g_2}^{g_1} |\partial_3 \boldsymbol{\phi}(\rho,\varphi,y_3)|^2 \, dy_3 \, \rho \, d\rho \right] \\
&\qquad \leq C\,(\dot{\delta}^t)^2 \, r^{2-\beta_1} \, \|\boldsymbol{\phi}\|^2_{2;\,\Gamma^t} + C\,(\dot{\delta}^t)^2 \, r^2 \, \|\nabla \boldsymbol{\phi}\|^2_{2;\,\Omega^t}.
\end{aligned}$$

Considering also the contribution from the component $a_3$ of function $\mathbf{a}$, we can derive in principle the same inequality. Thus, we have

$$\int_{\Omega^t_c} |\mathbf{a}(.\,,t)|^2 \, |\boldsymbol{\phi}|^2 \, d\mathbf{y} \leq c_6 \, (\dot{\delta}^t)^2 \, r^{2-\beta_1} \, \|\boldsymbol{\phi}\|^2_{2;\,\Gamma^t} + c_7 \, (\dot{\delta}^t)^2 \, r^2 \, \|\nabla \boldsymbol{\phi}\|^2_{2;\,\Omega^t}. \tag{44}$$

where $c_6 = c_6(\kappa_1, \kappa_2)$ and $c_7 = c_7(\kappa_1, \kappa_2)$. Substituting estimate (44) to (43), we obtain

$$\left| \int_{\Omega_c^t} \boldsymbol{\phi} \cdot \nabla \boldsymbol{\phi} \cdot \mathbf{a} \, \mathrm{d}\mathbf{y} \right| \leq |\dot{\delta}^t| \, \|\nabla \boldsymbol{\phi}\|_{2;\,\Omega^t} \left[ c_6 \, r^{2-\beta_1} \, \|\boldsymbol{\phi}\|_{2;\,\Gamma^t}^2 + c_7 \, r^2 \, \|\nabla \boldsymbol{\phi}\|_{2;\,\Omega^t} \right]^{1/2}$$

$$\leq |\dot{\delta}^t| \, \sqrt{c_6 r^{2-\beta_1}} \, \|\nabla \boldsymbol{\phi}\|_{2;\,\Omega^t} \, \|\boldsymbol{\phi}\|_{2;\,\Gamma^t} + |\dot{\delta}^t| \, \sqrt{c_7 r^2} \, \|\nabla \boldsymbol{\phi}\|_{2;\,\Omega^t}^2$$

$$\leq \frac{|\dot{\delta}^t|}{2} \left[ \sqrt{c_6 r^{2-\beta_1}} + \sqrt{c_7 r^2} \right] \|\nabla \boldsymbol{\phi}\|_{2;\,\Omega^t}^2 + \frac{|\dot{\delta}^t|}{2} \, \sqrt{c_6 r^{2-\beta_1}} \, \|\boldsymbol{\phi}\|_{2;\,\Gamma^t}^2. \tag{45}$$

Note that the integral in (7) at times $t$ such that $\tau \leq |t^c - t|$ can be estimated by means of the inequality following from the continuous imbedding $W^{1,2}(\Omega^t) \hookrightarrow L^6(\Omega^t)$. Since the cone parameters in the definition of the cone property of $\Omega^t$ (see [1, p. 66]) are independent of $t$ for $\tau \leq |t^c - t|$, the constant in the imbedding inequality is independent of $t$ as well. Hence we have

$$\left| \int_{\Omega^t} \boldsymbol{\phi} \cdot \nabla \mathbf{a} \cdot \boldsymbol{\phi} \, \mathrm{d}\mathbf{x} \right| \leq \|\nabla \mathbf{a}\|_{2;\,\Omega^t} \, \|\boldsymbol{\phi}\|_{2;\,\Omega^t}^{1/2} \, \|\boldsymbol{\phi}\|_{6;\,\Omega^t}^{3/2}$$

$$\leq C \, \|\nabla \mathbf{a}\|_{2;\,\Omega^t} \, \|\boldsymbol{\phi}\|_{2;\,\Omega^t}^{1/2} \left( \|\boldsymbol{\phi}\|_{2;\,\Omega^t}^{3/2} + \|\nabla \boldsymbol{\phi}\|_{2;\,\Omega^t}^{3/2} \right)$$

$$\leq \frac{\nu}{20} \, \|\nabla \boldsymbol{\phi}\|_{2;\,\Omega^t}^2 + C(\nu) \left( \|\mathbf{a}\|_{2;\,\Omega^t} + \|\mathbf{a}\|_{2;\,\Omega^t}^4 \right) \|\boldsymbol{\phi}\|_{2;\,\Omega^t}^2 \tag{46}$$

at times $t$ such that $\tau < |t - t^c|$, which implies the validity of (7) at these times. The same inequality can also be derived for the part of the integral on $\Omega^t \smallsetminus \Omega_c^t$ at times $t \in (t^c - \tau, t^c) \cup (t^c, t^c + \tau)$. (Then constant $C$ depends not only on $\nu$, but also on $r$.) Thus, considering $0 < |t - t^c| < \tau$ and summing (45) and (46) (with $\Omega^t \smallsetminus \Omega_c^t$ instead of $\Omega^t$ in (46)), we obtain

$$\left| \int_{\Omega^t} \boldsymbol{\phi} \cdot \nabla \mathbf{a} \cdot \boldsymbol{\phi} \, \mathrm{d}\mathbf{x} \right| \leq \frac{|\dot{\delta}^t|}{2} \left[ \sqrt{c_6 r^{2-\beta_1}} + \sqrt{c_7 r^2} \right] \|\nabla \boldsymbol{\phi}\|_{2;\,\Omega^t}^2 + \frac{\nu}{20} \, \|\nabla \boldsymbol{\phi}\|_{2;\,\Omega^t}^2$$

$$+ \frac{|\dot{\delta}^t|}{2} \, \sqrt{c_6 r^{2-\beta_1}} \, \|\boldsymbol{\phi}\|_{2;\,\Gamma^t}^2 + C(\nu, r) \left( \|\mathbf{a}\|_{2;\,\Omega^t} + \|\mathbf{a}\|_{2;\,\Omega^t}^4 \right) \|\boldsymbol{\phi}\|_{2;\,\Omega^t}^2. \tag{47}$$

This inequality implies (7) if

$$\frac{|\dot{\delta}^t|}{2} \left[ \sqrt{c_6 r^{2-\beta_1}} + \sqrt{c_7 r^2} \right] \leq \frac{\nu}{20} \qquad \text{and} \qquad \frac{|\dot{\delta}^t|}{2} \, \sqrt{c_6 r^{2-\beta_1}} \leq \frac{\gamma}{4}. \tag{48}$$

Both the inequalities can be fulfilled in the case $\beta_1 < 2$: it is sufficient to assume that $r$ has been "sufficiently small" from the beginning. In the case $\beta_1 = 2$, however, (48) is fulfilled only for $|\dot{\delta}^t|$ "sufficiently small" for $0 < |t - t^c| < \tau$, where $\tau > 0$ can be in fact chosen to be arbitrarily small. Since $r$ can also be considered to be arbitrarily small, we need $|\dot{\delta}^t|$ to satisfy

$$\frac{|\dot{\delta}^t|}{2} [\sqrt{c_6} < \frac{\nu}{20} \qquad \text{and} \qquad \frac{|\dot{\delta}^t|}{2} \sqrt{c_6} \leq \frac{\gamma}{4}. \tag{49}$$

Summarizing now the results of all the preceding subsections, we obtain:

**Theorem 7.1** *Suppose that domain $\Omega^t$ satisfies conditions (a1) and (a2) with the specifications described at the beginning of Sect. 7, including conditions (a8)–(a10). Suppose that the speed $|\dot{\delta}^t|$ satisfies the additional "conditions of smallness"* (49) *for $0 < |t - t^c| < \tau$ (where $\tau > 0$ can be arbitrarily small) in the case $\beta_1 = 2$. Then the weak solution of the problem* (1)–(5) *exists.*

*Remark 7.2* We recall that the weak solution does not exist in the case $\beta_1 = 2$ and $\dot{\delta}^t$ not approaching zero for $t \to t^c$ if Dirichlet's no-slip boundary condition is used instead of Navier's slip boundary condition (3), see Starovoitov [10]. Thus, boundary condition (3) enables us to consider a larger class of collisions of bodies, moving in the viscous incompressible fluid, than the traditional no-slip boundary condition.

*Remark 7.3* An analogous theorem can also be proven in the case when bodies $\mathcal{B}_1$, $\mathcal{B}_2$ strike with more general surfaces than rotationally symmetric ones, e.g. if their surfaces in the neighbourhood of points of collision coincide with graphs of appropriate concave up, respectively concave down, functions.

**Acknowledgements** The research has been supported by the Grant Agency of the Czech Republic (grant No. 13-00522S), by the Academy of Sciences of the Czech Republic (RVO 67985840) and by the University of Sud Toulon-Var.

# References

1. R. Adams, *Sobolev Spaces* (Academic, New York/San Francisco/London, 1975)
2. E. Feireisl, J. Neustupa, J. Stebel, Convergence of a Brinkman-type penalization for compressible fluid flows. J. Differ. Equ. **250**(1), 596–606 (2011)
3. E. Feireisl, O. Kreml, Š. Nečasová, J. Neustupa, J. Stebel, Weak solutions to the barotropic Navier-Stokes system with slip boundary conditions in time dependent domains. J. Differ. Equ. **254**, 125–140 (2013)
4. H. Fujita, N. Sauer, On existence of weak solutions of the Navier-Stokes equations in regions with moving boundaries. J. Fac. Sci. Univ. Tokyo Sec. 1A **17**, 403–420 (1970)
5. J. Leray, J.L. Lions, Quelques résultats de Višik sur les problèmes elliptiques non linèaires par les méthodes de Minty-Browder. Bull. Soc. Math. France **93**, 97–107 (1965)
6. J.L. Lions, *Quelques Méthodes de Résolution des Problémes aux Limites Non linéaires* (Dunod, Gauthier-Villars, Paris, 1969)
7. J. Neustupa, Existence of a weak solution to the Navier-Stokes equation in a general time-varying domain by the Rothe method. Math. Methods Appl. Sci. **32**, 653–683 (2009)
8. J. Neustupa, P. Penel, The Navier-Stokes equations with Navier's boundary condition around moving bodies at presence of collisions. C. R. Math. **347**, 685–690 (2009)

9. J. Neustupa, P. Penel, A weak solvability of the Navier-Stokes equation with Navier's boundary condition around a ball striking the wall, in *Advances in Mathematical Fluid Mechanics,* ed. by R. Rannacher, A. Sequeira (Springer, Berlin, 2010), pp. 385–408
10. V.N. Starovoitov, Behavior of a rigid body in an incompressible viscous fluid near a boundary. Int. Ser. Num. Math. **147**, 313–327 (2003)
11. R. Temam, *Navier-Stokes Equations* (North-Holland, Amsterdam/New York/Oxford, 1977)

# Effects of Fluid-Boundary Interaction on the Stability of Boundary Layers in Plasma Physics

**Masashi Ohnawa**

*Dedicated to Prof. Yoshihiro Shibata on the occasion of his 60th anniversary*

**Abstract** The present article addresses the asymptotic stability of stationary solutions to a system of the Euler-Poisson equations with an interaction between fluids and boundaries. The system describes motions of positive ions in a boundary layer called a sheath in a plasma flow, formed under the Bohm criterion, which requires the outflow velocity to exceed the ion acoustic velocity. In this model, charged particles accumulate as the plasma arrive at the boundary, which, at the same time affects fluid flows by changing the electric field over the entire domain through the change in the boundary condition. We show that the boundary layer is asymptotically stable under this fluid-boundary interactive setting.

**Keywords** Euler-Poisson system • Fluid-structure interaction • Weighted energy method

**Mathematics Subject Classification (2010).** Primary 74F10; Secondary 35B40, 76K05

## 1 Introduction

We study the initial boundary value problem to the following system in the one dimensional half space $\mathbb{R}_+ := \{x \in \mathbb{R} |\ x > 0\}$:

$$\rho_t + (\rho u)_x = 0, \tag{1a}$$

$$(\rho u)_t + (\rho u u)_x + K\rho_x + \rho\phi_x = 0, \tag{1b}$$

---

M. Ohnawa (✉)
Faculty of Marine Science, Department of Ocean Sciences, Tokyo University of Marine Science and Technology, 108-8477 Tokyo, Japan
e-mail: ohnawa@m.kaiyodai.ac.jp

 
H. Amann et al. (eds.), *Recent Developments of Mathematical Fluid Mechanics*, Advances in Mathematical Fluid Mechanics, DOI 10.1007/978-3-0348-0939-9_21

$$-\phi_{xx} = \rho - e^{\phi}, \tag{1c}$$

$$\phi_{xt}(t,0) = \left(\rho u + e^{\phi} u_e\right)(t,0), \tag{1d}$$

$$\lim_{x\to\infty} \phi(t,x) = 0, \tag{1e}$$

for the initial conditions prescribed as

$$(\rho, u)(0,x) = (\rho_0, u_0)(x)$$
$$\text{with } \inf_{x\in\mathbb{R}_+} \rho_0(x) > 0, \quad \lim_{x\to\infty} (\rho_0, u_0)(x) = (\rho_+, u_+) \in \mathbb{R}^2, \tag{2}$$

$$\phi_x(0,0) = q_0 \in \mathbb{R}, \tag{3}$$

where $\rho_+ > 0$, $u_+ < 0$, $u_e > 0$ and $q_0$ are constants.

This Euler-Poisson system describes the behavior of isothermal flows of positive ions. Unknown functions $\rho$ and $u$ represent the density and the velocity of positive ions respectively, while $\phi$ stands for the electrostatic potential. Positive constants $K$ and $u_e$ correspond to the absolute temperature of ions and the thermal velocity of electrons respectively. To construct a classical solution for the Poisson equation (1c), the condition

$$\rho_+ = 1 \tag{4}$$

is necessary. Equations (1a) and (1b) are the conservation of mass and momentum. Equation (1c) is the Gauss law, where we incorporate the Boltzmann relation which relates the electron density $\rho_e$ to the potential $\phi$ by $\rho_e = e^{\phi}$. A reference point of the potential $\phi$ is set as in (1e).

When a plasma is in contact with materials, a boundary layer called a sheath is formed. As sheath formation process is important for industrial applications, it has been intensively studied in plasma physics since I. Langmuir's works in 1920s. In 1949, D. Bohm derived from physical ansatz that the 'nondegenerate Bohm criterion'

$$u_+^2 > K + 1, \quad u_+ < 0 \tag{5}$$

or the 'degenerate Bohm criterion'

$$u_+^2 = K + 1, \quad u_+ < 0 \tag{6}$$

are necessary for sheath formation. For more physical backgrounds, see a comprehensive review [4]. In [5], the conditions for the existence of stationary solutions are given mathematically. The asymptotic stability of the stationary solution is shown in [2], in which the potential on the boundary is fixed as often seen in laboratory experiments.

Following these works, we define a sheath by a monotone stationary solution to (1)–(3), but to take accumulation of charged particles on the boundary into account, which corresponds to plasma flows outside laboratories, we adopt the boundary condition (1d) from [1]. Since $-\phi_x(t,0)$ is positively proportional to the net quantity of charged particles accumulated on the boundary, its temporal derivative is equal to the net flux of charged particles $[\rho(-u) + (-\rho_e)u_e](t,0)$, which (1d) claims. For the present problem, the stationary problem reads

$$(\tilde{\rho}\tilde{u})_x = 0, \tag{7a}$$

$$(\tilde{\rho}\tilde{u}\tilde{u})_x + K\tilde{\rho}_x + \tilde{\rho}\tilde{\phi}_x = 0, \tag{7b}$$

$$-\tilde{\phi}_{xx} = \tilde{\rho} - e^{\tilde{\phi}} \tag{7c}$$

with conditions corresponding to (1d), (1e) and (2), that is,

$$(\tilde{\rho}\tilde{u} + e^{\tilde{\phi}}u_e)(0) = 0, \tag{8}$$

$$\lim_{x\to\infty}(\tilde{\rho}, \tilde{u}, \tilde{\phi})(x) = (\rho_+, u_+, 0). \tag{9}$$

Note that

$$e^{\tilde{\phi}(0)}u_e = -\rho_+u_+ = |u_+| \quad \text{or} \quad \tilde{\phi}(0) = \log\frac{|u_+|}{u_e} =: \phi_b \tag{10}$$

follows from (4), (7a), (8) and the stationary problem is reduced to the Dirichlet problem. The conditions for the existence of stationary solutions obtained in [5] are summarized as follows.

**Proposition 1.1** *Consider a stationary problem* (7) *and* (9) *with the Dirichlet data*

$$\tilde{\phi}(0) = \phi_D.$$

(i) *When* (5) *is satisfied, the problem has a unique monotone solution if $\phi_D$ is smaller than a certain positive constant. The solution converges to the asymptotic state* (9) *exponentially fast as $x \to \infty$.*
(ii) *When* (6) *is satisfied, the problem has a unique monotone solution if and only if $\phi_D \leq 0$ is satisfied. The solution converges to the asymptotic state* (9) *algebraically fast as $x \to \infty$.*

To investigate the stability of the stationary solution, we define variables for the perturbation $(\psi, \eta, \sigma) = (\log\rho, u, \phi) - (\log\tilde{\rho}, \tilde{u}, \tilde{\phi})$ and subtract (7) from (1) to obtain

$$\psi_t + (\tilde{u} + \eta)\psi_x + \eta_x + \eta\tilde{v}_x = 0, \tag{11a}$$

$$\eta_t + (\tilde{u} + \eta)\eta_x + K\psi_x + \sigma_x + \eta\tilde{u}_x = 0, \tag{11b}$$

$$-\sigma_{xx} = \tilde{\rho}\,(e^{\psi} - 1) - e^{\tilde{\phi}}(e^{\sigma} - 1), \tag{11c}$$

$$\sigma_{xt}(t,0) = \left\{\left(\tilde{\rho} e^{\psi}(\tilde{u} + \eta) - \tilde{\rho}\tilde{u}\right) + |u_+|(e^{\sigma} - 1)\right\}(t,0), \tag{11d}$$

$$\lim_{x\to\infty} \sigma(t,x) = 0, \tag{11e}$$

where (10) is used. The initial conditions to (11) are

$$(\psi,\eta)(0,x) = (\psi_0,\eta_0)(x) := (\log\rho_0 - \log\tilde{\rho}, u_0 - \tilde{u})(x)$$
$$\text{with} \inf_{x\in\mathbb{R}_+} \psi_0(x) > -\infty, \quad \lim_{x\to\infty}(\psi_0,\eta_0)(x) = (0,0), \tag{12}$$

$$\sigma_x(0,0) = r_0 := q_0 - \tilde{\phi}_x(0). \tag{13}$$

One of the main obstacles to the stability analysis is the fact that real parts of all spectra of the system linearized around the asymptotic state are zero, if we consider the problem over the whole space. To overcome this difficulty, [2] makes use of the weighted energy method and proves the asymptotic stability of the stationary solution under the Dirichlet condition for both nondegenerate and degenerate problems. It also shows that the real parts of all spectra of exponentially weighted variables with a suitable index are negative if and only if the nondegenerate Bohm criterion (5) is fulfilled.

Another difficulty comes from the interaction between the fluid and the boundary. These kinds of problems have been actively studied in the last two decades especially in the numerical computation society. More familiar situations include interactions between viscous fluid flows and motions of the rigid bodies, elastic plates or deformable objects as bubbles, wherein fluids exert pressure or stresses on the bodies while the displacement or the deformation of the bodies affect the surrounding fluids. The existence of local solutions is not trivial for these settings. For the current problem, time dependent Neumann conditions makes the problem more difficult compared to that under the Dirichlet condition. To the best of the author's knowledge, this is the first mathematical work concerning the Euler-Poisson system with fluid-boundary interactions. Here are our main results.

**Theorem 1.2 (Nondegenerate Problem)** *Assume the condition* (5). *Suppose* $(1 + \beta_0 x)^{\lambda/2}\psi_0, (1+\beta_0 x)^{\lambda/2}\eta_0 \in H^2(\mathbb{R}_+)$ *for certain constants* $\lambda \geq 2$ *and* $\beta_0 > 0$. *Then the following statements hold:*

≪ *For an arbitrary* $\alpha \in (0,\lambda]$, *there exists a positive constant* $\delta$ *such that if*

$$|u_e/|u_+| - 1| + \|((1+\beta_0 x)^{\lambda/2}\psi_0, (1+\beta_0 x)^{\lambda/2}\eta_0)\|_{H^2} + |r_0| \leq \delta \tag{14}$$

*is satisfied, the initial boundary value problem* (11)–(13) *uniquely has a global solution* $(\psi,\eta,\sigma)$ *which verifies* $((1+\beta x)^{\alpha/2}\psi, (1+\beta x)^{\alpha/2}\eta, (1+\beta x)^{\alpha/2}\sigma) \in (\mathfrak{X}_2^0([0,\infty)))^2 \times \mathfrak{X}_2^2([0,\infty))$ *for a certain positive constant* $\beta$. ≫

*Moreover, it exhibits an algebraic decay*

$$\|((1+\beta x)^{\alpha/2}\psi,(1+\beta x)^{\alpha/2}\eta)\|_{H^2}^2(t)+\|(1+\beta x)^{\alpha/2}\sigma\|_{H^4}^2$$
$$\leq C(\|((1+\beta x)^{\lambda/2}\psi_0,(1+\beta x)^{\lambda/2}\eta_0)\|_{H^2}^2+r_0^2)\,(1+\beta t)^{-(\lambda-\alpha)}. \qquad (15)$$

**Theorem 1.3 (Degenerate Problem)** *Assume the condition* (6). *Let* $\lambda_0$ *be the unique real solution to the equation* $\lambda_0(\lambda_0-1)(\lambda_0-2)-12(\lambda_0+2)=0$ $(\lambda_0=5.5693\cdots)$ *and suppose* $(1+\beta_0 x)^{\lambda/2}\psi_0,(1+\beta_0 x)^{\lambda/2}\eta_0\in H^2(\mathbb{R}_+)$ *holds for certain constants* $\lambda\in[4,\lambda_0)$ *and* $\beta_0>0$. *Then the statements bracketed in Theorem 1.2 hold. Moreover, it exhibits an algebraic decay*

$$\|((1+\beta x)^{\alpha/2}\psi,(1+\beta x)^{\alpha/2}\eta)\|_{H^2}^2(t)+\|(1+\beta x)^{\alpha/2}\sigma\|_{H^4}^2(t)$$
$$\leq C\left(\|((1+\beta x)^{\lambda/2}\psi_0,(1+\beta x)^{\lambda/2}\eta_0)\|_{H^2}^2+r_0^2\right)(1+\beta t)^{-(\lambda-\alpha)/3}. \qquad (16)$$

**Notations** For a nonnegative integer $i$, we denote by $H^i(\mathbb{R}_+)$ the $i$-th order Sobolev space in the $L^2$ sense, equipped with the canonical norm $\|\cdot\|_{H^i}$. For arbitrary nonnegative integers $i$ and $j$, a function space $\mathfrak{X}_i^j$ is defined by

$$\mathfrak{X}_i^j([0,T]):=\bigcap_{k=0}^{i}C^k([0,T];H^{j+i-k}(\mathbb{R}_+)).$$

For arbitrary positive constants $\alpha$, $\beta$ and a nonnegative integer $i$, we denote an algebraically weighted norm by

$$\|f\|_{\alpha,\beta,i}(t):=\left(\int_{\mathbb{R}_+}(1+\beta x)^{\alpha}\sum_{0\leq j\leq i}(\partial_x^j f)^2(t,x)dx\right)^{1/2}.$$

In the next section, we present key ideas for proving main theorems. Readers interested in details are referred to [3], where isentropic cases are also treated. The proof of Theorem 1.3 is basically the same as Theorem 1.2 so we focus on the proof of Theorem 1.2. Necessary modifications for the degenerate problem are seen in [2].

## 2 Outline of the Proof

### 2.1 Local Solvability

In this subsection, the local solvability of the problem (11)–(13) is established. First, we discuss the unique existence of the solution to the nonlinear elliptic equation under a fixed Neumann boundary condition.

**Lemma 2.1** *Let $(\tilde{\rho}, \tilde{u}, \tilde{\phi})$ be a solution to (7)–(9). For a function $f \in L^2 \cap L^\infty(\mathbb{R}_+)$ with $\operatorname{ess\,inf}_{x\in\mathbb{R}_+}(f + \tilde{\rho})(x) > 0$ and an arbitrary $r \in \mathbb{R}$, the ordinary differential equation*

$$-\theta''(x) + e^{\tilde{\phi}}(e^{\theta} - 1)(x) = f(x), \quad x > 0, \tag{17a}$$

$$\theta'(0) = r, \tag{17b}$$

$$\lim_{x\to\infty} \theta(x) = 0, \tag{17c}$$

*has a solution $\theta$ uniquely in $\theta \in H^2(\mathbb{R}_+)$. This solution defines the Neumann-to-Dirichlet map $\mathfrak{D} : (f, \theta'(0)) \mapsto \theta(0)$.*

*Proof* Apply the Leray-Schauder fixed point theorem with the aid of truncation method by Stampacchia. □

Making use of this result, we obtain the local solution to the hyperbolic-elliptic coupled system with fluid-boundary interactions as follows.

**Proposition 2.2** *Suppose the initial data $(\psi_0, \eta_0)$ satisfies*

$$(\tilde{u} + \eta_0)(0) + \sqrt{K} < 0 \tag{18}$$

*and $(1 + \beta_0 x)^{\lambda/2}\psi_0(x)$, $(1 + \beta_0 x)^{\lambda/2}\eta_0(x) \in H^2(\mathbb{R}_+)$ for certain positive constants $\lambda$ and $\beta_0$ with $\lambda\beta_0 \ll 1$. Then there exists a constant $T > 0$ such that the initial boundary value problem (11), (12) and (13) admits a unique solution $(\psi, \eta, \sigma)$ over $t \in [0, T]$ satisfying*

$$(1 + \beta_0 x)^{\lambda/2}\psi, (1 + \beta_0 x)^{\lambda/2}\eta \in \mathfrak{X}_2^0([0, T]), \ (1 + \beta_0 x)^{\lambda/2}\sigma \in \mathfrak{X}_2^2([0, T]), \tag{19a}$$

$$\tilde{u}(0) + \eta(t, 0) + \sqrt{K} < 0, \tag{19b}$$

$$\inf_{x\in\mathbb{R}_+} \psi(t, x) > -\infty. \tag{19c}$$

*Proof* We construct the local solution to (11), (12) and (13) iteratively. To avoid inessential complexity in the argument, let us construct the solution in the non-weighted Sobolev space.

First, we set $(\psi^{(0)}, \eta^{(0)})(t, x) = (\psi_0, \eta_0)(x) \in (H^2(\mathbb{R}_+))^2$ for $t \geq 0$. Assuming $\psi^{(n)}, \eta^{(n)} \in \mathfrak{X}_1^1([0, T_0])$, $T_0 > 0$ for a nonnegative integer $n$, $(n + 1)$-th iteration consists of three steps as follows:

*Step 1* Set $(v^{(n)}, u^{(n)})(t, x) = (\psi^{(n)}, \eta^{(n)})(t, x) + (\tilde{v}, \tilde{u})(x)$, $\rho^{(n)}(t, x) = \exp v^{(n)}(t, x)$ and solve the ordinary differential equation

$$\begin{aligned}\frac{d}{dt}[\sigma_x]_{x=0}^{(n)}(t) &= [\rho^{(n)}u^{(n)} - \tilde{\rho}\tilde{u}]_{x=0}(t) + |u_+| \left\{\exp\left([\sigma]_{x=0}^{(n)}(t)\right) - 1\right\} \\ &= [\rho^{(n)}u^{(n)} - \tilde{\rho}\tilde{u}]_{x=0}(t)\end{aligned}$$

$$
\begin{aligned}
&\quad + |u_+| \left\{\exp\left(\mathfrak{D}\left(\rho^{(n)}(t,\cdot) - \tilde{\rho}(\cdot), [\sigma_x]_{x=0}^{(n)}(t)\right)\right) - 1\right\} \\
&=: F^{(n)}(t, [\sigma_x]_{x=0}^{(n)}(t)), \\
[\sigma_x]_{x=0}^{(n)}(0) &= r_0 \qquad (20)
\end{aligned}
$$

for $[\sigma_x]_{x=0}^{(n)}(t)$, where $\mathfrak{D}$ is the Neumann-to-Dirichlet map defined in Lemma 2.1. It is easy to check the Lipschitz continuity of $F^{(n)}$ in (20) and we have $[\sigma_x]_{x=0}^{(n)}(t)$ over a certain positive interval $[0, T_1]$.

*Step 2* For each $t \in [0, T_1]$, solve the nonlinear elliptic equation

$$
-\sigma_{xx}^{(n)}(t,x) + e^{\tilde{\phi}(x)}\left(\exp \sigma^{(n)}(t,x) - 1\right) = \rho^{(n)}(t,x) - \tilde{\rho}(x), \quad x \in \mathbb{R}_+,
$$
$$
\lim_{x\to\infty} \sigma^{(n)}(t,x) = 0
$$

for $\sigma^{(n)}(t,x)$ under the Neumann boundary condition obtained in Step 1. The solvability is assured by Lemma 2.1.

*Step 3* We solve the linear hyperbolic system for $\omega := (\sqrt{K}\psi, \eta)^{(n+1)}(t,x)$

$$
\omega_t + \begin{pmatrix} u^{(n)} & \sqrt{K} \\ \sqrt{K} & u^{(n)} \end{pmatrix} \omega_x + \eta^{(n+1)} \begin{pmatrix} \sqrt{K}\tilde{v}_x \\ \tilde{u}_x \end{pmatrix} + \begin{pmatrix} 0 \\ \sigma_x^{(n)} \end{pmatrix} = 0,
$$
$$
(\psi, \eta)^{(n+1)}(0,x) = (\psi_0, \eta_0)(x)
$$

over $[0, T_1] \times \mathbb{R}_+$ by applying the Kato theory noting (18).

The convergence of $\{(\psi^{(n)}, \eta^{(n)}, \sigma^{(n)})\}_{n=0}^{\infty}$ when restricted on a certain positive interval $[0, T]$ is shown by an energy method. □

## 2.2 A-Priori Estimates

Together with Proposition 2.2, a-priori estimates below complete the proof of Theorem 1.2.

**Proposition 2.3** *Under* (5), *suppose* $(\psi, \eta, \sigma)$ *is a solution to* (11), (12) *and* (13) *satisfying* (19), *with* $\lambda \geq 2$ *and* $0 < \beta_0 \ll 1$. *Then for an arbitrary* $\alpha \in (0, \lambda]$, *there exists a positive constant* $\delta_0$ *independent of* $T$, *such that if*

$$
|\phi_b| + \sup_{t\in[0,T]} \left(\|((1+\beta_0 x)\psi(t,x), (1+\beta_0 x)\eta(t,x))\|_{H^2}(t) + |\sigma_x(t,0)|\right) \leq \delta_0,
$$

*with $\phi_b$ defined in* (10) *is satisfied, the solution verifies*

$$\|(1+\beta x)^{\alpha/2}\psi\|_{H^2}^2(t) + \|(1+\beta x)^{\alpha/2}\eta\|_{H^2}^2(t) + \|(1+\beta x)^{\alpha/2}\sigma\|_{H^4}^2(t)$$
$$\le C(\|(1+\beta x)^{\lambda/2}\psi_0\|_{H^2}^2 + \|(1+\beta x)^{\lambda/2}\eta_0\|_{H^2}^2 + r_0^2)\,(1+\beta t)^{-(\lambda-\alpha)},$$

*for $t \in [0,T]$, where $C$ and $\beta(\le \beta_0)$ are certain positive constants.*

In the following two lemmata, we show only basic estimates to obtain Proposition 2.3. Higher order derivatives are estimated similarly.

**Lemma 2.4** *Assume the same conditions as in Proposition 2.3. Let $N_{\beta_0}(T) := \sup_{0\le t\le T}\left(\|(\psi,\eta)\|_{2,\beta_0,2}(t) + |\sigma_x(t,0)|\right)$.*

(i) *For an arbitrary $\alpha \in (0,\lambda]$, there exists a positive constant $\varepsilon_1$ independent of $T$ such that if*

$$0 < \beta_1 \le \varepsilon_1 \ \text{and}\ |\phi_b| + N_{\beta_0}(T) \le \varepsilon_1\beta_1 \tag{21}$$

*are satisfied, it holds for $t \in [0,T]$ that*

$$\frac{d}{dt}\int_{\mathbb{R}} W_1E_1dx + c_1\beta_1 \,\|\, (\psi,\eta,\sigma)\,\|_{\alpha-1,\beta_1,1}^2 \le C_1\Big[\sigma^2+\sigma_x^2\Big]_{x=0} \tag{22}$$

*for certain positive constants $c_1$ and $C_1$, where $W_1(x) := (1+\beta_1 x)^\alpha$ and*

$$E_1(t,x) := \frac{\rho}{2}\left\{e^{\tilde\phi}\left(K\psi^2+\eta^2\right) + K\psi_x^2 + \eta_x^2 + 2\tilde\rho(e^\psi - 1 - \psi)\right\}(t,x).$$

(ii) *For an arbitrary $\alpha \in (0,\lambda]$, there exists a positive constant $\varepsilon_2$ independent of $T$ such that if*

$$0 < \beta_2 \le \varepsilon_2 \ \text{and}\ |\phi_b| + N_{\beta_0}(T) \le \varepsilon_2\beta_2 \tag{23}$$

*are satisfied, it holds for $t \in [0,T]$ that*

$$\frac{d}{dt}\left(\int_{\mathbb{R}} W_2E_2dx + c_0\beta_2\Big[e^\sigma - 1 - \sigma\Big]_{x=0}\right) + c_2\Big[\psi^2+\eta^2+\sigma^2+\sigma_x^2\Big]_{x=0}$$
$$\le C_2\beta_2\|(\psi,\eta)\|_{\alpha-1,\beta_2,1}^2 \tag{24}$$

*for certain positive constants $c_0$, $c_2$ and $C_2$, where $W_2(x) := (1+\beta_2 x)^\alpha$ and*

$$E_2(t,x) := \frac{\rho}{2}\left(K\psi + \eta^2\right) + \frac{1}{2}\left(e^\phi\sigma^2 + \sigma_x^2\right)(t,x).$$

*Proof* Noting

$$\|\sigma\|_{\alpha,\beta}^2 + c\|\sigma_x\|_{\alpha,\beta}^2 \leq \big(1 + C(N_{\beta_0}(T) + |\phi_b|)\big)\|\psi\|_{\alpha,\beta}^2 + C\big[\sigma_x^2\big]_{x=0}, \tag{25}$$

$$c\|\sigma_t\|_{\alpha,\beta,1}^2 + \frac{d}{dt}\Big[e^{\sigma} - 1 - \sigma\Big]_{x=0}(t) \leq C\|(\psi,\eta)\|_{\alpha,\beta,1}^2, \tag{26}$$

hold if $|\phi_b| + N_{\beta_0}(T)$ is sufficiently small, the derivations are straightforward. □

**Lemma 2.5** *Assume the same conditions as in Lemma 2.4. If*

$$|\phi_b| + N_{\beta_0}(T) \leq \min\{\varepsilon_1^2, \varepsilon_2^2, \tilde{\gamma}\varepsilon_1\varepsilon_2\}, \quad \textit{where } \tilde{\gamma} := \frac{1}{2}\min\left\{1, \frac{c_1c_2}{4C_1C_2}\right\} \tag{27}$$

*is satisfied, there exist positive constants* $\beta_1, \beta_2 (\leq \beta_0)$ *which satisfy* (21), (23) *and*

$$\beta_2/\beta_1 \leq 2\tilde{\gamma}.$$

*Here* $\varepsilon_j, c_j, C_j (j = 1, 2)$ *are positive constants appeared in the statements of Lemma 2.4. Moreover, defining* $W_0(t) = (1 + \beta_1 t)^{\zeta}$ *for* $\zeta \geq 0$*, it holds*

$$\begin{aligned} &W_0(t)\,\|(\psi,\eta,\sigma)\|_{\alpha,\beta_1,1}^2(t) \\ &+ \beta_1\int_0^t W_0(s)\|(\psi,\eta,\sigma)\|_{\alpha-1,\beta_1,1}^2(s)ds + \int_0^t W_0(s)\Big[\psi^2 + \eta^2 + \sigma^2\Big]_{x=0}(s)ds \\ &\leq C\Big(\|(\psi_0,\eta_0)\|_{\alpha,\beta_1,1}^2 + r_0^2\Big) + C\int_0^t W_0'(s)\,\|(\psi,\eta,\sigma)\|_{\alpha,\beta_1,1}^2(s)ds, \quad t \in [0,T]. \end{aligned}$$

*Proof* Finding desired $\beta_1$ and $\beta_2$ under (27) is easy. The estimate (22) has an integral term on the left while having a boundary term on the right. The estimate (24) has an opposite property. By setting $1 \gg \gamma \gg \gamma\beta_1 \gg \beta_2$ roughly speaking, multiplication of (22) by $\gamma$ and addition to (24) followed by integration over $[0, t]$ with a weight of $W_0(t)$ yields the estimate above. □

## Comparison of the Current Problem to the Dirichlet Problem [2]

The most essential difference is seen in the appearance of boundary terms on the right hand sides of (15) and (16). These terms are traced back to (25) and (26). For the one dimensional Dirichlet problem, estimate (22) alone was enough to establish the basic estimate because no boundary terms appeared. In the Dirichlet problem, estimate (24) was used only for the multidimensional case but without the necessity to introduce secondary weight function as in Lemma 2.5. The linear stability analysis in [3] reveals that the fluid-boundary interaction does not act against stabilization, but its good effect is limited only on the boundary and does not stabilize the system in the interior domain.

## References

1. J.W. Cipolla, M.B. Silevitch, On the temporal development of a plasma sheath. J. Plasma Phys. **25**, 373–389 (1981)
2. S. Nishibata, M. Ohnawa, M. Suzuki, Asymptotic stability of boundary layers to the Euler-Poisson equations arising in plasma physics. SIAM J. Math. Anal. **44**, 761–790 (2012)
3. M. Ohnawa, Asymptotic stability of plasma boundary layers to the Euler-Poisson equations with fluid-boundary interaction. SIAM J. Math. Anal. **47**, 2795–2831 (2015)
4. K.-U. Riemann, The Bohm criterion and sheath formation. J. Phys. D. Appl. Phys. **24**, 493–518 (1991)
5. M. Suzuki, Asymptotic stability of stationary solutions to the Euler-Poisson equations arising in plasma physics. Kinet. Relat. Model. **4**, 569–588 (2011)

# On Incompressible Two-Phase Flows with Phase Transitions and Variable Surface Tension

**Jan Prüss, Senjo Shimizu, Gieri Simonett, and Mathias Wilke**

*Dedicated to Professor Yoshihro Shibata on the occasion of his 60th anniversary*

**Abstract** Our study of the basic model for incompressible two-phase flows with phase transitions consistent with thermodynamics (Prüss et al., Evol Equ Control Theory 1:171–194, 2012; Prüss and Shimizu, J Evol Equ 12:917–941, 2012; Prüss et al., Commun Part Differ Equ 39:1236–1283, 2014; see also Prüss et al., Interfaces Free Bound 15:405–428, 2013) is extended to the case of temperature-dependent surface tension. We prove well-posedness in an $L_p$-setting, study the stability of the equilibria of the problem, and show that a solution which does not develop singularities exists globally, and if its limit set contains a stable equilibrium it converges to this equilibrium in the natural state manifold for the problem as time goes to infinity.

**Keywords** Compactness • Convergence to equilibria • Entropy • Generalized principle of linearized stability • Kinetic undercooling • Marangoni forces • Phase transitions • Semiflow • Stability • Surface tension • Two-phase Navier-Stokes equations

## 1 Introduction

Let $\Omega \subset \mathbb{R}^n$ be a bounded domain of class $C^3$, $n \geq 2$. $\Omega$ contains two phases: at time $t$, phase $i$ occupies subdomain $\Omega_i(t)$ of $\Omega$. Assume $\partial\Omega_1(t) \cap \partial\Omega = \emptyset$; this means no *boundary intersection*. The closed compact hypersurface

J. Prüss (✉) • M. Wilke
Institut für Mathematik, Martin-Luther-Universität Halle-Wittenberg, 06099 Halle, Germany
e-mail: jan.pruess@mathematik.uni-halle.de; mathias.wilke@mathematik.uni-halle.de

S. Shimizu
Graduate School of Human and Environmental Studies, Kyoto University, Kyoto 606-8501, Japan
e-mail: shimizu.senjo.5s@kyoto-u.ac.jp

G. Simonett
Department of Mathematics, Vanderbilt University, Nashville, TN 37240, USA
e-mail: gieri.simonett@vanderbilt.edu



H. Amann et al. (eds.), *Recent Developments of Mathematical Fluid Mechanics*, Advances in Mathematical Fluid Mechanics, DOI 10.1007/978-3-0348-0939-9_22

$\Gamma(t) := \partial\Omega_1(t) \subset \Omega$ forms the interface between the phases, which is allowed to be disconnected.

We let $u$ denote the velocity field, $\pi$ the pressure field, $T$ the stress tensor, $D(u) = (\nabla u + [\nabla u]^{\mathsf{T}})/2$ the rate of strain tensor, $\theta$ the (absolute) temperature field, $\nu_\Gamma$ the outer normal of $\Omega_1$, $u_\Gamma$ the velocity field of the interface, $V_\Gamma = u_\Gamma \cdot \nu_\Gamma$ the normal velocity of $\Gamma(t)$, $H_\Gamma = H(\Gamma(t)) = -\mathrm{div}_\Gamma \nu_\Gamma$ the curvature of $\Gamma(t)$, $j_\Gamma$ the phase flux, and $[\![v]\!] = v_2 - v_1$ the jump of the quantities $v_i \in C(\overline{\Omega}_i(t))$ across $\Gamma(t)$.

Several quantities are derived from the specific *free energy densities* $\psi_i(\theta)$ as follows. $\epsilon_i(\theta) := \psi_i(\theta) + \theta\eta_i(\theta)$ means the specific internal energy density in phase $i$, $\eta_i(\theta) = -\psi_i'(\theta)$ the density of the specific entropy, $\kappa_i(\theta) := \epsilon_i'(\theta) > 0$ the heat capacity, and $l(\theta) = \theta[\![\psi'(\theta)]\!] = -\theta[\![\eta(\theta)]\!]$ the latent heat. Further $d_i(\theta) > 0$ denotes the coefficient of heat conduction in Fourier's law, $\mu_i(\theta) > 0$ the viscosity in Newton's law, $\rho_1, \rho_2 > 0$ the constant, positive densities of the phases, and $\sigma > 0$ the (coefficient of) surface tension. In the sequel we drop the index $i$, as there is no danger of confusion; we just keep in mind that the coefficients depend on the phases.

In the previous papers [14, 15] we have mathematically analyzed the following problem with sharp interface:

Find a family of closed compact hypersurfaces $\{\Gamma(t)\}_{t\geq 0}$ contained in $\Omega$ and appropriately smooth functions $u : \mathbb{R}_+ \times \bar{\Omega} \to \mathbb{R}^n$, and $\pi, \theta : \mathbb{R}_+ \times \bar{\Omega} \to \mathbb{R}$ such that

$$\begin{aligned}
\rho(\partial_t u + u\cdot\nabla u) - \operatorname{div} T &= 0 && \text{in } \Omega\setminus\Gamma(t),\\
T = 2\mu(\theta)D(u) - \pi I, \quad \operatorname{div} u &= 0 && \text{in } \Omega\setminus\Gamma(t),\\
[\![\tfrac{1}{\rho}]\!] j_\Gamma^2 \nu_\Gamma - [\![T\nu_\Gamma]\!] = \sigma H_\Gamma \nu_\Gamma, \quad [\![u]\!] &= [\![\tfrac{1}{\rho}]\!] j_\Gamma \nu_\Gamma && \text{on } \Gamma(t),\\
u = 0 \text{ on } \partial\Omega, \quad u(0) &= u_0 && \text{in } \Omega.
\end{aligned} \tag{1}$$

$$\begin{aligned}
\rho\kappa(\theta)(\partial_t\theta + u\cdot\nabla\theta) - \operatorname{div}(d(\theta)\nabla\theta) - 2\mu|D(u)|_2^2 &= 0 && \text{in } \Omega\setminus\Gamma(t),\\
-l(\theta)j_\Gamma - [\![d(\theta)\partial_{\nu_\Gamma}\theta]\!] = 0, \quad [\![\theta]\!] &= 0 && \text{on } \Gamma(t),\\
\partial_\nu\theta = 0 \text{ on } \partial\Omega, \quad \theta(0) &= \theta_0 && \text{in } \Omega.
\end{aligned} \tag{2}$$

$$\begin{aligned}
[\![\psi(\theta)]\!] + [\![\tfrac{1}{2\rho^2}]\!] j_\Gamma^2 - [\![\tfrac{T\nu_\Gamma\cdot\nu_\Gamma}{\rho}]\!] &= 0 && \text{on } \Gamma(t),\\
V_\Gamma = u_\Gamma\cdot\nu_\Gamma &= u\cdot\nu_\Gamma - \tfrac{1}{\rho} j_\Gamma && \text{on } \Gamma(t),\\
\Gamma(0) &= \Gamma_0.
\end{aligned} \tag{3}$$

This model is explained in more detail in [12]; see also [1, 6]. It is thermodynamically consistent in the sense that in absence of exterior forces and heat sources, the total mass and the total energy are preserved, and the total entropy is nondecreasing.

This model is in some sense the simplest non-trivial sharp interface model for incompressible Newtonian two-phase flows taking into account phase transitions driven by temperature.

Note that in this model neither *kinetic undercooling* nor *temperature dependence of the surface tension* $\sigma$, i.e. *Marangoni forces*, have been taken into account. It is the aim of the present paper to remove these shortcomings. Here we concentrate on the case of constant densities $\rho_i > 0$ which are not equal, i.e. $[\![\rho]\!] \neq 0$. To achieve this goal, the model has to be adjusted carefully. Surface tension $\sigma(\theta_\Gamma)$ here is precisely the *free surface energy density*. Therefore we define, in analogy to the bulk case, the *surface energy density* $\epsilon_\Gamma$, the *surface entropy density* $\eta_\Gamma$, the *surface heat capacity* $\kappa_\Gamma$, and *surface latent heat* $l_\Gamma$ by means of the relations

$$\epsilon_\Gamma(\theta_\Gamma) = \sigma(\theta_\Gamma) + \theta_\Gamma \eta_\Gamma(\theta_\Gamma), \quad \eta_\Gamma(\theta_\Gamma) = -\sigma'(\theta_\Gamma)$$
$$\kappa_\Gamma(\theta_\Gamma) = \epsilon'_\Gamma(\theta_\Gamma) = -\theta_\Gamma \sigma''(\theta_\Gamma), \quad l_\Gamma(\theta_\Gamma) = \theta_\Gamma \sigma'(\theta_\Gamma).$$

Then *total surface energy* will be

$$\mathsf{E}_\Gamma = \int_\Gamma \epsilon_\Gamma(\theta_\Gamma)d\Gamma,$$

and *total surface entropy* reads

$$\Phi_\Gamma = \int_\Gamma \eta_\Gamma(\theta_\Gamma)d\Gamma.$$

Note that in case $\sigma = const$ we have $\mathsf{E}_\Gamma = \sigma|\Gamma|$ and $\Phi_\Gamma = 0$. We point out that experiments have shown that in certain situations surface heat capacity cannot be neglected, see [2]. In case $\kappa_\Gamma$ is not identically zero (i.e. if $\sigma$ is not a linear function of $\theta_\Gamma$), there will also be a non-trivial *surface heat flux* $q_\Gamma$ which we assume to satisfy Fourier's law, that is

$$q_\Gamma = -d_\Gamma(\theta_\Gamma)\nabla_\Gamma\theta_\Gamma,$$

with coefficient of surface heat diffusivity $d_\Gamma(\theta_\Gamma) > 0$. In analogy to the bulk, the surface heat flux induces the contribution $\int_\Gamma -(q_\Gamma \cdot \nabla_\Gamma)/\theta_\Gamma^2\, d\Gamma$ to the production of surface entropy. *Kinetic undercooling* with coefficient $\gamma(\theta_\Gamma) > 0$ produces surface entropy $\int_\Gamma \gamma(\theta) j_\Gamma^2/\theta_\Gamma\, d\Gamma$. Following the derivation in [12], this leads to the following three modifications of the system (1)–(3). The momentum balance on the interface becomes

$$[\![\frac{1}{\rho}]\!] j_\Gamma^2 \nu_\Gamma - [\![T\nu_\Gamma]\!] = \sigma(\theta_\Gamma)H_\Gamma\nu_\Gamma + \sigma'(\theta_\Gamma)\nabla_\Gamma\theta_\Gamma.$$

The energy balance on the interface reads

$$\kappa_\Gamma(\theta_\Gamma)\frac{D}{Dt}\theta_\Gamma + \mathrm{div}_\Gamma q_\Gamma = [\![d(\theta)\partial_{\nu_\Gamma}\theta]\!] + l(\theta)j_\Gamma + l_\Gamma(\theta_\Gamma)\mathrm{div}_\Gamma u_\Gamma + \gamma(\theta_\Gamma)j_\Gamma^2,$$

and the Gibbs-Thomson law changes to

$$[\![\psi(\theta)]\!] + [\![\frac{1}{2\rho^2}]\!]j_\Gamma^2 - [\![\frac{T\nu_\Gamma\cdot\nu_\Gamma}{\rho}]\!] = -\gamma(\theta_\Gamma)j_\Gamma.$$

Here $D/Dt$ denotes the Lagrangian derivative coming from the velocity $u_\Gamma$ of the interface, and we employ the symbol $P_\Gamma$ for the orthogonal projection onto the tangent bundle of $\Gamma$. Note that $\theta_\Gamma = \theta|_\Gamma$ is the trace of $\theta$ on $\Gamma$ as $[\![\theta]\!] = 0$. We assume the tangential part of $u$ to be continuous across $\Gamma$, i.e. $[\![P_\Gamma u]\!] = 0$, and $P_\Gamma u_\Gamma = P_\Gamma u|_\Gamma$. Then the quantities $j_\Gamma$, $V_\Gamma$ and $u_\Gamma$ can be expressed as

$$j_\Gamma = [\![u\cdot\nu_\Gamma]\!]/[\![1/\rho]\!], \quad V_\Gamma = [\![\rho u\cdot\nu_\Gamma]\!]/[\![\rho]\!], \quad u_\Gamma = P_\Gamma u + V_\Gamma\nu_\Gamma.$$

The complete extended model now reads as follows.

In the bulk $\Omega\setminus\Gamma(t)$:

$$\begin{aligned} \rho(\partial_t u + u\cdot\nabla u) - 2\mathrm{div}(\mu(\theta)D(u)) + \nabla\pi &= 0,\\ 2D(u) = \nabla u + [\nabla u]^{\mathsf{T}}, \quad \mathrm{div}\,u &= 0, \\ \rho\kappa(\theta)(\partial_t\theta + u\cdot\nabla\theta) - \mathrm{div}(d(\theta)\nabla\theta) &= 2\mu(\theta)|D(u)|_2^2. \end{aligned} \tag{4}$$

On the interface $\Gamma(t)$:

$$\begin{aligned} &[\![P_\Gamma u]\!] = 0,\ P_\Gamma u_\Gamma = P_\Gamma u|_\Gamma,\ [\![u\cdot\nu_\Gamma]\!] = [\![1/\rho]\!]j_\Gamma,\ [\![\theta]\!] = 0,\ \theta_\Gamma = \theta|_\Gamma,\\ &[\![1/\rho]\!]j_\Gamma^2\nu_\Gamma - 2[\![\mu(\theta)D(u)\nu_\Gamma]\!] + [\![\pi]\!]\nu_\Gamma = \sigma(\theta_\Gamma)H_\Gamma\nu_\Gamma + \sigma'(\theta_\Gamma)\nabla_\Gamma\theta_\Gamma,\\ &\kappa_\Gamma\frac{D}{Dt}\theta_\Gamma - \mathrm{div}_\Gamma(d_\Gamma(\theta_\Gamma)\nabla_\Gamma\theta_\Gamma) = \\ &\qquad = [\![d(\theta)\partial_\nu\theta]\!] + l(\theta)j_\Gamma + \gamma(\theta_\Gamma)j_\Gamma^2 + l_\Gamma(\theta_\Gamma)\mathrm{div}_\Gamma u_\Gamma,\\ &[\![\psi(\theta)]\!] + [\![1/2\rho^2]\!]j_\Gamma^2 - 2[\![\mu(\theta)D(u)\nu_\Gamma\cdot\nu_\Gamma/\rho]\!] + [\![\pi/\rho]\!] = -\gamma(\theta_\Gamma)j_\Gamma,\\ &V_\Gamma = u\cdot\nu_\Gamma - j_\Gamma/\rho. \end{aligned} \tag{5}$$

On the outer boundary $\partial\Omega$:

$$u = 0, \quad \partial_\nu\theta = 0.$$

Initial conditions:

$$\Gamma(0) = \Gamma_0, \quad u(0) = u_0, \quad \theta(0) = \theta_0.$$

This model has conservation of total mass and total energy, and total entropy is non-decreasing. Indeed, along smooth solutions we have

$$\begin{aligned}\frac{d}{dt}(\Phi_b(t) + \Phi_\Gamma(t)) &= \int_\Omega [2\mu(\theta)|D(u)|_2^2/\theta + d(\theta)|\nabla\theta|^2/\theta^2]dx \\ &+ \int_\Gamma [d_\Gamma(\theta_\Gamma)|\nabla_\Gamma\theta_\Gamma|^2/\theta_\Gamma^2 + \gamma(\theta_\Gamma)j_\Gamma^2/\theta_\Gamma]d\Gamma \geq 0,\end{aligned}$$

where $\Phi_b = \int_\Omega \rho\eta(\theta)dx$.

For the sake of well-posedness we assume that $\psi_i$ and $\sigma$ are strictly concave. Experiments show that $\sigma$ is also strictly decreasing and positive for low temperatures. Therefore, $\sigma$ has precisely one zero $\theta_c > 0$ which we call the *critical temperature*. As the problem in the range $\theta > \theta_c$ is no longer well-posed, we restrict our attention to the interval $\theta \in (0, \theta_c)$. In all of the paper we impose the following assumptions.

**(a) Regularity.**

$$\mu_i, d_i, d_\Gamma, \gamma \in C^2(0, \theta_c), \quad \psi_i, \sigma \in C^3(0, \theta_c).$$

**(b) Well-posedness.**

$$\kappa_i, \kappa_\Gamma, \mu_i, d_i, d_\Gamma, \sigma > 0, \ \gamma \geq 0 \ \text{in} \ (0, \theta_c), \quad 0 < \theta_0 < \theta_c \ \text{in} \ \bar{\Omega}.$$

**(c) Compatibilities.**

$$\begin{aligned}&\operatorname{div} u_0 = 0 \text{ in } \Omega \setminus \Gamma_0,\\ &2P_{\Gamma_0}[\![\mu(\theta_0)D(u)\nu_{\Gamma_0}]\!] + \sigma'(\theta_0)\nabla_{\Gamma_0}\theta_0 = 0, \quad P_{\Gamma_0}u_\Gamma(0) = P_{\Gamma_0}u_0|_{\Gamma_0}\\ &P_{\Gamma_0}[\![u_0]\!] = 0, \ [\![\theta_0]\!] = 0, \ \theta_\Gamma(0) = \theta_0|_{\Gamma_0}, \ u_0|_{\partial\Omega} = 0, \ \partial_\nu\theta_0|_{\partial\Omega} = 0.\end{aligned}$$

Below we present a rather complete analysis of problem (4), (5) which parallels that in [15] where the simpler case $\sigma > 0$ constant and $\gamma \equiv 0$ has been studied. We obtain local well-posedness of the problem in an $L_p$-setting, prove that the stability properties of equilibria are the same as in [15], and we show that any bounded solution that does not develop singularities converges to an equilibrium as $t \to \infty$ in the state manifold $\mathcal{SM}$ which is the same as in [15]. We refer to [11, 13] for related results, and to [3, 4] for background material on maximal regularity for parabolic boundary value problems.

## 2 Total Entropy and Equilibria

**(a)** As we have seen in Sect. 1, the total mass $\mathsf{M} = \int_{\Omega\setminus\Gamma} \rho\, dx$ and the total energy

$$\mathsf{E} = \mathsf{E}(u, \theta, \theta_\Gamma, \Gamma) := \int_{\Omega\setminus\Gamma} \{(\rho/2)|u|^2 + \rho\varepsilon(\theta)\}\, dx + \int_\Gamma \epsilon_\Gamma(\theta_\Gamma)\, d\Gamma$$

are conserved, and the total entropy

$$\Phi = \Phi(\theta, \theta_\Gamma, \Gamma) := \int_{\Omega\setminus\Gamma} \rho\eta(\theta)\, dx + \int_\Gamma \eta_\Gamma(\theta_\Gamma)\, d\Gamma$$

is nondecreasing along smooth solutions. Even more, $-\Phi$ is a strict Lyapunov functional in the sense that it is strictly decreasing along smooth solutions which are non-constant in time. Indeed, if at some time $t_0 \geq 0$ we have

$$\frac{d}{dt}\Phi(u(t_0), \Gamma(t_0)) = 0,$$

then

$$\int_\Omega [2\mu(\theta)|D(u)|^2/\theta + d(\theta)|\nabla\theta|^2/\theta^2]dx + \int_\Gamma [d_\Gamma(\theta_\Gamma)|\nabla_\Gamma\theta_\Gamma|^2/\theta^2 + \gamma(\theta_\Gamma)j_\Gamma^2]\, d\Gamma = 0,$$

which yields $D(u(t_0)) = 0$ and $\nabla\theta(t_0) = 0$ in $\Omega$, as well as $\nabla_\Gamma\theta_\Gamma(t_0) = 0$ and $j_\Gamma(t_0) = 0$ on $\Gamma(t_0)$. As in [15] this implies $u(t_0) = 0$ and $\theta(t_0) = const = \theta_\Gamma(t_0)$ in $\Omega$. From the equations we see that $(V_\Gamma(t_0), j_\Gamma(t_0)) = (0, 0)$ and therefore $\pi$ is also constant in the components of the phases, and

$$\begin{aligned} [\![\pi]\!] &= \sigma(\theta_\Gamma)H_\Gamma, \\ [\![\psi(\theta_\Gamma)]\!] + [\![\pi/\rho]\!] &= 0. \end{aligned}$$

These relations show that the curvature $H_\Gamma$ is constant over all of $\Gamma$, and it determines the values of the pressures in the phases, in particular $\pi$ is constant in each phase, not only in its components. Since $\Omega$ is bounded, we may conclude that $\Gamma(t_0)$ is a union of finitely many, say $m$, disjoint spheres of equal radius, i.e. $(u(t_0), \theta(t_0), \Gamma(t_0))$ is an equilibrium. Therefore, the *limit sets* of solutions in the state manifold $\mathcal{SM}_\Gamma$, to be defined below, are contained in the $(mn+2)$-dimensional manifold of equilibria

$$\mathcal{E} = \Big\{\Big(0, \theta_*, \bigcup_{1\leq l\leq m} S_{R_*}(x_l)\Big) : \theta_* \in (0, \theta_c), \bar{B}_{R_*}(x_l) \subset \Omega, \qquad \bar{B}_{R_*}(x_l) \cap \bar{B}_{R_*}(x_k) = \emptyset, k \neq l\Big\}, \tag{6}$$

where $S_{R_*}(x_l)$ denotes the sphere with radius $R_*$ and center $x_l$. Here $R_* > 0$ is uniquely determined by the total mass and by the number $m$ of spheres, and $\theta_*$ is uniquely given by the total energy.

**(b)** Another interesting observation is the following. Consider the critical points of the functional $\Phi(u, \theta, \theta_\Gamma, \Gamma)$ with constraint $\mathsf{M} = \mathsf{M}_0$, $\mathsf{E}(u, \theta, \theta_\Gamma, \Gamma) = \mathsf{E}_0$, say on

$$\{(u, \theta, \theta_\Gamma, \Gamma) : (u, \theta) \in BUC(\bar{\Omega} \setminus \Gamma)^{n+1}, \Gamma \in \mathcal{MH}^2(\Omega), \theta_\Gamma \in C(\Gamma)\},$$

see below for the definition of $\mathcal{MH}^2(\Omega)$. So here we do not assume from the beginning that $\theta$ is continuous across $\Gamma$, and $\theta_\Gamma$ denotes surface temperature. Then by the method of Lagrange multipliers, there are constants $\lambda, \mu \in \mathbb{R}$ such that at a critical point $(u_*, \theta_*, \theta_{\Gamma*}, \Gamma_*)$ we have

$$\Phi'(u_*, \theta_*, \theta_{\Gamma*}, \Gamma_*) + \lambda \mathsf{M}'(\Gamma_*) + \mu \mathsf{E}'(u_*, \theta_*, \theta_{\Gamma*}, \Gamma_*) = 0. \tag{7}$$

The derivatives of the functionals are given by

$$\begin{aligned}\langle \Phi'(u, \theta, \theta_\Gamma, \Gamma) | (v, \vartheta, \vartheta_\Gamma, h)\rangle &= (\rho\eta'(\theta)|\vartheta)_\Omega + (\eta_\Gamma'(\theta_\Gamma)|\vartheta_\Gamma)_\Gamma \\ &\quad - ([\![\rho\eta(\theta)]\!] + \eta_\Gamma(\theta_\Gamma)H(\Gamma)|h)_\Gamma, \\ \langle \mathsf{M}'(\Gamma)|h\rangle &= -([\![\rho]\!]|h)_\Gamma,\end{aligned}$$

with $H(\Gamma) := H_\Gamma$, and

$$\begin{aligned}\langle \mathsf{E}'(u, \theta, \theta_\Gamma, \Gamma) | (v, \vartheta, \vartheta_\Gamma, h)\rangle &= (\rho u|v)_\Omega + (\rho\epsilon'(\theta)|\vartheta)_\Omega + (\epsilon_\Gamma'(\theta_\Gamma)|\vartheta_\Gamma)_\Gamma \\ &\quad - ([\![(\rho/2)|u|^2 + \rho\epsilon(\theta)]\!] + \epsilon_\Gamma(\theta_\Gamma)H(\Gamma)|h)_\Gamma.\end{aligned}$$

Setting first $(u, \vartheta_\Gamma, h) = (0, 0, 0)$ and varying $\vartheta$ in (7) we obtain

$$\eta'(\theta_*) + \mu\epsilon'(\theta_*) = 0 \quad \text{in } \Omega,$$

and similarly varying $\vartheta_\Gamma$ yields

$$\eta_\Gamma'(\theta_{\Gamma*}) + \mu\epsilon_\Gamma'(\theta_{\Gamma_*}) = 0 \text{ on } \Gamma_*.$$

The relations $\eta(\theta) = -\psi'(\theta)$ and $\epsilon(\theta) = \psi(\theta) - \theta\psi'(\theta)$ imply $0 = -\psi''(\theta_*)(1 + \mu\theta_*)$, and this shows that $\theta_* = -1/\mu$ is constant in $\Omega$, since $\kappa(\theta) = -\theta\psi''(\theta) > 0$ for all $\theta \in (0, \theta_c)$ by assumption. Similarly on $\Gamma_*$ we obtain $\theta_{\Gamma_*} = -1/\mu$ constant as well, provided $\kappa_\Gamma(\theta_\Gamma) > 0$, hence in particular $\theta_* \equiv \theta_{\Gamma_*}$.

Next varying $v$ implies $\mu u_* = 0$, and hence $u_* = 0$ as $\mu \neq 0$. Finally, varying $h$ we get

$$-([\![\rho\eta(\theta_*)]\!] + \eta_\Gamma(\theta_{\Gamma_*})H_\Gamma) - \lambda[\![\rho]\!] - \mu([\![\rho\epsilon(\theta_*)]\!] + \epsilon_\Gamma(\theta_{\Gamma_*})H(\Gamma_*)) = 0 \text{ on } \Gamma_*.$$

This implies with the above relations

$$[\![\rho\psi(\theta_*)]\!] + \sigma(\theta_*)H(\Gamma_*) = \lambda[\![\rho]\!]\theta_{\Gamma_*}.$$

Since $\theta_*$ is constant and assuming $\theta_* \in (0, \theta_c)$, we see with $\sigma_* > 0$ that $H(\Gamma_*)$ is constant. Therefore, as $\Omega$ is bounded, $\Gamma_*$ is a sphere whenever connected, and a union of finitely many disjoint spheres of equal size otherwise. Thus the critical points of the entropy functional for prescribed energy are precisely the equilibria of the problem (4), (5).

**(c)** Going further, suppose we have an equilibrium $e_* := (0, \theta_*, \theta_{\Gamma_*}, \Gamma_*)$ where the total entropy has a local maximum w.r.t. the constraints $\mathsf{M} = \mathsf{M}_0$ and $\mathsf{E} = \mathsf{E}_0$. Then $\mathcal{D}_* := [\Phi + \lambda\mathsf{M} + \mu\mathsf{E}]''(e_*)$ is negative semi-definite on the kernel of $(\mathsf{M}', \mathsf{E}')(e_*)$, where $\lambda$ and $\mu$ are the fixed Lagrange multipliers found above. The kernel of $\mathsf{M}'(e)$ is given by $(1|h)_\Gamma = 0$, as $[\![\rho]\!] \neq 0$, and that of $\mathsf{E}'(e)$ is determined by the identity

$$(\rho u|v)_\Omega + (\rho\kappa(\theta)|\vartheta)_\Omega + (\kappa_\Gamma(\theta_\Gamma)|\vartheta_\Gamma)_\Gamma - ([\![\rho\epsilon(\theta)]\!] + \epsilon_\Gamma(\theta_\Gamma)H(\Gamma)|h)_\Gamma = 0,$$

which at equilibrium yields

$$(\rho\kappa_*|\vartheta)_\Omega + (\kappa_{\Gamma_*}|\vartheta_\Gamma)_\Gamma = 0, \tag{8}$$

where $\kappa_* := \kappa(\theta_*)$, $\kappa_{\Gamma_*} := \kappa_\Gamma(\theta_*)$, and $\sigma_* = \sigma(\theta_{\Gamma_*})$. On the other hand, a straightforward calculation yields with $z = (v, \vartheta, \vartheta_\Gamma, h)$

$$-\langle\mathcal{D}_* z|z\rangle = (\rho v|v)_\Omega + \frac{1}{\theta_*}\big[(\rho\kappa_*\vartheta|\vartheta)_\Omega + (\kappa_{\Gamma*}\vartheta_\Gamma|\vartheta_\Gamma)_\Gamma - \sigma_*\theta_*(H'(\Gamma_*)h|h)_\Gamma\big]. \tag{9}$$

As $\kappa_*$ and $\kappa_{\Gamma_*}$ are positive, we see that the form $\langle\mathcal{D}z|z\rangle$ is negative semi-definite as soon as $H'(\Gamma_*)$ is negative semi-definite. We have

$$H'(\Gamma_*) = (n-1)/R_*^2 + \Delta_{\Gamma_*},$$

where $\Delta_{\Gamma_*}$ denotes the Laplace-Beltrami operator on $\Gamma_*$ and $R_*$ means the radius of an equilibrium sphere. To derive necessary conditions for an equilibrium $e_*$ to be a local maximum of entropy, we suppose that $\Gamma_*$ is not connected, i.e. $\Gamma_*$ is a finite union of spheres $\Gamma_*^k$. Set $\vartheta = \vartheta_\Gamma = 0$, and let $h = h_k$ be constant on $\Gamma_*^k$ with $\sum_k h_k = 0$. Then the constraint (8) holds, and with the volume $\omega_n$ of the unit sphere in $\mathbb{R}^n$

$$\langle \mathcal{D}z|z\rangle = (\sigma_*\theta_*)((n-1)/R_*^2)\omega_n R_*^{n-1}\sum_k h_k^2 > 0,$$

hence $\mathcal{D}$ cannot be negative semi-definite in this case, as $\sigma_* > 0$. Thus if $e_*$ is an equilibrium with locally maximal total entropy, then $\Gamma_*$ must be connected, and hence both phases are connected.

On the other hand, if $\Gamma_*$ is connected then $H'(\Gamma_*)$ is negative semi-definite on functions with mean zero, hence in this case $\mathcal{D}$ is in fact positive semi-definite. We will see below that connectedness of $\Gamma_*$ is precisely the condition for stability of the equilibrium $e_*$.

**(d)** Summarizing, we have shown

- The total energy is constant along smooth solutions of (4), (5).
- The negative total entropy is a strict Lyapunov functional for (4), (5).
- The equilibria of (4), (5) are precisely the critical points of the entropy functional with prescribed total energy and total mass.
- If the entropy functional with prescribed energy and total mass has a local maximum at $e_* = (0, \theta_*, \theta_{\Gamma_*}, \Gamma_*)$, then $\Gamma_*$ is connected.

It should be noted that we are using the term *equilibrium* to describe a *stationary solution* of the system, while *a thermodynamic equilibrium* would—by definition—require the entropy production to be zero. Our results show that stable equilibria are precisely those that guarantee the system to be in a thermodynamic equilibrium.

## 3 Local Well-Posedness and the Semiflow

$\theta_\Gamma$ is not a real system variable as it is the trace of $\theta$ on the interface. For analytical reasons it is a useful quantity while we consider the linear problem and the nonlinear problem on the reference manifold $\Sigma$ after a Hanzawa transform. Local well-posedness of Problem (4), (5) is based on maximal $L_p$-regularity of its principal linearization and on the contraction mapping principle.

### 3.1 Principal Linearization

The principal part of the linearized problem reads as follows

$$\begin{aligned}
\rho\partial_t u - \mu_0(x)\Delta u + \nabla\pi &= \rho f_u && \text{in } \Omega\setminus\Sigma,\\
\operatorname{div} u &= g_d && \text{in } \Omega\setminus\Sigma,\\
P_\Sigma[\![u]\!] + c(t,x)\nabla_\Sigma h &= P_\Sigma g_u && \text{on } \Sigma,\\
-2[\![\mu_0(x)D(u)\nu_\Sigma]\!] + [\![\pi]\!]\nu_\Sigma - \sigma_0(x)\Delta_\Sigma h\nu_\Sigma - \sigma_1(x)\nabla_\Sigma\theta_\Sigma &= g && \text{on } \Sigma,\\
u &= 0 && \text{on } \partial\Omega,\\
u(0) &= u_0 && \text{in } \Omega.
\end{aligned} \tag{10}$$

$$\begin{aligned}
\rho\kappa(x)\partial_t\theta - d(x)\Delta\theta &= \rho\kappa(x) f_\theta && \text{in } \Omega\setminus\Sigma,\\
\kappa_{\Gamma 0}(x)\partial_t\theta_\Sigma - d_{\Gamma 0}(x)\Delta_\Sigma\theta_\Sigma &= \kappa_{\Gamma 0}(x) g_\theta && \text{on } \Sigma,\\
[\![\theta]\!] = 0, \quad \theta_\Sigma &= \theta|_\Sigma && \text{on } \Sigma,\\
\partial_\nu\theta &= 0 && \text{on } \partial\Omega,\\
\theta(0) = \theta_0 \quad \text{in } \Omega, \qquad \theta_\Sigma(0) &= \theta_0|_\Sigma && \text{on } \Sigma,
\end{aligned} \tag{11}$$

$$\begin{aligned}
[\![\rho]\!]\partial_t h - [\![\rho u\cdot\nu_\Sigma]\!] + b(t,x)\cdot\nabla_\Sigma h &= [\![\rho]\!] f_h && \text{on } \Sigma,\\
-2[\![(\mu_0(x)/\rho)D(u)\nu_\Sigma\cdot\nu_\Sigma]\!] + [\![\pi/\rho]\!] &= g_h && \text{on } \Sigma,\\
h(0) &= h_0 && \text{on } \Sigma.
\end{aligned} \tag{12}$$

Here $\mu_0, \kappa_0, d_0, \kappa_{\Gamma 0}, d_{\Gamma 0}, \sigma_0$ and $\sigma_1$ are functions of $x$, which are realized by $\mu_0(x) = \mu(\theta_0(x))$ and so on. The difference between the linear problem for variable surface tension (10)–(12) and the linear problem for constant surface tension [15, Sect. 3.1] is the fourth equation of (10) and the second equation of (11).

Observe that (11) decouples from the remaining problems (10)–(12). Maximal $L_p$-regularity of (11) has been proved in [16], while maximal $L_p$-regularity of (10) and (12) has been considered in [15]. Therefore we obtain the maximal $L_p$-regularity of (10)–(12).

**Theorem 3.1** *Let $p > n+2$, $\rho_i > 0$, $\rho_2 \neq \rho_1$, $\mu_{0i}, \kappa_{0i}, d_{0i} \in C(\bar{\Omega}_i)$, $\kappa_{\Gamma 0}$, $d_{\Gamma 0}$, $\sigma_0$, $\sigma_1 \in C(\Sigma)$, $\mu_{0i}, \kappa_{0i}, d_{0i} > 0$, $\kappa_{\Gamma 0}, d_{\Gamma 0}, \sigma_0, \sigma_1 > 0$, $i = 1,2$,*

$$(b,c) \in W_p^{1-1/2p}(J; L_p(\Sigma))^{n+1} \cap L_p(J; W_p^{2-1/p}(\Sigma))^{n+1},$$

*where* $J = [0, a]$. *Then the coupled system* (10)–(12) *admits a unique solution* $(u, \pi, \theta, \theta_\Sigma, h)$ *with regularity*

$$
\begin{aligned}
&(u, \theta) \in H_p^1(J; L_p(\Omega))^{n+1} \cap L_p(J; H_p^2(\Omega \setminus \Sigma))^{n+1} =: \mathbb{E}_{u,\vartheta}(J),\\
&[\![u \cdot \nu_\Sigma]\!] \in H_p^1(J; \dot{W}_p^{-1/p}(\Sigma)), \quad \pi \in L_p(J; \dot{H}_p^1(\Omega \setminus \Sigma)) = \mathbb{E}_\pi(J),\\
&\theta_\Sigma \in H_p^1(J; W_p^{-1/p}(\Sigma)) \cap L_p(J; W_p^{2-1/p}(\Sigma)) =: \mathbb{E}_{tr\theta}(J),\\
&\pi_i := \pi_{|\partial\Omega_i} \in W_p^{1/2-1/2p}(J; L_p(\Sigma)) \cap L_p(J; W_p^{1-1/p}(\Sigma)) =: \mathbb{E}_{tr\pi}(J), \ i = 1, 2,\\
&h \in W_p^{2-1/2p}(J; L_p(\Sigma)) \cap H_p^1(J; W_p^{2-1/p}(\Sigma)) \cap L_p(J; W_p^{3-1/p}(\Sigma)) =: \mathbb{E}_h(J),
\end{aligned}
$$

*if and only if the data* $(f_u, f_\theta, g_\theta, g_d, P_\Sigma g_u, g, f_h, g_h, u_0, \theta_0, \theta_0|_\Sigma, h_0)$ *satisfy the following regularity*

(a) $(f_u, f_\theta) \in L_p(J; L_p(\Omega))^{n+1}$,
(b) $g_\theta \in L_p(J; W_p^{-1/p}(\Sigma))$,
(c) $g_d \in H_p^1(J; \dot{H}_p^{-1}(\Omega)) \cap L_p(J; H_p^1(\Omega \setminus \Sigma))$,
(d) $(g, g_h) \in W_p^{1/2-1/2p}(J; L_p(\Sigma))^{n+1} \cap L_p(J; W_p^{1-1/p}(\Sigma))^{n+1}$,
(e) $(P_\Sigma g_u, f_h) \in W_p^{1-1/2p}(J; L_p(\Sigma))^n \cap L_p(J; W_p^{2-1/p}(\Sigma))^n$,
(f) $(u_0, \theta_0, \theta_0|_\Sigma, h_0) \in X_\gamma := W_p^{2-2/p}(\Omega \setminus \Sigma)^{n+1} \times W_p^{2-3/p}(\Sigma) \times W_p^{3-2/p}(\Sigma)$,

*and compatibility conditions:*

(g) $\operatorname{div} u_0 = g_d(0)$ *in* $\Omega \setminus \Sigma$,
(h) $P_\Sigma[\![u_0]\!] + c(0, \cdot)\nabla_\Sigma h_0 = P_\Sigma g_u(0)$ *on* $\Sigma$,
(i) $-P_\Sigma[\![\mu_0(\cdot)(\nabla u_0 + [\nabla u_0]^{\mathsf{T}})\nu_\Sigma]\!] - \sigma_1(\cdot)\nabla_\Sigma \theta_\Sigma = P_\Sigma g(0)$ *on* $\Sigma$.

*The solution map*

$$[(f_u, f_\theta, g_\theta, g_d, P_\Sigma g_u, g, f_h, g_h, u_0, \theta_0, \theta_0|_\Sigma, h_0) \mapsto (u, \pi, \theta, \theta_\Sigma, h)]$$

*is continuous between the corresponding spaces.*

*Remark 3.2* $X_\gamma$ is the time trace space of the solution space $\mathbb{E}(J)$ with

$$\mathbb{E}(J) := \mathbb{E}_u(J) \times \mathbb{E}_\theta(J) \times \mathbb{E}_{tr\theta}(J) \times \mathbb{E}_h(J).$$

## 3.2 Local Existence

As in [15], the basic result for local well-posedness of Problem (4), (5) in an $L_p$-setting is the following theorem, which is proved by the contraction mapping principle.

**Theorem 3.3** *Let $p > n+2$, $\rho_1, \rho_2 > 0$, $\rho_1 \neq \rho_2$. We assume the conditions* **(a), (b), (c)** *in Sect. 1 and the regularity conditions*

$$(u_0, \theta_0) \in W_p^{2-2/p}(\Omega \setminus \Gamma_0)^{n+1}, \quad \Gamma_0 \in W_p^{3-2/p}.$$

*Then there exists a unique $L_p$-solution of Problem* (4), (5) *on some possibly small but nontrivial time interval $J = [0, \tau]$.*

Here the notation $\Gamma_0 \in W_p^{3-2/p}$ means that $\Gamma_0$ is a $C^2$-manifold, such that its (outer) normal field $\nu_{\Gamma_0}$ is of class $W_p^{2-2/p}(\Gamma_0)$. Therefore, the Weingarten tensor $L_{\Gamma_0} = -\nabla_{\Gamma_0}\nu_{\Gamma_0}$ of $\Gamma_0$ belongs to $W_p^{1-2/p}(\Gamma_0)$ which embeds into $C^{\alpha+1/p}(\Gamma_0)$, with $\alpha = 1-(n+2)/p > 0$ since $p > n+2$ by assumption. For the same reason we also have $u_0 \in C^{1+\alpha}(\bar{\Omega}_i(0))^n$, $\theta_0 \in C^{1+\alpha}(\bar{\Omega}_i(0))$, $i = 1, 2$, and $V_0 \in C^{1+\alpha}(\Gamma_0)$. The notion $L_p$-solution means that $(u, \pi, \theta, \theta_\Gamma, \Gamma)$ is obtained as the push-forward of an $L_p$-solution $(\bar{u}, \bar{\pi}, \bar{\theta}, \bar{\theta}_\Sigma, h)$ of the transformed problem, which means that $(\bar{u}, \bar{\theta}, \bar{\theta}_\Gamma, h)$ belongs to $\mathbb{E}(J)$. The regularity of the pressure is obtained from the equations.

### 3.3 *Time-Weights*

For later use we need an extension of the local existence result to spaces with time weights. For this purpose, given a UMD-Banach space $Y$ and $\mu \in (1/p, 1]$, we define for $J = (0, t_0)$

$$K^s_{p,\mu}(J; Y) := \{u \in L_{p,loc}(J; Y) : \; t^{1-\mu}u \in K^s_p(J; Y)\},$$

where $s \geq 0$ and $K \in \{H, W\}$. It has been shown in [10] that the operator $d/dt$ in $L_{p,\mu}(J; Y)$ with domain

$$\mathcal{D}(d/dt) = {}_0H^1_{p,\mu}(J; Y) = \{u \in H^1_{p,\mu}(J; Y) : \; u(0) = 0\}$$

is sectorial and admits an $H^\infty$-calculus with angle $\pi/2$. This is the main tool to extend Theorem 3.3 to the time weighted setting, where the solution space $\mathbb{E}(J)$ is replaced by $\mathbb{E}_\mu(J)$, and

$$z \in \mathbb{E}_\mu(J) \Leftrightarrow t^{1-\mu}z \in \mathbb{E}(J).$$

The trace spaces for $(u, \theta, h)$ for $p > 3$ are then given by

$$(u_0, \theta_0) \in W_p^{2\mu-2/p}(\Omega \setminus \Sigma)^{n+1}, \; \theta_0|_\Sigma \in W_p^{2\mu-3/p}(\Sigma), \; h_0 \in W_p^{2+\mu-2/p}(\Sigma),$$
$$h_1 := \partial_t h|_{t=0} \in W_p^{2\mu-3/p}(\Sigma), \tag{13}$$

where for the last trace we need in addition $\mu > 3/2p$. Note that the embeddings

$$\mathbb{E}_{\mu,u,\theta}(J) \hookrightarrow C(J; C^1(\bar{\Omega}_i))^{n+1}, \quad \mathbb{E}_{\mu,h}(J) \hookrightarrow C(J; C^{2+\alpha}(\Sigma)) \cap C^1(J; C^1(\Sigma))$$

with $\alpha = 1/2 - n/p > 0$ require $\mu > 1/2 + (n+2)/2p$, which is feasible since $p > n+2$ by assumption. This restriction is needed for the estimation of the nonlinearities.

For these time weighted spaces we have the following result.

**Corollary 3.4** *Let $p > n+2$, $\mu \in (1/2 + (n+2)/2p, 1]$, $\rho_1, \rho_2 > 0$, $\rho_1 \neq \rho_2$. We assume that the conditions* **(a), (b), (c)** *in Sect. 1 and the regularity conditions*

$$(u_0, \theta_0) \in W_p^{2\mu-2/p}(\Omega \setminus \Gamma_0)^{n+1}, \quad \Gamma_0 \in W_p^{2+\mu-2/p}$$

*are satisfied. Then the transformed problem admits a unique solution*

$$z = (u, \theta, \theta_\Sigma, h) \in \mathbb{E}_\mu(0, \tau)$$

*for some nontrivial time interval $J = [0, \tau]$. The solution depends continuously on the data. For each $\delta > 0$ the solution belongs to $\mathbb{E}(\delta, \tau)$, i.e. it regularizes instantly.*

## 4 Linear Stability of Equilibria

**1.** We call an equilibrium *non-degenerate* if the balls making up $\Omega_1(t)$ do neither touch each other nor the outer boundary; this set is denoted by $\mathcal{E}$. To derive the full linearization at a non-degenerate equilibrium $e_* := (0, \theta_*, \theta_{\Sigma*}, \Sigma) \in \mathcal{E}$, note that the quadratic terms $u \cdot \nabla u$, $u \cdot \nabla\theta$, $|D(u)|_2^2$, $[\![u]\!]j_\Gamma$, and $j_\Gamma^2$ give no contribution to the linearization. Therefore we obtain the following fully linearized problem for $(u, \pi, h)$, the relative temperature $\vartheta = (\theta - \theta_*)/\theta_*$ and $\vartheta_\Sigma = \vartheta|_\Sigma$.

$$\begin{aligned}
\rho\partial_t u - \mu_*\Delta u + \nabla\pi &= \rho f_u && \text{in } \Omega \setminus \Sigma,\\
\operatorname{div} u &= g_d && \text{in } \Omega \setminus \Sigma,\\
P_\Sigma[\![u]\!] &= P_\Sigma g_u && \text{on } \Sigma,\\
-2P_\Sigma[\![\mu_* D(u)\nu_\Sigma]\!] - \theta_*\sigma_*'\nabla_\Sigma\vartheta_\Sigma &= P_\Sigma g && \text{on } \Sigma,\\
-2[\![\mu_* D(u)\nu_\Sigma \cdot \nu_\Sigma]\!] + [\![\pi]\!] + \sigma_*\mathcal{A}_\Sigma h - \theta_*\sigma_*' H_*\vartheta_\Sigma &= g \cdot \nu_\Sigma && \text{on } \Sigma,\\
u &= 0 && \text{on } \partial\Omega,\\
u &= u_0 && \text{in } \Omega,
\end{aligned} \tag{14}$$

$$
\begin{aligned}
\rho\kappa_*\partial_t\vartheta - d_*\Delta\vartheta &= \rho\kappa_* f_\theta && \text{in } \Omega\setminus\Sigma,\\
[\![\vartheta]\!] &= 0 && \text{on } \Sigma,\\
\kappa_{\Gamma_*}\partial_t\vartheta_\Sigma - d_{\Gamma_*}\Delta_\Sigma\vartheta_\Sigma - (l_*/\theta_*)j_\Sigma - [\![d_*\partial_{\nu_\Sigma}\vartheta]\!] - \sigma_*'\mathrm{div}_\Sigma u_\Sigma &= \kappa_{\Gamma_*} g_\theta && \text{on } \Sigma,\\
\partial_\nu\vartheta &= 0 && \text{on } \partial\Omega,\\
\vartheta &= \vartheta_0 && \text{in } \Omega,
\end{aligned}
\tag{15}
$$

$$
\begin{aligned}
-2[\![\mu_* D(u)\nu_\Sigma\cdot\nu_\Sigma/\rho]\!] + [\![\pi/\rho]\!] + l_*\vartheta_\Sigma + \gamma_* j_\Sigma &= g_h && \text{on } \Sigma,\\
\partial_t h - [\![\rho u\cdot\nu_\Sigma]\!]/[\![\rho]\!] &= f_h && \text{on } \Sigma,\\
h(0) &= h_0 && \text{on } \Sigma,
\end{aligned}
\tag{16}
$$

where $\mu_* = \mu(\theta_*)$, $\kappa_* = \kappa(\theta_*)$, $d_* = d(\theta_*)$, $\sigma_* = \sigma(\theta_*)$, $l_* = l(\theta_*)$, $\gamma_* = \gamma(\theta_*)$, $l_{\Gamma_*} = l_\Gamma(\theta_*)$, $\kappa_{\Gamma_*} = \kappa_\Gamma(\theta_*)$, $d_{\Gamma_*} = d_\Gamma(\theta_*)$, and

$$\mathcal{A}_\Sigma = -H'(0) = -(n-1)/R_*^2 - \Delta_\Sigma, \quad H_* = (n-1)/R_*.$$

Here we used that $l_{\Gamma_*}/\theta_* = \sigma_*' = \sigma'(\theta_*)$. Finally, $u_\Sigma$ denotes the transformed velocity field $u_\Gamma$, and $j_\Sigma$ is given by

$$j_\Sigma := [\![u\cdot\nu_\Sigma]\!]/[\![1/\rho]\!].$$

The time-trace space $\mathbb{E}_\gamma$ of $\mathbb{E}(J)$ is given by

$$(u_0,\vartheta_0,\vartheta_0|_\Sigma,h_0) \in \mathbb{E}_\gamma = [W_p^{2-2/p}(\Omega\setminus\Sigma)]^{n+1} \times W_p^{2-3/p}(\Sigma) \times W_p^{3-2/p}(\Sigma),$$

and the space of right hand sides is

$$
\begin{aligned}
&((f_u,f_\theta), g_d, (f_h, P_\Sigma g_u), (g, g_\theta, g_h)) \in \mathbb{F}(J)\\
&\qquad := \mathbb{F}_{u,\theta}(J) \times \mathbb{F}_d(J) \times \mathbb{F}_h(J)^{n+1} \times \mathbb{F}_\theta(J)^{n+2},
\end{aligned}
$$

where

$$\mathbb{F}_{u,\theta}(J) = L_p(J\times\Omega)^{n+1}, \quad \mathbb{F}_d(J) = H_p^1(J;\dot{H}_p^{-1}(\Omega)) \cap L_p(J;H_p^1(\Omega)),$$

and

$$\mathbb{F}_\theta(J) = W_p^{1/2-1/2p}(J;L_p(\Sigma)) \cap L_p(J;W_p^{1-1/p}(\Sigma)),$$

$$\mathbb{F}_h(J) = W_p^{1-1/2p}(J;L_p(\Sigma)) \cap L_p(J;W_p^{2-1/p}(\Sigma)).$$

As the terms $(l_*/\theta_*)j_\Sigma$ and $\sigma_*'\mathrm{div}_\Sigma u_\Sigma$ are lower order, the remaining system is triangular, where the equations for $\theta$ decouple. Therefore, as in Sect. 3, it follows from the maximal regularity results in [5, 9, 15] and a standard perturbation argument that the operator $\mathbb{L}$ defined by the left hand side of (14)–(16) is an isomorphism from $\mathbb{E}$ into $\mathbb{F}\times\mathbb{E}_\gamma$. The range of $\mathbb{L}$ is determined by the natural compatibility conditions. If the time derivatives $\partial_t$ are replaced by $\partial_t+\omega$, $\omega>0$ sufficiently large, then this result is also true for $J=\mathbb{R}_+$.

**2.** We introduce a functional analytic setting as follows. Set

$$X_0 = L_{p,\sigma}(\Omega)\times L_p(\Omega)\times W_p^{-1/p}(\Sigma)\times W_p^{2-1/p}(\Sigma),$$

where the subscript $\sigma$ means solenoidal, and define the operator $L$ by

$$\begin{aligned}&L(u,\vartheta,\vartheta_\Sigma,h)=\\&\big(-(\mu_*/\rho)\Delta u+\nabla\pi/\rho,-(d_*/\rho\kappa_*)\Delta\vartheta,\\&-(1/\kappa_{\Gamma_*})(d_{\Gamma_*}\Delta_\Sigma\vartheta_\Sigma+(l_*/\theta_*)j_\Sigma+[\![d_*\partial_{\nu_\Sigma}\vartheta]\!]+\sigma_*'\mathrm{div}_\Sigma u_\Sigma),-[\![\rho u\cdot\nu_\Sigma]\!]/[\![\rho]\!]\big).\end{aligned}$$

To define the domain $\mathcal{D}(L)$ of $L$, we set

$$\begin{aligned}X_1=\{(u,\vartheta,\vartheta_\Sigma,h)\in H_p^2(\Omega\setminus\Sigma)^{n+1}\times W_p^{2-1/p}(\Sigma)\times W_p^{3-1/p}(\Sigma):\\ \mathrm{div}\,u=0\ \text{ in }\ \Omega\setminus\Sigma,\ P_\Sigma[\![u]\!]=0,\ [\![\vartheta]\!]=0\ \text{ on }\ \Sigma,\\ u=0,\ \partial_\nu\vartheta=0\ \text{ on }\ \partial\Omega\},\end{aligned}$$

and

$$\mathcal{D}(L)=\{(u,\vartheta,\vartheta_\Sigma,h)\in X_1: 2P_\Sigma[\![\mu_*D(u)\nu_\Sigma]\!]+\theta_*\sigma_*'\nabla_\Sigma\vartheta_\Sigma=0\ \text{ on }\ \Sigma\}.$$

$\pi$ is determined as the solution of the weak transmission problem

$$\begin{aligned}&(\nabla\pi|\nabla\phi/\rho)_2=((\mu_*/\rho)\Delta u|\nabla\phi)_2,\quad \phi\in H_{p'}^1(\Omega),\ \phi=0\text{ on }\Sigma,\\&[\![\pi]\!]=-\sigma_*\mathcal{A}_\Sigma h+\theta_*\sigma_*'\vartheta_\Sigma H_*+2[\![\mu_*(D(u)\nu_\Sigma|\nu_\Sigma)]\!],\qquad \text{on }\Sigma,\\&[\![\pi/\rho]\!]=2[\![(\mu_*/\rho)(D(u)\nu_\Sigma|\nu_\Sigma)]\!]-l_*\vartheta-\gamma_*[\![u\cdot\nu_\Sigma]\!]/[\![1/\rho]\!]\qquad\text{on }\Sigma.\end{aligned}$$

Let us introduce solution operators $T_k$, $k\in\{1,2,3\}$, as follows

$$\begin{aligned}\frac{1}{\rho}\nabla\pi&=T_1((\mu_*/\rho)\Delta u)+T_2(-\sigma_*\mathcal{A}_\Sigma h+\theta_*\sigma_*'\vartheta_\Sigma H_*+2[\![\mu_*(D(u)\nu_\Sigma|\nu_\Sigma)]\!])\\&\quad+T_3(2[\![(\mu_*/\rho)(D(u)\nu_\Sigma|\nu_\Sigma)]\!]-l_*\vartheta-\gamma_*[\![u\cdot\nu_\Sigma]\!]/[\![1/\rho]\!]).\end{aligned}$$

We refer to Köhne et al. [7] for the analysis of such transmission problems. The linearized problem can be rewritten as an abstract evolution problem in $X_0$.

$$\dot{z} + Lz = f, \quad t > 0, \quad z(0) = z_0, \tag{17}$$

where $z = (u, \vartheta, \vartheta_\Sigma, h)$, $f = (f_u, f_\theta, f_\Sigma, f_h)$, $z_0 = (u_0, \vartheta_0, \vartheta_0|_\Sigma, h_0)$, provided $(g_d, g_u, g, g_\theta, g_h) = 0$. The linearized problem has maximal $L_p$-regularity, hence (17) has this property as well. Therefore, by a well-known result, $-L$ generates an analytic $C_0$-semigroup in $X_0$; see for instance Proposition 1.1 in [8].

Since the embedding $X_1 \hookrightarrow X_0$ is compact, the semigroup $e^{-Lt}$ as well as the resolvent $(\lambda + L)^{-1}$ of $-L$ are compact as well. Therefore, the spectrum $\sigma(L)$ of $L$ consists of countably many eigenvalues of finite algebraic multiplicity, and it is independent of $p$.

**3.** Suppose that $\lambda$ with $\mathrm{Re}\,\lambda \geq 0$ is an eigenvalue of $-L$. This means

$$\begin{aligned}
\lambda\rho u - \mu_*\Delta u + \nabla\pi &= 0 && \text{in } \Omega\setminus\Sigma,\\
\operatorname{div} u &= 0 && \text{in } \Omega\setminus\Sigma,\\
P_\Sigma[\![u]\!] &= 0 && \text{on } \Sigma,\\
-2P_\Sigma[\![\mu_* D(u)\nu_\Sigma]\!] - \theta_*\sigma_*'\nabla_\Sigma\vartheta_\Sigma &= 0 && \text{on } \Sigma,\\
-2[\![\mu_* D(u)\nu_\Sigma\cdot\nu_\Sigma]\!] + [\![\pi]\!] + \sigma_*\mathcal{A}_\Sigma h - \theta_*\sigma_*' H_*\vartheta_\Sigma &= 0 && \text{on } \Sigma,\\
u &= 0 && \text{on } \partial\Omega,
\end{aligned} \tag{18}$$

$$\begin{aligned}
\lambda\rho\kappa_*\vartheta - d_*\Delta\vartheta &= 0 && \text{in } \Omega\setminus\Sigma,\\
[\![\vartheta]\!] = 0, \quad \vartheta &= \vartheta_\Sigma && \text{on } \Sigma,\\
\lambda\kappa_{\Gamma_*}\vartheta_\Sigma - d_{\Gamma_*}\Delta_\Sigma\vartheta_\Sigma - (l_*/\theta_*)j_\Sigma - [\![d_*\partial_{\nu_\Sigma}\vartheta]\!] - \sigma_*'\mathrm{div}_\Sigma u_\Sigma &= 0 && \text{on } \Sigma,\\
\partial_\nu\vartheta &= 0 && \text{on } \partial\Omega,
\end{aligned} \tag{19}$$

$$\begin{aligned}
-2[\![\mu_* D(u)\nu_\Sigma\cdot\nu_\Sigma/\rho]\!] + [\![\pi/\rho]\!] + l_*\vartheta_\Sigma + \gamma_* j_\Sigma &= 0 && \text{on } \Sigma,\\
\lambda[\![\rho]\!]h - [\![\rho u\cdot\nu_\Sigma]\!] &= 0 && \text{on } \Sigma.
\end{aligned} \tag{20}$$

Observe that on $\Sigma$ we may write

$$u_k = P_\Sigma u + \lambda h\nu_\Sigma + j_\Sigma\nu_\Sigma/\rho_k = u_\Sigma + j_\Sigma\nu_\Sigma/\rho_k, \quad k = 1, 2.$$

By this identity, taking the inner product of the problem for $u$ with $u$ and integrating by parts we get

$$\begin{aligned}
0 &= \lambda|\rho^{1/2}u|_2^2 - (\mathrm{div}\, T(u,\pi,\theta_*)|u)_2 \\
&= \lambda|\rho^{1/2}u|_2^2 + 2|\mu_*^{1/2}D(u)|_2^2 \\
&\quad + ([\![T(u,\pi,\theta_*)\nu_\Sigma]\!]|P_\Sigma u + \lambda h\nu_\Sigma)_\Sigma + ([\![T(u,\pi,\theta_*)\nu_\Sigma\cdot\nu_\Sigma/\rho]\!]|j_\Sigma)_\Sigma \\
&= \lambda|\rho^{1/2}u|_2^2 + 2|\mu_*^{1/2}D(u)|_2^2 + \sigma_*\bar\lambda(\mathcal{A}_\Sigma h|h)_\Sigma + l_*(\vartheta|j_\Sigma)_\Sigma \\
&\quad + \gamma_*|j_\Sigma|_\Sigma^2 - \theta_*\sigma_*'H_*\bar\lambda(\vartheta|h)_\Sigma - \theta_*\sigma_*'(\nabla_\Sigma\vartheta|P_\Sigma u)_\Sigma,
\end{aligned}$$

since $[\![T(u,\pi,\theta_*)\nu_\Sigma]\!] = \sigma_*\mathcal{A}_\Sigma h\nu_\Sigma - \theta_*\sigma_*'\vartheta_\Sigma H_*\nu_\Sigma - \theta_*\sigma_*'\nabla_\Sigma\vartheta_\Sigma$ and, moreover, $[\![T(u,\pi,\theta_*)\nu_\Sigma\cdot\nu_\Sigma/\rho]\!] = l_*\vartheta + \gamma_* j_\Sigma$. On the other hand, the inner product of the equation for $\vartheta$ with $\vartheta$ by an integration by parts and $[\![\vartheta]\!] = 0$ leads to

$$\begin{aligned}
0 &= \lambda|(\rho\kappa_*)^{1/2}\vartheta|_2^2 + |d_*^{1/2}\nabla\vartheta|_2^2 + ([\![d_*\partial_{\nu_\Sigma}\vartheta]\!]|\vartheta)_\Sigma \\
&= \lambda(|(\rho\kappa_*)^{1/2}\vartheta|_2^2 + |\kappa_{\Gamma_*}^{1/2}\vartheta_\Sigma|_\Sigma^2) + |d_*^{1/2}\nabla\vartheta|_2^2 + |d_{\Gamma_*}^{1/2}\nabla_\Sigma\vartheta_\Sigma|_\Sigma^2 \\
&\quad - l_*(j_\Sigma|\vartheta)_\Sigma/\theta_* + \sigma_*'(P_\Sigma u|\nabla_\Sigma\vartheta_\Sigma)_\Sigma + \lambda\sigma_*'H_*(h|\vartheta)
\end{aligned}$$

where we employed $[\![d_*\partial_{\nu_\Sigma}\vartheta]\!] = \kappa_{\Gamma_*}\lambda\vartheta_\Sigma - d_{\Gamma_*}\Delta_\Sigma\vartheta_\Sigma - (l_*/\theta_*)j_\Sigma - \sigma_*'\mathrm{div}_\Sigma u_\Sigma$ and $u_\Sigma = P_\Sigma u + (u_\Sigma\cdot\nu_\Sigma)\nu_\Sigma$. Adding the first identity to the second multiplied by $\theta_*$ and taking real parts yields the important relation

$$\begin{aligned}
0 &= \mathrm{Re}\,\lambda|\rho^{1/2}u|_2^2 + 2|\mu_*^{1/2}D(u)|_2^2 + \sigma_*\mathrm{Re}\,\lambda(\mathcal{A}_\Sigma h|h)_\Sigma \\
&\quad + \theta_*(\mathrm{Re}\,\lambda|(\rho\kappa_*)^{1/2}\vartheta|_2^2 + |d_*^{1/2}\nabla\vartheta|_2^2) \qquad (21)\\
&\quad + \gamma_*|j_\Sigma|_\Sigma^2 + \theta_*(\mathrm{Re}\,\lambda|\kappa_{\Gamma_*}^{1/2}\vartheta_\Sigma|_\Sigma^2 + |d_{\Gamma_*}^{1/2}\nabla_\Sigma\vartheta_\Sigma|_\Sigma^2).
\end{aligned}$$

On the other hand, if $\mathrm{Im}\,\lambda \neq 0$, taking imaginary parts separately we get

$$\begin{aligned}
0 &= \mathrm{Im}\,\lambda|\rho^{1/2}u|_2^2 - \mathrm{Im}\,\lambda\,\sigma_*(\mathcal{A}_\Sigma h|h)_\Sigma + \mathrm{Im}\,l_*(\vartheta|j_\Sigma)_\Sigma \\
&\quad - \mathrm{Im}\,\{\bar\lambda\theta_*\sigma_*'H_*(\vartheta|h)_\Sigma\} - \mathrm{Im}\,\theta_*\sigma_*'(\nabla_\Sigma\vartheta|P_\Sigma u)_\Sigma \\
0 &= \theta_*\mathrm{Im}\,\lambda(|(\rho\kappa_*)^{1/2}\vartheta|_2^2 + |\kappa_{\Gamma_*}^{1/2}\vartheta_\Sigma|_\Sigma^2) - \mathrm{Im}\,l_*(j_\Sigma|\vartheta)_\Sigma \\
&\quad + \mathrm{Im}\,\{\lambda\theta_*\sigma_*'H_*(h|\vartheta)_\Sigma\} + \mathrm{Im}\,\theta_*\sigma_*'(P_\Sigma u|\nabla_\Sigma\vartheta)_\Sigma,
\end{aligned}$$

hence

$$\sigma_*(\mathcal{A}_\Sigma h|h)_\Sigma = |\rho^{1/2}u|_2^2 - \theta_*(|(\rho\kappa_*)^{1/2}\vartheta|_2^2 + |\kappa_{\Gamma_*}^{1/2}\vartheta_\Sigma|_\Sigma^2).$$

Inserting this identity into (21) leads to

$$0 = 2\mathrm{Re}\,\lambda|\rho^{1/2}u|_2^2 + 2|\mu_*^{1/2}D(u)|_2^2 + \theta_*|d_*^{1/2}\nabla\vartheta|_2^2 + \gamma_*|j_\Sigma|_\Sigma^2 + \theta_*|d_{\Gamma_*}^{1/2}\nabla_\Sigma\vartheta_\Sigma|_\Sigma^2.$$

This shows that if $\lambda$ is an eigenvalue of $-L$ with $\mathrm{Re}\,\lambda \geq 0$ then $\lambda$ is real. In fact, otherwise this identity implies $\vartheta = const = \vartheta_\Sigma$, $D(u) = 0$ and $j_\Sigma = 0$, and then $u = 0$ by Korn's inequality and the no-slip condition on $\partial\Omega$, as well as $(\vartheta, \vartheta_\Sigma, h) = (0, 0, 0)$ by the equations for $\vartheta$ and $h$, since $\lambda \neq 0$.

**4.** Suppose now that $\lambda > 0$ is an eigenvalue of $-L$. Then we further have

$$\lambda \int_\Sigma h d\Sigma = \int_\Sigma (u_k \cdot \nu_\Sigma - j_\Sigma/\rho_k) d\Sigma = -\rho_k^{-1} \int_\Sigma j_\Sigma\, d\Sigma = 0,$$

as $\int_\Sigma u_k \cdot \nu_\Sigma\, d\Sigma = \int_{\Omega_k} \mathrm{div}\, u_k\, dx$. Hence the mean values of $h$ and $j_\Sigma$ both vanish since the densities are non-equal. Integrating the equations for $\vartheta$ and $\vartheta_\Sigma$, we obtain from this the relation

$$\kappa_{\Gamma_*} \int_\Sigma \vartheta_\Sigma\, d\Sigma + \int_\Omega \rho\kappa_*\vartheta\, dx = 0.$$

Since $\mathcal{A}_\Sigma$ is positive semidefinite on functions with mean zero in case $\Sigma$ is connected, by (21) we obtain $(u, \vartheta, h) = (0, 0, 0)$, i.e. in this case there are no positive eigenvalues. On the other hand, if $\Sigma$ is disconnected, there is at least one positive eigenvalue. To prove this we need some preparations.

**5.** First we consider the heat problem

$$\begin{aligned} \lambda\rho\kappa_*\vartheta - d_*\Delta\vartheta &= 0 && \text{in } \Omega \setminus \Sigma, \\ [\![\vartheta]\!] &= 0 && \text{on } \Sigma, \\ \vartheta_\Sigma = \vartheta|_\Sigma &= g && \text{on } \Sigma, \\ \partial_\nu\vartheta &= 0 && \text{on } \partial\Omega, \end{aligned} \tag{22}$$

and define $D_\lambda^H g = -[\![d_*\partial_{\nu_\Sigma}\vartheta]\!]$ on $\Sigma$, where $D_\lambda^H$ denotes the Dirichlet-to-Neumann operator for this heat problem. The properties of $D_\lambda^H$ are stated in [16]. Then the solution $\vartheta$ of (19) can be expressed by

$$(\kappa_{\Gamma_*}\lambda - d_{\Gamma_*}\Delta_\Sigma + D_\lambda^H)\vartheta_\Sigma - (l_*/\theta_*)j_\Sigma + \lambda\sigma_*' H_* h - \sigma_*' \mathrm{div}_\Sigma P_\Sigma u = 0, \tag{23}$$

where we made use of the identity

$$\mathrm{div}_\Sigma u_\Sigma = \mathrm{div}_\Sigma P_\Sigma u - H_* u_\Sigma \cdot \nu_\Sigma = \mathrm{div}_\Sigma P_\Sigma u - H_*\lambda h.$$

**6.** Next we solve the asymmetric Stokes problem

$$\begin{aligned} \lambda\rho u - \mu_*\Delta u + \nabla\pi &= 0 && \text{in } \Omega\setminus\Sigma,\\ \operatorname{div} u &= 0 && \text{in } \Omega\setminus\Sigma,\\ P_\Sigma[\![u]\!] &= 0 && \text{on } \Sigma,\\ -[\![T(u,\pi,\theta_*)\nu_\Sigma\cdot\nu_\Sigma]\!] &= g_1 && \text{on } \Sigma,\\ -[\![T(u,\pi,\theta_*)\nu_\Sigma\cdot\nu_\Sigma/\rho]\!] &= g_2 && \text{on } \Sigma,\\ P_\Sigma[\![T(u,\pi,\theta_*)\nu_\Sigma]\!] &= g_3 && \text{on } \Sigma,\\ u &= 0 && \text{on } \partial\Omega, \end{aligned} \tag{24}$$

to obtain the output

$$\begin{aligned} [\![\rho u\cdot\nu_\Sigma]\!]/[\![\rho]\!] &= S_\lambda^{11}g_1 + S_\lambda^{12}g_2 + S_\lambda^{13}g_3,\\ [\![u\cdot\nu_\Sigma]\!]/[\![1/\rho]\!] &= S_\lambda^{21}g_1 + S_\lambda^{22}g_2 + S_\lambda^{23}g_3,\\ P_\Sigma u &= S_\lambda^{31}g_1 + S_\lambda^{32}g_2 + S_\lambda^{33}g_3. \end{aligned}$$

Note that $g_1, g_2, (S_\lambda g)_1, (S_\lambda g)_2 \in L_2(\Sigma)$ are scalar functions, while $g_3$ and $(S_\lambda g)_3$ are vectors tangent to $\Sigma$, i.e., $g_3$ and $(S_\lambda g)_3 \in L_2(\Sigma; T\Sigma)$, with $T\Sigma$ being the tangent bundle of $\Sigma$. For this problem we have

**Proposition 4.1** *The operator $S_\lambda$ for the Stokes problem* (24) *admits a bounded extension to $L_2(\Sigma)^2 \times L_2(\Sigma; T\Sigma)$ for $\lambda \geq 0$ and has the following properties.*

**(i)** *If $u$ denotes the solution of* (24)*, then*

$$(S_\lambda g|g)_{L_2} = \lambda\int_\Omega \rho|u|^2\,dx + 2\int_\Omega \mu_*|D(u)|_2^2\,dx,\ \lambda\geq 0,\ g\in L_2(\Sigma)^2\times L_2(\Sigma;T\Sigma).$$

**(ii)** *$S_\lambda \in \mathcal{B}(L_2(\Sigma)^2 \times L_2(\Sigma; T\Sigma))$ is self-adjoint, positive semidefinite, and compact; in particular*

$$\begin{aligned} S_\lambda^{11} &= [S_\lambda^{11}]^*, & S_\lambda^{22} &= [S_\lambda^{22}]^*, & S_\lambda^{33} &= [S_\lambda^{33}]^*\\ S_\lambda^{12} &= [S_\lambda^{21}]^*, & S_\lambda^{13} &= [S_\lambda^{31}]^*, & S_\lambda^{23} &= [S_\lambda^{32}]^*. \end{aligned}$$

**(iii)** *For each $\beta \in (0, 1/2)$ there is a constant $C_\beta > 0$ such that*

$$|S_\lambda|_{\mathcal{B}(L_2)} \leq \frac{C_\beta}{(1+\lambda)^\beta}, \quad \lambda \geq 0.$$

**(iv)** $|S_\lambda|_{\mathcal{B}(L_2,H_2^1)} \le C$ *uniformly for* $\lambda \ge 0$.

**(v)** $S_\lambda^{11}, S_\lambda^{22} : L_{2,0}(\Sigma) \to H_2^1(\Sigma) \cap L_{2,0}(\Sigma)$ *are isomorphisms, for each* $\lambda \ge 0$, *where* $L_{2,0}(\Sigma) = \{u \in L_2(\Sigma) \mid \int_\Sigma u\, d\Sigma = 0\}$.

*Proof* The assertions follow from similar arguments as in the proof of [15, Proposition 4.3]. □

**7.** The following lemma is needed in the proof of the main result of this section.

**Lemma 4.2** *Let $H$, $V$ be Hilbert spaces. Let $B$ be a positive definite operator on $H$, $A : \mathcal{D}(A) \subset H \to V$ be a closed, densely defined operator such that $A\mathcal{D}(B^{1/2}) \subset \mathcal{D}(A^*)$. Then $A^*A + B$ is a self-adjoint positive definite operator with $\mathcal{D}(A^*A + B) = \mathcal{D}(B)$ and*

$$|A(A^*A + B)^{-1}A^*|_{\mathcal{B}(V)} \le 1.$$

*In addition, if $A(A^*A + B)^{-1}A^*v = v$, then $v = 0$.*

*Proof* By the closed graph theorem $A : \mathcal{D}(B^{1/2}) \to V$ is bounded. The closedness of $A^*$ in turn implies that the operator $A^*A : \mathcal{D}(B^{1/2}) \to H$ is closed. Another application of the closed graph theorem then shows that $A^*A : \mathcal{D}(B^{1/2}) \to H$ is bounded as well. This implies that $A^*A$ is a lower order perturbation of $B$ and also that $A^*A + B$ is self-adjoint and positive definite. Therefore, $(A^*A + B)^{-1} : H \to \mathcal{D}(B)$ exists and is bounded.

For $v \in \mathcal{D}(A^*)$ we have with $K := ACA^* := A(A^*A + B)^{-1}A^*$

$$\begin{aligned} |Kv|_V^2 &= (CA^*v \mid A^*ACA^*v)_H \\ &= (CA^*v \mid A^*v)_H - (CA^*v \mid BCA^*v)_H \\ &= (ACA^*v \mid v)_V - (BCA^*v \mid CA^*v)_H \\ &= (Kv \mid v)_V - (Bw \mid w)_H \end{aligned}$$

with $w = CA^*v = (A^*A + B)^{-1}A^*v$. Since $B$ is positive definite, there exists $\beta > 0$ such that

$$|Kv|_V^2 \le |Kv|_V|v|_V - \beta|w|_H^2 \le |Kv|_V|v|_V \tag{25}$$

which shows $|K|_{\mathcal{B}(V)} \le 1$. Moreover if $Kv = v$, then $|v|^2 \le |v|^2 - \beta|w|^2$ holds from (25). Hence $w = 0$, and consequently $v = Kv = Aw = 0$. □

Now suppose that $\lambda > 0$ is an eigenvalue of $-L$. We set

$$g = \begin{pmatrix} \theta_*\sigma_*' H_*\vartheta_\Sigma - \sigma_*\mathcal{A}_\Sigma h \\ -l_*\vartheta_\Sigma - \gamma_* j_\Sigma \\ \theta_*\sigma_*'\nabla_\Sigma\vartheta_\Sigma \end{pmatrix} = \begin{pmatrix} -\sigma_*\mathcal{A}_\Sigma h \\ -\gamma_* j_\Sigma \\ 0 \end{pmatrix} + \theta_* Q\vartheta_\Sigma \tag{26}$$

with $Q = (\sigma'_* H_*, -(l_*/\theta_*), \sigma'_* \nabla_\Sigma)^\mathsf{T}$ to obtain

$$(\lambda h, j_\Sigma, P_\Sigma u)^\mathsf{T} = S_\lambda g, \quad S_\lambda = (S_\lambda^{ij})_{1\le i,j\le 3}. \tag{27}$$

We recall (23). Since

$$\begin{aligned} -(l_*/\theta_*)j_\Sigma + \lambda\sigma'_* H_* h - \sigma'_* \mathrm{div}_\Sigma P_\Sigma u &= Q^* S_\lambda g \\ &= \theta_* Q^* S_\lambda Q \vartheta_\Sigma - Q^* S_\lambda (\sigma_* \mathcal{A}_\Sigma h, \gamma_* j_\Sigma, 0)^\mathsf{T}, \end{aligned}$$

(23) is equivalent to

$$(\kappa_{\Gamma_*}\lambda - d_{\Gamma_*}\Delta_\Sigma + D_\lambda^H + \theta_* Q^* S_\lambda Q)\vartheta_\Sigma = Q^* S_\lambda (\sigma_* \mathcal{A}_\Sigma h, \gamma_* j_\Sigma, 0)^\mathsf{T}. \tag{28}$$

Observing that $\kappa_{\Gamma_*}\lambda - d_{\Gamma_*}\Delta_\Sigma + D_\lambda^H + \theta_* Q^* S_\lambda Q$ is injective for $\lambda \ge 0$, we solve the equation above for $\vartheta_\Sigma$ to the result

$$\vartheta_\Sigma = L_\lambda Q^* S_\lambda (\sigma_* \mathcal{A}_\Sigma h, \gamma_* j_\Sigma, 0)^\mathsf{T}$$

with $L_\lambda = (\kappa_{\Gamma_*}\lambda - d_{\Gamma_*}\Delta_\Sigma + D_\lambda^H + \theta_* Q^* S_\lambda Q)^{-1}$. We set

$$S_\lambda Q = (u_1, u_2, u_3)^\mathsf{T}.$$

Combining (26)–(28), we obtain

$$\begin{aligned} \begin{pmatrix} \lambda h \\ j_\Sigma \\ P_\Sigma u \end{pmatrix} &= \theta_* S_\lambda Q L_\lambda (S_\lambda Q)^* \begin{pmatrix} \sigma_* \mathcal{A}_\Sigma h \\ \gamma_* j_\Sigma \\ 0 \end{pmatrix} - S_\lambda \begin{pmatrix} \sigma_* \mathcal{A}_\Sigma h \\ \gamma_* j_\Sigma \\ 0 \end{pmatrix} \\ &= -(S_\lambda - \theta_* S_\lambda Q L_\lambda Q^* S_\lambda) \begin{pmatrix} \sigma_* \mathcal{A}_\Sigma h \\ \gamma_* j_\Sigma \\ 0 \end{pmatrix}. \end{aligned} \tag{29}$$

In order to obtain the positivity of $S_\lambda - \theta_* S_\lambda Q L_\lambda Q^* S_\lambda$, we symmetrize it as

$$S_\lambda - \theta_* S_\lambda Q L_\lambda Q^* S_\lambda = S_\lambda^{1/2}(I - \theta_* S_\lambda^{1/2} Q L_\lambda Q^* S_\lambda^{1/2}) S_\lambda^{1/2} =: S_\lambda^{1/2}(I - K) S_\lambda^{1/2},$$

with

$$K = \theta_* S_\lambda^{1/2} Q L_\lambda Q^* S_\lambda^{1/2} = A(A^*A + B)^{-1} A^*.$$

Here we have set $A = \theta_*^{1/2} S_\lambda^{1/2} Q$ with domain $\mathcal{D}(A) = H_2^1(\Sigma)$ and $B = \kappa_{\Gamma_*}\lambda - d_{\Gamma_*}\Delta_\Sigma + D_\lambda^H$ with domain $\mathcal{D}(B) = H_2^2(\Sigma)$. Furthermore, $H = L_2(\Sigma)$ and $V = L_2(\Sigma)^2 \times L_2(\Sigma; T\Sigma)$.

By Lemma 4.2, we know $|K| \leq 1$, therefore it holds that

$$\begin{aligned}(S_\lambda v - \theta_* S_\lambda Q L_\lambda Q^* S_\lambda v \mid v) &= |S_\lambda^{1/2} v|^2 - (K S_\lambda^{1/2} v \mid S_\lambda^{1/2} v)\\ &\geq |S_\lambda^{1/2} v|^2 - |K||S_\lambda^{1/2} v|^2 \geq 0,\end{aligned}$$

which shows the positivity of $S_\lambda - \theta_* S_\lambda Q L_\lambda Q^* S_\lambda$.

Writing the upper left $2 \times 2$ block of $S_\lambda - \theta_* S_\lambda Q L_\lambda Q^* S_\lambda$ as

$$S_\lambda^0 := \begin{pmatrix} R_\lambda^1 & R_\lambda^* \\ R_\lambda & R_\lambda^2 \end{pmatrix},$$

$S_\lambda^0$ is also positive. Then by (29) it holds that

$$\begin{pmatrix} \lambda h \\ 0 \end{pmatrix} + S_\lambda^0 \begin{pmatrix} \sigma_* \mathcal{A}_\Sigma h \\ \gamma_* j_\Sigma \end{pmatrix} + \begin{pmatrix} 0 \\ j_\Sigma \end{pmatrix} = \begin{pmatrix} \lambda h \\ 0 \end{pmatrix} + \begin{pmatrix} R_\lambda^1 & R_\lambda^* \\ R_\lambda & R_\lambda^2 + 1/\gamma_* \end{pmatrix} \begin{pmatrix} \sigma_* \mathcal{A}_\Sigma h \\ \gamma_* j_\Sigma \end{pmatrix} = \begin{pmatrix} 0 \\ 0 \end{pmatrix}. \tag{30}$$

The following lemma is needed to solve (30).

**Lemma 4.3 (Schur)** *Let $H$ be a Hilbert space, $S, T, R \in \mathcal{B}(H)$, $S = S^*$, $T = T^*$ and suppose that $T$ is invertible. If*

$$\begin{pmatrix} S & R^* \\ R & T \end{pmatrix} \geq 0 \quad \text{on } H \times H,$$

*then $S - R^* T^{-1} R \geq 0$ on $H$.*

*Proof* For $x$ fixed, we set $(x, y)^\mathsf{T} = (x, -T^{-1} R x)^\mathsf{T}$. Then

$$0 \leq \left( \begin{pmatrix} S & R^* \\ R & T \end{pmatrix} \begin{pmatrix} x \\ y \end{pmatrix} \middle| \begin{pmatrix} x \\ y \end{pmatrix} \right) = ((S - R^* T^{-1} R)x \mid x),$$

and this proves the assertion. □

Since $S_\lambda^0$ is positive, $R_\lambda^1$ and $R_\lambda^2$ are positive as well. This implies, in particular, that $(R_\lambda^2 + \varepsilon)$ is positive definite for any $\varepsilon > 0$. Thus (30) is equivalent to the equation

$$\lambda h + (R_\lambda^1 - R_\lambda^* (R_\lambda^2 + \frac{1}{\gamma_*})^{-1} R_\lambda) \sigma_* \mathcal{A}_\Sigma h = 0. \tag{31}$$

We first show that

$$R_\lambda^1 - R_\lambda^* (R_\lambda^2 + \frac{1}{\gamma_*})^{-1} R_\lambda : L_{2,0}(\Sigma) \to L_{2,0}(\Sigma)$$

is injective. Let

$$\begin{pmatrix} S & R^* \\ R & T \end{pmatrix} := \begin{pmatrix} R_\lambda^1 & R_\lambda^* \\ R_\lambda & R_\lambda^2 + \frac{1}{2\gamma_*} \end{pmatrix}$$

and observe that the assertion of Lemma 4.3 holds true for this matrix with $H = L_{2,0}(\Sigma)$. Suppose $(R_\lambda^1 - R_\lambda^*(R_\lambda^2 + \frac{1}{\gamma_*})^{-1}R_\lambda)h = 0$ for some $h \in L_{2,0}(\Sigma)$. Then

$$\begin{aligned} 0 &= ((R_\lambda^1 - R_\lambda^*(R_\lambda^2 + \frac{1}{\gamma_*})^{-1}R_\lambda)h|h) \\ &= ((S - R^*T^{-1}R)h|h) + \frac{1}{2\gamma_*}(T^{-1}(T + \frac{1}{2\gamma_*})^{-1}Rh|Rh) \geq c|Rh|^2 \geq 0. \end{aligned}$$

Thus, $Rh = 0$, and then also $Sh = 0$. (That is, $R_\lambda^1 h = 0$). This implies

$$0 = (S_\lambda^0(h,0)|(h,0)) = ((I - K)S_\lambda^{1/2}(h,0,0)|S_\lambda^{1/2}(h,0,0)). \tag{32}$$

We conclude that $(I - K)S_\lambda^{1/2}(h,0,0) = (0,0,0)$, and Lemma 4.2 then yields $S_\lambda^{1/2}(h,0,0) = (0,0,0)$. Therefore, $S_\lambda(h,0,0) = (0,0,0)$, and in particular $S_\lambda^{11}h = 0$. We can now, at last, infer from Proposition 4.1(v) that $h = 0$.

An analogous argument shows that $R_\lambda^1 : L_{2,0}(\Sigma) \to L_{2,0}(\Sigma)$ is injective as well. Indeed, if $R_\lambda^1 h = 0$ for some $h \in L_{2,0}(\Sigma)$, then (32) holds, and the proof proceeds just as above. We note that $R_\lambda^1$ admits a representation

$$R_\lambda^1 = S_\lambda^{11}(I - C_\lambda)$$

on $L_{2,0}(\Sigma)$, with $C_\lambda$ a compact operator. Since $R_\lambda^1$ is injective on $L_{2,0}(\Sigma)$, $(I - C_\lambda)$ must be so as well. Consequently, $(I - C_\lambda)$ is a bijection as it has Fredholm index zero. Proposition 4.1(v) then implies

$$R_\lambda^1 \in \mathrm{Isom}(L_{2,0}(\Sigma), L_{2,0}(\Sigma) \cap H_2^1(\Sigma)),$$

i.e., $R_\lambda^1$ is an isomorphism between the indicated spaces.

A similar argument now shows that

$$R_\lambda^1 - R_\lambda^*(R_\lambda^2 + \frac{1}{\gamma_*})^{-1}R_\lambda \in \mathrm{Isom}(L_{2,0}(\Sigma), L_{2,0}(\Sigma) \cap H_2^1(\Sigma)).$$

Setting $T_\lambda := [R_\lambda^1 - R_\lambda^*(R_\lambda^2 + \frac{1}{\gamma_*})^{-1}R_\lambda]^{-1}$, Eq. (31) can be written as

$$\lambda T_\lambda h + \sigma_* \mathcal{A}_\Sigma h = 0.$$

This equation can be treated in the same way as in [15]. As a conclusion,

$$B_\lambda := \lambda T_\lambda + \sigma_* \mathcal{A}_\Sigma$$

has a nontrivial kernel for some $\lambda_0 > 0$, which implies that $-L$ has a positive eigenvalue. Even more is true. We have seen that $B_\lambda$ is positive definite for large $\lambda$ and $B_0 = \sigma_* \mathcal{A}_\Sigma$ has $-\sigma(n-1)/R_*^2$ as an eigenvalue of multiplicity $m-1$ in $L_{2,0}(\Sigma)$. Therefore, as $\lambda$ increases to infinity, $m-1$ eigenvalues $\mu_k(\lambda)$ of $B_\lambda$ must cross through zero, this way inducing $m-1$ positive eigenvalues of $-L$.

**8.** Next we look at the eigenvalue $\lambda = 0$. Then (21) yields

$$2|\mu_*^{1/2} D(u)|_2^2 + \gamma_* |j_\Sigma|_\Sigma^2 + \theta_*(|d_*^{1/2}\nabla\vartheta|_2^2 + |d_{\Gamma_*}^{1/2}\nabla_\Sigma\vartheta_\Sigma|_\Sigma^2) = 0,$$

hence $\vartheta$ is constant, $D(u) = 0$ and $j_\Sigma = 0$ by the flux condition for $\vartheta$. This further implies that $[\![u]\!] = 0$, and therefore Korn's inequality yields $\nabla u = 0$ and then we have $u = 0$ by the no-slip condition on $\partial\Omega$. This implies further that the pressures are constant in the phases and $[\![\pi]\!] = -\sigma_* \mathcal{A}_\Sigma h + \theta_* \sigma_*' H_* \vartheta_\Sigma$, as well as $[\![\pi/\rho]\!] = -l_* \vartheta$. Thus the dimension of the eigenspace for eigenvalue $\lambda = 0$ is the same as the dimension of the manifold of equilibria, namely $mn+2$ if $\Omega_1$ has $m \geq 1$ components. We set $\Sigma = \bigcup_{1\leq k\leq m}\Sigma_k$. Hence, $\Sigma$ consists of $m$ spheres $\Sigma_k$, $k = 1,\dots,m$, of equal radius with $\overline{\Sigma_k} \cap \overline{\Sigma_j} = \emptyset$, $l \neq j$, $\overline{\Sigma_k} \subset \Omega$, $k = 1,\dots,m$. The kernel of $L$ is spanned by $e_{\vartheta,\vartheta_\Sigma} = (0,1,1,0)$, $e_h = (0,0,0,1)$, $e_{ik} = (0,0,0,Y_k^i)$ with the spherical harmonics $Y_k^i$ of degree one for the spheres $\Sigma_k$, $i = 1,\dots,n$, $k = 1,\dots,m$.

To show that the equilibria are normally stable or normally hyperbolic, it remains to prove that $\lambda = 0$ is semi-simple. So suppose we have a solution of $L(u,\vartheta,h) = \sum_{i,k}\alpha_{ik}e_{ik} + \beta e_{\theta,\vartheta_\Sigma} + \delta e_h$. This means

$$\begin{aligned} -\mu_*\Delta u + \nabla\pi &= 0 \quad \text{in } \Omega\setminus\Sigma,\\ \operatorname{div} u &= 0 \quad \text{in } \Omega\setminus\Sigma,\\ P_\Sigma[\![u]\!] &= 0 \quad \text{on } \Sigma,\\ -[\![T(u,\pi,\theta_*)\nu_\Sigma]\!] + \sigma_*\mathcal{A}_\Sigma h\nu_\Sigma - \theta_*\sigma_*' H_*\vartheta_\Sigma\nu_\Sigma - \theta_*\sigma_*'\nabla_\Sigma\vartheta_\Sigma &= 0 \quad \text{in } \Sigma,\\ u &= 0 \quad \text{on } \partial\Omega. \end{aligned} \tag{33}$$

$$\begin{aligned} -d_*\Delta\vartheta &= \rho\kappa_*\beta \quad \text{in } \Omega\setminus\Sigma,\\ [\![\vartheta]\!] &= 0 \quad \text{on } \Sigma,\\ -d_{\Gamma_*}\Delta_\Sigma\vartheta_\Sigma - (l_*/\theta_*)j_\Sigma - [\![d_*\partial_{\nu_\Sigma}\vartheta]\!] - \sigma_*'\operatorname{div}_\Sigma u_\Sigma &= \kappa_{\Gamma_*}\beta \quad \text{on } \Sigma,\\ \partial_\nu\vartheta &= 0 \quad \text{on } \partial\Omega. \end{aligned} \tag{34}$$

$$\begin{aligned} -[\![T(u,\pi,\theta_*)\nu_\Sigma\cdot\nu_\Sigma/\rho]\!] + l_*\vartheta_\Sigma + \delta_* j_\Sigma &= 0 \quad \text{on } \Sigma,\\ -[\![\rho u\cdot\nu_\Sigma]\!]/[\![\rho]\!] &= \sum_{i,k}\alpha_{ik}Y_k^i + \delta \quad \text{on } \Sigma. \end{aligned} \tag{35}$$

We have to show $\alpha_{ik} = \beta = \delta = 0$ for all $i, k$. By $\operatorname{div} u = 0$, we have

$$\delta|\Sigma| = \delta|\Sigma| + \sum_{i,k} \alpha_{ik} \int_\Sigma Y_k^i \, d\Sigma = -\int_\Sigma [\![\rho u \cdot \nu_\Sigma]\!]/[\![\rho]\!] \, d\Sigma = 0,$$

which implies $\delta = 0$. Since $\int_\Sigma j_\Sigma \, d\Sigma = 0$ we have

$$\begin{aligned}\beta[(\rho\kappa_*|1)_\Omega + \kappa_{\Gamma_*}|\Sigma|] &= -\sigma_*' \int_\Sigma \operatorname{div}_\Sigma u_\Sigma \, d\Sigma = \sigma_*' H_* \int_\Sigma u_\Sigma \cdot \nu_\Sigma \, d\Sigma \\ &= \sigma_*' H_* \int_\Sigma [\![\rho u \cdot \nu_\Sigma]\!]/[\![\rho]\!] \, d\Sigma = 0,\end{aligned}$$

which implies $\beta = 0$.

Integrating and adding up the first equation of (33) multiplied by $u$, the fourth equation of (33) multiplied by $u_\Sigma$, the first equation of (34) multiplied by $\theta_*\vartheta$, the third equation of (34) multiplied by $\theta_*\vartheta_\Sigma$, and the first equation of (35) multiplied by $j_\Sigma$, we obtain

$$2|\mu_*^{1/2} D(u)|_2^2 + \gamma_* |j_\Sigma|_\Sigma^2 + \theta_*(|d_*^{1/2}\nabla\vartheta|_2^2 + |d_{\Gamma_*}^{1/2}\nabla_\Sigma\vartheta_\Sigma|_\Sigma^2) + \sigma_*(\mathcal{A}_\Sigma h|\nu_\Sigma \cdot u_\Sigma)_\Sigma = 0.$$

Next we observe that

$$(\mathcal{A}_\Sigma h \mid \nu_\Sigma \cdot u_\Sigma)_\Sigma = (\mathcal{A}_\Sigma h \mid [\![\rho u \cdot \nu_\Sigma]\!]/[\![\rho]\!])_\Sigma = -\sum_{i,k} \alpha_{ik} (\mathcal{A}_\Sigma h \mid Y_k^i)_\Sigma = 0,$$

since $\mathcal{A}_\Sigma$ is self-adjoint and the spherical harmonics $Y_k^i$ are in the kernel of $\mathcal{A}_\Sigma$. We can now conclude that $u_\Sigma = u - j_\Sigma \nu_\Sigma/\rho = 0$, and hence

$$0 = u_\Sigma \cdot \nu_\Sigma = -\sum_{i,k} \alpha_{ik} Y_k^i.$$

Thus $\alpha_{ik} = 0$ for $1 \le i \le n$, $1 \le k \le m$, as the spherical harmonics $Y_k^i$ are linearly independent. Therefore, the eigenvalue $\lambda = 0$ is semi-simple.

**9.** Let us summarize what we have proved.

**Theorem 4.4** *Let $L$ denote the linearization at $e_* := (0, \theta_*, \theta_*|_\Sigma, \Sigma) \in \mathcal{E}$ as defined above. Then $-L$ generates a compact analytic $C_0$-semigroup in $X_0$ which has maximal $L_p$-regularity. The spectrum of $L$ consists only of eigenvalues of finite algebraic multiplicity. Moreover, the following assertions are valid.*

(i) *The operator $-L$ has no eigenvalues $\lambda \neq 0$ with nonnegative real part if and only if $\Sigma$ is connected.*
(ii) *If $\Sigma$ is disconnected, then $-L$ has precisely $m - 1$ positive eigenvalues.*
(iii) *$\lambda = 0$ is an eigenvalue of $L$ and it is semi-simple.*

(iv) *The kernel $N(L)$ of $L$ is isomorphic to the tangent space $T_{e_*}\mathcal{E}$ of the manifold of equilibria $\mathcal{E}$ at $e_*$.*

*Consequently, $e_* = (0, \theta_*, \theta_*|_\Sigma, \Sigma) \in \mathcal{E}$ is normally stable if and only if $\Sigma$ is connected, and normally hyperbolic if and only if $\Sigma$ is disconnected.*

## 5 Nonlinear Stability of Equilibria

**1.** We look at Problem (4), (5) in the neighborhood of a non-degenerate equilibrium $e_* = (0, \theta_*, \Gamma_*) \in \mathcal{E}$. Employing a Hanzawa transform with reference manifold $\Sigma = \Gamma_*$ as in [15, Sect. 3], the transformed problem becomes

$$\begin{aligned}
\rho\partial_t u - \mu_*\Delta u + \nabla\pi &= F_u(u,\pi,\vartheta,h),\\
\operatorname{div} u &= G_d(u,h)\\
P_\Sigma[\![u]\!] &= G_u(u,h)\\
-2P_\Sigma[\![\mu_* D(u)\nu_\Sigma]\!] - \theta_*\sigma_*'\nabla_\Sigma\vartheta_\Sigma &= G_\tau(u,\vartheta,\vartheta_\Sigma,h),\\
-2[\![\mu_* D(u)\nu_\Sigma\cdot\nu_\Sigma]\!] + [\![\pi]\!] + \sigma_*\mathcal{A}_\Sigma h - \theta_*\sigma_*' H_*\vartheta_\Sigma &= G_\nu(u,\vartheta,h) + G_\gamma,\\
-2[\![\mu_* D(u)\nu_\Sigma\cdot\nu_\Sigma/\rho]\!] + [\![\pi/\rho]\!] + l_*\vartheta_\Sigma + \gamma_* j_\Sigma &= G_h(u,\vartheta,h),\\
u &= 0,\\
u(0) &= u_0,
\end{aligned} \tag{36}$$

with $G_\gamma = G_\gamma(\vartheta_\Sigma, h)$, where $\mu_* = \mu(\theta_*)$, $\sigma_* = \sigma(\theta_*)$, $\sigma_*' = \sigma'(\theta_*)$, $l_* = l(\theta_*)$, $\gamma_* = \gamma(\theta_*)$, and

$$\mathcal{A}_\Sigma = -H'(0) = -(n-1)/R_*^2 - \Delta_\Sigma.$$

For the relative temperature $\vartheta = (\theta - \theta_*)/\theta_*$ we obtain

$$\begin{aligned}
\rho\kappa_*\partial_t\vartheta - d_*\Delta\vartheta &= F_\theta(u,\vartheta,h) && \text{in } \Omega\setminus\Sigma,\\
\kappa_{\Gamma_*}\partial_t\vartheta_\Sigma - d_{\Gamma_*}\Delta_\Sigma\vartheta_\Sigma - (l_*/\theta_*)j_\Sigma - &&&\\
-[\![d_*\partial_{\nu_\Sigma}\vartheta]\!] - \sigma_*'\operatorname{div}_\Sigma u_\Sigma &= G_\theta(u,\vartheta,\vartheta_\Sigma,h) && \text{on } \Sigma,\\
[\![\vartheta]\!] &= 0 && \text{on } \Sigma,\\
\partial_\nu\vartheta &= 0 && \text{on } \partial\Omega,\\
\vartheta(0) &= \vartheta_0 && \text{in } \Omega,
\end{aligned} \tag{37}$$

with $\kappa_* = \kappa(\theta_*)$, $d_* = d(\theta_*)$, $\kappa_{\Gamma *} = \kappa_\Gamma(\theta_*)$, $d_{\Gamma *} = d_\Gamma(\theta_*)$. Finally, the evolution of $h$ is determined by

$$\begin{aligned} \partial_t h - [\![\rho u \cdot \nu_\Sigma]\!]/[\![\rho]\!] &= F_h(u,h) \quad \text{on } \Sigma, \\ h(0) &= h_0. \end{aligned} \tag{38}$$

Here $F_u$, $G_d$, $G_u$, $G_v$, $F_\theta$, $F_h$ are the same as in the case $\sigma$ where is constant; see [15, Sect. 3], and

$$\begin{aligned} G_\tau(u,\vartheta,\vartheta_\Sigma,h) &= 2P_\Sigma[\![(\mu(\theta)-\mu(\theta_*))D(u)\nu_\Sigma]\!] - 2P_\Sigma[\![\mu(\theta)D(u)M_0(h)\nabla_\Sigma h]\!] \\ &\quad - P_\Sigma[\![\mu(\theta)(M_1(h)\nabla u + [M_1(h)\nabla u]^{\mathsf{T}})(\nu_\Sigma - M_0(h)\nabla_\Sigma h)]\!] \\ &\quad + [\![\mu(\theta)((I-M_1)\nabla u + [(I-M_1)\nabla u]^T) \\ &\qquad\qquad (\nu_\Sigma - M_0\nabla_\Sigma h)\cdot\nu_\Sigma]\!] M_0(h)\nabla_\Sigma h \\ &\quad - \theta_*(\sigma'_* - (\sigma'(\theta_\Sigma)/\beta(h))\nabla_\Sigma\vartheta_\Sigma \\ &\quad - \theta_*(\sigma'(\theta_\Sigma)/\beta(h))P_\Sigma(I - P_\Gamma(h)M_0(h))\nabla_\Sigma\vartheta_\Sigma \\ &\quad + (\sigma'_*(\theta_\Sigma)/\beta(h))(P_\Gamma(h)M_0(h)\nabla_\Sigma\vartheta_\Sigma\cdot\nu_\Sigma)M_0(h)\nabla_\Sigma h, \end{aligned}$$

$$\begin{aligned} G_\gamma(\vartheta_\Sigma,h) &= \sigma(\theta_\Sigma)H(h) - \sigma(\theta_*)H'(0)h - \theta_*\sigma'_*H_*\vartheta_\Sigma \\ &\quad + \theta_*\sigma'(\theta_\Sigma)/\beta(h)(P_\Gamma(h)M_0(h)\nabla_\Sigma\vartheta_\Sigma\cdot M_0(h)\nabla_\Sigma h), \end{aligned}$$

$$\begin{aligned} G_\theta(u,\vartheta,\vartheta_\Sigma,h) &= (\kappa_{\Gamma_*} - \kappa_\Gamma(\theta_\Sigma))\partial_t\vartheta_\Sigma - \kappa_\Gamma(\theta_\Sigma)u_\Sigma\cdot P_\Gamma(h)M_0(h)\nabla_\Sigma\vartheta_\Sigma \\ &\quad - (d_{\Gamma_*} - d_\Gamma(\theta_\Sigma))\Delta_\Sigma\vartheta_\Sigma \\ &\quad - (d_\Gamma(\theta_\Sigma)\Delta_\Sigma\vartheta_\Sigma - \mathrm{tr}\{P_\Gamma(h)M_0(h)\nabla_\Sigma \\ &\qquad\qquad (d_\Gamma(\theta_\Sigma)P_\Gamma(h)M_0(h)\nabla_\Sigma\vartheta_\Sigma)\}) \\ &\quad - [\![d_*\partial\nu_\Sigma\vartheta]\!] + [\![d(\theta)(I-M_1(h))\nabla_\Sigma\vartheta_\Sigma\cdot\nu_\Gamma]\!] \\ &\quad - (\sigma'_*\mathrm{div}\,u_\Sigma - (\theta_\Sigma/\theta_*)\sigma'(\theta_\Sigma)\mathrm{tr}\{P_\Gamma(h)M_0(h)\nabla_\Sigma u_\Sigma\}) \\ &\quad - (1/\theta_*)(l(\theta_*)j_\Sigma - l(\theta_\Sigma)j_\Gamma) + (\gamma(\theta_\Sigma)/\theta_*)j_\Gamma^2, \end{aligned}$$

$$\begin{aligned} G_h(u,\vartheta,h) &= 2[\![(\mu(\theta)-\mu(\theta_*))D(u)\nu_\Sigma\cdot\nu_\Sigma/\rho]\!] \\ &\quad - 2([\![\mu(\theta)D(u)\nu_\Sigma\cdot\nu_\Sigma/\rho]\!] - [\![\mu(\theta)D(u)\nu_\Gamma\cdot\nu_\Gamma/\rho]\!]) \\ &\quad - [\![\mu(\theta)(M_1(h) + [M_1(h)]^{\mathsf{T}})\nu_\Gamma\cdot\nu_\Gamma/\rho]\!] \\ &\quad - ([\![\psi(\theta)]\!] + \theta_*[\![\psi'(\theta_*)]\!] - \theta_*[\![\psi'(\theta_*)]\!]\vartheta_\Sigma) \\ &\quad - (\gamma(\theta)j_\Gamma - \gamma(\theta_*)j_\Sigma) - [\![(1/2\rho^2)]\!]j_\Gamma^2, \end{aligned}$$

cf. [15, Sect. 2] for the definition of $M_0(h)$, $M_1(h)$, $P_\Gamma(h)$, and $\beta(h)$.

Moreover,

$$\nu_\Gamma = \beta(h)(\nu_\Sigma - M_0(h)\nabla_\Sigma h), \quad j_\Gamma = [\![u\cdot\nu_\Gamma]\!]/[\![1/\rho]\!], \quad j_\Sigma = [\![u\cdot\nu_\Sigma]\!]/[\![1/\rho]\!].$$

The nonlinearities are $C^1$ from $\mathbb{E}$ to $\mathbb{F}$, satisfying $(F_i'(0), G_k'(0)) = (0,0)$ for all $i \in \{u, \theta, h\}$ and $k \in \{d, u, \tau, \nu, \gamma, h, \theta\}$.

The state manifold locally near the equilibrium $e_* = (0, \theta_*, \Gamma_*)$ reads as

$$\begin{aligned} \mathcal{SM} := \Big\{&(u,\vartheta,h) \in L_p(\Omega)^{n+1} \times C^2(\Sigma) : \\ &(u,\vartheta) \in W_p^{2-2/p}(\Omega\setminus\Sigma)^{n+1},\ h \in W_p^{3-2/p}(\Sigma), \\ &\operatorname{div} u = G_d(u,h) \text{ in } \Omega\setminus\Sigma, \quad u = 0,\ \partial_\nu\theta = 0 \text{ on } \partial\Omega, \\ &P_\Sigma[\![u]\!] = G_u(u,\vartheta,h),\ [\![\vartheta]\!] = 0 \text{ on } \Sigma, \\ &-2P_\Sigma[\![\mu_* D(u)\nu_\Sigma]\!] - \theta_*\sigma_*'\nabla_\Sigma\vartheta = G_\tau(u,\vartheta,\vartheta_\Sigma,h) \text{ on } \Sigma\Big\}. \end{aligned} \tag{39}$$

Note that only $(u, \vartheta, h)$ need to be considered as state manifold for the flows, as $\vartheta_\Sigma$ is the trace of $\vartheta$, and the pressure $\pi$ can be recovered from the flow variables, as in [15]. Due to the compatibility conditions this is a nonlinear manifold. By parameterizing this manifold over its tangent space

$$\begin{aligned} \mathcal{SX} := \Big\{&(u,\vartheta,h) \in L_p(\Omega)^{n+1} \times C^2(\Sigma) : \\ &(u,\vartheta) \in W_p^{2-2/p}(\Omega\setminus\Sigma)^{n+1},\ h \in W_p^{3-2/p}(\Sigma), \\ &\operatorname{div} u = 0 \text{ in } \Omega\setminus\Sigma, \quad u = 0,\ \partial_\nu\theta = 0 \text{ on } \partial\Omega, \\ &P_\Sigma[\![u]\!] = 0,\ [\![\vartheta]\!] = 0,\ -2P_\Sigma[\![\mu_* D(u)\nu_\Sigma]\!] - \theta_*\sigma_*'\nabla_\Sigma\vartheta = 0 \text{ on } \Sigma\Big\}, \end{aligned}$$

the nonlinear problem (36)–(38) is written in the form studied in [15], where the generalized principle of linear stability is proved for the problem with constant surface tension. An adaption of this proof implies the following result.

**Theorem 5.1** *Let $p > n+2$, $\rho_1, \rho_2 > 0$, $\rho_1 \neq \rho_2$, and suppose the assumptions* **(a), (b), (c)** *in Sect. 1. Then in the topology of the state manifold $\mathcal{SM}$ we have:*

(i) $(0, \theta_*, \Gamma_*) \in \mathcal{E}$ *is stable if and only if* $\Gamma_*$ *is connected.*
(ii) *Any solution starting in a neighborhood of a stable equilibrium exists globally and converges to a possibly different stable equilibrium in the topology of $\mathcal{SM}$.*
(iii) *Any solution starting and* staying in *a neighborhood of an unstable equilibrium exists globally and converges to a possibly different unstable equilibrium in the topology of $\mathcal{SM}$.*

## 6 Qualitative Behaviour of the Semiflow

In this section we study the global properties of problem (4), (5), following the approach of [15, Sect. 6].

Recall that the closed $C^2$-hypersurfaces contained in $\Omega$ form a $C^2$-manifold, which we denote by $\mathcal{MH}^2(\Omega)$ as in [15, Sect. 2]. The charts are the parameterizations over a given hypersurface $\Sigma$, and the tangent space consists of the normal vector fields on $\Sigma$. We define a metric on $\mathcal{MH}^2(\Omega)$ by means of

$$d_{\mathcal{MH}^2}(\Sigma_1, \Sigma_2) := d_H(\mathcal{N}^2\Sigma_1, \mathcal{N}^2\Sigma_2),$$

where $d_H$ denotes the Hausdorff metric on the compact subsets of $\mathbb{R}^n$. This way $\mathcal{MH}^2(\Omega)$ becomes a Banach manifold of class $C^2$.

Let $d_\Sigma(x)$ denote the signed distance for $\Sigma$. We may then define the *level function* $\varphi_\Sigma$ by means of

$$\varphi_\Sigma(x) = \phi(d_\Sigma(x)), \quad x \in \mathbb{R}^n,$$

where

$$\phi(s) = s(1 - \chi(s/a)) + \operatorname{sgn} s\,\chi(s/a), \quad s \in \mathbb{R}.$$

It is easy to see that $\Sigma = \varphi_\Sigma^{-1}(0)$, and $\nabla\varphi_\Sigma(x) = \nu_\Sigma(x)$, for each $x \in \Sigma$. Moreover, $\kappa = 0$ is an eigenvalue of $\nabla^2\varphi_\Sigma(x)$, the remaining eigenvalues of $\nabla^2\varphi_\Sigma(x)$ are the principal curvatures $\kappa_j$ of $\Sigma$ at $x \in \Sigma$.

If we consider the subset $\mathcal{MH}^2(\Omega, r)$ of $\mathcal{MH}^2(\Omega)$ which consists of all closed hyper-surfaces $\Gamma \in \mathcal{MH}^2(\Omega)$ such that $\Gamma \subset \Omega$ satisfies the ball condition with fixed radius $r > 0$ then the map $\Phi : \mathcal{MH}^2(\Omega, r) \to C^2(\bar{\Omega})$ defined by $\Phi(\Gamma) = \varphi_\Gamma$ is an isomorphism of the metric space $\mathcal{MH}^2(\Omega, r)$ onto $\Phi(\mathcal{MH}^2(\Omega, r)) \subset C^2(\bar{\Omega})$.

Let $s - (n-1)/p > 2$; for $\Gamma \in \mathcal{MH}^2(\Omega, r)$, we define $\Gamma \in W^s_p(\Omega, r)$ if $\varphi_\Gamma \in W^s_p(\Omega)$. In this case the local charts for $\Gamma$ can be chosen of class $W^s_p$ as well. A subset $A \subset W^s_p(\Omega, r)$ is (relatively) compact, if and only if $\Phi(A) \subset W^s_p(\Omega)$ is (relatively) compact.

As an ambient space for the **state manifold** $\mathcal{SM}_\Gamma$ of Problem (1)–(3) we consider the product space $L_p(\Omega)^{n+1} \times \mathcal{MH}^2(\Omega)$ and set

$$\begin{aligned}
\mathcal{SM}_\Gamma := \Big\{&(u, \theta, \Gamma) \in L_p(\Omega)^{n+1} \times \mathcal{MH}^2(\Omega) : \\
&(u, \theta) \in W^{2-2/p}_p(\Omega \setminus \Gamma)^{n+1}, \ 0 < \theta < \theta_c, \ \Gamma \in W^{3-2/p}_p, \\
&\operatorname{div} u = 0 \text{ in } \Omega \setminus \Gamma, \quad u = \partial_\nu\theta = 0 \text{ on } \partial\Omega, \\
&P_\Gamma[\![u]\!] = 0, \ [\![\theta]\!] = 0 \text{ on } \Gamma, \\
&2P_\Gamma[\![\mu(\theta)D(u)\nu_\Gamma]\!] + \sigma'(\theta)\nabla_\Gamma\theta = 0 \text{ on } \Gamma\Big\}.
\end{aligned}$$

Charts for these manifolds are obtained by the charts induced by $\mathcal{MH}^2(\Omega)$, followed by a Hanzawa transformation.

Applying Theorem 3.3 and re-parameterizing the interface repeatedly, we see that (1)–(3) yields a local semiflow on $\mathcal{SM}_\Gamma$.

**Theorem 6.1** *Let $p > n+2$, $\sigma, \rho_1, \rho_2 > 0$, $\rho_1 \neq \rho_2$, and suppose the assumptions **(a), (b), (c)** in Sect. 1.*

*Then problem* (4), (5) *generates a local semiflow on the state manifold $\mathcal{SM}_\Gamma$. Each solution $(u, \theta, \Gamma)$ of the problem exists on a maximal time interval $[0, t_*)$, where $t_* = t_*(u_0, \theta_0, \Gamma_0)$.*

Again we note that the pressure $\pi$ as well as the phase flux $j_\Gamma$ and $\theta_\Gamma$ are dummy variables which are determined for each $t$ by the principal variables $(u, \theta, \Gamma)$. In fact,

$$j_\Gamma = [\![u \cdot \nu_\Gamma]\!]/[\![1/\rho]\!],$$

$\theta_\Gamma$ is the trace of $\theta$, and $\pi$ is determined by the weak transmission problem

$$\begin{aligned}
&(\nabla\pi|\nabla\phi/\rho)_{L_2(\Omega)} \\
&= (2\rho^{-1}\mathrm{div}(\mu(\theta)D(u)) - u\cdot\nabla u|\nabla\phi)_{L_2(\Omega)}, \quad \phi \in H^1_{p'}(\Omega),\ \phi = 0 \text{ on } \Gamma, \\
&[\![\pi]\!] = 2[\![\mu(\theta)D(u)\nu_\Gamma]\!]\nu_\Gamma + \sigma(\theta_\Gamma)H_\Gamma - [\![1/\rho]\!]j_\Gamma^2 \ \text{ on } \Gamma, \\
&[\![\pi/\rho]\!] = 2[\![(\mu(\theta)/\rho)D(u)\nu_\Gamma\cdot\nu_\Gamma]\!] - [\![1/(2\rho^2)]\!]j_\Gamma^2 - [\![\psi(\theta)]\!] - \gamma(\theta_\Gamma)j_\Gamma \ \text{ on } \Gamma,
\end{aligned}$$

Concerning such transmission problems we refer to [7].

## 6.1 Convergence

There are several obstructions against global existence:

- *regularity*: the norms of either $u(t)$, $\theta(t)$, or $\Gamma(t)$ become unbounded;
- *geometry*: the topology of the interface changes; or the interface touches the boundary of $\Omega$.
- *well-posedness*: the temperature leaves the range $0 < \theta(t) < \theta_c$.

Note that the compatibility conditions,

$$\begin{aligned}
&\mathrm{div}\, u(t) = 0 \text{ in } \Omega\setminus\Gamma(t), \quad u = 0,\ \partial_\nu\theta = 0 \text{ on } \partial\Omega, \\
&P_\Gamma[\![u(t)]\!] = 0,\ [\![\theta]\!] = 0,\ P_\Gamma[\![\mu(\theta)(\nabla u + [\nabla u]^{\mathsf{T}})\nu_\Gamma]\!] + \sigma'(\theta_\Gamma)\nabla_\Gamma\theta_\Gamma = 0 \text{ on } \Gamma(t),
\end{aligned}$$

are preserved by the semiflow.

Let $(u, \theta, \Gamma)$ be a solution in the state manifold $\mathcal{SM}_\Gamma$ with maximal interval $[0, t_*)$. By the *uniform ball condition* we mean the existence of a radius $r_0 > 0$

such that for each $t$, at each point $x \in \Gamma(t)$ there exists centers $x_i \in \Omega_i(t)$ such that $B_{r_0}(x_i) \subset \Omega_i$ and $\Gamma(t) \cap \bar{B}_{r_0}(x_i) = \{x\}$, $i = 1, 2$. Note that this condition bounds the curvature of $\Gamma(t)$, prevents parts of it to touch the outer boundary $\partial\Omega$, and to undergo topological changes. Hence if this condition holds, then the volumes of the phases are preserved.

With this property, combining the local semiflow for (4), (5) with the Lyapunov functional, i.e., the negative total entropy, and compactness we obtain the following result.

**Theorem 6.2** *Let $p > n+2$, $\rho_1, \rho_2 > 0$, $\rho_1 \neq \rho_2$, and suppose the assumptions **(a)**, **(b)**, **(c)** in Sect. 1. Suppose that $(u, \theta, \Gamma)$ is a solution of* (4), (5) *in the state manifold $\mathcal{SM}_\Gamma$ on its maximal time interval $[0, t_*)$. Assume that the following conditions hold on $[0, t_*)$:*

(i) $\sup_{0<t<t_*}(|u(t)|_{W_p^{2-2/p}}, |\theta(t)|_{W_p^{2-2/p}}, |\Gamma(t)|_{W_p^{3-2/p}}) < \infty$;
(ii) $\Gamma(t)$ *satisfies the uniform ball condition;*
(iii) $\inf_{0<t<t_*} \theta(t) > 0$ *and* $\sup_{0<t<t_*} \theta(t) < \theta_c$ *on* $\bar{\Omega}$.

*Then $t_* = \infty$, i.e. the solution exists globally, and its limit set $\omega_+(u, \theta, \Gamma) \subset \mathcal{E}$ is non-empty. If further $(0, \theta_\infty, \Gamma_\infty) \in \omega_+(u, \theta, \Gamma)$ with $\Gamma_\infty$ connected, then the solution converges in $\mathcal{SM}$ to this equilibrium.*

*Conversely, if $(u(t), \theta(t), \Gamma(t))$ is a global solution in $\mathcal{SM}$ which converges to an equilibrium $(0, \theta_*, \Gamma_*) \in \mathcal{E}$ in $\mathcal{SM}$ as $t \to \infty$, then (i), (ii) and (iii) are valid.*

*Proof* The assertions follow by similar arguments as in the proof of Theorem 6.2 in [15]. □

**Acknowledgements** The work of SS was partially supported by JSPS Grant-in-Aid for Scientific Research (B) #24340025 and Challenging Exploratory Research #23654048. The work of GS was partially supported by a grant from the Simons Foundation (#245959) and by a grant from NSF (DMS-1265579).

## References

1. D.M. Anderson, P. Cermelli, E. Fried, M.E. Gurtin, G.B. McFadden, General dynamical sharp-interface conditions for phase transformations in viscous heat-conducting fluids. J. Fluid Mech. **581**, 323–370 (2007)
2. K.S. Das, C.A. Ward, Surface thermal capacity and its effect on the boundary conditions of fluid-fluid interfaces. Phys. Rev. E **75**, 065303 (2007)
3. R. Denk, M. Hieber, J. Prüss, $\mathcal{R}$-boundedness, Fourier multipliers, and problems of elliptic and parabolic type, in *AMS Memoirs*, vol. 788 (American Mathematical Society, Providence, RI, 2003)
4. R. Denk, M. Hieber, J. Prüss, Optimal $L^p$-$L^q$-estimates for parabolic boundary value problems with inhomogeneous data. Math. Z. **257**, 193–224 (2007)
5. R. Denk, J. Prüss, R. Zacher, Maximal $L_p$-regularity of parabolic problems with boundary dynamics of relaxation type. J. Funct. Anal. **255**, 3149–3187 (2008)
6. M. Ishii, H. Takashi, *Thermo-Fluid Dynamics of Two-Phase Flow* (Springer, New York, 2006)

7. M. Köhne, J. Prüss, M. Wilke, Qualitative behaviour of solutions for the two-phase Navier-Stokes equations with surface tension. Math. Ann. **356**, 737–792 (2013)
8. J. Prüss, Maximal regularity for evolution equations in $L_p$-spaces. Conf. Sem. Mat. Univ. Bari **285**, 1–39 (2003)
9. J. Prüss, S. Shimizu, On well-posedness of incompressible two-phase flows with phase transitions: the case of non-equal densities. J. Evol. Equ. **12**, 917–941 (2012)
10. J. Prüss, G. Simonett, Maximal regularity for evolution equations in weighted $L_p$-spaces. Arch. Math. **82**, 415–431 (2004)
11. J. Prüss, G. Simonett, R. Zacher, Convergence of solutions to equilibria for nonlinear parabolic problems. J. Differ. Equ. **246**, 3902–3931 (2009)
12. J. Prüss, Y. Shibata, S. Shimizu, G. Simonett, On well-posedness of incompressible two-phase flows with phase transition: the case of equal densities. Evol. Equ. Control Theory **1**, 171–194 (2012)
13. J. Prüss, G. Simonett, R. Zacher, Qualitative behaviour of solutions for thermodynamically consistent Stefan problems with surface tension. Arch. Ration. Mech. Anal. **207**, 611–667 (2013)
14. J. Prüss, G. Simonett, R. Zacher, On the qualitative behaviour of incompressible two-phase flows with phase transition: the case of equal densities. Interfaces Free Bound. **15**, 405–428 (2013)
15. J. Prüss, S. Shimizu, M. Wilke, On the qualitative behaviour of incompressible two-phase flows with phase transition: the case of non-equal densities. Commun. Part. Differ. Equ. **39**, 1236–1283 (2014)
16. J. Prüss, G. Simonett, M. Wilke, On thermodynamically consistent Stefan problems with variable surface energy. Arch. Ration. Mech. Anal. (2015). doi:10.1007/s00205-015-0938-y

# On the Nash-Moser Iteration Technique

**Paolo Secchi**

*Dedicated to Professor Y. Shibata in occasion of his 60-th birthday*

**Abstract** The aim of this work is to provide a brief presentation of the Nash-Moser iteration technique for the resolution of nonlinear equations, where the linearized equations admit estimates with a loss of regularity with respect to the given data.

**Keywords** Characteristic free boundary • Compressible Euler equations • Implicit function theorems • Nash-Moser theorem • Nonlinear hyperbolic systems

## 1 Introduction

The aim of this work is to provide a brief presentation of the Nash-Moser iteration method for the resolution of nonlinear equations, where the linearized equations admit estimates with a loss of regularity with respect to the given data. This method was originally introduced by Nash in [16] for solving the isometric embedding problem. Moser in [14, 15] simplified the method at the expense of a loss of regularity, and showed how to apply it in a more general setting. Since then, this iteration technique is known as the Nash-Moser method. Hörmander, in his paper [10] on the boundary problem of physical geodesy, improved on Moser's scheme by reducing the loss of regularity, using a scheme more similar to Nash's original.

Our personal interest is motivated by the study of certain characteristic free-boundary problems for systems of nonlinear conservation laws that arise in fluid dynamics. Interesting and challenging problems arise when the unknown free-boundary is weakly but not strongly stable, i.e. the Kreiss-Lopatinskiĭ condition only holds in weak form. A typical difficulty in the analysis of weakly stable problems is the loss of regularity in the a priori estimates of solutions. Short-time existence results have been obtained for various weakly stable nonlinear problems, typically

P. Secchi (✉)
DICATAM, Section of Mathematics, University of Brescia, Via Valotti 9, 25133 Brescia, Italy
e-mail: paolo.secchi@unibs.it



H. Amann et al. (eds.), *Recent Developments of Mathematical Fluid Mechanics*, Advances in Mathematical Fluid Mechanics, DOI 10.1007/978-3-0348-0939-9_23

by the use of a Nash-Moser scheme to compensate for the loss of derivatives in the linearized energy estimates.

Alinhac [1] used a modified version of Hörmander's scheme to prove the short-time existence of rarefaction waves for a class of conservation laws. More recently, Coulombel and Secchi [6] introduced an additional modification to prove the existence of compressible vortex sheets for the two dimensional Euler equations, provided the Mach number is sufficiently large. A scheme similar to the one used in [6] is also considered by Trakhinin in the paper [19] on current-vortex sheets (see also [5]) and by Secchi and Trakhinin in the paper [18] on the plasma-vacuum interface problem.

In this paper we aim to provide a brief presentation of the Nash-Moser iteration method, whilst keeping in mind that our main interest is the application to PDE problems. We present a simplified version of the scheme of Coulombel and Secchi [6]; our exposition is also much indebted with [2]. We refer to [2, 9, 11] for a general description of the method. Other related classical references are [3, 4, 13, 17], see also the recent paper [7].

Finally, it is interesting to recall that the isometric embedding problem, originally solved by Nash with this method, was solved much later with an ordinary fixed-point argument, see [8]. Nevertheless, the Nash-Moser method remains a fundamental tool of nonlinear analysis for the study of perturbation problems.

Given $\mathcal{F} : X \to Y$, with $X, Y$ Banach spaces, suppose we wish to solve the nonlinear equation

$$\mathcal{F}(u) = f. \tag{1}$$

We assume $\mathcal{F}(0) = 0$; here $f$ is a given "small" perturbation and we look for a solution $u$ close to 0.

Assume $\mathcal{F}$ is continuously differentiable and the differential $d\mathcal{F}(\cdot)$ is invertible in a neighborhood of $u = 0$, so that $\mathcal{F}$ is locally invertible. One of the most classical methods for solving such a nonlinear equation via linearization is Newton iteration method, where the approximating sequence is defined by

$$\begin{aligned} u_0 &= 0, \\ u_{k+1} &= u_k + (d\mathcal{F}(u_k))^{-1}(f - \mathcal{F}(u_k)), \quad k \geq 1. \end{aligned} \tag{2}$$

It is well-known that Newton's method has a fast convergence rate:

$$\|u_{k+1} - u_k\|_X \leq C\|u_k - u_{k-1}\|_X^2.$$

However, for this scheme to make sense, we need the inverse $(d\mathcal{F}(u))^{-1}$. In fact, the linearized equation

$$d\mathcal{F}(u)v = g \tag{3}$$

may be difficult or impossible to solve for $v \in X$, hence we may not be able to define $(d\mathcal{F}(u))^{-1}$ in a neighborhood of $u = 0$.

In order to introduce the typical situations in which we may recourse to the Nash-Moser method, let us change the formulation as follows.

Instead of single spaces $X, Y$, suppose we are given scales of Banach spaces $X_0 \supset X_1 \supset \cdots \supset X_m \supset \ldots$ with increasing norms $\|\cdot\|_{X_m}$, $m \geq 0$, and spaces $Y_0 \supset Y_1 \supset \cdots \supset Y_m \supset \ldots$ with increasing norms $\|\cdot\|_{Y_m}$, $m \geq 0$. For instance $X_m = H^m$ (Sobolev spaces) or $X_s = C^s$ (Hölder spaces). Having in mind possible applications to nonlinear hyperbolic equations where the natural function spaces are the Sobolev spaces $H^m$, we will consider Banach scales $X_m, Y_m$ with discrete indices $m$ (instead of a continuous parameter $s \in \mathbb{R}$).

We again wish to solve (1) where now $\mathcal{F} : X_m \to Y_m$ for every $m \geq 0$, but $d\mathcal{F}(\cdot)$ is only invertible between $Y_m$ and $X_{m-s}$, with a loss of regularity of order $s$. To be more specific, let us suppose that we have a solution $v$ of (3) satisfying, for some given $u$, an estimate of the form

$$\|v\|_{X_m} \leq C\|g\|_{Y_{m+s}}$$

for all $m$ (in a finite interval), $s$ being a fixed number. In this case we say that the equation is solved with a "loss of $s$ derivatives"(clearly in arbitrary families of spaces $X_m, Y_m$ this expression cannot make sense).

Trying to apply again Newton's method (2) we would get

$$\begin{aligned} & u_k \in X_m, \\ & u_{k+1} = u_k + (d\mathcal{F}(u_k))^{-1}\underbrace{(f - \mathcal{F}(u_k))}_{\in Y_m} \in X_{m-s} \end{aligned}$$

with a finite loss of regularity at each step. Iteration is then impossible.

Furthermore, this loss $s$ may be doubled by a loss $s'$ due to the cost of solving (3) in terms of information about the coefficients of the equation, that is about $u$. Let us suppose, for example, that the solution $v$ satisfies an estimate of the form

$$\|v\|_{X_m} \leq C\left(\|g\|_{Y_{m+s}} + \|g\|_{Y_{m_0}}\|u\|_{X_{m+s'}}\right) \tag{4}$$

for all $m$, with $m_0, s$ and $s'$ fixed. In Newton's method, which uses $d\mathcal{F}(u_k)$ to calculate $u_{k+1}$, the solution to one step becomes the coefficient of the next, and the loss $s'$ is added to $s$.

We will see that the nature of this double loss of derivatives determines the applicability of the Nash-Moser technique. Roughly speaking, it is sufficient for the losses $s$ and $s'$ to be fixed, in which case (4) is said to be a "tame" estimate.

To overcome this difficulty, the key idea of Nash was to modify Newton's scheme (2) by including a smoothing operator at each step to compensate for the loss of regularity. Let us set $u_{k+1} = u_k + \delta u_k$ and write (2) as

$$\delta u_k = (d\mathcal{F}(u_k))^{-1}g_k, \qquad g_k = f - \mathcal{F}(u_k).$$

Now let us suppose we have a family of smoothing operators $\{S_X(\theta)\}_{\theta\geq 1}$

$$S_X(\theta) : X_0 \to X_\infty := \cap_{m\geq 0} X_m$$

satisfying $S_X(\theta) \to Id$ as $\theta \to \infty$ and other properties that will be detailed later on.[1] We modify the scheme by setting[2]

$$\delta u_k = (d\mathcal{F}(S_X(\theta_k)u_k))^{-1} S_Y(\theta_k) g_k, \tag{5}$$

where $\{\theta_k\}_{k\geq 1}$ is a sequence such that $\theta_k \to \infty$ as $k \to \infty$. Since $S_X(\theta_k) \to Id_X$ and $S_Y(\theta_k) \to Id_Y$ as $k \to \infty$, the scheme looks like Newton iteration for large $k$, so we might expect it to converge under certain conditions. In fact, balancing in appropriate way the fast convergence rate of Newton's scheme and loss of regularity gives the convergence of the approximating sequence.

In applications, very often one observes a rather big difference of regularity between the solution $u$ and the data $f$. That is, generally the regularity of the solution is not optimal compared with the regularity of the data that appear exaggerated. Thus, other modifications to the scheme are introduced in order to reduce this difference of regularity. In this paper, we will not insist on this feature of the method.

Depending on the particular problem under consideration, sometimes *after* having solved the equation by the Nash-Moser technique, one can try to get the optimal regularity of solution, see [12].

## 2 Statement of the Nash-Moser Theorem

We first give the main assumptions on the function $\mathcal{F}$.

**Assumption 2.1** *For all $u \in U \cap X_\infty$, where $U$ is a bounded open neighborhood of 0 in $X_{m_0}$ for some $m_0 \geq 0$, the function $\mathcal{F} : X_m \to Y_m$ is twice differentiable and satisfies the tame estimate*

$$\|d^2\mathcal{F}(u)(v_1, v_2)\|_{Y_m} \leq C\big(\|v_1\|_{X_{m+r}}\|v_2\|_{X_{m_0}} + \|v_1\|_{X_{m_0}}\|v_2\|_{X_{m+r}} + \|v_1\|_{X_{m_0}}\|v_2\|_{X_{m_0}}(1 + \|u\|_{X_{m+r'}})\big) \tag{6}$$

*for all $m \geq 0$ and for all $v_1, v_2 \in X_\infty$, for some fixed integers $r, r' \geq 0$. The constant $C$ is bounded for $m$ bounded.*

---

[1] Similar smoothing operators $S_Y(\theta) : Y_0 \to Y_\infty := \cap_{m\geq 0} Y_m$ are introduced as well.

[2] Following Hörmander's method [2, 11], in the sequel our iteration scheme will be a little more elaborated than in (5).

**Assumption 2.2** *For all $u \in U \cap X_\infty$ there exists a linear mapping $\Psi(u) : Y_\infty \to X_\infty$ such that $d\mathcal{F}(u)\Psi(u) = Id$, and satisfying the tame estimate*

$$\|\Psi(u)g\|_{X_m} \le C\left(\|g\|_{Y_{m+s}} + \|g\|_{Y_{m_0}}\|u\|_{X_{n+s'}}\right) \tag{7}$$

*for all $m \ge 0$ and some fixed integers $s, s' \ge 0$. The constant $C$ is bounded for $m$ bounded.*

The method requires a family of smoothing operators; for its construction in Sobolev and Hölder spaces we refer the reader to [1, 2].

**Definition 2.3** The decreasing family of Banach spaces $\{X_m\}_{m\ge 0}$ satisfies *the smoothing hypothesis* if there exists a family $\{S_\theta\}_{\theta\ge 1}$ of operators $S_\theta : X_0 \to X_\infty := \cap_{m\ge 0} X_m$ such that

$$\|S_\theta u\|_{X_\beta} \le C\,\theta^{(\beta-\alpha)_+}\,\|u\|_{X_\alpha} \qquad \forall\,\alpha,\beta \ge 0\,, \tag{8a}$$

$$\|S_\theta u - u\|_{X_\beta} \le C\,\theta^{\beta-\alpha}\,\|u\|_{X_\alpha} \qquad 0 \le \beta \le \alpha\,, \tag{8b}$$

$$\|\frac{d}{d\theta}S_\theta u\|_{X_\beta} \le C\,\theta^{\beta-\alpha-1}\,\|u\|_{X_\alpha} \qquad \forall\,\alpha,\beta \ge 0\,. \tag{8c}$$

Here we use the classical notation $(\beta-\alpha)_+ := \max(0, \beta-\alpha)$. The constants in the inequalities are uniform with respect to $\alpha, \beta$, when $\alpha, \beta$ belong to some bounded interval.

In the decreasing family of Banach spaces $\{Y_m\}_{m\ge 0}$ we will introduce similar smoothing operators $S_Y(\theta) : Y_0 \to Y_\infty := \cap_{m\ge 0} Y_m$. To not overload the notation such smoothing operators $S_Y(\theta)$ will be again denoted by $S_\theta$.

Now we can give our statement of the Nash-Moser theorem.

**Theorem 2.4** *Let $\{X_m\}_{m\ge 0}$ and $\{Y_m\}_{m\ge 0}$ be two decreasing family of Banach spaces, each satisfying the smoothing hypothesis, and assume that both Assumptions 6 and 7 hold. Let $m'$ be a positive integer such that $m' \ge m_0 + \max\{r, r'\} + \max\{s, s'\}$.*

*i) There exists a constant $0 < \epsilon \le 1$ such that if $f \in Y_{m'+s+1}$ with*

$$\|f\|_{Y_{m'+s+1}} \le \epsilon,$$

*the equation $\mathcal{F}(u) = f$ has a solution $u \in X_{m'}$, in the sense that there exists a sequence $\{u_n\} \subset X_\infty$ such that $u_n \to u$ in $X_{m'}$, $\mathcal{F}(u_n) \to f$ in $Y_{m'+s}$, as $n \to \infty$.*

*ii) Moreover, if there exists $m'' > m'$ such that $f \in Y_{m''+s+1}$, then the solution constructed $u \in X_{m''}$.*

As for the regularity of $u$ see Remark 3.10 at the end of Sect. 3.6.

# 3 Proof of Theorem 2.4

## 3.1 Description of the Iterative Scheme

The scheme starts from $u_0 = 0$. Assume that $u_k$ are already given for $k = 0, \dots, n$. We consider

$$u_{n+1} = u_n + \delta u_n \,, \tag{9}$$

where the differences $\delta u_n$ will be specified later on. Given $\theta_0 \geq 1$, let us set $\theta_n := (\theta_0^2 + n)^{1/2}$, and consider the smoothing operators $S_{\theta_n}$. We decompose

$$\mathcal{F}(u_{n+1}) - \mathcal{F}(u_n) = d\mathcal{F}(u_n)(\delta u_n) + e_n' = d\mathcal{F}(S_{\theta_n} u_n)(\delta u_n) + e_n' + e_n'' ,$$

where $e_n'$ denotes the usual "quadratic"error of Newton's scheme, and $e_n''$ the "substitution"error. Let us also set

$$e_n := e_n' + e_n''. \tag{10}$$

The iteration proceeds as follows. Assume that $f$ is as in the statement of the theorem. Given

$$\begin{array}{lll} u_0 := 0 \,, & f_0 := S_{\theta_0} f \,, & E_0 := 0 \,, \\ u_1, \dots, u_n \,, & f_1, \dots, f_{n-1} \,, & e_0, \dots, e_{n-1}, \end{array}$$

we first compute for $n \geq 1$

$$E_n := \sum_{k=0}^{n-1} e_k \,. \tag{11}$$

These are the accumulated errors at the step $n$. Then we compute $f_n$ from the equations:

$$\sum_{k=0}^{n} f_k + S_{\theta_n} E_n = S_{\theta_n} f \,, \tag{12}$$

and we solve the linear equation

$$d\mathcal{F}(S_{\theta_n} u_n) \, \delta u_n = f_n, \tag{13}$$

finding $\delta u_n$. Finally, we compute $e_n$ from

$$\mathcal{F}(u_{n+1}) - \mathcal{F}(u_n) = f_n + e_n \,. \tag{14}$$

For $n = 0$ we only consider (13), (14) and compute $u_1, e_0$. Adding (14) from 0 to $N$, and combining with (12) gives

$$\mathcal{F}(u_{N+1}) - f = (S_{\theta_N} - I)f + (I - S_{\theta_N})E_N - e_N \,. \tag{15}$$

Because $S_{\theta_N} \to I$ as $N \to +\infty$, and since we expect $e_N \to 0$, we will formally obtain the solution of the problem (1) from

$$\mathcal{F}(u_{N+1}) \to f.$$

## 3.2 Introduction of the Iterative Scheme

We recall that the sequence $\{\theta_n\}_{n\geq 0}$ is defined by $\theta_n := (\theta_0^2 + n)^{1/2}$, for some $\theta_0 \geq 1$. Let us denote $\Delta_n := \theta_{n+1} - \theta_n$. In particular, the sequence $\{\Delta_n\}$ is decreasing, and tends to zero. Moreover, one has

$$\forall n \in \mathbb{N}\,, \quad \frac{1}{3\theta_n} \leq \Delta_n = \sqrt{\theta_n^2 + 1} - \theta_n \leq \frac{1}{2\theta_n}\,.$$

Let us take an integer $\alpha \geq m_0 + 1$, a small number $0 < \delta < 1$, and an integer $\tilde{\alpha} > \alpha$ that will be chosen later on. Our inductive assumption reads:

$$(H_{n-1}) \qquad \begin{cases} \forall\, k = 0, \dots, n-1\,, & \forall\, m \in [m_0, \tilde{\alpha}] \cap \mathbb{N}\,, \\ \|\delta u_k\|_{X_m} \leq \delta\, \theta_k^{m-\alpha-1}\, \Delta_k\,. \end{cases}$$

The next task is to prove that for a suitable choice of the parameters $\theta_0 \geq 1$, and $\delta > 0$, and for $f$ small enough, $(H_{n-1})$ implies $(H_n)$. In the end, we shall prove that $(H_0)$ holds for $f$ sufficiently small.

From now on, we assume that $(H_{n-1})$ holds. Let us show some basic consequences:

**Lemma 3.1** *If $\theta_0$ is big enough, chosen independently of $\alpha$, then for every $k = 0, \dots, n$, and for every integer $m \in [m_0, \tilde{\alpha}]$, we have*

$$\|u_k\|_{X_m} \leq \delta\, \theta_k^{(m-\alpha)_+}\,, \qquad m \neq \alpha\,, \tag{16a}$$

$$\|u_k\|_{X_\alpha} \leq \delta\, \log \theta_k\,. \tag{16b}$$

The proof follows from the triangle inequality, and from the classical comparisons between series and integrals. The choice of how large should be $\theta_0$ is independent of $\alpha$.

**Lemma 3.2** *If $\theta_0$ is big enough, chosen independently of $\alpha$, then for every $k = 0, \dots, n$, and for every integer $m \in [m_0, \tilde{\alpha} + \max\{r', s'\}]$, we have*

$$\|S_{\theta_k} u_k\|_{X_m} \le C\,\delta\,\theta_k^{(m-\alpha)+}\,, \qquad m \neq \alpha, \tag{17a}$$

$$\|S_{\theta_k} u_k\|_{X_\alpha} \le C\,\delta\,\log\theta_k\,. \tag{17b}$$

*For every $k = 0, \dots, n$, and for every integer $m \in [m_0, \tilde{\alpha}]$, we have*

$$\|(I - S_{\theta_k}) u_k\|_{X_m} \le C\,\delta\,\theta_k^{m-\alpha}\,. \tag{18}$$

The proof follows from Lemma 3.1 and the properties of the smoothing operators, respectively (8a) for (17) and (8b) for (18). We remark that the choice of how large should be $\theta_0$ is independent of $\alpha$.

## 3.3 Estimate of the Errors

### 3.3.1 Estimate of the Quadratic Errors

We start by proving an estimate for the quadratic error $e'_k$ of the iterative scheme. Recall that this error is defined by

$$e'_k := \mathcal{F}(u_{k+1}) - \mathcal{F}(u_k) - d\mathcal{F}(u_k)\delta u_k\,, \tag{19}$$

**Lemma 3.3** *Assume that $\alpha \ge m_0 + 1$ also satisfies $\alpha \ge m_0 + r' - r + 1$. There exist $\delta > 0$ sufficiently small, and $\theta_0 \ge 1$ sufficiently large, both chosen independently of $\alpha$, such that for all $k = 0, \dots, n-1$, and for all integer $m \in [m_0, \tilde{\alpha} - \max\{r, r'\}]$, one has*

$$\|e'_k\|_{Y_m} \le C\,\delta^2\,\theta_k^{L_1(m)-1}\,\Delta_k\,, \tag{20}$$

*where $L_1(m) := \max\{m + m_0 + r - 2\alpha - 2;\, (m + r' - \alpha)_+ + 2m_0 - 2\alpha - 2\}$.*

*Proof* The quadratic error given in (19) may be written as

$$e'_k = \int_0^1 (1-\tau)\, d^2\mathcal{F}(u_k + \tau\,\delta u_k)(\delta u_k, \delta u_k)\, d\tau\,.$$

From $(H_{n-1})$ and (16a), we have

$$\sup_{\tau\in[0,1]} \|u_k + \tau\,\delta u_k\|_{X_{m_0}} \le 2\,\delta\,,$$

so for $\delta$ sufficiently small we can apply the tame estimate (6). Using $(H_{n-1})$ and (16) we obtain (20). If $m + r' = \alpha$ we use (16b) with $\log \theta_k \le \theta_k$; it yields an estimate of the form (20) if $\alpha \ge m_0 + r' - r + 1$. □

### 3.3.2 Estimate of the Substitution Errors

Now we estimate the substitution error $e''_k$ of the iterative scheme, defined by

$$e''_k := d\mathcal{F}(u_k)\delta u_k - d\mathcal{F}(S_{\theta_k} u_k)\delta u_k \,. \tag{21}$$

**Lemma 3.4** *Assume that $\alpha \ge m_0 + 1$ also satisfies $\alpha \ge m_0 + r' - r + 1$. There exist $\delta > 0$ sufficiently small, and $\theta_0 \ge 1$ sufficiently large, both chosen independently of $\alpha$, such that for all $k = 0, \dots, n-1$, and for all integer $m \in [m_0, \tilde{\alpha} - \max\{r, r'\}]$, one has*

$$\|e''_k\|_{Y_m} \le C\,\delta^2\,\theta_k^{L(m)-1}\,\Delta_k \,, \tag{22}$$

*where* $L(m) := \max\{m + m_0 + r - 2\alpha; (m + r' - \alpha)_+ + 2m_0 - 2\alpha\}$.

*Proof* The substitution error given in (21) may be written as

$$e''_k = \int_0^1 d^2\mathcal{F}(S_{\theta_k} u_k + \tau(I - S_{\theta_k})u_k)(\delta u_k, (I - S_{\theta_k})u_k)\,d\tau \,.$$

As in the calculation for the quadratic error, we first show that we can apply (6) for $\delta$ sufficiently small. Then, the estimate (22) follows from $(H_{n-1})$, (16)–(18). □

Adding (20), (22) gives the estimate for the sum of errors defined in (10):

**Lemma 3.5** *Assume that $\alpha \ge m_0 + 1$ also satisfies $\alpha \ge m_0 - r' - r + 1$. There exist $\delta > 0$ sufficiently small, and $\theta_0 \ge 1$ sufficiently large, both chosen independently of $\alpha$, such that for all $k = 0, \dots, n-1$ and all integer $m \in [m_0, \tilde{\alpha} - \max\{r, r'\}]$, one has*

$$\|e_k\|_{Y_m} \le C\,\delta^2\,\theta_k^{L(m)-1}\,\Delta_k \,, \tag{23}$$

*where $L(m)$ is defined in Lemma 3.4.*

The preceding lemma immediately yields the estimate of the accumulated error $E_n$ defined in (11):

**Lemma 3.6** *Assume that $\alpha \ge m_0 + 1$ also satisfies $\alpha \ge m_0 + r' - r + 1$. Let $\tilde{\alpha} = 2\alpha + \max\{r, r'\} + 1 - m_0 - r$. There exist $\delta > 0$ sufficiently small, $\theta_0 \ge 1$ sufficiently large, both chosen independently of $\alpha$, such that*

$$\|E_n\|_{Y_p} \le C\,\delta^2\,\theta_n^{L(p)} \,, \tag{24}$$

*where we have set $p := \tilde{\alpha} - \max\{r, r'\}$.*

*Proof* For the estimate in $Y_p$ of the accumulated error we choose $p$ to be as large as possible, namely $p = \tilde{\alpha} - \max\{r, r'\}$. Moreover $\tilde{\alpha}$ is taken sufficiently large so that $L(p) \geq 1$. Then it follows from (23) that

$$\|E_n\|_{Y_p} \leq C\,\delta^2 \sum_{k=0}^{n-1} \theta_k^{L(p)-1}\,\Delta_k \leq C\,\delta^2\,\theta_n^{L(p)},$$

which gives (24). We can check that $L(p) \geq 1$ if

$$\begin{aligned}\tilde{\alpha} \geq \min\{2\alpha + \max\{r, r'\} + 1 - m_0 - r, 3\alpha + \max\{r, r'\} + 1 - 2m_0 - r'\}\\ = 2\alpha + \max\{r, r'\} + 1 - m_0 - r\,,\end{aligned}$$

which explains our choice for $\tilde{\alpha}$. □

## 3.4 *Estimate of $f_n$*

Going on with the iteration scheme, the next lemma gives the estimates of the source term $f_n$, defined by Eq. (12):

**Lemma 3.7** *Let $\alpha$ and $\tilde{\alpha}$ be given as in Lemma 3.6. There exist $\delta > 0$ sufficiently small and $\theta_0 \geq 1$ sufficiently large, both chosen independently of $\alpha$, such that for all integer $m \in [m_0, \tilde{\alpha} + s]$, one has*

$$\|f_n\|_{Y_m} \leq C\,\Delta_n\{\theta_n^{m-\alpha-s-1}\,\|f\|_{Y_{\alpha+s}} + \delta^2\,\theta_n^{L(m)-1}\}\,. \tag{25}$$

*Proof* From (12) we have

$$f_n = (S_{\theta_n} - S_{\theta_{n-1}})f - (S_{\theta_n} - S_{\theta_{n-1}})E_{n-1} - S_{\theta_n}e_{n-1}\,.$$

Using (8c) gives

$$\|(S_{\theta_n} - S_{\theta_{n-1}})f\|_{Y_m} \leq C\,\Delta_{n-1}\theta_{n-1}^{m-\alpha-s-1}\,\|f\|_{Y_{\alpha+s}} \tag{26}$$

for all $m \geq 0$. Using (8c), (24) gives

$$\|(S_{\theta_n} - S_{\theta_{n-1}})E_{n-1}\|_{Y_m} \leq C\,\Delta_{n-1}\delta^2\,\theta_{n-1}^{m-p-1+L(p)} \leq C\,\Delta_{n-1}\delta^2\,\theta_{n-1}^{L(m)-1}\,, \tag{27}$$

because $m - p + L(p) \leq L(m)$ for all $m \geq 0$. Moreover, from (8a), (23) we get

$$\|S_{\theta_n}e_{n-1}\|_{Y_m} \leq C\,\Delta_{n-1}\delta^2\,\theta_{n-1}^{L(m)-1}\,, \tag{28}$$

for all $m \geq m_0$. Finally, using $\theta_{n-1} \leq \theta_n \leq \sqrt{2}\theta_{n-1}$ and $\Delta_{n-1} \leq 3\,\Delta_n$ in (26)–(28) yields (25). □

## 3.5 Proof of Induction

We now consider problem (13), that gives the solution $\delta u_n$.

**Lemma 3.8** *Let $\alpha \geq m_0 + \max\{r, r'\} + \max\{s, s'\} + 1$, and let $\tilde{\alpha}$ be given as in Lemma 3.6. If $\delta > 0$ is sufficiently small, $\theta_0 \geq 1$ is sufficiently large, both chosen independently of $\alpha$, and $\|f\|_{Y_{\alpha+s}}/\delta$ is sufficiently small, then for all $m \in [m_0, \tilde{\alpha}]$, one has*

$$\|\delta u_n\|_{X_m} \leq \delta\,\theta_n^{m-\alpha-1}\,\Delta_n\,. \tag{29}$$

*Proof* Let us consider problem (13). By (17a) $S_{\theta_n} u_n$ satisfies

$$\|S_{\theta_n} u_n\|_{X_{m_0}} \leq C\delta\,.$$

So for $\delta$ sufficiently small we may apply (7) in order to obtain

$$\|\delta u_n\|_{X_m} \leq C\,\left(\|f_n\|_{Y_{m+s}} + \|f_n\|_{Y_{m_0}}\,\|S_{\theta_n} u_n\|_{X_{m+s'}}\right). \tag{30}$$

Estimating the right-hand side of (30) by Lemma 3.7 and (17a) yields

$$\begin{aligned}\|\delta u_n\|_{X_m} \leq{}& C\,\Delta_n\{\theta_n^{m-\alpha-1}\,\|f\|_{Y_{\alpha+s}} + \delta^2\,\theta_n^{L(m+s)-1}\}\\ &+ C\,\Delta_n\,\{\theta_n^{m_0-\alpha-s-1}\,\|f\|_{Y_{\alpha+s}} + \delta^2\,\theta_n^{L(m_0)-1}\}\,\delta\,\vartheta_n^{(m+s'-\alpha)_+}\,.\end{aligned} \tag{31}$$

One checks that, for $\alpha \geq m_0 + \max\{r, r'\} + \max\{s, s'\} + 1$, and $m \in [m_0, \tilde{\alpha}]$, the following inequalities hold true:

$$\begin{aligned}&L(m+s) < m-\alpha\,,\\ &m_0-\alpha-s+(m+s'-\alpha)_+ \leq m-\alpha\,,\\ &L(m_0)+(m+s'-\alpha)_+ < m-\alpha\,.\end{aligned} \tag{32}$$

From (31), we thus obtain

$$\|\delta u_n\|_{X_m} \leq C\left(\|f\|_{Y_{\alpha+s}} + \delta^2\right)\theta_n^{m-\alpha-1}\,\Delta_n\,, \tag{33}$$

and (29) follows, for $\delta > 0$ and $\|f\|_{Y_{\alpha+s}}/\delta$ sufficiently small. □

The crucial point of the method is seen in (32): the quadratic nature of the errors is reflected in the estimate (23) by the presence of the term "$-2\alpha$", while the tame nature of the estimates contributes linearly in $m$ (with $|L'(m)| \le 1$). It is the "$-2\alpha$" term which allows (for $\alpha$ sufficiently large) to get (32) and close the induction.

Lemma 3.8 shows that $(H_{n-1})$ implies $(H_n)$ provided that $\alpha \ge m_0 + \max\{r, r'\} + \max\{s, s'\} + 1$, $\tilde{\alpha} = 2\alpha + \max\{r, r'\} + 1 - m_0 - r$, $\delta > 0$ is small enough, $\|f\|_{Y_{\alpha+s}}/\delta$ is small enough, and $\theta_0 \ge 1$ is large enough. We fix $\alpha$, $\tilde{\alpha}$, $\delta > 0$, and $\theta_0 \ge 1$, and we finally prove $(H_0)$.

**Lemma 3.9** *If $\|f\|_{Y_{\alpha+s}}/\delta$ is sufficiently small, then property $(H_0)$ holds.*

*Proof* Let us consider problem (13) for $n = 0$:

$$d\mathcal{F}(0)\,\delta u_0 = S_{\theta_0} f .$$

Applying (7) gives

$$\|\delta u_0\|_{X_m} \le C\, \|S_{\theta_0} f\|_{Y_{m+s}} \le C\, \theta_0^{(m-\alpha)+}\, \|f\|_{Y_{\alpha+s}} .$$

Then

$$\|\delta u_0\|_{X_m} \le \delta\, \theta_0^{m-\alpha-1} \Delta_0 , \qquad m_0 \le m \le \tilde{\alpha},$$

provided $\|f\|_{Y_{\alpha+s}}/\delta$ is taken sufficiently small. □

### 3.6 Conclusion of the Proof of Theorem 2.4 i)

Given an integer $\alpha \ge m_0 + \max\{r, r'\} + \max\{s, s'\} + 1$, in agreement with the requirements of Lemma 3.8, we take $\tilde{\alpha} = 2\alpha + \max\{r, r'\} + 1 - m_0 - r$ as in Lemma 3.6. If $\delta > 0$ and $\|f\|_{Y_{\alpha+s}}/\delta$ are sufficiently small, $\theta_0 \ge 1$ is sufficiently large, then $(H_n)$ holds true for all $n$. Let us set $m' = \alpha - 1$. In particular, from $(H_n)$ we obtain

$$\sum_{n\ge 0} \|\delta u_n\|_{X_{m'}} < +\infty , \tag{34}$$

so the sequence $\{u_n\}$ converges in $X_{m'}$ towards some limit $u \in X_{m'}$. From (15) we have

$$\mathcal{F}(u_{n+1}) - f = (S_{\theta_n} - I)f + (I - S_{\theta_n})E_n + e_n .$$

Using (8b), (23), (24) we can pass to the limit in the right-hand side in $Y_{m'+s}$ and get

$$\lim_{n\to\infty} \mathcal{F}(u_{n+1}) = \mathcal{F}(u) = f .$$

Therefore $u$ is a solution of (1), and the proof of Theorem 2.4 i) is complete.

*Remark 3.10* In view of (7) with a loss of regularity of order $s$ from $g$, given $f \in Y_{m'+s+1}$ we could wish to find a solution $u \in X_{m'+1}$ instead of $u \in X_{m'}$ as above. The regularity of $u$ follows from the condition $m' < \alpha$ for the convergence of the series (34). Working with spaces $X_m$ with integer index the condition yields $m' \leq \alpha - 1$; in spaces with real index it would be enough $m' \leq \alpha - \epsilon$, for all $\epsilon > 0$, and we would get $u \in X_{m'+1-\epsilon}$.

## 3.7 Additional Regularity of the Solution Constructed

Let us now prove assertion ii) of the Nash-Moser theorem. Let us assume that $f \in Y_{m''+s+1}$, with $m'' > m'$. Let us set $\alpha' = m'' + 1$ and define $\tilde{\alpha}'$ accordingly, $\tilde{\alpha}' = 2\alpha' + \max\{r, r'\} + 1 - m_0 - r$. The proof is obtained by finite induction. For it we shall use the estimate $(H_n)$ which is now true for all $n$, and the estimates that can be obtained from it.

We consider again (31) and remark that the exponents of $\theta_n$ of the terms not involving $f$ are strictly less than $m - \alpha - 1$, as shown in (32). On the other hand, the terms in (31) involving $f$ come from (25), or more precisely from (26). Using the fact that $f$ is now more regular, we can substitute (26) by

$$\|(S_{\theta_n} - S_{\theta_{n-1}})f\|_{Y_m} \leq C\,\Delta_n \theta_n^{m-\alpha-s-2}\,\|f\|_{Y_{\alpha+s+1}},$$

and, accordingly, instead of (33) we find

$$\|\delta u_n\|_{X_m} \leq C\left(\|f\|_{Y_{\alpha+s+1}} + \delta^2\right)\theta_n^{m-\alpha-2}\,\Delta_n \leq C\,\theta_n^{m-\alpha-2}\,\Delta_n, \qquad \forall n \geq 0. \tag{35}$$

Starting from these new estimates instead of $(H_n)$, we can revisit the proof of assertion i). Note that in $e'_k, e''_k$ there is at least one factor involving $\delta u_n$ in each term. Estimating this factor by (35) gives

$$\|e_k\|_{Y_m} \leq C\,\delta\,\theta_k^{L(m)-2}\,\Delta_k\,.$$

Going on with the repetition of the proof we obtain

$$\|E_n\|_{Y_{p+1}} \leq C\,\delta\,\theta_n^{L(p)}\,,$$

$$\|f_n\|_{Y_m+1} \leq C\,\Delta_n\{\theta_n^{m-\alpha-s-1}\,\|f\|_{Y_{\alpha+s+1}} + \delta^2\,\theta_n^{L(m)-1}\}\,,$$

$$\|\delta u_n\|_{X_{m+1}} \leq C\left(\|f\|_{Y_{\alpha+s+1}} + \delta\right)\theta_n^{m-\alpha-1}\,\Delta_n\,.$$

This gives the gain of one order. After a finite number of iterations of the same procedure we find

$$\|\delta u_n\|_{X_m} \leq C\,\theta_n^{m-\alpha'-1}\,\Delta_n \qquad \text{for all } m \in [m_0, \tilde{\alpha}'].$$

The conclusion of the proof of assertion ii) follows as for (34).

## 4 Simplified Case

To understand better the role of parameters in the induction of the proof, let us assume for simplicity that $m_0 = 0,\ r = r' = s = s' = 1$. Then estimate (23) holds with $L(m) = m + 1 - 2\alpha$. The number $p = \tilde{\alpha} - 1$ in (24) is chosen such that $L(p) = 1$ which yields $p = 2\alpha,\ \tilde{\alpha} = 2\alpha + 1$. To close the induction we choose $\alpha$ from (32) that now reads

$$\begin{aligned} &m + 2 - 2\alpha < m - \alpha\,, \\ &-\alpha - 1 + (m + 1 - \alpha)_+ \leq m - \alpha\,, \\ &1 - 2\alpha + (m + 1 - \alpha)_+ < m - \alpha\,. \end{aligned}$$

Here it is sufficient to take $\alpha > 2$, i.e. $\alpha \geq 3$, and $(H_n)$ will hold for all $m \in [0, 2\alpha + 1]$. The quadratic nature of the errors with the presence of the term "$-2\alpha$"allows (for $\alpha$ sufficiently large) to close the induction. Thus, the same nonlinearity of the equation is exploited for the convergence of the approximating sequence.

## References

1. S. Alinhac, Existence d'ondes de raréfaction pour des systèmes quasi-linéaires hyperboliques multidimensionnels. Commun. Partial Differ. Equ. **14**, 173–230 (1989)
2. S. Alinhac, P. Gérard, *Opérateurs pseudo-différentiels et théorème de Nash-Moser* (InterEditions, Paris, 1991)
3. V.I. Arnol'd, Small denominators. I. Mapping the circle onto itself. Izv. Akad. Nauk SSSR Ser. Mat. **25**, 21–86 (1961)
4. V.I. Arnol'd, Small denominators and problems of stability of motion in classical and celestial mechanics. Usp. Mat. Nauk **18**, 91–192 (1963)
5. G.-Q. Chen, Y.-G. Wang, Existence and stability of compressible current-vortex sheets in three-dimensional magnetohydrodynamics. Arch. Ration. Mech. Anal. **187**, 369–408 (2008)
6. J.F. Coulombel, P. Secchi, Nonlinear compressible vortex sheets in two space dimensions. Ann. Sci. École Norm. Sup. **41**(4), 85–139 (2008)
7. I. Ekeland, An inverse function theorem in Fréchet spaces. Ann. Inst. Henri Poincaré Anal. Non Linéaire **28**, 91–105 (2011)
8. M. Günther, On the perturbation problem associated to isometric embeddings of Riemannian manifolds. Ann. Glob. Anal. Geom. **7**, 69–77 (1989)
9. R.S. Hamilton, The inverse function theorem of Nash and Moser. Bull. Am. Math. Soc. (N.S.) **7**, 65–222 (1982)

10. L. Hörmander, The boundary problems of physical geodesy. Arch. Ration. Mech. Anal. **62**, 1–52 (1976)
11. L. Hörmander, Implicit function theorems. Stanford University Lecture Notes (1977)
12. T. Kato, *Abstract Differential Equations and Nonlinear Mixed Problems*. Lezioni Fermiane [Fermi Lectures] (Scuola Normale Superiore, Pisa, 1985)
13. A.N. Kolmogorov, On conservation of conditionally periodic motions for a small change in Hamilton's function. Dokl. Akad. Nauk SSSR (N.S.) **98**, 527–530 (1954)
14. J. Moser, A new technique for the construction of solutions of nonlinear differential equations. Proc. Natl. Acad. Sci. U.S.A. **47**, 1824–1831 (1961)
15. J. Moser, A rapidly convergent iteration method and non-linear differential equations II. Ann. Scuola Norm. Sup. Pisa **20**(3), 499–535 (1966)
16. J. Nash, The imbedding problem for Riemannian manifolds. Ann. Math. **63**(2), 20–63 (1956)
17. J. Schwartz, On Nash's implicit functional theorem. Commun. Pure Appl. Math. **13**, 509–530 (1960)
18. P. Secchi, Y. Trakhinin, Well-posedness of the plasma-vacuum interface problem. Nonlinearity **27**, 105–169 (2014)
19. Y. Trakhinin, The existence of current-vortex sheets in ideal compressible magnetohydrodynamics. Arch. Ration. Mech. Anal. **191**, 245–310 (2009)

# Rate of Convergence to the Stationary Solution of the Navier-Stokes Exterior Problem

**Masao Yamazaki**

*Dedicated to Professor Yoshihiro Shibata on the occasion of his sixtieth birthday*

**Abstract** This paper is concerned with the nonstationary Navier-Stokes equation in two-dimensional exterior domains with stationary external forces, and provides the rate of convergence of solutions to the stationary solution under the smallness condition of the stationary solution.

**Keywords** Exterior problem • Navier-Stokes equations • Rate of convergence • Stationary solutions

## 1 Introduction

Let $\Omega$ be an exterior domain in the two-dimensional space $\mathbb{R}^2$ with $C^{3+\gamma}$-boundary $\Gamma$ with some $\gamma > 0$. We are concerned with stability of the stationary solution of the following nonstationary Navier-Stokes equation on $\Omega$ with stationary external force and stationary inhomogeneous boundary data:

$$\frac{\partial u}{\partial t}(x,t) - \Delta u(x,t) + \big(u(x,t)\cdot\nabla\big)u(x,t) + \nabla\tilde{p}(x,t) = f(x) \quad \text{in } \Omega\times(0,+\infty), \tag{1}$$

$$\nabla\cdot u(x,t) = 0 \qquad \text{in } \Omega\times(0,+\infty), \tag{2}$$

$$u(x,t) = a(x) \quad \text{on } \Gamma\times(0,+\infty), \tag{3}$$

M. Yamazaki (✉)
Department of Mathematics, Faculty of Science and Engineering, Waseda University, Shinjuku, Tokyo 169-8555, Japan
e-mail: masao.yamazaki@waseda.jp



H. Amann et al. (eds.), *Recent Developments of Mathematical Fluid Mechanics*, Advances in Mathematical Fluid Mechanics, DOI 10.1007/978-3-0348-0939-9_24

$$u(x,t) \to 0 \qquad \text{as } |x| \to \infty, \tag{4}$$

$$u(x,0) = u_0(x) \text{ in } \Omega. \tag{5}$$

where the vector-valued unknown function $u(x,t)$, the scalar-valued unknown function $\tilde{p}(x,t)$, the vector-valued given function $a(x)$, $u_0(x)$ and $f(x)$ stand for the velocity, the pressure, the stationary boundary data, the initial data and the stationary external force satisfying the outflow condition

$$\int_\Gamma a(x) \cdot n(x)\, ds(x) = 0 \tag{6}$$

respectively, where $n(x)$ denotes the outer normal vector at $x \in \Gamma$.

Finn and Smith, in the pioneering work [9], started the study on the stationary Navier-Stokes equation

$$-\Delta w(x) + \big(w(x) \cdot \nabla\big) w(x) + \nabla \pi(x) = f(x) \text{ in } \Omega, \tag{7}$$

$$\nabla \cdot w(x) = 0 \qquad \text{in } \Omega, \tag{8}$$

$$w(x) = a(x) \text{ on } \Gamma, \tag{9}$$

$$w(x) \to w_\infty \text{ as } |x| \to \infty, \tag{10}$$

in two-dimensional exterior domains, first in the case with definite velocity $w_\infty \neq 0$ at infinity with no external force. This problem was considered by Gilbarg and Weinberger [15, 16] and Amick [1, 2]. In particular, Amick [1, 2] considered the case where the domain is symmetric with respect to the direction of $w_\infty$. Then Galdi and Simader [12] considered the problem for external force $f(x)$ with little regularity, and Galdi and Sohr [13], Sazonov [24] and Russo [22] obtained precise asymptotic behavior of $w(x)$ and $\pi(x)$. (For more complete references, see Galdi [11].)

In order to consider the case $w_\infty = 0$, Galdi [10, Sect. 3] and Pileckas and Russo [21] posed the assumption

$$\begin{gathered}\text{The domain } \Omega \text{ is invariant under the mappings}\\ (x_1, x_2) \mapsto (-x_1, x_2), \quad (x_1, x_2) \mapsto (x_1, -x_2)\end{gathered} \tag{D4}$$

with a specific coordinate variables $(x_1, x_2)$, and assumed that the external force $f(x) = \big(f_1(x), f_2(x)\big)$ satisfies the condition

$$\begin{cases} f_1(-x_1, x_2) = -f_1(x_1, x_2), & f_2(-x_1, x_2) = f_2(x_1, x_2), \\ f_1(x_1, -x_2) = f_1(x_1, x_2), & f_2(x_1, -x_2) = -f_2(x_1, x_2), \end{cases} \tag{U4}$$

and find a solution $w(x) = \big(w_1(x), w_2(x)\big)$ satisfying (U4), which is exactly the same as [10, (3.18)]. In particular, Pileckas and Russo [21] proved the existence of a weak solution with finite Dirichlet integral satisfying (U4) for external force satisfying (U4) belonging to a suitable class. For related works, see Russo [23] and the references cited therein.

On the other hand, in this case there seems to be few results on strong solutions corresponding to the physically reasonable solutions in the three-dimensional case. There are two points of difficulty. For one thing, typical decay order of stationary solutions of exterior problems are the same as that of the fundamental solutions of the Laplacian due to the net force exerted to the obstacle. (See Novotny and Padula [20], Borchers and Miyakawa [4], Kozono et al. [18].) Here we limit ourselves to the domains satisfying (D4) above, and the external force and the boundary data satisfying (U4) above in order to eliminate the net force. In this case the typical decay order of the solutions is the same as the derivative of the fundamental solution, which we call critically decaying solutions.

However, in the case $n = 2$, the product of these functions belongs only in the weak-$L^1$ space, in which almost no functional analytic tools are available in finding function spaces where we can apply the Banach fixed point theorem, of which fact the second difficulty consists. Hence we must rely on pointwise estimates for the time being. By this method Yamazaki [26, 27] showed the existence of solutions of the nonlinear equation, for $\Omega = \mathbb{R}^2$ and for exterior domains respectively, under assumptions on symmetry stronger than (D4) and (U4). Yamazaki [27] also showed that, if the external force decays faster, then the solutions decay faster than the derivative of the fundamental solution, which we call supercritically decaying solutions. Nakatsuka [19] recently showed that, if the solution constructed in [27] is sufficiently small, then it coincides with the on in [21] under symmetry condition weaker than (U4).

In this article we are concerned with the stability of the stationary solutions above. Putting $v(x,t) = u(x,t) - w(x)$ and $p(x,t) = \tilde{p}(x,t) - \pi(x)$, we see that $\big(v(x,t), p(x,t)\big)$ enjoys the system:

$$\frac{\partial v}{\partial t}(x,t) - \Delta v(x,t) + \big(w(x)\cdot\nabla\big)v(x,t) + \big(v(x,t)\cdot\nabla\big)w(x) + \big(v(x,t)\cdot\nabla\big)v(x,t) + \nabla p(x,t) = 0 \quad \text{in } \Omega\times(0,+\infty), \tag{11}$$

$$\nabla\cdot v(x,t) = 0 \quad \text{in } \Omega\times(0,+\infty), \tag{12}$$

$$v(x,t) = 0 \quad \text{on } \Gamma\times(0,+\infty), \tag{13}$$

$$v(x,t)\to 0 \quad \text{as } |x|\to\infty, \tag{14}$$

$$v(x,0) = v_0(x) \text{ in } \Omega, \tag{15}$$

where $v_0(x) = u_0(x) - w(x)$.

Then Galdi and Yamazaki [14] showed the global stability of the stationary solutions above if they are sufficiently small. Namely, we consider the following two cases:

(C) $\Omega$ is an exterior domain satisfying (D4), $w(x)$ is a sufficiently small critically decaying stationary solution satisfying (U4), and $v_0(x) \in L^2_\sigma(\Omega)$ is a function satisfying (U4).
(S) $\Omega$ is an arbitrary exterior domain, $w(x)$ is a sufficiently small supercritically decaying stationary solution, and $v_0(x) \in L^2_\sigma(\Omega)$ is arbitrary.

Then there uniquely exists a time-global solution $v(t)$ of (11)–(15) belonging to a suitable class. Moreover, $v(\cdot, t)$ tends to 0 in $L^q_\sigma(\Omega)$ for every $q \in [2, \infty)$ and in $\left(H^1_0(\Omega)\right)^2$ as $t \to \infty$. Precise conditions, statement and notations are introduced in the next section.

The purpose of this paper is to show that $v(\cdot, t)$ tends to 0 also in $\left(L^\infty(\Omega)\right)^2$, and to obtain the decay rate of $\|v(\cdot, t)\|_q$ for $2 < q \leq \infty$ and $\|\nabla v(\cdot, t)\|_2$ as $t \to \infty$. For this purpose we show some estimates of the resolvent of the perturbed operator by employing the representation of the Neumann series expansion in Kozono and Yamazaki [17] as well as the resolvent estimate by Borchers and Varnhorn [5] and Dan and Shibata [7, 8]. We need also the estimate for fractional powers of the resolvent as well, due to the fact that the coerciveness holds only for more limited exponents.

This paper is organized as follows. In Sect. 2 we introduce some function spaces and state the main result. In Sect. 3 we recall some preliminary results concerning inequalities and the Stokes semigroup. Section 4 is devoted to the treatment of the semigroup generated by perturbed operators by the aforementioned manner. Finally the main result is proved in Sect. 5.

## 2 Notations and Main Result

We first introduce some function spaces. For a domain $\Omega \subset \mathbb{R}^2$, let $C_0^\infty(\Omega)$ denote the set of infinitely differentiable functions on $\Omega$ supported by a compact subset of $\Omega$, and let $C_{0,\sigma}^\infty(\Omega)$ denote the set of vector-valued functions $\varphi(x) = \left(\varphi_1(x), \varphi_2(x)\right) \in \left(C_0^\infty(\Omega)\right)^2$ such that $\operatorname{div} \varphi \equiv 0$. Next, for a domain $\Omega$ in $\mathbb{R}^2$ and $q \in [1, \infty]$, let $L^q(\Omega)$ denote the standard Lebesgue spaces equipped with the norm $\|\cdot\|_q$. Next, for $q \in (1, \infty)$ and $s > 0$, let $H^s_q(\mathbb{R}^2)$ denote the Sobolev space consisting of the functions satisfying

$$\|u\|_{H^s_q(\mathbb{R}^2)} = \left\|\mathcal{F}^{-1}\left[\langle\xi\rangle^s F[u](\xi)\right]\right\|_q < \infty,$$

where $\langle\xi\rangle = \sqrt{1 + |\xi|^2}$, and let $\dot{H}^s_q(\mathbb{R}^2)$ denote the set of functions modulo polynomials such that

$$\|f\|_{\dot{H}^s_q(\mathbb{R}^2)} = \left\|\mathcal{F}^{-1}\left[|\xi|^s F[u](\xi)\right]\right\|_q < \infty.$$

Every modulo class in $\dot{H}^2_q(\mathbb{R}^2)$ has a unique canonical representative in the space $L^{2q/(2-sq)}(\mathbb{R}^2)$ if $s < 2/q$, and has a unique canonical representative modulo constants in the space $\dot{H}^1_{2q/(2-sq)}(\mathbb{R}^2)$ if $2/q \le s < 1 + 2/q$.

We next introduce the Littlewood-Paley decomposition. Let $\chi(t)$ be a monotone-decreasing smooth function on $\mathbb{R}$ such that $\chi(t) \equiv 1$ holds for $t < 4/3$ and that $\chi(t) \equiv 0$ holds for $t > 5/3$. Then we put

$$\Phi(\xi) = \chi(|\xi|) \qquad \text{for } \xi \in \mathbb{R}^2,$$

$$\varphi_j(\xi) = \chi\left(2^{-j}|\xi|\right) - \chi\left(2^{1-j}|\xi|\right) \text{ for } j \in \mathbb{Z} \text{ and } \xi \in \mathbb{R}^2 \setminus \{0\}.$$

Suppose that $q, r \in [1, \infty]$ and $s \in \mathbb{R}$. For a tempered distribution $f \in \mathcal{S}'(\mathbb{R}^2)$, put

$$\|f\|_{B^s_{q,r}} = \left( \left\| \mathcal{F}^{-1}\left[\Phi(\xi)\mathcal{F}[f](\xi)\right] \right\|_q^r + \sum_{j=1}^{\infty} 2^{sjr} \left\| \mathcal{F}^{-1}\left[\varphi_j(\xi)\mathcal{F}[f](\xi)\right] \right\|_q^r \right)^{1/r},$$

$$\|f\|_{\dot{B}^s_{q,r}} = \left( \sum_{j=-\infty}^{\infty} 2^{sjr} \left\| \mathcal{F}^{-1}\left[\varphi_j(\xi)\mathcal{F}[f](\xi)\right] \right\|_q^r \right)^{1/r}$$

for $1 \le r < \infty$, and

$$\|f\|_{B^s_{q,\infty}} = \max\left\{ \left\| \mathcal{F}^{-1}\left[\Phi(\xi)\mathcal{F}[f](\xi)\right] \right\|_q, \sup_{j\ge 1} 2^{sj} \left\| \mathcal{F}^{-1}\left[\varphi_j(\xi)\mathcal{F}[f](\xi)\right] \right\|_q \right\}$$

$$\|f\|_{\dot{B}^s_{q,\infty}} = \sup_{j\in\mathbb{Z}} 2^{sj} \left\| \mathcal{F}^{-1}\left[\varphi_j(\xi)\mathcal{F}[f](\xi)\right] \right\|_q .$$

Then we introduce the Besov space by $B^s_{q,r}(\mathbb{R}^2) = \{f \in \mathcal{S}' \mid \|f\|_{\dot{B}^s_{q,r}} < \infty\}$ and the homogeneous Besov space by $\dot{B}^s_{q,r}(\mathbb{R}^2) = \{f \in \mathcal{S}'/\mathcal{P} \mid \|f\|_{\dot{B}^s_{q,r}} < \infty\}$, where $\mathcal{S}'$ and $\mathcal{P}$ denote the set of tempered distributions on $\mathbb{R}^2$ and the set of polynomials of $x_1$ and $x_2$ respectively. The spaces $B^0_{q,r}(\mathbb{R}^2)$ and $\dot{B}^0_{q,r}(\mathbb{R}^2)$ are contained in $L^{q_1}(\mathbb{R}^2)+L^{q_2}(\mathbb{R}^2)$ for every $q_1, q_2$ such that $q_1 < q < q_2$. In particular, if $r \le \min\{q, 2\}$, these spaces are contained in $L^q(\mathbb{R}^2)$. If either $0 \le s < 2/q$ or $s = 2/q$ and $r = 1$, every modulo class of $\dot{B}^s_{q,r}(\mathbb{R}^2)$ possesses a canonical representative in $\dot{B}^0_{2q/(2+q-sq),r}(\mathbb{R}^2)$, and if $2/q \le s < 2/q + 1$ and $1 < r \in \infty$ or $2/q < s \le 2/q + 1$ and $r = 1$, every modulo class in $\dot{B}^s_{q,r}(\mathbb{R}^2)$ possesses a canonical representative modulo constants in

$\dot{B}^1_{2q/(2+q-sq),r}(\mathbb{R}^2)$. (See Bourdaud [6].) We also observe that the spaces $B^s_{2,2}(\mathbb{R}^2)$ and $\dot{B}^s_{2,2}(\mathbb{R}^2)$ coincide with $H^s_2(\mathbb{R}^2)$ and $\dot{H}^s_2(\mathbb{R}^2)$ respectively for $s \geq 0$, and the spaces $B^s_{\infty,\infty}(\mathbb{R}^2)$ and $\dot{B}^s_{\infty,\infty}(\mathbb{R}^2)$ coincide with the Hölder space $C^s(\mathbb{R}^2)$ and the homogeneous Hölder space $\dot{C}^s(\mathbb{R}^2)$ respectively for $s > 0$ and $s \notin \mathbb{N}$.

For a domain $\Omega \subset \mathbb{R}^2$ with $C^{3+\gamma}$-boundary with some $\gamma > 0$ and for $X = H^s_q, \dot{H}^s_q, B^s_{q,r}$ and $\dot{B}^s_{q,r}$ with $s$ such that $0 < s \leq 3$, let $X(\Omega)$ denote the set of the restrictions of the elements of $X(\mathbb{R}^2)$ equipped with the norm

$$\|u\|_{X(\Omega)} = \inf\left\{ \|\tilde{u}\|_{X(\mathbb{R}^2)} \;\middle|\; \tilde{u}|_\Omega = u \right\},$$

and let $X_0(\Omega)$ denote the closure of the space $C_0^\infty(\Omega)$ in $X(\Omega)$. If $q = 2$, we abbreviate $H^s_q(\Omega)$, $\dot{H}^s_q(\Omega)$, $H^s_{q,0}(\Omega)$ and $\dot{H}^s_{q,0}(\Omega)$ to $H^s(\Omega)$, $\dot{H}^s(\Omega)$, $H^s_0(\Omega)$ and $\dot{H}^s_0(\Omega)$ respectively. Then we have the following properties: Suppose that $\Omega$ is a domain with compact boundary $\Gamma$ of class $C^{3+\gamma}$ with some $\gamma > 0$. Then every function $f \in H^1_q(\Omega)$ has the trace $f|_\Gamma \in B^{1-1/q}_{q,q}(\Gamma)$, and there exists a constant $C$ such that the trace estimate

$$\|f|_\Gamma\|_{B^{1-1/q}_{q,q}} \leq C\|f\|_{H^1_q}, \tag{16}$$

where the Besov space $B^{1-1/q}_{q,q}(\Gamma)$ is defined by coordinate patch form one-dimensional Besov spaces. (See Bergh-Löfström [3] or Triebel [25].)

We next introduce the Helmholtz decomposition. Let $\Omega$ be either the whole plane, the half plane, a bounded domain or an exterior domain with $C^{3+\gamma}$ boundary $\Gamma$. For every $q \in (1, \infty)$, the space $\left(L^q(\Omega)\right)^2$ admits the direct sum decomposition $L^q_\sigma(\Omega) \oplus G^q(\Omega)$, where

$$L^q_\sigma(\Omega) = \left\{ u \in \left(L^q(\Omega)\right)^2 \;\middle|\; \operatorname{div} u = 0 \text{ in } \Omega,\, n \cdot u = 0 \text{ on } \Gamma \right\}$$

and

$$G^q(\Omega) = \left\{ \operatorname{grad} f \in \left(L^q(\Omega)\right)^2 \;\middle|\; f \in L^q_{\mathrm{loc}}(\Omega) \right\}.$$

Let $P_q$ denote the projection from $\left(L^q(\Omega)\right)^2$ onto $L^q_\sigma(\Omega)$ with respect to the direct sum decomposition above. Then we have $P_q = P_r$ on $\left(L^q(\Omega) \cap L^r(\Omega)\right)^2$. From this fact we can abbreviate $P_q$ to $P$. Define the Stokes operator $A$ by $-P\Delta$.

In order to state our main results, we introduce some function spaces. For a positive number $b$, let $\mathcal{X}(b)$ denote the set of continuous functions $w(x)$ on $\Omega$ satisfying $\|w\|_{\mathcal{X}(b)} = \sup_{x\in\Omega}(1+|x|)^b|w(x)| < \infty$. Then the space $\mathcal{X}(b)$ becomes a Banach space with norm $\|\cdot\|_{\mathcal{X}(b)}$. In the case that $\Omega$ satisfies the condition (D4), let $\mathcal{X}_4(b)$ denote the set of vector-valued functions $w \in \left(\mathcal{X}(b)\right)^2$ satisfying the condition (U4).

*Remark 2.1* The following assertion is proved in [27]. Suppose that $1 \le b \le 2$ and that the exterior domain $\Omega$ satisfies a symmetry condition stronger than (D4). If the boundary datum $a(x)$ is sufficiently small in $B^{1-1/q}_{q,q}(\Omega)$ with some $q > 2$ and satisfies the outflow condition (6), and a symmetry condition stronger than (U4), and if the external force $f(x)$ is given by a finite sum of the derivatives of sufficiently small potential functions and satisfies the same symmetry condition as is posed on $a(x)$, then there uniquely exists a stationary solution $\big(w(x), \pi(x)\big)$ of (7)–(10) such that $w \in \big(\mathcal{X}(b)\big)^2$ and that $\|w\|_{\mathcal{X}(b)}$ is sufficiently small, and $\nabla w$ is sufficiently small in $\big(L^2(\Omega)\big)^4$.

Suppose that $\big(w(x), \pi(x)\big)$ is a solution of the stationary equations (7)–(10) such that $w \in \big(\mathcal{X}(b) \cap \dot{H}^1(\Omega)\big)^2$ holds with some $b \ge 1$. We consider the stability of $w(x)$ by considering the system (11)–(15) for $v_0 \in L^2_\sigma(\Omega)$. Applying the operator $P$ to (11) and making use of (12)–(14), we have

$$\frac{dv}{dt}(t) = -Av(t) - P\big[(w \cdot \nabla)v(t) + \big(v(t) \cdot \nabla\big)w + \big(v(t) \cdot \nabla\big)v(t)\big]. \tag{17}$$

Then (17) with (15) is formally equivalent to the integral equation

$$v(t) = \exp(-tA)v_0 - \int_0^t \exp\big(-(t-\tau)A\big)$$
$$P\big[(w \cdot \nabla)v(\tau) + \big(v(\tau) \cdot \nabla\big)w + \big(v(\tau) \cdot \nabla\big)v(\tau)\big]\, d\tau. \tag{18}$$

In the following we assume either one of the following assumptions, which are precise statement of those given in the introduction.

(C) $\Omega$ is an exterior domain with $C^{3+\gamma}$ boundary satisfying (D4), the pair of functions $\big(w(x), \pi(x)\big)$ is a solution of (7)–(10) such that $w \in \mathcal{X}_4(1)$ and that $\|w\|_{\mathcal{X}(1)}$ and $\|\nabla w\|_2$ are sufficiently small, and $v_0(x) \in L^2_\sigma(\Omega)$ satisfies (U4).

(S) $\Omega$ is an arbitrary exterior domain with $C^{3+\gamma}$ boundary, the pair of functions $\big(w(x), \pi(x)\big)$ is a solution of (7)–(10) such that $w \in \big(\mathcal{X}(b)\big)^2$ with some $b > 1$ such that $\|w\|_{\mathcal{X}(b)}$ and $\|\nabla w\|_2$ are sufficiently small, and $v_0(x) \in L^2_\sigma(\Omega)$ is arbitrary.

Then Galdi and Yamazaki [14] showed the following theorem.

**Theorem 2.2** *Under the assumption* (C) *or* (S), *there uniquely exists a solution* $v(\cdot, t) \in BC\big([0,\infty), L^2_\sigma(\Omega)\big)$ *of the integral equation* (18) *such that* $v(x, 0) = v_0(x)$ *and that* $t^{1/2}v(\cdot, t) \in BC\left((0,\infty), \big(H^1_0(\Omega)\big)^2\right)$. *This solution belongs to the class*

$$C\big([0,\infty), L^2_\sigma(\Omega)\big) \cap C^1\big((0,\infty), L^2_\sigma(\Omega)\big) \cap C\left((0,\infty), \big(H^2(\Omega)\big)^2\right),$$

*and is a solution of the abstract differential equation* (17). *Furthermore, the function* $\|v(\cdot,t)\|_2$ *is monotone-decreasing with respect to* $t$, *and* $v(\cdot,t)$ *enjoys the decay properties* $\|v(\cdot,t)\|_q \to 0$ *as* $t \to \infty$ *for every* $q \in [2,\infty)$ *and* $\|\nabla v(\cdot,t)\|_2 \to 0$ *as* $t \to \infty$.

*Remark 2.3* The condition (6) is necessary for the existence of the stationary solution $w(x) \in \mathcal{X}(b)$ with $b > 1$. Indeed, for sufficiently large $R > 0$, we have

$$\int_{|x|=R} u(x) \cdot \frac{x}{R}\, ds(x) = O(R^{1-b}),$$

which tends to 0 if $b > 1$. Hence the divergence theorem yields (6).

Our main result is the following theorem, which provides the decay rate of $v(t)$ with various norms. In particular, we give the decay of $\|v(t)\|_\infty$. Note that even its boundedness was not shown in [14].

**Theorem 2.4** *In the same assumptions as in Theorem 2.2, we have* $\|v(t)\|_q = o\left(t^{1/q-1/2}\right)$ *for* $q \in [2,\infty)$, $\|\nabla v(t)\|_2 = o\left(t^{-1/2}\right)$ *and* $\|v(t)\|_\infty = o\left(t^{-1/2}\sqrt{\log t}\right)$ *as* $t \to \infty$.

## 3 Preliminary Results on the Stokes Semigroup

In this section we list some results needed in the proof of the main results. The first one is the Sobolev embedding theorem and the Gagliardo-Nirenberg inequality.

**Lemma 3.1** *Suppose that* $q, r, s \in (1,\infty)$ *satisfy* $r > s$ *and*

$$0 < \theta = \frac{2(r-s)q}{(2q-2s+qs)r} \leq 1.$$

*Then we have the inclusion relation* $\dot{H}^1_{q,0}(\Omega) \cap L^s(\Omega) \subset L^r(\Omega)$, *and the inequality* $\|u\|_r \leq C\|\nabla u\|_q{}^\theta \|u\|_s{}^{1-\theta}$ *holds with a constant* $C$. *Furthermore, if* $\theta < 1$, *this result holds also for* $r = \infty$, *and if* $\theta = 1$, *the inclusion relation* $\dot{H}^1_q(\Omega) \subset L^r(\Omega)$ *holds.*

The second one is Hardy's inequality.

**Lemma 3.2** *Suppose that* $\Omega$ *is an exterior domain in* $\mathbb{R}^2$. *Then there exists a constant* $C$ *such that, for every* $u(x) \in \dot{H}^1_0(\Omega)$, *we have*

$$\int_\Omega \frac{|u(x)|^2}{|x|^2\left(1 + \left|\log|x|\right|\right)^2}\, dx \leq C\|\nabla u\|_2{}^2.$$

If the domain and the function satisfy symmetry condition, the lemma above is substantially improved. Namely, we have the following lemma. (See [10, Lemma 3.1].)

**Lemma 3.3** *Suppose that* $\Omega$ *is an exterior domain in* $\mathbb{R}^2$ *satisfying the condition* (D4). *Then there exists a constant* $C$ *such that, for every* $u(x) \in \dot{H}_0^1(\Omega)$ *satisfying* (U4), *we have*

$$\int_\Omega \frac{|u(x)|^2}{|x|^2}\, dx \le C\|\nabla u\|_2{}^2.$$

From Lemmas 3.2 and 3.3 we have the following estimate, which plays the key role later.

**Proposition 3.4** *Suppose that* $w(x) \in \left(\mathcal{X}(b)\right)^2$ *with some* $b \ge 1$. *If* $b = 1$, *we also suppose that* $\Omega$ *satisfies* (D4). *Then there exists a constant* $C_b$ *such that the following assertion holds. For every* $u(x) = \left(u_1(x), u_2(x)\right) \in \left(H_0^1(\Omega)\right)^2$ *in the case* $b > 1$, *and every* $u(x) \in \left(H_0^1(\Omega)\right)^2$ *satisfying* (U4) *in the case* $b = 1$, *we have* $w(x) \cdot u(x) \in L^2(\Omega)$ *with the estimate*

$$\|w(x) \cdot u(x)\|_2 \le C_b \|w\|_{\mathcal{X}(b)} \|\nabla u\|_2 \, .$$

We next recall some properties of the Stokes operator and Stokes semigroup proved by Borchers and Varnhorn [5] and Dan and Shibata [7, 8], and add some observations needed later.

**Proposition 3.5** *Put* $D = \{\zeta \in \mathbb{C} \mid \zeta \ne 0, |\arg\zeta| \le 3\pi/4\}$. *Then the Stokes operator* $A$ *satisfies the following assertions:*

(1) *For every* $q$ *and* $r$ *such that* $1 < q \le r \le \infty$, *there exists a positive constant* $C_{q,r}$ *such that, for every* $\zeta \in D$, *the operator* $(\zeta + A)^{-1}$ *is a bounded operator from* $L^q_\sigma(\Omega)$ *to* $\left(L^r(\Omega)\right)^2$ *satisfying the estimate*

$$\left\|(\zeta + A)^{-1} u\right\|_r \le C_{q,r} |\zeta|^{-1+1/q-1/r} \|u\|_q$$

*for every* $u \in L^q_\sigma(\Omega)$ *and every* $\zeta \in D$. *In particular, if* $q \le r < \infty$, *the function* $(\zeta + A)^{-1}u$ *belongs to* $L^r_\sigma(\Omega)$.

(2) *For every* $q$ *and* $r$ *such that* $1 < q \le r \le 2$, *There exists a positive constant* $C_{q,r}$ *such that, for every* $\zeta \in D$, *the operator* $\nabla(\zeta + A)^{-1}$ *is a bounded operator from* $L^q_\sigma(\Omega)$ *to* $\left(L^r(\Omega)\right)^4$ *satisfying the estimate*

$$\left\|\nabla(\zeta + A)^{-1} u\right\|_r \le C_{q,r} |\zeta|^{-1/2+1/q-1/r} \|u\|_q$$

*for every* $u \in L^q_\sigma(\Omega)$ *and every* $\zeta \in D$.

This proposition immediately implies the following theorem. (See [5, 7, 8].)

**Theorem 3.6** *For every* $q \in (1, \infty)$, *the operator* $-A$ *generates a bounded analytic* $C^0$*-semigroup* $\exp(-tA)$ *on the space* $L^q_\sigma(\Omega)$, *and it satisfies the following estimates:*

(1) *Suppose that* $q \le r \le \infty$ *and that* $\alpha \ge 0$. *Then there exists a positive constant* $C_{q,r,\alpha}$ *such that the estimate*

$$\|A^{\alpha} \exp(-tA)u\|_r \le C_{q,r,\alpha} t^{1/r-1/q-\alpha} \|u\|_q$$

*holds for every* $u \in L^q_\sigma(\Omega)$.

(2) *Suppose that* $q \le r \le 2$. *Then we can choose* $C_{q,r,1/2}$ *so large that the estimate*

$$\|\nabla \exp(-tA)u\|_r \le C_{q,r,1/2} t^{1/r-1/q-1/2} \|u\|_q$$

*holds for every* $u \in L^q_\sigma(\Omega)$ *as well.*

*Proof* Put $\Gamma = \Gamma_1 + \Gamma_2 + \Gamma_3$, where

$$\Gamma_1 = \left\{ re^{3\pi i/4} \;\middle|\; \infty > r \ge \frac{1}{|z|} \right\}, \quad \Gamma_2 = \left\{ \frac{e^{i\theta}}{|z|} \;\middle|\; \frac{3\pi}{4} \ge \theta \ge -\frac{3\pi}{4} \right\},$$
$$\Gamma_3 = \left\{ re^{-3\pi i/4} \;\middle|\; \frac{1}{|z|} \le r < \infty \right\}.$$

Then for $z \in \mathbb{C}$ such that $z \ne 0$ and $|\arg z| \le \pi/12$, we can write

$$\exp(-tA)u = \frac{1}{2\pi i} \int_\Gamma e^{t\zeta} (\zeta + A)^{-1} u \, d\zeta.$$

Hence Assertion (1) with $q \le r < \infty$ and $\alpha \ge 0$, Assertion (1) with $r = \infty, \alpha = 0$ and Assertion (2) follow from Proposition 3.5. Assertion (2) for $r = \infty$ and $\alpha > 0$ follows from the estimate

$$\begin{aligned}
\|A^{\alpha} \exp(-tA)u\|_\infty &\le \left\| \exp\left(-\frac{t}{2}A\right) A^{\alpha} \exp\left(-\frac{t}{2}A\right) u \right\|_\infty \\
&\le C_{2q,\infty,0} \left(\frac{t}{2}\right)^{-1/2q} \left\| A^{\alpha} \exp\left(-\frac{t}{2}A\right) u \right\|_{2q} \\
&\le C_{2q,\infty,0} C_{q,2q,\alpha} \left(\frac{t}{2}\right)^{-1/2q+(-1/q+1/2q-\alpha)} \|u\|_q .
\end{aligned}$$

This completes the proof. □

In order to obtain the estimate for higher order derivative, we make use of operator we have the following lemma.

**Lemma 3.7** *We have the following assertions:*

1. *For* $v \in D(A^{1/2}) = L^2_\sigma(\Omega) \cap \left(H^1_0(\Omega)\right)^2$, *we have the identity* $\|\nabla v\|_2 = \left\|A^{1/2} v\right\|_2$.
2. *For* $v \in D(A) = L^2_\sigma(\Omega) \cap \left(H^1_0(\Omega) \cap H^2(\Omega)\right)^2$, *there exists a constant* $C$ *such that we have the estimate* $\left\|\nabla^2 v\right\|_2 \le C \left(\|Av\|_2 + \left\|A^{1/2} v\right\|_2\right)$.

*Proof* We first prove Assertion (1). For $v \in D(A)$, we have

$$\|\nabla v\|_2{}^2 = (-\Delta v, v) = (Av, v) = \left\|A^{1/2}v\right\|_2{}^2.$$

Since $D(A)$ is dense in $D(A^{1/2})$ and both sides are well-defined for $v \in D(A^{1/2})$, we obtain the conclusion.

We turn to the proof of Assertion (2). For $v \in L^2_\sigma(\Omega) \cap \left(H^1_0(\Omega) \cap H^3(\Omega)\right)^2$ and $R > 0$ such that $\partial\Omega \subset B(0, R)$, we have the identity

$$\begin{aligned}
&\sum_{j,k=1}^{2} (\nabla_j \nabla_k v, \nabla_j \nabla_k v) \\
&= \sum_{j,k=1}^{2} \left( \oint_{\partial\Omega} (\nabla_j \nabla_k v, n_j \nabla_k v)\, ds(x) + \int_1^2 \oint_{|x|=rR} (\nabla_j \nabla_k v, n_j \nabla_k v)\, ds(x) dr \right) \qquad (19)\\
&\quad + \sum_{k=1}^{2} (-\Delta \nabla_k v, \nabla_k v).
\end{aligned}$$

Since $\nabla_j \nabla_k v \in \left(L^2(\Omega)\right)^8$ and $\nabla_k v \in \left(L^2(\Omega)\right)^4$, the Schwarz inequality yields

$$\begin{aligned}
&\left| \int_1^2 \oint_{|x|=rR} (\nabla_j \nabla_k v, n_j \nabla_k v)\, ds(x) dr \right| \\
&\qquad \le \left( \int_{R \le |x| \le 2R} |\nabla_j \nabla_k v(x)|^2 dx \right)^{1/2} \left( \int_{R \le |x| \le 2R} |\nabla_k v(x)|^2 dx \right)^{1/2}.
\end{aligned}$$

The right-hand side of this inequality tends to 0 as $R \to \infty$. Hence, substituting this fact and into (19), we obtain

$$\sum_{j,k=1}^{2} (\nabla_j \nabla_k v, \nabla_j \nabla_k v) = \sum_{j,k=1}^{2} \oint_{\partial\Omega} (\nabla_j \nabla_k v, n_j \nabla_k v)\, ds(x) + \sum_{k=1}^{2} (-\Delta \nabla_k v, \nabla_k v).$$

Since the identity $-\Delta \nabla_k v = -\nabla_k \Delta v = \nabla_k A v$ holds, it follows that

$$\left\|\nabla^2 v\right\|_2{}^2 = \sum_{j,k=1}^{2} \oint_{\partial\Omega} (\nabla_j \nabla_k v, n_j \nabla_k v)\, ds(x) + \sum_{k=1}^{2} (\nabla_k A v, \nabla_k v). \qquad (20)$$

In the same way we have

$$(Av, -\Delta v) = \sum_{k=1}^{2} \left( \oint_{\partial\Omega} (Av, n_k \nabla_k v)\, ds(x) + (\nabla_k A v, \nabla_k v) \right). \qquad (21)$$

Subtracting (21) from (20) and making use of the trace estimate (16) and Assertion (1), we obtain

$$\begin{aligned}
&\left\|\nabla^2 v\right\|_2^{\,2}\\
&\le |(Av, -\Delta v)| + C\left\|\nabla^2 v|_{\partial\Omega}\right\|_{H^{-1/2}} \|\nabla v|_{\partial\Omega}\|_{H^{1/2}}\\
&\quad + C\|Av|_{\partial\Omega}\|_{H^{-1/2}} \|\nabla v|_{\partial\Omega}\|_{H^{1/2}}\\
&\le \|Av\|_2 \left\|\nabla^2 v\right\|_2 + \frac{1}{4}\left\|\nabla^2 v\right\|_2^{\,2} + 4C^2\|\nabla v\|_2^{\,2} + \frac{1}{4}\|Av\|_2^{\,2} + 4C^2\|\nabla v\|_2^{\,2}\\
&\le \frac{3}{4}\left\|\nabla^2 v\right\|_2^{\,2} + \frac{3}{4}\|Av\|_2^{\,2} + 8C^2\left\|A^{1/2} v\right\|_2^{\,2}.
\end{aligned}$$

We thus conclude

$$\left\|\nabla^2 v\right\|_2^{\,2} \le 3\|Av\|_2^{\,2} + 32C^2\left\|A^{1/2} v\right\|_2^{\,2}$$

for $v \in L^2_\sigma(\Omega) \cap \left(H^1_0(\Omega) \cap H^3(\Omega)\right)^2$. Since this space is dense in $D(A)$ and both sides are well-defined for $v \in D(A)$, we obtain the conclusion. □

From this estimate we have the following proposition.

**Proposition 3.8** *We have the following assertions:*

*(1) Suppose that* $1 < q \le 2$. *Then there exists a constant* $C'_q$ *such that, for every* $u \in L^q_\sigma(\Omega)$ *and every* $t > 0$, *the function* $\exp(-tA)u$ *belongs to the space* $\left(H^1_0(\Omega) \cap H^2(\Omega)\right)^2$, *and satisfies the estimate*

$$\left\|\nabla^2 \exp(-tA)u\right\|_2 \le C'_{q,s} t^{-1/q}(1 + t^{-1/2})\|u\|_q.$$

*(2) There exists a constant* $C''_s$ *such that, for every* $u \in L^2_\sigma(\Omega) \cap \left(H^1_0(\Omega)\right)^2$, *the function* $\exp(-tA)u$ *satisfies the estimate*

$$\left\|\nabla^2 \exp(-tA)u\right\|_2 \le C''_s(1 + t^{-1/2})\|\nabla u\|_2.$$

*Proof* We first show Assertion (1). Since $\exp(-tA)$ is a bounded analytic $C^0$-semigroup on $L^2_\sigma(\Omega)$, we have

$$\begin{aligned}
\left\|A^{j/2} \exp(-tA)u\right\|_2 &\le C\left(\frac{t}{2}\right)^{-j/2} \left\|\exp\left(-\frac{t}{2}\right)u\right\|_2\\
&\le C\left(\frac{t}{2}\right)^{-j/2+1/2-1/q} \|u\|_q
\end{aligned}$$

for $j = 1, 2$. Now the conclusion follows from Lemma 3.7, (2).

Assertion (2) follows from the estimate

$$\left\|A^{j/2}\exp(-tA)u\right\|_2 \le Ct^{-(j-1)/2}\left\|A^{1/2}u\right\|_2$$

for $j = 1, 2$ and Lemma 3.7, (1). □

This proposition immediately implies the following corollary.

**Corollary 3.9** *Suppose that* $1 \le s < 3/2$. *Then we have the following assertions:*

*(1) Suppose that* $1 < q \le 2$. *Then there exists a constant* $C'_{q,s}$ *such that, for every* $u \in L^q_\sigma(\Omega)$ *and every* $t > 0$, *the function* $\exp(-tA)u$ *belongs to the space* $\left(H^s_0(\Omega)\right)^2$, *and satisfies the estimate*

$$\|\exp(-tA)u\|_{\dot{H}^s} \le C'_{q,s}t^{-1/q}(1 + t^{(s-1)/2})\|u\|_q.$$

*(2) There exists a constant* $C''$ *such that, for every* $u \in L^2_\sigma(\Omega) \cap \left(H^1_0(\Omega)\right)^2$, *the function* $\exp(-tA)u$ *satisfies the estimate*

$$\|\exp(-tA)u\|_{\dot{H}^s} \le C''_s(1 + t^{-(s-1)/2})\|\nabla u\|_2.$$

*Proof* Assertion (1) with $s = 1$ coincides with Theorem 3.6 with $r = 2$, and Assertion (2) with $s = 1$ follows immediately from Lemma 3.7, (1).

For every $\theta \in (0, 1)$, let $X_\theta$ denote the complex interpolation space $\left(\dot{H}^1_0(\Omega), \dot{H}^1_0(\Omega) \cap \dot{H}^2(\Omega)\right)_\theta$. Then it follows from Assertion (1) and Proposition 3.8 that

$$\|\exp(-tA)u\|_{X_\theta} \le C'_{q,s}t^{-1/q}(1 + t^{-\theta/2})\|u\,|_q \tag{22}$$

for $u \in L^q_\sigma(\Omega)$, and

$$\|\exp(-tA)u\|_{X_\theta} \le C''_s(1 + t^{-\theta/2})\|\nabla u\|_2 \tag{23}$$

for $u \in L^2_\sigma(\Omega) \cap \left(H^1_0(\Omega)\right)^2$. If $0 < \theta < 1/2$,

For every $\theta \in (0, 1)$, we see that $X_\theta = \dot{H}^1_0(\Omega) \cap H^{1+\theta}(\Omega)$, which coincides with $\dot{H}^1_0(\Omega) \cap \dot{H}^{1+\theta}_0(\Omega)$ if $0 < \theta < 1/2$. Hence the conclusion follows from (22) and (23). □

# 4 Perturbation of the Stokes Operator

In this section we introduce a perturbation of the operator $A$, and show some properties. Suppose that $w$ satisfies $w \in (\mathcal{X}(b))^2$ with some $b \geq 1$ and $\nabla w \in (L^2(\Omega))^4$, and put

$$B[u] = P\{(w \cdot \nabla)u + (u \cdot \nabla)w\}.$$

Then, for every $u \in D(A) = L^q_\sigma(\Omega) \cap H^1_{q,0}(\Omega) \cap H^2_q(\Omega)$ with $1 < q \leq 2$, we have $\nabla u \in (L^q(\Omega))^4$, which implies $(w \cdot \nabla)u \in (L^q(\Omega))^2$. We moreover have $u \in L^{2q/(2-q)}_\sigma(\Omega)$ if $1 < q < 2$ and $u \in (L^\infty(\Omega))^2$ if $q = 2$, which imply $(u \cdot \nabla)w \in (L^q(\Omega))^2$ in both cases. Hence the operator $L_w[u] = Au + B[u]$ is well-defined on $u \in D(A)$.

In the sequel we obtain the resolvent estimate of this operator. For this end we first define the fractional power $(\zeta + A)^{-1/2}$. Let $\mu(\lambda)$ denote the spectral measure associated with the operator $A$ on $L^2_\sigma(\Omega)$. Then we can write

$$(\zeta + A)^{-1} = \int_0^\infty \frac{1}{\zeta + \lambda}\, d\mu(\lambda), \quad (\zeta + A)^{-1/2} = \int_0^\infty \frac{1}{\sqrt{\zeta + \lambda}}\, d\mu(\lambda)$$

for $\zeta \in D$ on $L^2_\sigma(\Omega)$. Then the operator $(\zeta + A)^{-1/2}$ is holomorphic in the interior of $D$ with values in bounded linear operators on $L^2_\sigma(\Omega)$. Here we note that $\zeta \in D$ implies $\zeta + \lambda \in D$ for every $\lambda \geq 0$, and hence the branch of $\sqrt{\zeta + \lambda}$ is well-defined. It is easy to see that $\{(\zeta + A)^{-1/2}\}^2 = (\zeta + A)^{-1}$. For the operator $(\zeta + A)^{-1/2}$ we have the following lemma.

**Lemma 4.1** *We have the following assertions:*

(1) *There exists a positive constant $C_2$ such that, for every $\zeta \in D$ and every $u \in L^2_\sigma(\Omega)$, we have $(\zeta + A)^{-1/2}u \in D(A^{1/2})$ with the estimates*

$$\|(\zeta + A)^{-1/2}u\|_2 \leq C_2|\zeta|^{-1/2}\|u\|_2, \quad \|\nabla(\zeta + A)^{-1/2}u\|_2 \leq C_2\|u\|_2.$$

(2) *For every $r$ satisfying $2 < r < \infty$, there exists a constant $C_r$ such that we have the estimate*

$$\|(\zeta + A)^{-1/2}u\|_r \leq C_r|\zeta|^{-1/r}\|u\|_2$$

*for every $\zeta \in D$ and every $u \in L^2_\sigma(\Omega)$.*

(3) *For every $q$ satisfying $1 < q < 2$, there exists a constant $C_q$ such that we have the estimate*

$$\|(\zeta + A)^{-1/2}u\|_2 \leq C_q|\zeta|^{-1+1/q}\|u\|_q$$

*for every $\zeta \in D$ and every $u \in L^2_\sigma(\Omega) \cap L^q_\sigma(\Omega)$.*

*Proof* We first show Assertion (1). Fix $u \in L^2_\sigma(\Omega)$. Then, for every $\zeta \in D$, we have

$$\left\|(\zeta + A)^{-1/2}u\right\|_2^2 = \left((\bar{\zeta} + A)^{-1/2}(\zeta + A)^{-1/2}u, u\right)$$
$$= \left(\int_0^\infty \frac{1}{\sqrt{(\bar{\zeta} + \lambda)(\zeta + \lambda)}}\, d\mu(\lambda)u, u\right).$$

Since the inequality $|\bar{\zeta} + \lambda| = |\zeta + \lambda| \geq |\zeta|/\sqrt{2}$ holds for every $\zeta \in D$ and $\lambda \geq 0$, we can estimate

$$\left\|(\zeta + A)^{-1/2}u\right\|_2^2 \leq \frac{2}{|\zeta|}\|u\|_2^2.$$

Next, for $u \in D(A^{1/2})$ and $\zeta \in D$, Lemma 3.7, (1) yields

$$\left\|\nabla(\zeta + A)^{-1/2}u\right\|_2^2 = \left\|A^{1/2}(\zeta + A)^{-1/2}u\right\|_2^2$$
$$= \left(\int_0^\infty \frac{\lambda}{\sqrt{(\bar{\zeta} + \lambda)(\zeta + \lambda)}}\, d\mu(\lambda), u\right).$$

Since the inequality $|\bar{\zeta} + \lambda| = |\zeta + \lambda| \geq \lambda/\sqrt{2}$ holds for every $\zeta \in D$ and $\lambda \geq 0$, we can estimate

$$\left\|A^{1/2}(\zeta + A)^{-1/2}u\right\|_2^2 \leq 2\|u\|_2^2$$

for $u \in D(A^{1/2})$ and $\zeta \in D$. From these inequalities we see that the required inequalities hold for $u \in D(A^{1/2})$. Since $D(A^{1/2})$ is dense in $L^2_\sigma(\Omega)$, we obtain the conclusion with $C_2 = \sqrt{2}$.

Assertion (2) follows from Assertion (1) and Lemma 3.1, and Assertion (3) follows from Assertion (2) and duality argument. □

From this lemma we can prove the following estimate.

**Lemma 4.2** *Suppose that* $w \in (\mathcal{X}(b))^2$ *with some* $b \geq 1$ *and* $\nabla w \in (L^2(\Omega))^4$. *Suppose also that* $\zeta \in \mathbb{C} \setminus \{0\}$ *satisfies* $|\arg \zeta| \leq 3\pi/4$. *Then the operator* $(\zeta + A)^{-1/2}B(\zeta + A)^{-1/2}$ *is bounded in* $L^2_\sigma(\Omega)$, *and it satisfies the estimate*

$$\left\|(\zeta + A)^{-1/2}B\left[(\zeta + A)^{-1/2}u\right]\right\|_2 \leq C\|w\|_{\mathcal{X}(b)}\|u\|_2,$$

*where* $C$ *is a constant depending only on* $\Omega$.

*Proof* Suppose that $\varphi \in C_{0,\sigma}^{\infty}(\Omega)$. In view of the equalities $\nabla \cdot w = 0$ and $\nabla \cdot (\zeta + A)^{-1/2}u = 0$, we have

$$\begin{aligned} &\left|\left(\varphi, (\zeta + A)^{-1/2}P\left\{(w\cdot\nabla)(\zeta + A)^{-1/2}u + \left((\zeta + A)^{-1/2}u\cdot\nabla\right)w\right\}\right)\right| \\ &= \left|-\left(\nabla(\zeta + A)^{-1/2}\varphi, w(\zeta + A)^{-1/2}u\right)\right| \\ &\leq \left\|\nabla(\zeta + A)^{-1/2}\varphi\right\|_2 \left\|w(\zeta + A)^{-1/2}u\right\|_2. \end{aligned} \tag{24}$$

Then Lemma 4.1 implies

$$\left\|\nabla(\zeta + A)^{-1/2}\varphi\right\|_2 \leq C\|\varphi\|_2 \tag{25}$$

with a constant $C$ depending only on $\Omega$. On the other hand, in view of the fact

$$(\zeta + A)^{-1/2}u \in D(A^{1/2}) = L_\sigma^2(\Omega) \cap \left(H_0^1(\Omega)\right)^2,$$

Proposition 3.4 and Lemma 4.1 imply

$$\left\|w(\zeta + A)^{-1/2}u\right\|_2 \leq C\|w\|_{\mathcal{X}(b)}\left\|\nabla(\zeta + A)^{-1/2}u\right\|_2 \leq C\|w\|_{\mathcal{X}(b)}\|u\|_2, \tag{26}$$

where the constant $C$ depends only on $\Omega$. Substituting (25) and (26) into (24) and observing the denseness of $C_{0,\sigma}^{\infty}(\Omega)$ in $L_\sigma^2(\Omega)$, we obtain the conclusion. □

For this operator we have the following proposition.

**Proposition 4.3** *For every $q$, $r$ such that $1 < q \leq 2 \leq r < \infty$, there exist positive numbers $A$ and $A_{q,r}$ such that, for every $w \in \left(\mathcal{X}(b)\right)^2$ satisfying $\nabla w \in \left(L^2(\Omega)\right)^4$ and $\|w\|_{\mathcal{X}(b)} \leq A$, we have the estimates*

$$\begin{aligned} \left\|(\zeta + L_w)^{-1}u\right\|_r &\leq A_{q,r}|\zeta|^{-1+1/q-1/r}\|u\|_q, \\ \left\|\nabla(\zeta + L_w)^{-1}u\right\|_2 &\leq A_{q,2}|\zeta|^{-1+1/q}\|u\|_q \end{aligned}$$

*for every $u \in L_\sigma^q(\Omega)$ and every $\zeta \in D$.*

*Proof* Suppose that $\|w\|_{\mathcal{X}(b)} \leq 1/2C$. Then Lemma 4.2 implies that the series

$$\sum_{j=0}^{\infty}\{-(\zeta + A)^{-1/2}B(\zeta + A)^{-1/2}\}^j$$

converges to an operator $T$. This operator is bounded on $L_\sigma^2(\Omega)$ uniformly in $\zeta \in D$, and it satisfies

$$(\zeta + A)^{-1/2}T(\zeta + A)^{-1/2} = (\zeta + A)^{-1}\sum_{j=0}^{\infty}\{-B(\zeta + A)^{-1}\}^j.$$

It follows that

$$(\zeta + A)^{-1/2} T (\zeta + A)^{-1/2} = (\zeta + A + B)^{-1} = (\zeta + L_w)^{-1}. \tag{27}$$

For $q$ and $r$ as in the assumption, Lemma 4.1 implies

$$\left\|(\zeta + A)^{-1} u\right\|_2 \leq C_q |\zeta|^{-1+1/q} \|u\|_q \tag{28}$$

for $u \in L^q_\sigma(\Omega) \cap L^2_\sigma(\Omega)$, and

$$\left\|(\zeta + A)^{-1} u\right\|_r \leq C_r |\zeta|^{-1/r} \|u\|_2, \quad \left\|\nabla(\zeta + A)^{-1} u\right\|_2 \leq C_2 \|u\|_2 \tag{29}$$

for $u \in L^2_\sigma(\Omega)$. Hence the required estimates follow from (27)–(29). □

From this proposition we have the following theorem.

**Theorem 4.4** *Let $w$ be the same as in Proposition 4.3. Then the operator $-L_w$ generates a bounded analytic $C^0$-semigroup $\exp(-tL_w)$ on $L^2_\sigma(\Omega)$. Furthermore, for every $q$ and $r$ such that $1 < q \leq 2 \leq r < \infty$, there exists a constant $B_{q,r}$ such that we have the estimates*

$$\|\exp(-tL_w)u\|_r \leq B_{q,r} t^{-1/q+1/r} \|u\|_q, \quad \|\nabla \exp(-tL_w)u\|_2 \leq B_{q,2} t^{-1/q} \|u\|_q$$

*for every $u \in L^q_\sigma(\Omega)$ and $t > 0$.*

*Proof* Putting $q = r = 2$, we have $\left\|(\zeta + L_w)^{-1} u\right\|_2 \leq C|\zeta|^{-1} \|u\|_2$ for $u \in L^2_\sigma(\Omega)$ and $\zeta \in D$. Let $\Gamma_1$, $\Gamma_2$, $\Gamma_3$ and $\Gamma$ be the same as in the proof of Theorem 3.6. Then, for $z \in \mathbb{C}$ such that $z \neq 0$ and $|\arg z| \leq \pi/12$, we introduce the operator $T(z)$ on $L^2_\sigma(\Omega)$ by

$$T(z)u = \frac{1}{2\pi i} \int_\Gamma e^{z\zeta} (\zeta + L_w)^{-1} u \, d\zeta.$$

Then we have $2\pi/3 \leq \arg z\zeta \leq 5\pi/6$ for $\zeta \in \Gamma_1$. It follows that $|e^{z\zeta}| \leq e^{-|z||\zeta|/2}$. Hence we have

$$\begin{aligned} \left\| \frac{1}{2\pi i} \int_{\Gamma_1} e^{z\zeta} (\zeta + L_w)^{-1} u \, d\zeta \right\|_2 &\leq \frac{1}{2\pi} \int_{1/|z|}^\infty e^{-r|z|/2} \frac{A_{2,2}}{r} \|u\|_2 \, dr \\ &\leq \frac{A_{2,2}\|u\|_2}{2\pi} \int_1^\infty \frac{e^{-\theta/2}}{\theta} \, d\theta \leq C A_{2,2} \|u\|_2. \end{aligned} \tag{30}$$

In the same way we have

$$\left\| \frac{1}{2\pi i} \int_{\Gamma_3} e^{z\zeta} (\zeta + L_w)^{-1} u \, d\zeta \right\|_2 \leq C A_{2,2} \|u\|_2. \tag{31}$$

We finally have

$$\left\| \frac{1}{2\pi i} \int_{\Gamma_2} e^{z\zeta} (\zeta + L_w)^{-1} u \, d\zeta \right\|_2 \le \frac{A_{2,2} \|u\|_2}{2\pi} \int_{-3\pi/4}^{3\pi/4} d\theta \le A_{2,2} \|u\|_2. \tag{32}$$

It follows from (30)–(32) that $T(z)$ defines a bounded analytic semigroup on $L^2_\sigma(\Omega)$. Since $D(L_w) = D(A)$ is dense in $L^2_\sigma(\Omega)$, the family of operators $T(t)$ coincides with the bounded analytic $C^0$-semigroup $\exp(-tL_w)$ on $L^2_\sigma(\Omega)$.

Next, fix $q$ such that $1 < q \le 2$, and consider a function $u \in L^2_\sigma(\Omega) \cap L^q_\sigma(\Omega)$. Then, for every $r$ such that $2 \le r < \infty$, we have

$$\begin{aligned}
\left\| \frac{1}{2\pi i} \int_{\Gamma_1} e^{z\zeta} (\zeta + L_w)^{-1} u \, d\zeta \right\|_r &\le \frac{1}{2\pi} \int_{1/|z|}^{\infty} e^{-r|z|/2} A_{2,2} \|u\|_q r^{-1+1/q-1/r} \, dr \\
&\le \frac{A_{2,2} \|u\|_q}{2\pi} |z|^{1/r-1/q} \int_1^\infty \frac{e^{-\theta/2}}{\theta} \, d\theta \le C A_{2,2} |z|^{1/r-1/q} \|u\|_2.
\end{aligned}$$

Arguing for $\Gamma_2$ and $\Gamma_3$ in the same way, we conclude

$$\|\exp(-tL_w) u\|_r \le C A_{2,2} |z|^{1/r-1/q} \|u\|_2.$$

In the same way we obtain

$$\|\nabla \exp(-tL_w) u\|_2 \le C A_{2,2} |z|^{-1/q} \|u\|_2.$$

Since $L^2_\sigma(\Omega) \cap L^q_\sigma(\Omega)$ is dense in $L^q_\sigma(\Omega)$, we obtain the conclusion. □

## 5 Asymptotic Behavior: Proof of Theorem 2.4

Theorem 2.2 implies that, for every $\varepsilon > 0$, there exists a positive number $T_0$ such that, for every $t \ge T_0$ we have $\|v(t)\|_2 < \varepsilon$, $\|v(t)\|_4 < \varepsilon$ and $\|\nabla v(t)\|_2 < \varepsilon$.

Next, for $T_1$ such that $T_0 < T_1 < \infty$, we put

$$\alpha(T_1) = \sup_{T_0 \le t \le T_1} \max \left\{ (t - T_0)^{1/4} \|v(t)\|_4, (t - T_0)^{1/2} \|\nabla v(t)\|_2 \right\}.$$

Then the function $\alpha(T_1)$ is continuous and monotone-increasing. For $t \in [T_0, T_1]$, we can write

$$v(t) = \exp(-(t - T_0)L_w) v(T_0) + \int_{T_0}^t \exp(-(t - \tau)L_w) P \left[ \left( v(\tau) \cdot \nabla \right) v(\tau) \right] d\tau.$$

From this we can estimate

$$\|v(t)\|_4 \le B_{2,4}(t-T_0)^{-1/4}\|v(T_0)\|_2 + C_{4/3}\int_{T_0}^{t} B_{4/3,4}(t-\tau)^{-1/2}\|v(\tau)\|_4\|\nabla v(\tau)\|_2\,d\tau$$
$$\le B_{2,4}(t-T_0)^{-1/4}\varepsilon + C_{4/3}\alpha(t)^2\int_{T_0}^{t} B_{4/3,4}(t-\tau)^{-1/2}\tau^{-3/4}\,d\tau$$

where $C_{4/3}$ denotes the operator norm of the projection $P$ from $\left(L^{4/3}(\Omega)\right)^2$ to $L_\sigma^{4/3}(\Omega)$. This implies

$$(t-T_0)^{1/4}\|v(t)\|_4 \le B_{2,4}\varepsilon + C_{4/3}B_{4/3,4}B\left(\frac{1}{2},\frac{1}{4}\right)\alpha(T_1)^2. \tag{33}$$

In the same way, from the estimate

$$\|\nabla v(t)\|_2 \le B_{2,2}(t-T_0)^{-1/2}\varepsilon + C_{4/3}\alpha(t)^2\int_{T_0}^{t} B_{4/3,2}(t-\tau)^{-3/4}\tau^{-3/4}\,d\tau$$

it follows that

$$(t-T_0)^{1/2}\|\nabla v(t)\|_2 \le B_{2,2}\varepsilon + C_{4/3}B_{4/3,2}B\left(\frac{1}{4},\frac{1}{4}\right)\alpha(T_1)^2. \tag{34}$$

Hence, putting

$$C_1 = \max\left\{C_{4/3}B_{4/3,4}B\left(\frac{1}{2},\frac{1}{4}\right), C_{4/3}B_{4/3,2}B\left(\frac{1}{4},\frac{1}{4}\right)\right\},$$
$$C_2 = \max\{B_{2,4}, B_{2,2}, 1\}$$

and taking the maximum of (33) and (34), we see that

$$\max\left\{(t-T_0)^{1/4}\|v(t)\|_4, (t-T_0)^{1/2}\|\nabla v(t)\|_2\right\} \le C_1\alpha(T_1)^2 + C_2\varepsilon.$$

Taking the supremum for $t \in [T_0, T_1]$, we see that $\alpha(T_1)$ satisfies

$$\alpha(T_1) \le C_1\alpha(T_1)^2 + C_2\varepsilon. \tag{35}$$

We suppose that $\varepsilon < 1/4C_1C_2$. Then there exists two distinct roots of the equation $C_1X^2 - X + C_2\varepsilon = 0$. Let $f(\varepsilon)$ denote the smaller one; namely,

$$f(\varepsilon) = \frac{1-\sqrt{1-4C_1C_2\varepsilon}}{2C_1} = \frac{2C_2\varepsilon}{1+\sqrt{1-4C_1C_2\varepsilon}}.$$

Then we have $f(\varepsilon) > \epsilon$. Hence we have $\alpha(T_1) < f(\varepsilon)$ if we take $T_1$ sufficiently close to $T_0$. From this we see $\alpha(T_1) \leq f(\varepsilon)$ for every $T_1 > T_0$. Indeed, if $\alpha(T_1) > f(\varepsilon)$ holds for some $T_1$, the intermediate theorem implies that

$$f(\varepsilon) < \alpha(T_2) < \frac{1 + \sqrt{1 - 4C_1C_2\varepsilon}}{2C_1}$$

holds with some $T_2 \in (T_0, T_1]$. However, for every $\beta$ such that

$$f(\varepsilon) < \beta < \frac{1 + \sqrt{1 - 4C_1C_2\varepsilon}}{2C_1},$$

we have $\beta > C_1\beta^2 + C_2\varepsilon$. This contradicts the estimate (35) with $T_1$ replaced by $T_2$.

It follows that

$$\|v(T_1)\|_4 \leq f(\varepsilon)(T_1 - T_0)^{-1/4} \leq \sqrt[4]{2} f(\varepsilon) T_1^{-1/4}$$

and

$$\|\nabla v(T_1)\|_2 \leq f(\varepsilon)(T_1 - T_0)^{-1/2} \leq \sqrt{2} f(\varepsilon) T_1^{-1/2}$$

for every $T_1 \geq 2T_0$. On the other hand, we have $\|v(T_1)\|_2 \leq \varepsilon \leq f(\varepsilon)$. Hence Lemma 3.1 implies that the estimate

$$\|v(T_1)\|_q \leq C_q f(\varepsilon) T_1^{-1/2+1/q}$$

holds for every $T_1 > 2T_0$ and $q \in [2, \infty)$. Since we have $f(\varepsilon) \to +0$ as $\varepsilon \to +0$, we conclude that

$$\|v(t)\|_q = o(t^{1/q-1/2}) \text{ for } q \in [2, \infty) \text{ and } \|\nabla v(t)\|_2 = o(t^{-1/2}).$$

It remains only the estimate for $\|v(t)\|_\infty$, First, since $v(t) \in H_0^1(\Omega)$ and since $H_0^1(\Omega)$ can be regarded as a closed subset of $H_0^1(\mathbb{R}^2)$, we have

$$\begin{aligned} \|v(t)\|_{\dot{B}^0_{\infty,2}(\Omega)} &\leq \|v(t)\|_{\dot{B}^0_{\infty,2}(\mathbb{R}^2)} \\ &\leq C\|\nabla v(t)\|_{L^2(\mathbb{R}^2)} = C\|\nabla v(t)\|_{L^2(\Omega)} = o(t^{-1/2}). \end{aligned}$$

We next recall the Littlewood-Paley decomposition. For every $k \in \mathbb{Z}$ we have

$$\Phi(2^{-k}\xi) + \sum_{j=k+1}^{\infty} \varphi_j(\xi) = 1.$$

Then, for every fixed $\varepsilon \in (0, 1]$, choose $T \geq 2$ so large that

$$\sup_{t \geq T-1} \max \left\{ t^{3/8} \|v(t)\|_8, t^{1/2} \|\nabla v(t)\|_2, t^{1/2} \|v(t)\|_{\dot{B}^0_{\infty,2}} \right\} \leq \varepsilon.$$

Suppose that $t \geq T$. We remark that, for every $\tau \in [t-1, t]$, we have $\tau \geq t-1 \geq t/2$.

Next, for a fixed $t \geq T$, choose $k$ as the smallest positive integer such that $t \leq 2^{2k}$; namely, $k \geq (\log_2 t)/2$. We then put

$$v^{(1)}(t) = \mathcal{F}^{-1}\left[\Phi(2^k \xi)\mathcal{F}[v(t)]\right], \quad v^{(2)}(t) = \sum_{j=-k+1}^{k-1} \left[\varphi_j(\xi)\mathcal{F}[v(t)]\right],$$

$$v^{(3)}(t) = \sum_{j=k}^{\infty} \left[\varphi_j(\xi)\mathcal{F}[v(t)]\right].$$

Then we have $v(t) = v^{(1)}(t) + v^{(2)}(t) + v^{(3)}(t)$.

We first have

$$\left\| v^{(1)}(t) \right\|_\infty \leq \|v(t)\|_8 \left\| 2^{-2k} \left(\mathcal{F}^{-1}[\Phi]\right) (2^{-k} \cdot) \right\|_{8/7} = C 2^{-k/4} \|v(t)\|_8 \leq C t^{-1/2} \varepsilon. \tag{36}$$

Next, in order to estimate $\left\| v^{(3)}(t) \right\|_\infty$, we employ another representation

$$v(t) = \exp(-A) v(t-1) + g(t), \tag{37}$$

where

$$g(t) = \int_{t-1}^{t} \exp(-(t-\tau)A) P\left[(w \cdot \nabla) v(\tau) + \left(v(\tau) \cdot \nabla\right) w + \left(v(\tau) \cdot \nabla\right) v(\tau)\right] d\tau.$$

Then Lemma 3.1 and Corollary 3.9, (2) imply that

$$\begin{aligned} \|\exp(-A) v(t-1)\|_{\dot{C}^{1/3}} &\leq C \|\exp(-A) v(t-1)\|_{\dot{H}^{4/3}} \\ &\leq C \|\nabla v(t-1)\|_2 \leq C\varepsilon (t-1)^{-1/2} \leq 2C\varepsilon t^{-1/2} \end{aligned} \tag{38}$$

Next, for $\tau \in [t-1, t]$, we have

$$\|(w \cdot \nabla) v(\tau)\|_{8/5} \leq \|w\|_8 \|\nabla v(\tau)\|_2 \leq C \|w\|_{\mathcal{X}(b)} \varepsilon t^{-1/2},$$

$$\left\| \left(v(\tau) \cdot \nabla\right) w \right\|_{8/5} \leq \|v(\tau)\|_8 \|\nabla w\|_2 \leq C \|\nabla w\|_2 \varepsilon t^{-3/8},$$

and

$$\big\|\big(v(\tau)\cdot\nabla\big)v(\tau)\big\|_{8/5} \le C\varepsilon^2 t^{-7/8}.$$

Summing up these estimates we conclude that

$$\begin{aligned}&\big\|P\big[(w\cdot\nabla)v(\tau)+\big(v(\tau)\cdot\nabla\big)w+\big(v(\tau)\cdot\nabla\big)v(\tau)\big]\big\|_{8/5}\\&\qquad\qquad\qquad\le C(\|w\|_{\mathcal{X}(b)}+\|\nabla w\|_2+1)t^{-3/8}.\end{aligned}$$

Hence Lemma 3.1 and Corollary 3.9, (2) imply

$$\begin{aligned}\|g(t)\|_{\dot{C}^{1/3}} &\le C\int_{t-1}^{t}\big\|\exp\big(-(t-\tau)A\big)\\&\qquad P\big[(w\cdot\nabla)v(\tau)+\big(v(\tau)\cdot\nabla\big)w+\big(v(\tau)\cdot\nabla\big)v(\tau)\big]\big\|_{\dot{H}^{4/3}}\,d\tau\\&\le C\int_{t-1}^{t}(t-\tau)^{-2/3-5/4+1}\\&\qquad\big\|P\big[(w\cdot\nabla)v(\tau)+\big(v(\tau)\cdot\nabla\big)w+\big(v(\tau)\cdot\nabla\big)v(\tau)\big]\big\|_{8/5}\,d\tau\\&\le C\varepsilon(\|w\|_{\mathcal{X}(b)}+\|\nabla w\|_2+1)t^{-3/8}\int_{t-1}^{t}(t-\tau)^{-11/12}\,d\tau\\&\le C\varepsilon(\|w\|_{\mathcal{X}(b)}+\|\nabla w\|_2+1)t^{-3/8}.\end{aligned}\tag{39}$$

It follows from (37)–(39) that

$$\|v(t)\|_{\dot{C}^{1/3}} \le C\varepsilon(\|w\|_{\mathcal{X}(b)}+\|\nabla w\|_2+1)t^{-3/8}.$$

Since $\dot{C}^{1/3}$ coincides with $\dot{B}^{1/3}_{\infty,\infty}$, we have

$$\begin{aligned}\big\|\mathcal{F}^{-1}[\varphi_j\mathcal{F}[v(t)]]\big\|_{\infty} &\le C2^{-j/3}\|v(t)\|_{\dot{C}^{1/3}}\\&\le C2^{-j/3}\varepsilon(\|w\|_{\mathcal{X}(b)}+\|\nabla w\|_2+1)t^{-3/8}.\end{aligned}$$

Summing up we obtain

$$\begin{aligned}\big\|v^{(3)}(t)\big\|_{\infty} &\le \sum_{j=k}^{\infty}\big\|\mathcal{F}^{-1}[\varphi_j\mathcal{F}[v(t)]]\big\|_{\infty}\\&\le C2^{-k/3}\varepsilon(\|w\|_{\mathcal{X}(b)}+\|\nabla w\|_2+1)t^{-3/8}\\&\le C\varepsilon(\|w\|_{\mathcal{X}(b)}+\|\nabla w\|_2+1)t^{-3/8-1/6}\\&\le C\varepsilon(\|w\|_{\mathcal{X}(b)}+\|\nabla w\|_2+1)t^{-1/2}.\end{aligned}\tag{40}$$

We finally observe that

$$\begin{aligned}\|v^{(2)}(t)\|_\infty &\le \sum_{j=-k+1}^{k-1} \|\mathcal{F}^{-1}[\varphi_j \mathcal{F}[v(t)]]\|_\infty \\ &\le \left( \sum_{j=-k+1}^{k-1} \|\mathcal{F}^{-1}[\varphi_j \mathcal{F}[v(t)]]\|_\infty^{\ 2} \right)^{1/2} \sqrt{2k-1} \\ &\le C\sqrt{\log t}\|v(t)\|_{\dot{B}^0_{\infty,2}} \le C\varepsilon t^{-1/2}\sqrt{\log t}.\end{aligned} \tag{41}$$

Summing up (36), (40) and (41) we conclude that

$$\|v(t)\|_\infty = o\left(t^{-1/2}\sqrt{\log t}\right).$$

**Acknowledgements** The author is very grateful to the referee for pointing out an important mistake. Partly supported by the International Research Training Group (IGK 1529) on Mathematical Fluid Dynamics funded by DFG and JSPS and associated with TU Darmstadt, Waseda University in Tokyo and the University of Tokyo, and by Grant-in-Aid for Scientific Research (C) 25400185, Ministry of Education, Culture, Sports, Science and Technology, Japan

# References

1. C.J. Amick, On Leray's problem of steady Navier-Stokes flow past a body in the plane. Acta Math. **161**, 71–130 (1988)
2. C.J. Amick, On the asymptotic form of steady Navier-Stokes flow past a body in the plane. J. Differ. Equ. **91**, 149–167 (1991)
3. J. Bergh, J. Löfström, *Interpolation Spaces* (Springer, Berlin, 1976)
4. W. Borchers, T. Miyakawa, On stability of exterior stationary Navier-Stokes flows. Acta Math. **174**, 311–382 (1995)
5. W. Borchers, W. Varnhorn, On the boundedness of the Stokes semigroup in two-dimensional exterior domains. Math. Z. **213**, 275–299 (1993)
6. G. Bourdaud, Réalisations des espaces de Besov homogènes. Ark. Mat. **26**, 41–54 (1988)
7. W. Dan, Y. Shibata, On the $L_q$-$L_r$ estimates of the Stokes semigroup in a two dimensional exterior domain. J. Math. Soc. Jpn. **51**, 181–207 (1999)
8. W. Dan, Y. Shibata, Remark on the $L_q$-$L_\infty$ estimate of the Stokes semigroup in a 2-dimensional exterior domain. Pac. J. Math. **189**, 223–239 (1999)
9. R. Finn, D.R. Smith, On the stationary solution of the Navier-Stokes equations in two dimensions. Arch. Ration. Mech. Anal. **25**, 26–39 (1967)
10. G.P. Galdi, Stationary Navier-Stokes problem in a two-dimensional exterior domains, in *Handbook of Differential Equations, Stationary Partial Differential Equations*, vol. I, ed. by M. Chipot, P. Quittner (North-Holland, Amsterdam, 2004), pp. 71–155
11. G.P. Galdi, *An Introduction to the Mathematical Theory of the Navier-Stokes Equations, Steady-State Problems*. Springer Monographs in Mathematics, 2nd edn. (Springer, New York, 2011)
12. G.P. Galdi, C.G. Simader, Existence, uniqueness and $L^q$-estimates for the Stokes problem in an exterior domain. Arch. Ration. Mech. Anal. **112**, 291–318 (1990)

13. G.P. Galdi, H. Sohr, On the asymptotic structure of plane steady flow of a viscous fluid in exterior domains, Arch. Ration. Mech. Anal. **131**, 101–119 (1995)
14. G.P. Galdi, M. Yamazaki, Stability of stationary solutions of two-dimensional Navier-Stokes exterior problem, in *Proceedings of the Workshop Mathematical Fluid Dynamics and Nonlinear Wave*. GAKUTO International Series of Mathematical Sciences and Applications (Gakkotosho, Tokyo, to appear)
15. D. Gilbarg, H.F. Weinberger, Asymptotic properties of Leray's solution of the stationary two-dimensional Navier-Stokes equations. Uspekhi Mat. Nauk **29**, 109–122 (1974); English translation Russ. Math. Surv. **29**, 109–123 (1974)
16. D. Gilbarg, H.F. Weinberger, Asymptotic properties of steady plane solutions of the Navier-Stokes equations with bounded Dirichlet integral. Ann. Sci. Norm. Super. Pisa **5**(4), 381–404 (1978)
17. H. Kozono, M. Yamazaki, On a larger class of stable solutions to the Navier-Stokes equations in exterior domains. Math. Z. **228**, 751–785 (1998)
18. H. Kozono, H. Sohr, M. Yamazaki, Representation formula, net force and energy relation of the stationary Navier-Stokes equations in 3-dimensional exterior domains. Kyushu J. Math. **51** 239–260 (1997)
19. T. Nakatsuka, On uniqueness of symmetric Navier-Stokes flows around a body in the plane. Adv. Differ. Equ. **20**, 193–212 (2015)
20. A. Novotny, M. Padula, Note on decay of solutions of steady Navier-Stokes equations in 3-D exterior domains. Differ. Integral Equ. **8**, 1833–1842 (1995)
21. K. Pileckas, R. Russo, On the existence of vanishing at infinity symmetric solutions to the plane stationary exterior Navier-Stokes problem. Math. Ann. **352**, 643–658 (2012)
22. A. Russo, On the asymptotic behavior of $D$-solutions of the plane steady-state Navier-Stokes equations. Pac. J. Math. **246**, 253–256 (2010)
23. R. Russo, On Stokes' problem, in *Advances in Mathematical Fluid Mechanics*, ed. by R. Rannacher, A. Sequeira (Springer, Heidelberg, 2010), pp. 473–511
24. L.I. Sazonov, Asymptotic behavior of the solution to the two-dimensional stationary problem of flow past a body far from it. Mat. Zametki **65**, 246–253 (1999). English translation Math. Notes **65**, 202–206 (1999)
25. H. Triebel, *Theory of Function Spaces* (Birkhäuser, Basel, 1983)
26. M. Yamazaki, The stationary Navier-Stokes equation on the whole plane with external force with antisymmetry. Ann. Univ. Ferrara **55**, 407–423 (2009)
27. M. Yamazaki, Unique existence of stationary solutions to the two-dimensional Navier-Stokes equations on exterior domains, in *Mathematical Analysis on the Navier-Stokes Equations and Related Topics, Past and Future–in Memory of Professor Tetsuro Miyakawa*. Gakuto International Series in Mathematical Sciences and Applications, vol. 25 (Gakkotosho, Tokyo, 2011), pp. 220–241

MIX
Papier aus verantwortungsvollen Quellen
Paper from responsible sources
FSC® C105338

If you have any concerns about our products,
you can contact us on
**ProductSafety@springernature.com**

In case Publisher is established outside the EU,
the EU authorized representative is:
**Springer Nature Customer Service Center GmbH**
**Europaplatz 3, 69115 Heidelberg, Germany**

Printed by Libri Plureos GmbH
in Hamburg, Germany